*Electron–Molecule Interactions
and Their Applications*
Volume 1

Electron–Molecule Interactions and Their Applications

VOLUME 1

EDITED BY

L. G. Christophorou

Atomic, Molecular and High Voltage Physics Group
Health and Safety Research Division
Oak Ridge National Laboratory
Oak Ridge, Tennessee
and
Department of Physics
The University of Tennessee
Knoxville, Tennessee

1984

ACADEMIC PRESS, INC.

(Harcourt Brace Jovanovich, Publishers)

Orlando   San Diego   San Francisco   New York   London
Toronto   Montreal   Sydney   Tokyo   São Paulo

COPYRIGHT © 1984, BY ACADEMIC PRESS, INC.
ALL RIGHTS RESERVED.
NO PART OF THIS PUBLICATION MAY BE REPRODUCED OR
TRANSMITTED IN ANY FORM OR BY ANY MEANS, ELECTRONIC
OR MECHANICAL, INCLUDING PHOTOCOPY, RECORDING, OR ANY
INFORMATION STORAGE AND RETRIEVAL SYSTEM, WITHOUT
PERMISSION IN WRITING FROM THE PUBLISHER.

ACADEMIC PRESS, INC.
Orlando, Florida 32887

United Kingdom Edition published by
ACADEMIC PRESS, INC. (LONDON) LTD.
24/28 Oval Road, London NW1 7DX

Library of Congress Cataloging in Publication Data

Main entry under title:

Electron-molecule interactions and their applications.

    Includes bibliographical references and indexes.
    1. Electron-molecule collisions.
I. Christophorou, L.G.
QC794.6.C6E374   1983   539.7'2112   83-7648
ISBN 0-12-174401-9 (v. 1 : alk. paper)

PRINTED IN THE UNITED STATES OF AMERICA

84 85 86 87   9 8 7 6 5 4 3 2 1

Contents

Contributors ix
Preface xi
Contents of Volume 2 xiii

1. Elastic Scattering of Electrons by Molecules
G. Csanak, D. C. Cartwright, S. K. Srivastava, and S. Trajmar

    I. Introduction 2
   II. Experimental Techniques 5
  III. Theoretical Methods 41
  IV. Comparison of Experimental and Theoretical Results 89
   V. Applications 119
      References 142

2. Excitation of Molecules by Electron Impact
S. Trajmar and D. C. Cartwright

     I. Introduction 156
    II. Definition of Electron Collision Cross Sections 158
   III. Electron-Impact Spectroscopy 160
   IV. Experimental Techniques for Measuring Electron-Impact Excitation Cross Sections 166
    V. Electron-Impact Excitation Cross Section Data 171
   VI. Generation of Optical Data by Electron-Impact Techniques 181
  VII. Recent Developments 190
 VIII. Theory of Electron–Molecule Collisions 193
   IX. Comparison of Theory and Experiment 202
    X. Applications of Electron Collision Data 226
      References 242

3. *Ionization of Molecules by Electron Impact*
   T. D. Märk

   I. Introduction   251
   II. Ionization Mechanisms and Types of Ions Produced   252
   III. Total and Partial Ionization Cross Sections   279
   IV. Differential Ionization Cross Sections   306
   References   315

4. *Dissociation of Molecules by Electron Impact*
   E. C. Zipf

   I. Introduction   335
   II. Models   338
   III. Experimental Techniques   346
   IV. Measurements   363
   References   397

5. *Electron–Molecule Resonances*
   J. B. Hasted and D. Mathur

   I. Introduction   403
   II. Theory of Resonances   405
   III. Experimental Techniques for the Study of Resonances   424
   IV. Experimental Results   432
   V. Conclusions   471
   References   471

6. *Electron Attachment Processes*
   L. G. Christophorou, D. L. McCorkle, and A. A. Christodoulides

   I. Introduction   478
   II. Modes of Production of Negative Ions   479
   III. Techniques for the Study of Electron Attachment Processes   495
   IV. Dissociative Electron Attachment to Ground-State Molecules   504
   V. Dissociative Electron Attachment to "Hot" Molecules (Effects of Temperature on Dissociative Electron Attachment)   558
   VI. Dissociative Electron Attachment to Electronically Excited Molecules   568
   VII. Molecular Parent Negative Ions   569

VIII.	Negative Ions Formed by Ion Pair Processes and by Collisions of Molecules with Ground-State and Rydberg Atoms	593
IX.	Doubly Charged Negative Ions	605
	References	606

7. Electron Detachment Processes
R. L. Champion and L. D. Doverspike

I.	Introduction	619
II.	Nomenclature and Experimental Techniques	621
III.	Atomic Reactants	623
IV.	Molecular Targets	639
V.	Detachment Accompanied by Rearrangement	656
VI.	Molecular Negative Ions	673
VII.	Detachment Rate Constants	677
VIII.	Summary	678
	References	679

Index   683

Contributors

Numbers in parentheses indicate the pages on which the authors' contributions begin.

D. C. CARTWRIGHT (1, 155), University of California, Los Alamos National Laboratory, Los Alamos, New Mexico 87545

R. L. CHAMPION (619), Department of Physics, The College of William and Mary, Williamsburg, Virginia 23185

A. A. CHRISTODOULIDES (477), Department of Physics, University of Ioannina, Ioannina, Greece, and Department of Physics, The University of Tennessee, Knoxville, Tennessee 37916

L. G. CHRISTOPHOROU (477), Atomic, Molecular and High Voltage Physics Group, Health and Safety Research Division, Oak Ridge National Laboratory, Oak Ridge, Tennessee 37830, and Department of Physics, The University of Tennessee, Knoxville, Tennessee 37916

G. CSANAK (1), University of California, Los Alamos National Laboratory, Los Alamos, New Mexico 87545

L. D. DOVERSPIKE (619), Department of Physics, The College of William and Mary, Williamsburg, Virginia 23185

J. B. HASTED (403), Department of Physics, Birkbeck College, University of London, London, England WC1E 7HX

T. D. MÄRK (251), Institut für Experimentalphysik, Leopold Franzens Universität, A-6020 Innsbruck, Österreich

D. MATHUR (403), Tata Institute of Fundamental Research, Bombay 400 005, India

D. L. McCORKLE (477), Department of Physics, The University of Tennessee, Knoxville, Tennessee 37916

S. K. SRIVASTAVA (1), California Institute of Technology, Jet Propulsion Laboratory, Pasadena, California 91109

S. TRAJMAR (1, 155), California Institute of Technology, Jet Propulsion Laboratory, Pasadena, California 91109

E. C. ZIPF (335), Department of Physics and Astronomy, University of Pittsburgh, Pittsburgh, Pennsylvania 15260

Preface

In this two-volume treatise a balanced and comprehensive account is given of *electron–molecule interactions* in dilute and dense gases and (briefly) in liquid media. The electron—especially the *slow* electron—is chased patiently, and the intricate ways in which it and molecules interact in their encounters—both in very-low-pressure gases and in high- and ultra-high-pressure gases and liquids—are described. Electron–atom interactions are covered only to the extent that they balance and enhance the understanding of the interactions of electrons with simple and complex molecules.

Basic recent knowledge on electron–molecule interactions illuminates, through its transdisciplinary nature, wide areas of pure and applied science and constitutes a fertile soil from which many modern technologies draw their strength, growth, and innovation. This is amply demonstrated by the discussions in this work and by the voluminous literature on electron–molecule interactions, which originates rather evenly from a wide spectrum of pure and applied science.

In Volume 1 the first four chapters deal with the primary electron–molecule interactions of elastic electron scattering by molecules and rotational excitation of molecules by electrons (Chapter 1), vibrational and electronic excitation (Chapter 2), ionization (Chapter 3), and dissociation (Chapter 4) of molecules by electrons. Chapter 5 focuses on electron–molecule resonances, Chapter 6 on electron attachment processes, and Chapter 7 on electron detachment (principally collisional). With the exception of Chapter 6, the experimental information discussed in Volume 1 has been obtained at low pressures using beam techniques.

In Volume 2 the first chapter deals with electron transfer reactions and the second with electron-molecular positive-ion recombination. Chapter 3 focuses on electron motion in high-pressure gases and shifts the discussion on electron–molecule interactions from single- to multiple-collision conditions, which is the domain of electron swarm studies. In Chapter 4 knowledge on electron–molecule interactions in gases is linked to that on similar processes in the liquid state.

Selected examples on the translation of the results of basic research on electron–molecule interactions to application are discussed in Chapter 5, and a rather exhaustive account of the electron affinity of molecules, atoms, and radicals is given in the last chapter (6).

I am deeply grateful to my colleagues who contributed to this work and to my collaborators at Oak Ridge National Laboratory and The University of Tennessee for their fine work, advice, cooperation, and patience. The encouragement received from the management of Oak Ridge National Laboratory and The University of Tennessee is acknowledged with gratitude.

Special thanks are due to Mrs. Lana Melton, Mrs. Waldean Richardson, and Mrs. Amy Bliss for their invaluable assistance with the manuscripts and to the staff of Academic Press for their excellent cooperation.

To my wife, Eratoula, and daughters, Penelope and Yianna, a debt of gratitude is due for their sacrifices and support.

L. G. Christophorou
Oak Ridge, Tennessee

Contents of Volume 2

1. *Electron Transfer Reactions*
   T. F. Moran

2. *Electron–Molecular Positive-Ion Recombination*
   J. Wm. McGowan and J. B. A. Mitchell

3. *Electron Motion in Low- and High-Pressure Gases*
   S. R. Hunter and L. G. Christophorou

4. *Interphase Physics: Linking Knowledge on Electron–Molecule Interactions in Gases to Knowledge on Such Processes in Condensed Matter*
   L. G. Christophorou and K. Siomos

5. *From Basic Research to Application*
   L. G. Christophorou and S. R. Hunter

6. *Electron Affinities of Atoms, Molecules, and Radicals*
   A. A. Christodoulides, D. L. McCorkle, and L. G. Christophorou

*Electron–Molecule Interactions
and Their Applications*
Volume 1

1 Elastic Scattering of Electrons by Molecules

G. Csanak and D. C. Cartwright
University of California
Los Alamos National Laboratory
Los Alamos, New Mexico

and

S. K. Srivastava and S. Trajmar
California Institute of Technology
Jet Propulsion Laboratory
Pasadena, California

I.	Introduction	2
II.	Experimental Techniques	5
	A. Instrumentation	5
	B. Methods	21
	C. Summary of Experimental Data	27
III.	Theoretical Methods	41
	A. Historical Perspective	41
	B. Details of the Theoretical Methods	54
	C. Summary of Theoretical Data	89
IV.	Comparison of Experimental and Theoretical Results	89
	A. Historical Perspective	89
	B. The Validity of Certain Physical Approximations	100
	C. Specific Comparisons between Theory and Experiment	104
V.	Applications	119
	A. Phenomena Involving Elastic Scattering and Rotational Excitation	119
	B. Planetary Atmospheres	127
	C. Gas Lasers	137
	Appendix A. Summary of Review Articles on Electron Scattering by Molecules (Elastic and Rotational)	137
	References	142

I. Introduction

Whenever free electrons collide with molecules, a wide variety of kinetic processes may take place. These processes can be divided into two general categories: those in which the electron loses a portion of its kinetic energy to the excitation of internal degrees of freedom of the molecule (inelastic collisions) and those in which no energy is transferred to the internal motion of the molecule (elastic collisions). During elastic collisions the electron loses some energy owing to momentum transfer, but since this energy loss is proportional to the ratio of the mass of the electron to the mass of the molecule, it is generally small compared with the energy lost to excitation of internal molecular degrees of freedom. In this chapter we discuss nonresonant elastic scattering and rotational excitation whereas nonresonant vibrational and electronic excitation of molecules are the subject of Chapter 2. Resonance effects in electron scattering are treated in Chapter 5.

Electron collision processes, in general, play important roles in both manmade and naturally occurring plasmas. The quantities that are needed for determining the energy balance and transport properties of electrons in plasmas are the various electron-impact cross sections, which are defined in the following manner. Assume that a beam of N electrons per second impinge on a molecular target consisting of N_M molecules (at low density). Then the number of electrons deflected N_d per second by the molecules, into a solid angle $d\Omega$, defined by polar angles θ and ϕ, is given by

$$N_d = N_M N \frac{d\sigma}{d\Omega}(E_o, \theta, \phi) \, d\Omega. \tag{1}$$

Here E_o is the kinetic energy of the incident electrons and $d\Omega = \sin\theta \, d\theta \, d\phi$. The quantity $d\sigma/d\Omega(E_o, \theta, \phi)$, having the dimensions of area per unit solid angle, is called the *differential cross section* (DCS) for the particular elastic or inelastic process. In a gaseous target, the molecules freely rotate and the cross section becomes independent of the azimuthal polar angle ϕ. The total number of particles per second elastically scattered from an incident beam of unit flux and energy E_o is obtained by integrating the differential cross section over all polar angles. This integration yields the *integral cross section*, which is defined by

$$\sigma(E_o) = 2\pi \int_0^\pi \frac{d\sigma}{d\Omega}(E_o, \theta) \sin\theta \, d\theta. \tag{2}$$

The cross section for momentum transfer is given by the expression

$$\sigma^M(E_o) = 2\pi \int_0^\pi \frac{d\sigma}{d\Omega}(E_o, \theta) \left(1 - \frac{k_f}{k_i} \cos\theta\right) \sin\theta \, d\theta \, d\phi, \tag{3}$$

where k_i and k_f are the initial and final momenta of the electron, respectively.

The above definitions of cross sections apply to "ideal" measurements of electron–molecule collision events. In practice, electron-impact cross sections are measured for an ensemble of target molecules with instruments of finite energy and angular resolutions and represent averages over ensemble and the instrumental ranges of the parameters. Furthermore, with the exception of H_2, rotational excitation (and in some cases even vibrational excitation) cannot be resolved from elastic scattering with present experimental techniques. The elastic cross sections measured with today's techniques, in most cases, represent simultaneous elastic scattering and rotational excitation (and deexcitation). As a consequence, the measured elastic DCSs, unless stated otherwise represent

$$\mathrm{DCS}(E_o, \theta) = \overline{\frac{d\sigma(E_o, \theta)}{d\theta}} = \sum_j \sum_{j'} N_j \sigma(E_o, \theta; j \to j'), \qquad (4)$$

where j and j' are the initial and final rotational state quantum numbers, respectively, and N_j the population fraction, or weighting factor, of the state j. The various theoretical results that have been reported vary from no averaging having been performed to complete averaging over the initial rotational states, so the individual papers must be read with care. In some theoretical treatments of elastic scattering, the $\Delta j = 0$ condition is enforced to simplify the work; that is,

$$d\sigma(E_o, \theta)/d\theta = \sum_j N_j \sigma(E_o, \theta; j). \qquad (5)$$

In other theoretical models used for rotational excitation, the elastic scattering is considered to be independent of j, so the summation in Eq. (5) is not needed. It should be emphasized, therefore, that in comparing experiment and theory these differences in the theoretical approaches need to be taken into account.

The first reported measurements of elastic DCSs date back to around 1930, e.g., Harnwell (1929a,b) McMillan (1930), Arnot (1929, 1931), Bullard and Massey (1931), Hughes and McMillan (1932), and Mohr and Nicoll (1932). During the same period Ramsauer and Kollath (1930, 1931b,c, 1932) developed instrumentation which was suitable for the measurement of *total* scattering cross sections (the sum of the cross sections for elastic and all inelastic scattering processes) down to very low electron impact energies. For electron-impact energies less than the lowest vibrational and electronic excitation thresholds such total cross sections are equivalent to integral elastic cross sections (averaged over rotational processes). At very low electron energies, the swarm methods, originally introduced by Townsend and Bailey (1921, 1922), have been utilized to obtain momentum transfer and total scattering cross sections. These early efforts have been well reviewed by Massey and Burhop (1969) and Massey (1969).

The physical principles invoked in electron–atom or electron–molecule elastic scattering were identified in the early days of quantum mechanics (Born, 1926; Oppenheimer, 1928; Massey and Mohr, 1932a and b) and have been summarized in the classic volumes "Electronic and Ionic Impact Phenomena" (Massey and Burhop 1969; Massey 1969). Both theoretical and experimental methods for studying electron–molecule collisions have undergone rapid development in the last 15 years. This progress can be attributed to the availability of high-speed computers and to the motivation provided by the practical need of cross section data for laser development, various plasma and atomic fusion schemes, magnetohydrodynamic power generation, and astrophysical and atmospheric modeling purposes.

This chapter is intended to provide an introduction to the fundamental physics of electron–molecule elastic scattering and rotational excitation and a summary of the recent developments. A brief discussion of experimental and theoretical methods is given and references to work that have reported cross section data are listed in tabular form. The main emphasis in discussing the recent experimental work will be placed on e-beam/molecular-beam technique yielding DCS data, since this technique is relatively new and provides information not available from swarm data. Electron swarm and total cross section measurements, and the question of resonances, will be discussed only briefly because they have been the topics of recent reviews (e.g., Huxley and Crompton, 1974; Schulz, 1973a,b; Trajmar and Register, 1984; Trajmar et al., 1983). In discussing the theoretical methods we attempt to describe the collision physics contained in the various theoretical models rather than the details of the numerical methods, and comparisons between experimental and theoretical results are made to emphasize these aspects of the various theoretical models. A few applications involving the use of elastic and rotational excitation data are given at the end of this chapter. Only the recent developments that pertain to elastic electron scattering will be discussed in some detail and only data published after 1970 will be covered. The list of references in this chapter is therefore not exhaustive but rather augments earlier reviews on this subject.

Earlier reviews of the experimental techniques that have been used to study electron scattering are contained in Massey and Burhop (1969), Massey (1969), Kollath (1958), Fite (1962), McDaniel (1964), Hasted (1964), Klemperer (1965), Kuyatt (1968), Golden et al. (1971), Sevier (1972), Bonham and Fink (1974), Srivastava (1979) and Trajmar and Register (1984). A summary of recent theoretical reviews is given in Appendix A. The most comprehensive review of the theory of electron–molecule scattering is that of Lane (1980).

II. Experimental Techniques

A. Instrumentation

The apparatus for the study of elastic scattering of electrons by molecules is known as an *electron scattering spectrometer*. Various designs for such a spectrometer have been published (e.g., Simpson, 1964; Preston *et al.*, 1973; Brunt *et al.*, 1977; Roy and Carette, 1977; Brewer *et al.* 1980a and 1980b), and the basic components are illustrated in Fig. 1. A spectrometer consists of an electron monochromator, an electron energy analyzer, a target, and peripheral electronic devices. The first three components are placed inside a vacuum chamber, which generally can be evacuated to about 10^{-8} Torr. When a target gas is introduced into the chamber for the purpose of cross section measurements, the pressure rises to the range 10^{-5}–10^{-6} Torr. In differential (in angle) scattering measurements, magnetic fields are undesirable since they influence the electron trajectories and, hence, the collection characteristics. Therefore, collimation, transport, and energy selection of electrons are usually achieved by electrostatic techniques and the vacuum chamber is lined with magnetic shielding material to reduce the stray magnetic field to a value of a few milliGauss along the electron path and in the scattering region.

In the following we shall describe these components in detail.

1. *Electron Monochromator*

The electron monochromator produces an electron beam with a well-defined angular divergence and energy distribution. The energy distribution is

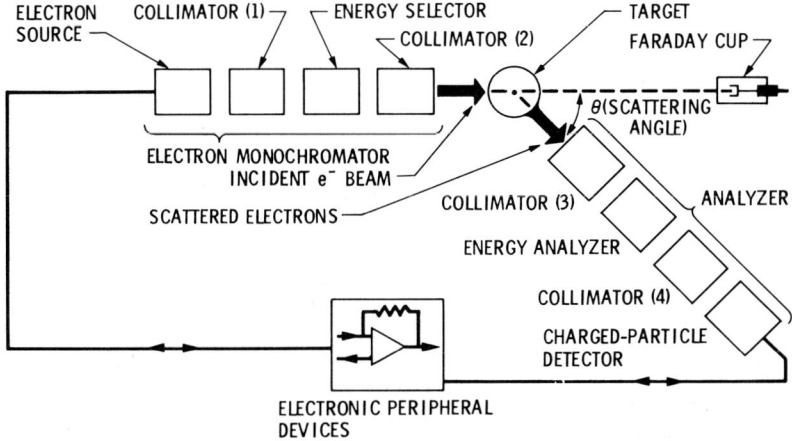

Fig 1. Basic components of an electron scattering spectrometer.

usually narrow (a few hundredths of an electronvolt). This beam is focused into the interaction region with the desired impact energy. For optimum performance several design factors have to be optimized as discussed in the papers by Kuyatt and Simpson (1967) and Read et al. (1974). The main components of the electron monochromator are the electron source, the collimator, and the energy selector. They are discussed below.

a. Electron Sources

The electron source generates a beam of electrons that is transported to the target by two collimators and an energy selector (see Fig. 1). It is designed to produce a maximum electron current in a beam with minimum angular divergence. The key component of the electron source is the electron emitter for which two different designs have been used. One utilizes the photoelectric effect, or photoionization, of atoms or molecules, and the other employs thermionic emission.

Generation of electrons by the photoelectric effect for collision studies was used in the 1930s by Ramsauer and Kollath (1930, 1932). Recently, van Brunt and Gallagher (1977) produced electron beams of very narrow energy spread by laser photoionization. The main drawback with the photoelectric (or photo-ionization) electron emitters is the low electron current, which, because of additional losses associated with the various subcomponents in the energy selector, is too low to be useful for collision studies at the present time.

Thermionic emission has most generally been used as the source of electrons. These devices are based on the fact that, when a metal is heated to a sufficiently high temperature, free electrons are produced with an energy distribution given by

$$dN(E) \propto \exp\left[-\frac{W + E}{kT}\right] dE, \tag{6}$$

where $dN(E)$ is the number of electrons emitted per second between energies E and $E + dE$, k the Boltzmann constant, T the absolute temperature, and W the work function of the metal. Pure metals need to be heated to about $T = 2000$ K for electron emission, which is somewhat of a disadvantage because it produces electrons with a broad energy width (~ 0.5 eV) and also presents heat dissipation problems. A number of techniques have been tried to lower the cathode temperature: (a) thoriated-tungsten cathode, which has a work function much less than pure tungsten; (b) oxide-coated cathode (BaSrO), which has a metal surface (generally tungsten) on which barium oxide or alkaline earth oxides are applied; and (c) dispenser-type cathodes, which consist of porous sintered-tungsten, molybdenum, or nickel into which barium is absorbed. These cathodes are heated indirectly, but because the operating temperatures are lower than for pure metals, the energy width of the emitted electrons is smaller (~ 0.3 eV). However, they are easily "poisoned" by molecular gases and lose their electron emission efficiency. Another area that occasionally presents a serious problem is the maintenance of a uniform cathode surface potential. In electron collision experiments, the cathode potential is taken as the reference in measuring the kinetic energy of the electrons. Any spatial nonuniformity in the potential at the surfaces spreads the energy distribution of the electron beam and, therefore, introduces an uncertainty in the value of the reference energy. In present-day electron scattering experiments, hairpin filaments of pure tungsten are used. They are advantageous because they are inert to most gases, present a point source for electron emission, and prevent uncertainties caused by surface nonuniformity of the potential. In addition, since they are physically small devices, their heat capacities are correspondingly small and heat dissipation is not a serious problem.

There are several designs for extraction of the electrons from the filament (or cathode) and for forming them into a beam. The most commonly used design is that of Pierce (1954), shown schematically in Fig. 2, which has properties that are well understood. It consists of a tungsten filament, an equipotential surface (Pierce element) inclined at an angle of $67.5°$ to the axis of the source, and an anode with an aperture of diameter smaller than the anode–cathode separation. The Pierce element is (in principle) at the filament potential, but in practice, a small potential difference is applied between the two to maximize the electron current. The maximum (space-charge-limited) current J_{max} A/m^2 emitted from this diode can be calculated from the relation (Child, 1911)

$$J_{max} = a(V_A^{3/2}/d^2), \qquad (7)$$

where the perveance a is equal to 2.33×10^{-6}; V_A is the voltage across the diode and d the separation between the cathode and the anode. A more

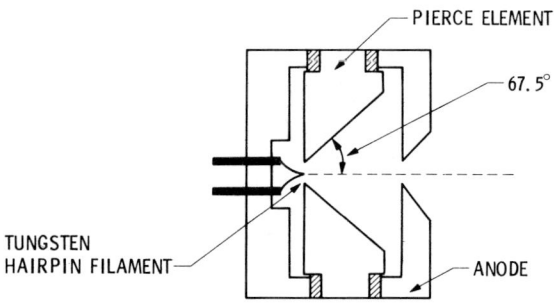

Fig. 2. Electron source based on the Pierce design.

detailed treatment for the maximum current emitted for the various source geometries is given by Read *et al.* (1974).

The divergence of the electron beam emerging from the anode hole must be known to design the collimator properly, and it is determined by the cathode–anode spacing d and the anode aperture radius r. The cathode–anode diode forms a weak lens (Calbick lens) whose focal length f is equal to $-3d$. If an empirical correction of 10% is applied for the distortion of the field near the aperture, then the divergence angle α (rad) is given by

$$\alpha = r/2.7d. \tag{8}$$

The divergence characteristics associated with the various electron sources are different, but in general, α should be as small as possible, which means that the ratio r/d should be small. Harting and Burrows (1970) have treated this problem in greater detail for a Pierce diode and have derived conditions for optimum values of the anode voltage and d. It is to be noted that, although the Pierce design is strictly applicable to a flat cathode, not to a pointed filament, this restriction is usually disregarded in practice.

Another type of electron source utilizes a "Wehnelt cylinder" and an anode to extract electrons from a tunsten filament. A schematic diagram illustrating a typical design of this source is shown in Fig. 3. This source consists of a pointed (normally V-shaped) tungsten filament, which is placed inside a "Wehnelt" cylinder. When the filament is heated, electrons are thermionically emitted and a variable voltage (negative with respect to the tip of the filament) is then applied on the Wehnelt cylinder, which causes an electron cloud formation in front of the filament. A high positive voltage on the anode extracts the electrons from this cloud and the hole in the cylinder forms the electron beam. The intensity of this beam is controlled by the variable voltage on the Wehnelt cylinder. The charge cloud has the advantage that it "shields" the filament, thereby reducing any effect that may be caused by the irregularities of the filament surface and the temperature variation on the wire.

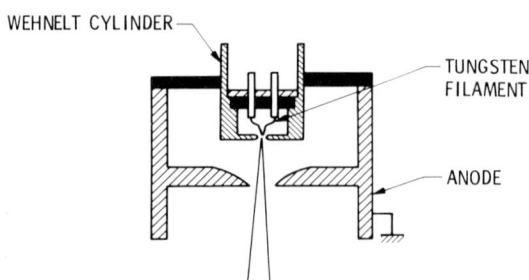

Fig. 3. A Wehnelt-type electron source.

This type of electron source has a very large field near the hole on the Wehnelt cylinder, which acts as a lens and focuses the beam of electrons very close to the cylinder hole. As a consequence, the electron beam emerging from the source has a high angle of divergence, which is undesirable for an ideal electron source. A more detailed description of this source is given by Bonham and Fink (1974). Other types of sources have been reviewed by Simpson and Kuyatt (1963).

b. Collimators

Electrons extracted from the electron source are collimated and focused at the entrance aperture of the energy selector by a collimator with the desired divergence angle (Fig. 1). The simplest collimator generally consists of electrostatic lenses and apertures. Details on the properties and design of electrostatic lenses can be found in the books by Spangenberg (1948), El-Kareh and El-Kareh (1970), and Harting and Read (1976). The collimator between the electron source and the energy selector usually consists of two electrostatic lenses; the first accelerates the electrons and the second decelerates and focuses them at the entrance aperture (real or virtual) of the selector. A field-free region is formed between the two lenses where one or more apertures and beam deflectors are placed to define the window and pupil and to steer the electron beam.

Collimator (2) of Fig. 1 accelerates the electrons exiting from the energy selector and focuses them on to the target. The task of this collimator is slightly more difficult than for collimator (1) because it is required to focus the electron beam on the target for all incident electron energies with the same magnification. This is usually accomplished by adding a variable focal length, or "zoom," electrostatic lens in tandem with the accelerator lens. A zoom lens utilizes three or more cylindrical (or aperture) electrostatic elements to accomplish the variable focusing. Details on these lenses can be found in the publications by Heddle (1969), Read (1970), DiChio et al. (1974), Harting and Read (1976), Chutjian (1979), and the references therein.

c. Energy Selectors

The energy spread in the electron beam, which is focused at the entrance of the energy selector by collimator (1) of Fig. 1, is large (~ 0.5 eV FWHM). For the study of elastic and inelastic scattering of electrons by molecules, a much narrower energy spread is required, and this is achieved by the energy selector. Electrostatic energy selectors can be divided into two categories: the retarding potential difference type and the electrostatic deflection type.

i. Retarding Potential Selector. The retarding potential difference (RPD) energy selector works on the principle that if a parallel beam of electrons of energy E_o enters a retarding field with potential V, then only electrons with energy greater than V will pass through the field. This principle is illustrated in Figs. 4 (a–c). Figure 4a shows the distribution of an energy unselected beam of electrons with peak energy E_o. If the retarding potential V is equal to $E_1 < E_o$, then only the electrons shown by the hatched areas in 4b pass through selector. However, if the retarding potential E_1 is increased by a small amount ΔE (so that $V = E_1 + \Delta E > E_o$), then the electrons shown by the hatched area in 4c will pass through the selector. If the two electron intensities, represented by 4b and 4c are subtracted from each other, then the number of electrons contained in the small energy interval ΔE can be measured. Thus, in a scattering experiment, if the scattered electron signals are recorded for the two cases shown in 4b and 4c and subtracted from each other, then the final data will give results for an electron beam whose apparent energy spread is ΔE. There are several methods of achieving this (Fox et al., 1951; Schulz and Fox, 1957; Schulz, 1964; Chantry, 1969; Knoop et al., 1970; Golden and Zecca, 1971). To obtain the best energy resolution, the retarding field must act in the direction of the velocity of the electrons. Simpson (1961), in a detailed study of RPD

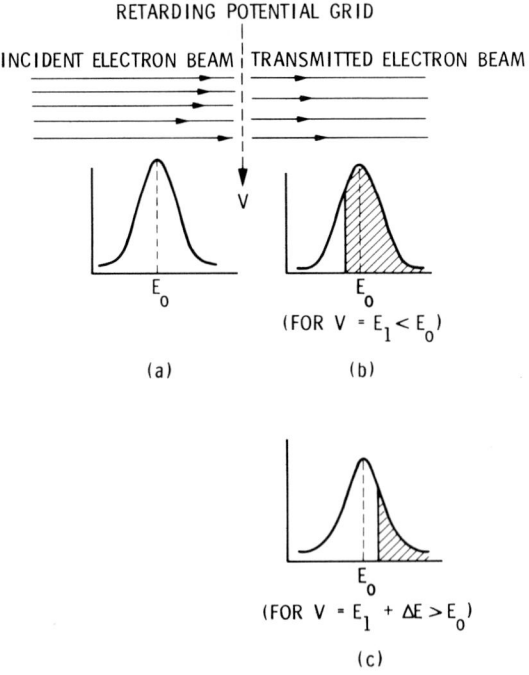

Fig. 4. Principle of RPD method. Hatched areas represent the electron flux of the incident e beam that passes through the retarding potential.

selectors, showed that the nonaxial components of velocity present in a beam of electrons are responsible for the poor resolution of these energy selectors. In general, where a beam is collimated by a magnetic field, a resolution of 0.1 eV can be achieved. Golden and Zecca (1971) reported a resolution of 8 meV with a carefully designed electrostatic collimation system. Heddle (1968) pointed out that this method can introduce spurious structure under certain experimental conditions, and more recently, Wiesemann (1981) discussed the properties of RPD energy selectors.

ii. Electrostatic Deflection Selectors. When high energy resolution is required in electron scattering experiments, electrostatic deflection selectors are most frequently used. Among them the most commonly used ones are (1) parallel plate, (2) cylindrical, (3) hemispherical, and (4) coaxial cone velocity selectors. In an excellent review article, Rudd (1972) discussed the properties of the first three selectors, and the fourth one is described by Brewer *et al.* (1980a and 1980b). Many other types of selectors have been proposed but have not gained wide acceptance.

(1) *Parallel-plate selector.* A schematic diagram of this selector is shown in Fig. 5a. It is the simplest selector from the point of view of construction as well

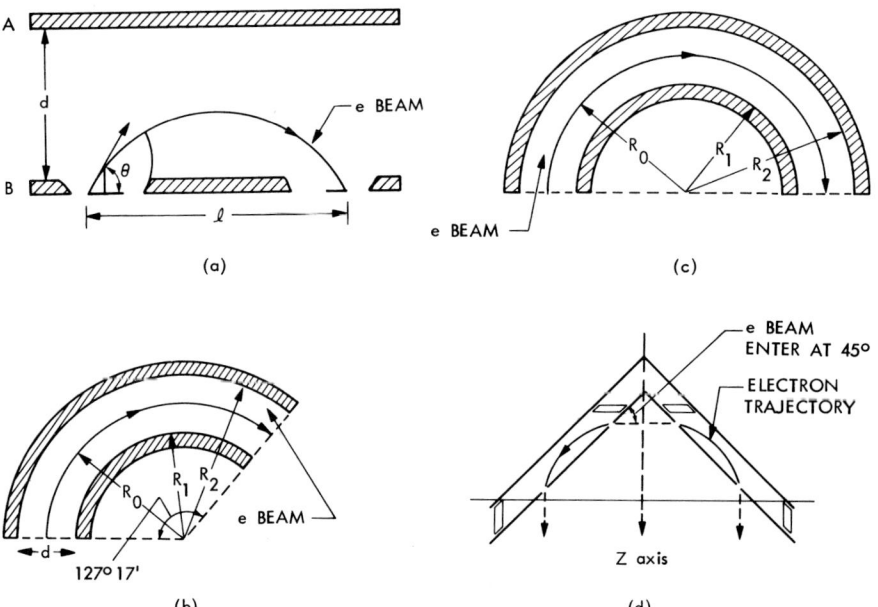

Fig. 5. Various types of electron energy selectors: (a) parallel plate selector, (b) cylindrical selector, (c) spherical selector, and (d) coaxial cone selector.

as for theoretical treatment. It consists of two metal plates (A and B) separated by a distance d. One of these plates (B) has two slits situated at a distance l from each other, and a voltage difference is applied between the two plates. For electrons plate B is kept positive with respect to plate A. A slit in plate B allows the electrons to enter the space between the two plates at an angle θ. Under these conditions, the electrons describe a parabolic trajectory (similar to one traversed by a stone thrown in the gravitational field of earth) and exit through the second slit. For a given velocity of incident electrons, the maximum range occurs when $\theta = 45°$. The energy selection is achieved by narrowing the slits and by varying the voltage difference between the two plates. Further details on this type of selector can be found in the papers by Yarnold and Bolton (1955), Harrower (1955), and Eland and Danby (1968).

(2) *Cylindrical selector.* A schematic diagram of this type of selector is shown in Fig. 5b. It consists of two concentric cylinders separated by a distance d and having a sector angle of $127°17'$. It is based on the principle, initially shown by Hughes and Rojansky (1929), that focusing of a charged-particle beam occurs at every $127°17'$ when it travels in a radial, inverse first-power electric field. The focusing occurs only in one dimension (in a line perpendicular to the length of the slit). In operation, a voltage difference ΔV is applied between the inner and outer cylinders. The polarity of the voltages on the two cylinders depends on the charge of the particle being energy selected. The beam enters through a slit located at the center of the gap and is focused on a slit at the other end of the gap. If $E_o = qV_o$ is the energy of the central beam of electrons entering the selector, then the voltages on the inner V_1 and outer V_2 cylinderical plates for refocusing the beam at the center of exit gap are given by the following expressions:

$$V_2 = 2V_o \ln\left(\frac{2R_2}{R_1 + R_2}\right) \tag{9}$$

and

$$V_1 = 2V_o \ln\left(\frac{2R_1}{R_1 + R_2}\right), \tag{10}$$

where R_1 and R_2 are the radii of the inner and outer cylinders, respectively. The voltage difference ΔV between the two cylindrical plates is given by

$$\Delta V = 2V_o \ln(R_1/R_2). \tag{11}$$

The resolution of the selector depends on the energy of the charged particle being energy selected and the dimensions of the two slits. This selector performs well for producing low-energy (~ 0–10 eV) beams of electrons and has been successfully used for generating electron-impact spectra, for studying resonances, and for energy analysis of negative ions generated in dissociative attachment processes. Typically, it operates at an energy resolution of about 0.04 eV and e-beam currents of about 10^{-8} A.

To improve the performance of this selector, several changes have been made. Marmet and Kerwin (1960) added grids between the cylindrical plates of the selector, which improved the signal-to-background ratio. Matsuda (1961) introduced additional electrodes to improve the focusing properties. Froitzheim et al. (1975) showed that spurious structures occasionally seen in the transmitted spectrum of a 127° cylindrical analyzer are caused by specular reflections of electrons from the inner and outer cylinders towards the exit slit. They found that this structure can be reduced by a factor of 10 if the cylinders are appropriately corrugated.

(3) *Hemispherical selector.* This type of selector has been widely used in electron scattering experiments. It was first introduced by Purcell (1938), who provided the theoretical analysis and built an experimental model. It consists of two concentric hemispheres of radii R_1 and R_2 between which a voltage ΔV is applied (Fig. 5c). This voltage difference produces a $1/r^2$ electrostatic field between the two hemispheres. The charged particles enter the selector at the center of the gap between the two hemispheres and, after suffering a deflection of 180°, are focused at the other end of the gap. The focusing is two-dimensional, with unit magnification, and occurs along a line at the exit of the gap according to the energy of the electron. An aperture placed in this plane then can achieve the energy selection. In practice it is more desirable to put the physical aperture itself in an element of the electrostatic lens system where the kinetic energy of the electrons are higher and, therefore, space-charge and surface field effects are not so serious. This aperture is imaged as a virtual aperture at the exit plane of the selector and performs the energy selection. If $E_o = qV_o$ is the energy of the charged particles in electron volts entering the gap at R_o [$R_o = 1/2(R_1 + R_2)$], then the voltage difference ΔV required between the two hemispheres to refocus the beam at R_o as it leaves the gap is given by

$$\Delta V = V_o \left[\frac{R_2}{R_1} - \frac{R_1}{R_2} \right]. \tag{12}$$

The voltages on the inner V_1 and outer V_2 spheres with respect to the voltage at the center of the gap are given by

$$V_1 = \pm V_o \left[1 - \frac{R_2}{R_1} \right] = \pm 2 V_o \left[1 - \frac{R_o}{R_1} \right] \tag{13}$$

and

$$V_2 = \pm V_o \left[1 - \frac{R_1}{R_2} \right] = \pm 2 V_o \left[1 - \frac{R_o}{R_2} \right]. \tag{14}$$

The ± signs depend on the sign of the charge of the particles, and for electrons the minus sign should be used. Typically, such a selector can provide an energy resolution of 0.040 eV and an e-beam current of about 10^{-8} A.

There are a number of variations that have been employed in the past in the design of hemispherical selectors. For example, since it is difficult to manufacture two concentric hemispheres with great precision, Jost (1979) developed a design of a "simulated" electron energy selector. In this device the spherical equipotentials are produced in the region of electron-beam trajectories by a set of nonspherical electrodes. It was found that the electron optical properties of this device were very similar to those of a spherical condenser. In an another variation two selectors are put in tandem (Register *et al.*, 1980a), which results in a great improvement in the signal-to-background ratio. It is also possible to put the object and image at some distance away from the entrance and exit apertures if the sector angle of a hemispherical analyzer is reduced below 180°. A discussion regarding this can be found in publications by Dempster (1937), Browne *et al.* (1951), and Wollnik (1967).

(4) *Coaxial-cone selector.* It is also possible to select the energy of a charged particle by passing it through an electrostatic field generated between two coaxial cones when a potential difference is applied between them. Brewer *et al.* (1980a, 1980b) made both analytical and experimental determinations of the properties of this type of energy selector. They found that the coaxial-cone geometry has several advantages for applications in scattering experiments. First, the deflection angle is 90°, which offers some geometrical advantages. Second, the focusing of charged particles takes place in a plane perpendicular to the deflection plane, which enhances the transmission properties. Third, the selector is easy to fabricate.

2. *Electron Energy Analyzer*

The function of an electron energy analyzer is to determine the energy distribution of the scattered electrons. The components of a typical electron energy analyzer are shown in Fig. 1. A collimator (3) focuses the scattered electrons at the entrance slit of an energy analyzer and another collimator (4) accelerates and focuses the electrons on the cathode of a charged-particle detector. A channel electron multiplier, or an open-window-type electron multiplier, has generally been used in the past as a particle detector. Each electron generates a pulse as the output signal of the electron multiplier, which is then sent to peripheral electronics. The design considerations for the various components of the analyzer are similar to those of the monochromator as discussed in Section II. A.1. Because of the low signal level in the analyzer, space-charge problems do not arise and some of the difficulties encountered in the electron energy selector are removed. Details on the design parameters for a typical electron energy analyzer can be found in a paper by Read *et al.* (1974).

3. Target

In the past, two different types of target geometries have been used. One is a static gas geometry and the other is a molecular-beam geometry. They will be described in the following paragraphs.

a. Static Gas Target.

A schematic diagram showing a typical experimental arrangement is shown in Fig. 6. The target molecules are confined in a scattering chamber whose pressure and temperature can be accurately measured. There are several types of scattering chambers that have been discussed in detail by Kuyatt (1968). The earliest, and most often used, arrangement (see, e.g., Bromberg, 1975) employs a big vacuum chamber filled with the target gas, within which the monochromator and energy analyzer are placed. Either the electron energy selector, the analyzer, or both can be differentially pumped. In another widely used arrangement, the target gas is contained in a small chamber. The electrons enter and exit this chamber through small apertures. Angular coverage is achieved by making the chamber wall flexible or moving the exit aperture by some mechanism.

The angular distribution of scattered electrons is usually measured by rotating the analyzer around the static target. The scattered current $I_S(E_o, \theta)_n$ associated with excitation process n at a nominal scattering angle θ and nominal impact energy E_o is given by

$$I_S(E_o, \theta)_n = \int_0^\infty \int_{\Delta\Omega} I_o f(E_o) \frac{d^2\sigma(E_o', \theta')_n}{d\theta' \, dE_o'} \rho l \, d\Omega' \, dE_o', \tag{15}$$

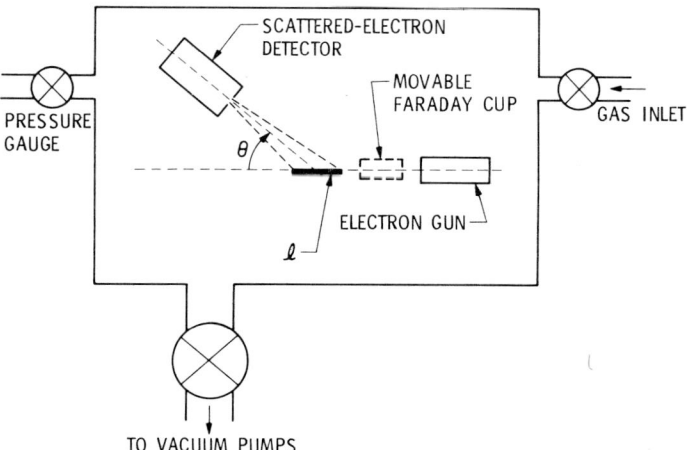

Fig. 6. Schematic diagram of a typical static gas scattering arrangement.

where I_o is the electron-beam current with energy distribution function $f(E_o)$ incident on the target, $d^2\sigma/d\theta'\,dE_o'$ the doubly differential cross section, ρ the target density, $\Delta\Omega$ the solid angle of detection, and l the path length of scattering. The prime refers to the variable actual energy and angle within the range of detection. In Eq. (15) $l\,d\Omega$ is known as the "effective path length." In electron-beam experiments, the elastic cross section can be taken as independent of impact energy over the energy width of the incident electron beam (with the exception of resonance and threshold regions), and the integration over the energy distribution can be performed using

$$\int_0^\infty f(E_o)\,dE_o = 1.$$

Under these assumptions Eq. (15) reduces to

$$I_S(E_o, \theta) = I_o\,\text{DCS}(E_o, \theta)_n \rho \int_{\Delta\Omega} l\,d\Omega, \tag{16}$$

where Eq. (4) defines the instrumentally averaged differential cross section $\text{DCS}(E_o, \theta)_n$ at nominal impact energy impact energy E_o and scattering angle θ. It was assumed in this analysis that the electron beam can be represented by a line with negligible tangential dimensions. For most static gas geometries and for high incident electron-beam energies, the effective path length is found to be proportional to $1/\sin\theta$. In this case, the quantities relating the scattering intensity to the cross section in Eq. (16) can be measured and the absolute value of the differential cross section (DCS) can be obtained. However, if the energy of the incident electron beam is low (less than about 100 eV), the calculation of the effective path length becomes more difficult and uncertain because of the complicated electron trajectory. This question has been discussed by Kuyatt (1968), Vuškovič et al. (1970), Bederson and Kieffer (1971), Bonham and Wellenstein (1973), Shuttleworth et al. (1977), Von Busch (1975), and Brinkmann and Trajmar (1981). In practice, at low impact energies one measures the relative DCS and the absolute value is determined by some type of normalization procedure. The various normalization procedures will be discussed in the following section.

Although the static gas target offers the possibility of determining the absolute values of DCS at high incident electron energies, it has a number of drawbacks. First, for most measurements the electron energy selector and analyzer have to be differentially pumped, which introduces experimental problems. Second, for molecular species that are produced by heating (such as CsCl, LiF) and that can be formed into a beam but are difficult to confine in the scattering chamber, the static gas target geometry is unsuitable. Third, the Doppler broadening of the scattered e-beam caused by the thermal motion of the molecules in a scattering chamber can be large, particularly for light molecules and low incident-electron energies. For example, Celotta and Huebner (1979) estimate that a full width at a half maximum (FWHM) of 0.028 eV will be added to a scattering spectrometer's overall resolution when 20 eV electrons are scattered by a static helium target at 300 K.

b. Molecular Beam Target.

The method of forming a molecular beam depends on the species to be studied. For those that exist in gaseous form at normal temperature, the beam is formed by flowing the gas through an orifice, a hypodermic needle, a capillary array, or a supersonic nozzle. The spatial density and velocity distributions of the molecules in the beam depend on the source structure and conditions. The best collimation and lowest pumping requirements are achieved with capillary arrays or by pulsed supersonic beams.

Figure 7 shows a typical experimental arrangement for producing a beam by flowing a gas through a capillary array. An effusive flow results when the molecular beams are formed by flowing the gas through a tube or capillary array at a very low source pressure (e.g., pressure in chamber F_3 of Fig. 7). For

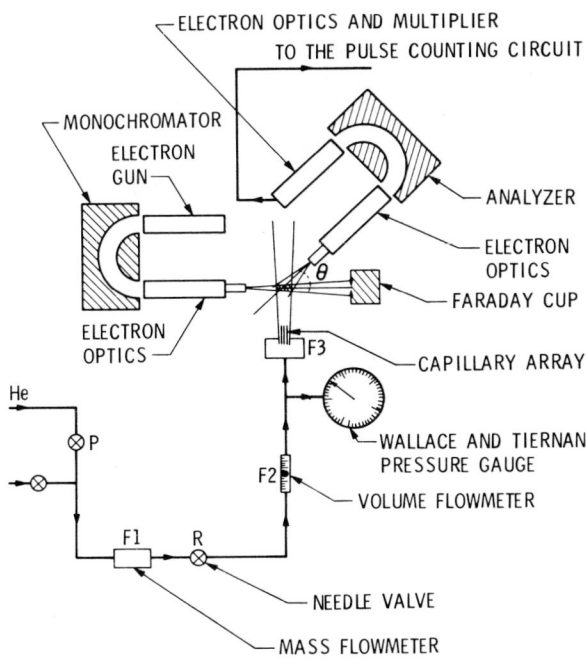

Fig. 7. Schematic diagram of a crossed electron-beam–molecular-beam arrangement.

such a flow ("molecular flow" condition), molecules travel in straight lines from the source orifice without collisions and the distribution of densities and velocities of molecules in the beam can be accurately computed. Unfortunately, the pressures at which molecular flow conditions prevail are very low for most molecules and the beam densities obtained under this condition are not large enough for differential scattering experiments. Higher beam densities ($\sim 10^{13}$ cm^{-3}) are obtained when the pressure behind the tube or capillary array is increased and a "transition flow" regime is reached. In a series of papers Olander and his associates (Jones *et al.* 1969; Olander, 1969; Olander and Kruger, 1970; Olander *et al.* 1970) presented theoretical details on the distributions of molecules emanating from a single capillary and a capillary array under a variety of source conditions. The question of utilizing effusive sources in the "transition flow" region has recently been discussed by Brinkmann and Trajmar (1981) and Register and Trajmar (1984).

Kantrowitz and Gray (1957) first suggested that, if an effusive source is replaced by a supersonic jet, then the beam intensities could be increased by many orders of magnitude. The problem was that to keep a certain level of vacuum in a collision chamber, a high pumping speed was needed, even for a small supersonic jet (Kistiakowsky and Slichter, 1951). Since then, many advances in vacuum techniques and designs of supersonic nozzles have taken place. Today, supersonic beams with intensities of the order of 10^{19} mol s^{-1} cm^{-2} at the collimating orifice can be produced by relatively low pumping speeds (Campargue, 1964). The progress in this field has been reviewed by Aderson *et al.* (1965) and by Anderson (1974). A supersonic beam is formed by raising the source pressure to give a source Knudsen number less than unity (Anderson, 1974). A molecular free-jet then expands into the vacuum and is collimated by a skimmer. The resulting beam is of high intensity and has a narrow velocity distribution. The rapid expansion of the beam also results in cooling and may cause nonequilibrium between the translational, rotational, and vibrational degrees of freedom of a diatomic or polyatomic gas. This effect has either been suggested or used in the past by several workers (Milne and Greene, 1969; Dyke *et al.*, 1972; Sinha *et al.*, 1973; Kukolich *et al.*, 1974; Gallagher and Fenn, 1974; Smalley *et al.*, 1976) to prepare molecules in their lowest rotational and vibrational levels. It has also been shown that the cooling causes cluster formation and may be employed to prepare weakly bound molecular complexes (Hagena (1974); Dyke and Muenter 1974; Harris *et al.*, 1974; Freeman *et al.*, 1974; Janda *et al.*, 1975; Nguyen *et al.*, 1975; Smalley *et al.*, 1976). In addition, the large velocity gradient present in supersonic beams causes molecular alignment. This polarization of the angular momentum of the molecules in the target has been studied by Sinha *et al.* (1974) for the case of alkali dimers produced in a supersonic beam. No electron scattering experiment using pulsed supersonic beams has been reported yet.

Experimental techniques for forming molecular beams of substances that exist either in solid or liquid phase at room temperature are different from those mentioned above. Beams of such substances are generally formed by heating them in a crucible and then allowing the vapor to effuse through a hole or tube from the crucible. The material for the crucible depends on the temperature requirement and on the sample to be studied. For high-temperature work, tantalum crucibles were used by Trajmar et al. (1977). The crucible was heated either by wrapping a resistive wire around it or by electron bombardment, the latter being more suitable for producing high temperatures. One such experimental arrangement is shown in Fig. 8. Ramsey, in his book (1956), described several methods of producing molecular beams by heating various substances in a crucible.

When an electron beam crosses a beam of molecules at 90° ("crossed-beam" geometry), the expression for the scattered intensity of electrons is not as simple as for the static gas target geometry [Eq. (16)]. The most important difference is that the target density varies appreciably over the scattering volume (i.e., the volume where e-beam and molecular beam overlap as viewed by the detector). Brinkmann and Trajmar (1981) derived an expression for the scattered intensity of electrons for a crossed electron-beam–molecular-beam geometry. They have taken into account the effect of the variation of target density, electron-beam flux, detector solid angle from scattering point to scattering point, the dependence of cross section with angle within the solid angle of detection, and the overall "instrument function" including the

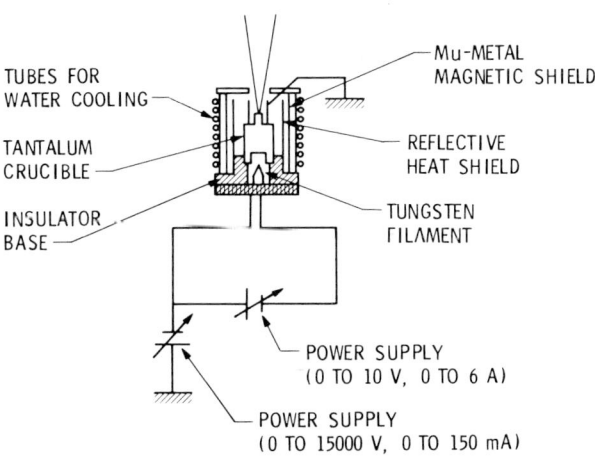

Fig. 8. Schematic diagram showing the arrangement for producing molecular beams of substances found in solid phase at normal temperature and pressure.

detector response and transmission efficiency of the electron optics. The resulting expression is given as

$$I(E_o, \theta) = F(E_o)\text{DCS}(E_o, \theta) \int_V \rho(\mathbf{r}) f_e(\mathbf{r}) F_\sigma(\mathbf{r}, \theta') \Delta\Omega(\mathbf{r}) \, d\mathbf{r} \, d\theta', \quad (17)$$

where $I(E_o, \theta)$ is the number of elastically scattered electrons detected per second for incident electron energy E_o and scattering angle θ, $F(E_o)$ the energy-dependent efficiency factor of the analyzer; $\text{DCS}(E_o, \theta)$ the differential cross section at nominal E_o and θ averaged over the energy resolution of the apparatus, $\rho(\mathbf{r})$ and $f_e(\mathbf{r})$ are the target density and electron flux spatial distribution, respectively, $\Delta\Omega(\mathbf{r})$ is the solid angle of detection for a scattering point located at $\mathbf{r}$, and $F_\sigma(\mathbf{r}, \theta')$ represents the deviation of the scattering cross section, at scattering angle θ', from the average value corresponding to nominal scattering angle θ. Here θ' is the variable scattering angle within the view cone associated with the scattering point located at position $\mathbf{r}$. The integration is performed over the whole scattering volume V and it represents the "effective scattering volume," or the "effective path length."

Brinkmann and Trajmar (1981) carried out a detailed study for commonly used targets and geometries with assumed characteristics. They calculated the effective scattering volume and the numerical values of the volume correction factors, which are the reciprocals of the effective scattering volumes normalized to unity at $\theta = 90°$.

To obtain absolute values of the DCS, one has to calculate the actual values of the volume integral in Eq. (17). This can be done by an iterative procedure if the electron- and target-beam characteristics, scattering geometry, and instrumental functions are known. The problem is that these quantities are not generally known and are difficult to determine and to hold constant throughout an experiment at low electron energies. It is more convenient, therefore, to apply the volume correction factor $F(\theta)$, which is defined as

$$F(\theta) = \frac{\left[\int_V \rho(\mathbf{r}) f_e(\mathbf{r}) F_\sigma(\mathbf{r}, \theta') \Delta\Omega(\mathbf{r}) \, d\theta' \, d\mathbf{r}\right]_{90°}}{\left[\int_V \rho(\mathbf{r}) f_e(\mathbf{r}) F_\sigma(\mathbf{r}, \theta') \Delta\Omega(\mathbf{r}) \, d\theta' \, d\mathbf{r}\right]_\theta} \quad (18)$$

to transform the measured scattering intensities to relative cross sections and then normalize the cross sections by other means or to apply a calibration technique, where the effective scattering volumes cancel out (e.g., relative flow technique).

4. Peripheral Electronic Devices

The main role of these devices is to amplify the electrical pulse generated by the analyzer for each detected electron, count these pulses, store the counts in some type of memory as a function of some experimental variable, control the entire operation of the scattering spectrometer in such a way that it acquires data automatically or semiautomatically, and finally convert the signal spectrum into the desired form (e.g., cross sections). With the recent advancements in the field of digital electronics, such devices are becoming easily accessible.

In a typical elastic scattering cross section measurement, the electrical pulses generated by the electron detector (of the order of 15 to 20 mV) are amplified by a charge-sensitive amplifier to about 3 to 5 V. These pulses are fed to a multichannel scaler (MCS) that records the signal as a function of channel number. The channel number corresponds to the variable parameter and the number of channels employed depends on the details of the experiment. The MCS also controls the electron-impact spectrometer (Fig. 1) and varies the voltage on the monochromator, and/or analyzer, depending on the mode of operation. Each channel is thus set to collect the scattered signal for a particular value of the desired parameter and for a specified amount of dwell time. The data are acquired by repetitive scanning by the MCS (sometimes for several hours for weak scattered signals) to average out drifts in the experimental conditions and to reduce the statistical errors.

B. Methods

The two most common ways of operating an electron-impact spectrometer are the "energy loss" and the "impact energy" modes. In the first case the impact energy and scattering angle are fixed and the signal is recorded as the function of the electron energy loss (ΔE). One thus obtains an "electron energy-loss spectrum." A typical electron energy-loss spectrum is shown in Fig. 9 for which the target molecule is H_2. The peak locations correspond to the energy levels of the molecule, and the intensities are related to the excitation cross sections. In general, the elastic plus rotational spectral features and the nearby vibrational excitation features (which do not appear in this particular spectrum) are well separated from the features corresponding to electronic excitation. The variations of the intensity of these features (area) as a function of scattering angle gives the relative values of differential cross section. In the second mode ("impact energy") the dependence of the intensity of a particular energy-loss feature (that is the cross section) on the impact energy is measured. Here the scattering angle and electron energy loss are constant and the incident electron energy is the variable parameter. Such an example is shown in Fig. 10.

For elastic scattering cross section measurements, one operates in the energy-loss mode, and the energy and angular dependence of the elastic scattering signal constitute the basic information. To generate the elastic cross

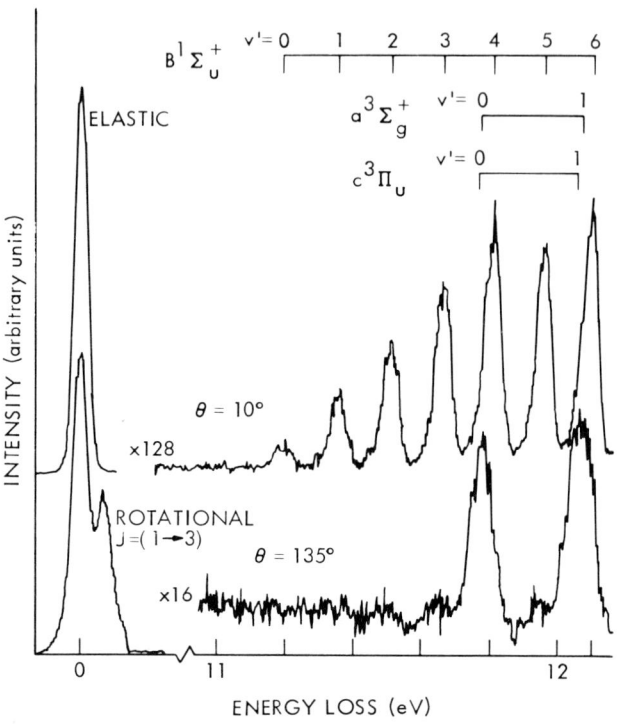

Fig. 9. Typical energy loss spectra showing elastic peak and various inelastic features for H_2 with $E_0 = 20$ eV.

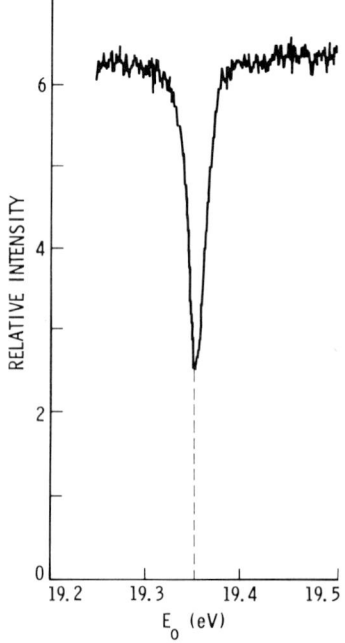

Fig. 10. The 19.35-eV resonance in He at 90° scattering angle.

section data from this information, accurate knowledge of the impact energy and scattering angle is necessary and the relative scattering intensities have to be converted to absolute cross sections. We shall briefly discuss these problems in the following paragraphs.

1. Energy Calibration of the Incident Electron Beam

The incident energy of the electron beam is usually estimated from the potential difference V between the filament and the target region. However, because of contact potentials between the various components of the electron energy selector, patch fields present on the surfaces, the work function of the filament, and the space charge produced by the electron cloud in front of the filament, the actual value of E_0 is different from V. Therefore, the value of E_0 obtained by reading the potential difference V needs to be corrected. This correction can sometimes be as large as 1 eV. Several methods have been employed to estimate the magnitude of this correction, some of which have been reviewed by Kuyatt (1968) and Heddle (1968).

Perhaps the oldest method of energy calibration is the retarding potential method. A retarding potential is applied between the last electrode of the monochromator and a Faraday cup which measures the current. In an ideal case, a plot of the retarding potential against the Faraday cup current will show a very sharp drop from which the energy of the e-beam can be deduced. In practice, the beams of electrons are not strictly monoenergetic and perfectly collimated, and even if we assume that the width of the electron energy distribution is negligibly small, there is always a transverse component of velocity. The retarding potential, being axial, does not have any effect on this component, so a sharp cutoff for the Faraday cup current does not occur and the method will provide only a crude estimate of E_o. Simpson (1961) considered this problem in detail and gave expressions for the energy resolution that can be achieved by this method.

A second method of calibration employs threshold values of inelastic processes such as fluorescence from the excited states of atoms and molecules, generation of metastable species, excitation of vibrational and electronic states of molecules observed by the energy-loss spectroscopy, and the appearance potential for ionization of a gas. In most cases, the thresholds are not very sharply defined and an uncertainty, amounting to several electron volts, can result. Furthermore, it is well known that the excitation functions near threshold have structures that may introduce error into the calibration.

The most widely used method for the energy calibration in the field of electron scattering spectroscopy is to measure the energy positions of well-known resonances observed in electron–atom/molecule collision processes. Generally, the resonance in the 90° elastic differential cross section of helium at 19.367 ± 0.07 eV (Brunt et al., 1977; Fig. 10) has been employed for this purpose. Resonances found in other atoms and molecules were reviewed by Schulz (1973a,b). Because the space- and surface-charge effects depend on the gas used, it is advisable to employ the resonances of the gas under investigation for energy calibration. For gases whose resonance intensities are either weak or their energy locations are not well known, one can mix a small amount of helium with the gas and then use the helium resonance for calibration. Since the energy location of this resonance is higher than the ionization potentials of most gases, great care should be taken to avoid the errors caused by the presence of ions. The effect of ions can be detected by varying the ratio of the gas–helium mixture and extrapolating to zero He concentration.

2. Calibration of the Scattering Angle

It is important to know accurately the value of the scattering angle, particularly if $DCS(E_o, \theta)$ varies rapidly with the angle (e.g., for low scattering angles). The scattering angle is generally varied by rotating the analyzer around the target (Fig. 1) and is measured with respect to the incident electron beam. Usually the turntable on which the analyzer is mounted is mechanically calibrated for the angle. The angular position of the analyzer axis with respect to the axis of the electron energy selector is read either mechanically or electrically. In practice, because of the various mechanical and electrical errors that appear after repeated operation of the spectrometer, the angular calibration may not hold. Furthermore, the mechanical axes of both the electron energy selector and analyzer may not coincide with their optical axes, since the former employ deflectors for steering the electron beam. Since the optical axes may change from experiment to experiment, angular calibration is required each time the spectrometer is operated.

A simple method that has been employed is to determine the symmetry of a strong inelastic scattering feature around the nominal zero angle. The intensity of this signal is recorded as a function of the nominal scattering angle on both sides of the mechanically determined nominal zero-degree scattering angle. The scattered intensity should be symmetrical on both sides of the actual zero degree. Although measurements very near to actual zero degree sometimes are not feasible (because of interference by the incident electron beam), an extrapolation to zero from both sides can be done, and thus, the actual zero degree can be determined.

3. Methods of Normalization of Relative Cross Section Values

As has already been discussed previously, it is possible to measure absolute values of cross sections when a static gas geometry is employed. However, the accuracy in the determination of $DCS(E_o, \theta)$ depends on the energy of the incident electron beam. At high e-beam energies, the scattering geometry can be calculated accurately, but at low energies this calculation becomes difficult and inaccurate. The situation is more complicated in the case of electron-beam–molecule-beam geometry, and the determination of the absolute value of the effective scattering volume is difficult.

In practice, relative values of DCSs are usually determined and subsequently normalized by some method to obtain the absolute values. For electron–molecule scattering there are three reliable methods that can be employed for this purpose: (a) normalization by a theory in the Born limit, (b) normalization by using the relative flow technique, (c) normalization by using integral elastic cross sections. These methods will be described in the following sections.

a. Normalization by Theory in the Born Limit

If the energy of the incident electron beam is large ($E_o \geq 500$ eV for elastic scattering or for rotational, vibrational, and valence shell electron excitation) and the scattering angle is small, then one can calculate the value of $DCS(E_o, \theta)$ by utilizing either a Born or a Born-type approximation. This value can then be used to normalize the relative experimental data. The method employs the determination of the differential excitation function, which is obtained by fixing the scattering angle and recording the elastically scattered electron intensity as a function of energy E_o. Then, according to Eq. (17), this intensity will be directly proportional to $DCS(E_o, \theta)$. If the experimental curve is corrected for the variation of electron-beam current and flux distribution as E_o changes and is recorded from low to high incident electron energies (~ 1–500 eV), then the high-energy end of the excitation function can be normalized to a theoretical value, thereby providing the absolute value to the whole curve. Although the method appears simple, there are several difficulties associated with it. The variation of electron beam with E_o cannot be recorded with great accuracy, and with this variation, the effective scattering volume also changes [i.e., the integral in (Eq. 17)]. In addition, the transmission efficiency of the analyzer system usually changes with the energy of the scattered electron, and the calibration of this efficiency over a wide energy range is difficult.

b. Normalization Using the Relative Flow Technique

This technique was described by Srivastava et al. (1975a) and is capable of giving reliable results for molecular species that are gas phase at normal temperature and pressure. The technique utilizes as a secondary standard the helium elastic scattering cross section which have recently been measured with high accuracy (Register et al., 1980b). The experimental arrangement employed is shown in Fig. 7. The experimental procedure is as follows. First, the gas (M_2), whose cross section is not known, is flowed through a capillary array. The flow rate, pressure behind the capillary array, and angular distribution $I(E_o, \theta, M_2)$ [defined by Eq. (17)] for the elastically scattered electrons are measured for M_2. The gas M_2 is then replaced by He in such a way that the e-beam current and other experimental conditions remain unchanged. The flow rate, the pressure behind the capillary array, and the elastic scattering signal $I(E_o, \theta, \text{He})$ for He are now measured as functions of scattering angle. The ratio of the respective differential cross sections is given by

$$\frac{\text{DCS}(E_o, \theta, M_2)}{\text{DCS}(E_o, \theta, \text{He})} = \frac{I(E_o, \theta, M_2)}{I(E_o, \theta, \text{He})} \frac{\left[\int_V \rho(\mathbf{r}) f_e(\mathbf{r}) \Delta\Omega(\mathbf{r}) dr\right]_{\text{He}}}{\left[\int_V \rho(\mathbf{r}) f_e(\mathbf{r}) \Delta\Omega(\mathbf{r}) dr\right]_{M_2}}. \quad (19)$$

If the two gases flow under conditions of molecular flow (details are given by Srivastava *et al.*, 1975a), the above equation reduces to

$$\frac{\text{DCS}(E_o, \theta, M_2)}{\text{DCS}(E_o, \theta, \text{He})} = \frac{I(E_o, \theta, M_2)}{I(E_o, \theta, \text{He})} \left(\frac{m(\text{He})}{m(M_2)}\right)^{\frac{1}{2}} \frac{F(\text{He})}{F(M_2)}, \quad (20)$$

and

$$\frac{\text{DCS}(E_o, \theta, M_2)}{\text{DCS}(E_o, \theta, \text{He})} = \frac{I(E_o, \theta, M_2)}{I(E_o, \theta, \text{He})} \frac{p(\text{He})}{p(M_2)}, \quad (21)$$

where $m(\text{He})$ and $m(M_2)$ are the atomic and the molecular weights of He and M_2, respectively, $F(\text{He})$ and $F(M_2)$ the flow rates, and $p(\text{He})$ and $p(M_2)$ the pressures behind the capillary array for the two gases. More details on this technique and extension to the nonmolecular flow regime can be found in a review paper by Trajmar and Register (1984). This technique has yielded reliable data for a number of molecules listed in Table I.

c. Normalization Using Integral Elastic Cross Sections

Equation (2) can be written in the following form for the case where only the relative values of the elastic DCSs are known:

$$\sigma(E_o) = 2\pi P(E_o) \int_0^\pi \text{DCS}_o(E_o, \theta_{\text{rel}}) \sin\theta \, d\theta, \quad (22)$$

where $\sigma(E_o)$ is the absolute integral cross section, $\text{DCS}(E_o, \theta)_{\text{rel}}$ the relative differential cross section, and $P(E_o)$ a proportionality constant that relates the relative values of the DCSs to their absolute values by

$$\text{DCS}(E_o, \theta) = P(E_o)[\text{DCS}(E_o, \theta)]_{\text{rel}}. \quad (23)$$

If experimental values of relative cross sections between 0 and 180° are known and if the absolute values of $\sigma(E_o)$ are also available, then from Eq. (22) one can determine the value of $P(E_o)$. This value of $P(E_o)$ can then be used in Eq. (23) to determine the absolute value of the elastic DCSs. However, experimental (geometric) limitations do not permit the measurement of $[DCS(E_o, \theta)]_{rel}$ over the *entire* angular range. For example, in most experiments carried out in our laboratory, elastic scattering measurements can be made only for angles between 10 and 135°. For scattering angles less than about 10° the incident electron beam begins to interfere with the elastic scattered signal, and for angles greater than 135°, mechanical limitations hinder the measurements. Therefore, in the low and high angular regions the relative values of the DCSs are obtained by some type of extrapolation method. Usually, if a theoretical calculation is found to agree with the shape of the experimentally determined angular distribution, then the shape predicted by the theory in the low and high angular region can be used for extrapolation of the experimental data. One also has to consider that effective scattering volume corrections may have to be applied to the measured scattering signals to obtain the relative DCS.

The absolute values of $\sigma(E_o)$ may be available from the theoretical calculations or can be obtained from *total* electron scattering cross sections $\sigma_T(E_o)$. Total scattering cross sections can be accurately determined by e-beam attenuation (or time-of-flight or recoil techniques) (Bederson and Kieffer, 1971). In this method, a beam of electrons is passed through the gas under study and its attenuation is measured as a function of energy. From the knowledge of the attenuation, the *total* scattering cross section $\sigma_T(E_o)$ can be determined. At low impact energies, where the vibrational excitation channels are not open, the total scattering cross section is identical to the "vibrationally elastic" integral cross section. At higher energies the total scattering cross section is related to $\sigma(E_o)$ by the relation

$$\sigma(E_o) = \sigma_T(E_o) - \sigma_{ion}(E_o) - \sigma_{exc}(E_o), \qquad (24)$$

where $\sigma_{ion}(E_o)$ and $\sigma_{exc}(E_o)$ are the cross sections for ionizations and excitation of the molecule, respectively. Experimental values of $\sigma_{ion}(E_o)$ are available for most molecules with acceptable accuracy, and $\sigma_{exc}(E_o)$ can sometimes be estimated. In principle then, $\sigma(E_o)$ can be obtained from Eq. (24), but in practice, $\sigma_{exc}(E_o)$ is currently known for only a very few molecules and only over a limited energy range.

C. Summary of Experimental Data

1. Tabulation of Data

Tables I and II summarize the experimental elastic scattering and rotational excitation cross sections for electron–molecule collisions published since 1970.

(Text continues on p. 37)

TABLE I
A summary of elastic cross section measurements[a,b,c]

Reference	Energy range (eV) Angular range (deg)	Method of Normalization	Type of Data
H_2			
Weingartshofer et al. (1970)	11-13 20°, 50° and 110°	to σ_T	DCS
Linder and Schmidt (1971a)	0.6-10.8 20°-120°	to σ_T	DCS, σ (Rot. Res.)
Lloyd et al. (1974)	30, 50, 100 and 200 20-130°	to Born theory at 60°	DCS
Srivastava et al. (1975a)	3-75 20-135°	to He elastic	DCS, σ, σ^M
Fink et al. (1975a)	100-1000 3-130°	to theory at 90°	DCS, σ
van Wingerden et al. (1977)	100-2000 5-50°	to N_2 elastic	DCS, σ
Shyn and Sharp (1981)	2-200 6-156°	to He elastic	DCS, σ, σ^M

1 Elastic Scattering of Electrons by Molecules

N_2

Reference	Energy / Angle	Notes	Data
Bromberg (1970)	300, 400, 500 / 2–110°	abs. meas.	DCS
Truhlar et al. (1972a)	20 / 20–85°	to σ_T	DCS, σ
Srivastava et al. (1976a)	5–75 / 20–135°	to He elastic	DCS, σ, σ^M
DuBois and Rudd (1976)	20–800 / 2–150°	abs. meas.	DCS, σ
Jansen et al. (1976)	100–3000 / 5–50°	abs. meas.	DCS
Shyn and Carignan (1980)	1.5–400 / 6–168°	to calc.	DCS, σ, σ^M
Jung et al. (1982)	2.22 and 2.47 / 15–105	to Ne and Ar elastic	DCS (Rot. Unfolded)
Tanaka et al. (1980)	5 / 30–130°		DCS (Rot. Unfolded)

O_2

Reference	Energy / Angle	Notes	Data
Linder and Schmidt (1971b)	0.74–4 / 20–110°	to σ_T	DCS, σ
Trajmar et al. (1971)	4–45 / 10–90°	to σ_T	DCS, σ
Bromberg (1974a)	300, 400 and 500 / 2–40°	abs. meas.	DCS
Wakiya (1978a,b)	20–500 / 5–130°	to σ_T and to DCS (30°) of Bromberg (1974a)	DCS, σ

(table continues)

TABLE I (cont.)

Reference	Energy range (eV) Angular range (deg)	Method of Normalization	Type of Data
CO			
Bromberg (1970)	300, 400 and 500 2-110°	abs. meas.	DCS
Truhlar et al. (1972b)	6-80 15-85°	to theory	DCS, σ
Truhlar et al. (1972a)	20 20-85°	to σ	DCS, σ
DuBois and Rudd (1976)	200, 500 and 800 3-150°	abs. meas.	DCS
Tanaka et al. (1978)	3-100	to He	DCS, σ, σ^M
Jung et al. (1982)	1.8 and 2.1 15°-120°	to Ne and Ar elastic	DCS (Rot. Unfolded)
NO			
Kubo et al. (1981)	3-30 30-140°	to He elastic	DCS, σ, σ^M
HF			
Rohr and Linder (1976)	2 80°	to HCl	DCS

HCl			
Rohr and Linder (1976)	3 20, 60 and 100°	to σ_T	DCS
HBr			
Rohr (1977, 1978a)	0.2–10 20–120°	to He elastic	DCS, σ
LiF			
Vuskovic et al. (1978)	5.4 and 20 20–130°	to theory	DCS, σ, σ^M
KI			
Slater et al. (1974b)	0.81–15.7	recoil method	DCS, σ, σ^M
Rudge et al. (1976)	6.7, 15.7 and 60 15–130°	to theory	DCS, σ, σ^M
CsF			
Slater et al. (1972, 1974a)	0.69–6.81	recoil method	DCS, σ, σ^M

(*table continues*)

Table I (cont.)

Reference	Energy range (eV) Angular range (deg)	Method of Normalization	Type of Data
CsCℓ			
Becker et al. (1974)	0.47–15.7	recoil method	DCS, σ, σ^M
Vušković and Srivastava (1981)	5 and 20 10–120°	to theory	DCS, σ, σ^M
H$_2$O			
Jung et al. (1982)	2.14 and 6.0 15–105°	to Ar and Ne elastic	DCS (Rot. Unfolded)
Lassettre and White (1974)	500 and 516 4–77.5°	to Bromberg (1974b)	DCS
Nishimura (1979)	30–90 10–135°	to He elastic	DCS, σ, σ^M
Bromberg (1974b)	500 4–30°	abs. meas.	DCS
H$_2$S			
Rohr (1978b)	0.3–10 10–120°	to He elastic	DCS
HCN			
Srivastava et al. (1978)	3–50 20–130°	to He elastic	DCS, σ, σ^M

1 Elastic Scattering of Electrons by Molecules   33

CO_2

Danner (1970)	1.6-10 5-145°	to DCS sum of Ramsauer and Kollath (1932)	DCS, σ
Bromberg (1974a)	300, 400 and 500 2-45°	abs. meas.	DCS
Shyn et al. (1978)	3-90 -150°	to theory	DCS, σ, σ^M
Register et al. (1980a)	4-50 15-140°	to He elastic	DCS, σ, σ^M

N_2O

Kubo et al. (1981)	5, 10 and 20 30-140°	to He elastic	DCS, σ, σ^M

SO_2

Orient et al. (1982)	12-200 15-145°	to He elastic	DCS, σ, σ^M

NH_3

Harshbarger et al. (1979)	300, 400 and 500 2-10°	abs. meas.	DCS
Bennani et al. (1979)	35,000 3 x 10^{-3}-7°	to theory	DCS

As_4

Daimon et al. (1979)	70 and 500 3-138°	to theory	DCS

(table continues)

Table I (cont.)

Reference	Energy range (eV) Angular range (deg)	Method of Normalization	Type of Data
CH_4			
Rohr (1979a and 1980)	0.1-10 10-120°	to σ_T	DCS
Newell et al. (1979)	7.5-50 20-130°	to He elastic	DCS
Sohn et al. (1983)	0.1-1.8 35-110°	to He elastic	DCS
Tanaka et al. (1982)	3-20 30-140°	to He elastic	DCS, σ, σ^M
C_2H_2			
Fink et al. (1975b)	100-1000 3-130°	to theory	DCS
C_2H_4			
Fink et al. (1975b)	100-1000 3-130°	to theory	DCS
C_2H_6			
Fink et al. (1975b)	100-1000 3-130°	to theory	DCS
Matsunaga et al. (1981)	3-20 30-140°	to He elastic	DCS

1 Elastic Scattering of Electrons by Molecules

Molecule	Reference	Energy; Angle	Normalization	Type
C_3H_8	Matsunaga et al. (1981)	3–20 30–140°	to He elastic	DCS
CCl_2F_2	Rohr (1979b)	1 eV 20°	to He elastic	DCS
CCl_3F	Rohr (1979b)	2 eV; 20° 4 eV; 60°	to He elastic	DCS
CCl_4	Daimon et al. (1979)	70, 300 and 500 5–135°	to theory	DCS
SF_6	Srivastava et al. (1976b)	5–75 20–135°	to He elastic	DCS, σ, σ^M
UF_6	Srivastava et al. (1976b)	10–75 20–135°	to He elastic	DCS, σ, σ^M

[a] *Elastic* here means elastic plus rotational excitation and occasionally may also include unresolved vibrational excitation. In cases when the rotational excitation was separated from elastic scattering, this is indicated in the last column.

[b] Momentum transfer cross sections are included only when they have been deduced from beam–beam differential cross section measurements.

[c] The symbols DCS, σ, σ^M, and σ_T refer to differential, integral elastic, elastic momentum transfer, and total scattering cross sections, respectively.

TABLE II
Rotational excitation cross section measurements[a]

Reference	Energy range (eV) Angular range (deg)	Method of Normalization	Type of Data
H_2			
Linder and Schmidt (1971o)	1.5–10 20–120°	to H_2 elastic	$j = 1 \to 3$ DCS, σ
Wong and Schulz (1974)	4.5 20–100°	to H_2 elastic	$j = 1 \to 3$ DCS
Srivastava et al. (1975b)	40 eV; 10–135° 3–40 eV; 20° 15–100 eV; 115°	to H_2 elastic	$j = 1 \to 3$
N_2			
Wong and Dube (1978)	1.1–3.9 60°	to He elastic	$\Delta j = 4$ DCS
Tanaka et al. (1980)	5 30–130°	N_2 elastic	$j = 0 \to 2, 4$ DCS
Jung et al. (1982)	2.22 and 2.47 15–105°	to Ar and Ne elastic	$\Delta j = +2, +4$; DCS $\Delta j = \pm 2, \pm 4$; σ
CO			
Jung et al. (1982)	1.8 and 2.1 eV 15–120°	to Ar and Ne elastic	$\Delta j = +1, +2, +3, +4$ DCS
H_2O			
Jung et al. (1982)	2.14 and 6.0 15–105°	to Ar and Ne elastic	$\Delta j = \pm 1$, DCS

[a] The rotational structure in the case of H_2 was experimentally resolved. In the other cases the combined elastic and

1 Elastic Scattering of Electrons by Molecules   37

Reference to cross section data published before 1970 can be found in NBS Publication 426 (Kieffer, 1976). In the tables given here, the cross section information is grouped according to individual molecules, and the references are presented along with the energy and angular ranges covered, the method of normalization, and the type of data. As mentioned previously, for most molecules it is not possible to resolve rotational structure from the elastic scattering feature. Therefore, unless specified by "Rot. Res." or "Rot. Unfolded," the data given by the references cited here include the contribution of rotational excitation to the elastic scattering (i.e., they are "vibrationally elastic" cross sections).

2. *Systematics in the Data*

Figure 11 shows DCSs for vibrationally elastic scattering of electrons by eight different molecules and three different rare gas atoms to illustrate the similarities and differences in the magnitudes and shapes in their scattering characteristics. As shown in Fig. 11a, the DCSs for CO_2, N_2, CO, O_2, and H_2 at 10 eV (for $\theta \lesssim 100°$) are strikingly similar in both shape and magnitude but differ from those for helium and neon at 12 eV. The DCS for scattering by H_2 demonstrates the weakest scattering into the backward direction at this incident energy of all the diatomic molecules for which data exists. In Fig. 11b the DCSs for vibrationally elastic scattering by CH_4, SF_6, and UF_6 are shown along with that for Kr and H_2 (to serve as a reference curve), again for 10 eV incident energy. Note the difference in the shapes of the DCSs for the "complex" targets (Kr, CH_4, SF_6, UF_6) from those for the simple targets (diatomics and CO_2). The double minima (and deep single minimum) shown in Fig. 11b appear to be characteristic of targets with "complex" electronic structure and are present for atomic (e.g., Kr) as well as molecular targets. Although it has been suggested that these multiple minima are associated with the electronic shell structure of the target, no quantitative explanation has been reported yet to explain these multiple minima. Inspection of the data in Fig. 11 also shows a *weak* correlation between the size of the target and the magnitude of the DCS in the forward scattering direction.

The electron energy and scattering angle dependence of vibrationally elastic electron scattering is illustrated in Fig. 12 for the case of N_2. This figure (Siegel *et al.*, 1978) displays the following important characteristics of elastic scattering by N_2, the qualitative aspects of which are generally true for all atoms and molecules. At very low incident electron energy ($E_o \lesssim 0.01$ eV), the DCS for elastic scattering is nearly isotropic because the centrifugal barrier excludes all but S-wave ($l = 0$) scattering. As the incident electron energy increases, the DCS generally becomes more and more forward peaked, which is a manifestation of the fact that more and more partial waves contribute to the scattering process and the fact that the electron transfers less and less

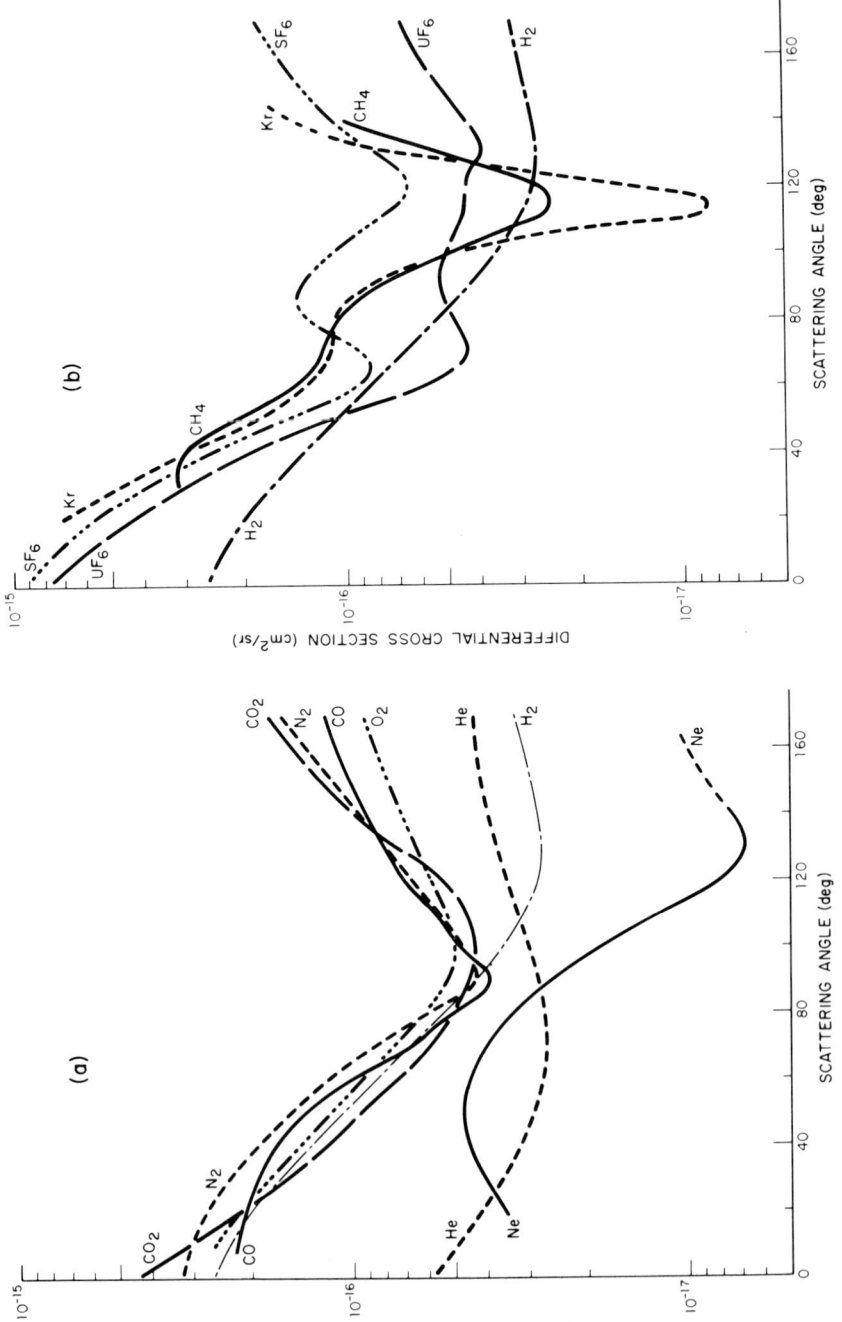

Fig. 11. DCS for elastic scattering of electrons: (a) for H_2, CO_2, N_2, CO, O_2 (10 eV) and He and Ne (12 eV). [The data for Neon and Helium were taken from Register and Trajmar (1984) and from Register et al. (1980b), respectively, and the data for the other molecules were taken from Trajmar et al. (1983).]; (b) for H_2, Kr, CH_4, SF_6, and UF_6 (all 10 eV). [The data for scattering by Kr was taken from Srivastava et al. (1981) and the other data from the review by Trajmar et al. (1983).]

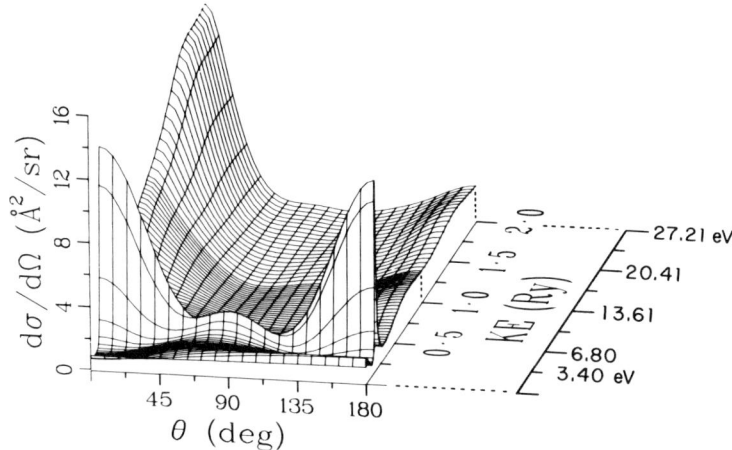

Fig. 12. Three-dimensional plot of the DCS for elastic scattering of electrons by N_2 as a function of the incident electron energy 0.001–2.0 Ry as well as the scattering angle (Siegel et al., 1978). The energy axis is shown in both Rydberg atomic units (13.6058 eV/Ry) and electron volts.

momentum to the molecule. N_2 has an interesting d-wave shape resonance centered near 2.4 eV which dramatically alters the general characteristics just described and which stands out in this figure. In the case of resonances, the shape of the DCS is determined by other physical processes (e.g., see Chang, 1977, 1983) and does not follow the qualitative arguments that apply in the nonresonant region.

Figure 13 displays the incident-electron energy dependence of the measured integral vibrationally elastic scattering cross section for a variety of molecules ranging from H_2 to SF_6. The purpose of this figure is to illustrate the substantial variation in both the magnitude and electron energy dependence of the integral vibrationally elastic scattering cross sections. Note that the integral cross sections shown in Fig. 13 are always equal to or less than the *total* scattering cross section for the same molecule because the latter contains contributions from inelastic scattering.

As a final example of the general behavior of "elastic" scattering cross sections, we show in Fig. 14 the energy-weighted integral momentum transfer cross section for "elastic" scattering by polar molecules as a function of the dipole moment (from Collins and Norcross, 1978). The purpose of this figure is to emphasize the very large cross sections for electron scattering by polar molecules for electron energies less than 1 eV. As a reference point for comparison, the corresponding cross sections for H_2 and N_2 are about $100 \text{ eV} - a_0^2$, or less, for $E_0 \lesssim 1$ eV.

Detailed comparisons among the various theoretical results and selected experimental data will be made in Section IV.

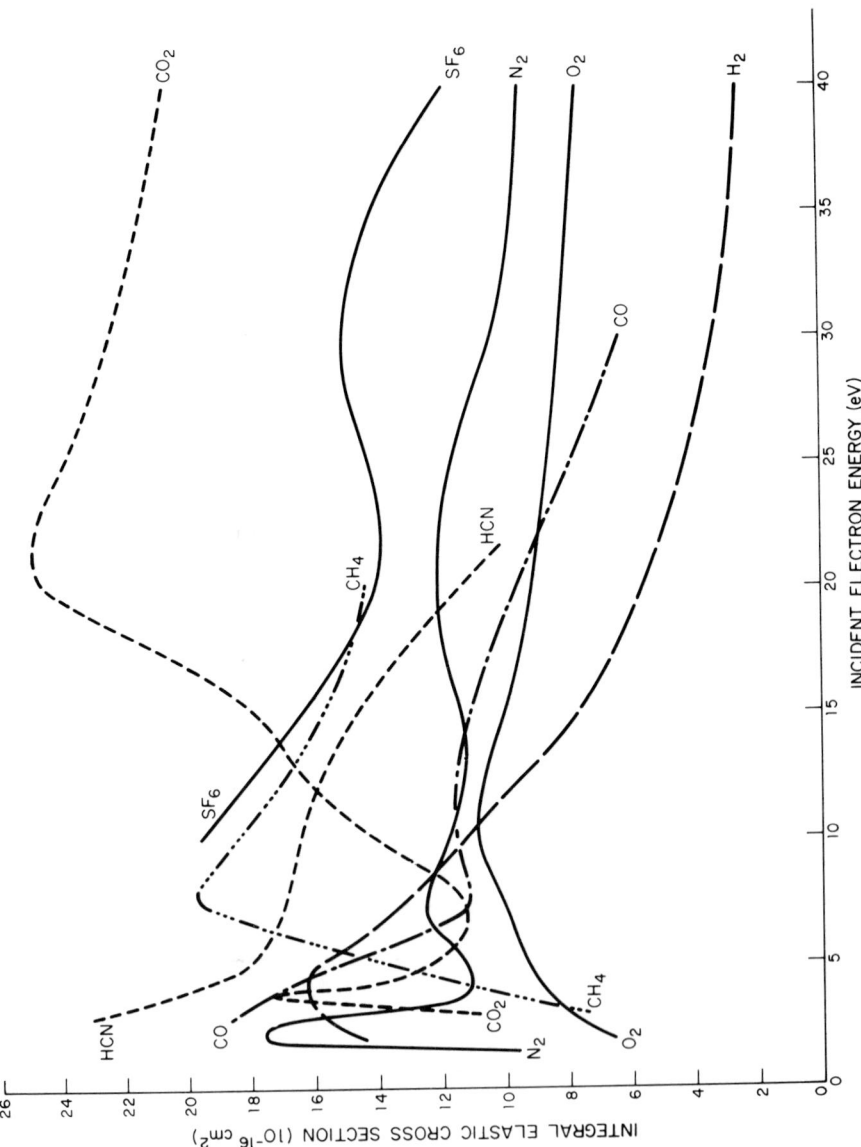

Fig. 13. Integral elastic scattering cross section as a function of incident electron energy for a variety of molecules. [All the data shown here were taken from the review of Trajmar et al. (1983).]

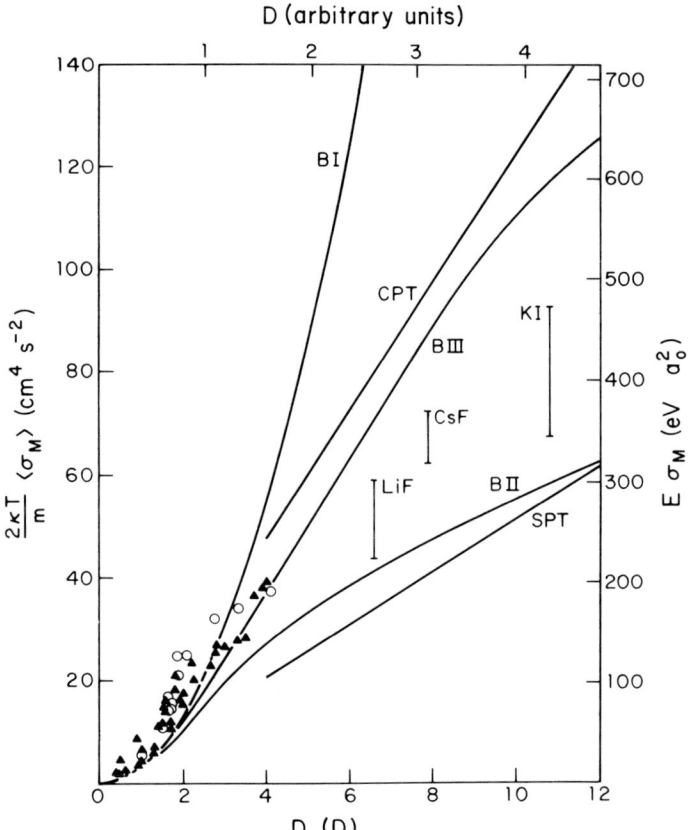

Fig. 14. Energy-weighted integral momentum transfer cross section (right ordinate) for electron scattering as a function of the dipole moment (from Collins and Norcross, 1978). The data are from thermal energy electron swarm measurements and the various theoretical curves are discussed in detail by Collins and Norcross (1978).

III. Theoretical Methods

A. Historical Perspective

1. Fixed-Nuclei Model*

The first electron–molecule scattering calculations were reported by Sir Harrie Massey and co-workers in the 1930s. In his pioneering work on the elastic electron scattering by molecular hydrogen, Massey (1930) stated that

---

* We make a distinction between *approximations* to exact theories and *models* that replace physical reality and are more easily solvable than the complete problem.

In an actual case the axes of the molecules may be taken as in random directions, the hydrogen molecule having no permanent dipole moment and so being only slightly oriented by the electric field of the incident beam. This permits us to average the scattering over all orientation to obtain experimental conditions.

Thus Massey assumed that the molecular axes remained fixed during the electron collision process and then averaged the cross section over random orientations of these axes, with the internuclear distance fixed. This intuitive approach is called the "fixed-nuclei" (FN) model of electron–molecule scattering. [See Table III for a description of the terms and a definition of the symbols used in this chapter.] Massey (1930) and Massey and Mohr (1932a) used the Born approximation (Born, 1926) within the FN model for $e^- - H_2$

TABLE III

Descriptions of Terms and Definitions of Symbols

Term	Description
AN approximation	Scattering amplitude and cross section information obtained in the adiabatic-nuclei (AN) approximation.
FN model	Scattering amplitude and cross section information obtained in the fixed-nuclei (FN) model for which the *cross section* is obtained for fixed nuclear orientation and *then* averaged over all orientations.
CN approximation	The electron–molecule Schrödinger equation is solved with the clamped nuclei (CN) Hamiltonian which contains only the electron kinetic operator, electron–electron, and electron–nuclei interaction terms (*no nuclear kinetic energy and nucleus–nucleus interaction term included*).
CC approximation	An approximation scheme in which the "close-coupling" (CC) equations are solved with a potential (static, model-exchange, polarization, etc.) obtained using an eigenstate expansion, generally leading to a system of integro-differential equations coupling the eigenstates.
LAB frame	The molecule as viewed from a coordinate system (frame) fixed in the laboratory (LAB).
BODY frame	The scattering process is described with the coordinate system (frame) attached to the nuclei (BODY).
S	Solution obtained using a static (S) potential.
SME	Solution obtained using the static and model exchange (SME) (i.e., local exchange) potential.
SP	Solution obtained using a potential with static and polarization (SP) terms.
SMEP	Solution obtained using a potential with static plus model exchange and polarization (SMEP) terms.
SE	Solution obtained using the exact static exchange (SE) potential (containing a nonlocal exchange potential).
SEP	Solution obtained using the (exact) static exchange plus (model) polarization (SEP) potential.

1 Elastic Scattering of Electrons by Molecules

TABLE III (cont.)

Term	Description
SOE	Solution obtained using a model where the static potential is used and an orthogonality condition between the scattering orbital and the occupied orbitals is used to simulate the effect of exchange (SOE).
SOEP	Solution obtained for the static plus polarization (SOEP) problem using the orthogonality restriction.
DCS_E, σ_E	The elastic differential (integral) cross section calculated according to the AN approximation.
DCS_R, σ_R	The rotational excitation differential (integral) cross section calculated from the CN scattering amplitude according to the AN approximation.

Symbols	Definition
σ_E	Integral cross section for elastic scattering[a]
σ_E^M	Momentum transfer cross section for elastic scattering[a]
DCS_E	Differential cross section for elastic scattering[a]
σ_R	Integral cross section for rotational excitation[b]
σ_R^M	Momentum transfer cross section for rotational excitation[b]
DCS_R	Differential cross section for rotational excitation
σ_{ER}	Integral cross section for elastic scattering and rotational excitation (i.e., for vibrationally elastic scattering)[c]
σ_{ER}^M	Momentum transfer cross section for elastic scattering and rotational excitation (i.e., for vibrationally elastic scattering)
DCS_{ER}	Differential cross section for elastic scattering and rotational excitation (i.e., for vibrationally elastic scattering)
$\langle \sigma \rangle$	Integral cross section calculated in the FN model[d]
$\langle DCS \rangle$	Differential cross section calculated in the FN model[d]
$\langle \sigma^M \rangle$	Momentum transfer cross section calculated in the FN model[b]

[a] Elastic scattering cross sections may refer to the $\sigma_{j \to j}$ integral cross section with $j = 0, 1, 2 \ldots$ or a weighted average of these cross sections:

$$\sigma_E = \sum_j P_j \sigma_{j \to j} \quad \text{with} \quad \sum_j P_j = 1.$$

For the specific definition used see the quoted works.

[b] Cross sections for rotational excitation may refer to one $\sigma_{j \to j'}$ cross section or for a whole series of them with $j = 0, 1, 2, \ldots$ and $j' = 0, 1, 2, \ldots$.

[c] Cross sections for elastic and rotational excitation may be the simple sum of cross sections, e.g., $\sigma_{ER} = \sigma_{11} + \sigma_{13}$ or a weighted sum, e.g.,

$$\sigma_{ER} = 0.88(\sigma_{11} + \sigma_{13}) + 0.12(\sigma_{33} + \sigma_{31}).$$

For the specific definition used of σ_{ER} see the quoted works.

[d] This cross section is conveniently interpreted as the combined elastic and rotational excitation cross section (see text for details).

elastic scattering as did Massey and Mohr (1932a) and Bullard and Massey (1931) for e^--N_2 scattering. The calculations were performed using a coordinate system attached to the nuclei, called the "BODY frame," and the resulting cross sections were then averaged over all orientations of the nuclear axes in the space-fixed, laboratory frame ("LAB frame"). They obtained good agreement with experiment for the elastic DCS at high incident-electron energies (energies greater than 100 eV) but poor agreement in the low-energy region. Subsequently, Massey and co-workers (Massey and Mohr, 1931, 1932b, 1934) identified the fundamental physical effects required to explain the elastic scattering process at all incident electron energies. (Here we discuss only *nonresonant* electron scattering processes.)

Those physical effects, which are not included in the Born approximation, are

(1) *distortion*, which is the effect of the electrostatic field of the target molecule upon the incident electron wave, distorting it from its asymptotic plane-wave character;

(2) *exchange*, which refers to the fact that the incident and target electrons are truly indistinguishable during the collision process and requires that the total (electron plus molecule) wave function be antisymmetric with respect to the interchange of the coordinates of the incident electron with those of the target;

(3) *polarization*, which is a "backcoupling effect" upon the incident electron wave caused by the distortion of the electron charge distribution of the target, which in turn was caused by the incident electron.

In a series of papers, Massey and co-workers (Massey and Mohr, 1931, 1932b, 1934; Bullard and Massey, 1931) attempted to demonstrate that these effects are capable of correctly explaining elastic electron scattering by molecules. In particular, Massey and Mohr (1934) obtained good agreement for e^--H_2 elastic scattering when the distortion and polarization effects were incorporated (in a very approximate way) into the calculation and showed that polarization is particularly important for small-angle scattering. Because of computational limitations, these authors could not provide detailed numerical proof of the importance of these physical effects at that time and it had to wait for the relatively recent development of high-speed computers to give a complete proof of the fundamental validity of their concepts. Stier (1932) and Fisk (1936, 1937), the first investigators (besides Massey and co-workers) to apply the FN model, discussed e^--N_2 and $e^--N_2,-O_2,-H_2$, and $-Cl_2$ scattering, respectively, assuming a *semiempirical static potential* and solving this scattering model exactly. Subsequently Nagahara (1953, 1954) improved upon the calculations using an *ab initio* static potential in his work on e^--H_2 scatterings. Massey and Ridely (1956), who first included exchange into the treatment, demonstrated its great significance for low-energy scattering.

2. LAB-Frame Close-Coupling (LF–CC) Approach

Quite interestingly, in his work on electron–polar molecule scattering (representing the molecule by a point dipole), Massey (1932) introduced the rotation of the molecule (i.e., of the dipole) explicitly into the treatment viewed from a laboratory-fixed coordinate system (called "LAB frame"). By expanding the complete scattering function in terms of the quantum-mechanical eigenstates of the rotor, he obtained a coupled system of partial differential equations for the coefficient functions of the expansion (also called channel functions), which depend on the coordinates of the scattered electron. By observing that there is no first-order distortion for the incident electron wave in the case of a dipole field, he concluded that the *rotational excitation of weakly polar molecules* can be treated at all energies in the Born approximation. This is the result of the anisotropic and long-range nature of the dipole interaction. This concept was extended by Gerjuoy and Stein (1955a, b) to low-energy rotational excitation by homonuclear diatomic molecules. These latter authors argued that, at low energies, the long-range quadrupole potential dominates and the various partial waves do not become distorted (except the s wave in the internal region, which is, however, not too important for rotational excitation). Massey's general treatment of the "rotating dipole" was further extended and generalized in the classic work of Arthurs and Dalgarno (1960), who considered particle scattering by a rigid rotor with an arbitrary interaction potential as a model for electron–molecule scattering. By expanding the scattering functions in terms of total angular momentum eigenfunctions of the coupled particle–rigid rotor system, coupled differential equations were obtained for the "coefficient functions" that depend on the *radial* coordinate of the scattered electron. These coupled differential equations are referred to as the rotational *close-coupling* (CC) equations in which individual equations, representing rotational channels, are connected by various coupling terms. In practice, only a finite number of equations are retained, and this is then called the rotational *CC approximation*. The rotational CC equation representation of Massey (1932) and of Arthurs and Dalgarno (1960) was subsequently generalized to include the effects of exchange and polarization and to allow for vibrational excitation. In the most general form, the $(N + 1)$-electron wave function is expanded in terms of the rotational, vibrational, and electronic states of the N-electron target molecular wave functions. Within this rotational, vibrational, electronic CC representation, it can be shown that the close coupling of electronic channels can be replaced exactly by a nonlocal and energy-dependent potential (which, when added to the static and exchange potentials, the sum is usually referred to as an "optical potential") that enters the rotational–vibrational CC equations and further complicates them. This "optical potential" usually consists of static, exchange, polarization, and absorption terms. The nonlocal exchange potential is sometimes simulated by orthogonalizing the scattering electron orbital to the bound orbitals with equal symmetry or substituted by a local exchange potential. The energy-dependent nonlocal polarization potential is usually replaced by a *local adiabatic* polarization potential, sometimes with local *nonadiabataic* corrections. In the weak-coupling limit (i.e., when coupling between all channels except two is ignored), the *distorted-wave* approximation (DWA) is obtained [see, e.g., Mott and Massey, (1965), pp. 349–351]. With the advent of high-speed computers, first the DWA and later the solution of the rotational CC-equations (with approximations for the exchange and polarization potentials) became practical and results appeared in the literature for realistic systems. The first reported solution of the rotational CC equations was made by Lane and Geltman (1967) on electron–molecular hydrogen elastic scattering and rotational excitation. In this work, the effect of exchange was simulated by a local potential. Shortly thereafter, DWA results were reported by Ardill and Davison (1968) with the exact exchange included. The work of Henry and Lane (1969) treated exchange and adiabatic polarization in an *ab initio* way within the rotational CC representation and their results, which are in excellent agreement with experiment, show clearly the great importance of both exchange and polarization in low-energy electron–molecule scattering.

3. *The Adiabatic-Nuclei Approximation*

Subsequently, it was realized that cross sections for rotational and vibrational excitation can be obtained from a generalization of the FN model by allowing for nuclear motion. Massey (1930) wrote

> A molecule is effectively rigid to a 20,000-volt electron, the latter moving across the molecular diameter before any appreciable non-rigid motion has taken place. However, when necessary, the effects of such motions may always be brought in by calculating the scattered amplitude as a function $f(\theta, \phi, d)$ of the orientation angles θ, ϕ, of the molecule and the distance d between the nuclei. The observed cross section will be obtained calculating
>
> $$\left| \int_0^\pi \int_0^{2\pi} \int_{-\infty}^{\infty} f(\theta, \phi, d) \chi(\theta, \phi) \bar{\chi}(\theta, \phi) \xi(d) \bar{\xi}(d) \sin\theta \, d\theta \, d\phi \, dd \right|^2 \qquad (25)$$
>
> where $\chi(\theta, \phi)$ is the rotational wave-function of the molecule and $\xi(d)$ the vibrational function. The fact that the scattering probability does depend on θ, ϕ, d leads to a finite probability for energy interchange between the incident electron and molecular rotation and vibration, an effect of importance in the case of slow electrons.

With these considerations, Massey (1930) introduced what will be referred to here as the "adiabatic-nuclei" (AN) approximation. As is clear from Massey's description, the AN approximation proceeds in two steps. In the first step, the scattering problem is solved in the "BODY frame" (i.e., in a coordinate system attached to the nuclei) and the scattering amplitude, $f(\theta, \phi, d)$ is calculated at fixed nuclear orientation and distance. The calculation of $f(\theta, \phi, d)$ involves the solution of the electron scattering problem with fixed nuclei, that is,

1 Elastic Scattering of Electrons by Molecules

a solution of the Schrödinger equation with fixed nuclear orientation and distance. We shall refer to this as the "clamped-nuclei" problem and the appropriate Schrödinger equation as the "clamped-nuclei" (CN) Schrödinger equation. (see Table III). In the second step, the transformation to the "LAB frame" is made and the nuclear wave functions $\chi(\theta, \phi)$ and $\xi(d)$ are introduced. Thus the AN approximation refers to the "LAB frame" *before* the cross section calculation is completed. Figure 15 gives a schematic representation of the relationship among the various "frames" and is helpful in visualizing the connections between the various approximations.

The cross section formula in the AN approximation given by Massey (1930) was obtained on an intuitive basis, and no quantitative criteria were given for the limits of its applicability. The AN approximation for electron–molecule scattering was subsequently placed on a firm quantitative foundation by Altshuler (1957), who applied the general "adiabatic approximation" of Chase (1956) to this problem and discussed the quantitative criteria for its applicability. He found that the AN approximation holds if

(i) the period of target motion (i.e., rotation and vibration) is much greater than the time required for the electron to cross the region of interaction (i.e., the collision time) and

(ii) the process involves the exchange of only a small number of quanta with the target (i.e., that the number of excited states of the target that contribute significantly to the total wave function is limited).

Under "normal conditions" the rotational ($\tau_r \simeq 10^{-12}$ s) and the vibrational periods ($\tau_v \simeq 10^{-14}$ s) are long compared with the collision time for nonpolar molecules ($t_c \gtrsim 10^{-16}$ s), and the adiabatic criteria are expected to be satisfied. However, the collision time is unusually long close to threshold and at the energy of a resonance for which condition (i) does not hold and the adiabatic-nuclei approximation is not applicable. As mentioned earlier, for polar molecules, the collision time is not clearly defined because of the long-range dipole potential. Indeed, Altshuler (1957) found that the Born approximation for the fixed dipole gave an infinite DCS in the forward direction (and consequently the integral cross section diverged), and Mittleman and von Holdt (1965) showed that this characteristic also exists for the exact solution of scattering by a point dipole in the AN approximation. Later, it was shown by Garrett (1971) that even an *exact* AN approximation would lead to divergencies for any real polar molecule. It can be shown, however, that, except in the vicinity of the forward direction, "adiabatic conditions" hold even for polar molecules (Collins and Norcross, 1978). For zero-angle scattering and its immediate vicinity, however, the "LAB frame" approach has to be used. [Luckily, the "LAB frame" Born approximation is satisfactory (in most cases) for these angles.] This result shows that the AN approximation is useful even

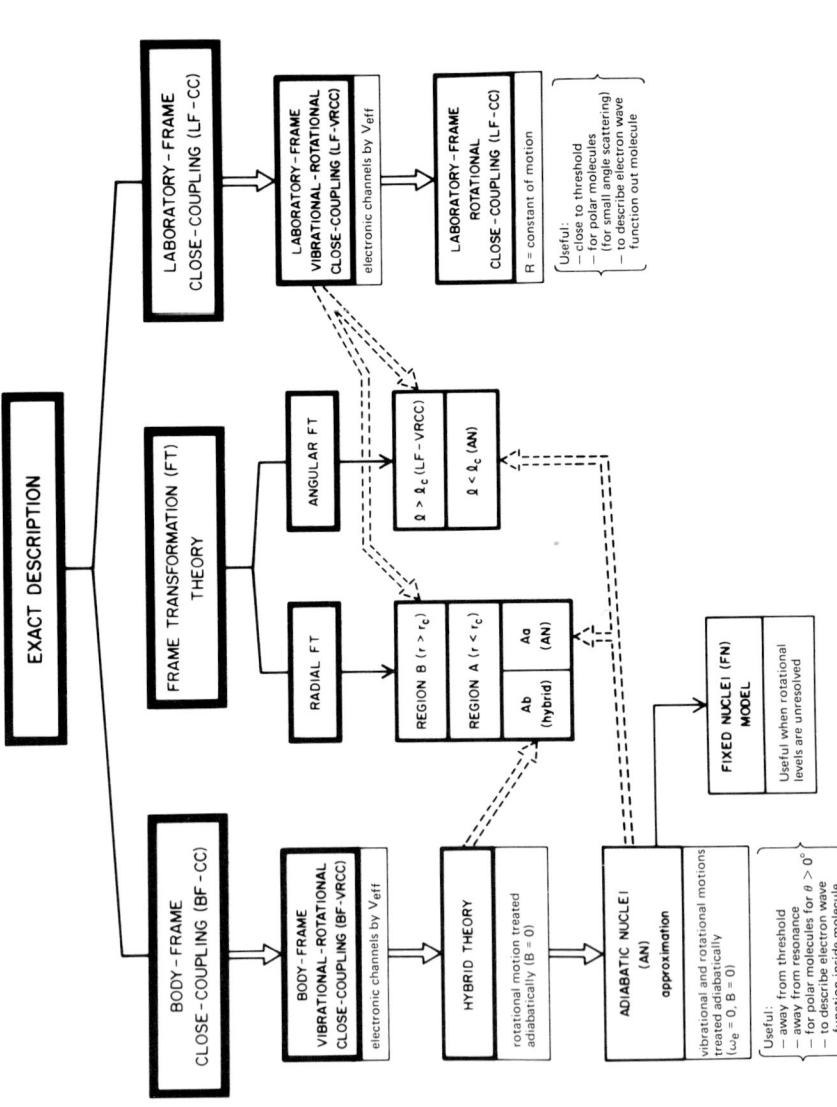

Fig. 15. Schematic diagram of the relationships between the various frame representations and the exact descriptions used to treat elastic scattering and rotational and vibrational excitation of molecules. The quantities B and ω_e denote the rotational and vibrational constants for diatomic molecules See text for detailed discussion

TABLE IV

Summary of Clamped-Nuclei (CN) Approximation Methods

Fundamental assumptions

(a) Electron kinetic energy is greater than the vibrational and rotational energy spacing.
(b) Calculation is made for homonuclear scattering *or* for the wave function inside of the molecule for electron–polar molecule scatterings *or* for small incident electron angular momenta for electron–polar molecule scatterings.

Scattering approximations

A. *Plane-wave-type theories*
   Born
   Born–Oppenheimer
   Bonham–Ochkur–Rudge
   Polarized Born
B. *Glauber-type theories*
C. *Distorted wave theories*
D. *Close-coupling approximations*
   Static potential (S)
   Static plus model exchange potential (SME)
   Static plus model exchange plus polarization potential (SMEP)
   Static exchange approximation (SE)
   Static exchange plus polarization (SEP)
   Static orthoganalized exchange (SOE)
   Static orthoganalized exchange plus polarization (SOEP)

---

for electron scattering from polar molecules but not without the limitations stated above. The conditions for applicability of the AN approximation were reinvestigated by Oksyuk (1966), Hara (1969c), and Chang and Temkin (1970) and have recently been discussed in detail by Shimamura (1984). The breakdown of the AN approximation close to threshold has been recently investigated numerically for $e^- - H_2$ rotational excitation by Feldt and Morrison (1982) and Morrison *et al.* (1983). A summary of the clamped-nuclei approximation methods is given in Table IV.

4. *Relationship between FN Model and AN Approximation Results*

In the FN model only elastic electron scattering can be calculated, while the AN approximation allows for rotational and vibrational excitation. An important result was first identified by Oksyuk (1966) and Takayanagi (1967) [and discussed in detail by Lane (1980)] namely, that

$$d\sigma(\text{FN})/d\Omega = \sum_{v'J'M'} (d\sigma/d\Omega)_{v'J'M'}(\text{AN}), \qquad (26)$$

where $(d\sigma/d\Omega)_{v'J'M'}(AN)$ is the differential cross section for the excitation of the $v'J'M'$ state (where $v'J'M'$ are the final state vibrational and rotational quantum numbers) in the adiabatic nuclei (AN) approximation and $d\sigma(FN)/d\Omega$ is the elastic differential cross section in the fixed nuclei (FN) model. This result, [Eq. (26)] has the following interesting practical consequence. If one wishes to compare the results of AN calculations with (elastic) experimental results for which rotational and vibrational transitions are unresolved, then it follows from Eq. (26) that theoretical results obtained using the FN model are adequate and generally easier to obtain than summing detailed state-to-state results using the AN approximation to obtain the same result. The situation was very aptly summarized by Golden et al. (1971):

> It is in this way, then, that one finally learns precisely what is meant physically by elastic scattering in the fixed-nuclei approximation: it is the sum over all rotational states [starting from *any* fixed initial rotational state] of the scattering cross sections computed in the adiabatic-nuclei approximation.*

We would like to mention that quite often an approximation scheme is used that includes "adiabatic rotation" but excludes the possibility of vibration. This scheme can be called the adiabatic nuclei–rotation, fixed nuclei–vibration approximation. In this scheme the internuclear distance is fixed $(R = R_o)$, and hence no vibration is allowed, but adiabatic rotation is taken into consideration.

After the FN model and AN approximation were placed on a solid quantitative foundation and its regions of applicability were well defined, it became clear that its implementation is much easier than the solution of the "LAB frame" CC equations. In fact, calculations were soon reported for e^-–H_2 scattering using the AN approximation or the FN model and employing the *exact* static exchange potential. The first accurate static exchange (or continuum Hartree–Fock) calculations using the FN model were reported by Tully and Berry (1969), using a partial differential equation approach, and by Hara (1969a) using spheroidal coordinates. Temkin and co-workers (Temkin and Vasavada, 1967; Temkin et al., 1969) used single-center expansion for e^-–H_2^+ scattering in the FN model. Hara (1969a,c) also completed calculations with *adiabatic* polarization included in the calculation for electron–molecular hydrogen scattering and obtained excellent agreement with experiment, for both elastic and rotational excitation DCSs.

5. Relationship between "LAB Frame" and "BODY Frame" Representations

It was observed in analyzing the results of the "LAB frame" CC calculations, and of the AN ("BODY frame") approximation for e^-–H_2

---
* Here the phrase "fixed nuclei approximation" refers to what we call the FN model.

1 Elastic Scattering of Electrons by Molecules

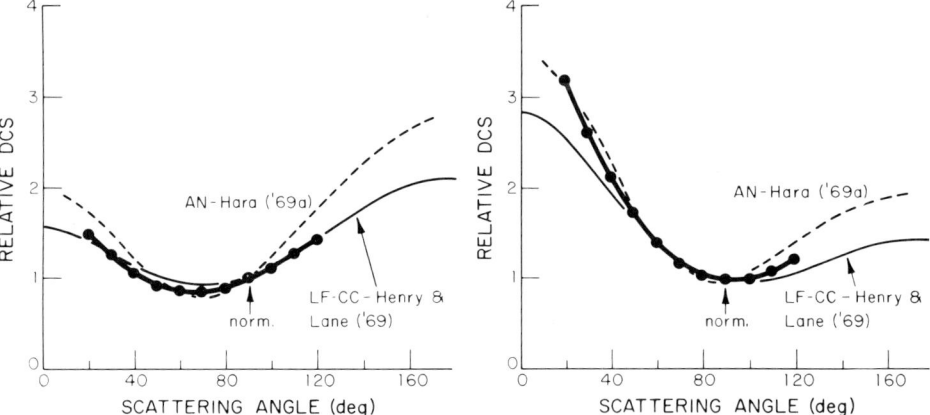

Fig. 16. Relative DCSs for elastic scattering of: (a) 2.5-eV incident energy electrons by H_2; (b) same as (a) except for 4.5-eV incident energy. [The experimental data are from Linder and Schmidt (1971a) and the theoretical curves are the adiabatic nuclei (AN) results of Hara (1969a) and the laboratory frame-close coupling (LF–CC) results of Henry and Lane (1969). The experimental data have been normalized to the theoretical results at a scattering angle of $90°$.]

scattering that, if static exchange effects were included in both treatments, they gave practically identical results. This fact is illustrated in Fig. 16, which compares results from AN (Hara, 1969a) and LF–CC (Henry and Lane, 1969) calculations to each other and to experiment at two different incident electron energies for elastic scattering by H_2. A similar agreement is also found for rotational excitation from Hara (1969c) and Henry and Lane (1969). There, except at low incident energy and small scattering angles, results from AN approximation and LF–CC calculations are in good agreement. This agreement raised questions as to the general relationship between the two approaches. This problem was discussed by Bottcher (1969) who first introduced the exact "BODY frame" description for the electron–molecule scattering process. By expanding the *exact* electron + rotor wave function in the rotating "BODY frame" he obtained coupled differential equations for the "coefficient functions" of the expansions which describe the state of the scattered electron relative to the rotating "BODY frame." By introducing the approximation that neglects dynamical effects (of the rotation), he showed that the usual form of AN approximation can be recovered and identified clearly the precise conditions of applicability for the AN approximation. These results implied that, under conditions for which the AN approximation is valid, the "LAB frame" CC equations and the scattering equation in the AN approximation will lead to essentially identical results (See Figs. 16–18). Subsequently, Fano (1970a) analyzed the advantages and disadvantages of the two approaches. He pointed out that, when the electron is far away from the

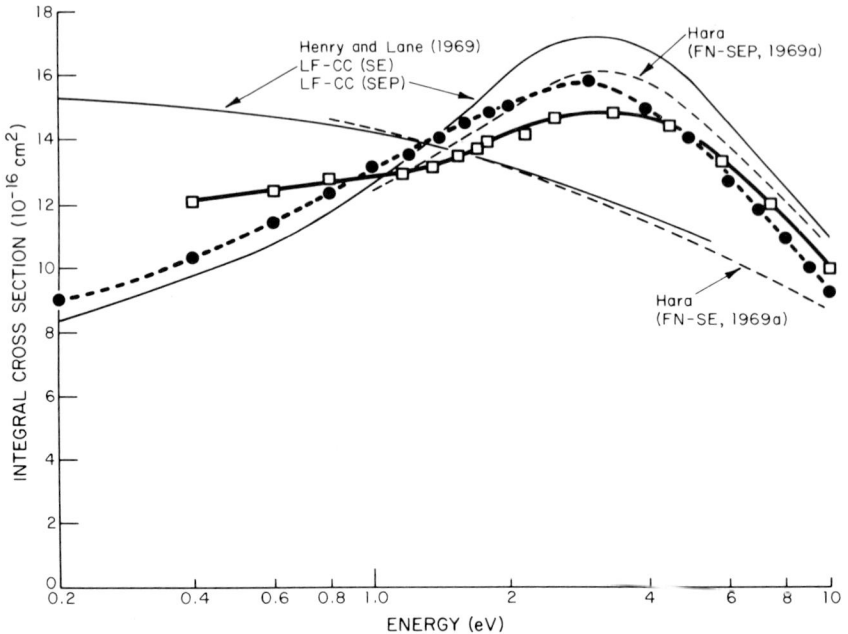

Fig. 17. Integral cross section for "elastic" scattering of electrons by H_2 in the energy range from 0.2 to 10 eV. The experimental data are from Ramsauer and Kollath (1930, □) and from Golden et al. (1966, ●). The theoretical results shown are from the LF–CC calculations of Henry and Lane (1969) and from the FN-SEP calculations of Hara (1969a).

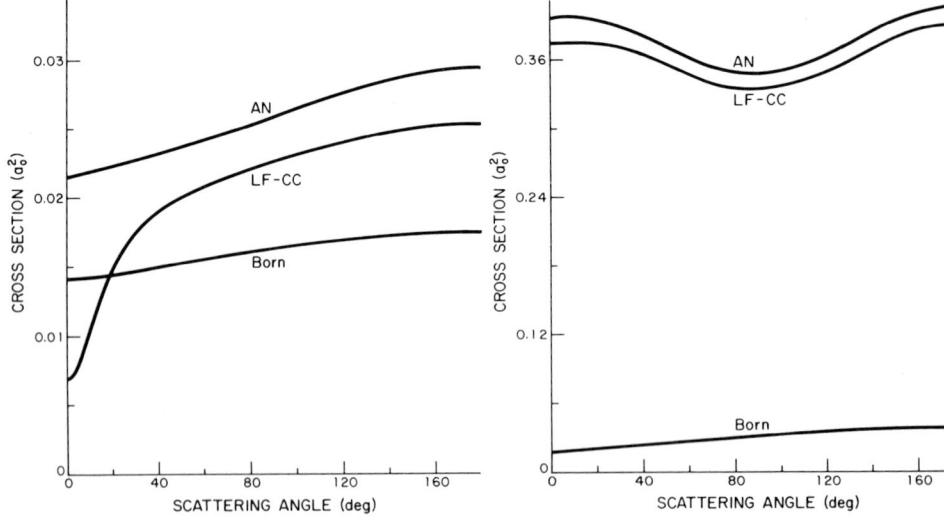

Fig. 18. (a) DCSs for the $J = 0 \to 2$ rotational excitation of H_2 at an incident electron energy of 0.1 eV (Feldt and Morrison, 1982). The results obtained from AN and laboratory frame–close coupling (LF–CC) calculations, obtained using the same model potential, are compared to show the differences between the two methods for treating nuclear motion. The Born results (Feldt and Morrison, 1982) are also shown for comparison. As evident, the results from the AN and the LF–CC calculations are in reasonable agreement except for scattering angles less than 40°. (b) Same as (a) except for an incident electron energy of 1 eV. At this energy and above, the results from the AN and the LF–CC calculations are in good agreement at all scattering angles but the Born results are too low by a factor of about three.

molecule, the angular momentum of the incident electron l and the molecule j (describing its rotational state), as well as the vibrational quantum number v of the molecule, are approximately constants of the motion and the "LAB frame" approach is advantageous. On the other hand, when the electron is inside (or very close to) the molecule and forms a "collision complex" with it, the projection of the electron angular momentum along the internuclear axis l_z and the internuclear distance R are approximately good quantum numbers. The total angular momentum $\mathbf{J} = \mathbf{l} + \mathbf{j}$ is a good quantum number in both cases. Fano (1970a) emphasized that both approaches are exact and are connected by a unitary transformation, what he called the *"frame transformation."* In a practical calculation, when truncations of the expansions are made, Fano (1970a) and Chang and Fano (1972) recommended that the "BODY frame" approach be used for the electron inside the molecule and the "LAB frame" approach be used when the electron is outside (see Fig. 15). This scheme will be called the *"frame transformation theory."* The wave functions obtained in the two approaches should then be connected at a convenient intermediate electron–molecule distance. In some fortunate situations this "connecting point" may lie at infinity, and in that case, no "LAB frame" CC equations need to be solved and only the transformation of the "BODY frame" T matrices into the LAB frame need to be performed. (The scattered electron always "leaves" the molecule in a well-defined rotational and vibrational state.) The greatest utility of the frame transformation theory is near the rotational threshold [and so it was used by Fano (1970b) to interpret H_2 photoionization spectra] and for electron–polar-molecule scattering.

6. *Recent Developments*

Under normal conditions the AN approximation appeared adequate for homonuclear molecules, and major efforts have been made in the last 10 years to solve specific problems as accurately as possible. The exact treatment of exchange for a nonspherical system represents, even today, a major computational effort, and numerical techniques are still being developed to deal with this problem. On the one hand, accurate noniterative and iterative techniques are being developed to solve the complex coupled intergrodifferential equations, while, on the other hand, the so-called L^2 methods (R-matrix and T-matrix techniques) have been introduced to treat the static exchange problem in the AN approximation. These techniques are referred to as L^2 methods because they take full advantage of matrix techniques developed in quantum chemistry and utilize L^2 functions (i.e., square integrable functions) to solve the scattering problem. Computationally, the L^2 functions are used to handle the internal part of the problem, which is the most complex one.

Substantial efforts have also been made to develop equivalent local potentials that describe the exchange effect accurately. These potentials are usually referred to as "local exchange" potentials and were first used by Hara (1967) to approximate the exact nonlocal exchange potential in a scattering problem. Methods have also been developed to replace exchange by orthogonalization of the scattering orbitals to the bound-state orbitals, a technique referred to as the "pseudopotential approach" (Burke and Chandra, 1972). The combination of local exchange potential and orthogonalization has recently been attempted with success (Morrison and Collins, 1981).

The inclusion of target polarization is a more difficult problem. As mentioned earlier, detailed calculations including adiabatic polarization were reported by Temkin and co-workers (Temkin and Vasavada, 1967, Temkin *et al.*, 1969) for $e^- - H_2^+$ scattering and by Hara (1969a,b) for electron-H_2 scattering, but the majority of calculations in both the low- and intermediate-energy region used semiempirical polarization potentials with adjustable parameters. Recently, however, results have been reported using *ab initio* polarization potentials by Klonover and Kaldor (1978) and by Schneider (1977), using many-body perturbation theory and pseudostate expansion methods, respectively. The most recent work of Schneider and Collins (1982, 1983) treats both exchange and polarization effects in a very complete manner, and it is quite likely that further work will proceed along these lines.

As mentioned above, the majority of calculations on electron scattering by diatomic molecules utilized the AN approximation. A notable exception is the work of Truhlar and co-workers at intermediate energies (reviewed by Truhlar *et al.*, 1979), where the LAB frame CC approach was used.

The calculations discussed above have generally ignored the vibrational degrees of freedom of the molecule. Recently, however, Choi and Poe (1977a,b) reported an extensive analysis of the CC equations in the rotating frame where vibrational coupling was included. It was pointed out earlier by Chandra and Temkin (1976a) that vibrational coupling is important to explain resonances (where the AN approximation does not work), and they introduced the so-called hybrid theory, where vibrational CC was explicitly introduced but rotation was treated adiabatically. Their treatment has been generalized by Choi and Poe (1977a,b) who introduced rotational–vibrational interactions into the treatment of the scattering problem.

B. Details of the Theoretical Methods

1. *LAB Frame Treatment*

As discussed above, the LAB frame treatment describes the collision process (including the rotation and vibration of the molecule) from a coordinate system attached to the space-fixed "laboratory." The simplest model to investigate the LAB frame treatment is one in which the internal structure and vibration of the molecule is neglected and the molecule is replaced by a rigid linear rotor from which an electron is scattered. The electron interacts with this rigid rotor via an interaction potential which depends on the coordinates of the electron in the LAB frame and the orientation of the rigid linear rotor. The various physical interactions of the electron with the molecule can be built into the particle–rotor interaction potential, including (approximately) the electrostatic potential, the adiabatic- and nonadiabatic polarization potentials, as well as a local exchange potential. This model will be discussed in some detail in Section III.B.1.a.

A slightly more complex model is that of the vibrating rotor for which the vibrational degrees of freedom are introduced but in which the molecule is still approximated by a structureless model system. The interaction potential now depends on the extension, as well as the orientation, of the vibrating rotor and permits a description of both rotational and vibrational excitation. This model will be discussed in Section III.B.1.b.

The next level of sophistication is a description that introduces the internal structure of the molecule by allowing "exchange" between the incident and the target electrons. In this treatment, the wave function of the molecule is introduced explicitly to provide for the possibility of antisymmetrization of the total (electron plus molecule) wave function and is discussed in Section III.B.1.c.

The next and final step of sophistication is to consider the possibility of a real (or virtual) electronic excitation of the molecule along with the vibrational degrees of freedom, and this procedure is briefly outlined in Section III.B.1.d. A summary of the LF–CC equations is given in Table V to help guide the reader through the various approximations that are usually employed.

a. Electron Scattering by a Rigid Rotor

i. Close-Coupling (CC) Equations We elaborate on the case of electron scattering by a linear rigid rotor following the work of Arthurs and Dalgarno (1960). The more general case of a symmetric top rotor has been discussed by Massey (1932), Bottcher (1969), Golden *et al.* (1971), and Abusalbi *et al.* (1983). The Hamiltonian of an interacting particle–linear rigid rotor system can be written as

$$H = H_{\text{rot}} - \frac{\hbar^2}{2m} \nabla_r^2 + V(\mathbf{r}, \hat{R}), \tag{27}$$

where $\mathbf{r}$ refers to the coordinates of the electron, $\hat{R}$ specifies the direction of the rigid linear rotor in the LAB frame, and m is the mass of the electron.

TABLE V

Summary of Laboratory Frame–Close Coupled (LF–CC) Equations

Complete LF–CC treatment	Rotational–vibrational–electronic CC equations (exact treatment). No calculations reported.
Rotational–vibrational LF–CC equations:	Only ground electronic state is explicitly included; coupling to open and closed channel electronic states are approximated via absorption and polarization potentials.
Rotational LF–CC equations	Vibrational degrees of freedom omitted; exchange effects properly included; coupling to open and closed electronic channels are approximated via absorption and polarization potentials.
Particle–vib–rotor scattering model, CC equation	Target molecule approximated by vib–rotor; particle–vibrator interaction potential may include static, polarization, absorption and model exchange potentials.
Particle rotor scattering model, CC equations	Target molecule approximated by rigid rotor. No vibrational excitation is possible. Interaction potential may include static, polarization, absorption, local exchange terms.

The eigenfunctions of H_{rot} are (Herzberg, 1950: p. 66)

$$H_{\text{rot}} Y_{jm}(\hat{R}) = (\hbar^2/2I)j(j+1)Y_{jm}(\hat{R}), \tag{28}$$

where j is the rotational quantum number, I the moment of inertia of the rotor, and $Y_{jm}(\hat{R})$ the usual spherical harmonics. If Ψ refers to the scattering wave function of the electron plus rotor system, then Ψ is the solution of the Schrödinger equation

$$H\Psi = E\Psi, \tag{29}$$

with appropriate boundary conditions that describe the incident electron and the rotor in its initial state. Since the electron and rotor form a closed system, the total angular momentum of this system is a conserved quantity. In the following discussion, $\tilde{J}^2$ and $\tilde{J}_z$ will refer to the magnitude of the *total* angular momentum and Z-component operators, with $\hbar^2 J(J+1)$ and $\hbar M_Z$ eigenvalues, respectively. Arthurs and Dalgarno (1960) expanded Ψ in terms of the eigenfunctions of the total angular momentum defined by

$$\mathscr{Y}_{Jjl}^M(\hat{r},\hat{R}) = \sum_{m_l=-l}^{l}\sum_{m_j=-j}^{j}(jlm_jm_l|jlJM)Y_{lm_l}(\hat{r})Y_{jm_j}(\hat{R}), \tag{30}$$

where $(jlm_jm_l|jlJM)$ is the Clebsch–Gordan coefficient as defined by Edmonds (1957). Assuming (Arthurs and Dalgarno, 1960) (i) that Ψ_{jl}^{JM} is a solution of the Schrödinger equation, Eq. (29), with energy eigenvalue $E = E_j$,

1 Elastic Scattering of Electrons by Molecules

(ii) that it is simultaneously an eigenfunction of the $\tilde{J}^2$ and $\tilde{J}_z$ operators with eigenvalues $\hbar^2 J(J+1)$ and $\hbar M_z$, respectively, (iii) that it represents a state for which the incoming electron enters with angular momentum l, and (iv) that the initial state of the rotor is characterized by the quantum number j, then Ψ_{jl}^{JM} can be expanded in terms of the functions $\mathscr{Y}_{Jjl}^M$, as,

$$\Psi_{jl}^{JM}(\mathbf{r}, \hat{R}) = \sum_{j'} \sum_{l'} r^{-1} u_{j'l'}^{Jjl}(r) \mathscr{Y}_{Jj'l'}^M(\hat{r}, \hat{R}). \tag{31}$$

By substituting Ψ_{jl}^{JM} in this form into Eq. (29), the following coupled system of differential equations, called *rotational close coupling (CC) equations*, are obtained for the $u_{j'l'}^{Jjl}(r)$ functions:

$$\frac{\hbar^2}{2m}\left(-\frac{d^2}{dr^2} + \frac{l'(l'+1)}{r^2} - k_{j'j}^2\right) u_{j'l'}^{Jjl}(r) + \sum_{j''}\sum_{l''} \langle j''l''; J | V | j'l'; J \rangle u_{j''l''}^{Jjl}(r) = 0, \tag{32}$$

where

$$k_{j'j}^2 = \frac{2m}{\hbar^2}\left[E_j - \frac{\hbar^2}{2I} j'(j'+1)\right], \tag{33}$$

and

$$\langle j''l''; J | V | j'l'; J \rangle = \int\int \mathscr{Y}_{Jj''l''}^M(\hat{r}, \hat{R}) V(\hat{r}, \hat{R}) \mathscr{Y}_{Jj'l'}^M(\hat{r}, \hat{R}) \, d\hat{r} \, d\hat{R}. \tag{34}$$

From the definition of Ψ_{jl}^{JM} one obtains the following asymptotic form for $u_{j'l'}^{Jjl}(r)$:

$$u_{j'l'}^{Jjl}(r) \xrightarrow[r \to \infty]{} \delta_{jj'}\delta_{ll'} \exp[-i(k_{jj}r - \tfrac{1}{2}l'\pi)] - (k_{jj}/k_{j'j})^{1/2}$$
$$\times \mathbf{S}^J(jl; j'l') \exp[i(k_{j'j}r - \tfrac{1}{2}l'\pi)], \tag{35}$$

which defines the scattering matrix $\mathbf{S}^J(jl; j'l')$. We are interested in that solution of Eq. (29) that corresponds to an incident electron with linear momentum k_{jj} and initial rotor state characterized by the quantum number j. The corresponding total wave function Ψ_j can be given then in the form

$$\Psi_j(\mathbf{r}, \hat{R}) = \frac{i(\pi)^{1/2}}{k_{jj}} \sum_{J=0}^{\infty} \sum_{M=-J}^{J} \sum_{l=|J-j|}^{J+j} (jlm_j 0 | jlJM) i^l (2l+1)^{1/2} \Psi_{jl}^{JM}(\mathbf{r}, \hat{R}), \tag{36}$$

which has the desired asymptotic form

$$\Psi_j(\mathbf{r}, \hat{R}) \xrightarrow[r \to \infty]{} \exp(i\mathbf{k}_{jj} \cdot \mathbf{r}) Y_{jm_j}(\hat{R})$$
$$+ \sum_{j'} \frac{i}{k_{jj}} \left(\frac{k_{jj}}{k_{j'j}}\right)^{1/2} \frac{\exp(ik_{j'j}r)}{r} \sum_{m_{j'}=-j'}^{j'} q(j'm_{j'}; jm_j | \hat{r}) Y_{jm_j}(\hat{R}), \tag{37}$$

where $q(j'm_{j'}; jm_j|\hat{r})$ is the reaction amplitude defined by

$$q(j'm_{j'}; jm_j|\hat{r}) = \sum_{J=0}^{\infty} \sum_{M=-J}^{J} \sum_{l=|J-j|}^{J+j} \sum_{l'=|J-j'|}^{J+j'} \sum_{ml=-l'}^{l'} l - l'$$
$$\times (\pi)^{1/2}(2l+1)^{1/2}(jlm_j 0|jlJM)(j'l'm_{j'}m_{l'}|j'l'JM)$$
$$\times \mathbf{T}^J(jl; j'l') Y_{l'm_{l'}}(\hat{r}), \qquad (38)$$

where

$$\mathbf{T}^J(jl; j'l') = \delta_{jj'}\delta_{ll'} - \mathbf{S}^J(jl; j'l'). \qquad (39)$$

The integral cross section for the $j \to j'$ transition can now be written in the form

$$\sigma(j'; j) = \frac{\pi}{(2j+1)k_{jj}^2} \sum_{J=0}^{\infty} \sum_{l=|J-j|}^{J+j} \sum_{l'=|J-j'|}^{J+j'} (2J+1)|\mathbf{T}^J(jl; j'l')|^2$$
$$\equiv \sum_{J=0}^{\infty} \sigma_J(j; j'). \qquad (40)$$

On the basis of Eq. (40), we can refer to the $J = 1, l = l' = 1$ contribution to the integral cross section as the $J = 1, pp$ term in the sum, and these contributions are additive. It should be noted that the CC equations for the $u_{j'l'}^{Jj}(r)$ functions are uncoupled for different J values, reflecting the fact that the total angular momentum J is a good quantum number. This fact is often referred to as the "block diagonal" feature of the CC equations. We also note that the linear rotor can model diatomic (in $^1\Sigma$ electronic states) or linear polyatomic molecules [see, e.g., Shimamura (1984)] and that these considerations can be generalized to a "symmetric top" rotor, which can similarly model symmetric polyatomic molecules (see Abusalbi et al. 1983). The $V(\mathbf{r}, \hat{R})$ potential can be made to incorporate all the many-body effects of the problem in an approximate way) including the electrostatic field, adiabatic and non-adiabatic polarization terms and a local exchange potential.

In actual calculations, the rotational CC equations [Eq. (32)] can be solved for only a finite number of J values and hence only a limited number of l and j values. This limitation will be referred to here as the rotational CC *approximation*. That is, the coupled equations have been truncated to a manageable number of equations. The first solutions in the rotational CC approximation for electron–molecule scattering were reported by Lane and Geltman (1967) for e^-–H_2 scattering. The interaction potential included the short- and long-range part of the electrostatic potential and a long-range adiabatic dipole polarization potential with a cut off. The effect of exchange was simulated by making the electrostatic potential more attractive inside the molecule. Results from this calculation are shown in Fig. 19 in which the CC results are compared with results from DW and Born calculations, as well as with experimental data.

Subsequently, Itikawa and Takayanagi (1969) and Itikawa (1969) reported results from rotational CC calculations for e^-–CN and e^-–HCl elastic scattering and rotational excitation, and Crawford and Dalgarno (1971) for e^-–CO scattering. Truhlar and co-workers have done extensive work using the LAB frame rotational CC equations with static, local exchange, adiabatic, and nonadiabatic polarization potentials and have reported results for e^-–H_2 (Truhlar and Brandt, 1976), for e^-–N_2 (Truhlar et al., 1976, Brandt et al., 1976, Onda and Truhlar, 1978; Onda and Truhlar, 1979a–c; Onda and Truhlar 1980a,b), for e^-–CO (Onda and Truhlar, 1980c), for e^-–CO_2 (Thirumalai et al., 1981a, Thirumalai and Truhlar, 1981, Onda and Truhlar, 1979d), e^-–C_2H_2 (Thirumalai et al., 1981b), and for e^-–CH_4 (Abusalbi et al., 1983) scattering. This work was summarized by Truhlar et al. (1979) and by Truhlar (1981), and the results are compared with other theoretical and experimental results in Figs. 23–31 of this chapter. Varracchio (1979) reported an interesting numerical study comparing CC, DW, and field-theoretical results for elastic scattering and rotational excitation of H_2. Recent LAB-frame rotational CC calculations were reported for strongly polar molecule targets by Allison (1975), by Itikawa (1976), by Rudge (1978a), by Gianturco and Rahman (1978), and by Collins and Norcross (1978). These and other works for polar molecules were reviewed by Itikawa (1978a), Norcross and Collins (1982), and Shimamura (1984) to which the reader is referred for further details.

ii. Distorted Wave (DW) Approximation. The solution of Eq. (32) simplifies tremendously if all the coupling terms except one are neglected. In this case the entire problem of calculating the scattering matrix reduces to the solution of the uncoupled equations

$$\left[\frac{\hbar^2}{2m}\left(-\frac{d^2}{dr^2} + \frac{l'(l'+1)}{r^2} - k_{j'j}^2\right) + \langle j'l'; J|V|j'l'; J\rangle\right]u_{j'l'}^{Jjl}(r) = 0. \quad (41)$$

This approximation is called the *LAB frame–distorted wave (LF–DW) approximation* and the solution of this equation can be used to construct the scattering matrix for elastic scattering and rotational excitation. The LF–DW approximation was first used by Mjolsness and Sampson (1964), Dalgarno and Henry (1965), and Sampson and Mjolsness (1965) for electron–molecule elastic scattering and rotational excitation. Lane and Geltman (1967) also reported results of DW calculations and compared them with CC equation results and experiment and that comparison is shown in Fig. 19. Recently, the DW approximation was used by Rudge et al. (1976) and by Rudge (1978a,b) for electron scattering by numerous polar molecules. As is shown in Figs. 19 and 26 and will be discussed later, the DW approximation works reasonably well near threshold but breaks down as the electron energy increases.

Finally, we note (Takayanagi and Itikawa, 1970) that, in the LF–DW approximation, the electron–molecule interaction potential is averaged over

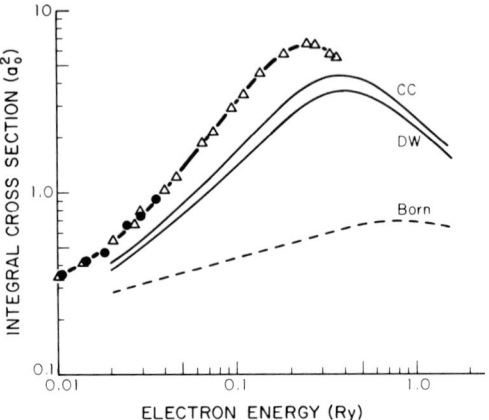

Fig. 19. Integral cross section for rotational excitation ($J = 0 \to 2$) of H_2 as a function of the incident electron energy. The data points indicate results obtained from the analysis of electron swarm data (Crompton et al. 1969, ●; Gibson, 1970, △). The solid curves are close coupling (CC) and distorted wave (DW) results of Lane and Geltman (1967). The dashed wave is Born approximation results of Lane and Geltman (1967).

the initial and final rotational states of the molecule, thereby reducing considerably the nonspherical part of the interaction. This approximation appears to be justified on physical grounds, only if the electron moves slowly towards the molecule.

iii. Born Approximation If the interaction potential V is completely neglected in calculating the wave function, then the LAB-frame Born approximation (LAB–Born) is obtained. In this case, there is no need for angular momentum decomposition and a closed-form T matrix can be obtained. The scattering amplitude can then be written as

$$f(jm_j, j'm_{j'} | \hat{r}') = -\frac{m}{2\pi\hbar^2} \int d\mathbf{r} \exp(i\mathbf{K} \cdot \mathbf{r}) \int Y^*_{j'm_{j'}}(\hat{R}) V(\mathbf{r}, \hat{R}) Y_{jm_j}(\hat{R}) d\hat{R}, \quad (42)$$

where $\mathbf{K}$ is the momentum-transfer vector defined by

$$\mathbf{K} = \mathbf{k} - \mathbf{k}', \quad (43)$$

where $\mathbf{k}$ and $\mathbf{k}'$ refer to the initial and final momentum of the electron, respectively, and $\hat{r}'$ refers to the direction of the outgoing electron. The DCS is obtained in the form

$$\frac{d\sigma}{d\Omega}(j,j'|\hat{r}') = \frac{k'}{k(2j+1)} \sum_{m_j=-j}^{j} \sum_{m_{j'}=-j'}^{j'} |f(jm_j j'm_{j'}|\hat{r}')|^2. \quad (44)$$

1 Elastic Scattering of Electrons by Molecules

In the case of a pure dipole potential (Massey, 1932), i.e., when

$$V(\mathbf{r}, \hat{R}) = -e\mathbf{D} \cdot \mathbf{r}/r^2, \tag{45}$$

where **D** is the dipole moment (with magnitude D) directed along **R**, one obtains

$$\frac{d\sigma}{d\Omega}(j,j'|\hat{r}') = \frac{4}{9}\left(\frac{Dem}{\hbar^2}\right)^2 (2j'+1)(j0j'0|jj'10)^2 \frac{k'}{k} \cdot \frac{1}{K^2}, \tag{46}$$

which reduces to the following form in the special case of $j' = j \pm 1$:

$$\frac{d\sigma}{d\Omega}(j,j+1|\hat{r}') = \frac{4}{3}\left(\frac{Dem}{\hbar^2}\right)^2 \frac{j+1}{2j+1} \frac{k'}{k} \frac{1}{k^2 + k'^2 - 2kk'\cos\theta} \tag{47}$$

and

$$\frac{d\sigma}{d\Omega}(j,j-1|\hat{r}') = \frac{4}{3}\left(\frac{Dem}{\hbar^2}\right)^2 \frac{j}{2j+1} \frac{k'}{k} \frac{1}{k^2 + k'^2 - 2kk'\cos\theta}. \tag{48}$$

For the integral cross sections, we obtain for these cases

$$\sigma(j,j+1) = \frac{8\pi}{3k^2}\left(\frac{Dem}{\hbar^2}\right)^2 \frac{j+1}{2j+1} \ln\frac{k+k'}{|k-k'|}, \tag{49}$$

$$\sigma(j,j-1) = \frac{8\pi}{3k^2}\left(\frac{Dem}{\hbar^2}\right)^2 \frac{j}{2j+1} \ln\frac{k+k'}{|k-k'|}. \tag{50}$$

As mentioned above, the first calculation in the LAB-frame Born approximation for the particle–rigid rotor model was reported by Massey (1932) for the rotating dipole. Subsequently, Gerjuoy and Stein (1955a,b) used the LF–Born approximation and a long-range quadrupole potential to show that rotational excitation is more probable at low energies than previously thought [based on previous Born calculations (Morse, 1953) which used a short-range potential for V]. These results also showed that the Born approximation is accurate for rotational excitation close to threshold where the weak, long-range potential determines the cross section. Subsequently, Takayanagi (1966) analyzed the Born approximation for polar molecules and discussed its limitations [see also the review of Takayanagi and Itikawa (1970) and the works of Crawford (1967), Chang (1970), and Itikawa (1971a, 1972) for a comprehensive discussion on the Born approximation]. Dalgarno and Moffett (1963) emphasized the importance of the polarization potential in conjunction with the Born approximation. Lane and Geltman (1967) also completed Born calculations for elastic scattering and rotational excitation of H_2 in which they compared the results with those from CC and DW approximations using the same electron interaction potential in all three cases (see Fig. 19). They found large deviation from the CC results except close to threshold where all approximation schemes (CC, DW, and Born) agreed (see Fig. 19). A similar study was made by Itikawa and Takayanagi (1969) for the case of rotational excitation of CN and HCl for which they obtained agreement among the three calculations to within about 10% and a practically identical energy dependence. The agreement between the Born and CC calculations were better for HCl (smaller dipole moment) than for CN in accordance with the original prediction of Massey (1932). Truhlar and Rice (1970, 1974), Trajmar et al. (1970), and Truhlar et al. (1971) used the LF–Born approximation, with adiabatic and nonadiabatic polarization effects included in the potential, for elastic scattering by H_2, and the same scheme was used by Truhlar (1972) and Onda and Truhlar (1979c) for elastic scattering by N_2.

b. Particle Scattering by a Vibrating Rotor

In this model, vibration is allowed for the rotor, which is still considered structureless. The total Hamiltonian for a particle plus vibrating linear rotor can be written as

$$H = H_{\text{rot}} + H_{\text{vibr}} - \frac{\hbar^2}{2m}\nabla_r^2 + V(\mathbf{r}, \hat{R}), \tag{51}$$

where H_{vibr} is the vibrational Hamiltonian which can be taken, for simplicity, to be the harmonic-oscillator Hamiltonian, whose eigenvalue problem can be written as

$$H_{\text{vibr}}\chi_v(R) = \hbar\omega(v + \tfrac{1}{2})\chi_v(R), \tag{52}$$

where $v = 0, 1, 2, \ldots$ and $\chi_v(R)$ are the usual harmonic-oscillator wave functions. The electron scattering wave function $\Psi^{JM}(\mathbf{r}, \mathbf{R})$ (a simultaneous eigenfunction of the H, $\tilde{J}^2$, and $\tilde{J}$ operators) is now expanded in terms of the basis functions

$$\Phi^{(lj)}_{vJM}(\hat{r}, \mathbf{R}) = \chi_v(R)\mathscr{Y}^M_{Jlj}(\hat{r}, \hat{R}) \tag{53}$$

in the form

$$\Psi^{JM}_{ljv}(\mathbf{r}, \mathbf{R}) = \sum_{l'j'v'} r^{-1} F^{Jljv}_{l'j'v'}(r)\chi_{v'}(R)\Phi^{(l'j')}_{v'JM}(\hat{r}, \hat{R}). \tag{54}$$

If this expression is substituted into the Schrödinger equation, the following coupled system of differential equations is obtained for the $F^{J,ljv}_{l'j'v'}(r)$ functions:

$$\left[-\frac{\hbar^2}{2m}\frac{d^2}{dr^2} + \frac{\hbar^2 l'(l'+1)}{2mr^2} + \hbar Bj'(j'+1) + \hbar\omega\left(v' + \frac{1}{2}\right) - E \right] F^{J,ljv}_{l'j'v'}(r)$$

$$+ \sum_{l''j''v''} (\Phi^{(l'j')}_{JM}\chi_{v'}|V|\Phi^{(l''j'')}_{JM}\chi_{v''}) F^{J,ljv}_{l''j''v''}(r) = 0, \tag{55}$$

1 Elastic Scattering of Electrons by Molecules

where

$$(\Phi_{JM}^{(l'j')}\chi_{v'}|V|\Phi_{JM}^{(l''j'')}\chi_{v''}) = \int \Phi_{JM}^{(l'j')}(\hat{r},\hat{R})\chi_{v'}(R)V(\hat{r},\hat{R})\phi_{JM}^{(l''j'')}(\hat{r},\hat{R})\chi_{v''}(R)\,d\hat{r}\,d\mathbf{R}. \tag{56}$$

The system of equations defined by Eq. (55) are the *vibrational–rotational CC equations*. The principal use of these equations is for calculating vibrational excitation, which obviously cannot happen without allowing for vibrational degrees of freedom. The conceptual significance of the expansion in Eq. (54) is that the asymptotic conditions can be conveniently formulated in terms of these functions since, for large r, the expansion reduces to one term. This is a manifestation of the fact that, when the electron is far away from the molecule, the molecule is in well-defined rotational and vibrational states. Accordingly j, J, and v are approximately good quantum numbers for large r(Fano, 1970a). Numerical solution of the rotational–vibrational CC equations was reported by Brandt *et al.* (1976) and Truhlar *et al.* (1976) for elastic scattering and rotational excitation of N_2 and by Rudge (1980) for scattering by HF. Brandt *et al.* (1976) included the following (v, j) "states": (0,0), (0,2), (1,0), and (1,2), whereas Rudge (1980) included the $j = 0, 1, 2$ and $v = 0, 1, 2$ quantum states. The vibrating rotor model was also used earlier in conjunction with the Born approximation for rotational and vibrational excitation of H_2 by Carson (1954).

c. Ab Initio Treatment: Electrostatic and Exchange Effects

The internal structure of the molecule and the exchange effects were first introduced into the LAB frame treatment of electron–molecule scattering by Ardill and Davison (1968). They wrote the Hamiltonian of the electron plus molecule system as

$$H = H_{el} + H_{rot}, \tag{57}$$

where H_{rot} is the rotational Hamiltonian [with eigenfunctions given by Eq. (28)] and

$$H_{el} = H_T - (\hbar^2/2m)\nabla_{N+1}^2 + V \tag{58}$$

and where H_T is the target (i.e., molecular) Hamiltonian at fixed internuclear conditions, $-(\hbar^2/2m)\nabla_{N+1}^2$ is the kinetic-energy operator of the incident electron, and

$$V = \sum_{i=1}^{n} \left\{ \frac{e^2}{|\vec{r}_{N+1} - \vec{r}_i|} - \frac{Z_A e^2}{|\vec{r}_{N+1} - \vec{r}_A|} - \frac{Z_B e^2}{|\vec{r}_{N+1} - \vec{r}_B|} \right\} \tag{59}$$

is the electron–molecule interaction potential (given here for a diatomic molecule); $\mathbf{r}_{N+1}\sigma_{N+1}$ refers to the spatial and spin coordinate of the incident electron, respectively, $\mathbf{r}_i\sigma_i$ ($i = 1, 2, \ldots, N$) are those of the target electrons, and $\mathbf{r}_A$ and $\mathbf{r}_B$ are the spatial coordinates of the nuclei with nuclear charge Z_A and Z_B, respectively. Again $\Psi^{JM}(\mathbf{r}_1\sigma_1, \mathbf{r}_2\sigma_2, \ldots, \mathbf{r}_N\sigma_N, \mathbf{r}_{N+1}, \sigma_{N+1})$ denotes the scattering function that is a simultaneous eigenfunction of the $\tilde{J}^2$ and $\tilde{J}_z$ operators. The scattering function can be sought in the form (for simplicity we assume a $^1\Sigma$ ground state)

$$\Psi_{jl}^{JM}(1, \ldots, N, N+1)$$
$$= \frac{1}{(2\pi)^{1/2}} \mathscr{A}\left[\Phi_o(1, 2, \ldots, N) \sum_{j'l'} r^{-1} u_{j'l'}^{Jjl}(r_{N+1}) \mathscr{Y}_{Jj'l'}^M(\hat{r}_{N+1}, \hat{R}) \chi_{1/2}^m(\sigma_{N+1})\right], \tag{60}$$

where $\mathscr{A}$ is the antisymmetrization operator, $\Phi_o(1, 2, \ldots, N)$ the ground-state wave function of the molecule with fixed (equilibrium) internuclear distance (with $1 \equiv \mathbf{r}_1\sigma_1, \ldots, N \equiv \mathbf{r}_N\sigma_n$), and $\chi_{1/2}^m(i)$ the one-electron spin function with spin projection m ($m = \frac{1}{2}$ or $-\frac{1}{2}$). Substituting Ψ_{jl}^{JM}, as given by Eq. (60), into the Schrödinger equation and projecting it onto $\phi_o(r_1, \ldots, N) \mathscr{Y}_{Jjl}^M(\hat{r}_{N+1}, \hat{R})$, the following coupled system of integrodifferential equations is obtained.

$$\frac{\hbar^2}{2m}\left[-\frac{d^2}{dr^2} + \frac{l'(l'+1)}{r_{N+1}^2} - k_{j'j}^2\right] u_{j'l'}^{Jjl}(r_{N+1}) + \sum_{j''}\sum_{l''} \langle j'l'; J|V|j''l''; J\rangle u_{j''l''}^{Jjl}(r_{N+1})$$
$$= \sum_{j''}\sum_{l''} \int_0^\infty K(j'l'; j''l''; J|r, r_{N+1}) u_{j''l''}^{Jjl}(r_{N+1}), \tag{61}$$

where K, the exchange kernel, is a complicated expression depending on the target. In the case of $N = 2$ (hydrogen molecule), K was given by Henry and Lane (1969) in the following simple form:

$$K(j'l'; j''l''; J|r_1, r_3)$$
$$= r_1 r_3 \int \mathscr{Y}_{j'l'}^{JM*}(\hat{r}, \hat{R}) \Phi_o(\mathbf{r}_1\mathbf{r}_2, \hat{R}) r_{13}^{-1} \Phi_o(\mathbf{r}_2\mathbf{r}, \hat{R}) \mathscr{Y}_{j''l''}^{JM}(\hat{r}, \hat{R}) d\hat{r}_1 d\vec{r}_2 dr_2 d\hat{R} \tag{62}$$

where $\phi_o(\mathbf{r}_1\mathbf{r}_2, \hat{R})$ is the spatial part of the electronic wave function of H_2 with $\hat{R}$ internuclear orientation (i.e., this is a "rotated" wave function). In comparing Eq. (61) with Eq. (32), the appearance of the integral operator with kernel $K(j'l', j''l'', J/r_1, r)$ is evident, and it describes the exchange effect in each particular channel, as well as in the coupling terms. The numerical solution of Eq. (61) is an exceedingly arduous task, even when only a small number of terms are included into the sum for j'' and l''. Quite often, therefore, the *non-*

local exchange potential represented by $K(j'l', j''l'', j/r_1, r)$ is replaced by a *local exchange potential*. For scattering problems, nonlocal exchange potentials were first discussed by Hara (1967) and later by Riley and Truhlar (1975). Local potential approximations for exchange coupling potentials have been discussed by Truhlar and Mullaney (1978). The only results reported in the literature that solved Eq. (61) numerically are those of Henry and Lane (1969) for elastic scattering and rotational excitation by H_2 and by Itikawa and Ashihara (1971) for e^-–HCl scattering. The results from the calculations by Henry and Lane (1969) are shown in Figs. 16, 17, 21, and 22, where they are compared with results from other theories and with the available experimental data.

If all but one coupling term (both direct and exchange) is neglected in Eq. (61), one obtains what is called the *exchange distorted wave* (EDW) *approximation*, which was used by Ardill and Davison (1968) in a calculation for rotational excitation of H_2. They found that exchange is important in the pp contribution to the rotational excitation cross section at low incident electron energies ($E < 0.5$ eV).

In this discussion the vibrational motion of the nuclei has been ignored and a more complete treatment, which also includes vibration, is presented by Choi and Poe (1977a,b).

d. Complete LAB-frame Treatment: The Introduction of Electronic Channels

In the complete LAB-frame formulation for the electron plus molecule scattering function $\psi_{jlv\alpha_o}^{JM}$ (where the indices $jlv\alpha_o$ refer to the incident channel, α is the electronic quantum number of the Born–Oppenheimer state, and it is assumed that the target states are well described within the Born–Oppenheimer approximation), v and j are the vibrational and rotational quantum numbers, respectively, and J is the total angular momentum of the electron (with angular momentum l) plus the target molecule (with rotational angular momentum j), and S the total spin of the electron plus molecule system. This function is expanded in the form

$$\Psi_{jlv\alpha o}^{JMSM_S} = \mathscr{A}\left[\sum_{p'} r^{-1} u_p^{p'}(r_{N+1}) \Phi_{p'}(1, 2, \ldots, N, N+1)\right], \qquad (63)$$

where the LAB-frame channel functions are defined by

$$\Phi'_p(1, 2, \ldots, N, N+1) = \sum_{m_S(M_S)\alpha} \mathscr{Y}_{jl}^M(\hat{r}_{n+1}, \hat{R}) \chi_v(R) \Psi_\alpha(\mathbf{r}_1 \sigma_1, \cdots, \mathbf{r}_N \sigma_N; \mathbf{R})$$

$$\times \chi_{m_S}^S(\sigma_{N+1}) \tfrac{1}{2} S_\alpha S_j m_S(M_S)_\alpha \qquad (64)$$

and where the channel index p refers to the collection of quantum numbers $p = (\alpha, v, j, l, J, S)$. By substituting formula (63) into the Schrödinger equation for $\Psi_{jlv\alpha o}^{JMSMs}$ and performing the usual integrations, a coupled system of integrodifferential equations is obtained for the $u_{p'}^{P}(r)$ functions. Since the channel index now includes rotational, vibrational, and electronic quantum numbers, these integrodifferential equations are of the utmost complexity and no solution has yet been reported for these equations. Since it is extremely difficult to incorporate all channels into an actual calculation, the effect of the electronic channels is usually approximated by the inclusion of *adiabatic* and *nonadiabatic* polarization potentials to account for virtual excitation of the electronic channels and by "absorption" potentials to account for real electronic excitation. In effect the "complete" treatment described here justifies the use of effective potentials in the models already described in this section. When the electronic channels are converted into effective potentials, only rotational and vibrational degrees of freedom remain and one obtains the so-called vibrational–rotational CC equations. Vibrational–rotational CC equations have been solved numerically by Henry (1970), who took nonlocal exchange effects into consideration. As will be seen later, a CC calculation for the *vibrational* degrees of freedom is significant only in the case of a resonance, and in most cases, the adiabatic approximation suffices for describing the nonresonant scattering processes.

2. BODY-Frame Treatment

As discussed above, the coordinate system attached to the nuclei (called the BODY-frame) may be a better choice as a frame of reference for description of the electron–molecule scattering problem since, for a certain incident energy range, the nuclei can be considered stationary during the scattering process (the AN approximation). Historically, the first approach (based on intuition) was the FN model (Massey, 1930, Stier, 1932) where the nuclei were considered fixed during the electron scattering process, the elastic scattering problem was solved in the BODY-frame, the cross section was calculated, and an averaging was then performed for all possible molecular orientations. These ideas were put on a firm mathematical basis by Chase (1956) and Altshuler (1957) who generalized the Born–Oppenheimer approximation to scattering problems and introduced the AN approximation. Again the nuclei were considered fixed during the electron scattering process, but the nuclear motion was introduced using the adiabatic nuclei hypothesis, the scattering matrix was transformed to the LAB-frame, and only then was the cross section calculated. This latter procedure allowed the calculation of rotational and vibrational cross sections, as well as those for elastic scattering. Chase (1956), Altshuler (1957), Okysuk (1966), Hara (1969c), and Chang and Temkin (1970) discussed the limits of applicability of the AN approximation. The most recent development in using the adiabatic hypothesis was the introduction of the so-called rotating frame by Bottcher (1969). This method describes the complete scattering problem in an *ab initio* way and takes the coordinate system that rotates with the electron plus molecule system (i.e., with the "collision complex"). This approach is an exact description of the scattering problem, and Bottcher (1969) showed that the usual form of the AN approximation can be recovered. It can also be shown that this description is completely equivalent to the LAB-frame treatment. While Bottcher (1969) discussed the particle–rotor problem in the rotating frame, the same approach was used by Chang and Fano (1972) and Choi and Poe (1977a,b) to treat the particle–vibrating rotor problem. Exchange effects were considered in an *ab initio* manner in the BODY-frame by Henry and Chang (1972). The complete scattering problem from the rotating frame point of view was reviewed by Lane (1980).

a. Particle Scattering by a Rigid Rotor

To depict the essential features of the BODY-frame treatment and its relationship to the LAB-frame approach, the rigid-rotor model will be treated first. In the LAB-frame treatment, Ψ^{JM} is expanded in terms of the $\mathscr{Y}_{jjl}^M$ functions [see Eq. (31)], while in the case of the BODY-frame treatment, it will be expanded in terms of functions that describe the rigid rotation of the collision complex and the electron rotation around the molecular axis (Chang and Fano, 1972; Harter *et al.*, 1978). These basis functions are of the form

$$X_{JM}^{(l\Lambda)}(\hat{r}', \hat{R}) = Y_{l\lambda}(\hat{r}')D_{\lambda+\bar{\Lambda},M}^{(J)}(\hat{R}), \qquad \Lambda = \lambda + \bar{\Lambda} \tag{65}$$

where J is the total angular momentum of the particle plus rotor system, with projection M onto the LAB-frame z axis, λ the projection of the electronic angular momentum onto the molecular axis (the BODY z axis), $\bar{\Lambda}$ the molecular angular momentum quantum number (for Σ states, $\bar{\Lambda} = 0$) and $\Lambda = \lambda + \bar{\Lambda}$, $\hat{R}$ refers to the polar coordinates of the BODY-frame z axis in the LAB-frame, and $\hat{r}' \equiv \vartheta', \phi'$ are the electron polar coordinates in the BODY-frame. $D_{\Lambda,M}^{(J)}(\hat{R})$ is the Wigner function [Edmonds (1957)]. $X_{JM}^{(l\Lambda)}(\hat{r}', \hat{R})$ describes the state of the collision complex with total angular momentum quantum numbers J and M for which the electron circles around the BODY axis with angular momentum projection λ.

This basis is related to the $\mathscr{Y}_{Jjl}^M(\hat{r}, \hat{R})$ basis defined in the LAB-frame treatment via a unitary transformation

$$X_{JM}^{(l\Lambda)}(\hat{r}', \hat{R}) = \sum_j \mathscr{Y}_{Jjl}^M(\hat{r}, \hat{R}) U_{j\Lambda}^{(lJ)} \tag{66}$$

called a *frame transformation* by Chang and Fano (1972) for which $U_{j\Lambda}^{(lJ)}$ are the expansion coefficients. If the complete scattering function Ψ^{JM} is expanded in terms of the functions $X_{JM}^{(l\Lambda)}$ in the form

$$\Psi^{JM}(\mathbf{r}, \hat{R}) = \sum_{l,\Lambda} r^{-1} G_{l\Lambda}^{(J)}(r) X_{JM}^{(l\Lambda)}(\hat{r}', \hat{R}), \tag{67}$$

then the following coupled system of differential equations is obtained for the $G_{l\Lambda}^{(J)}(r)$ functions,

$$\left(-\frac{\hbar^2}{2m}\frac{d^2}{dr^2} + \frac{\hbar^2 l(l+1)}{2mr^2} - E\right) G_{l\Lambda}^{(J)}(r) + \sum_{\Lambda'} (X_{JM}^{(l\Lambda)}|H_{\text{rot}}|X_{JM}^{(l\Lambda')}) G_{l\Lambda'}^{(J)}(r)$$

$$+ \sum_{l'} (X_{JM}^{(l\Lambda)}|V|X_{JM}^{(l'\Lambda)}) G_{l'\Lambda}^{(J)}(r) = 0, \tag{68}$$

where

$$(X_{JM}^{(l\Lambda)}|H_{\text{rot}}|X_{JM}^{(l\Lambda')}) = \int X_{JM}^{(l\Lambda)}(\hat{r}, \hat{R}) H_{\text{rot}}(\hat{R}) X_{JM}^{(l\Lambda')}(\hat{r}, \hat{R}) d\hat{r} d\hat{R}$$

and

$$(X_{JM}^{(l\Lambda)}|V|X_{JM}^{(l'\Lambda)}) = \int X_{JM}^{(l\Lambda)}(\hat{r}, \hat{R}) V(\mathbf{r}, \hat{R}) X_{JM}^{(l'\Lambda)}(\hat{r}, \hat{R}) d\hat{r} d\hat{R}.$$

The boundary conditions to be imposed on the $G_{l\Lambda}^{(J)}(r)$ functions are too complicated to be discussed here and the interested reader is referred to Choi and Poe (1977a,b) for a complete discussion. It can be easily seen that

$$(X_{JM}^{(l\Lambda)}|H_{\text{rot}}|X^{(l\Lambda')}) = B \sum_j \tilde{U}_{\Lambda j}^{(lJ)} j(j+1) U_{j\Lambda'}^{(lJ)}, \tag{69}$$

where B is the rotational constant and a measure of the rotational level separation. If this level separation B is much smaller than E and the $(X_{JM}^{(l\Lambda)}|V|X_{JM}^{(l\Lambda)})$ matrix elements, then it can be neglected relative to them; and Eq. (68) reduces to the form

$$\left(-\frac{\hbar^2}{2m}\frac{d^2}{dr^2} + \frac{\hbar^2 l(l+1)}{2mr^2} - E\right) G_{l\Lambda}^{(J)}(r) + \sum_{l'} (X_{JM}^{(l\Lambda)}|V|X_{JM}^{(l'\Lambda)}) G_{l'\Lambda}^{(J)}(r) = 0. \tag{70}$$

This is exactly the adiabatic nuclei (AN) approximation obtained in this form if the electron scattering function Ψ_J^{ad} is expanded, at fixed R, in terms of the $Y_{lm}(\hat{r}')$ spherical harmonics in the BODY-frame:

$$\Psi_J^{\text{ad}}(\mathbf{r}, \hat{R}) = \sum_{l,\Lambda} r^{-1} G_{l\Lambda}^{(J)}(r) Y_{l\Lambda}(\hat{r}'), \tag{71}$$

and this expansion is substituted into the clamped-nuclei Schrödinger equation. Thus it is clear that the exact rotating coordinate system equations give the usual form for the AN approximation under the conditions just discussed, and the exact BODY-frame treatment reveals the conditions under which the AN approximation holds (see Bottcher, 1969).

b. Particle Scattering by Vibrating Rotor

The vibrational degrees of freedom in the rotating frame approach were discussed by Chang and Fano (1972) in which they wrote the complete Hamiltonian in the form

$$H = H_{rot} + H_{vibr} - (\hbar^2/2m)\Delta^2 + V(r, \vartheta', R), \tag{72}$$

where H_{vibr} is the vibrational Hamiltonian and now the interaction potential V depends on R also. As discussed above the LAB system used the $\Phi_{vlj}^{JM}(\hat{r}, \hat{R}) = \chi_v(R)\mathcal{Y}_{jlj}^M(\hat{r}, \hat{R})$ basis set, and in the BODY-frame, one can use

$$X_{JM}^{l\Lambda v}(\hat{r}', \mathbf{R}) = \chi_v(R) X_{JM}^{(l\Lambda)}(\hat{r}', \hat{R}) \tag{73}$$

as a basis set and expand Ψ^{JM} in the form

$$\Psi^{JM}(\mathbf{r}, \mathbf{R}) = \sum_{l\Lambda v} r^{-1} G_{l\Lambda v}^{(J)}(r) \chi_v(R) X_{JM}^{(l\Lambda)}(\hat{r}', \hat{R}), \tag{74}$$

obtaining the following equations for the $G_{l\Lambda v}^{(J)}(r)$ functions:

$$\left(-\frac{\hbar^2}{2m} \frac{d^2}{dr^2} + \frac{\hbar^2 l(l+1)}{2mr^2} + \hbar\omega\left(v + \frac{1}{2}\right) - E \right) G_{l\Lambda v}^{(J)}(r)$$

$$+ B \sum_{\Lambda} \sum_{j} \tilde{U}_{\Lambda j}^{(lJ)} \hbar^2 j(j+1) U_{j\Lambda'}^{(lJ)} G_{l\Lambda' v}^{(J)}(r) + \sum_{l'v'} (\Lambda lv |V| \Lambda l'v') G_{l'\Lambda v'}^{(J)}(r) = 0, \tag{75}$$

where

$$(\Lambda lv |V| \Lambda l'v') = \int X_{JM}^{l\Lambda v}(\hat{r}', \mathbf{R}) V(\hat{r}, \hat{R}) X_{JM}^{l\Lambda v'}(\hat{r}', \mathbf{R}) d\hat{r} \, d\mathbf{R}.$$

These equations constitute the *vibrational–rotational CC equations in the BODY-frame*. However, it has been pointed out by Fano (1970a) and by Chang and Fano (1972) that, under adiabatic conditions, the internuclear distance R is a good quantum number and is a complementary quantum number to v (the vibrational quantum number). Thus, to recover the adiabatic limit, it is desirable to introduce the quantity R as a quantum number in place of v through the transformation

$$H_{l\Lambda}^{(J)}(r, R) = \sum_v G_{l\Lambda v}^{(J)}(r) \chi_v(R) \tag{76}$$

and express Ψ^{JM} in terms of these functions in the form

$$\Psi^{JM}(\mathbf{r}, \mathbf{R}) = \sum_{l,\Lambda} r^{-1} H^{(J)}_{l\Lambda}(r, R) X^{(l,\Lambda)}_{JM}(\hat{r}, \hat{R}). \tag{77}$$

From Eq. (75) the following equations can be obtained for the $H^{J}_{l,\Lambda}(r, R)$ functions:

$$\left(-\frac{\hbar^2}{2m}\frac{d^2}{dr^2} + \frac{\hbar^2 l(l+1)}{2mr^2} - E\right)H^{(J)}_{l\Lambda}(r, R)$$

$$+ B\sum_{\Lambda'}\left[\sum_{J} U^{(lJ)}_{\Lambda j} \hbar^2 j(j+1) U^{(lJ)}_{j\Lambda'}\right]H^{(J)}_{l\Lambda}(r, R)$$

$$+ \hbar\omega \int_0^\infty dR' \left[\sum_v \chi_v(R)\left(v + \frac{1}{2}\right)\chi_v(R')\right]H^{(J)}_{l\Lambda}(r, R')$$

$$+ \sum_{l'} (l\Lambda|V|l'\Lambda) H^{(J)}_{l'\Lambda}(r, R) = 0. \tag{78}$$

In the exact scheme, described by the above equation, the $H^{(J)}_{l\Lambda}(r, R)$ functions are connected for different values of R and Λ. However, the AN approximation assumes that $\hbar\omega \ll E$ and $\hbar^2 B \ll E$ and the above equations decouple because the second and third terms can be neglected, and separate equations are obtained for each value of R and Λ, showing that they indeed are good quantum numbers (Fano, 1970a, Chang and Fano, 1972). Since the vibrational spacing ω is larger than the rotational spacing B, adiabatic conditions may hold for rotation (i.e., $\hbar^2 B \ll E$) but not for vibration (i.e., $\hbar\omega \simeq E$). In this case, a hybrid method can be obtained by substituting $B = 0$ into Eq. (75) or Eq. (78). In this hybrid theory, rotational motion is treated adiabatically (i.e., Λ has become a good quantum number) but vibrational motion is dynamically coupled. The hybrid method was first introduced by Chandra and Temkin (1976a) to treat $e^- - N_2$ scattering and to explain the vibrational structure of the $E = 2.4$ eV resonance and was later derived from the general BODY-frame CC equations by Choi and Poe (1977b). Choi and Poe (1977a,b) generalized the hybrid method of Chandra and Temkin (1976a) by introducing vibrational–rotational interaction into the Hamiltonian and they applied this technique (Poe and Choi, 1979, 1981) to the calculation of elastic scattering and rotational and vibrational excitation of CO, H_2, D_2, and O_2 molecules by electrons. Some of their results for e^-–CO scattering are shown in Figs. 29 and 30. In these calculations, the electron–vibrator potential $V(r, \vartheta', R)$ was taken as a sum of static, local exchange, and semiempirical adiabatic polarization potentials. Since these calculations were designed principally for vibrational excitation, they will not be discussed here in detail.

c. *Ab Initio Treatment: Electrostatic and Exchange Effects*

The electronic structure of the molecule and associated electrostatic and exchange effects were introduced into the BODY-frame treatment by Henry and Chang (1972). They have shown that, if the electron-plus-molecule scattering wave function is written in the form

$$\Psi^{JM}(1,\ldots,N,N+1)$$

$$= \frac{1}{(2\pi)^{1/2}} \mathscr{A}\left(\Phi_o(1,2,\ldots,N) \sum_{l,\Lambda,v} r_{N+1}^{-1} G_{l\Lambda v}^{(J)}(r_{N+1}) \chi_v(R) X_{JM}^{(l\Lambda)}(\hat{r},\hat{R})\right)$$

$$= \frac{1}{(2\pi)^{1/2}} \mathscr{A}\left(\Phi_o(1,2,\ldots,N) \sum_{l,\Lambda} r_{N+1}^{-1} H_{l\Lambda}^{(J)}(r_{N+1},R) X_{JM}^{(l\Lambda)}(\hat{r}_{N+1},\hat{R})\right), \quad (79)$$

then the following equations are obtained for the $H_{l\Lambda}^{J}(r,R)$ functions:

$$\left(-\frac{\hbar^2}{2m}\frac{d^2}{dr^2} + \frac{\hbar^2 l(l+1)}{2mr^2} - E\right) H_{l\Lambda}^{(J)}(r,R) + \sum_{l'}\left(X_{JM}^{(l\Lambda)}|V|X_{JM}^{(l'\Lambda)}\right) H_{l'\Lambda}^{(J)}(r,R)$$

$$+ \sum_{\Lambda}\left(X_{JM}^{(l\Lambda)}|H_{\text{rot}}|X_{JM}^{(l\Lambda')}\right) H_{l\Lambda'}^{(J)}(r,R)$$

$$+ \hbar\omega \sum_v \left(v + \frac{1}{2}\right)\chi_v(R)\int_0^\infty dR' H_{l\Lambda}^{(J)}(r,R')\chi_v(R')$$

$$= \sum_{l'\Lambda'} \int_0^\infty dr_1\, K(l\Lambda,l'\Lambda';JM|r_1,r;R) H_{l'\Lambda'}^{(J)}(r,R), \quad (80)$$

where K is the exchange kernel, which for H_2 can be written in the form
$K(l\Lambda,l'\Lambda';JM|r_1,r;R)$

$$= \int d\hat{r}\int d\mathbf{r}_2\, X_{JM}^{(l\Lambda)*}(\hat{r},\hat{R})\Phi_o(\mathbf{r}_1\mathbf{r}_2;R)\frac{1}{|\mathbf{r}-\mathbf{r}_2|}\Phi_o(\hat{r}_1\hat{r}_2;R)X_{JM}^{(l'\Lambda')}(\hat{r},\hat{R}). \quad (81)$$

In the adiabatic limit [i.e., when H_{rot} is ignored and $\omega = 0$ is substituted into Eq. (80)] the *static exchange AN approximation* is obtained. If exchange effects are also neglected, i.e., $K = 0$ is used, then the *static AN approximation* is obtained. In the last 10 years, the majority of calculations for electron–molecule elastic scattering used one of these approximations. These calculations will be reviewed in some detail in a subsequent section.

d. *Complete BODY-Frame Treatment*

The complete BODY-frame treatment has been described in detail by Lane (1980). Much as for the complete LAB frame treatment, the electron-plus-molecule scattering function $\Psi^{JM}(1,2,\ldots,N,N+1)$ is expanded in terms of eigenstates that are generalizations of those given by Eq. (65) and that represent ground and excited states of the target molecule. They can be written in the form

$$\Phi_p(\mathbf{r}_1\sigma_1,\ldots,\mathbf{r}_N\sigma_N,\mathbf{R},\hat{\mathbf{r}}_{N+1},\sigma_{N+1})$$
$$= Y_{lm}(\hat{\mathbf{r}}_{N+1})[(2J+1)/8\pi]^{1/2}D^{J}_{M\Lambda}(\hat{R})\chi_{m_s}(\sigma_{N+1})\psi_\alpha(\mathbf{r}_1\sigma_1,\ldots,\mathbf{r}_N\sigma_N,\mathbf{R}), \tag{82}$$

where p, the channel index, refers to the collection of quantum numbers $p = (l, m, J, M, \Lambda, m_s, \alpha)$. The BODY-frame expansion then takes the form

$$\Psi^{JM}(1,2,\ldots,N,N+1)$$
$$= \mathscr{A}\left\{\sum_p r^{-1}U_p(r_{N+1},R)\Phi_p(\mathbf{r}_1\sigma_1,\mathbf{r}_2\sigma_2,\ldots,\mathbf{r}_N\sigma_N,\mathbf{R},\hat{\mathbf{r}}_{N+1},\sigma_{N+1})\right\}. \tag{83}$$

Substituting this expansion into the Schrödinger equation for the scattering function Ψ^{JM} and projecting it onto the channel states $\Phi_{p'}$, a coupled system of integrodifferential equations is obtained for the $U_p(r_{N+1}, R)$ functions (see Lane, 1980). These equations have channel and coupling potentials that are both local and nonlocal (owing to exchange), and through the vibrational Hamiltonian, $U_p(r_{N+1}, R)$ are connected to $U_p(r_{N+1}, R')$ with $R' \neq R$ (R is not a good quantum number). The rotational Hamiltonian connects the $U_p(r_{N+1}, R)$ functions with different Λ's, and Λ is not a good quantum number in the exact scheme.

If the AN approximation is introduced (i.e., terms containing the vibrational and rotational Hamiltonian are neglected), then Λ and R *become* good quantum numbers and the problem reduces to solving the Schrödinger equation for the motion of electrons with fixed nuclei, i.e., the clamped-nuclei Schrödinger equation. Furthermore, if only the ground-state wave function of the molecule, i.e., only terms with $\alpha = \alpha_0$ referring to the ground state, is included in the sum for p, then the *static exchange approximation* is recovered (see the previous section). Additional electronic states can be approximately taken into account by introducing *polarization and absorption potentials*, and the majority of the calculations that have been reported used semiempirical polarization potentials (see below). The few CC calculations (Chung and Lin, 1978, Weatherford, 1980, Holly *et al.*, 1981) reported to date were designed to calculate electronic excitation cross sections and will be discussed in Chapter 2. The conversion of the set of equations involving coupled electronic channels to a one-electron equation involving an effective (optical) potential goes beyond the scope of this work, and the interested reader is referred to the reviews of Fetter and Watson (1965), Truhlar *et al.* (1971), Truhlar *et al.* (1979) and Truhlar (1981), and references therein.

e. Adiabatic-Nuclei (AN) Calculations for Electron–Molecule Scattering

i. Low-Energy Theories   After it became clear that the AN approximation (when valid) gives practically identical results to those obtained by solving the LAB-frame CC equations, the majority of calculations reported were made using the AN approximation because of its greater simplicity than the LAB-frame CC equation technique. Recall that the AN approximation is expected to be valid *except*

(a)   close to the threshold,
(b)   at a resonance energy,
(c)   for a polar molecules (at small scattering angles).

The AN approximation requires solution of the Schrödinger equation in the continuum for the electron-plus-molecule system at fixed internuclear distances and orientation. Thus the electron–molecule scattering problem in the AN approximation is analogous to the problem of electron scattering by atoms (except that in the molecular case the target is not spherically symmetric). Consequently, in applying the AN approximation to electron–molecule scattering, the same terminology and very similar computational techniques can be used as for electron–atom scattering. For example, one can consider static (S), static exchange (SE), static exchange polarization (SEP) approximation schemes [see, e.g., Mott and Massey (1965) for the definition of these terms in electron–atom scattering theory]. The only difference is computational in that spherical symmetry cannot be utilized to solve the scattering equations. We discuss below the static, static exchange, and static exchange polarization scattering approximations that have been employed in the AN approximation.

Static (S) approximation.   The simplest approximation of electron scattering (atom or molecule) is when the electron is allowed to interact only with the electrostatic field of the target. In this case, the many-body Schrödinger equation reduces to a one-electron scattering problem characterized by a *local* potential. In the case of electron–atom scattering, the potential is spherically symmetric and the powerful partial-wave analysis technique can be used to reduce the problem to a series of differential equations. In the case of *electron–molecule* scattering (with fixed nuclei), the potential is *not* spherically symmetric and if a single-centered partial-wave analysis is performed, the various partial waves will be *coupled to each other* because of the anisotropy of the molecular electrostatic force field. One then obtains a system of differential equations coupled by the free-electron angular momentum. For diatomic molecular targets, where cylindrical symmetry holds, the dependence on the azimuthal angle can be factored out. The early calculations of Stier (1932) for

$e^- - N_2$ scattering and of Fisk (1936, 1937) for $e^- - N_2$, $-O_2$, $-H_2$, and $-Cl_2$ scattering were performed using a semiempirical form for the electrostatic interaction potential. The mathematical form for the electrostatic potential was chosen so that the Schrödinger equation was separable in prolate spheroidal coordinates. By adjusting free parameters in the interaction potential, reasonably good results were obtained for $e^- - N_2$ and $-O_2$ scattering integral cross sections, but poor results for those of electron–Cl_2 scattering. Subsequently, Nagahara (1953, 1954) treated $e^- - H_2$ scattering using an *ab initio* electrostatic potential, prolate spheroidal coordinates, and an expansion in terms of Legendre polynomials depending on one of these coordinates. He obtained a coupled system of differential equations (which he solved numerically in a drastically truncated form) and obtained substantially improved results over those of Fisk for the $e^- - H_2$ intergral elastic scattering cross section. All of these authors used the FN model for calculation of the cross sections. Takayanagi and Geltman (1965) used the distorted-wave approximation to solve the coupled radial equations (angular-momentum coupling) and reported FN elastic scattering and AN rotational excitation cross sections for $e^- - H_2$ and $e^- - N_2$ scattering (see Fig. 26). Recently, Crees and Moores (1975, 1977) used prolate spheroidal coordinates to solve the $e^- - N_2$ static problem, but they did not report any cross section values. Finally, we mention that the clamped-nuclei (CN) Schrödinger equation has been solved for a dipole potential (to model a polar molecule) by Mittleman and Holdt (1965) and for a hard sphere–dipole potential by Onda (1976).

Static exchange (SE) approximation. The first quantitative indication as to the importance of exchange in low-energy electron–molecule scattering came from Massey and Ridley (1956), who used the Hulthen–Kohn variational principle, along with a properly antisymmetrized trial function, to calculate the contribution of exchange. They obtained good agreement with the integral elastic scattering cross section data for $e^- - H_2$ scattering when exchange was added to the lowest symmetry partial wave ($l = 0$, $m = 0$). To solve the static exchange problem in general, however, is an arduous task, even for elastic $e^- - H_2$ scattering, because of the nonlocal and anisotropic nature of the exchange potential for molecules. The problem was initially attacked using three different approaches:

(a) using the single-center expansion technique (Wilkins and Taylor, 1967, Burke and Sinfailam, 1970);
(b) using spheroidal prolate coordinates along with an expansion technique (Hara, 1969a); and
(c) solving the partial differential equations in polar coordinates numerically (Tully and Berry, 1969).

When the single-center expansion is used [method (a)], a coupled system of integrodifferential equations is obtained that can be solved (in a truncated form) by

(i) noniterative techniques (Smith and Henry, 1973; Buckley and Burke, 1977; Raseev et al. (1978),

(ii) algebraic techniques (Crees and Moores, 1975; Schneider and Collins, 1981; Collins and Schneider, 1981), and

(iii) iterative techniques (Collins et al., 1978, 1980).

Various other numerical techniques, generally referred to as L^2 *techniques*, have been developed to handle the serious difficulties associated with the exchange terms. These include the R-matrix techniques of Schneider and co-workers (Schneider, 1975; Morrison and Schneider, 1977; Schneider and Hay, 1976) and of Burke and co-workers (Burke et al., 1977; Buckley et al., 1979; and Noble et al., 1982), the T-matrix technique of McKoy, Rescigno, and co-workers (Rescigno et al., 1974a,b, 1975; McCurdy et al., 1976; Rescigno et al., 1976; Fliflet and McKoy, 1978; Fliflet et al., 1978; Levin et al., 1980), the application of the Kohn variational method (Collins and Robb, 1980), and the application of the Schwinger variational method and extensions of it to this problem by McKoy and co-workers (Watson and McKoy, 1979; Lucchese and McKoy, 1979; Watson et al., 1980; Lucchese et al., 1980; Lucchese and McKoy, 1981; Watson et al., 1981; Takatsuka and McKoy, 1980, 1981a,b,c; Lee et al., 1981a,b). These various numerical techniques have been discussed in detail in contributions to the volume, "Electron–Molecule and Photon–Molecule Collisions" (eds. T. N. Rescigno et al., 1979) and are systematically reviewed in the comprehensive article by Buckley et al. (1984) on computational methods.

Since the numerical solution of the *exact* static exchange AN approximation is a difficult task, concerted efforts were made to develop local potentials that could be used as good approximations to the nonlocal exchange potential. These types of potentials are called *local exchange potentials* and are known to be useful for bound-state problems where their use originated with Slater (1951). The first reported successful derivation of a local exchange potential for scattering problems, a generalization and adoptation of Slater's procedure, was given by Hara (1967). Subsequently, other local exchange potentials, and modifications of the one introduced by Hara (1967), were recommended by Riley and Truhlar (1975), Baille and Darewych (1977), and Morrison and Collins (1978). This latest work reported results of both exact static exchange and static plus various types of local exchange potentials for $e^- - H_2$ and $-N_2$ scattering. This allowed an assessment to be made of the utility of the local exchange approximation for these systems. Although still approximate, the inclusion of exchange by these methods significantly improves the results over those employing purely static potentials.

An alternative, somewhat heuristic, but interesting technique was introduced by Burke and Chandra (1972) and applied to e^-–CO scattering by Chandra (1975). They argue that the major physical effect of exchange is to keep the incident electron away from the target electrons of the same spin. They accomplished this restriction computationally by retaining the static potential but replacing the exchange potential by a requirement of orthogonality between scattering orbital and the bound-state orbitals of the same spin. (Here we will refer to this approximation scheme as the static-orthogonalized exchange (SOE) approximation.) This technique, although it gives some good results, also showed fundamental deficiencies. A very comprehensive study has recently been reported by Collins *et al.* (1980) and Morrison and Collins (1981) in which they compared exact static exchange results with static local exchange and static orthogonalization results, as well as with the results of those calculations that combine the local exchange approximation with an orthogonalization requirement. They studied electron scattering by H_2, N_2, LiH, CO, HCl, and LiF, which represents scattering by a wide range of representative polar and nonpolar systems. From this study, they concluded that "orthogonalization alone is an insufficient representation of the exchange interaction for electron–molecule collisions." In some cases [e.g., e^-–HCl in Collins *et al.* (1980)] they obtained excellent results with a local exchange potential. However, in every case studied they obtained the best results by combining a local exchange potential with an orthogonalization procedure. The greatest utility of local exchange potentials and orthogonalization techniques is evidently for polyatomic molecules where exact static exchange calculations are extremely arduous and time consuming. There have been several studies of this type (Gianturco and Thompson, 1976, 1977; Salvini and Thompson, 1971; Morrison *et al.*, 1977; Morrison and Lane, 1977; Collins and Morrison, 1982; and Jain and Thompson, 1982) using local exchange potentials and/or orthogonalization, with encouraging results.

Static exchange plus polarization (SEP) approximation. The first accurate static exchange calculations (Hara, 1969a, Tully and Berry, 1969) showed that good agreement with experimental data can be obtained *only if* "polarization effects," as well as exchange, are incorporated into the scattering model and this fact is well illustrated in Fig. 22. In this figure, the integral cross section for rotational excitation of H_2 is shown for static (S), static model exchange (SME), static plus polarization (SP), and static model exchange plus polarization (SEP) in the AN approximation and compared with the LF–CC results of Henry and Lane (1969) and with experiment (Gibson, 1970). When both exchange and polarization are included, the results agree well with both the LF–CC and experimental data. As mentioned above, polarization is the back-coupling effect exerted on the electron by the electrostatic molecular field distorted by the incident electron. Mathematically, it was shown by Castillejo *et al.* (1960) that polarization originates from converting the coupling of electronic channels into an effective potential. The polarization potential is, in principle, a nonlocal, energy-dependent, complex potential, and it can be introduced by using a wide variety of different scattering formalisms [e.g., the Feshbach (1958) projection-operator formalism; the many-body Green's function formalism of Schneider et al. (1970)]. For a review and further references, see Fetter and Watson (1965), Truhlar et al. (1971), and Csanak and Taylor (1972). In its simplest form, called the adiabatic polarization potential, it reduces to the form

$$V(\mathbf{r}, \mathbf{R}) = -(1/r^4)[\alpha_0(R) + \alpha_2(R)P_2(\hat{R} \cdot \hat{r})]. \qquad (84)$$

Since this potential is singular at the origin and is certainly not applicable there (this form is correct only when $r \to \infty$), it is customary to introduce a cutoff function or parameter which makes it regular at the origin. The most frequently used form is written as

$$V(\mathbf{r}, \mathbf{R}) = -(1/r^4)[1 - \exp(-r/r_o)^6][\alpha_0(R) + \alpha_2(R)P_2(\hat{R} \cdot \hat{r})], \qquad (85)$$

where r_σ is an empirically chosen parameter. The cutoff function allows computations to be done at all r values and yet simulates the most important nonadiabatic polarization effect (Truhlar et al., 1971). The adiabatic polarization potential can be calculated in a more *ab initio* manner via first-order perturbation theory in which the molecular field is perturbed by a charged particle at a fixed distance. The difference in energy between the perturbed and unperturbed molecule depends on the location of the charged particle and provides the polarization potential (or more accurately *induced* polarization potential). This technique was employed by Lane and Henry (1968), Hara (1969b), Truhlar and Van Catledge (1978), Morrison and Hay (1977), Dixon et al. (1979), and Eades et al. (1979) to construct adiabatic polarization potentials for use in electron–molecule scattering calculations. The only polarization potential of this type that has been used in an AN approximation is by Hara (1969a,c) for H_2. The results are in agreement with experiment (see Figs. 16, 17, 22) and show that, for low incident energies, both exchange and polarization are important in order to obtain results that agree with experiment.

A somewhat more sophisticated approach, called the *polarized orbital approximation*, initially developed for electron–atom scattering by Temkin (1957), was applied in the AN approximation by Temkin and Vasavada (1967) and by Temkin et al. (1969) for e^-–H_2^+ scattering. In this approach the adiabatic polarization effect is introduced into the wave function and the Kohn–Hulthen variational principle is then used to derive a coupled system of differential-equations. It can be shown (see, e.g., Csanak and Taylor, 1972) that, in effect, the polarized orbital approximation introduces exchange polarization effects (i.e., second-order exchange) into the treatment within the adiabatic (with respect to the electron speed) approximation. The only calculation reported so far using this method is that of Temkin et al. (1969).

The polarization potential (adiabatic, semiempirical, nonadiabatic, etc.) can be added to any of the local exchange potentials or to a combination of the static potential and an orthogonalization procedure to simulate exchange (see, e.g., Morrison and Collins, 1978, 1981). For a complete review of calculations that belong to this class of approximations, the reader should consult Lane (1980) and Buckley et al. (1984).

Although the approaches discussed above introduce polarization in a semiempirical (or heuristic) manner, results of calculations have recently been reported that treat polarization in an *ab initio* way. Schneider (1977) reported results for $e^- - H_2$ scattering using the "pseudostate expansion" method, whereas Klonover and Kaldor (1978) reported $e^- - H_2$ elastic scattering and rotational excitation cross section results using the "many-body perturbation" theory along with a "T-matrix" technique. Recently, detailed calculations have been reported by Schneider and Collins (1982, 1983) that employed the optical potential method of Feshbach (1958) and an algebraic–equation technique for the integral cross section for $e^- - H_2$ and $e^- - N_2$ elastic scattering. Their results are in very good agreement with the experimental values. These calculations can be considered the first, complete *ab initio*, calculations in the AN-approximation for electron-molecule scattering that takes into consideration all physically important effects.

3. Frame Transformation Theory

Some of the most penetrating insight into the electron–molecule scattering process in recent times came from the observations of Fano (1970a) that the LAB-frame is appropriate for electron–molecule distances r that are much larger than the dimensions of the molecule and that the BODY-frame is appropriate for small values of r for which the electron is located inside of the molecule and forms a collision complex with it.

These observations led to the formulation by Fano (1970b) and by Chang and Fano (1972) of the *frame transformation* (FT) theory. This theory uses the BODY-frame at small electron–molecule distances r along with the AN approximation to calculate the inner part of the solution and the LAB-frame with neglect of exchange potentials at large r to calculate the outer part of the solution. These two parts of the solution are connected at a conveniently (and hence somewhat arbitrarily) chosen intermediate value, $r = r_o$. In its first application, Fano (1970b) used FT theory for interpretation of the photoionization spectra of H_2 (i.e., electron plus H_2^+ scattering). Chang and Fano (1972) also discussed application of the FT theory for the particle–vibrator model as follows. They divided the $0 < r < \infty$ range into regions A and B, separated at $r = r_o$, and region A was further subdivided into subregions Aa and Ab, separated from each other at $r = r_1$ (see Fig. 15). In region Aa (i.e., for $r < r_1$) the AN approximation is assumed to hold with respect to both rotation and vibration, and hence both R and Λ are good quantum numbers and Eq. (78) is used with $B = 0$ and $\omega = 0$. In region Ab (i.e., for $r_1 < r < r_o$), rotation is assumed to be adiabatic but vibration is treated dynamically and consequently Eq. (78) is used with $B = 0$. The two regions (i.e., Aa and Ab) are connected at $r = r_1$ in such a way that the solution $H_{l\Lambda}^J(r, R)$ (valid in the Aa region) is distributed according to the vibrational functions $\chi_v(R)$ at $r = r_1$ (the boundary of Aa and Ab) according to the formula

$$G_{l\Lambda v}^J(r) = \int_0^\infty dR\, \chi_v(R) H_{l\Lambda}^J(r, R), \qquad (86)$$

which is the inverse of formula (76). The $G_{l\Lambda v}^J(r_1)$ values so obtained are used at $r = r_1$ to initialize the $G_{l\Lambda v}^J(r)$ functions to be solved for in the Ab region. The integration of Eq. (78) with $B = 0$ then proceeds to the $r = r_o$ point. At $r = r_o$, the BODY-frame is changed to the LAB-frame and the

$$F_{ljv}^{J,l'j'v'}(r) = \sum_\Lambda U_{j\Lambda}^{lJ} G_{l\Lambda v}^J(r) \qquad (87)$$

relation is used at $r = r_o$ to initialize the $F_{jv}^{J,l'j'v'}(r)$ functions and Eq. (55) is integrated outward. In this asymptotic region it is not necessary that the electron–molecule interaction potential include an exchange contribution.

The FT theory for a molecule with internal structure has been described in detail by Lane (1980) and has been applied numerically to e^-–CO scattering by Chandra (1977) and to the electron–polar molecule scattering problem with a dipole potential by Clark (1979).

FT theory has also been generalized to polyatomic molecules by Harter et al. (1978) and reviewed in detail by Le Dourneuf et al. (1983).

An alternative method, called angular frame transformation (AFT), was introduced by Collins and Norcross (1978), principally for electron–polar molecule scattering. These authors argue that it is the electronic angular momentum that determines which system to use. If l refers to the electronic angular momentum quantum number, then for a range of angular momenta $l \leq l_t$ (l_t to be determined on physical grounds), the AN approximation should be used for any value of r, and for $l > l_t$ the LAB frame treatment is needed for all r values. This technique has been successfully applied by Collins et al. (1980) and Siegel et al. (1980a,b, 1981a,b), and the reader can find further details in the review of Norcross and Collins (1982).

Finally, we again note that electron scattering by strongly polar molecules poses a special problem for which the AN approximation is useful but can never be used exclusively. For small-angle scattering (i.e., for large values of l), the AN approximation has to be replaced by the LAB-frame treatment (i.e., to use AFT theory), or with a cutoff in the potential, and in some cases (e.g., for very large values of l), the LAB-frame Born approximation suffices. A comparison of theoretical results (Itikawa, 1976) obtained from LF–CC calculations and in the AN approximation is shown in Fig. 20 for which the LF–CC results are probably the most accurate. The results in this figure illustrate the inadequacy of the Born approximation and the difficulty with the Glauber model at intermediate and large scattering angles. For the special problems of electron–polar molecule scattering, the reader is referred to the work of Collins and Norcross (1978), the review of Norcross and Collins (1982), and to the recent work of Padial *et al.* (1983). These references also discuss the semiclassical and "scattering-length" methods that have been applied to electron–polar molecule scattering and are not discussed here.

Shimamura (1979) has made an interesting comparison between the "reduced" DCSs ($\theta \leq 70°$) predicted by various LF–CC calculations for vibrationally elastic scattering and that predicted by the Glauber theory by

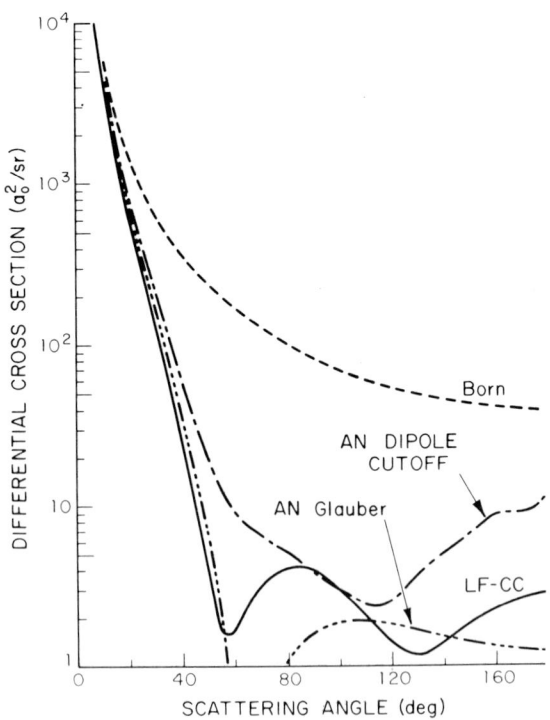

Fig. 20. Calculated DCSs for the $J = 0 \to 1$ rotational excitation cross section of CsF by electrons of 1-eV incident energy. The solid curve is the LF–CC results of Itikawa (1976). The ———— curve is the exact AN calculation results of Onda (1976) using dipole cut-off potential. The —–—–— curve is AN Glauber results of Ashihara *et al.* (1975). The dashed curve is LF–Born results of Itakawa (1976).

1 Elastic Scattering of Electrons by Molecules

plotting the data in terms of the "reduced" momentum transfer as shown in Fig. 34. The "reduced" DCS is defined as DCS(reduced) = (E_o/D^4)DCS, where E_o is the incident electron energy, D the dipole moment, and the "reduced" momentum transfer is $q_R = 2D(q/k_i)$, where q is the momentum transfer and k_i the incident electron wave number. The comparison in Fig. 34 shows that, for scattering angles from 0 out to some critical angle (depending on the dipole moment), the DCS for vibrationally elastic scattering is well described by the Glauber theory treating only the electron–dipole interaction. For the DCS beyond this critical angle, the shape and magnitude of the DCS depends on the details of the particular molecule.

4. High- and Intermediate-Energy Theories

In the high- and intermediate-energy region, it is natural to use the AN approximation for homonuclear molecules since the collision time is clearly short compared to the rotational and vibrational periods. On the other hand, in this energy region there are a large number of open electronic channels whose effect cannot be ignored. In actual calculations, only the ground electronic state is included explicitly and the effect of the open and closed electronic channels is approximately represented by absorption and polarization potentials. The physical description is simplified by the fact that, for these electron energies, perturbative approximations (which assume that the effect of the target upon the incident electron is small) work reasonably well and the accurate solution of the CC equations is, consequently, not necessary.

In the very high-energy region (i.e., above 1 keV incident electron energy) it is expected that both the incident and scattered electron can be described well by a plane wave (i.e., the distortion effect of the target is completely negligible), and the electron scattering process is well characterized by the Born approximation (Born, 1926). At somewhat lower energies (in the 100 eV < E_o < 1 keV region), distortion of the incident and scattered electron wave functions by the target molecule can no longer be ignored. Incorporation of this distortion effect into calculations was made principally along two lines: the Glauber approximation and multiple scattering theory combined with the partial wave method. The various approximations used to describe high-energy electron–molecule scattering have been discussed by Craggs and Massey (1956), Massey (1969) Davis (1971), and Bonham and Fink (1974) to which the reader is referred to for further details.

a. Born-Type Approximations

In Born-type calculations, both the incident and scattered electron are represented by plane-waves. The Born T matrix in the AN approximation is given in the following form for an N-electron diatomic molecule composed of atoms of nuclear charge Z_A and Z_B (Massey, 1930, Massey and Mohr, 1932a,b):

$$T^{Born}_{k_1 k_2}(\mathbf{R}) = \int \exp(-i\mathbf{k}_1 \cdot \mathbf{r}) \left\{ \sum_{i=1}^{N} \frac{e^2}{|\mathbf{r} - \mathbf{r}_i|} - \frac{Z_A e^2}{|\mathbf{r} - \mathbf{r}_A|} - \frac{Z_B e^2}{|\mathbf{r} - \mathbf{r}_B|} \right\} \exp(i\mathbf{k}_2 \cdot \mathbf{r})$$
$$\times \Psi_o(\mathbf{r}_1, \mathbf{r}_2, \ldots, \mathbf{r}_N; \mathbf{R}) \Psi_o(\mathbf{r}_1, \mathbf{r}_2, \ldots, \mathbf{r}_N; \mathbf{R}) \, d\mathbf{r} \, d\mathbf{r}_1 \cdots d\mathbf{r}_N$$
$$\equiv \int \exp(-\mathbf{k}_1 \cdot \mathbf{r}) V_{stat}(\mathbf{r}, \mathbf{R}) \exp(\mathbf{k}_2 \cdot \mathbf{r}), \tag{88}$$

where $V_{stat}(\mathbf{r}, \mathbf{R})$ is defined by the formula

$$V_{stat}(\mathbf{r}, \mathbf{R}) = -\frac{Z_A e^2}{|\mathbf{r} - \mathbf{r}_A|} - \frac{Z_B e^2}{|\mathbf{r} - \mathbf{r}_B|}$$
$$+ \sum_{i=1}^{N} \int \Psi_o^*(\mathbf{r}_1, \ldots, \mathbf{r}_N; \mathbf{R}) \frac{e^2}{|\mathbf{r} - \mathbf{r}_i|} \Psi_o(\mathbf{r}_1, \ldots, \mathbf{r}_N; \mathbf{R}) \, d\mathbf{r}_1 \cdots d\mathbf{r}_N \tag{89}$$

and represents the electrostatic potential of the ground-state molecule at fixed internuclear distance R and orientation $\hat{R}$.

b. *The Born–Oppenheimer and First-Order Approximations for Exchange*

The Born approximation does not take into account the possibility of the electron exchange between the incident electron and one of those of the target molecule. Exchange was introduced within the plane-wave approximation scheme by Oppenheimer (1928) and the resulting approximation is called the *Born–Oppenheimer scattering approximation*. Its application to electron–molecule scattering problem was first discussed by Massey and Mohr (1931). The Born–Oppenheimer (BO) approximation allows exchange but still assumes that both incident and scattered electrons are described by plane waves. For example, the T matrix for $e^- - H_2$ scattering can be given in the following form in the BO approximation (in atomic units):

$$T^{BO}_{k_1 k_2}(\mathbf{R}) = T^{Born}_{k_1 k_2}(\mathbf{R}) + T^{Opp}_{k_1 k_2}(\mathbf{R}), \tag{90}$$

where

$$T^{Opp}_{k_1 k_2}(\mathbf{R}) = \int \exp(-i\mathbf{k}_1 \cdot \mathbf{r}) \left\{ \sum_{i=1}^{2} \frac{1}{|\mathbf{r} - \mathbf{r}_i|} - \frac{1}{|\mathbf{r} - \mathbf{r}_A|} - \frac{1}{|\mathbf{r} - \mathbf{r}_B|} \right\}$$
$$\times \exp(i\mathbf{k}_2 \cdot \mathbf{r}_1) \Psi_o(\mathbf{r}_1, \mathbf{r}_2) \Psi_o(\mathbf{r}, \mathbf{r}_2) \, d\mathbf{r} \, d\mathbf{r}_1 \, d\mathbf{r}_2, \tag{91}$$

where i is an electron index and the subscripts A, B denote the nuclei.

The BO approximation was improved by Bell and Moiseiwitsch (1963) via the introduction of the so-called first-order exchange (FOE) approximation. In this approximation the T matrix is calculated from the formula

$$T^{BFOE}_{k_1 k_2}(\mathbf{R}) = T^{Born}_{k_1 k_2}(\mathbf{R}) + T^{FOE}_{k_1 k_2}(\mathbf{R}), \tag{92}$$

where

$$T^{FOE}_{k_1 k_2} = \int \exp(-i\mathbf{k}_1 \cdot \mathbf{r}) \Psi_o(\mathbf{r}_1, \mathbf{r}_2) \left\{ \left[\sum_{i=1}^{2} \frac{1}{|\mathbf{r} - \mathbf{r}_i|} - \frac{1}{|\mathbf{r} - \mathbf{r}_A|} - \frac{1}{|\mathbf{r} - \mathbf{r}_B|} \right] \right.$$
$$\left. - V_o(\mathbf{r}, \mathbf{r}_A, \mathbf{r}_B) \right\} \exp(+i\mathbf{k}_2 \cdot \mathbf{r}) \Psi_o(\mathbf{r}, \mathbf{r}_2) \, d\mathbf{r} \, d\mathbf{r}_2, \tag{93}$$

with

$$V_o(\mathbf{r}, \mathbf{r}_A, \mathbf{r}_B) = \int \left[\sum_{i=1}^{2} \frac{e^2}{|\mathbf{r} - \mathbf{r}'_i|} - \frac{Z_A e^2}{|\mathbf{r} - \mathbf{r}_A|} - \frac{Z_B e^2}{|\mathbf{r} - \mathbf{r}_B|} \right] |\Psi_o(\mathbf{r}'_1 \mathbf{r}'_2; \mathbf{R})|^2 \, d\mathbf{r}'_1 \, d\mathbf{r}'_2. \tag{94}$$

c. *The Bonham–Ochkur–Rudge Approximations for Exchange*

Further improvements upon the BO and FOE approximations were made by Bonham (1962), Ochkur (1963), and Rudge (1965) who introduced what is frequently referred to as the *Ochkur (O) or Ochkur–Rudge (OR) approximation*. To be historically accurate, we shall refer to it as the Bonham–Ochkur–Rudge (BOR) approximation. In this approximation, the T matrix is given by the formula

$$T^{BBOR}_{k_1 k_2}(\mathbf{R}) = T^{Born}_{k_1 k_2}(\mathbf{R}) + T^{BOR}_{k_1 k_2}(\mathbf{R}), \tag{95}$$

where (in the Ochkur form)

$$T^{BOR}_{k_1 k_2}(\mathbf{R}) = \frac{1}{k_1^2} T^{BORN}_{k_1 k_2}(\mathbf{R}). \tag{96}$$

d. *Incorporation of Absorption and Polarization*

The next step in the improvement beyond the plane-wave theories was the (approximate) incorporation of coupling between open and closed electronic channels via absorption and polarization potentials. As mentioned above, the electron–molecule scattering problem can be reformulated, using optical (or effective) potentials (Feshbach, 1958, Fetter and Watson, 1965, Truhlar *et al.* 1971, Csanak and Taylor, 1972), as a one-electron scattering problem with an energy-dependent, nonlocal, complex potential. The optical potential, and the plane-wave approximation, can be introduced into the optical potential formulation. In this scheme (sometimes referred to as the polarized Born (PB) approximation), the T matrix is given in the form

$$T^{PB}_{k_1 k_2}(\mathbf{R}) = \int \exp(-i\mathbf{k}_1 \cdot \mathbf{r}) V_{opt}(\mathbf{r}, \mathbf{r}'; \mathbf{R}) \exp(i\mathbf{k}_2 \cdot \mathbf{r}') d\mathbf{r} \, d\mathbf{r}', \quad (97)$$

where $V_{opt}(\mathbf{r}, \mathbf{r}'; \mathbf{R})$ is the (approximate) optical potential and may include static, exchange, polarization, and absorption terms; i.e.,

$$V_{opt}(\mathbf{r}, \mathbf{r}'; \mathbf{R}) = V_{stat}(\mathbf{r}; \mathbf{R}) \delta(\mathbf{r} - \mathbf{r}') + V_{exch}(\mathbf{r}, \mathbf{r}'; \mathbf{R}) + V_{pol}(\mathbf{r}, \mathbf{r}'; \mathbf{R}) + V_{abs}(\mathbf{r}, \mathbf{r}'; \mathbf{R}). \quad (98)$$

The static part of V_{opt}, V_{stat}, is a *local* potential. The exchange part, by nature, is *nonlocal* and its contribution to the T matrix can be approximated, e.g., by the Bonham–Ochkur–Rudge (BOR) approximation to the T matrix: $T^{BOR}_{k_1 k_2}(R)$. The quantity $V_{pol}(\mathbf{r}, \mathbf{r}'; \mathbf{R})$ is energy-dependent and nonlocal but usually approximated by an energy-independent (adiabatic) polarization potential according to

$$V_{pol}(\mathbf{r}, \mathbf{r}'; \mathbf{R}) = V_{pol}(\mathbf{r}, \mathbf{R}) \delta(\mathbf{r} - \mathbf{r}'), \quad (99)$$

which is also used in calculations that consider absorption effects, e.g.,

$$V_{abs}(\mathbf{r}, \mathbf{r}'; \mathbf{R}) = V_{abs}(\mathbf{r}; \mathbf{R}) \delta(\mathbf{r} - \mathbf{r}'). \quad (100)$$

e. Summary of Born-type Calculations

The first calculation reported for any electron–molecule scattering process was by Massey (1930) who used the Born approximation, as given by Eq. (88), to study high-energy electron scattering by H_2. Massey (1930) pointed out that at high energies the molecular scattering can be approximated by the coherent scattering of two separated "atomic targets," resulting in a diffraction-type of pattern superimposed on the "atomic" differential cross section. This observation led to the field of gas-phase electron diffraction studies, which will not be discussed here; see, for example, Massey (1969), Davis (1971) and Bonham and Fink (1974). The approximation scheme in which the molecular scattering is approximated by the coherent scattering of two "atomic" (i.e., spherically symmetric) targets is called the *separated atom (SA) approximation*. The atomic scattering amplitude need not be calculated in the Born approximation, and the atomic targets need not be identical to the free atoms. If the atomic targets are identified with the free atoms then the SA approximation is called the *independent atom model* (IAM). The work of Massey (1930) on $e^- - H_2$ scattering was completed by Massey and Mohr (1932a). These authors showed that the orientation-averaged elastic scattering Born cross section can be written in the form

$$\sigma^{Born}(\vartheta) = \sigma^{Born, SA}(\vartheta) + \delta\sigma, \quad (101)$$

1 Elastic Scattering of Electrons by Molecules

where $\sigma^{\text{Born,SA}}(\vartheta)$ is the Born–SA approximation and has the form

$$\sigma^{\text{Born,SA}}(\vartheta) = [1 + \sin(KR)/KR]\sigma_{\text{at}}(\vartheta), \tag{102}$$

where $\sigma_{\text{at}}(\vartheta)$ is the Born cross section for scattering from the atomic target, $K = |\mathbf{k}_2 - \mathbf{k}_1| = 2k \sin \vartheta/2$ the magnitude of the momentum transfer, $\delta\sigma$ is called the multiple-scattering contribution.

It was shown by Massey that $\delta\sigma$ in Eq. (101) can be neglected for $KR \geq \pi$, and Massey and Mohr (1932a) showed that at high incident energies $\delta\sigma$ is always small. At high incident energies, when the electron de Broglie wavelength is small relative to the interatomic distance, the multiple-scattering description is useful. [For a more detailed discussion of the multiple-scattering approach see, e.g., Craggs and Massey (1956) and Massey (1969).]

The SA approximation in the Born approximation was applied by Massey and Mohr (1932a) to $e^- - N_2$ scattering, and they obtained good agreement for the relative DCS in the intermediate-energy region.

Following these initial reports, few intermediate- and high-energy calculations were reported until 1965, probably because of a lack of new experimental data. Bonham and Iijima (1963) studied the influence of various molecular wave functions upon the Born approximation for $e^- - H_2$ scattering. The next significant report on this field was that of Khare and Moiseivitsch (1965), who investigated the accuracy of the SA approximation using the Born approximation and introduced exchange effects using the FOE approximation of Bell and Moiseiwitsch (1963). They showed that for $e^- - H_2$ scattering the SA approximation is accurate to within 4% (i.e., $\delta\sigma$ is less than 4% of σ^{Born}). They also studied the effect of exchange and showed that incorporation of exchange improved agreement with experiment for electron energies less than 200 eV. However, the only experimental data available at that time were still from the 1930s. It was not until 1969 that new experimental data became available for intermediate energy $e^- - H_2$ scattering (William, 1969). The appearance of these new data in the literature was soon followed by the Born-type calculations of Khare and Shobha (1972, 1974). They showed that results for the DCS improve dramatically for small angles ($\theta < 20°$) if polarization is included into the optical potential. With the inclusion of these two physical effects, reasonable agreement is obtained with experiment for an incident electron energy as low energy as 30 eV.

Beginning in the early 1970s, the quality and quantity of experimental data improved significantly, and more detailed comparisons could be made with existing theories. This was done in great detail for the H_2 target by van Wingerden et al. (1977) who concluded that the experimental data lie closest to the theoretical results of Khare and Shobha (1972). However, some deviation was detected between theory and experiment at low scattering angles for impact energies greater than 400 eV. Subsequently, Gupta and Khare (1978) introduced an energy-dependent polarization potential that improved the agreement with experiment, but for an unknown reason, theory and experiment still disagree for $E > 2$ keV and small scattering angles.

f. Glauber (Eikonal) Approximations

After the work of Massey and collaborators in the 1930s (Massey, 1930, Massey and Mohr, 1932a,b, 1934, Massey and Bullard, 1933), it became obvious that the distortion effect the molecule exerts upon the incident and scattered electron has to be taken into consideration to obtain highly accurate results in the high-energy region and good accuracy in the intermediate energy region. One way of doing this is by using the Glauber, or eikonal, approximation. The eikonal approximation (also called geometrical optics or semiclassical approximation) was introduced by Moliérc (1947) for potential-scattering problems and later generalized by Glauber (1954, 1959) for many-particle systems.

In its simplest application, for potential scattering problems with potential $U(\mathbf{r})$, the eikonal scattering amplitude is given by the formula (Glauber, 1954; 1959)

$$f_E = \frac{k}{2\pi i} \int d\mathbf{b} \exp(i\mathbf{K} \cdot \mathbf{b}) \{\exp[i\chi(\mathbf{b},\mathbf{k})]\} - 1, \qquad (103)$$

where the *eikonal phase shift* is given by

$$\chi(\mathbf{b},\mathbf{k}) = -\frac{1}{2k} \int_{-\infty}^{\infty} U(\mathbf{b},z) \, dz \qquad (104)$$

and the z axis is chosen along the direction of the incident particle and $\mathbf{b}$ is the impact parameter. The general radius vector $\mathbf{r}$ of the electron is given in the form $\mathbf{r} = \mathbf{b} + z \cdot \mathbf{v}$, where $\mathbf{v}$ is the unit vector perpendicular to the momentum transfer vector $\mathbf{K} = \mathbf{k}_1 - \mathbf{k}_2$.

In the generalization of the eikonal formula by Glauber (1954, 1959), $U(\mathbf{r})$ became the interaction potential of the electron with the multielectron system $U(\mathbf{r}) \to U(\mathbf{r}, \mathbf{r}_1, \ldots, \mathbf{r}_N)$, $\chi(\mathbf{b}, k)$ became an operator

$$\chi^G(\mathbf{b},\mathbf{k}) = -\tfrac{1}{2} \int_{-\infty}^{\infty} U(\mathbf{b}, z; \mathbf{r}, \ldots, \mathbf{r}_N) \, dz, \qquad (105)$$

and the Glauber scattering amplitude is defined by

$$f_G(i \to f) = \frac{k}{2\pi i} \int d\mathbf{b} \exp(i\mathbf{q} \cdot \mathbf{b}) \langle f | \exp[i\chi^G(\mathbf{b},\mathbf{k})] - 1 | i \rangle, \qquad (106)$$

where $|f\rangle$ and $|i\rangle$ are the final and initial target states, respectively, of the N-electron system. In the case of elastic scattering by an N-electron molecule, we obtain

$$f_G(0 \to 0) = \frac{k}{2\pi i} \int d\mathbf{b} \exp(i\mathbf{q} \cdot \mathbf{b}) \langle 0|\exp[i\chi^G(\mathbf{b},\mathbf{k})] - 1|0\rangle. \quad (107)$$

Under certain circumstances [discussed in detail by Yates and Tenney (1972)], the integral can be approximated in the following way:

$$\langle 0|\exp[i\chi^G(\mathbf{b},\mathbf{k})] - 1|0\rangle \simeq \exp[i\langle 0|\chi^G(\mathbf{b},\mathbf{k})|0\rangle] - 1$$
$$\equiv \exp[i\bar{\chi}^G(\mathbf{b},\mathbf{k})] - 1, \quad (108)$$

where

$$\bar{\chi}^G(\mathbf{b},\mathbf{k}) = -\frac{i}{k} \int_{-\infty}^{\infty} dz \langle 0|V(\mathbf{r},\mathbf{r}_1,\ldots,\mathbf{r}_N)|0\rangle \quad (109)$$

and

$$f_G^{\text{stat}}(0 \to 0) = \frac{k}{2\pi i} \int d\mathbf{b} \exp(i\mathbf{q} \cdot \mathbf{b}) \{\exp[i\bar{\chi}^G(\mathbf{b},\mathbf{k})] - 1\}. \quad (110)$$

This is called the Glauber static approximation, which is equivalent to the eikonal approximation for a static potential U [see Eqs. (103) and (104)]. A natural generalization, in the framework of the Glauber approximation, is obtained by including static, exchange, polarization, and absorption terms into the potential U.

The first application of the Glauber (eikonal) approximation to electron–molecule scattering was reported by Yates and Tenney (1972), who studied double-scattering effects for e^-–N_2 scattering at 500 eV and 40 keV incident energies, and for I_2 and U_2 (a hypothetical molecule) for 40-keV incident energies. At 500-eV incident energy, agreement with experiment was poor for e^-–N_2 scattering, and they concluded that polarization, exchange, and chemical binding effects are important.

Subsequently, Ashihara et al. (1975) treated electron–polar molecule scattering using the Glauber (eikonal) approximation in which the point-dipole potential was included within the Glauber approximation and the quadrupole potential was treated within the Born approximation. Bhattacharyya and Ghosh (1975) used an optical-potential eikonal theory to calculate the elastic scattering cross section for e^-–H_2 scattering in which the optical potential $V_{\text{opt}}(\mathbf{r})$ was given in the form

$$V_{\text{opt}}(\mathbf{r}) = V_{\text{stat}}(\mathbf{r}) + V_{\text{pol}}(\mathbf{r}) + V_{\text{exch}}(\mathbf{r}), \quad (111)$$

where $V_{\text{stat}}(\mathbf{r})$ represents static, $V_{\text{pol}}(\mathbf{r})$ polarization, and $V_{\text{exch}}(\mathbf{r})$ local exchange potentials, respectively. Bhattacharyya and Ghosh (1976) used the same scheme within the AN approximation to calculate the rotational excitation cross section for H_2. A very similar calculation, with the inclusion of an absorption potential, was reported by Khare et al. (1977) for electron–H_2 elastic scattering. The incorporation of exchange within the eikonal approximation and the SA approximation was discussed by Huang and Chan (1977), M. K. Srivastava et al. (1978), and Jhanwar et al. (1980), and a detailed study on static, exchange, polarization, and absorption effects was reported by Dhal et al. (1979). Gianturco and Lamanna (1980) reported calculations in the Glauber approximation for e^--H_2 and e^--N_2 scattering which included the long-range quadrupole and the polarization potential.

g. *Separated-Atom Distorted-Wave (SA–DW) Theories*

Based on an investigation of gas-phase electron diffraction data, Schomaker and Glauber (1952) observed that the Born approximation breaks down for molecules containing heavy atoms at high energies ($E > 1$ keV), even though the "separated-atom" approximation is known to be valid [see, e.g., Massey (1969: pp. 678–682)] at this energy. The reason for this breakdown is the absence of consideration of "distortion effects" in the Born approximation which are more important for heavy than light atoms at this energy. These effects can be introduced by solving the electron scattering Schrödinger equation for the electron–atom problem (using the electrostatic field of the atom along with the partial-wave expansion) and introducing this atomic scattering amplitude into the separated-atom formula. The first application of this technique was made by Massey and Bullard (1933), who calculated the DCS for elastic e^--N_2 scattering in the energy range 30–780 eV and found reasonable agreement with experiment. They found that, even at 780 eV, there is substantial deviation from the Born approximation. At intermediate energies ($E_o \lesssim 200$ eV), the SA approximation does not hold because long-range electrostatic and polarization effects become important.

A detailed multiple-scattering scheme was developed by Hayashi and Kuchitsu (1976) to account for both short- and long-range effects. The short-range potentials are assumed to be centered on the atoms, and the long-range potential is assumed to be centered in the "center of the molecule." Scattering from each short-range potential, and from the long-range potential, is determined by the partial wave method. The total effect of all potentials is considered within the multiple-scattering series [see, e.g., Rodberg and Thaler (1967)] and further approximations were introduced to make the calculations practical. This theory was applied by Hayashi and Kuchitsu (1977) to e^--H_2 elastic scattering, and excellent agreement with experiment was obtained at 200- and 400-eV energies but somewhat less satisfactory agreement at 75-, 81.6- and 100-eV energies.

A similar, but somewhat simpler, two-potential approch has been used by Choi et al. (1979) for e^-–N_2 scattering. The scattering amplitudes for the separated atoms were calculated with the partial-wave technique and the cross section in the SA approximation was calculated from those amplitudes. The cross section associated with the long-range potential was calculated independently using the Born approximation. The two cross sections were then added to obtain the molecular cross section (i.e., it was assumed that the short- and long-range scatterings combine "incoherently"). The same model was successfully applied by Lee and Freitas (1981a,b, 1983) for e^-–H_2, e^-–N_2, and e^-–CO scatterings. A still different multiple-scattering method was introduced by Dill and Dehmer (1977) based on the "muffin-tin" potential method used in solid-state physics. In spite of its great successes the discussion of this method goes beyond the scope of the present work. It was reviewed in detail by Lane (1980).

C. Summary of Theoretical Results

Tables VI and VII contain summaries of theoretical work that has been reported for DCS and ICS results on elastic electron scattering by molecules and of rotational excitation of molecules by electron impact. (Calculations reporting eigenphase sums, resonance energies, and widths are *not* included into these tables.) Table VI contains a summary of the work employing LAB frame approximations and Table VII a summary of the clamped-nuclei results. Each table is further subdivided according to the molecules.

IV. Comparison of Experimental and Theoretical Results

A. Historical Perspective

The theory of electron–molecule scattering was motivated, from the very beginning, by experimental data that needed to be interpreted. The original experimental data were *relative differential cross sections* (*DCS*), which included elastic scattering, rotational excitation (and sometimes vibrational excitation), *integral cross sections* (*ICS*) and *momentum transfer cross sections*. Since the greatest number of experiments were for H_2 and N_2 targets, the first calculations (Massey, 1930; Massey and Mohr, 1931; Massey and Bullard, 1933; Stier, 1932; Fisk, 1936, 1937; Roscoe, 1938) were performed to interpret these data. Massey and collaborators concentrated on the high- and intermediate-energy region and used the Born, Born–Oppenheimer, or

(*Text continues on p. 100*)

TABLE VI
LAB Frame Approximations

	Author(s)	Energy Range^a (eV)	Quantities Reported	Comments
	Close-coupling			
H$_2$	Lane and Geltman (1967)	0–20.4	σ_E, σ_R	SMEP
	Lane and Henry (1968)	0.03–20.0	σ_R	SP
	Lane and Geltman (1969)	0–20.4	σ_{ER}, σ_R, DCS_E, DCS_R, σ_{ER}^M, σ_E	SMEP
	Henry and Lane (1969)	0–13.6	σ_E, σ_R, DCS_E, DCS_R, σ^M, σ_{ER}	S, SE, SEP
	Truhlar and Brandt (1976)	10, 40	σ_E, σ_R, σ_{ER}, σ_E^M, σ_R^M, DCS_{ER}, DCS_R, DCS_E	SMEP
	Feldt and Morrison (1982)	0–3.8	σ_R, DCS_R	SMEP
N$_2$	Brandt et al (1976)	30–75	DCS_E, DCS_R, σ_E^M, σ_R^M, σ_{ER}^M, σ_{ER}^M	SMEP
	Truhlar et al (1976)	5, 10	σ_E, σ_R, DCS_{ER}^M	SMEP
	Onda and Truhlar (1978)	30–50	DCS_E, σ_E, σ_E^M, σ_R, DCS_R, DCS_{ER}^M, σ_R^M	SP, SMEP
	Onda and Truhlar (1979a)	30	DCS_{ER}, σ_E, σ_E^M, DCS_E, DCS_R, σ_E, σ_R, σ_E^M, σ_R^M	SMEP
	Onda and Truhlar (1979b)	50	DCS_E, DCS_R, DCS_{ER}	SMEP
	Onda and Truhlar (1979c)	10, 50	DCS_E, DCS_R, σ_E, σ_R, σ_E^M, σ_R^M	
	Onda and Truhlar (1980a)	20–35	DCS_E, DCS_R, σ_E, σ_E^M	SMEP
	Onda and Truhlar (1980b)	30	DCS_E, DCS_R	SMEP

1 Elastic Scattering of Electrons by Molecules

	Reference	Energy range	Quantities
CO	Crawford and Dalgarno (1971)	0.005–0.1	$\sigma_E^M, \sigma_R^M, \sigma_E, \sigma_R, DCS_E, DCS_R$
	Onda and Truhlar (1980c)	10	$DCS_E, DCS_R, \sigma_E, \sigma_R, \sigma_E^M, \sigma_R^M$, SMEP
CsF	Allison (1975)	1, 2	$\sigma_R, \sigma_R^M, DCS_R$
	Itikawa (1976)	1–5	$DCS_E, DCS_R, \sigma_E, \sigma_R, \sigma_E^M, \sigma_R^M$
	Collins and Norcross (1978)	1	$\sigma_{ER}, \sigma_{ER}^M$
	Rudge (1978b)	0.25–8.0	$\sigma_{ER}, \sigma_{ER}^M, DCS_E, DCS_R, DCS_{ER}$
HCl	Itikawa and Takayanagi (1969)	0.01–20.0	$\sigma_E, \sigma_R, \sigma_{ER}$
	Itikawa (1969)	0.01–5.0	$DCS_E, DCS_R, \sigma_E^M, \sigma_R^M$
	Itikawa and Ashihara (1971)	1	DCS_R
	Gianturco and Rahman (1978)	0–0.4	σ_E, σ_R
LiF	Collins and Norcross (1977)	0.3–7.0	$\sigma_{ER}^M, \sigma_E^M, \sigma_R^M$
	Collins and Norcross (1978)	0.5–3.0	$\sigma_E^M, \sigma_R^M, \sigma_E, \sigma_R, DCS_{ER}$
	Rudge (1978a)	0.5–20.0	$\sigma_{ER}, DCS_{ER}, \sigma_{ER}^M, \sigma_E, \sigma_R, \sigma_E^M, \sigma_R^M$
KOH	Collins et al. (1979)	0.01–10.0	σ_{ER}^M, σ^E
CsOH			
KI	Collins and Norcross (1978)	6.74	$DCS_{ER}, \sigma_{ER}^M, \sigma_E^M, \sigma_R^M, \sigma_E, \sigma_R, \sigma_{ER}$
	Rudge (1978b)	0.25–8.0	$\sigma_E, \sigma_R, \sigma_E^M, \sigma_R^M$

(table continues)

TABLE VI (cont.)

	Author(s)	Energy Range[a] (eV)	Quantities Reported	Comments
HF	Rudge (1980)	0.46-1.5	σ_R, σ_R, DCS_{ER}	
CO₂	Onda and Truhlar (1979d)	1-20	DCS_{ER}, σ_E, σ_R, DCS_E, DCS_R	SMEP
	Thirumalai et al (1981a)	10	DCS_E, σ_E, DCS_R, σ_R	SMEP
C₂H₂	Thirumalai et al (1981b)	10	DCS_E, DCS_R, DCS_{ER}, σ_E, σ_R, σ_{ER}, σ_E^M, σ_R^M, σ_{ER}^M	SMEP
CH₄	Abusalbi et al (1983)	10	DCS_E, DCS_R, σ_E, σ_R, σ_{ER}	SMEP
	Distorted Wave			
H₂	Dalgarno and Henry (1965)	0.05-1.0	σ_R	
	Sampson and Mjolsness (1965)	0.05-1.0	σ_R	
	Lane and Geltman (1967)	0.04-25.0	σ_R	
	Varracchio (1979)	0.8-40	σ_R, σ_E	
N₂	Mjolsness and Sampson (1964)	0.007-1.0	σ_R	
	Sampson and Mjolsness (1965)	0.007-1.0	σ_R	
KI	Rudge et al (1976)	6.74, 15.7, 60.0	DCS_{ER}, σ_{ER}^M, σ_{ER}^M	
	Rudge	0.5-20.0	σ_{ER}, σ_{ER}^M, DCS_{ER}	

1 Elastic Scattering of Electrons by Molecules

Born			
H_2	Carson (1954)	0-250	σ_R
	Gerjuoy and Stein (1955b)	0.03-0.2	σ_{ER}, σ_R
	Dalgarno and Moffett (1963)		
	Dalgarno and Henry (1965)	0.005-1.0	σ_R
	Lane and Geltman (1967)	0.04-25.0	σ_R
	Truhlar and Rice (1970)	1-100	σ_E, DCS_E
	Trajmar et al (1970)	7-100	DCS_E
	Truhlar et al (1971)	45	DCS_E
	Truhlar and Rice (1974)	30-200	DCS_E
	Feldt and Morrison (1982)	0.005	DCS_R, σ_R
N_2	Gerjuoy and Stein (1955a)	0.02-0.6	σ_{ER}, σ_E
	Truhlar (1972)	30, 83	DCS_E
	Brandt and Truhlar (1973)	10, 30	DCS_E, σ_{ER}, σ_{ER}^M
H_2O	Itikawa (1972)	0.01-0.08	σ_R
NH_3	Itikawa (1971a)	0.01-0.1	σ_R
	Itikawa (1971b)	0.01-0.1	σ_R

(table continues)

TABLE VI (cont.)

	Author(s)	Energy Range[a] (eV)	Quantities Reported	Comments
	Hybrid Theory			
N_2	Chandra and Temkin (1976a)	0-5	σ_{ER}, DCS_{ER}, σ_{ER}^M	
	Chandra and Temkin (1976b)	5, 10	DCS_{ER}	
CO	Poe and Choi (1979)	0.1-10.0	σ_{ER}, σ_{ER}^M, DCS_{ER}	
H_2	Poe and Choi (1981)	0.75-10.0	σ_{ER}, DCS_{ER}	
D_2	Poe and Choi (1981)	1-10	σ_{ER}	
	Frame-Transformation			
CO	Chandra (1977)	0.005-10.0	σ_E^M, σ_R^M, σ_{ER}^M, σ_E, σ_R, σ_{ER}, DCS_E, DCS_R	
	Multiple-Frame			
LiF	Collins and Norcross (1978)	0.544, 2.0, 3.0, 5.44, 20	σ_E, σ_R, σ_E^M, σ_R^M, DCS_{ER}	
	Siegel et al (1980)	5-20	DCS_{ER}	
	Siegel et al (1981a)	1.0-20.0	σ_{ER}, σ_{ER}^M, σ_E, σ_R	
CsCl	Siegel et al (1981b)	5, 20	DCS_E, DCS_R, DCS_{ER}	
HCl	Padial et al (1983)	0.01-12.0	σ_{ER}, σ_E, σ_R	

[a] Energy range usually refers to *the broadest range* for which *any* of the cross sections were reported. (It does not mean than all cross sections are available in that range.)

1 Elastic Scattering of Electrons by Molecules

TABLE VII

Clamped Nuclei (CN) Approximation

H₂

Author(s)	Energy Range* (eV)	Quantites Reported	Comments
Close-Coupling			
Fisk (1936)	0-25	$\langle\sigma\rangle$	S (semi-empirical)
Nagahara (1954)	1-25	$\langle\sigma\rangle$	S
Carter et al (1958)	1.2-23.0	$\langle\sigma\rangle$	S, SE (spherical approximation)
Oksyuk (1966)	0-25	$\sigma_E, \sigma_R, \sigma_{ER}$	S (semi-empirical)
Hara (1967)	0-11	$\langle\sigma\rangle, \langle\sigma^M\rangle, \langle DCS\rangle$	SME
Wilkins and Taylor (1967)	0.1-13.6	$\langle\sigma\rangle$	SE
Tully and Berry (1969)	0-4.9	$\langle\sigma\rangle, \langle DCS\rangle$	SE
Hara (1969a)	0.3-27.8	$\langle\sigma\rangle, \langle\sigma^M\rangle, \langle DCS\rangle$	SE
Hara (1969c)	0.5-13.6	σ_R, DCS_R	SE
Rescigno et al (1975)	1.2-6.7	$\langle\sigma\rangle, \langle DCS\rangle$	SE
Schneider (1975)	0-13.6	$\langle\sigma\rangle, \langle DCS\rangle$	SE
Schneider (1977)	0-11	$\langle\sigma\rangle$	SEP
Baille and Darewych (1977)	0-20	σ_E, DCS_E, DCS_R	SME
Morrison and Collins (1978)	0.3-20.4	$\langle\sigma\rangle$	S, SME, SE
Klonover and Kaldor (1978)	1-10	$\sigma_E, DCS_E, \sigma_R, DCS_R$	SE, SEP
Levin et al (1979)	3.4-6.7	DCS_R	SE
Bell (1981)	0.1-60	$\langle\sigma\rangle$	SEP
Feldt and Morrison (1982)	0-4	DCS_R, σ_R	SMEP
Schneider and Collins (1982)	0.1-10.0	$\langle\sigma\rangle$	SEP
Gibson and Morrison (1983)	0-10	σ_R, DCS_R	SMEP

TABLE VII (cont.)

	Author(s)	Energy Range (eV)	Quantities Reported	Comments
N_2	Stier (1932)	0-10	$\langle\sigma\rangle$, $\langle DCS\rangle$	S (semi-empirical)
	Fisk (1936)	0-25	$\langle\sigma\rangle$	S (semi-empirical)
	Oksyuk (1966)	0-25	σ_E, σ_R, σ_{ER}	S (semi-empirical)
	Burke and Sinfailam (1970)	0-27.2	$\langle\sigma\rangle$, σ_R	SE
	Burke and Chandra (1972)	0-10	$\langle\sigma\rangle$, $\langle\sigma^M\rangle$, σ_R, $\langle DCS\rangle$	SOE
	Brandt and Truhlar (1973)	10, 30	$\langle DCS\rangle$	SP (spherical approximation)
	Sawada et al (1974)	0-500	$\langle\sigma\rangle$, $\langle DCS\rangle$	SP (semi-empirical)
	Chandra (1975)	0-10	DCS_E, DCS_R, σ_E^M, σ_R^M	SOE
	Morrison and Schneider (1977)	0-13.6	$\langle\sigma\rangle$, $\langle DCS\rangle$	SE
	Buckley and Burke (1977)	1-5	$\langle\sigma\rangle$, $\langle\sigma^M\rangle$, σ_R, $\langle DCS\rangle$	SE
	Flifet et al (1978)	0.5-10.0	$\langle\sigma\rangle$, $\langle\sigma^M\rangle$, σ_R^M, σ_E	SE
	Morrison and Collins (1978)	0.1-13.6	$\langle\sigma\rangle$, $\langle\sigma^M\rangle$, $\langle DCS\rangle$	SMEP
NH_3	Massey and Mohr (1932 a)	10, 30, 60	$\langle DCS\rangle$	
	Altshuler (1957)	0.2-1.0	σ_{ER}	
H_2O	Altshuler (1957)	0.36	σ_{ER}	
$CsCl$	Ashihara et al (1975)	3-15	$\langle DCS\rangle$, $\langle\sigma^M\rangle$	

1 Elastic Scattering of Electrons by Molecules

		Glauber		
H_2	Bhattacharyya and Ghosh (1975)	9.4–100.0		$\langle DCS \rangle$, $\langle \sigma \rangle$
	Bhattacharyya and Ghosh (1976)	45		$\langle DCS \rangle$, DCS_R
	Huang and Chan (1977)	30–200		$\langle DCS \rangle$, $\langle \sigma \rangle$
	Khare et al (1977)	30–200		$\langle DCS \rangle$
	Srivastava et al (1978)	30, 75, 200		$\langle DCS \rangle$
	Bhattacharyya et al (1978)	20–200		DCS_E, DCS_R, $\langle DCS \rangle$, σ_E, σ_R, $\langle \sigma \rangle$, σ_E^M, σ_R^M, $\langle \sigma^M \rangle$
	Gien (1978)	100		$\langle DCS \rangle$, $\langle \sigma \rangle$
	Jain et al (1978)	75–700		$\langle DCS \rangle$
	Dhal et al (1979)	50, 100, 200		$\langle DCS \rangle$
	Jhanwar et al (1980)	50–1000		$\langle DCS \rangle$
	Gianturco and Lamanna (1980)	10–50		DCS_E, DCS_R, σ_E, σ_R
	La Gattuta (1980)	40, 100, 200		DCS_E, DCS_R
	Jhanwar et al (1982)	50–1000		$\langle DCS \rangle$, $\langle \sigma \rangle$
N_2	Yates and Tenney (1972)	400, 40000		$\langle DCS \rangle$
	Bhattacharyya and Goswami (1982)	20–200		DCS_E, DCS_R, $\langle DCS \rangle$, σ_E, σ_R, $\langle \sigma \rangle$, σ_E^M, σ_R^M, $\langle \sigma^M \rangle$
CsCl	Ashihara et al (1975)	4.5–15.0		DCS_E, DCS_R, $\langle DCS \rangle$, $\langle \sigma \rangle$, $\langle \sigma^M \rangle$
	Takayanagi (1974)	4.47		DCS_R
KI	Shimamura (1976)	6.74, 15.7, 60.0		$\langle DCS \rangle$

(table continues)

TABLE VII (cont.)

	Author(s)	Energy Range (eV)	Quantites Reported	Comments
	Multiple-Scattering Partial-Wave			
N_2	Massey and Bullard (1933)	30, 60, 83, 205, 410, 780	$\langle DCS \rangle$	
	Wedde and Strand (1974)	40-1000	$\langle \sigma \rangle$, $\langle DCS \rangle$	
	Hayashi and Kuchitsu (1976)	100-500	$\langle DCS \rangle$	
	Dill and Dehmer (1977)	0-1000	$\langle \sigma \rangle$	
	Rumble and Truhlar (1979)	13.6	$\langle \sigma \rangle$	S, SP, SME
	Burke et al (1983)	1-31.6	$\langle \sigma \rangle$	SEP
	Rumble et al (1983)	5-50	$\langle \sigma \rangle$, $\langle \sigma^M \rangle$, c_R, $\langle DCS \rangle$, DCS_E, DCS_R	SMEP
CO	Chandra (1975)	0.1-10.0	$\langle \sigma^M \rangle$	SOE
	Levin et al (1980)	1.2-6.7	$\langle \sigma^M \rangle$	SE
F_2	Schneider and Hay (1976)	0-12	$\langle \sigma \rangle$, $\langle DCS \rangle$	
	Rescigno et al (1976a)	0-14	$\langle \sigma \rangle$	SE
HCl	Collins et al (1980)	3	$\langle DCS \rangle$	SMEP
CO_2	Morrison et al (1977)	0.05-10.0	$\langle \sigma^M \rangle$, $\langle DCS \rangle$, $\langle \sigma \rangle$	S, SME, SMEP
	Morrison and Lane (1977)	0-10	σ_E, σ_R	SMEP
	Lucchese and McKoy (1982)	0-13.6	$\langle \sigma \rangle$, $\langle DCS \rangle$	SE
	Collins and Morrison (1982)	0.2-2.0	$\langle \sigma \rangle$	SE
CsCl	Onda (1976)	4.77	DCS_E, DCS_R	
O_2	Fisk (1936)	0-25	$\langle \sigma \rangle$	S (semi-empirical)
	Oksyuk (1966)	0-25	σ_E, σ_R, σ_{ER}	S (semi-empirical)
CH_4	Gianturco and Thompson (1976)	0-13.6	$\langle \sigma \rangle$	SOEP
	Jain and Thompson (1982)	0-15	$\langle \sigma \rangle$, $\langle DCS \rangle$	SMEP
H_2O	Jain and Thompson (1982)	1-10	$\langle \sigma^M \rangle$	

1 Elastic Scattering of Electrons by Molecules

	Reference	Energy range	Cross sections
H_2	Massey and Mohr (1932a)	30, 80, 400	⟨DCS⟩
	Massey and Mohr (1934)	29, 83, 205	⟨DCS⟩
	Roscoe (1938)	400	⟨DCS⟩
	Khare and Moiseiwitsch (1965)	50-912	⟨DCS⟩
	Rozsnyai (1967)	50-400	⟨DCS⟩, ⟨σ⟩
	Khare and Shobha (1972)	30-912	⟨DCS⟩
	Ford and Browne (1973)	arbitrary	⟨DCS⟩, ⟨σ⟩*	⟨DCS⟩ is given as a function of K valid for any E
	Liu and Smith (1973)	arbitrary	⟨DCS⟩
	Khare and Shobha (1974)	30-912	⟨DCS⟩, ⟨σ⟩
	Gupta and Khare (1978)	100-2000	⟨DCS⟩, ⟨σ⟩
	*⟨σ⟩ results were published by van Wingerden et al (1977).		
	Siegel et al (1978)	0-30	⟨DCS⟩
	Choi et al (1979)	50-500	⟨σ⟩, ⟨σ^M⟩, ⟨DCS⟩
	Siegel et al (1980b)	0-1000	⟨σ⟩, ⟨DCS⟩
	Lee and Freitas (1983)	50-800	⟨σ⟩, ⟨DCS⟩
H_2	Hayashi and Kuchitsu (1977)	75-400	⟨DCS⟩
	Lee and Freitas (1981a)	100-1000	⟨DCS⟩
	Lee and Freitas (1981b)	20-81.6	⟨DCS⟩
O_2	Wedde and Strand (1974)	40-1000	⟨σ⟩
CO	Khare and Raj (1979)	400	⟨DCS⟩
	Lee and Freitas (1983)	500-800	⟨σ⟩, ⟨DCS⟩

* Energy range usually refers to the broadest energy range for which any of the cross sections when reported. (It does not mean that all cross sections are available in that range.)

DW–SA approximations to interpret the experimental data. These studies led to the introduction of static, exchange, and polarization effects, as well as the separated-atom (SA) and independent-atom (IA) models for high-energy scattering. Stier (1932) and Fisk (1936, 1937) concentrated on low-energy ICS data, used the FN model, and solved the CN Schrödinger equation with a static potential (with mixed success). Massey (1932) used the LF–Born approximation to interpret electron transport data in HCl gas and introduced the LF–CC scheme.

A significant development in interpreting low-energy swarm data was made by Gerjuoy and Stein (1955a,b) who proved that, close to threshold, results from the LF–Born approximation can explain the behavior of electron swarms in H_2 *if* the long-range quadrupole interaction potential is included. Their model was further improved by including polarization effects (Dalgarno and Moffet, 1963) and by allowing for distortion (Sampson and Mjolness, 1965).

Another significant development came when Lane and Geltman (1967) reported the first LF–CC calculation results for $e^- - H_2$ scattering, and subsequently, Wilkins and Taylor (1967), Hara (1969a,c), and Tully and Berry (1969) reported accurate AN–static exchange results for $e^- - H_2$ scattering. This was the first time that a quantitative assessment could be made as to the individual contributions to the elastic scattering that is produced by the static and the exchange interactions. The polarization effect was also introduced but was treated semiempirically (Hara, 1969a,c; Henry and Lane, 1969). These quantitative results verified Massey's prediction that static, exchange, and polarization effects are necessary to explain elastic electron–molecule scattering.

The accurate treatment of polar molecules also began in 1969 with the LF treatment of Itikawa and Takayanagi (1969) and Crawford and Dalgarno (1971). Simultaneously, it became clear that, because of divergencies, the AN approximation cannot be applied to describe elastic electron scattering by polar molecules (Altschuler, 1957, Mittleman and von Holdt, 1965, Garrett, 1971). A general formalism was needed to encompass all electron—molecule scattering and such a formalism was introduced by Chang and Fano (1972) via the frame transformation theory. This theory was later modified by Collins and Norcross (1978) to make it more applicable to electron–polar molecule scattering.

B. The Validity of Certain Physical Approximations

Comparison of experimental data with the results of numerical calculations provides information about the validity of certain approximation methods (for the energy range and molecule considered) and the answers provide physical insight into the electron scattering process. There are a number of questions that one should ask about the physics of the scattering process.

(1) What is the role of the nuclear motion during the scattering process? That is, under what conditions is it necessary to use the LF–CC equations and under what conditions does the AN approximation hold?

(2) How "rigorous" must the calculation be? That is, is it necessary to solve the complete LF–CC or BF–CC equations, or does the DW, Glauber, or Born approximation suffice? The answer to this question depends critically on the incident electron energy.

(3) What is the importance of the various parts of the interaction potentials such as static, exchange, polarization, and absorption? The answer to this is also energy dependent.

(4) What are the different physical effects that determine electron scattering by heteronuclear compared with homonuclear molecules?

(5) What is the dominant physical interaction that determines the integral, differential, and momentum transfer cross sections? What are the physical interactions that determine the elastic scattering cross sections as opposed to those for rotational excitation?

We shall try to give examples as partial answers to these questions. A detailed discussion on the comparison of experimental data with theoretical results was given, molecule by molecule, by Lane (1980), and very detailed comparisons of theory and experiment for polar molecules was made by Itikawa (1978a,b), Norcross and Collins (1982), and Shimamura (1983). The influence of various static, exchange, and polarization interaction potentials on the DCS was discussed, in the context of the LF–CC approximation, by Truhlar (1981).

1. LF–CC versus AN Approximation

Except close to threshold (i.e., for $E_o < 1$ eV), the rotational CC calculations are expected to give essentially identical results for homonuclear molecules to those from AN calculations if the physical effects considered in the two schemes are identical (see Sections II.A and II.B). This conclusion is supported by the results shown in Fig. 16 and Fig. 17, where the LF–CC calculations of Henry and Lane (1969) are compared with the results from AN calculations of Hara (1969a) for the differential and integral cross sections of $e^- - H_2$ scattering, respectively. The relationships between LF–CC and AN calculations for rotational excitation of H_2 close to threshold were recently investigated by Feldt and Morrison (1982) and Morrison et al. (1983). As shown in Fig. 18 the results from LF and AN calculations are in good agreement at 1.0 eV but generally disagree at 0.1 eV in the forward scattering direction. The Born results generally *underestimate* the cross section at all incident energies.

In a series of papers Ashihara *et al.* (1975), Onda (1976), and Itikawa (1976) made a similar study for rotational excitation of the *strongly polar molecule*, CsF, and the results from these calculations for rotational excitation are shown in Fig. 20. The results presented in this figure show the disagreement between the AN and LF frame representations for such targets at intermediate and large scattering angles, which is probably due to the breakdown in the AN approximation for strongly polar molecules. It also illustrates the breakdown of the Born approximation for strongly polar molecules, which, in this case, *overestimates* the cross section.

2. *Approximations: CC, DW, and Born*

Figure 19 shows results from the LAB frame calculations of Lane and Geltman (1967) for rotational excitation of H_2 using the CC, DW, and Born approximations and compares them with experiment. The results presented in this figure show that, for a particular frame representation (LF in this case), the DW approximation gives results very similar to those from CC calculations and in reasonably good agreement with experiment. As shown in the previous figures, the Born approximation *underestimates* the cross section. Detailed studies were made by Rudge (1978a) for LiF, in which several scattering approximations were compared and from which it was concluded that the DW approximation does not work well for strongly polar molecules because of the importance of channel coupling [see also Norcross and Collins (1982)].

3. *Different Physical Effects in the Interaction Potential*

Massey (1930) introduced the fundamental physical effects for elastic scattering as static, exchange, and polarization. The importance of these various physical interactions is shown in Figs. 17, 21, 22, and 23. Figure 17 contains a comparison of results from static exchange (SE) and static exchange polarization (SEP) calculations for the H_2 integral "elastic" cross section in both the LF–CC and the FN representations. It is clear from Fig. 17 that polarization effects must be included to obtain even qualitative agreement with experimental data. The results in Fig. 21 show that, if *only* polarization is combined with the static potential, the results (SP) are also in poor agreement with the experimental data. The results in Fig. 22 show the importance of the same potential terms for the rotational excitation integral cross section of H_2. It is clear that if static, exchange, and polarization effects are included (SEP), then good agreement is obtained with experiment.

1 Elastic Scattering of Electrons by Molecules

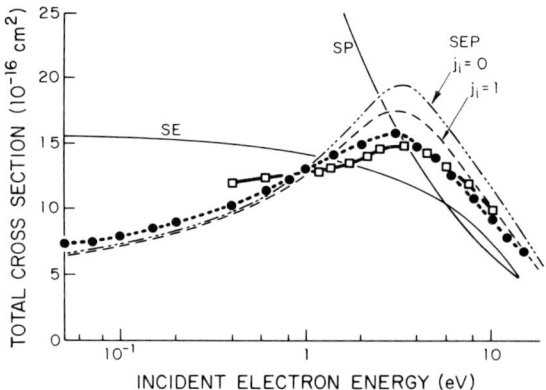

Fig. 21. Total electron scattering cross section for H_2 as a function of the incident electron energy. The experimental data are from Golden *et al.* (1966, ●) and Ramsauer and Kollath (1930, □) and the theoretical results are from the LF–CC calculations of Henry and Lane (1969). The SEP curves labeled $j_i = 0$ and 1 denote results obtained using only those initial rotational states.

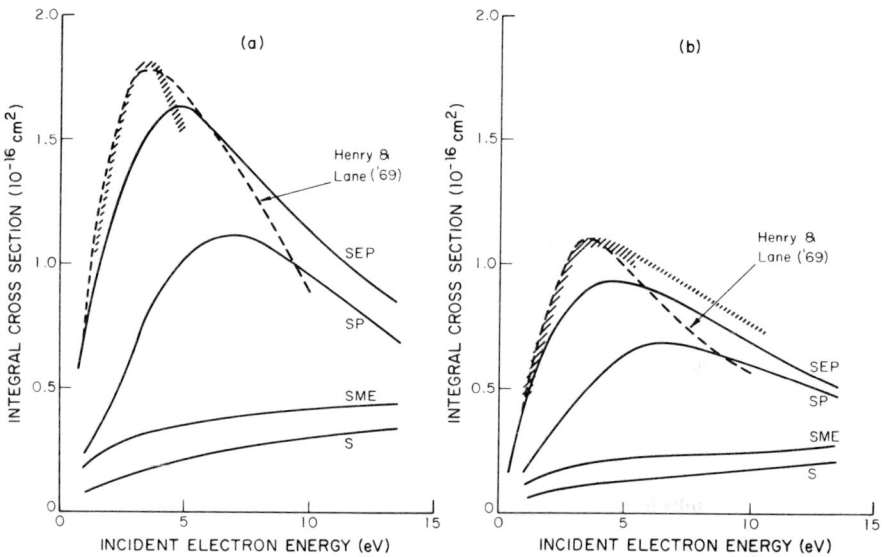

Fig. 22. (a) Integral cross section for rotational excitation ($j = 0 \to 2$) of H_2 by electrons in the energy range 0–15 eV. The shaded region denotes the cross section obtained by Gibson (1970) from an analysis of electron swarm data. The solid curves denote the AN approximation results of Hara (1969c) using static (S), static model exchange (SME), static plus polarization (SP), and static (model) exchange plus polarization (SEP) for the electron dynamics. The dashed curve denotes the LF–CC results by Henry and Lane (1969). (b) Same as (a) except for the $j = 1 \to 3$ rotational excitation process, with the shaded area denoting data obtained by Gibson (1970) and Linder and Schmidt 1971a.

C. Specific Comparisons between Theory and Experiment

Having discussed the most important physical interactions that need to be included to obtain good agreement between theory and experiment in the previous section, we now turn to some general comparisons. The purpose of the figures included in this section is to give the reader a feeling as to the current level of understanding of these scattering processes. The interested reader is also directed to the excellent summaries given by Lane (1980) and by Norcross and Collins (1982).

1. Molecular Hydrogen

Figure 23 contains a comparison of the available experimental data with results from various theoretical models for elastic scattering of 7 and 10 eV

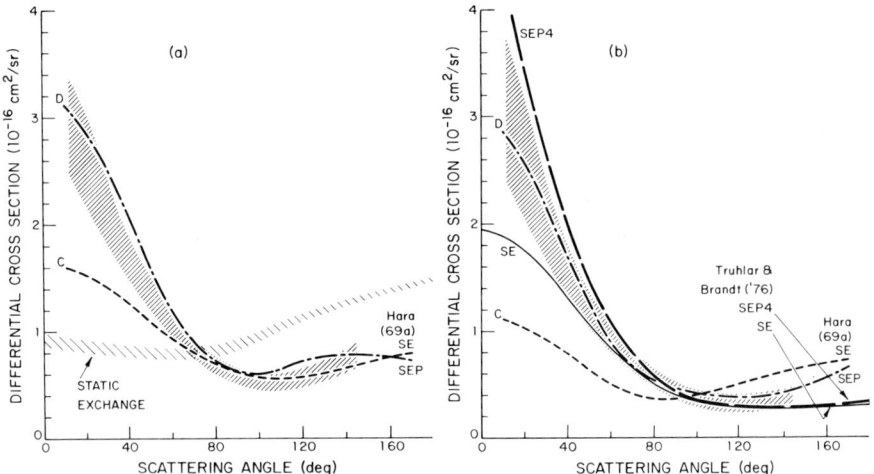

Fig. 23. (a) DCSs for "elastic" scattering of 7-eV incident energy electrons by H_2. The more densely shaded region contains the experimental data of Shyn and Sharp (1981), Linder and Schmidt (1971a), and Srivastava *et al.* (1975a). Linder and Schmidt resolved rotational excitation from elastic scattering and their results represent pure elastic scattering. The data of Shyn and Sharp lie along the upper edge of this shaded region for all scattering angles. The less densely shaded region contains the results of the static exchange body frame calculations of Schneider (1975), Rescigno *et al.* (1975), and Tully and Berry (1969). The two individual lines denote the FN approximation results of Hara (1969a), the specific model being indicated at the left end of the curves. (b) Same as (a) except for an incident electron energy of 10 eV. Again, the experimental results from Shyn and Sharp (1981) lie along the upper boundary of the shaded region. As is the case in (a), the results from Hara's model D (with polarization) are in excellent agreement with experiment. The results from two of the model calculations by Truhlar and Brandt (1976) are also shown by individual lines [SE is static plus model exchange; SEP4 is static plus model exchange plus polarization (model 4).]

electrons by H_2. To accentuate the experimental–theoretical comparison, a densely shaded region is used in Fig. 23 to denote the data from three different experiments. Similarly, the static exchange results of Schneider (1975) and of Rescigno et al. (1975), which agree well with each other, are shown in Fig. 23a as the lightly shaded region. The static exchange (with and without polarization) results of Hara (1969a), obtained using the AN approximation, are also shown. In Fig. 23b the LF–CC results of Truhlar and Brandt (1976) for two different scattering potentials are also shown. The results in Fig. 23 demonstrate again the necessity of including target polarization in the theory if quantitative agreement with experiment is to be obtained. It is worth noting that the comparisons in Fig. 23 show that polarization is *relatively* less important at 10 than it is at 7 eV.

A similar comparison is made at higher incident electron energies, 40 and 100 eV, in Fig. 24. Again the experimental data are given as a shaded region, and in Fig. 24a the LF–CC results of Truhlar and Brandt (1976) for SE and SE plus a model polarization potential are shown, whereas in Fig. 24b results from Born (Khare and Shoba, 1974) and from Glauber (Bhattacharyya and Ghosh, 1976) calculations are shown. The comparisons contained in Fig. 24 indicates that (i) target polarization is unimportant except for scattering angles

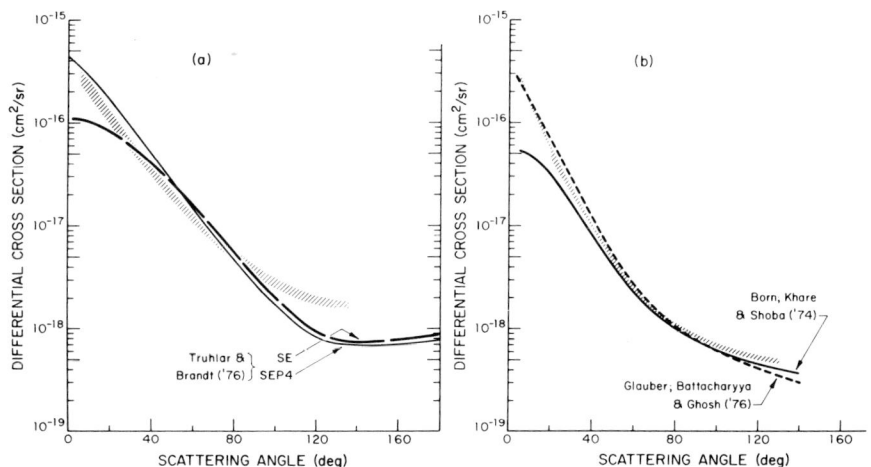

Fig. 24. (a) DCSs for elastic scattering of 40-eV incident energy electrons by H_2. The experimental data are from Shyn and Sharp (1981) and from Srivastava et al. (1975a) and are denoted by shading. The individual curves denote the static model exchange (SE) and static model exchange polarization (SEP4) lab frame results of Truhlar and Brandt (1976). (b) Same as (a) except for an incident electron energy of 100 eV. The shaded region denotes the range of the experimental data summarized by van Wingerden et al. (1977). Also shown are the Born results of Khare and Shoba (1974) and the Glauber model results of Battacharyya and Ghosh (1976).

less than 20° for electron energies greater than 40 eV and (ii) that the Glauber, and to a lesser extent, the Born models do reasonably well in describing the elastic scattering DCS at 100 eV and above, out to scattering angles of about 100°. All the theoretical models appear to underestimate the DCSs for scattering angles greater than 100°.

Figure 25 contains a comparison between measured and calculated DCSs for rotational excitation of H_2 at 4.5 and 10 eV incident electron energy. The experimental data are a Boltzmann-weighted average of the $j = 0 \to 2$ and $1 \to 3$ transitions. The AN approximation (SEP) results of Hara (1969c) are shown by the heavy solid and dashed lines, and the LF–CC model potential results of Truhlar and Brandt (1976) at 10 eV are shown by the light dashed

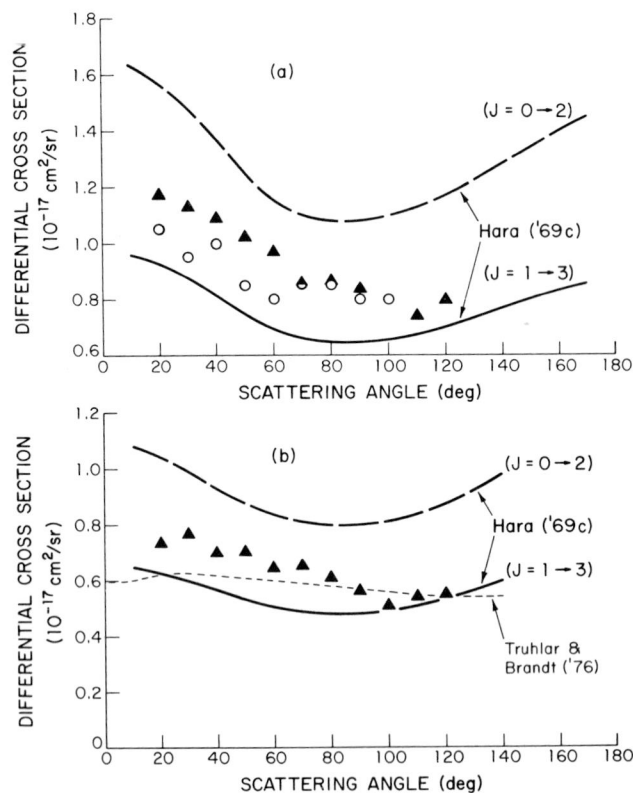

Fig. 25. (a) DCSs for rotational excitation of H_2 for an incident electron energy of 4.5 eV. The experimental data (○) are from Wong and Schulz (1974) and (△) from Linder and Schmidt (1971a) for the $J = 1 \to 3$ transition. The results from the AN–SEP calculations of Hara (1969c) are shown by the heavy lines. (b) Same as (a) except for an incident electron energy of 10 eV. In addition to the theoretical results of Hara (1969c), the LF–CC model potential results of Truhlar and Brandt (1976) are also shown.

1 Elastic Scattering of Electrons by Molecules

line. Both of the theoretical models predict results that are in reasonable agreement with experiment.

The experimental and theoretical integral cross sections for elastic scattering and for rotational excitation of H_2 are compared in Fig. 26 over the incident electron energy range from 0.001 to 1000 eV. The upper group of

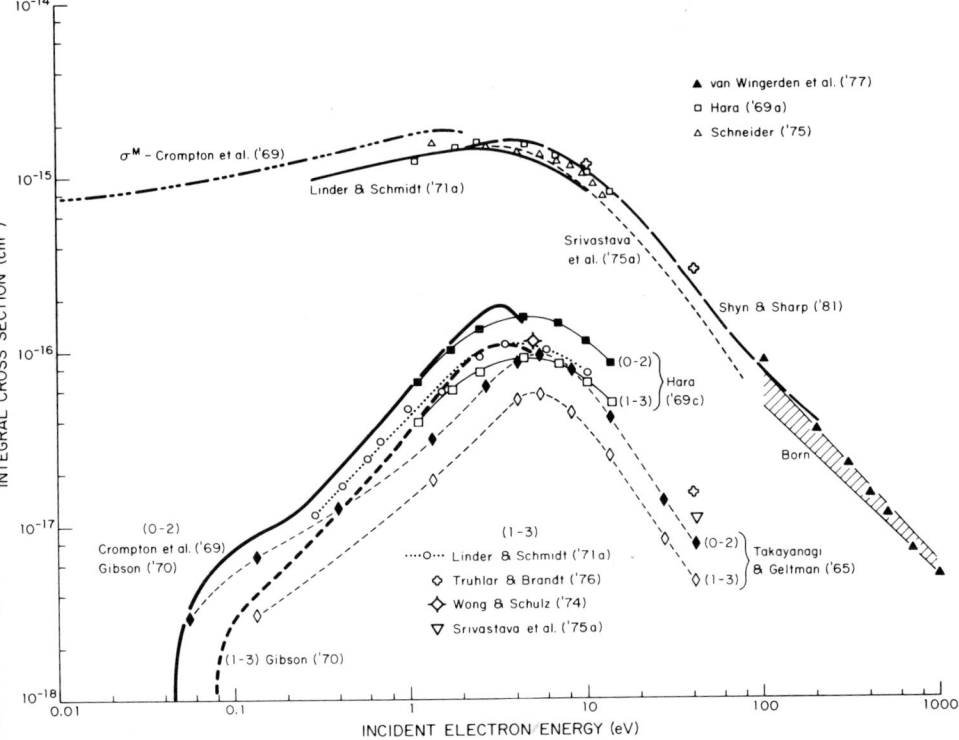

Fig. 26. Integral cross sections as a function of incident electron energy from 0.01 to 1000 eV for elastic electron scattering and rotational excitation of H_2. For "elastic" scattering, the experimental results are from Linder and Schmidt (1971a), Srivastava et al. (1975a), from Shyn and Sharp (1981) and from van Wingerden et al. (1977) at the higher energies. The results from Linder and Schmidt represent pure elastic scattering without rotational excitation. The integral total momentum transfer cross section deduced from electron swarm measurements (Crompton et al., 1969) is also shown for comparison. Theoretical results from Hara (1969c), Schneider (1975), Truhlar and Brandt (1976) and various Born model results from Gupta and Khare (1978) are shown. The lower grouping of curves are from rotational excitation and include experimental results for the (1-3) transition from Linder and Schmidt (1971a), Wong and Schulz (1974), Srivastava et al. (1975a) and Gibson (1970) (from swarm data). For the (0 → 2) transition, only the results from the analysis of swarm data by Crompton et al. (1969) and by Gibson (1970) are available. The corresponding theoretical results of Hara (1969c) and of Takayanagi and Geltman (1965) and of Truhlar and Brandt (1976) are shown for comparison.

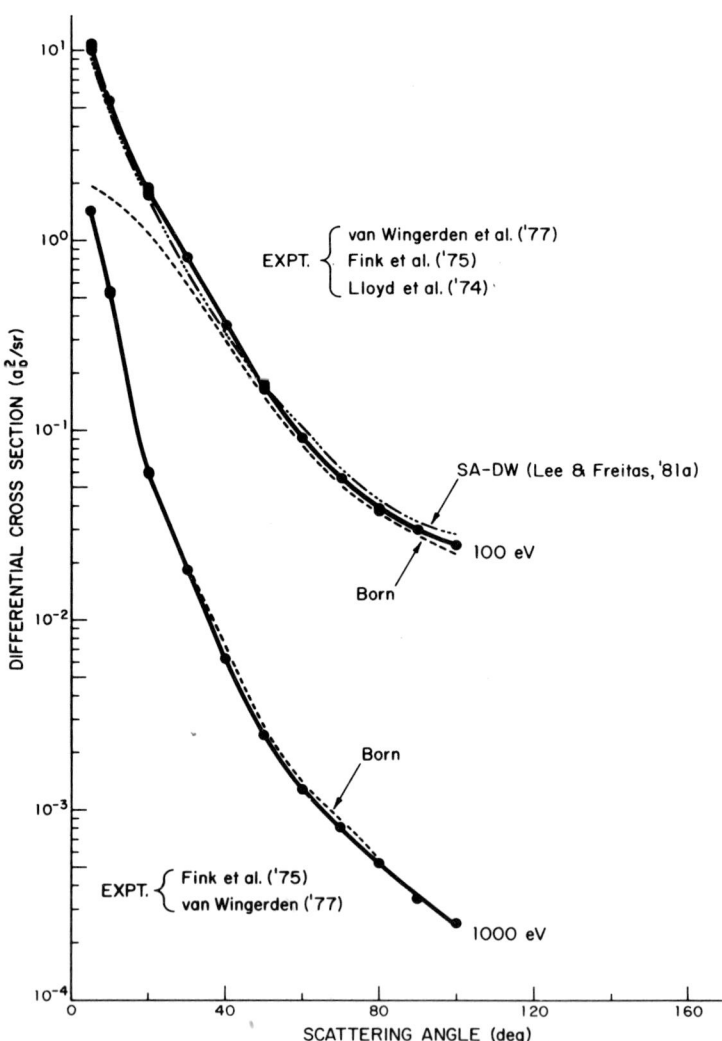

Fig. 27. DCSs for vibrationally elastic scattering by H_2 at 100 and 1000 eV. The experimental data, which are in excellent agreement with each other, were taken from van Wingerden *et al.* (1977), Fink *et al.* (1975a), and Lloyd *et al.* (1974). The Born results (Ford and Browne, 1973) and the separated atom–distorted wave (SA–DW) results of Lee and Freitas (1981a) are also shown.

curves and data points are for elastic scattering and the lower set is for rotational excitation. For reference purposes the momentum transfer cross section determined by Crompton et al. (1969) is also shown. The agreement between theory and experiment for elastic scattering, over the entire energy range, has to be considered as very good and that for rotational excitation it is reasonably good. One concludes from these comparisons that the physics governing elastic scattering and rotational excitation of H_2 is now reasonably well understood.

Figure 27 shows the DCS for vibrationally "elastic" electron scattering of 100- and 1000-eV electrons by H_2. As shown in this figure, the first-order-type theories (except Born at 100 eV, $\theta < 40°$) agree very well with the experimental data. At these energies, particularly 1000 eV, the effects of exchange and polarization are very small and "first order" theories accurately describe the scattering process.

2. Molecular Nitrogen

After H_2, N_2 is the next most frequently studied molecule, both experimentally and theoretically, for elastic scattering and rotational excitation. Figure 28 shows comparisons of the DCSs from theory and experiment for incident electron energies of 5, 30, and 400 eV. In Fig. 28a the BF-static exchange calculations of Siegel et al. (1978, 1980b) and the hybrid SEP results of Chandra and Temkin (1976a,b) are combined in the shaded region. The LF–CC results of Truhlar et al. (1976) are shown as the dot–dash line, and the relatively poor agreement with experiment is attributed to incomplete convergence of the rotational state expansion. The agreement between theory and experiment at 30 eV (Fig. 28b) is generally good except for the AN–Glauber model at scattering angles greater than 100° and is a result of the *absence* of exchange in the Glauber theory. As shown in Fig. 28c, the agreement between theory and experiment is excellent at electron energies of 400 eV and higher, indicating that the physics for the scattering process is well understood at these electron energies.

The integral cross section for "elastic" scattering of electrons by N_2 is shown in Fig. 29, for incident electron energies from 0.1 to 1000 eV. Data from three different molecular beam experiments are shown in the figure as well as the momentum transfer cross section of Engelhardt et al. (1964) for purposes of comparison. The results of SE calculations, which agree well with each other and are shown as the shaded band, predict the 2.4-eV resonance at too high an electron energy. The BF results of Choi et al. (1979) agree well with experiment for energies greater than 50 eV. The vibrational structure known to be present in the elastic cross section around the 2.4-eV resonance is not reproduced by the data shown here.

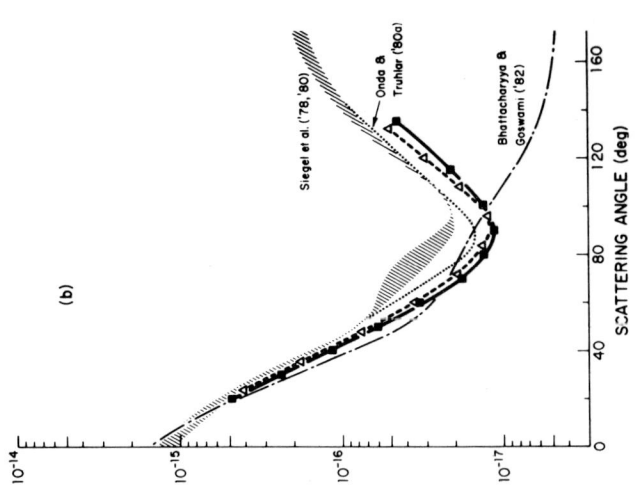
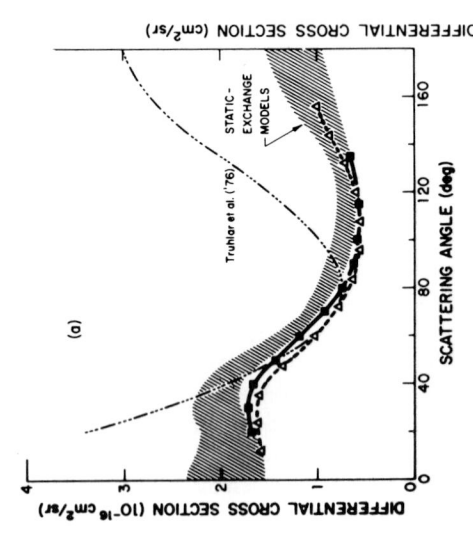

1 Elastic Scattering of Electrons by Molecules

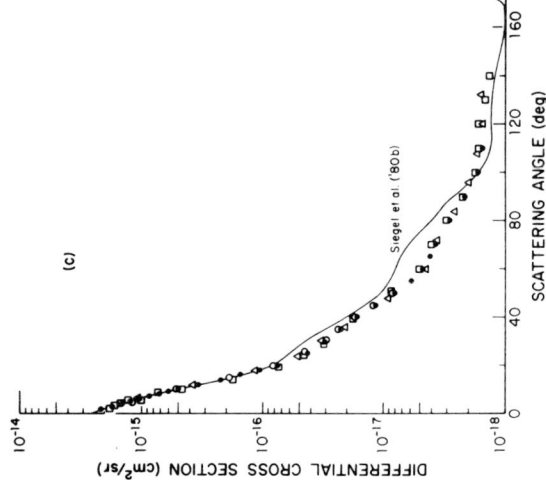

Fig. 28. (a) DCSs for "elastic" scattering of 5-eV energy electrons by N_2. The experimental data are from Srivastava *et al.* (1976, ■) and Shyn and Carignan (1980, △). The shaded area defines the region containing the results of the BF–SME calculations of Siegel *et al.* (1978, 1980b) and the static exchange–polarization (hybrid) calculations of Chandra and Temkin (1976b). The results of Truhlar *et al.* (1976), obtained from a LF–CC calculation using a four-state rotational–vibrational basis (without exchange), is shown as the dot–dash line. (b) Same as (a) except for 30-eV incident electron energy. The BF–SME results of Siegel *et al.* (1978, 1980b) are shown by the shaded region. The individual lines denote the AN Glauber model results of Bhattacharyya and Goswami (1982) and the LF–CC calculations employing model potentials of Onda and Truhlar (1980a). (c) Some as (a) and (b) except for an incident electron energy of 400 eV. The data of Bromberg (1970, ●), Dubois and Rudd (1976, □), Jansen *et al.* (1976, ○), and Shyn and Carignan (1980, △) are in excellent agreement with each other and with the BF–SME results of Siegel *et al.* (1980b).

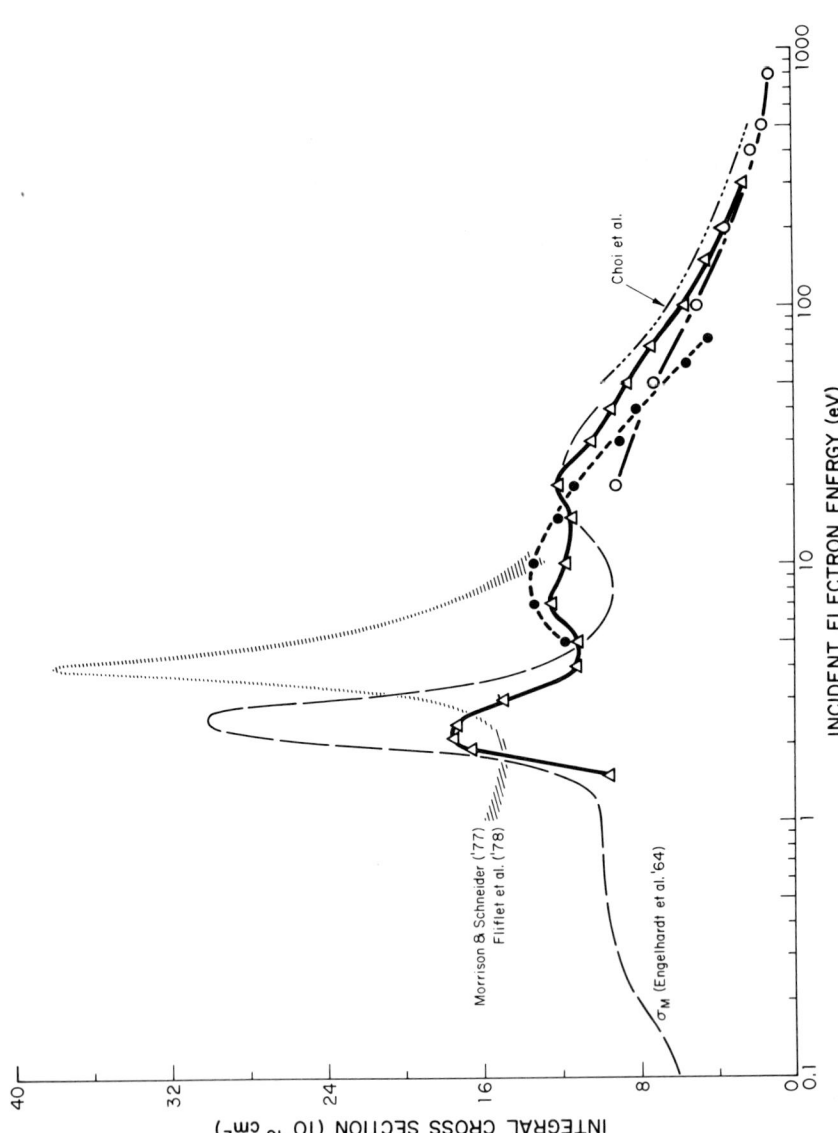

Fig. 29. Integral elastic electron scattering cross sections for N_2 in the 0.1–1000-eV impact energy region. The results from molecular-beam experiments are from Dubois and Rudd (1976, ○), Srivastava *et al.* (1976, ●), and Shyn and Carignan (1980, △). For purposes of comparison, the total *momentum* transfer cross section deduced from electron swarm data by Engelhardt *et al.* (1964) is also shown. The shaded region contains the results of static exchange models by Morrison and Schneider (1977)

3. Carbon Monoxide

Relative to H_2 and N_2 only a few experimental and theoretical results have been reported for "elastic" electron scattering by CO. Figure 30 compares some of the available DCS data at 10, 50, 500, and 800 eV. The results at 10 eV from the BF–CC hybrid and LF–CC SME calculations of Poe and Choi (1979) and of Onda and Truhlar (1980c) agree reasonably well with the experimental data. The agreement between theory and experiment is not as good at 50 eV.

Figure 31 compares various experimental and theoretical integral cross sections for "elastic" electron scattering of electrons by CO in the 0.1–1000-eV energy range. For purposes of comparison, the *momentum* transfer cross section deduced from electron swarm data by Land (1978) is also known. The "elastic" integral σ_E and the momentum transfer σ^M cross sections calculated

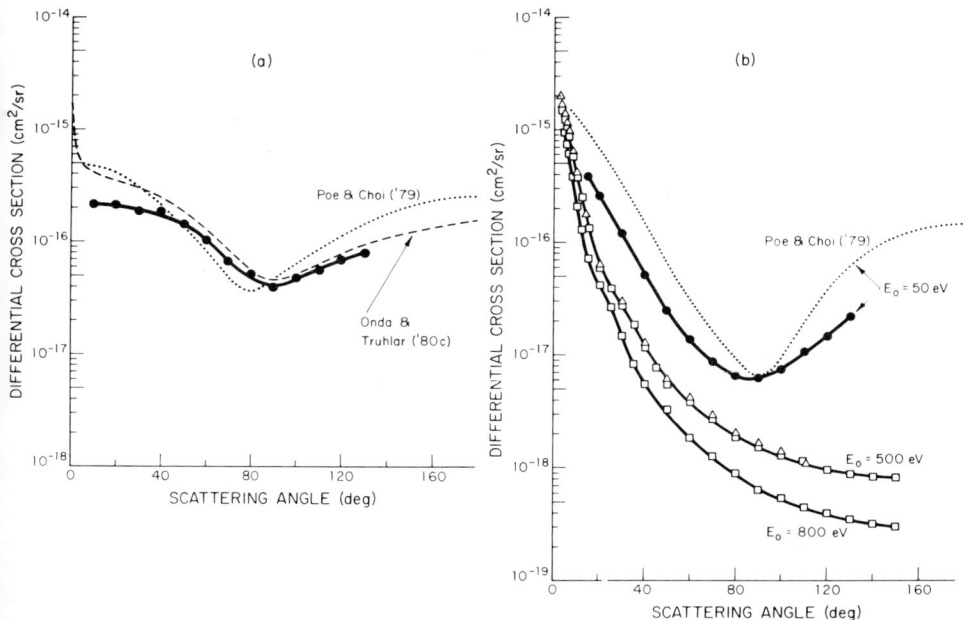

Fig. 30. (a) DCSs for "elastic" scattering of 10 eV electrons by CO. The experimental data are from Tanaka *et al.* (1978, ●) and the theoretical results are from the BF–CC hybrid calculations, using model exchange potentials (SME), of Poe and Choi (1979) and LF–CC calculations, again using model exchange potentials, of Onda and Truhlar (1980c). (b) Same as (a) except for 50 eV (Tanaka *et al.*, 1978, ●), 500 eV (Bromberg, 1970, △; Dubois and Rudd, 1976, □), and 800 eV [Dubois and Rudd, 1976, □]. Results from the BF–CC SME hybrid calculations of Poe and Choi (1979) for 50-eV energy electrons are shown for comparison.

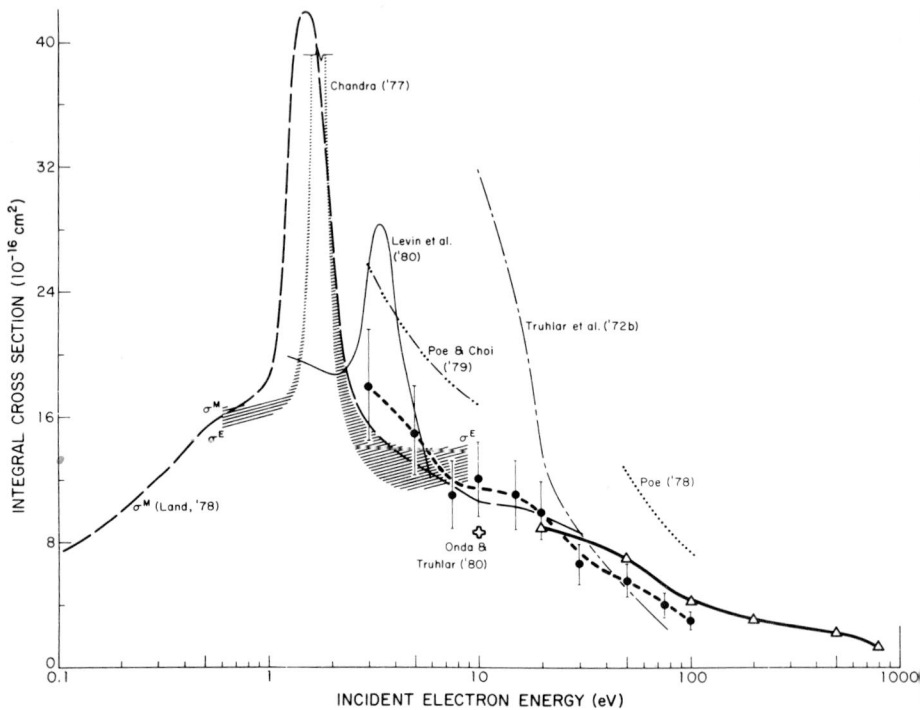

Fig. 31. Integral cross section for elastic scattering of electrons by CO for incident electron energies from 0.1 to 1000 eV. The results from molecular-beam experiments of DuBois and Rudd (1976, △) and Tanaka *et al.* (1978, ●) are shown as well as the integral momentum transfer cross section deduced from electron swarm data by Land (1978). The theoretical results shown are from a variety of models that are discussed in the text. The results from Poe (1978) are unpublished but were reported by Tanaka *et al.* (1978).

by Chandra (1977), using the (radial) frame transformation theory, define the boundaries of the shaded region. Other theoretical results shown are the AN–SE results of Levin *et al.* (1980), the LF–CC results of Truhlar *et al.* (1972a,b), the LF–CC SMEP results of Onda and Truhlar (1980c), the two-potential results of Poe [1978, see Tanaka *et al.* (1978)] and the "hybrid" theory results of Poe and Choi (1979). Except for the single value at 10 eV by Onda and Truhlar (1980c), the theories seem to overestimate the integral cross section for e^-–CO elastic scattering.

4. Methane

Relatively few experimental and theoretical results have been reported for "elastic" scattering of electrons by CH_4. Figure 32 contains a comparison of

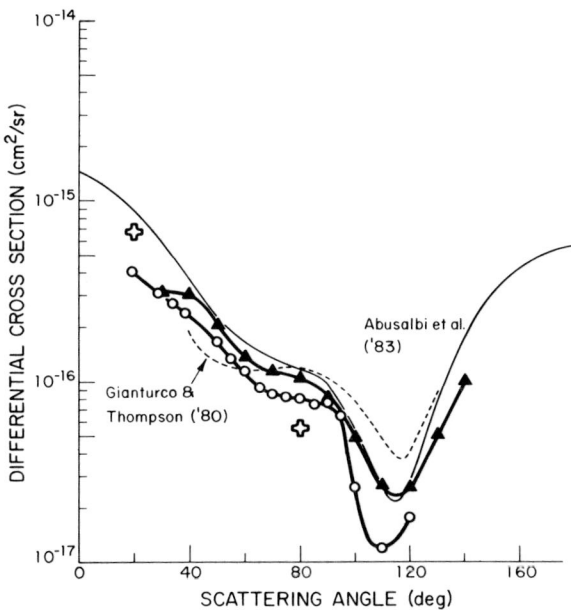

Fig. 32. DCS for vibrationally elastic scattering of 10-eV electrons by CH_4. The experimental data are from molecular-beam experiments of Newell *et al.* (1979, ○) Rohr (1980, ✢), and Tanaka *et al.* (1982, ▲). The theoretical results shown are from the AN–SOEP calculations of Gianturco and Thompson (1980) and the LF–CC SMEP calculations of Abusalbi *et al.* (1983).

the available data for an incident electron energy of 10 eV. All the data shown are from molecular-beam experiments and were recently acquired. Also shown are the AN static orthogonalized exchanged plus polarization (SOEP) results of Gianturco and Thompson (1980) and the LF–CC SMEP results of Abusalbi *et al.* (1983). The agreement between theory and experiment is reasonably good at this incident electron energy. Jain (1983) recently reported results in the 25–800-eV energy range obtained using a spherically symmetric analytical static potential along with model local exchange and polarization potentials. His results (not shown) are in reasonably good agreement with the limited experimental data at the higher energies.

5. *Polar Molecules*

There has been a great deal of experimental and theoretical interest in electron scattering by polar molecules the past 10 years which has led to new insight into the physics of electron–polar molecule scattering. Excellent

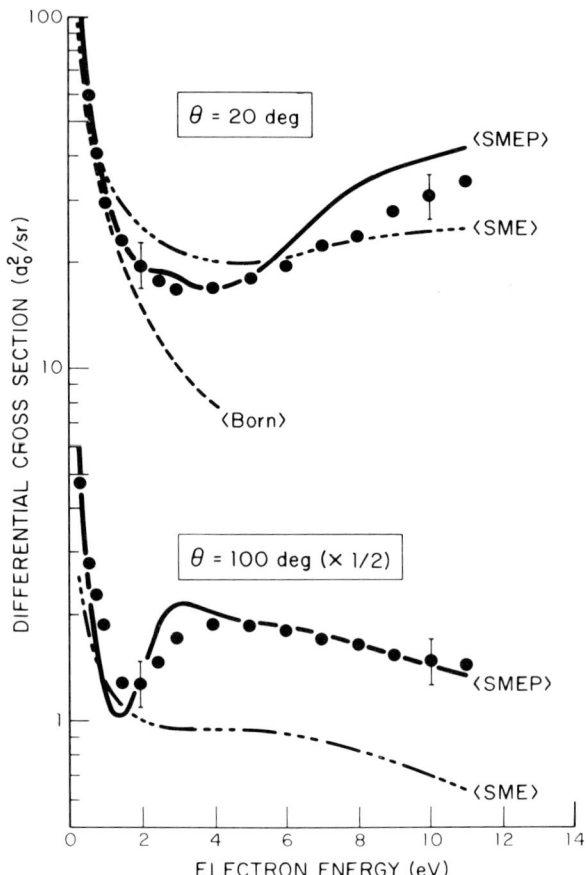

Fig. 33. Energy-dependence of the DCS for vibrationally elastic scattering of electrons by HCl at the fixed scattering angles of 20 and 100° (reduced by a factor of 2) (from Padial et al. 1983). The data points are from Rohr and Linder (1976) and the lines denote the results of CC calculations using various models for the electron–molecule interaction potential: SME is static plus model exchange; SMEP is static plus model exchange plus polarization. The brackets indicate that the DCS has been averaged over a thermal distribution of rotational states at T = 300 K.

reviews of this field have just been published by Norcross and Collins (1982) and Shimamura (1983) and, consequently, only the essential features will be summarized here.

Figure 33 shows the DCS for elastic electron scattering by HCl, at the fixed scattering angles of 20 and 100°, as a function of the incident electron energy. The results in this figure again show clearly the importance of including target polarization to obtain agreement with experiment in the low-energy region.

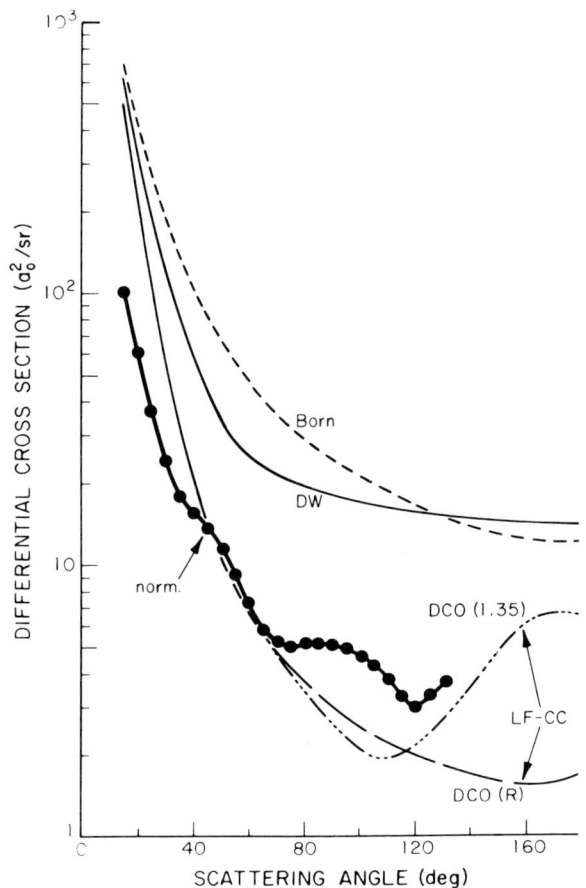

Fig. 34. DCS for "elastic" scattering of 6.74-eV electrons by KI (from Collins and Norcross, 1978). All the calculations are LF, and DCO stands for dipole cutoff at the indicated value for the cutoff parameter. The experimental data of Rudge et al. (1976, ●) have been normalized to the DCO(R) results at 40°.

Figure 34 compares various theoretical cross sections for "elastic" electron scattering by KI ($E_o = 6.74$ eV) with the available experimental data. The comparisons in this figure show that the Born and DW scattering approximations are generally inaccurate and that quantitative agreement with experiment generally requires a more sophisticated theory. The dominant role of the dipole potential is also evident from this figure.

Figure 35 shows an interesting comparison between the measured "reduced" DCS, as a function of the "reduced" momentum transfer, for

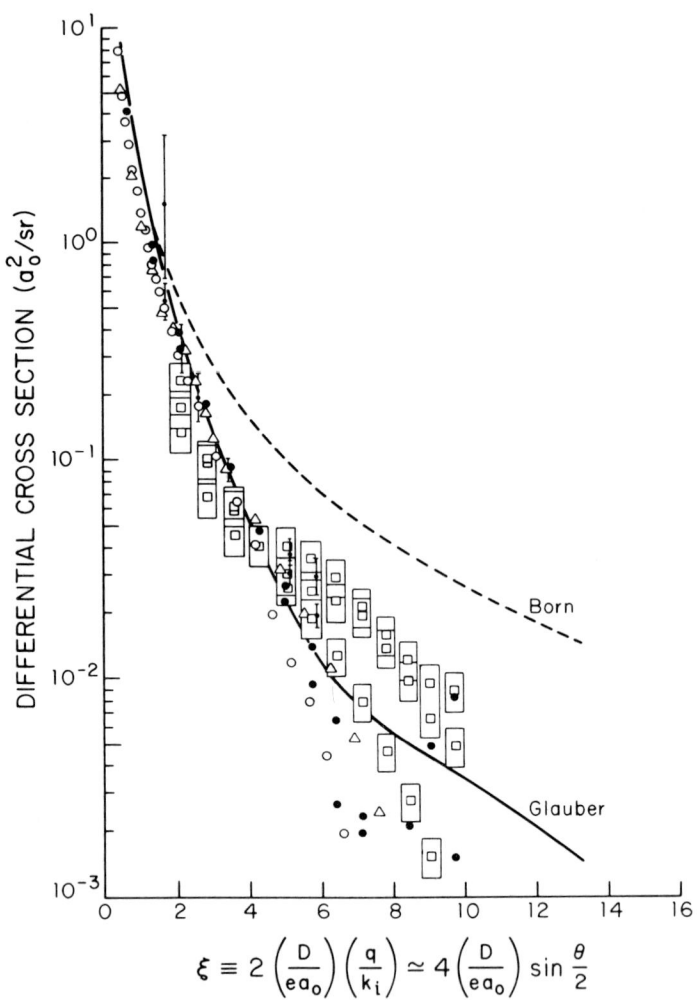

Fig. 35. "Reduced" differential cross section [DCS (reduced) = (E_o/D^4) DCS] (for scattering angles $\theta \leq 70°$) plotted versus "reduced" momentum transfer $\bar{q} = 2D(q/k) \cong 4D \sin(\theta/2)$; D being the dipole moment, q the momentum transfer, E_o the collision energy, and k the corresponding wave number of the electron (from Shimamura, 1979). Molecular-beam recoil measurements: ○, CsF (1.80 eV); △, CsCl (4.77 eV); ●, KI (6.74 eV, 15.7 eV). Energy-loss measurements: ⌽, LiF (5.4 eV, 20.0 eV); ⊟, KI (6.74 eV), 15.7 eV, 60 eV). The procedure for normalizing the measured cross section is nonunique, especially for the energy-loss measurements having large error bars at small θ where simple theories are accurate.

vibrationally elastic scattering by various polar molecules with the result predicted by the Glauber and Born approximations (from Shimamura, 1979). This comparison is interesting because the Glauber approximation for a point dipole depends only on the reduced momentum transfer and hence represents a universal curve (Shimamura, 1979). The agreement between the data and the universal curve is surprisingly good and the agreement between the shapes is independent of any normalization procedure. This comparison is discussed more completely by Shimamura (1984).

V. Applications

A. Phenomena Involving Elastic Scattering and Rotational Excitation

Knowledge of the elastic electron scattering properties of various gases is of central importance in the study of a wide variety of plasma phenomena because the effective momentum transfer cross section of electrons in gases is usually dominated by the elastic scattering process for electron energies below the threshold for electronic excitation (and outside any vibrational resonance region). It is the variation of the momentum transfer cross section with incident electron energy that produces corresponding variations in the expressions for the electron conductivity, diffusion coefficients, and other transport properties of electrons in gases. Consequently, the magnitude and energy dependence of the elastic scattering differential cross sections for the components of the gaseous media strongly influence the passage of electrons through the gas. Finally, excitation of rotational, vibrational, and electronic degrees of freedom frequently plays a major role in determining the energy thermalization properties of the electrons in the particular molecular gas, depending on the particular electron energy region of interest.

In this section the role of elastic scattering and rotational excitation for a few selected applications will be discussed. The role of vibrational and electronic excitation will be discussed in the following chapter. As pointed out by Shkarofsky *et al.* (1968), strictly speaking there is no such phenomenon as an elastic collision involving electrons. The deceleration of the free electron during the collision and the resulting creation of photons (*bremsstrahlung*) makes *every* electron collision inelastic even though the energy lost by the free electron to the radiation field is extremely small at the energies ($E \lesssim 1000$ eV) of interest to this review. Because the energy radiated by the free electron is so small, it is customary to call any collision process for which the *target* stays in its initial quantum states an elastic one. This terminology will also be followed in this section.

1. Kinetic Considerations

To understand the role of elastic scattering in various phenomena, a brief review of kinetic processes is helpful. The general properties of particle collisions in the kinetic theory of gases have been discussed in a very large number of excellent textbooks (e.g., Chapman and Cowling, 1958), so only those properties specific to the understanding of electron processes will be summarized here.

The concept of the *mean free path* is central in describing the role of collisions in all gas kinetic processes and originates from a classical hard-sphere formulation of the collision event. That is, the mean free path λ of a particle moving through other particles is defined as the average distance traveled between collisions. Assuming that the particles behave as solid spheres when colliding, that is, they suffer a collision when the distance between centers is less than

$$r_{12} = r_1 + r_2, \tag{112}$$

the distance a given particle travels between collisions when traversing a cylindrical volume of radius $2r_{12}$ containing a gas of density n is

$$\lambda = 1/\pi r_{12}^2 n. \tag{113}$$

For the case of electrons, $r_e \equiv r_1 \ll r_M \equiv r_2 \equiv r$, so

$$\lambda = 1/n\pi r^2 \equiv 1/n\sigma, \tag{114}$$

in which the quantity σ is defined as the cross-sectional area πr^2 of the atom or molecule with respect to an electron collision. In interpreting the macroscopic properties of electrons in a plasma environment, the relationship between λ and σ is retained, but since the real cross sections are always energy dependent, only quantities that have been *averaged* over the distribution of electron energies are really determined. Similarly, the collision frequency, which depends on the velocity of the electrons [for the usual case the gas velocity v is much less than the electron velocity v_e] and the collision cross section, can be *defined* as the appropriate average over the distribution of electron velocities.

$$\langle v_e \rangle = n_e n \langle v_e \sigma \rangle, \tag{115}$$

where the angular brackets denote averaging over the distribution of electron velocities (assuming the molecules are stationary compared with the electron). It is important to note that these definitions have evolved from the development of plasma theory and corresponding efforts to extract a wide variety of electron transport properties from various plasma measurements. During the past 50 years, a very large number of elegant experiments have been performed that accurately measured the electron drift velocity and diffusion coefficients in many gases. To extract the fundamental energy-dependent cross section information from such "averaged" data, sophisticated numerical techniques for solving the appropriate Boltzmann equation are being developed. There is an extremely rich literature (e.g., Huxley and Crompton, 1974; McDaniel, 1964; Shkarofaky et al., 1968) describing both the measurements themselves and the interpretation of the resulting data.

With the relatively recent development (last 15 years) of improved instrumentation described in the preceding sections, it is now possible to make direct measurements of fundamental collision processes (inelastic as well as elastic) for which the key independent variables characterizing the collision process (e.g., velocity and scattering angle) have been "averaged" over only the very narrow ranges of instrumental resolution. That is, the crossed electron-beam–molecular-beam experiments that are now conducted in many laboratories around the world represent "classic" scattering experiments that provide fundamental differential cross sections (DCSs) for each type of scattering process at a particular incident electron energy and for which the incident electron is scattered into various solid angles. This type of information is substantially more detailed than can be extracted from the measured transport properties of electron swarms in the same molecular gas. However, an interesting situation has developed as a result of these new "beam–beam" measurements, in that they are turning out to complement, rather than replace, the older electron swarm techniques. In fact the availability of DCS information for both elastic and inelastic processes has rekindled an interest in the theory of electron transport in molecular gases. The reason for this renewed interest is that the magnitudes and shapes of the DCSs show that customary two-term expansion solutions to the Boltzmann equation used for interpreting the electron swarm measurements are usually insufficient to account for the magnitudes of the inelastic cross sections and the different anisotropies for elastic and inelastic scattering processes. Some interesting aspects of this synergism between the results from beam–beam and from transport experiments will be discussed in the following sections.

2. The Momentum Transfer Cross Section

As described in Section I, the fundamental characteristic of the elastic scattering of an electron by a molecule is the associated DCS for the process. However, in transport phenomena in which the scattering of electrons by atoms and molecules plays an important role, it is a special form of angular weighting of the elastic DCS, the *elastic momentum transfer* (sometimes called "diffusion" or just "momentum transfer") cross section that is necessary. The definition of the momentum transfer cross section is obtained directly from the mathematical form in which the electron scattering DCS appears in the

Boltzmann equation (Shkarofsky et al., 1968: p. 82) for the diffusion of electrons through an atomic or molecular gas. This momentum transfer cross section σ^M is a measure of the *loss* of forward momentum suffered by the electrons as they drift through the gas and is defined as

$$\sigma^M = 2\pi \int_0^\pi \frac{d\sigma}{d\Omega}(\theta, E_o)[1 - \cos\theta] \sin\theta \, d\theta. \tag{116}$$

The physical interpretation (Hasted, 1972: p. 19) can be obtained as follows. As is well known (i.e., Shkarofsky et al., 1968: p. 75), an electron of mass m scattered by an atom or molecule of mass M ($M \gg m$) into the scattering angle θ by an elastic collision loses a portion ΔE of its initial kinetic energy E given by

$$\frac{\Delta E}{E_o} = \frac{2m}{M}(1 - \cos\theta). \tag{117}$$

The mean (averaged over all scattering angles) fractional energy loss per elastic collision associated with the momentum transferred is therefore given by (Hasted, 1972)

$$\frac{\Delta E}{E_o} = \frac{2m}{M} \frac{2\pi}{\sigma} \int_0^\pi \frac{d\sigma}{d\Omega}(\theta, E_o)[1 - \cos\theta] \sin\theta \, d\theta = \frac{2m}{M} \frac{\sigma^M}{\sigma}, \tag{118}$$

where σ denotes the usual integral cross section [Eq. (2)] for elastic scattering. The angle-dependent weighting of the elastic scattering DCS for the momentum transfer cross section emphasizes large-angle (backward) scattering because the weighting factor $(1 - \cos\theta)$ is zero for forward scattering, unity for 90°, and two for backward scattering. Although it is not particularly useful except to help visualize the effect of the special angular weighting, it is straightforward to use Eq. (116) to define the DCS for momentum transfer as

$$\text{DCS}^M = \frac{d\sigma^M}{d\Omega}(\theta, E) \equiv \frac{d\sigma}{d\Omega}(\theta, E)[1 - \cos\theta]. \tag{119}$$

The above discussion shows that the most important property of electron–molecule scattering for determining the transport of the electrons through the molecular gas is a special weighting of the elastic scattering DCS; i.e., the momentum (or diffusion) cross section [Eq. (118)]. The results discussed above further emphasize the point that, if the transport of electrons through a particular gaseous medium is to be predicted, then the DCS for elastic scattering needs to be measured at each incident electron energy of interest.

Figure 36 contains comparisons of elastic and momentum transfer DCSs (at selected incident electron energies) for N_2 to illustrate some of the characteristics described above. As noted in Section IV.C.2, the DCSs for

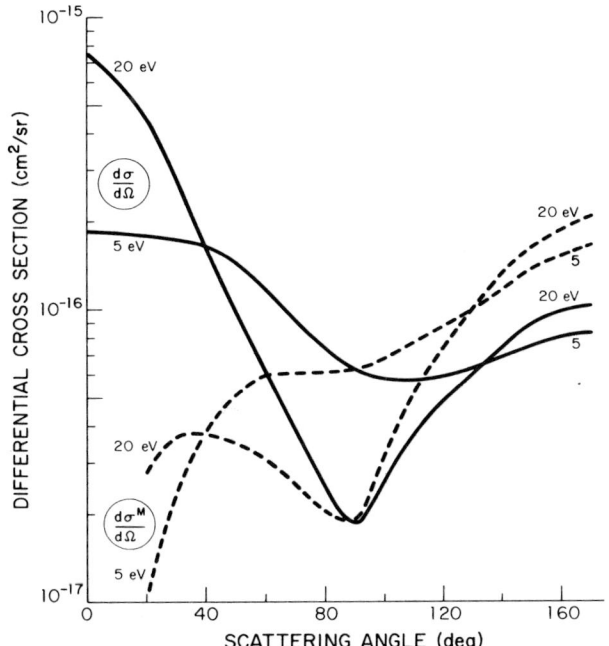

Fig. 36. DCS (solid line) and "momentum transfer" DCS (dashed lines) as defined by Eq. (119) for "vibrationally elastic" scattering of 5- and 20-eV electrons by N_2. The integral of the dashed curves over all scattering solid angles gives the integral momentum transfer cross section. The data used to construct this figure were taken from Trajmar et al. (1983).

elastic scattering by these molecules possess the characteristic strong forward-scattering peak but show substantial differences at intermediate and large scattering angles. As also shown in Fig. 36, the DCSs for elastic momentum transfer differ substantially in both shape and magnitude for different electron energies. The origin of these difference is, of course, very complicated, currently not completely understood, and is the reason why these DCSs need to be measured or predicted theoretically. The corresponding measured integral momentum transfer cross sections for a variety of molecules is shown in Fig. 37 to illustrate the differences in their electron energy dependence, as well as their different magnitudes.

In preparing these comparisons, the emphasis has been placed on the beam–beam data rather than those obtained from electron swarm measurements because the former are more recent and a number of excellent reviews of the swarm data already exist (McDaniel, 1964; Shkarofsky et al., 1968; Huxley and Crompton, 1974). However, as mentioned above, the swarm data

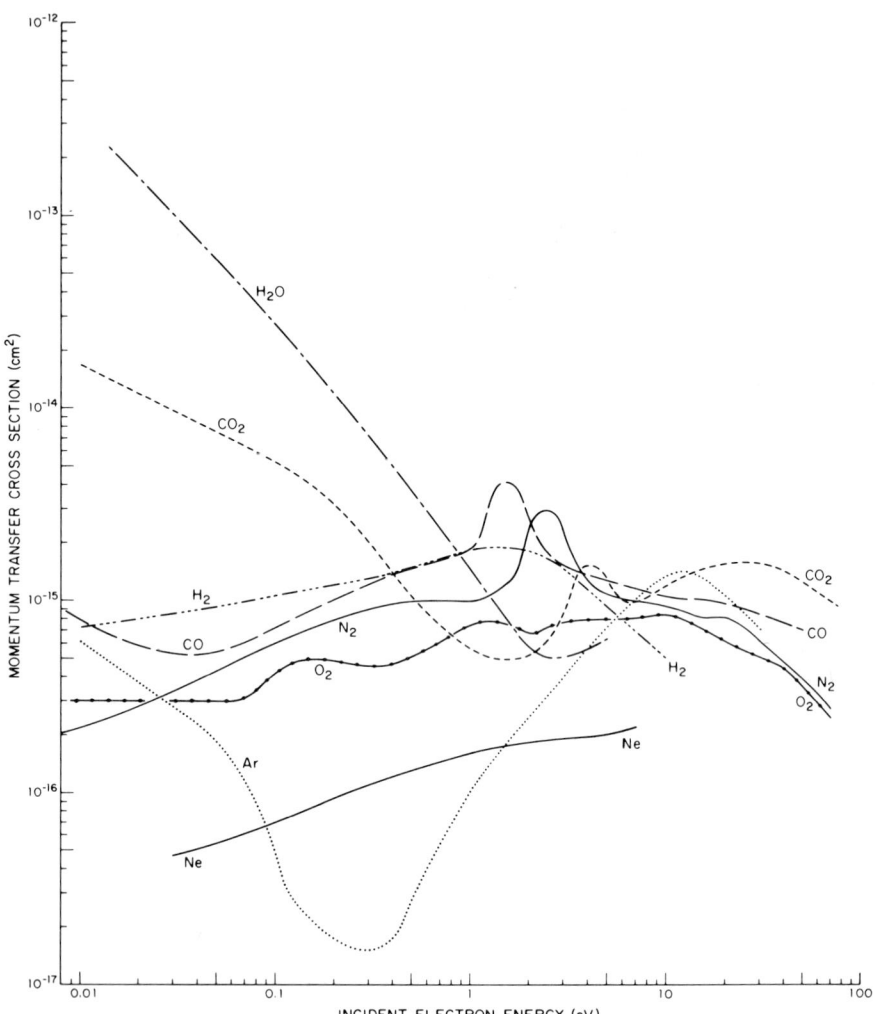

Fig. 37. Integral momentum transfer cross section as a function of incident electron energy from 0.01 to 100 eV for a variety of atoms and molecules. Data taken from Trajmar *et al.* (1983) and from Kieffer (1973). The curve for neon has been included to illustrate the well-known Ramsauer minimum. Note that the measured cross sections vary by three orders of magnitude at some electron energies.

are very important for at least two reasons. First, they are complementary to the beam–beam results because they extend to much lower electron energies (i.e., $E \lesssim 0.001$ eV) than is currently possible with beam–beam experiments (i.e., $E \gtrsim 0.1$ eV). Swarm techniques can also measure the various transport coefficients themselves more accurately (e.g., 1% or less) than can be currently obtained by integrating an elastic DCS. Secondly, in the electron energy region in which the results obtained by the two different methods can be directly compared ($\sim 0.1 - 5$ eV), such a comparison provides valuable insight into the validity of the method used to solve the Boltzmann equation applied to the swarm data. After a brief discussion of the role of rotational excitation, we shall return to a discussion of the solution to the Boltzmann equation since that is currently a topic of considerable interest (see Chapter 3 of Volume 2).

3. *Rotational Excitation of Molecules*

For the study of electron processes in molecular gases, the effect of rotational excitation usually needs to be included if those channels are energetically accessible. Although rotational excitation involves only a small energy transfer per collision, the energy transferred is still much greater than that associated with what we are calling elastic collisions (the rotational-excitation cross sections are nearly as large as elastic scattering cross sections) and this electron energy loss process has a major effect on the shape of the electron distribution for electron energies less than a few tenths of an electron volt.

The importance of the rotational excitation process in electron swarm studies in molecular gases has been known for a long time (Huxley and Crompton, 1974), and the first experimental estimates of rotational excitation cross sections came from the analysis of swarm data. For example, *ortho*-hydrogen and *para*-hydrogen have different electron transport properties for gas temperatures less than 100 K (Huxley and Crompton, 1974) which, because of their different rotational populations, is a manifestation of the effects of rotational excitation on the electron energy distribution.

Mozumder (1982) recently completed detailed calculations on the thermalization of electron energy in molecular hydrogen, as a function of the gas temperature (200–1000 K), including rotational and vibrational inelastic processes as well as the energy loss due to elastic collisions. To perform these calculations, Mozumder used published energy-dependent cross sections for momentum transfer and rotational and vibrational excitation; included microreversibility and detailed balancing; and took the long-time limit to determine the characteristic relaxation time for each process. The relaxation (or thermalization) time for a given process is then defined (from the asymptotic form of the relevant equations) as that time that characterizes the exponential decay of the excess electron temperature (Mozumder, 1982). Each process contributes independently, and additively, to the total relaxation time τ according to

$$\frac{1}{\tau} = \frac{1}{\tau_M} + \frac{1}{\tau_R} + \frac{1}{\tau_V}, \tag{120}$$

where the subscripts M, R, and V denote momentum transfer, rotational

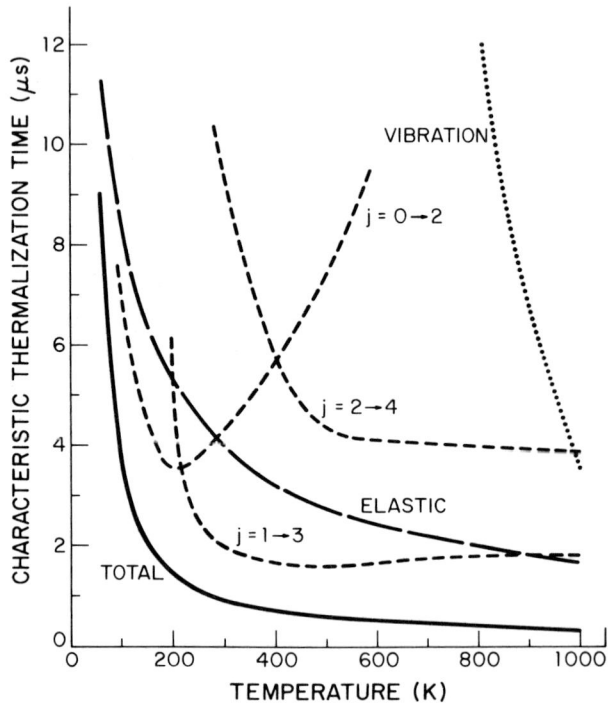

Fig. 38. Characteristic thermalization time (μsec) of electrons in H_2 at a density of 3.31×10^{16} molecules/cm^3 as a function of the gas temperature (K), constructed from the data of Mozumder (1982). The total relaxation time is shown by the solid curve, that due to elastic scattering by the long-dashed line, that due to various rotational transitions by the short-dashed curves, and that due to vibration as the dotted curve.

excitation, and vibrational excitation, respectively. The characteristic relaxation times obtained from these calculations are shown in Fig. 38 (for an H_2 density of 3.31×10^{16} cm^{-3}) as a function of the gas temperature. Not surprisingly, the results in Fig. 38 show that the relative importance of the elastic and various inelastic processes depends on the gas temperature and that the effect of elastic momentum transfer is not negligible at any temperature. For temperatures above 80 K, energy loss due to rotational excitation becomes the dominant process for H_2 and vibrational excitation begins to become important above about 900 K. Although this example considers H_2, which has the largest molecular rotational-level energy spacing, the central importance of rotational excitation in electron thermalization is a general characteristic of most molecules, as will be discussed further in the next section.

4. *Electron Collision Frequency*

The concept of an electron collision frequency is essential in any discussion of electron transport through a neutral gas or partially ionized plasma. However, there has been some ambiguity in the literature concerning the definition of an electron collision frequency, the particular form chosen depending a great deal on the specifics of the application. The reason for this is that the usual formulations for electron transport require an *effective* collision frequency, that is, the microscopic collision frequency *averaged* over the distribution of electron energies applicable to the specific transport phenomena being studied. Itikawa (1971b) has proposed a particular form for the effective collision frequency in gases such that the electrical conductivity due to the electron current is

$$\text{conductivity} = N_e e^2 / M_e \langle v_{\text{eff}} \rangle, \tag{121}$$

where N_e is the electron density (cm^{-3}), e and M_e are the electron charge and mass; and $\langle v_{\text{eff}} \rangle$ is the effective electron collision frequency. Since any definition of the effective collision frequency must be applicable to a wide range of plasma phenomena, Itikawa showed how various transport coefficients can be expressed in terms of $\langle v_{\text{eff}} \rangle$ for the case of a slightly ionized gas. He has also used this definition to determine the effective collision frequency for a broad range of atmospheric constituents and has updated them as improved microscopic cross section data became available (Itikawa, 1971b, 1974, 1978b). Aggarwal *et al.* (1979) extended the work of Itikawa (1971b) to determine the effective collision frequency in the earth's ionosphere and a comparison with the measured data will be made in the next section.

B. Planetary Atmospheres

All the planets in our solar system possess a gaseous atmosphere of one kind or another, and because they are exposed to solar radiation and because they have a magnetosphere, they also have an ionosphere (and possibly aurora) analogous to that associated with Earth. The solar radiation, and possibly precipitating charged particles, ionize the planet's atmospheric constituents to produce this ionosphere. The properties of the ionosphere will, therefore, be directly related to the incident solar and (if appropriate) particle spectra, and associated planetary magnetic field and to the particular atmospheric constituents and kinetic processes in which they participate. Understanding the complex relationships among our planet's atmosphere, the sun, and the magnetosphere has fascinated and intrigued the world's scientific minds for more than 100 years (e.g., Eather, 1980). Although tremendous progress has been made in the past 30 years in understanding the intricacies of the solar–terrestrial relationship from the judicious use of rockets and satellites, many major questions still remain to be answered. Since interest in understanding Earth's upper atmosphere remains a primary motivation for studying electron–molecule impact processes, we shall discuss the role of elastic scattering and rotational excitation in Earth's ionospheric and auroral processes in the next section and in other planetary atmospheres in the section after that.

1. Earth's Atmosphere

a. Composition

There are a large number of excellent textbooks and monographs that discuss various characteristics of Earth's upper atmosphere, and they will be drawn upon to summarize those aspects necessary to understand the role played by electron–molecule collisions in Earth's atmosphere. Typical altitude profiles of the major atomic and molecular constituents (Whitten and Poppoff, 1965: p. 70) of Earth's atmosphere are shown in Fig. 39. One of the reasons that there has been so much research done on electron-impact and

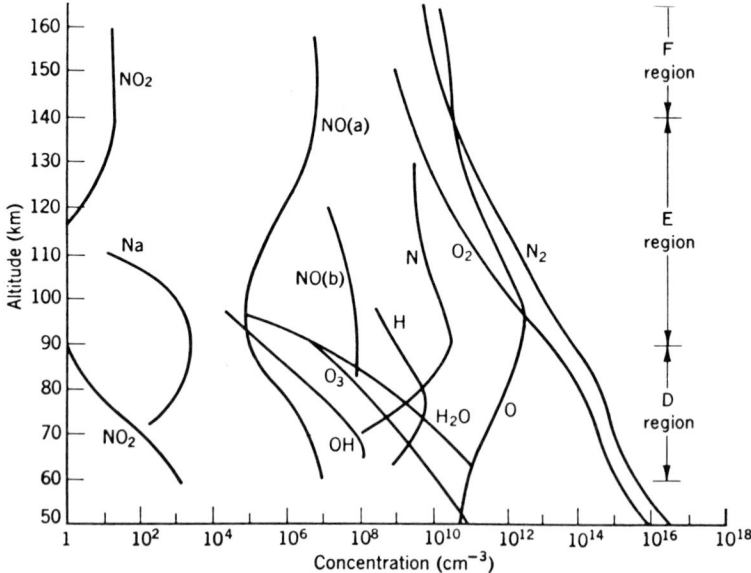

Fig. 39. Typical altitude (km) profiles as a function of the concentration (cm^{-3}) for a number of atomic and molecular constituents in the normal Earth's atmosphere. Two different curves for NO are shown to account for two different sets of data accumulated in the early 1960s (see Whitten and Poppoff, 1965: pp. 68–72). The altitude range for the D and E regions of the normal ionosphere are indicated in the figure. (R. C. Whitten and I. G. Poppoff, "Physics of the Lower Ionosphere," © 1965, p. 70. Reprinted by permission of Prentice-Hall, Inc., Englewood Cliffs, New Jersey.)

photoabsorption processes in O_2 and N_2 is because they, and their associated atomic fragments, are the dominant constituents throughout the ionospheric D and E regions as illustrated in Fig. 39. Although other molecular species are substantially less abundant except in the D region, they play very important roles in the atmospheric chemistry (Whitten and Popoff, 1965; Banks and Kockarts, 1973). Two different altitude profiles for NO are shown in Fig. 39 because there remains some controversy as to the NO concentration under various atmospheric conditions (e.g., daytime, nighttime, auroral excitation).

b. The Ionosphere

The first successful transmission of radio signals across the Atlantic by Marconi in 1901 probably marks the birth of ionospheric science (Giraud and Petit, 1978). The large number of practical applications associated with radio transmission, combined with the general scientific interest in understanding the complex solar–terrestrial relationship, has fostered active research programs around the world that continue to flourish today. A number of excellent monographs (e.g., Giraud and Petit, 1978; Bauer, 1973; Whitten and Popoff, 1968), textbooks (Banks and Kockarts, 1973), and review articles (e.g., Takayanagi and Itikawa, 1970) have been written dealing with various aspects of the very complex physical and chemical phenomena taking place in Earth's ionosphere and we draw selectively on their work to identify the role of electron-impact processes. In the absence of auroral events, the electron–ion pairs that characterize the ionosphere are produced by a combination of solar radiation, cosmic-ray particles, and secondary ionizations by the primary photoelectrons. These electrons, whose "initial" energy distribution ranges from a fraction of an electron volt to over 100 eV, lose their energy in molecular collisions as they move through the atmosphere until they either recombine or participate in the complex negative-ion chemistry. In spite of the voluminous laboratory work that has been done on recombination processes, it is still not clear (Giraud and Petit, 1978) how the laboratory data should be used to interpret the wide (neutral and electron) temperature range present in the ionosphere.

The rate of participation of the ionospheric electrons, positive and negative ions, and neutral species in the various physical and chemical processes is strongly temperature- (and hence altitude-) dependent; therefore the thermalization (energy balance) of the ionospheric plasma plays a major role in the composition and physical properties of the ionosphere. Detailed studies (Dalgarno, 1969; Takayanagi and Itikawa, 1970; Stubbe, 1971; Giraud and Petit, 1978: p. 231) reveal that the major mechanisms for electron thermalization in the earth's ionosphere are electron–electron (Coulomb) collisions, excitation of the O-atom fine-structure levels, rotational excitation of O_2 and

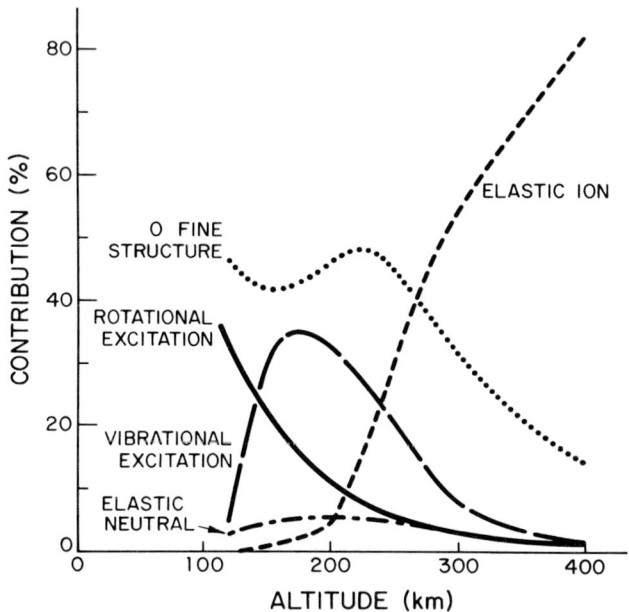

Fig. 40. Contribution (percent) of various electron collision processes to the total enery exchange rate as a function of altitude (km) in Earth's normal ionosphere, as constructed from data compiled by Stubbe (1971).

N_2, vibrational excitation of O_2 and N_2, and elastic collisions with the ions and neutrals. The rates for all these processes (except the first) depend (Giraud and Petit, 1978) on the difference between the electron and neutral kinetic temperatures as well as the electron temperature itself. The relative importance of the various electron–heavy-particle energy-transfer terms are summarized in Fig. 40 for a midlatitude atmospheric–ionospheric model at low-to-medium solar activity (Stubbe, 1971). As shown in Fig. 40, rotational excitation of the atmospheric molecules is an important electron thermalization process for the D, E, and lower F regions of the ionosphere. An excellent discussion of the thermodynamics of the energy balance in the ionosphere, including the electron, ion, and neutral-particle temperatures and transport effects is contained in the recent monograph by Giraud and Petit (1978).

c. Electron–Molecule Collision Data Needed

A good summary of the cross section information needed to better understand atmospheric processes has recently been given by Takayanagi (1982) who points out that the momentum transfer cross sections for the ma-

1 Elastic Scattering of Electrons by Molecules

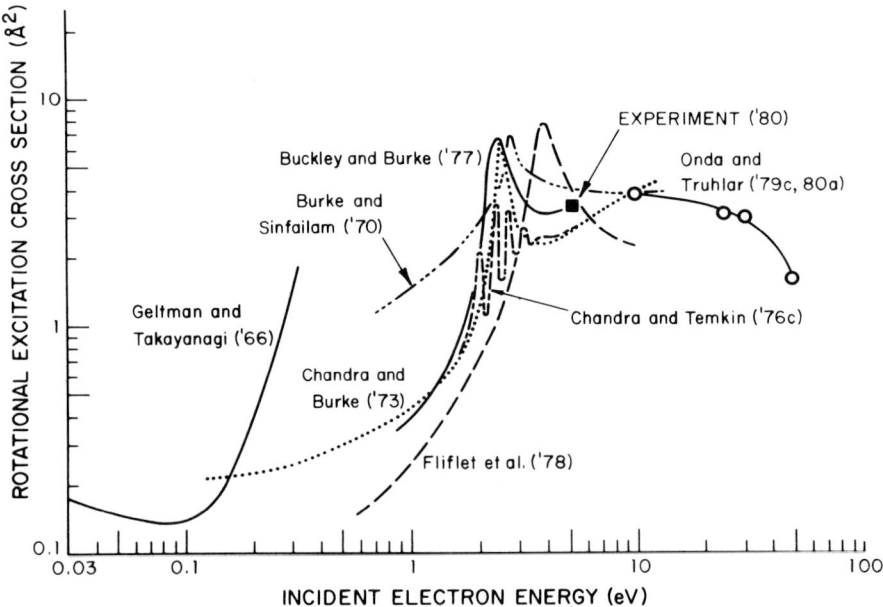

Fig. 41. Rotational excitation cross section (Å^2) for the $J = 0 \to 2$ transition in N_2 as a function of the incident electron energy, from 0.03 to 100 eV, as compiled by Takayanagi (1982). All the data shown in this figure are theoretical except for the single data point of Tanaka *et al.* (1980). (Adapted with permission from V. McKoy, H. Suzuki, K. Takayanagi, S. Trajmar (eds.). "Electron–Molecule Collisions and Photoionization Processes." Dearfield Beach, Florida. Verlag Chemie, 1983.)

jor molecular constituents in the atmosphere are reasonably well known. However, there remains a controversy as to the magnitude and energy dependence of the rotational-excitation cross section of O_2, and the rotational cross section for N_2 is uncertain to more than a factor of two for electron energies less than 2 eV. The situation for N_2, as compiled by Takayanagi (1982), is illustrated in Fig. 41, and the comparisons there suggest that considerable theoretical work and higher resolution experimental work remain to be done. In addition, cross sections for both elastic scattering and rotational excitation of NO, NH_3, OH, NO_2, and N_2O are currently poorly known but are needed for atmospheric studies.

Although the excellent experimental and theoretical work that has been done over the years to characterize the various electron collision processes has been very successful in helping geophysicists to better understand the ionospheric behavior, there remain a number of characteristics that are not yet fully explained. Figure 42, constructed from the results of Aggarwal *et al.* (1979), compares theoretical and experimental results for the altitude dependence of electron–neutral particle (v_{en}), electron–ion (v_{ei}), and total (v_g)

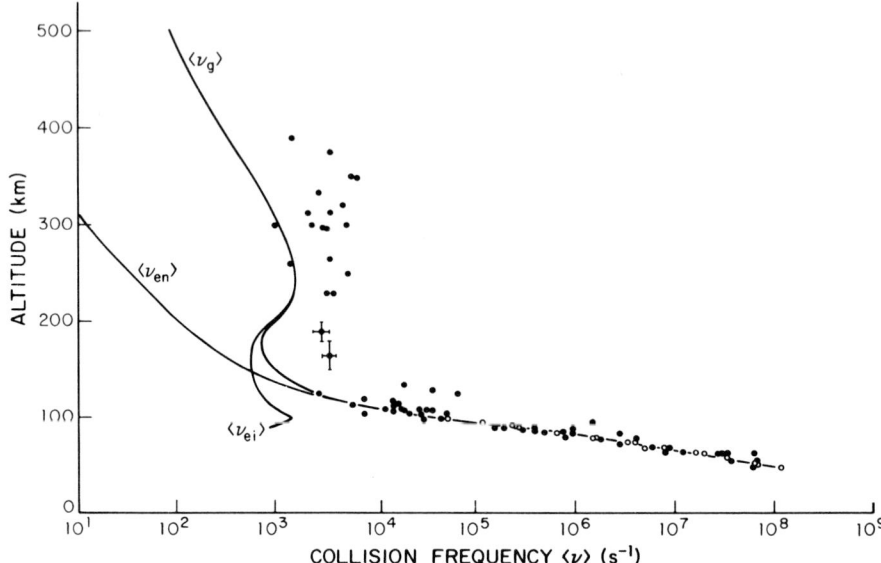

Fig. 42. Comparison between the measured (●) and calculated (○) altitude (km) profiles for electron–neutral particle (ν_{en}), electron–ion (ν_{ei}), and total–electron (ν_g) collision frequency (s^{-1}), as compiled by Aggarwal *et al.* (1979).

collision frequency for Earth's ionosphere from 50 to 500 km. As shown in the figure, the agreement between calculated and measured collision frequencies is excellent below 120 km but the calculated total collision frequency is about a factor of four below the measured results for altitudes greater than 120 km. At this writing, no quantitative explanation for this discrepancy has been offered. Clearly, either some, as yet unknown, species have large electron scattering cross sections in this altitude range or some of the cross section values used in the theoretical work are too small.

2. *The Aurora*

Any discussion of electron–molecule collision processes in Earth's atmosphere would be incomplete without at least some discussion of auroral phenomena. Documented observation of the natural auroral phenomena goes back more than 300 years (Eather, 1980) but the origin of the spectacular displays defied quantitative explanation until the early twentieth century. An idea as to the size of the "normal" (as opposed to "intense") aurora can be gathered from Fig. 43 [taken from the historical review of Eather 1980], which is a composite of seven successive satellite photographs showing the auroral

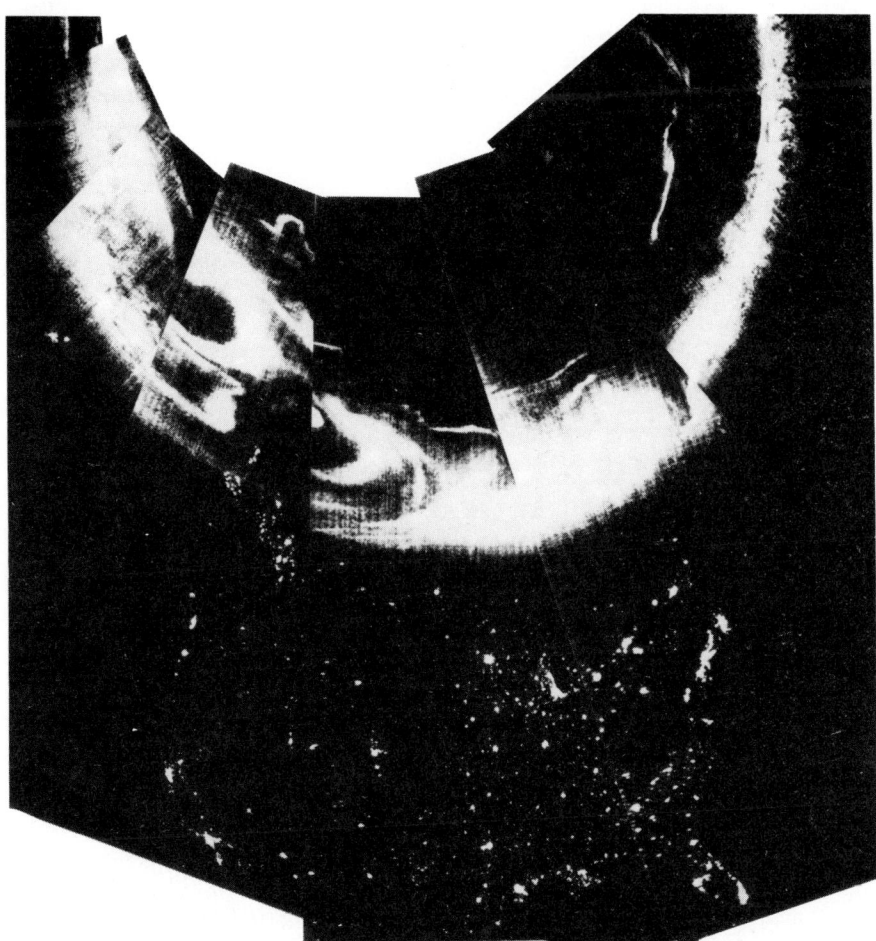

Fig. 43. Composite of seven consecutive satellite (infrared) photographs of the North American continent, showing a "normal" aurora across the top of the continent. The bright spots and regions below the aurora are from the lights of the major U.S. and Canadian cities. This picture gives an idea of the width and breadth of the "normal" (i.e., not particularly intense) aurora by utilizing the continental U.S. as a yardstick (Eather, 1980; courtesy of the American Geophysical Union).

oval spreading across the North American continent. An outline of the United States, as well as the locations of major U. S. cities, is produced by the bright spots below the auroral oval. The natural auroral phenomena are of interest to the scientific community because they represent a coupling between the sun, the magnetosphere, and Earth's ionosphere. This coupling produces not only the well-known spectacular optical displays following intense solar activity but also has a variety of effects on Earth's ionosphere that affect the transmission and reception of radio waves. A schematic diagram of a "normal" aurora and its relationship to the altitude profile of the ionosphere is

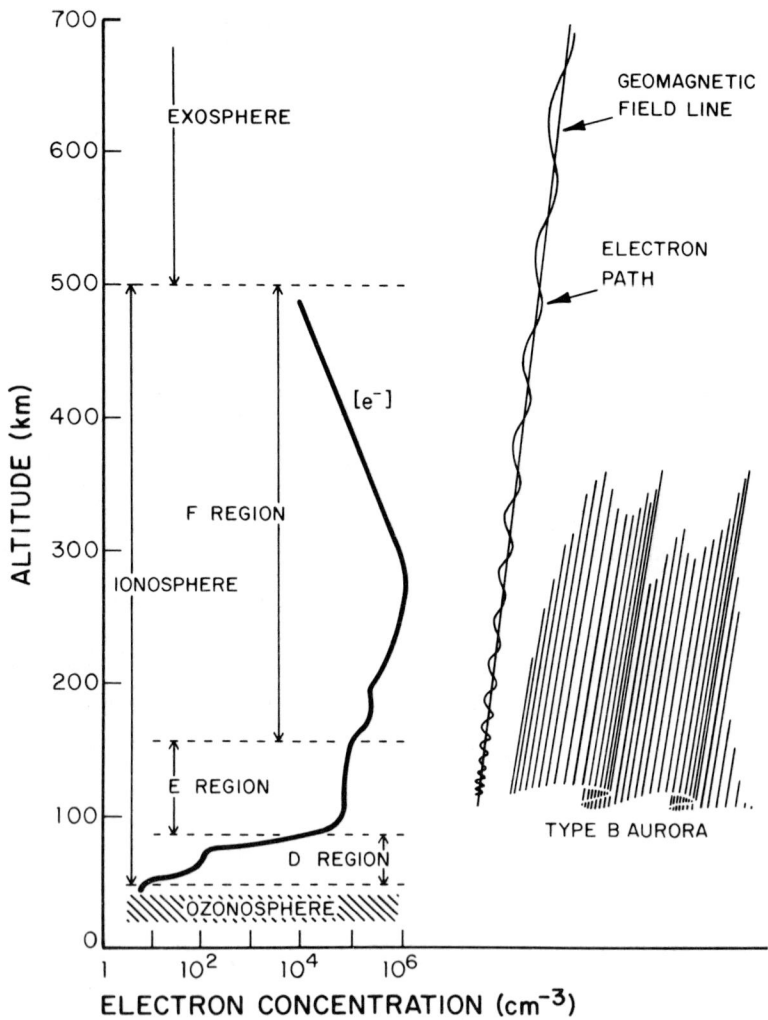

Fig 44. The left portion of this figure illustrates schematically the altitude (km) profile of the electron concentration (cm^{-3}) in Earth's "normal" ionosphere and the extent of the D, E, and F regions. The right portion is a schematic of Earth's geomagnetic lines, auroral electron trajectories, and a "normal" aurora that occur in Earth's polar region to show how the aurora penetrate into the ionosphere.

given in Fig. 44. As shown in this figure, the "normal" aurora terminates near the lower boundary of the ionosphere E region while intense aurora penetrate well into the D region of Earth's ionosphere.

An excellent review of all aspects of auroral phenomena was given by Vallance-Jones (1974), including the effects of aurora on the polar ionospheres. The role of electron-impact processes in the auroral phenomena is basically the same as outlined above for ionospheric processes, except for important differences involving the incident electron energy spectrum. That is, the gaseous species are the same, but the initial electron energy is 1–10 keV for auroras as contrasted to ~ 10 eV for ionospheric photo-electrons. As a result, the auroral electron energy distribution below about 200 km is characterized by a higher average energy than that produced by photoelectrons. Electron transport and energy thermalization are therefore determined by the relevant electron–molecule collisions cross sections for higher electron energies than in the case of ionospheric processes. An additional complexity associated with the aurora, not particularly important in Earth's ionosphere, is the generation of molecular species not found in the undisturbed atmosphere. For example, the absolute number densities for vibrationally excited N_2 and O_2 and metastable electronic states [e.g., $N_2(A\ ^3\Sigma_u^+)$] may be so high as to affect both the transport and energy balance of the electrons. In addition, the higher-energy characteristic of the auroral electron spectrum is apparently responsible for an observed enhanced concentration of NO (see Vallance-Jones, 1974: p. 193).

In summary then, the auroral phenomena requires the same basic cross section data, but for higher electron energies and possibly for additional molecules such as NO and vibrationally and electronically excited species.

3. Other Planetary Atmospheres

A great deal has been learned about the composition of the atmospheres of the other planets in our solar system as a result of both the U.S. and U.S.S.R. Explorer satellite systems and the U.S. Voyager spacecrafts. Although the information is still far from complete, a combination of all the available evidence suggests that the primary constituents are as summarized in Table VIII (Hunt and Bartlett, 1972; Hunt, 1974; Hunt, 1976).

Any detailed consideration of the ionospheric structure or interaction of a planet with the magnetosphere will require the same types of fundamental cross section data as required to model Earth's atmosphere. That is, the momentum transfer and rotational- and vibrational-excitation cross section for these gases will be needed, as a minimum, to determine the secondary electron distribution for the planetary atmosphere. Some work has been done on CO_2 because of its importance in the atmospheres of Venus and Mars. Morrison and Green (1978) looked at the thermalization of electrons in CO_2,

TABLE VIII

Components of Various Planetary Atmospheres

Planet	
Mercury	(H, He, O),?
Venus	Ar, N_2, HCl, HF, H_2O, CO_2, SO_2
Mars	Ne, Ar, Kr, Xe, N_2, O_2, CO, O_3, CO_2, H_2O
Jupiter/Io	He, K, Na, H_2, CO, CO_2, H_2O, SO_2, HCN, NH_3, CH_4, C_2H_2, C_2H_6, PH_3, GeH_4
Saturn	He, H_2, CH_4, NH_3, C_2H_2, C_2H_6, C_3H_4, C_3H_8
Uranus	H_2, CH_4, (He, NH_3), ?
Neptune	H_2, CH_4, (He, NH_3),?
Pluto	CH_4, (Ne) ?

including all relevant electron-impact processes and determined that vibrational and rotational excitation were both more efficient than momentum transfer for electron cooling for electron temperatures less than 7000 K. They pointed out that the magnitude of the effect due to rotational excitation is uncertain because no experimental data are available to calibrate the theoretical estimates used in the modeling. Fox and Dalgarno (1979) examined the processes by which energetic electrons lose energy in a weakly ionized gas of carbon dioxide and identified the deficiency of excitation, ionization, and dissociation cross section data.

Of the other molecular constituents of the planetary atmospheres listed in Table VIII, H_2, N_2, O_2, and CO have by far the most cross section data available, but as pointed-out above, even these data sets are not complete (e.g., rotational-excitation cross sections are available only for H_2). Fragmentary information on elastic electron scattering by H_2O, CH_4, HD, HCl, and NH_3 is available today, but very little on vibrational excitation, and no data on rotational excitation is available for these molecules. There is currently no electron scattering information available for O_3 or other somewhat esoteric molecules (Hunt, 1976) such as NH_3SH_2 that are believed to be important in the atmospheres of the heavy planets. Clearly, an enormous amount of basic experimental and theoretical research remains to be done on these planetary constituents before enough electron-impact cross section data are available to model these planetary atmospheres in any detail.

The Jupiter environment represents a unique scenario from the point of view of electron collision processes. This is due to volcanic activities of Io and the large charged particle fluxes in the Jovian magnetosphere. Volcanic eruptions are estimated to generate 10^8 g s^{-1} SO_2 on Io (Cheng, 1980). The interaction of this SO_2 with the magnetospheric electrons is mainly responsible for penetrating the Io plasma form and flux tube. Ultraviolet spectral observations indicate that a large portion of the excitation process is due to electrons. Analysis of this system and the aurora, airglow, and lightning in the atmospheres of Jupiter and Saturn is only in the beginning stage. It is clear, however, that a large number of electron collision processes are important in the atmospheres of the major planets.

C. Gas Lasers

Elastic electron scattering and rotational excitation of molecules by electrons certainly play an important role in the operation of any molecular gas discharge laser, and therefore these processes must be understood for each specific laser system of interest. However, since gas lasers are treated in some detail in Chapter 2 of Volume 1 and Chapter 5 of Volume 2 of this series, they will not be discussed here.

Appendix A

Summary of Review Articles on
the Theory of Electron Scattering by Molecules
(Elastic Scattering and Rotational Excitation)

J. D. Craggs and H. S. W. Massey, in *Handbuch der Physik*, edited by S. Flügge (Springer, Berlin, 1956).

The Collisions of Electrons with Molecules

> Comprehensive review of high-, intermediate-, and low-energy scattering of electrons by molecules. Detailed discussion of high-energy electron scattering: separated-atom approximation, independent-atom model, and multiple-scattering effects. Detailed presentation of low-energy elastic scattering calculations up to 1956 including discussion of prolate spherical coordinates and early low-energy calculations for electron–H_2, –N_2, –O_2, and –Cl_2 and their comparison with experimental data. Detailed review of swarm experiments and wave interaction studies summarizing experimental data obtained for H_2, D_2, and N_2. Review of the theory of vibrational and rotational excitation. (*Introductory*)

O. H. Crawford, A. Dalgarno, and P. B. Hays, *Mol. Phys.* **13**, 181–192 (1967).

Electron Collision Frequencies in Polar Gases

> The electron scattering by a fixed dipole and rotating dipole are considered in the Born approximation. The fixed-dipole Born-approximation results show reasonably good agreement with the experimental data for the momentum transfer cross section. The interpretation of swarm experiments for molecules with internal structure is discussed briefly. (*Introductory*)

K. Takayanagi, *Suppl. Progr. Theor. Phys.* **40**, 216–248 (1967).

Scattering of Slow Electrons by Molecules

Discusses electron–molecule interaction potentials and the adiabatic approximation. Detailed discussion of single-center and two-center expansion techniques. Born approximation calculations are reviewed for rotational excitation of homonuclear and polar molecules. (*Introductory*)

K. Takayanagi and Y. Itikawa, *Adv. At. Mol. Phys.* **6**, 105–153 (1970).

The Rotational Excitation of Molecules by Slow Electrons

Discusses electron–molecule interaction potentials and LAB frame CC equations. Detailed discussion is given of rotational excitation of molecules modeled by spherical, symmetrical, and assymetrical rotors in the Born approximation. Partial-wave analysis of the Born approximation is presented and detailed discussion is given on its validity. Adiabatic approximation and distorted wave calculations are described. Close-coupling calculations for homonuclear and heteronuclear molecules are described. (*Introductory*)

N. Chandra and S. K. Joshi, *Adv. Astron. Astrophys.* **7**, 1 (1970).

Scattering of Electrons by Diatomic Molecules

General review of electron–molecule scattering with detailed discussion of the fundamental equations of the subject. Born, Born–Oppenheimer, first-order exchange, Ochkur, and Ochkur–Rudge approximations are elaborated. Detailed presentation is given of electron scattering by H_2^+, H_2, and N_2 up to 1970. A short summary of electron scattering by polar molecules is also given. (*Introductory*)

D. E. Golden, N. F. Lane, A. Temkin, and E. Gerjuoy, *Rev. Mod. Phys.* **43**, 642–678 (1971).

Low-Energy Electron–Molecule Scattering Experiments and the Theory of Rotational Excitation

Complete review of experimental and theoretical work on low-energy electron–molecule scattering up to 1970. Discusses experimental techniques for the measurement of total, momentum transfer, and differential cross sections. Reviews experimental results obtained for important molecules. Describes LAB frame CC formalism for electron–H_2 scattering in complete form. Reviews CC calculations as well as distorted-wave and Born calculations. Compares theoretical results with experimental data. Reviews fixed-nuclei calculations and describes in detail the adiabatic-nuclei theory along with the single-center expansion. Reviews AN theory for rotational excitation. Compares results of AN calculations with experiment and with LAB frame close-coupling calculations. A classic of electron–molecule collision physics. (*Introductory*)

W. R. Garrett, *Mol. Phys.* **24**, 465–487 (1972).

Low-Energy Electron Scattering by Polar Molecules

All important aspects of electron–polar molecule scattering are reviewed. Scattering by a stationary dipole and rotating dipole are discussed in detail. The dipole-moment dependence of momentum transfer and elastic and rotational excitation cross sections are analyzed in detail. The problem of so-called critical binding is discussed in detail throughout the review. (*Introductory*)

K. Takayanagi, "Atomic Physics 4," edited by G. Zu Putlitz, E. W. Weber, and A. Winnacker (Plenum, New York, 1974).
Low-Energy Electron–Molecule Scattering

Short review of low-energy elastic and rotational excitation work. Physics is emphasized, only few equations are given. Both LAB frame CC and Born approximations and AN–CC and Glauber calculations are reviewed. (*Introductory*)

Y. Itikawa, *Phys. Rep.* **46**, 117–164 (1978).
Electron Scattering by Polar Molecules

Comprehensive and thorough review. After a short historical introduction the basic experimental techniques relevant to electron–polar molecule scattering are discussed and the theoretical methods are reviewed. The prominent role of the dipole potential is mentioned. The LAB frame CC equations are described and LAB frame–Born approximation is presented. Fixed-nuclei methods are surveyed, along with semiclassical methods. A separate section discusses the comparison of theoretical results with experimental data where the roles of dipole potential and short-range potentials are highlighted. A separate section discusses the "critical binding" problem and rotational resonances, and another section is dedicated to the applied aspects of this subject. (*Intermediate*)

D. G. Truhlar, K. Onda, R. A. Eades, and D. A. Dixon, *Int. J. Quantum Chem. Symp.* **13**, 601–632 (1979).
Effective Potential Approach to Electron–Molecule Scattering Theory

After a short presentation of the effective (optical) potential formalism of electron–molecule scattering theory, the authors review in detail the various numerical procedures used for the actual calculation of static, polarization, and model (local) exchange potentials. They briefly review the AN and LAB–CC calculations in which these effective potentials were utilized. A special section is dedicated to the analysis of results using these effective potentials for electron–N_2, –H_2, and –CO_2 scattering calculations. Complete reference list is given on the subject. (*Introductory*)

P. G. Burke, *Adv. At. Mol. Phys.* **15**, 471–506 (1979).
Theory of Low-Energy Electron–Molecule Collisions

Brief and coherent review touching on all important aspects of the problem. Discusses LAB frame and molecular frame representations and frame transformation theory. Separate section on L^2 methods (*T* matrix method, *R* matrix method). (*Intermediate*)

N. F. Lane, *Rev. Mod. Phys.* **52**, 29–119 (1980).
The Theory of Electron–Molecule Collisions

The most comprehensive review of this field with a complete list of references to 1979. Every aspect of (principally low-energy) electron–molecule scattering problem is discussed in great detail. Excellent summary of all important works covering in great depth the 1969–1979 period. Fundamental principles and essential points of physics are summarized with great lucidity, computational aspects are briefly mentioned but not detailed. A detailed discussion with complete derivations of the most general LAB frame and BODY frame CC equations, frame transformation theory, and AN approximation is given. Transformation formulas connecting BODY frame T matrix and LAB frame T matrix are discussed in detail with insight provided. R matrix and T matrix expansion methods are summarized. Fundamental equations of the static exchange, static model exchange approximations are given as well as the various polarization potentials used in AN calculations. Pseudopotential methods and weak scattering approximations (distorted-wave method, Born approximation) are briefly discussed. The application of the various theoretical methods are given molecule by molecule. Thorough discussion of electron–H_2, –N_2, –CO, and –CO_2 scattering with highlights of all significant calculations. Electron–polar molecule scattering is discussed in a very concentrated manner in a coherent summary of all significant works. (*Intermediate*)

P. G. Burke, in "Quantum Dynamics of Molecules," edited by R. G. Woolly (Plenum, New York, 1980).

Theory of Electron–Molecule Collisions

Complete review of theoretical methods used for low-energy electron–molecule calculations, including detailed presentation of CN–CC equations, LAB–CC equations, and frame transformation theory, with examples for calculational results and its comparison with experiment. Detailed presentation of L^2 methods (R matrix and T matrix method) is given. Vibrational excitation and dissociative attachment is also discussed. Resonance theory is detailed. Separate section on scattering by polar molecules. Short summary on scattering at intermediate and high energies. (*Introductory*)

D. G. Truhlar, in "Chemical Applications of Atomic and Molecular Electrostatic Potentials," edited by P. Politzer and D. G. Truhlar (Plenum, New York, 1981).

Effective Potentials for Intermediate-Energy Electron Scattering: Testing Theoretical Methods

After short introduction on effective potentials, detailed description is given of the calculation of static, local exchange, and polarization potentials to be used in intermediate-energy electron–molecule scattering (AN or LAB–CC) calculations. The results of scattering calculations with these effective potentials are compared with experimental data in the last section. N_2, H_2, CO, and C_2H_2 targets are discussed in detail. Very complete set of references are given. (*Introductory*)

B. D. Buckley, P. G. Burke, and C. J. Noble, in "Electron Molecule Collisions," edited by I. Shimamura and K. Takayanagi (Plenum Press, New York, 1984).

Computational Methods for Low-Energy Electron–Molecule Scattering

Very comprehensive review on the derivation and solution of LAB frame CC and fixed nuclei CC equations. Detailed review of computational techniques used for fixed nuclei calculations including integral equation method, iterative methods, linear algebraic equations method, and variational methods. Discussion on model exchange potential and polarization potentials to be used in these calculations. Discussion of continuum multiple-scattering method, R matrix, and T matrix methods. Short review of hybrid theory and resonance theory. Discussion on electronic excitation. (*Intermediate*)

D. W. Norcross and L. A. Collins, *Adv. At. Mol. Phys.* **18**, 341–397 (1982).

Recent Developments in the Theory of Electron Scattering by Highly Polar Molecules

Very complete review on all aspects of electron–polar molecule scattering with great physical insight and very detailed comparison between theoretical results and experimental data. After a short discussion on the Born approximation, the fixed nuclei approximation is discussed and the problem of divergencies is introduced along with the "critical dipole" problem. Subsequently the LAB frame and BODY frame formulation is summarized and frame transformation theory is discussed along with the multiple-frame formulation (called "angular frame transformation" by the authors and "hybrid formulation" by others). In the section "Approaches and Approximations" a very systematic presentation of all electron–polar molecule scattering calculation is given. First the exact static exchange approximation and the associated numerical techniques are reviewed, then approximations and representations of the interaction potential are discussed: polarization, exchange, and static potential along with model potentials, the dipole cutoff potential and the point dipole potential. Collisional approximations are discussed: close coupling, the multiple-scattering method, distorted-wave method, classical and semiclassical methods, and various forms of the Born approximation. In a section on cross section it is shown how the simple LAB frame approximations (e.g., Born) can be combined with accurate AN calculations to obtain accurate results. The "Applications" section contains one of the most detailed and comprehensive presentations of theory and experiment for a few targets (principally LiF) ever done for any molecule. In this section the physics and the complexity of the problem become very transparent. The authors analyze separately the various physical features which are manifested in integrated cross section, momentum transfer cross section, and the differential cross section. A masterful presentation of the subject. (*Advanced*)

M. A. Morrison, *Austr. J. Phys.* (1983).

The Physics of Low-Energy Electron–Molecule Collisions: A Guide for the Perplexed and the Uninitiated

A delightful review on the theory of low-energy electron–molecule scattering. It is an attempt to provide three things: (1) a qualitative introduction to the principal physical features of electron–molecule collisions, (2) a description of the essential ideas behind several approaches to this problem that are currently under investigation, and (3) comments on a few outstanding problems that seem to be particularly important and on work in the field since 1980. A detailed discussion is given on close-coupling techniques and the AN approximation and their relationships (frame transformation theory). Elaborate presentation is given on the calculation of the electron–molecule interaction potentials: static, model exchange, and polarization potentials. A short summary of current problems and approaches to solve them, emphasizing physical concepts, is also given. (*Introductory*)

I. Shimamura, in "Symposium on Electron–Molecule Collisions — Invited Papers," edited by I. Shimamura and M. Matsuzawa, pp. 13–30. (University of Tokyo Press, 1979); and in "Electron–Molecule Collisions," edited by I. Shimamura and K. Takayanagi (Plenum Press, New York, 1983).

Rotational Excitation of Molecules by Slow Electrons

Examines systematic in DCSs for scattering of electrons by different molecules by using scaling laws for highly polar molecules and reduced cross sections for the others. Investigates the region of validity for semiclassical approximations and scaling laws and the applicability of sum rules for scattering by polar molecules. The author discusses in detail the AN approximation and frame transformation theory and presents excellent summary of data on scattering by polar molecules. *(Advanced)*

Acknowledgments

The authors would like to thank all their colleagues who have generously provided results prior to publication and helpful comments on this review. They are especially indebted to Dr. Lee Collins for valuable comments and suggestions. Special thanks also go to Howard Shirley (LANL), who expertly prepared the illustrations, and to Patrick Romero (LANL), who patiently typed, and retyped, the manuscript.

References

Abusalbi, N., Eades, R. A., Nam, T., Thirumalai, D., Dixon, D. A., and Truhlar, D. G. (1983). *J. Chem. Phys.* **78**, 1213–1227.
Aggarwal, K. M., Nath, N., and Setty, C.S.G.K. (1979). *Planet. Space Sci.* **27**, 754–768.
Allison, A. C. (1975). *J. Phys. B* **8**, 325–330.
Altshuler, S. (1957). *Phys. Rev.* **107**, 114–117.
Anderson, J. B. (1974). "Molecular Beams and Low Density Gas Dynamics" (P. P. Wagner, ed.), Vol. 4. Dekker, New York.
Anderson, J. B., Andres, R. P., and Fenn, J. B. (1965). *Adv. At. Mol. Phys.* **1**, 275–315.
Ardill, R. W. B., and Davison, W. D. (1968). *Proc. R. Soc. London* **A304**, 465–477.
Arnot, F. L. (1929). *Proc. R. Soc. London* **A125**, 660–669.
Arnot, F. L. (1931). *Proc. R. Soc. London* **A133**, 615–636.
Arthurs, A. M., and Dalgarno, A. (1960). *Proc. R. Soc. London* **A256**, 540–551.
Ashihara, O., Shimamura, I., and Takayanagi, K. (1975). *J. Phys. Soc. Japan* **38**, 1732–1741.
Baille, P., and Darewych, J. W. (1977). *J. Chem. Phys.* 67, 3399–3400.
Banks, P. M., and Kockarts (1973). "Aeronomy." Academic Press, New York.
Bauer, S. J. (1973). "Physics of Planetary Ionospheres." Springer-Verlag, Heidelberg.
Becker, W. G., Frickes, M. G., Slater, R. C., and Stern, R. C. (1974). *J. Chem. Phys.* **61**, 2283–2289.
Bederson, B., and Kieffer, L. J. (1971). *Rev. Mod. Phys.* **43**, 601–640.
Bell, K. L. (1981). *J. Phys. B* **14**, 2895–2900.
Bell, K. L., and Moiseiwitsch, B. L. (1963). *Proc. R. Soc. London* **A276**, 346–353.
Bennani, A. L., Duguet, A., and Wellenstein, H. F., *J. Phys. B* **12** (1979) 461–472.
Bhattacharyya, P. K., and Ghosh, A. S. (1975). *Phys. Rev. A* **12**, 480–485.
Bhattacharyya, P. K., and Ghosh, A. S. (1976). *Phys. Rev. A* **14**, 1587–1590.
Bhattacharyya, P. K., and Goswami, K. K. (1982). *Phys. Rev. A* **26**, 2592–2605.
Bhattacharyya, P. K., Goswami, K. K., and Ghosh, A. S. (1978). *Phys. Rev. A* **18**, 1865–1877.
Bonham, R. A. (1962). *J. Chem. Phys.* **36**, 3260–3269.
Bonham, R. A. and Fink, M. (1974). "High Energy Electron Scattering." von Nostrand-Reinhold, New York.
Bonham, R. A., and Iijima, T. (1963), *J. Phys. Chem. Ithaca* 67, 2266–2272.
Bonham, R. A., and Wellenstein, H. F. (1973). *J. Appl. Phys.* **44**, 2631–2634.
Born, M. (1926). *Z. Phys.* **38**, 803–827.

Bottcher, C. (1969). *Chem. Phys. Lett.* **4**, 320–322.
Brandt, M. A., and Truhlar, D. G. (1973). *Chem. Phys. Lett.* **23**, 48–52.
Brandt, M. A., Truhlar, D. G., and Van-Catledge, F. A. (1976). *J. Chem. Phys.* **64**, 4957–4967.
Brewer, D. F. C., Newell, W. R., and Smith, A. C. H. (1980a). *J. Phys. E* **13**, 114–122.
Brewer, D. F. C., Newell, W. R., and Smith, A. C. H. (1980b). *J. Phys. E* **13**, 123–127.
Brinkmann, R. T., and Trajmar, S. (1981). *J. Phys. E* **14**, 245–255.
Bromberg, J. P. (n.d.), Unpublished results cited by Lassettre and White, 1974b.
Bromberg, J. P. (1970). *J. Chem. Phys.* **52**, 1243–1247.
Bromberg, J. P. (1974a). *J. Chem. Phys.* **60**, 1717–1721.
Bromberg, J. P. (1975). "The Physics of Electronic and Atomic Collisions" (J. S. Risley and R. Geballe, eds.), pp. 98–111. University of Washington Press, Seattle.
Browne, C. P., Craig, D. S., and Williamson, R. M. (1951). *Rev. Sci. Instrum.* **22**, 952–965.
Brunt, J. N. H., Read, F. H., and King, G. C. (1977). *J. Phys. E* **10**, 134–139.
Buckley, B. D., and Burke, P. G. (1977). *J. Phys. B* **10**, 725–739.
Buckley, B. D., Burke, P. G., and Voky Lan (1979). *Comput. Phys. Commun.* **17**, 175–179.
Buckley, B. D., Burke, P. G., and Noble, C. J. (1984). Computational methods for low-energy electron molecule scattering. In "Electron Molecule Collisions" (I. Shimamura and K. Takayanagi, eds.). Plenum Press, New York.
Bullard, E. C., and Massey, H. S. W. (1931). *Proc. R. Soc. London* **A133**, 637–651.
Burke, P. G. (1971/72). *Comments At. Mol. Phys.* **III**, 31–45.
Burke, P. G. (1972). *In* "Atoms and Molecules in Astrophysics" (T. R. Carson and M. J. Roberts, eds.), p. 1. Academic Press, London.
Burke, P. G., and Chandra, N. (1972). *J. Phys. B* **5**, 1696–1711.
Burke, P. G., and Sinfailam, A. L. (1970). *J. Phys. B* **3**, 641–659.
Burke, P. G., Mackey, I., and Shimamura, I. (1977). *J. Phys. B* **10**, 2497–2512.
Burke, P. G., Noble, C. J., and Salvini, S. (1983). *J. Phys. B* **16**, L113–L120.
Campargue, R. (1964). *Rev. Sci. Instrum.* **35**, 111–112.
Carson, T. R. (1954). *Proc. Phys. Soc. London* **A67**, 909–916.
Carter, C., March, N. H., and Vincent, D. (1958). *Proc. Phys. Soc. London* **71**, 2–16.
Castillejo, L., Percival, I. C., and Seaton, M. J. (1960). *Proc. R. Soc. London* **A254**, 259–271.
Celotta, R. J., and Huebner, R. H. (1979). *In* "Electron Spectroscopy: Theory Techniques, and Applications" (C. R. Brundle and A. D. Baker, eds.), Vol. 3, pp. 41–125. Academic Press, New York.
Chandra, N. (1975). *Phys. Rev. A* **12**, 2342–2352.
Chandra, N. (1977). *Phys. Rev. A* **16**, 80–108.
Chandra, N., and Burke, P. G. (1973), *J. Phys. B* **6**, 2355–2357.
Chandra, N., and Temkin, A. (1976a). *Phys. Rev. A* **13**, 188–203.
Chandra, N., and Temkin, A. (1976b). *J. Chem. Phys.* **65**, 4537–4537.
Chandra, N., and Temkin, A. (1976c). *Phys. Rev. A* **14**, 507–511; and NASA TN D-8347.
Chang, E. S. (1970). *Phys. Rev. A* **2**, 1403–1406.
Chang, E. S. (1977). *Phys. Rev. A* **16**, 1841–1849, 1850–1853.
Chang, E. S. (1983). *Phys. Rev. A* **27**, 709–716.
Chang, E. S., and Fano, U. (1972). *Phys. Rev. A* **6**, 173–185.
Chang, E. S. and Temkin, A. (1970). *J. Phys. Soc. Japan* **29**, 172–179.
Chantry, P. J. (1969). *Rev. Sci. Instrum.* **40**, 884–889.
Chapman, S., and Cowling, T. G. (1958). "Mathematical Theory of Non-Uniform Gases," Cambridge Univ. Press, New York.
Chase, D. M. (1956). *Phys. Rev.* **104**, 838–842.
Cheng, A. F. (1980). *Astrophys. J.* **242**, 812–827.
Child, C. D. (1911). *Phys. Rev.* **32**, 492–511.

Choi, B. H., and Poe, R. T. (1977a). *Phys. Rev. A* **16**, 1321–1830.
Choi, B. H. and Poe, R. T. (1977b). *Phys. Rev. A* **16**, 1831–1840.
Choi, B. H., Poe, R. T., Sun, J. C., and Shan, Y. (1979). *Phys. Rev. A* **19**, 116–124.
Chung, S., and Lin, C. C. (1978). *Phys. Rev. A* **17**, 1874–1891.
Chutjian, A. (1979). *Rev. Sci. Instrum.* **50**, 347–355.
Clark, C. W. (1979). *Phys. Rev. A* **20**, 1875–1889.
Collins, L. A., and Morrison, M. A. (1982). *Phys. Rev. A* **25**, 1764–1767.
Collins, L. A., and Norcross, D. W. (1977). *Phys. Rev. Lett.* **38**, 1208–1211.
Collins, L. A., and Norcross, D. W. (1978). *Phys. Rev. A* **18**, 467–498.
Collins, L. A., and Robb, W D. (1980). *J. Phys. B* **13**, 1637–1649.
Collins, L. A., and Schneider, B. I. (1981). *Phys. Rev. A* **24**, 2387–2401.
Collins, L. A., Norcross, D. W., and Schmid, G. B. (1979). *J. Phys. B* **12**, 1019–1030.
Collins, L. A., Henry, R. J. W., and Norcross, D. W. (1980a). *J. Phys. B* **13**, 2229–2307.
Collins, L. A., Robb, W. D., and Morrison, M. A. (1980b). *Phys. Rev. A* **21**, 488–495.
Craggs, J. D., and Massey, H. S. W. (1956). *In* "Handbuch der Physik", Vol. 37/1 (S. Flugge, ed.). Springer-Verlag, Berlin, 314–428.
Crawford, O. H. (1967). *J. Chem Phys.* **47**, 1100–1104.
Crawford, O. H., and Dalgarno, A. (1971). *J. Phys. B* **4**, 494–502.
Crees, M. A., and Moores, D. L. (1975). *J. Phys. B* **8**, L195–L199.
Crees, M. A., and Moores, D. L. (1977). *J. Phys. B* **10**, L225–L227.
Crompton, R. W., Gibson, D. K., and McIntosh, A. I. (1969). *Austral. J. Phys.* **22**, 715–731.
Csanak, G., and Taylor, H. S. (1972). *Phys. Rev. A* **6**, 1843–1855.
Daimon, H., Kondow, T., and Kuchitsu, K. (1979). In "Abstracts of Contributed Papers XI, Int. Conf. on the Physics of Electronic and Atomic Collisions, Kyoto," pp. 316–317.
Dalgarno, A. (1969). *Can. J. Chem.* **47**, 1723–1731.
Dalgarno, A., and Henry, R. J. W. (1965). *Proc. Phys. Soc.* **85**, 679–684.
Dalgarno, A., and Moffet, R. J. (1963). *Proc. Natl. Acad. Sci. India* **A33**, 511–521.
Danner, D. (1970). Ph.D. Thesis, Physikalisches Institute der Universitat Freiburg, West Germany.
Davis, M. I. (1971). "Electron Diffraction in Gases." Dekker, New York.
Dempster, A. J. (1937). *Phys. Rev.* **51**, 67–69.
Dhal, S. S., Srivastava, B. B., and Shingal (1979). *J. Phys. B* **12**, 2211–2216.
DiChio, D., Natali, S. V., Kuyatt, C. E., and Galejs, A. (1974). *Rev. Sci. Instrum.* **45**, 566–569.
Dill, D., and Dehmer, J. L. (1977). *Phys. Rev. A* **16**, 1423–1431.
Dixon, D. A., Eades, R. A., and Truhlar, D. G. (1979). *J. Phys. B* **12**, 2741–2753.
DuBois, R. D. and Rudd, M. E. (1976). *J. Phys. B* **15**, 2657–2667.
Dyke, T. R., and Muenter, J. S. (1974). *J. Chem. Phys.* **60**, 2929–2931.
Dyke, T. R., Romasevich, G. R., Klemperer, W., and Falconer, E. (1972). *J. Chem. Phys.* **57**, 2277–2284.
Eades, R. A., Truhlar, D. G., and Dixon, D. A. (1979). *Phys. Rev. A* **26**, 867–878.
Eather, R. H. (1980). "Majestic Lights, the Aurora in Science, History, and the Arts." American Geophysical Union, Washington, D.C.
Edmonds, A. R. (1957). "Angular Momentum in Quantum Mechanics." Princeton University Press, Princeton New Jersey.
Eland, J. D. H., and Danby, J. C. (1968). *J. Phys. E* **1**, 406–408.
El-Kareh, A. B., and El-Kareh, J. C. J. (1970), "Electron Beams, Lenses and Optics," Vols. I and II. Academic Press, New York.
Englehardt, A. G., Phelps, A. V., and Risk, C. G. (1964). *Phys. Rev.* **135**, A1566–A1574.
Fano, U. (1970a). *Comments At. Mol. Phys.* **2**, 47–52.
Fano, U. (1970b). *Phys. Rev. A* **2**, 353–365.

Feldt, A. N., and Morrison, M. A. (1982). *J. Phys. B* **15**, 301–308.
Feshbach, H. (1958). *Ann. Phys.* **5**, 357–390.
Fetter, A. L., and Watson, K. M. (1965). *Adv. Theor. Phys.* **1**, 115–194.
Fink, J., and Kisker, E. (1980). *Rev. Sci. Instrum.* **51**, 918–920.
Fink, M., Jost, K., and Hermann, D. (1975a). *Phys. Rev. A* **12**, 1374–1382.
Fink, M., Jost, K., and Hermann, D. (1975b). *J. Chem. Phys.* **63**, 1985–1987.
Fisk, J. B. (1936). *Phys. Rev.* **49**, 167–173.
Fisk, J. B. (1937). *Phys. Rev.* **51**, 25–28.
Fite, W. L. (1962). "Atomic and Molecular Processes" (D. R. Bates, ed.), p. 421. Academic Press, New York.
Fliflet, A. W., and McKoy, V. (1978). *Phys. Rev. A* **18**, 1048–1054.
Fliflet, A. W., Levin, D. A., Ma, M., and McKoy, V. (1978). *Phys. Rev. A* **17**, 160–169.
Ford, A. L., and Browne, J. C. (1973). *Chem. Phys. Lett.* **20**, 284–290.
Fox, J. L., and Dalgarno, A. (1979). *Planet. Space Sci.* **27**, 491–502.
Fox, R. E., Hickman, W. M., Grove, D. J., and Kheldass, T. (1951). *Phys. Rev.* **84**, 859–860.
Freeman, R. R., Matthison, E. M., Pritchard, D. E., and Klepper, D. (1974). *Phys. Rev. Lett.* **33**, 397–399.
Froitzheim, H., Ibach, H., and Lehwald, S. (1975). *Rev. Sci Instrum.* **46**, 1325–1328.
Gallagher, R. J., and Fenn, J. B. (1974). *J. Chem. Phys.* **60**, 3487–3491.
Garrett, W. R. (1971). *Phys. Rev. A* **4**, 2229–2235.
Geltman, S., and Takayanagi, K. (1966), *Phys. Rev.* **143**, 25–30.
Gerjuoy, E., and Stein, S. (1955a). *Phys. Rev.* **97**, 1671–1679.
Gerjuoy, E., and Stein, S. (1955b). *Phys. Rev.* **98**, 1848–1851.
Gianturco, F. A., and Lamanna, U. T. (1980). *Mol. Phys.* **40**, 793–804.
Gianturco, F. A., and Rahman, N. K. (1978). *J. Phys. B* **11**, 721–740.
Gianturco, F. A., and Thompson, D. G. (1976). *J. Phys. B* **9**, L383–L385.
Gianturco, F. A., and Thompson, D. C. (1977). *J. Phys. B* **10**, L21–L26.
Gianturco, F. A., and Thompson, D. G. (1980). *J. Phys. B* **13**, 613–625.
Gibson, D. K. (1970). *Austral. J. Phys.* **23**, 683–696.
Gibson, T. L., and Morrison, M. A. (1983). *J. Phys. B* to be published.
Gien, T. T. (1978). *Phys. Lett.* **68A**, 33–35.
Gilardini, A. (1972). "Los Energy Electron Collisions in Gases; Swarm and Plasma Methods Applied to their Study." Wiley, New York.
Giraud, A. and Petit, M. (1978). Ionospheric techniques and phenomena. *In* "Geophysics and Astrophysics Monographs," Vol. 13. Reidel, Dordrecht, Holland.
Glauber, R. J. (1955). *Phys. Rev.* **100**, 242–248.
Glauber, R. J. (1959). *In* "Lectures in Theoretical Physics" (W. E. Brittin and L. G. Dunham, eds.), pp. 315–414. Wiley (Interscience), New York.
Golden, D., and Zecca, A. (1971). *Rev. Sci. Instrum.* **42**, 210–216.
Golden, D. E., Bandel, H. W., and Salerno, J. A. (1966). *Phys. Rev.* **146**, 40–42.
Golden, D. E., Lane, N. F., Temkin, A., and Gerjuoy, E. (1971). *Rev. Mod. Phys.* **43**, 642–678.
Gupta, P. and Khare, S. P. (1978). *J. Chem. Phys.* **68**, 2193–2198.
Hagena, D. F. (1974). Cluster beams in nozzle sources. *In* "Molecular Beams and Low Density Gasdynamics," pp. 93–181. Mardel-Decker, New York.
Hara, S. (1967). *J. Phys. Soc. Japan* **22**, 710–718.
Hara, S. (1969a). *J. Phys. Soc. Japan* **27**, 1009–1019.
Hara, S. (1969b). *J. Phys. Soc. Japan* **27**, 1262–1267.
Hara, S. (1969c). *J. Phys. Soc. Japan* **27**, 1592–1597.
Harnwell, G. P. (1929a). *Phys. Rev.* **33**, 559–571.
Harnwell, G. P. (1929b). *Phys. Rev.* **34**, 661–672.

Harris, S. J., Novich, S. E., Winn, J. S., and Klemperer, W. (1974). *J. Chem. Phys.* **61**, 3866–3867.
Harrower, G. A. (1955). *Rev. Sci. Instrum.* **26**, 850–854.
Harshbarger, W. R., Skerbele, A., and Lassettre, E. N. (1979). *J. Chem. Phys.* **54**, 3784–3789.
Harter, W. G., Patterson, C. W., and da Paixâo, F. J. (1978). *Rev. Mod. Phys.* **50**, 37–83.
Harting E., and Burrows, K.M. (1970). *Rev. Sci. Instrum.* **41**, 97–101.
Harting, E., and Read, F. H. (1976). "Electrostatic Lenses." Elsevier, New York.
Hasted, J. B. (1964). "Physics of Atomic Collisions." Butterworth, London, and Washington, D.C.
Hasted, J. B. (1972). "Physics of Atomic Collisions." Elsevier, New York.
Hayashi, S., and Kuchitsu, K. (1976). *Chem. Phys. Lett.* **41**, 575–579.
Hayashi, S., and Kuchitsu, K. (1977). *J. Phys. Soc. Japan* **42**, 1319–1326.
Heddle, D. W. O. (1968). "Atomic Interactions, Part A" (B. Bederson and W. L. Fite, eds.), pp. 43, Academic Press, New York.
Heddle, D. W. O. (1969). *J. Sci. Instrum.* **2**, 1046–1050.
Henry, R. J. W. (1970). *Phys. Rev. A* **2**, 1349–1358.
Henry, R. J. W., and Chang, E. S. (1972). *Phys. Rev. A* **5**, 276–284.
Henry, R. J. W., and Lane, N. F. (1969). *Phys. Rev.* **183**, 221–231.
Herzberg, G. (1950). "Molecular Spectra and Molecular Structure, I Spectra of Diatomic Molecules." van Nostrand-Reinhold, New York.
Holley, T. K., Chung, S., Lin, C. C., and Lee, E. T. P. (1981). *Phys. Rev. A* **24**, 2946–2952.
Huang, J. T. J., and Chan, F. T. (1977). *Phys. Rev. A* **15**, 1782–1784.
Hughes, A. L., and McMillen, J. H. (1932). *Phys. Rev.* **41**, 39–48.
Hughes, A. L., and Rojansky, V. (1929). *Phys. Rev.* **34**, 284–290.
Hunt, G. E. (1974). *Proc. R. Soc. London* **A341**, 317–330.
Hunt, G. E. (1976). *Adv. Phys.* **25**, 455–487.
Hunt, G. E., and Bartlett, J. T. (1972). *Endeavour* **32**, 39–43.
Huxley, L. G. H., and Crompton, R. W. (1974). "The Diffusion and Drift of Electrons in Gases." Wiley, New York.
Itikawa, Y. (1969). *J. Phys. Soc. Japan* **27**, 444–452.
Itikawa, Y. (1971a). *J. Phys. Soc. Japan* **30**, 835–842.
Itikawa, Y. (1971b). *Planet. Space Sci.* **19**, 993–1007.
Itikawa, Y. (1971c). *J. Phys. Soc. Japan* **31**, 1532–1535.
Itikawa, Y. (1972). *J. Phys. Soc. Japan* **32**, 217–226.
Itikawa, Y (1974). *Atom. Data Nuc. Data Tables* **14**, 1–10.
Itikawa, Y. (1976). *J. Phys. Soc. Japan* **41**, 619–624.
Itikawa, Y. (1978a). *Phys. Rep.* **46**, 117–164.
Itikawa, Y. (1978b). *At. Data and Nucl. Data Tables* **21**, 69–75.
Itikawa, Y., and Ashihara, O. (1971). *J. Phys. Soc. Japan* **30**, 1461–1466.
Itikawa, Y., and Takayanagi, K. (1969). *J. Phys. Soc. Japan* **26**, 1254–1264.
Jain, A. (1983). *J. Chem. Phys.* **78**, 6579–6583.
Jain, D. K., and Khare, S. P. (1977). *Phys. Lett.* **63A**, 237–238.
Jain, A., and Thompson, D. G. (1982). *J. Phys. B* **15**, L631–L637.
Jain, A., Tripathi, A. N., and Srivastava, M. K. (1979). *Phys. Rev. A* **20**, 2352–2355.
Janda, K. C., Hemminger, J. C., Winn, J. S., Novick, S. E., Harris, S. J., and Klemperer, W. (1975). *J. Chem. Phys.* **63**, 1419–1421.
Jansen, R. H. J., DeHeer, F. J., Luyken, H. J., van Wingerden, B., and Blaauw, H. J. (1976). *J. Phys. B* **9**, 185–212.
Jhanwar, B. L., Khare, S. P., and Sharma, M. K. (1980). *Phys. Rev. A* **22**, 2451–2459.
Jhanwar, B. L., Khare, S. P., and Sharma, M. K. (1982). *Phys. Rev. A* **26**, 1392–1400.
Jones, R. H., Olander, D. R., and Kruger, V. R. (1969) *J. Appl. Phys.* **40**, 4641–4649.
Jost, K. (1979). *J. Phys. E* **12**, 1006–1012.

Jung, K., Antoni, T., Mullerr, R., Kochem, K. -H., and Ehrhardt, H. (1982). *J. Phys. B* **15**, 3535–3555; private communication (1982).
Kantrowitz, A. and Grey, J. (1957). *Rev. Sci. Instrum.* **22**, 328–332.
Khare, S. P., and Moiseiwitsch (1965). *Proc. Phys. Soc.* **85**, 821–839.
Khare, S. P., and Raj, D. (1979). *J. Phys. B* **12**, L351–L354.
Khare, S. P., Shingal, R., and Srivastava, B. B. (1977). *J. Phys. Soc. Japan* **43**, 2036–2041.
Khare, S. P., and Shobha, P. (1972). *J. Phys. B* **5**, 1938–1949.
Khare, S. P. and Shobha, P. (1974). *J. Phys. B* **7**, 420–427.
Kieffer, L. J. (1976). "Bibliography of Low Energy Electron and Photon Cross Section Data," NBS Special Publication 426, Natl. Bur. Stand. (U.S.), Washington, D. C.
Kieffer, L. J. (1973). "A Compilation of Electron Collision Cross Section Data for Modeling Gas Discharge Lasers," JILA Information Center Report 13, p. 139. University of Colorado, Boulder, Colorado.
Kistiakowsky, G. B., and Slichter, W. P. (1951). *Rev. Sci. Instrum.* **22**, 333–337.
Klemperer, O. (1965). *Rep. Prog. Phys.* **28**, 77–112.
Klonover, A., and Kaldor, U. (1978). *J. Phys. B* **11**, 1623–1632.
Knoop, F. W. E., Brongersma, H. H., and Boerboom, A. J. H. (1970). *Chem. Phys. Lett.* **5**, 450–452.
Kollath, R. (1958). "Handbusch der Physik" (S. Flugge, ed.), Vol. 34, p. 1. Springer-Verlag, Berlin.
Kubo, M., Matsunaga, D., Suzuki, T., and Tanaka, H. (1981). "XIIIth Int. Conf. Physics of Electronic and Atomic Collisions, Gatlinburg," Abstracts of Contributed Papers, pp. 360–361.
Kukolich, S. G., Oates, D. E., and Wang, J. H. S. (1974). *J. Chem. Phys.* **61**, 4686–4689.
Kuyatt, C. E. (1968). "Methods of Experimental Physics" (B. Bederson and W. L. Fite, eds.), Vol. 7, Part A. pp. 1–42, Academic Press, New York.
Kuyatt, C. E., and Simpson, J. A. (1967). *Rev. Sci. Instrum.* **38**, 103–111.
LaGattuta, K. J. (1980). *Phys. Rev. A* **21**, 547–555.
Land, J. E. (1978). *J. Appl. Phys.* **49**, 5716–5721.
Lane, N. F. (1980). *Rev. Mod. Phys.* **52**, 29–119.
Lane, N. F., and Geltman, S. (1967). *Phys. Rev.* **160**, 53–67.
Lane, N. F., and Geltman, S. (1969). *Phys. Rev.* **184**, 46–51.
Lane, N. F., and Henry, R. J. W. (1968). *Phys. Rev.* **173**, 183–190.
Lassettre, E. N., and White, E. R. (1974). *J. Chem. Phys.* **60**, 2460–2463.
Le Dourneuf, M., Lan, V. K., and Schneider, B. I. (1983). *In* "Electron-Atom and Electron-Molecule Collisions" (J. Nenze, ed.), p. 135–160. Plenum, New York.
Lee, M.-T., and Freitas, L. C. G. (1981a). *J. Phys. B* **14**, 1053–1064.
Lee, M.-T., and Freitas, L. G. C. (1981b). *J. Phys. B* **14**, 4691–4700.
Lee, M.-T., and Freitas, C. G. (1983). *J. Phys. B* **16**, 233–243.
Levin, D. A., Fliflet, A. W., and McKoy, V. (1979). *Phys. Rev. A* **20**, 491–498.
Levin, D. A., Fliflet, A. W., and McKoy, V. (1980). *Phys. Rev. A* **21**, 1202–1209.
Linder, F., and Schmidt, H. (1971a). *Z. Naturforsch.* **26a**, 1603–1617.
Linder, F., and Schmidt, H. (1971b). *Z. Naturforsch.* **26a**, 1617–1625.
Liu, J. W., and Smith, V. H. (1973). *J. Phys. B* **6**, L275–L279.
Lloyd, C. R., Teubner, P. J. O., Weigold, E., and Lewis, B. R. (1974). *Phys. Rev. A* **10**, 175–181.
Loeffler, K. H. (1969). *Z. Phys.* **27**, 145–149.
Lucchese, R. R., and McKoy, V. (1979). *J. Phys. B* **12**, L421–L424.
Lucchese, R. R., and McKoy, V. (1981). *Phys. Rev. A* **24**, 770–776.
Lucchese, R. R., and McKoy, V. (1982). *Phys. Rev. A* **25**, 1963–1968.
Lucchese, R. R., Watson, D. K., and McKoy, V. (1980). *Phys. Rev.* **22**, 421–426.
Lynch, M. G., Dill, D., Siegel, J., and Dehmer, J. (1979). *J. Chem. Phys.* **71**, 4249–4254.
McConnell, J. R., Holberg, J. B., Smith, G. R., Sandel, B. R., Shemansky, D. E., and Broadfoot, A. L. (1982). *Plant. Space Sci.* **30**, 151–167.

McCurdy, C. W., Rescigno, T. N., and McKoy, V. (1976). *J. Phys. B* **9**, 691–698.
McDaniel, E. W. (1964). "Collision Phenomena in Ionized Gases." Wiley, New York.
McMillan, J. H. (1930). *Phys. Rev.* **36**, 1034–1043.
Marmet, P., and Kerwin, L. (1960). *Can. J. Phys.* **38**, 787–796.
Massey, H. S. W. (1930). *Proc. R. Soc. London* **A129**, 616–627.
Massey, H. S. W. (1932). *Proc. Cambridge Philos. Soc.* **28**, 99–105.
Massey, H. S. W. (1969). *In* "Electronic and Ionic Impact Phenomena," Vol. II. Clarendum, Oxford.
Massey, H. S. W., and Bullard, E. C. (1933). *Proc. Cambridge Philos. Soc.* **29**, 511–521.
Massey, H. S. W., and Burhop, E. H. S. (1969). *In* "Electronic and Ionic Impact Phenomena," Vol. I. Clarendum, Oxford.
Massey, H. S. W., and Mohr, C. B. O. (1931). *Proc. R. Soc. London* **A132**, 605–630.
Massey, H. S. W., and Mohr, C. B. O. (1932a). *Proc. R. Soc. London* **A135**, 258–275.
Massey, H. S. W., and Mohr, C. B. O. (1932b). *Proc. R. Soc. London* **A136**, 289–311.
Massey, H. S. W., and Mohr, C. B. O. (1934). *Proc. R. Soc. London* **A146**, 880–899.
Massey, H. S. W., and Ridley, R. O. (1956). *Proc. Phys. Soc.* **A69**, 659–667.
Matsuda, H. (1961). *Rev. Sci. Instrum.* **32**, 850–852.
Matsunaga, D., Kubo, M., and Tanaka, H. (1981). "XIIth Int Conf. Physics of Electronic and Atomic Collisions, Gatlinburg," Abstracts of Contributed Paper, pp. 358–359.
Milne, T. A., and Greene, F. T. (1969). *Adv. High Temp. Chem.* **2**, 107–150.
Mittleman, M. H., and von Holdt, R. E. (1965). *Phys. Rev.* **140**, A726–A729.
Mjolsness, R. C., and Sampson, D. H. (1964). *Phys. Rev. Lett.* **13**, 812–815.
Mohr, C. B. O., and Nicoll, F. H. (1932). *Proc. R. Soc. London* **A138**, 469–478.
Moilére, G. (1947). *Z. Naturforsch.* **29**, 133–145.
Morrison, M. A., and Collins, L. A. (1978). *Phys. Rev. A* **17**, 918–938.
Morrison, M. A., and Collins, L. A. (1981). *Phys. Rev. A* **23**, 127–138.
Morrison, M. A., and Green, A. E. (1978). *J. Geophys. Res.* **83**, 1172–1174.
Morrison, M. A., and Hay, P. J. (1977). *J. Phys. B* **10**, L647–L652.
Morrison, M. A., and Lane, N. F. (1977). *Phys. Rev. A* **16**, 975–980.
Morrison, M. A., and Schneider, B. I. (1977). *Phys. Rev. A* **16**, 1003–1011.
Morrison, M. A., Lane, F. N., and Collins, L. A. (1977). *Phys. Rev. A* **15**, 2186–2201.
Morrison, M. A., Feldt, A. N., and Austin, D. (1983). *Phys. Rev. A*, to be published.
Morse, P. M. (1953). *Phys. Rev.* **90**, 51–55.
Mott, N. F., and Massey, H. S. W. (1965). "Theory of Atomic Collisions," 3rd ed., Oxford University Press.
Mozumder, A. (1982). *J. Chem. Phys.* **76**, 3277–3284.
Nagahara, S. (1953). *J. Phys. Soc. Japan* **8**, 165–168.
Nagahara, S. (1954). *J. Phys. Soc. Japan* **9**, 52–55.
Newell, W. R., Brewer, D. F. C., and Smith, A. C. H. (1979). *In* "XIth Int. Conf. Physics of Electronic and Atomic Collisions, Kyoto and Private Communications," Abstracts of Contributed Papers, pp. 308–309.
Nguyen, B., Bennani, A. L., and Rouault, M. (1975). *J. Phys. E* **8**, 909–912.
Nishimura, H. (1979). "XIth Int. Conf. Physics of Electronic and Atomic Collisions, Kyoto," Abstracts of Contributed Papers, p. 314.
Noble, C. J., Burke, P. G., and Salvini, S. (1982). *J. Phys. B* **15**, 3799–3786.
Norcross, D. W., and Collins, L. A. (1982). *Adv. At. Mol. Phys.* **13**, 341–397.
Ochkur, V. I. (1963). *Sov. Phys. JETP* **18**, 503–508 (1964). [*J. Exp. Theoret. Phys.* (*U.S.S.R.*) **45**, 734–741].
Oksyuk, Yu. D. (1966). *Sov. Phys. JETP* **22**, 373–881 [Original: *J. Exp. Theoret. Phys.* (*U.S.S.R.*) Original: **49**, 1261–1273 (1965)].

Olander, D. R. (1969). *J. Appl. Phys.* **40**, 4650–4657.
Olander, D. R., and Kruger, J. (1970). *J. Appl. Phys.* **41**, 2769–2776.
Olander, D. R., Jones, R. H., and Siekhaus, W. J. (1970). *J. Appl. Phys.* **41**, 4388–4391.
Onda, K. (1976). *J. Phys. Soc. Japan* **40**, 1437–1445.
Onda, K., and Truhlar, D. G. (1978). *J. Chem. Phys.* **69**, 1361–1373.
Onda, K., and Truhlar, D. G. (1979a). *J. Chem. Phys.* **70**, 1681–1689.
Onda, K., and Truhlar, D. G. (1979b). *J. Chem. Phys.* **71**, 5097–5106.
Onda, K., and Truhlar, D. G. (1979c). *J. Chem. Phys.* **71**, 5107–5123.
Onda, K., and Truhlar, D. G. (1979d). *J. Phys. B* **12**, 283–290.
Onda, K., and Truhlar, D. G. (1980a). *J. Chem. Phys.* **72**, 5249–5262.
Onda, K., and Truhlar, D. G. (1980b). *J. Chem. Phys.* **72**, 1415–1417.
Onda, K., and Truhlar, D. G. (1980c). *J. Chem. Phys.* **73**, 2688–2695.
Oppenheimer, J. R. (1928). *Phys. Rev.* **32**, 361–376.
Orient, O. J., Iga, I., and Srivastava, S. K. (1982). *J. Chem. Phys.* **77**, 3523–3526.
Padial, N. T., Norcross, D. W., and Collins, L. A. (1983). *Phys. Rev. A* **27**, 141–148.
Pierce, J. R. (1954). "Theory and Design of Electron Beams." Van Nostrand, Princeton, New Jersey.
Poe, R. T., and Choi, B. H. (1979). "Calculation of Electron-Carbon Monoxide Vibrational Transition Cross Sections," AFAPL-TR-79-2022.
Poe, R. T., and Choi, B. H. (1981). "Calculation of Electron-Hydrogen, Deuterium and Oxygen Molecules," AFWAL-TR-81-2051.
Preston, J. A., Hender, M. A., and McConkey, J. W. (1973). *J. Phys. E* **6**, 661–666.
Purcell, E. M. (1938). *Phys. Rev.* **15**, 818–826.
Ramsauer, C., and Kollath, R. (1930). *Ann. Phys.* **4**, 91–108.
Ramsauer, C., and Kollath, R. (1931b). *Ann. Phys.* **9**, 756–768.
Ramsauer, C., and Kollath, R. (1931c). *Ann. Phys.* **10**, 143–154.
Ramsauer, C. and Kollath, R. (1932). *Ann. Phys.* **12**, 529–561.
Ramsey, N. F. (1956). "Molecular Beams." Clarendon, Oxford.
Raseev, G., Guiusti-Suzor, A., and Lefebvre-Brion, H. (1978). *J. Phys. B* **11**, 2735–47.
Read, F. H. (1970). *J. Phys. C* **3**, 127–131.
Read, F. H., Comer, J., Imhof, R. E., Brunt, J. N. H., and Harting, E. (1974). *J. Electron Spectrosc. and Relat. Phenom.* **4**, 293–312.
Register, D. F., and Trajmar, S. (1983). *Phys. Rev.*
Register, D. F., Nishimura, H., and Trajmar, S. (1980a). *J. Phys. B* **13**, 1651–1662.
Register, D. F., Trajmar, S., and Srivastava, S. K. (1980b). *Phys. Rev. A* **21**, 1134–1151.
Rescigno, T. N., McCurdy, C. W., and McKoy, V. (1974a). *Chem. Phys. Lett.* **27**, 401–404.
Rescigno, T. N., McCurdy, C. W., and McKoy, V. (1974b). *Phys. Rev. A* **10**, 2240–2245.
Rescigno, T. N., McCurdy, C. W., and McKoy, V. (1975). *Phys. Rev. A* **11**, 825–829.
Rescigno, T. N., McCurdy, C. W., McKoy, V., and Bender, C. F. (1976b). *Phys. Rev. A* **13**, 216–223.
Rescigno, T. N., Bender, C. F., McCurdy, C. W., and McKoy, V. (1976a). *J. Phys. B* **9**, 2141–2146.
Rescigno, T., McKoy, V., and Schneider B. (1979). "Electron-Molecule and Photon-Molecule Collisions." Plenum, New York.
Riley, M. E., and Truhlar, D. G. (1975). *J. Chem. Phys.* **63**, 2181–2191.
Rodberg, L. S., and Thaler, R. M. (1967). "Introduction to the Quantum Theory of Scattering." Academic Press, New York.
Rohr, K. (1977). *J. Phys. B* **10**, L399–L401.
Rohr, K. (1978a). *J. Phys. B* **11**, 1849–1860.
Rohr, K. (1978b). *J. Phys. B* **11**, 4109–4117.

Rohr, K. (1979a). "XIth Int. Conf. Physics of Electronic and Atomic Collisions, Kyoto," Abstracts of Contributed Papers, pp. 324–325.
Rohr, K. (1979b). "XIth Int. Conf. Physics of Electronic and Atomic Collisions, Kyoto," Abstracts of Contributed Papers, pp. 322–323.
Rohr, K. (1980). *J. Phys. B* **13**, 4897–4905.
Rohr, K., and Linder, F. (1976). *J. Phys. B* **9**, 2521–2537.
Roscoe, R. (1938). *Philos. Mag.* **26**, 32–41.
Roy, D., and Carette, J. D. (1977). "Topics in Current Physics" (H. Ibach, ed.), Vol. 4, pp. 13–58. Springer-Verlag, Berlin.
Rozsnyai, B. F. (1967). *J. Chem. Phys.* **47**, 4102–4112.
Rudd, M. E. (1972). Electrostatic analyzers. *In* "Low-Energy Electron Spectrometry" (K. D. Sevier, ed.). Wiley (Interscience), New York.
Rudge, M. R. H. (1965). *Proc. Phys. Soc.* **86**, 763–772.
Rudge, M. R. H. (1978a). *J. Phys. B* **11**, 1503–1513.
Rudge, M. R. H. (1978b). *J. Phys. B* **11**, 2221–2228.
Rudge, M. R. H. (1980). *J. Phys. B* **13**, 1269–1279.
Rudge, M R. H., Trajmar, S., and Williams, W. (1976). *Phys. Rev. A* **13**, 2074–2086.
Rumble, J. R., and Truhlar, D. G. (1979). *J. Chem. Phys.* **70**, 4101–4107.
Rumble, J. R., Truhlar, D. G., and Morrison, M. A. (1983). To be published.
Salvini, S. and Thompson, D. G. (1971). *J. Phys. B* **4**, 461–467.
Sampson, D. H., and Mjolsness, R. C. (1965). *Phys. Rev.* **140**, A1466–A1476.
Sawada, T., Ganas, P. S., and Green, A. E. S. (1974). *Phys. Rev. A* **9**, 1130–1135.
Schneider, B. I. (1975). *Phys. Rev. A* **11**, 1957–1962.
Schneider, B. I. (1977). *Chem. Phys. Lett.* **51**, 578–581.
Schneider, B. I., and Collins, L. A. (1981). *Phys. Rev. A* **24**, 1264–1266.
Schneider, B. I., and Collins, L. A. (1982). *J. Phys. B* **15**, L335–L340.
Schneider, B. I., and Collins, L. A. (1983). *Phys. Rev. A* **27**, 2847–2857.
Schneider, B. I., and Hay, P. J. (1976). *Phys. Rev. A* **13**, 2049–2056.
Schneider, B. I., Taylor, H. S., and Yaris, R. (1970). *Phys. Rev. A* **1**, 855–867.
Schomaker, V., and Glauber, R. J. (1953). *Nature (London)* **170**, 290–
Schulz, G. J. (1964). *Phys. Rev.* **136**, A650–A656.
Schulz, G. J. (1973a). *Rev. Mod. Phys.* **45**, 378–422.
Schulz, G. J. (1973b). *Rev. Mod. Phys.* **45**, 423–486.
Schulz, G. J., and Fox, R. E. (1957). *Phys. Rev.* **106**, 1179–1181.
Sevier, K. D. (1972). "Low Energy Electron Spectrometry." Wiley (Interscience, New York.
Shimamura, I. (1976). Calculation quoted by Rudge *et al.* (1978b).
Shimamura, I. (1979). "Symposium on Electron-Molecule Collisions-Invited Papers" (I. Shimamura and M. Matsuzawa, eds.), pp. 13–30, University of Tokyo Press.
Shimamura, I. (1984). Rotational excitation of molecules by slow electrons. *In* "Electron-Molecule Collisions" (I. Shimamura and K. Takayanagi, eds.). Plenum, New York.
Shkarofsky, I. P., Johnston, T. W., and Bachynski, M. P. (1968). "The Particle Kinetics of Plasmas." Addison-Wesley, Reading, Massachusetts.
Shuttleworth, T., Newell, W. R., and Smith, A. C. H. (1977). *J. Phys. B* **10**, 1641–1651.
Shyn, T. W., and Carignan, G. R. (1980). *Phys. Rev. A* **22**, 923–929.
Shyn, T. W., and Sharp, W. E. (1981). *Phys. Rev. A* **24**, 1734–1740.
Shyn, T. W., Sharp, W. E., and Carignan, G. R. (1978). *Phys. Rev. A* **17**, 1855–1861.
Siegel, J., Dill, D., and Dehmer, J. L. (1978). *Phys. Rev. A* **17**, 2106–2109.
Siegel, J., Dehmer, J. L., and Dill, D. (1980a). *J. Phys. B* **13**, L215–L219.
Siegel, J., Dehmer, J L., and Dill, D. (1980b). *Phys. Rev. A* **21**, 85–94.
Siegel, J., Dehmer, J. L., and Dill, D. (1981a). *Phys. Rev. A* **23**, 632–640.

Siegel, J., Dehmer, J. L., and Dill, D. (1981b). *J. Phys. B* **14**, L441–L446.
Simpson, J. A. (1961). *Rev. Sci. Instrum.* **32**, 1283–1293.
Simpson, J. A. (1964). *Rev. Sci. Instrum.* **35**, 1698–1704.
Simpson, J. A., and Kuyatt, C. E. (1963). *J. Res. Natl. Bur. Stand. Sect. C* **67**, 279–281.
Sinha, M. P., Schultz, A., and Zare, R. N. (1973). *J. Chem. Phys.* **58**, 549–556.
Sinha, M. P., Caldwell, C. D., and Zare, R. N. (1974). *J. Chem. Phys.* **61**, 491–503.
Slater, J. C. (1951). *Phys. Rev.* **81**, 385–390.
Slater, R. C., Fickes, M. G., and Stern, R. C. (1972). *Phys. Rev. Lett.* **29**, 333–336.
Slater, R. C., Fickes, M. G., Becker, W. G., and Stern, R. C. (1974a). *J. Chem. Phys.* **60**, 4697–4709.
Slater, R. C., Fickes, M. G., Becker, W. G., and Stern, R. C. (1974b). *J. Chem. Phys.* **61**, 2290–2293.
Smalley, R. E., Levy, D. H., and Wharton, L. (1976). *J. Chem. Phys.* **64**, 3266–3276.
Smith, E. R., and Henry, R. J. W. (1973). *Phys. Rev. A* **7**, 1585–1590.
Sohn, W., Jung, K., and Ehrhardt, H. (1983). *J. Phys. B* **16**, 891–901.
Spangenberg, K. R. (1948). "Vacuum Tubes." McGraw-Hill, New York.
Srivastava, M. K., Tripathi, A. N., and Lal, M. (1978). *Phys. Rev. A* **18**, 2377–2378.
Srivastava, S. K. (1979). "Symposium on Electron-Molecule Collisions, Invited Papers" (I. Shimamura and M. Matsuzawa, eds.), pp. 1–12. University of Tokyo Press.
Srivastava, S. K., Chutjian, A., and Trajmar, S. (1975a). *J. Chem. Phys.* **63**, 2659–2665.
Srivastava, S. K., Hall, R. I., Trajmar, S., and Chutjian, A. (1975b). *Phys. Rev. A* **12**, 1399–1401.
Srivastava, S. K., Chutjian, A., and Trajmar, S. (1976a). *J. Chem. Phys.* **64**, 1340–1344.
Srivastava, S. K., Trajmar, S., Chutjian, A., and Williams, W. (1976b). *J. Chem. Phys.* **64**, 2767–2771.
Srivastava, S. K., Tanaka, H., and Chutjian, A. (1978). *J. Chem. Phys.* **69**, 1493–1497.
Srivastava, S. K., Tanaka, H., Chutjian, A., and Trajmar, S. (1981). *Phys. Rev. A* **23**, 2156–2166.
Stier, H. C. (1932). *Z. Phys.* **76**, 439–470.
Strobel, D. F., and Davis, J. (1980). *Astrophys. J.* **238**, L49–L52.
Stubbe, P. (1971). *J. Sci. Indion Res.* **30**, 379–387.
Takatsuka, K., and McKoy, V. (1980). *Phys. Rev. Lett.* **45**, 1396–1399.
Takatsuka, K., and McKoy, V. (1981a). *Phys. Rev. A* **23**, 2352–2358.
Takatsuka, K., and McKoy, V. (1981b). *Phys. Rev. A* **23**, 2359–2364.
Takatsuka, K., and McKoy, V. (1981c). *Phys. Rev. A* **24**, 2473–2480.
Takayanagi, K. (1966). *J. Phys. Soc. Japan* **21**, 507–514.
Takayanagi, K. (1967). *Progr. Theoret. Phys.* **40**, 216–248.
Takayanagi, K. (1974). *In* "Atomic Physics 4" (G. Zu Putlitz, E. W. Weber, and A. Winnacek, eds.). Plenum, New York.
Takayanagi, K. (1982). "First US–Japan Seminar on Electron-Molecule Collisions and Photoionization Processes (B. V. McKoy and S. Trajmar, eds.), pp. 108–117. Verlag Chemic International, Deerfield, Florida.
Takayanagi, K., and Geltman, S. (1965). *Phys. Rev.* **138**, A1003–A1010.
Takayanagi, K., and Itikawa, Y. (1970a). *Adv. At. Mol. Phys.* **6**, 105–153.
Takayanagi, K., and Itikawa, Y. (1970b). *Space Sci. Rev.* **11**, 380–450.
Tanaka, H., Srivastava, S. K., and Chutjian, A. (1978). *J. Chem. Phys.* **69**, 5329–5333.
Tanaka, H., Boesten, L., and Shimamura, I. (1980). 7th Int. Conf. Atomic Physics, Abstracts, p. 43. MIT Press.
Tanaka, H., Okada, T., Boesten, L., Suzuki, T., Yamamoto, T., and Kubo, M. (1982). *J. Phys. B* **15**, 3305–3319; private communication.
Temkin, A. (1957). *Phys. Rev.* **107**, 1004–1012.
Temkin, A., and Vasavada, K. V. (1967). *Phys. Rev.* **160**, 109–117.
Temkin, A., Vasavada, K. V., Chang, E. S., and Silver, A. (1969). *Phys. Rev.* **186**, 56–66.
Thirumalai, D., and Truhlar, D. G. (1981). *J. Chem. Phys.* **75**, 5207–5209.

Thirumalai, D., Onda, K., and Truhlar, D. G. (1981a). *J. Chem. Phys.* **74**, 6792–6805.
Thirumalai, D., Onda, K., and Truhlar, D. G. (1981b). *J. Chem. Phys.* **74**, 526–534.
Townsend, J. S., and Baily, V. A. (1921). *Philos. Mag.* **42**, 873–891.
Townsend, J. S., and Bailey, A. V. (1922). *Philos. Mag.* **44**, 1033–1052.
Trajmar, S., and Register, D. F. (1984). Experimental techniques for cross section measurements. *In* "Electron-Molecule Collisions (K. Takayanagi, and I. Shimamura, eds.). Plenum, New York.
Trajmar, S., Truhlar, D. G., and Rice, J. K. (1970). *J. Chem. Phys.* **52**, 4502–4515.
Trajmar, S., Cartwright, D. C., and Williams, W. (1971). *Phys. Rev. A* **4**, 1482–1492.
Trajmar, S., Williams, W., and Srivastava, S. K. (1977). *J. Phys. B* **10**, 3323–3333.
Trajmar, S., Register, D. F., and Chutjian, A. (1983). Physics Report **97**, 219–356.
Truhlar, D. G. (1972). *J. Chem. Phys.* **57**, 3260–3263.
Truhlar, D. G. (1981). *In* "Chemical Applications of Atomic and Molecular Electro-static Potentials (P. Politzer and D. G. Truhlar, eds.), pp. 123–172. Plenum, New York.
Truhlar, D. G., and Brandt, M. A. (1976). *J. Chem. Phys.* **65**, 3092–3101.
Truhlar, D. G., and Mullaney, N. A. (1978). *J. Chem. Phys.* **68**, 1574–1584.
Truhlar, D. G., and Rice, J. K. (1970). *J. Chem. Phys.* **52**, 4480–4501.
Truhlar, D. G., and Rice, J. K. (1974). *Phys. Lett.* **47A**, 327–374.
Truhlar, D. G., and Van-Catledge, F. A. (1978). *J. Chem. Phys.* **69**, 3575–3578.
Truhlar, D. G., Rice, J. K., Trajmar, S., and Cartwright, D. C. (1971). *Chem. Phys. Lett.* **91**, 299–305.
Truhlar, D. G., Trajmar, S., and Williams, W. (1972a). *J. Chem. Phys.* **57**, 3250–3259.
Truhlar, D. G., Williams, W., and Trajmar, S. (1972b). *J. Chem. Phys.* **57**, 4307–4313.
Truhlar, D. G., Brandt, M. A., Chutjian, A., Srivastava, S. K., and Trajmar, S. (1976). *J. Chem. Phys.* **65**, 2962–2969.
Truhlar, D. G., Onda, K., Eades, R. A., and Dixon, D. A. (1979a). *Int. J. Quantum Chem. Symp.* **13**, 601–632.
Truhlar, D. G., Onda, K., Eades, R. A., and Dixon, D. A. (1979b). *Int. J. Quantum Chem. Symp.* **13**, 601–632.
Tully, J. C., and Berry, R. S. (1969). *J. Chem. Phys.* **51**, 2056–2075.
Vallance-Jones, A. (1974). Aurora. *In* "Geophysics and Astrophysics Monographs," Vol. 9. Reidel, Dordrecht, Holland.
van Brunt, R. J., and Gallagher, A. (1977). Invited papers and progress reports. *In* "Xth ICPEAC" (G. Watel, ed.), p. 129. Paris.
van Wingerden, B., Weigold, E., de Heer, F. J., and Nygaard, K. J. (1977). *J. Phys. B* **10**, 1345–1362.
Varracchio, E. F. (1979). *J. Phys. B* **12**, 3427–3440.
Von Busch, F. (1975). *J. Phys. B* **8**, 1440–1452.
Vuškovič, L., Cvejanovic, S., and Kurepa, M. (1970). *Fizika* **2** (Suppl. 1), 26–31.
Vuškovič, L., Srivastava, S. K., and Trajmar, S. (1978). *J. Phys. B* **11**, 1643–1652.
Vuškovič, L., Srivastava, S. K. (1981). *J. Phys. B* **14**, 2677–2685.
Wakiya, K. (1978a). *J. Phys. B.* **11**, 3913–3930.
Wakiya, K. (1978b). *J. Phys. B.* **11**, 3931–3938.
Watson (1927). *Philos. Mag.* **3**, 849–853.
Watson, D. K., and McKoy, B. (1979). *Phys. Rev. A* **20**, 1474–1483.
Watson, D. K., Lucchese, R. R., McKoy, V., and Rescigno, T. N. (1980). *Phys. Rev. A* **21**, 738–744.
Watson, D. K., Rescigno, T. N., McKoy, V. (1981). *J. Phys. B.* **14**, 1875–1882.
Weatherford, C. A. (1980). *Phys. Rev. A* **22**, 2519–2528.
Wedde, T., and Strand, T. G. (1974). *J. Phys. B* **7**, 1091–1100.
Weingartshofer, A., Ehrhardt, H., Hermann, V., and Linder F. (1970). *Phys. Rev. A* **2**, 294–304.
Whitten, R. C., and Poppoff, I. G. (1965). "The Physics of the Lower Ionosphere." Prentice Hall, Englewood Cliffs, New Jersey.

Wiesemann, K. (1981). *J. Phys. E* **14**, 1404–1406.
Wilkins, K. L., and Taylor, H. S. (1967). *J. Chem. Phys.* **47**, 3532–3539.
William, K. G. (1969). "VI Int. Conf. on the Physics of Electronic Atomic Collisions," Abstracts, pp. 735–737, MIT Press, Boston.
Wollnik, H. (1967). Electrostatic Prism. *In* "Focusing of Charged Particles" (A. Septier, ed.), Vol. II, pp. 163. Academic Press, New York.
Wong, S. F., and Dube, L. (1978). *Phys. Rev. A* **17**, 570–576.
Wong, S. F., and Schulz, G. J. (1974). *Phys. Rev. Lett.* **32**, 1089–1092.
Yarnold, G. D., and Bolton, H. C. (1955). *J. Sci. Instrum.* **26**, 850.
Yates, A. C., and Tenney, A. (1972). *Phys. Rev. A* **5**, 2474–2481.
Yung, Y. L., Gladstone, G. R., and Chang, K. M. (1982). *Astrophys. J.* **254**, L65–L69.

2 Excitation of Molecules by Electron Impact

S. Trajmar
California Institute of Technology
Jet Propulsion Laboratory
Pasadena, California

and

D. C. Cartwright
University of California
Los Alamos National Laboratory
Los Alamos, New Mexico

I.	Introduction	156
II.	Definition of Electron Collision Cross Sections	158
III.	Electron-Impact Spectroscopy	160
IV.	Experimental Techniques for Measuring Electron-Impact Excitation Cross Sections	166
	A. Total Scattering Cross Sections	166
	B. Integral Excitation Cross Sections	167
	C. Differential Excitation Cross Sections	168
	D. Experimental Uncertainties: Error Estimation	170
V.	Electron-Impact Excitation Cross Section Data	171
	A. General Remarks	171
	B. Vibrational Excitation	172
	C. Electronic Excitation	174
	D. Total Electron Scattering Cross Sections	181
VI.	Generation of Optical Data by Electron-Impact Techniques	181
VII.	Recent Developments	190
	A. Study of Superexcited States	190
	B. Coincidence Measurements	191
	C. Electron Collisions in Laser Fields	191
	D. Electron Scattering with Spin-Selected Electrons	193
VIII.	Theory of Electron–Molecule Collisions	193
	A. Vibrational Excitation	195
	B. Electronic Excitation	197

IX.	Comparison of Theory and Experiment	202
	A. Vibrational Excitation	202
	B. Electronic Excitation	207
X.	Applications of Electron Collision Data	226
	A. Atmospheric Processes	226
	B. Electron-Beam and Discharge Processes	231
	C. Astrophysics	241
	References	242

I. Introduction

The first inelastic electron collision measurements were conducted by Lenard (1902) just five years after the discovery of the electron by Thomson. Early works included the study of energy losses suffered by an electron swarm in gases by Franck and Hertz (1914), the measurements of the angular distribution of inelastically scattered electrons by Dymond (1927) and by Mohr and Nicoll (1932), the introduction of the Ramsauer technique by Ramsauer and co-workers, and the development of the swarm technique for quantitative measurements by Townsend and co-workers, and represented the birth of the field of electron–molecule (atom) collison studies. The development of the quantum-mechanical theory of scattering provided the basis for understanding the experimental observations of electron scattering events, and the early electron scattering studies made substantial contributions to our understanding of the quantum states of atoms and molecules and to the formulation of the appropriate theory of these quantum states. The problem of an infinite differential cross section in the forward scattering direction associated with electron scattering described by classical mechanics was eliminated by the quantum-theoretical description. The possibility of spin exchange in electron-impact excitation, as pointed out by Oppenheimer (1928), made possible the study of electronic states not accessible by optical excitation from the ground state. A comprehensive review of these early works has been presented by Massey et al. (1969).

The early experiments had many limitations such as insufficient energy resolution, inefficient electron detection, and inferior vacuum techniques, but new technological developments after the 1940s brought about the modern age of electron collision studies. Introduction of the cylindrical analyzer by Marmet and Kerwin (1960), the hemispherical analyzer by Lassettre and co-workers (Skerbele and Lassettre, 1964) and by Simpson and co-workers (Simpson, 1964; Kuyatt and Simpson, 1967), the utilization of electron-multiplying methods in signal detectors, and general progress in vacuum techniques and electronics opened up new vistas for the study of electron collision processes over a wide energy range. The application of coincidence and laser excitation techniques to electron collision studies and the utilization of spin selected electron beams have enabled the experimentalist to approach the "perfect experiment" in which all quantum numbers

2 Excitation of Molecules by Electron Impact

associated with the system are well defined and no averaging over unresolved states is required.

From the theoretical point of view, the detailed study of electron–molecule (atom) scattering processes began with the development of quantum mechanics and was motivated by the data from the early electron scattering experiments involving the rare gases. What is now known as the Bethe–Born approximation (Bethe, 1930; Massey and Mohr, 1931) was first applied to treat electron scattering at high incident energies in 1930 and has been used extensively to better understand atomic and molecular structure as well as the interaction (Inokuti, 1971) of energetic particles with atomic and molecular gases. The pioneering work of Mohr and Nicoll (1932) showed that, in the medium- and low-impact-energy regions, distortion of the incident and scattered electron waves by the target needs to be included in order to explain the observed differential cross sections in mercury and argon. In the intervening 50 years since this first theoretical work, a variety of theories to treat inelastic electron–atom scattering evolved which range from modifications of plane-wave and semiclassical models to elaborate close-coupling and variational methods. Because the electron–molecule force field is more difficult to treat computationally than that for the analogous electron–atom process, and because much of the early experimental work was for rare-gas targets, the application of theory to treat electron-impact excitation of molecules received relatively little attention except for the early work on molecular hydrogen by Massey and Mohr (1932) and by Roscoe (1941). Renewed interest in the early 1960s in processes involving H_2^+, H_2, and N_2, combined with new computational methods for treating the nonspherical symmetry in the electron–molecule interaction, resulted in the application of plane-wave theories to treat excitation of specific molecular band systems in these three molecules. During the past five years, theoretical results obtained from the application of theories more advanced than plane-wave models for the description of electron-impact excitation of diatomic molecules have been reported. However, as will be discussed in a later section, there remains a genuine need for improvement in the theoretical methods because there are significant disagreements between the results obtained by the various theoretical methods themselves, as well as with the experimental data.

Electron-impact processes with atoms and molecules play a central role in a wide variety of naturally occurring and laboratory-produced phenomena and, as a result, have been the subject of considerable research, particularly in the last ten years. The importance of electron-impact excitation processes involving ground-state molecules, and molecules in metastable excited states, has been recognized for many years in the study of ionospheric (Whitten and Poppoff, 1971) and auroral (Vallance-Jones, 1974) processes in planetary atmospheres. The discovery of high-power (electron-beam and discharge-pumped) gas lasers has resulted in an increased interest in electron collision processes because of the important role they play in creating the population inversion (Willett, 1974). This renewed interest in electron scattering has resulted in the measurement of many of the electron-impact excitation cross sections of importance during the past ten years. The need for cross section information involving targets that are presently difficult to study experimentally (e.g., molecular free radicals, etc.), and recent results from electron–photon coincidence measurements (Blum and Kleinpoppen, 1979) which probe the details of the scattering process, have further stimulated theoretical studies. In spite of this renewed activity, however, much of the electron-impact excitation cross section information needed to model a variety of plasma and planetary processes is still not known and the absence of these basic cross section data hinders progress toward a more thorough understanding of various phenomena.

Electron collision with molecules can lead to a variety of processes (elastic scattering, excitation, ionization, dissociation, electron capture, etc., and to various combinations of these basic processes). In the present chapter we are concerned only with those processes which cause excitation of the molecule. For reasons stated in the first chapter, pure rotational excitation is discussed in connection with elastic scattering and we concern ourselves here *only with nonresonant vibrational and electronic excitations*. Two aspects of the electron collision phenomena will be treated: electron-impact spectroscopy and quantitative cross section measurements. Electron-impact spectroscopy yields information concerning the energy levels of the molecule and the variety of excitation paths. It represents the qualitative aspect. The measurement of specific cross sections is the quantitative aspect. In this review, the similarities and differences between optical and electron-impact spectroscopy and the utilization of electron spectroscopy will be briefly discussed. A brief summary of experimental techniques aimed at the determination of cross sections and of the available cross section data will also be given. [More detailed accounts of these matters have been given by Trajmar (1980), Trajmar *et al.* (1983), and Trajmar and Register (1984).] The various theoretical methods currently used to treat electron-impact excitation of molecules will be summarized and their results compared against available experimental data. The application of electron-impact excitation cross section data will be demonstrated through several examples related to laser and plasma systems, radiation chemistry, ionospheric phenomena, and astrophysical processes.

II. Definition of Electron Collision Cross Sections

The quantities that characterize a particular electron collision process, and are most important when trying to understand and model partially ionized systems, are the differential and integral cross sections. They represent the

2 Excitation of Molecules by Electron Impact

time-independent probability for the occurrence of a particular collision process and are well defined for most electron scattering experiments. One can encounter situations, however, when the definition of scattering cross sections is not feasible and only time-dependent probabilities can be considered. (For example, in the case of electron scattering in a pulsed high-intensity laser field.)

The notion of constant collision differential cross section, as used in gas kinetics, is not appropriate for electron collisions because the effective interaction potential between the electron and molecule depends strongly on the electron velocity (or impact energy E_0) and spherical angles of scattering $\Omega(\theta, \phi)$. The differential cross section is therefore a function of these parameters and the details of the scattering process. For the electron-impact excitation to quantum state n, in general, the double differential cross section is written as

$$\delta^2 \sigma_n(E_0, E_n, \Omega)/\delta\Omega \delta E_n,$$

where E_n refers to the excitation energy. In the case of excitation to a discrete final state, one can integrate over the excitation energy and obtain the customary differential (in angle) cross section,

$$d\sigma_n(E_0, \Omega)/d\Omega.$$

In most scattering experiments the target molecules are randomly oriented and one can, therefore, determine only cross sections averaged over molecular orientation. For this reason, the differential cross section becomes independent of the azimuthal scattering angle ϕ. Furthermore, the measuring apparatus has associated with it a finite energy and angular resolution and, therefore, the measured cross section represents a value averaged over these experimental variables. This measured differential cross section, at a given impact energy, is denoted as

$$(\text{DCS}_n(\theta))_{E_0} = \overline{(d\sigma_n(\theta)/d\Omega)}_{E_0}. \tag{1}$$

This is the quantity determined in most electron beam–gas beam scattering experiments and it has the dimension of area per unit solid angle.

In practice it is seldom possible to measure the cross section for excitation between two completely defined quantum states and, in most cases, one measures the average cross section for a number of experimentally indistinguishable processes. For example, transitions between individual hyperfine and magnetic sublevel quantum states can be measured only in specially designed scattering experiments (utilizing laser excitation or electron–photon coincidence techniques). Similarly, measured cross sections (in most cases) represent averages over various target and electron spin orientations. With the present state-of-the-art energy resolution, individual rotational transitions can be resolved only in H_2 and individual vibrational transitions can only be partially solved in complex molecules. For a comparison of experimental results with theory, a procedure which averages over unresolved initial states (i) and sums over unresolved final states (f) has to be carried out:

$$\left(\frac{d\sigma_n(\theta)}{d\Omega}\right)_{E_0} = \sum_i \sum_f N_i \left(\frac{d\sigma_n^{if}(\theta)}{d\Omega}\right)_{E_0}, \tag{2}$$

where N_i is the fractional population in initial state i.

Integration of the differential cross section over all scattering angles yields the integral $[\sigma_n(E_0)]$ and momentum transfer $[\sigma_n^M(E_0)]$ cross sections:

$$\sigma_n(E_0) = 2\pi \int_0^\pi \left(\frac{d\sigma_n(\theta)}{d\Omega}\right)_{E_0} \sin\theta \, d\theta, \tag{3}$$

$$\sigma_n^M(E_0) = 2\pi \int_0^\pi \left(\frac{d\sigma_n(\theta)}{d\Omega}\right)_{E_0} \left(1 - \frac{k_f}{k_i}\cos\theta\right) \sin\theta \, d\theta, \tag{4}$$

where k_i and k_f are the magnitude of the initial and final momentum of the electron, respectively. The total (or as sometimes called "grand total") electron scattering cross section is the sum of the integral cross sections for all energetically accessible processes:

$$\sigma_T(E_0) = \sum_n \sigma_n(E_0). \tag{5}$$

For experiments in which the scattered electron is detected in coincidence with other secondary particles (electrons, photons, ions, neutral fragments), cross sections differential in the energy and angle of these secondary species have to be correspondingly defined.

The determination of the differential cross section for electron-impact excitation between a specific pair of quantum states, as would be obtained by any of a large variety of theoretical methods, has been very well reviewed by Lane (1980) and therefore is not repeated here. In a subsequent section of this chapter, a brief summary of each theoretical method that has been used to obtain differential or integral cross sections will be presented in the context of comparing each method with the others and with experiments. Readers interested in specific details should consult Lane (1980) or the original references.

III. Electron-Impact Spectroscopy

Electron-impact spectroscopy refers to spectroscopic investigations in which the excitation is caused by electron impact and the information is obtained by determining the energy and angular distribution of the scattered electrons. Sometimes quantitative cross section measurements and detection

2 Excitation of Molecules by Electron Impact   161

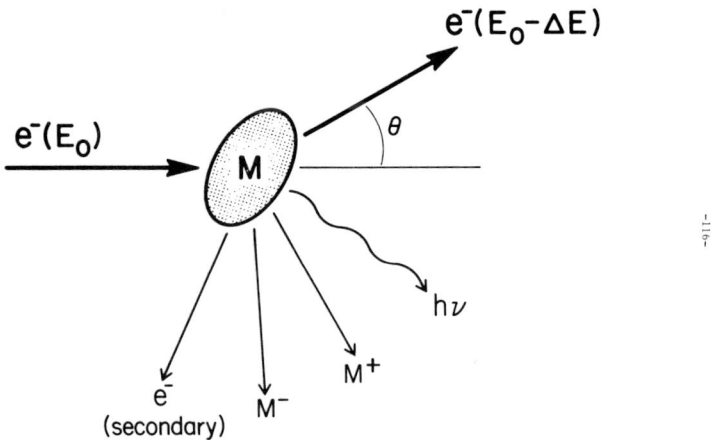

Fig. 1. Schematic diagram of a general electron scattering process in which incident electrons of known energy E_0 and momentum strike a target, lose energy ΔE to the target, and are scattered into the direction θ with respect to the incident electron beam. In general, the resulting excited target may emit a photon ($h\nu$), may have been ionized by the collision process to become a positive ion, or may have captured the electron to become a negative ion. In this chapter, only those processes involving electronic and vibrational excitation will be considered; the other possible processes are the subject of other chapters of this review.

of photons and particles other than electrons are also covered under this same topic. Here we discuss electron-impact spectroscopy to give a general background for electron beam–molecular beam collision studies. The techniques for generating absolute cross sections will be described in Section IV.

The experiments for spectroscopic studies are conceptually simple (see Fig. 1). A nearly monoenergetic electron beam collides with the gaseous molecular target under conditions where single collisions dominate. The electrons interact with the molecules and may lose (or gain) kinetic energy to the (from the) internal excitation of the target. The energy distribution of the scattered electrons, therefore, is related to the energy-level scheme of the target. The scattered intensity is related to the corresponding cross section and is the subject of measurement in spectrometric investigations. The angular distribution is determined by the nature of interaction and contains information concerning the properties of the states involved in the excitation process. Typical electron energy loss spectra are shown in Figs. 2 and 3. In a superficial way they resemble optical absorption spectra, but although there are a number of similarities, there are also significant differences between the two cases.

At high impact energies and small scattering angles (small momentum transfer) electrons excite in the same way as photons and the two spectra should be identical in the zero-momentum-transfer limit. This so-called "limit

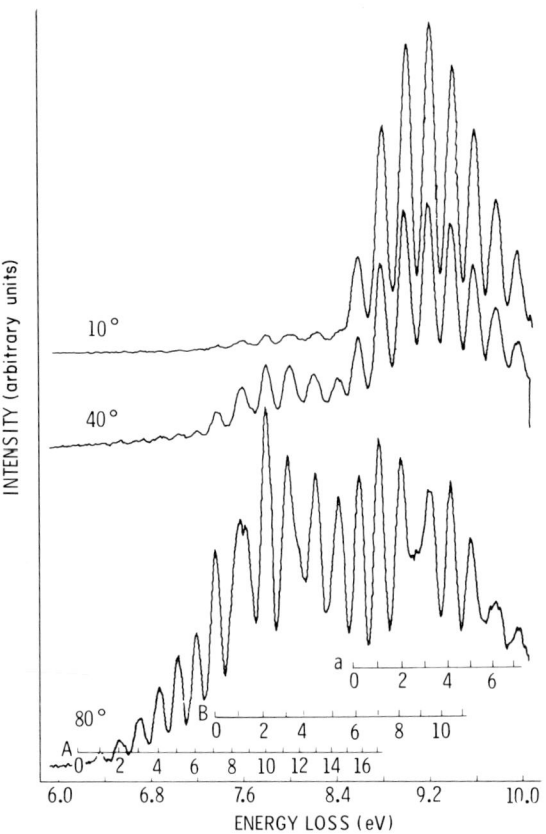

Fig. 2. Electron energy loss spectra (about 100 meV resolution) in molecular nitrogen for an incident electron energy of 20 eV, and scattering angles of 10°, 40°, and 80°. The locations of the vibrational levels associated with the A $^3\Sigma_u^+$, B $^3\Pi_g$, and a $^1\Pi_g$ electronic states are indicated in the figure. [From Cartwright et al. (1977a).]

theorem" makes possible the utilization of electrons as "pseudo" photons and the determination of optical absorption and ionization cross sections by electron-impact techniques. This application is especially important in the extreme UV and x-ray regions, where light sources are not readily available (with the exception of the rather expensive synchrotron). In high-energy electron scattering the target experiences a sharply pulsed (in time) electric field. The frequency components of this field can be obtained by Fourier transform and they represent a wide range of frequencies of essentially equal intensity. The fast electron, therefore, represents an ideal continuum light source and the energy-loss spectrum generated by it resembles an optical absorption spectrum.

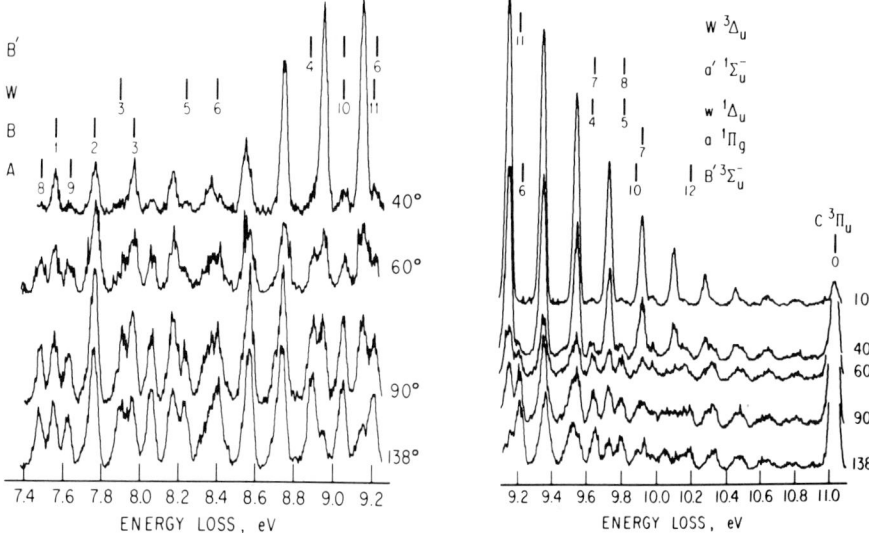

Fig. 3. Electron energy loss spectra (about 25 meV resolution) of molecular nitrogen for incident electron energy of 20.6 eV and scattering angles of 10°, 40°, 60°, 90°, and 138°. The spectra in the figure have been divided into two overlapping sections near an electron energy loss of 9.2 eV and certain resolved vibrational levels of the A $^3\Sigma_u^+$, B $^3\Pi_g$, W $^3\Delta_u$, B' $^3\Sigma_u^-$, C $^3\Pi_u$, a $^1\Pi_g$, w $^1\Delta_u$, and a' $^1\Sigma_u^-$ electronic states are indicated. [From Cartwright et al. (1977a).]

For slow electrons optical selection rules are not obeyed. Transitions to symmetry- and/or spin-forbidden states are easily achieved, especially at large scattering angles (large momentum transfer). The reason for these deviations from the optical case is the basic difference between the electric dipole interaction (in the optical case) and the more general electron interaction potential (for the electron case). The latter can be represented by all orders of multipoles and also contains electron exchange terms. This relaxation of optical selection rules makes possible the study of energy levels of molecules which are not observable by optical methods and has been extensively utilized for generating metastable species. A large number of molecular electronic states have been discovered by this method of low-energy, high-angle electron-impact excitation.

As an example for the application of electron-impact spectroscopy, we show the results of a recent investigation (Trajmar et al., 1976) of molecular oxygen in Fig. 4. In the 8.2–10.2 eV energy loss region, the strong Schumann–Runge continuum dominates for high-energy electron-impact excitation as indicated by the curve marked OPT. (to denote optical). As the scattering conditions are changed to low impact energies and high scattering angles, the continuum becomes weak and underlying forbidden transitions become apparent. Figure 4 clearly shows the singlet and triplet Π_g excitations as well as additional, still unassigned transitions. Similar studies have revealed new electronic states in other molecules (see, e.g., Kuppermann et al., 1968, 1979; King et al., 1977; Vušković et al., 1978; Wilden et al., 1979; Spence, 1981; Doering and McDiarmid, 1980 and 1982).

Let us now discuss briefly the principles underlying the "limit theorem" and the question of selection rules in connection with electron-impact excitation.

The concept of generalized oscillator strength, $f_{0n}^G(K)$ for electron scattering was introduced by Bethe (1930) and recently discussed in detail by Inokuti (1971);

$$f_{0n}^G(K) = \frac{\Delta E}{2} \frac{k_0}{k_n} K^2 \frac{d\sigma_{0n}^B}{d\Omega}(K). \tag{6}$$

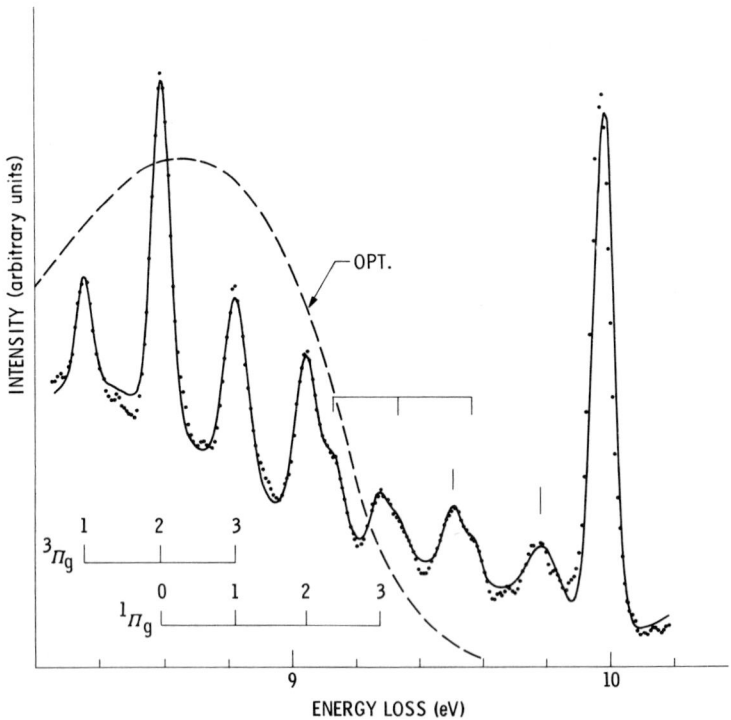

Fig. 4. Electron energy loss spectrum in molecular oxygen for a residual $(E_0 - \Delta E \equiv E_R)$ electron energy of 1 eV and a scattering angle of 90°. The experimental data are shown as the points and the synthetic spectrum is denoted by the solid line. Peaks associated with specific vibrational levels of various electronic states are indicated in the figure along with six features that have not been identified. The dashed curve marked OPT. denotes the well-known Schumann–Runge continuum seen in photoabsorption studies. [From Trajmar et al. (1976).]

2 Excitation of Molecules by Electron Impact

Here $d\sigma_{0n}^B(K)$ is the Born differential cross section for excitation of the target from state $|\psi_0\rangle$ to state $|\psi_n\rangle$, k_n and k_0 are the magnitudes of the final and initial momentum of the electron, K is the magnitude of the momentum transfer ($\mathbf{K} = \mathbf{k}_n - \mathbf{k}_0$ and $k_n^2 = k_0^2 - \Delta E$). The cross section is proportional to the square of the transition-matrix element:

$$\frac{d\sigma_{0n}^B}{d\Omega}(K) \propto |\langle \psi_n | \exp(i\mathbf{K}\cdot\mathbf{r}) | \psi_0 \rangle|^2. \tag{7}$$

The exponential is the Born operator. It is easy to show by power-series expansion (Lassettre, 1965) that in the limit $K \to 0$ this leads to

$$f_{0n}^G(K) \xrightarrow[K\to 0]{} f_{0n}^{\text{opt}}, \tag{8}$$

where f_{0n}^{opt} is the optical f value.

Lassettre et al. (1969) extended the use of Eq. (6) to cases in which the Born approximation does not hold and introduced the "apparent generalized oscillator strength" $f_{0n}^{AG}(K)$ in place of $f_{0n}^G(K)$. He was able to show that even in these cases

$$f_{0n}^{AG}(K) \xrightarrow[K\to 0]{} f_{0n}^{\text{opt}}, \tag{9}$$

which is called the "limit theorem." Equation (9) is equivalent to a selection rule which states that in the zero-momentum-transfer limit, optical selection rules apply to electron-impact excitation. The problems with the limit theorem, from the practical point of view, are that it represents a nonphysical limit (even for zero scattering angle the momentum transfer for inelastic scattering is different from zero) and that the extrapolation to zero momentum transfer involves arbitrariness.

Selection rules for electron-impact excitations based on group-theoretical arguments have been derived and summarized by Cartwright et al. (1971) and by Goddard et al. (1971). In the case of molecules, selection rules can be derived under two special conditions: (a) rules concerning a given scattering angle but arbitrary orientation of the molecule, and (b) rules concerning arbitrary scattering angle but specific orientation of the molecule. As an example of case (a), we mention the selection rule $\Sigma^- \not\leftrightarrow \Sigma^+$ for $0°$ and $180°$ scattering angle for diatomic molecules. Trajmar et al. (1971) first observed case (a) for the $X^3\Sigma_g^- \to b^1\Sigma_g^+$ excitation in O_2, and a simple symmetry argument can be used to derive this selection rule. The differential cross section is proportional to the square of the transition-matrix element

$$\frac{d\sigma}{d\Omega} \propto |\langle \psi_f | T | \psi_i \rangle|^2, \tag{10}$$

where T is the transition operator and ψ_i and ψ_f are the initial and final wave functions for the $(N+1)$-electron system. The transformation properties of the molecular wave functions under the symmetry operations of the point group of the electron–molecule system ($C_{\infty v}$ for diatomic molecules) are known and the operator T is known to be totally symmetric. In order for the matrix element in Eq. (10) to be nonvanishing, the product representation $[\psi_f \psi_i]$ also has to be totally symmetric. The $(N + 1)$-electron wave function can be factored as

$$\psi = \psi^e \psi^v \psi^r f_e, \qquad (11)$$

where $\psi^e,$ $\psi^v,$ and ψ^r are the electronic, vibrational, and rotational wave functions, respectively, and f_e is the scattered-electron wave function. (Here the Born–Oppenheimer adiabatic approximation, which has been found to be valid in general for electron-impact excitations, independent of E_0 and θ, is applied.) The combined vibrational wave functions in Eq. (10) yield the Franck–Condon factor just as in optical spectroscopy. Neither the Franck–Condon factor nor the rotational wave functions will influence the present symmetry arguments. For 0° or 180° scattering, the scattered-electron wave function must be totally symmetric under the operations of the point group of the diatomic molecule. The selection rule is, therefore, determined by the symmetry properties of the product representation $[\psi_f^e \psi_i^e]$. The electronic wave functions for the Σ^+ and Σ^- molecular states are symmetric and antisymmetric, respectively, with respect to reflection in the plane of the molecule. The product representation is therefore antisymmetric and requires a zero value for the transition-matrix element in Eq. (10) for any $\Sigma^+ \leftrightarrow \Sigma^-$ transition for which the electron is scattered into 0° or 180° (where f_e is symmetric).

IV. Experimental Techniques for Measuring Electron-Impact Excitation Cross Sections

Experimental techniques are only briefly discussed here because a more detailed treatment has been given by Trajmar and Register (1984).

A. Total Scattering Cross Sections

There are three methods commonly used for total scattering cross section measurements: the transmission method, the recoil method, and the transmission method with time-of-flight discrimination.

In the transmission method the attenuation of an electron beam passing through a gas is measured and is related to the total scattering cross section. One encounters some problems in dealing with the questions of (1) forward-direction scattering into the view cone of the detector, (2) treating electrons which undergo multiple collisions and are scattered back into the view cone of the detector, (3) defining the proper path length. In addition, pressure- and current-dependent problems have to be checked and minimized. Since a large fraction of the forward scattering arises from inelastic processes, this problem may be greatly reduced by the application of a retarding potential analyzer. These questions need to be carefully addressed in each measurement since they determine the error limits for the cross section data. The transmission technique has been applied to a large number of molecules.

The recoil technique is also basically a transmission technique but instead of the attenuation of the electron beam, the attenuation of the molecular beam is measured. This method has been applied extensively to atomic species and can be applied equally well (in principle) to molecules. The major difficulty is the accurate quantitative detection of neutral molecules. This difficulty must account for the absence of application of the recoil techniques to molecules so far.

In the time-of-flight method, the time distribution of the electron-beam pulse is converted to an energy distribution with an empty gas cell. When gas is present in the cell, the attenuation of the electron beam as a function of energy (time distribution) can be determined and related to the total scattering cross section as in other transmission experiments. The time-of-flight approach avoids some of the problems common to conventional transmission methods. It is limited at low impact energies (below a few tenths of an electron volt) by flux and at high impact energies (above 50 eV) by time resolution capabilities. Since inelastic forward direction scattering can not be discriminated against energetically in the time-of-flight measurements, caution must be used on molecular systems, especially those with large polarizabilities or permanent dipole moments. The method has been applied to about a dozen molecular species.

The total scattering cross sections measured by these techniques are generally accurate to within a few percent and the data obtained by various investigators agree with each other within this error limit.

B. Integral Excitation Cross Sections

Integral electron-impact excitation cross sections for molecules have been primarily obtained in the past by measuring optical excitation functions (or line excitation cross sections); that is, an "apparent" electron-impact excitation cross section for emission at a particular wavelength as a function of impact energy. Photoemission from a particular state excited by electron impact is monitored in these experiments. The optical excitation function can be corrected to various degrees of approximation to yield electron-impact excitation cross sections. The experimental determination of the electron-impact excitation cross section requires absolute measurement of the photon emission rate and knowledge of branching ratios for decay paths of the excited state and the scattering geometry. Further complications arise because of cascade contributions to the population of the excited state under study, radiation trapping for high target densities, and superelastic scattering. A large number of molecular transitions have been studied in the visible and the measurements are now being extended into the vacuum-UV region at several laboratories.

At low impact energies the trapped-electron method developed by Schulz (1958) can be utilized to obtain electron-impact excitation functions. A number of relative measurements were carried out but normalization of the data to the absolute scale is difficult and no cross section data have been generated by this method.

C. Differential Excitation Cross Sections

Differential (in angle) scattering cross sections (DCSs) represent information one level more detailed than integral cross sections. This type of data is needed for modeling of electron energy degradation processes for cases in which the angular dependence of the scattering process is explicitly considered, for stringent testing of theoretical models and for obtaining inelastic momentum transfer cross sections. Integral cross sections obtained by direct integration of differential cross sections are free from many of the deficiencies encountered in measurement of the integral cross section (e.g., cascade effects) but are subject to some errors due to the need for extrapolation of the measured data beyond experimentally accessible regions (from $0°$ to $180°$ scattering angles).

The techniques of elastic DCS measurements have been discussed in Chapter 1. The same techniques are applicable to inelastic DCS measurements except the scattering signal has to be determined in the appropriate energy loss channel. The first step is obtaining DCSs is to generate the energy loss spectra at various scattering angles and incident energies. (Typical energy loss spectra for N_2 are shown in Figs. 2 and 3.) The location of the spectral features characterizes the energy-level scheme of the target and the scattering intensities are related to the corresponding DCSs. Special care has to be exercised in transporting inelastically scattered electrons of varying residual energy with equal efficiency to the detectors. This requires an "achromatic" lens system (Chutjian, 1979). The general relationship between the scattered electron signal and the cross section is rather complex and the extraction of the desired cross section from this expression is complicated (Brinkmann and Trajmar, 1981; Trajmar and Register, 1984). It requires detailed knowledge of the target density and electron-beam flux distributions and the overall "instrumental function." The situation becomes much more manageable if a DCS averaged over the energy and angular resolution of the apparatus is removed from the integrand. Then the relationship for scattering process n becomes

$$I_n(E_0, \theta) = \overline{DCS_n(E_0, \theta)} V_{\text{eff}}(\theta)_{E_0} \qquad (12)$$

Here the flux, geometrical, and instrumental functions are included in the "effective scattering path length" $V_{\text{eff}}(\theta)$, which is obtained by the appropriate integrations. The values of the effective scattering volume (or "path length") can be calculated at each impact energy as a function of the scattering angle. Therefore, the absolute values of the DCS can, in principle, be obtained. It is, however, an easier and more reliable procedure to calculate relative path-length corrections and normalize the relative cross sections by other methods (Brinkmann and Trajmar, 1981; Trajmar and Register, 1984).

One approach to normalization utilizes total electron scattering, excitation, and ionization cross sections. Total cross sections can be measured with 1 to 3% accuracy and have recently become available for a large number of molecules. The integral elastic cross section σ_{elas} is obtained from the expression

$$\sigma_{\text{elas}} = \sigma_T - \sigma_{\text{ion}} - \sigma_{\text{exc}}, \qquad (13)$$

and is utilized to normalize the relative experimental elastic data, which in turn can be used to normalize the inelastic data. The ionization cross section σ_{ion} is usually available to an accuracy of about 5%. The total excitation cross section σ_{exc} is usually not available at low energies, but can be estimated with acceptable accuracy. At higher impact energies, σ_{exc} is comparable to the other quantities in Eq. (13) and the estimation accuracy will greatly influence the calibration.

High-energy and low-angle electron-impact measurements have been used to generate optical f values by extrapolation of the generalized oscillator strength $f^G(K)$ to zero momentum transfer. One can, however, reverse the procedure and utilize available optical f values to normalize relative generalized oscillator strengths which were obtained from, and are equivalent to, the experimentally measured DCS. The main difficulty with this procedure is the requirement of extrapolating the generalized oscillator strength (obtained from the relative DCS) to zero momentum transfer. This is usually performed by plotting the relative $f^G(K)$ against K^2 and observing the small K^2 tendency in the curve. The problem, as mentioned, is that for an inelastic process the limit of zero momentum transfer is unphysical. The other problem is that the most important points in this extrapolation are the low-angle DCS points, which are experimentally the most uncertain. This uncertainty in the low-angle data arises mainly from the difficulties of accurate effective path-length corrections at low scattering angles, the interference of the parent electron beam, and the rapid change of the DCS with scattering angle.

Optical excitation functions, if properly corrected for cascade, can also be used to normalize selective electron-impact integral cross sections. Since absolute optical excitation function measurements face most of the problems one encounters in absolute electron-impact cross section measurements and more, this normalization procedure is not very practical.

In principle, theoretical calculations could also serve as the basis for normalizing relative cross section data. In general, however, the theoretical calculations have not been accurate enough for this purpose although for simple species (H, He), highly accurate theoretical results have been obtained recently which certainly satisfy the criteria for a standard. Similar developments for simple molecular species can be expected in the future. One needs to consider, however, the possible limitations of the theoretical model. It may well describe a process at a particular energy or angular range but not in other ranges and may be reliable for certain types of transitions but not for others. At the present time, theoretical calculations are most useful in cases in which it is difficult to obtain quantitative experimental data (for corrosive materials, free-radical or excited species, etc.).

The most practical and reliable method of generating absolute electron-impact excitation cross sections is to normalize the measured relative cross sections to elastic scattering cross sections, which in turn can be accurately determined by normalization to elastic scattering by He. The procedure for obtaining the absolute elastic cross sections for gaseous molecular species is discussed in detail in Chapter 1 of this volume and by Trajmar and Register (1984) and Srivastava et al. (1975). In normalizing the inelastic cross sections to the elastic cross section special care must be taken of the electron energy sensitivity of the detector.

D. Experimental Uncertainties: Error Estimation

Perhaps as significant as any reported cross section is the accuracy that may be associated with that cross section. It is imperative, therefore, that error estimates associated with experimental results be as realistic as possible. In this section we will attempt to point out some of the types of experimental uncertainties associated with cross section measurements.

Experimental uncertainties may be classified as arising from either systematic or random sources. Systematic errors arise from known or unknown corrections which, if not applied, give a systematic bias to the results. Random errors are uncertainties which are due to the statistical nature of any measurement and may be reduced by repeating the experiment a large number of times. To cite the example mentioned previously, an effective path-length correction may be considered, a systematic error. If this geometry correction is not included in the analysis, the "correct" angular distribution will *never* be obtained from such measurements. The count rate in such an experiment represents a random error whose stastical uncertainty is $1/N^{1/2}$ where N is the accumulated number of counts. This error may be reduced by either increasing the count rate or the accumulation time. The background count rate, on the other hand, represents a systematic error with an associated statistical uncertainty.

In many experiments, systematic errors may be either avoided or corrected. Temperature dependence of the pressure reading, temporal drift in electron optics, timing calibrations in coincidence measurements, geometry and background corrections in crossed-beam experiments, and path-length corrections in time-of-flight measurements are all examples of correctable systematic errors. As most of these corrections are also subject to statistical uncertainties, it is clearly desirable to reduce the magnitude of the systematic effects within the constraint of retaining sufficient count rate to perform meaningful measurements. Once systematic corrections have been applied, one is still left with a lingering concern over unknown systematic errors. This question can only be answered by comparing the results against indepedent measurements of the same quantities that claim comparable accuracy.

Once the measurement errors have been reduced to statistical uncertainties, standard techniques for data analysis as described in most elementary statistics texts (Alder and Roessler, 1968; Bevington, 1969) may be employed (Trajmar and Register, 1984).

V. Electron-Impact Excitation Cross Section Data

A. General Remarks

The experimental methods described in Section IV are suitable for measuring cross sections for a variety of electron collision processes (e.g., elastic scattering, dissociation, ionization, etc.). In this chapter we deal only with vibrational (and associated rotational) and discrete-electronic-state excitations. Furthermore, with the exception of a few general remarks, we restrict our discussion to direct (nonresonant) excitation cross sections.

Experimental work prior to 1969 has been summarized by Massey *et al.* (1969) and Kieffer (1969a,b, 1971). More recently, Kieffer (1973) and Schulz (1976) compiled the cross section data necessary for modeling gas-discharge lasers. Very recently, Trajmar *et al.* (1983) reviewed the cross section data for electron-impact excitation of molecules. It is clear from these surveys that only fragmentary cross section data exist for vibrational and electronic excitations even for the most common molecular species. Most of the data are for total electron scattering cross sections and for excitation functions which have been measured by optical techniques and represent, in most cases, "apparent" excitation cross sections. The only molecule for which a reasonably extensive set of differential and integral cross sections is available is N_2. Even this case is far from complete, however, and the accuracy of these data is only about $\pm 30\%$.

The main reasons for the current lack of extensive experimental data are

(1) the experimental difficulties associated with normalization (and/or direct absolute measurement);

(2) problems encountered in the resolution and/or deconvolution of molecular electron energy loss spectra;

(3) excessive time requirements for carrying out the measurements for a large number of processes, at a large number of incident electron energies (and scattering angles) for each molecule; and

(4) handling of a very large body of experimental data points.

Significant progress has been made in recent years in all of these areas. The availability of accurate He cross sections (Register et al., 1980b) and application of the relative flow technique (Srivastava et al., 1975; Trajmar and Register, 1984) make possible the normalization of relative experimental results in a convenient and efficient way. Energy resolution of the order of 20 meV is now available for routine measurements. With this resolution, and with the help of computer unfolding codes, analysis and evaluation of the complex overlapping structures of energy loss spectra can be achieved. Computer control of the experiments and handling of the data make it possible to collect and manipulate extensive sets of data. One can expect that a considerable body of high-quality ($\pm 10\%$ accuracy) cross section data will become available in coming years as a result of these advances.

Here we give a brief overview of the available cross section data. A detailed compilation and discussion of the data is given by Trajmar et al. (1983).

B. Vibrational Excitation

Electrons (at nonrelativistic velocities) are very light compared to nuclei and one might expect electron-impact excitation of nuclear motions (rotation, vibration, recoil) to be relatively unimportant. The fact is that electrons are very effective in producing vibrational excitation, particularly at low impact energies, by interacting with the molecular electron distribution, which is coupled with the nuclear vibrational motion. This is a resonance mechanism for vibrational excitation which involves a temporary electron capture. The increased electromagnetic interaction causes an efficient distortion of the molecular charge distribution and this distortion leads to efficient energy transfer into vibrational excitation. (At low energies typical cross sections are

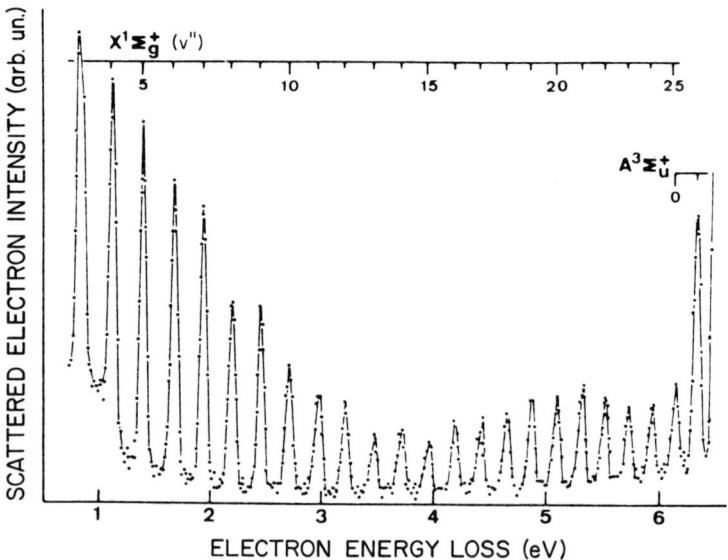

Fig. 5. Electron energy loss spectra in molecular nitrogen for incident electron energy of 10 eV and a scattering angle of 30°. These data show many vibrational levels of the ground electronic state being excited under these conditions. [From Huetz et al. (1980).]

10^{-16} cm^2.) Resonance processes can also lead to excitation of high vibrational levels. This occurs, for example, when vibrational excitation proceeds through a repulsive intermediate state (competing with dissociation or dissociative attachment) in connection with core-excited shape resonances. This, however, occurs with rather small probability. An interesting example for this type of excitation is shown in Fig. 5 (Huetz et al., 1980).

The behavior of the vibrational-excitation cross section in the resonance region depends critically on the lifetime of the negative ion state. Resonances with lifetimes which are short compared to the vibrational period result in broad (few electron volt) resonances and, therefore, in enhancements of the cross sections. When the lifetime of the compound state is long compared to the vibrational period, the vibrational structure in the negative ion is well developed and resonance excitation through these states can take place. The vibrational-excitation cross sections, as a function of impact energy, then exhibit sharp, narrow (few milli-electron-volt) spikes. When the lifetime is comparable to the vibrational period ($\sim 10^{-14}$ s), the vibrational-excitation cross sections oscillate (as a function of impact energy) and the peak cross sections appear at different energies depending on the final decay channel of excitation. The angular behavior of the cross sections in the resonance case depends on the symmetry of the resonance state and on the participating partial waves.

The direct electron-impact vibrational excitation of molecular vibration, especially at high impact energies, tends to approximate the $\Delta v = 1$ selection rule. These-excitation cross sections are of the order of 10^{-18} cm^2 and change smoothly with impact energy. Excitation to overtone or combination bands decrease by about an order of magnitude with increasing vibrational quantum numbers. The angular behavior of the cross sections depends on the relative importance of short- and long-range intereaction terms and is usually similar for the fundamental and overtone bands.

Recent vibrational-excitation cross section measurements are summarized in Table I and an excellent review on vibrational excitation in the resonance-dominated low-energy region is given by Schulz (1976). Recently Rohr (1979c) discussed the question of threshold resonances and cross section behavior in vibrational excitation. A detailed compilation of the recent cross section data was made by Trajmar et al. (1983). It should be noted that the rotational structure of the vibrational transition can be resolved only for H_2 at the present time and the vibrational-excitation cross sections, in general, represent averaging over initial, and summation over final, rotational states.

C. Electronic Excitation

In excitation of electronic states, resonance mechanisms do not appear to play a significant role. Only core-excited shape resonances cause significantly increased cross sections but only over an energy region which is small (a few electron volts) compared to the energy region over which direct excitation is effective (Huetz et al., 1980).

The largest cross sections for electronic excitation are associated with optically allowed transitions at intermediate impact energies (and small scattering angles). The value of these cross sections increase gradually with increasing impact energy from threshold to about ten times threshold energy and then slowly decrease at high energies. The angular distributions are forward peaked and this character becomes more enhanced with increasing impact energy.

At low impact energies (within a few electron volts of threshold) forbidden transitions dominate the energy loss spectra and, therefore, represent the important cross sections. Particularly significant are the spin-forbidden processes that readily occur by electron exchange and result in metastable species. This property of electron-impact excitation has been widely utilized to generate metastable atoms and molecules. Integral cross sections for spin-forbidden processes rise steeply near threshold, reach their peak value within a few electron volts of threshold, and then decrease sharply with increasing energy. The DCSs associated with these processes are nearly isotropic, reflecting the short-range nature of the spin exchange reaction.

TABLE I

Summary of Electron-Impact Vibrational-Excitation Cross Section Measurements since about 1970[a,b,c]

Reference	Energy (eV) and angular range	Excitation	Method, type of cross section	Estimated error (%)
H_2				
Ehrhardt et al. (1968)	5°–100° 0.027–10.23	$v = 1$	Beam–beam DCS, σ_v	
	1.03–8.01	$v = 2$		±15
	1.86–7.21	$v = 3$		
Trajmar et al. (1970)	7–81.6 10°–80°	$v = 1, 2, 3$	Beam–beam DCS, σ_v	30–50
Crompton et al. (1970)	0.7–1.2	$v = 1$	Swarm, σ_v	±5
Gibson (1970)	0.588–4.92	$v = 1$	Swarm, σ_v	±5
Weingartshofer et al. (1970)	11.0–13.0 10°, 40°, 90° 10°, 90° 10°, 70°, 100°, 120° 100°	$v = 1$ $v = 2$ $v = 3$ $v = 4$	Beam–beam Differential excitation functions	±20
Linder and Schmidt (1971a)	20°–120° 1.5–10.8	$v = 1, \Delta J = 0$ $v = 1, (1 \to 3)$	Beam–beam DCS, σ_v	±10
Wong and Schulz (1974)	4.5, 20°–100° 4.5, 20°–100° 4.5, 20°–100° 4.5, 20°–100° 4.5, 20°–80° 4.5, 20°–80°	$v = 1, \Delta J = 0$ $v = 1, (1 \to 3)$ $v = 2, \Delta J = 0$ $v = 2, (1 \to 3)$ $v = 3, (\Delta J = 0)$ $v = 3, (1 \to 3)$	Beam–beam DCS	±10

TABLE I (*cont.*)

Reference	Energy (eV) and angular range	Excitation	Method, type of cross section
N₂			
Pavlovic et al. (1972)	19–35 25°–90°	$v = 1, 2, 3$	Beam–beam DCS
D. G. Truhlar et al. (1976)	5, 10 20°–135°	$v = 1$	Beam–beam DCS, σ_v
D. G. Truhlar et al. (1977)	20–75 20°–135°	$v = 1$	Beam–beam DCS
Huetz et al. (1980)	9 30°	$v = 1$–25	Beam–beam DCS
Tanaka et al. (1981)	3–30 20°–130°	$v = 1$	Beam–beam DCS, σ_v
Jung et al. (1982)	2.47 15°–105° 1.8–2.8 15°	$v = 1$ $J = 0, \pm 2, \pm 4$ $v = 1$ $J = 0, +2, +4$	Beam–beam DCS, σ_v
O₂			
Linder and Schmidt (1971b)	0.1–1.5 60°	$v = 1$–4	Beam–beam DCS, σ_v
CO			
Ehrhardt et al. (1968)	Threshold to 4 eV	$v = 1$–7	Beam–beam, σ_v

Reference	Range	State	Method
Chutjian et al. (1972)	10–80	$v = 1$	Beam–beam DCS, σ_v
Land (1978)	40°, 80°		Swarm
Zubek and Szmytkowski (1979)	0.266–3.816	$v = 1$	Beam–beam DCS
	1–4	$v = 1$–6	
Chutjian and Tanaka (1980)	3–100	$v' = 1; v = 0, 2$–6	Beam–beam DCS, σ_v
	10°–135°	$v = 1$	
Jung et al. (1982)	1.8	$v = 1$	Beam–beam DCS
	15°–1°	$J = 0, 1, 2, 3, 4$	
HX (X = F, Cl, Br)			
Rohr and Linder (1976)	Threshold to 3.5 eV	HF; $v = 1, 2$	Beam–beam DCS, σ_v
	20°–120°		
	Threshold to 3.5 eV	HCl; $v = 1, 2, 3$	Beam–beam DCS, σ_v
	20°–120°		
Rohr (1977a, 1978a)	Threshold to 10 eV	HBr; $v = 1$–5	Beam–beam DCS, σ_v
	20°–120°		
H$_2$O			
Seng and Linder (1974) and (1976)	Threshold to 10 eV	$(1, 0, 0; 0, 0, 1)$	Beam–beam DCS, σ_v
	20°–110°		
Rohr (1977b)	Threshold to 3 eV	$(v_1; v_3)$	Beam–beam DCS
	20°–120°		

TABLE I (cont.)

Reference	Energy (eV) and angular range	Excitation	Method, type of cross section
H₂S			
Rohr (1978b)	2 60°	$n,0,1$ $1,0,n$ $n,1,0$ $0,1,n$	Beam–beam DCS
	Threshold to 4 eV 120°	$(1,0,0;0,0,1)$	Beam–beam DCS
	2 20°–120°	$(1,0,0;0,0,1)$ $0,1,0$	Beam–beam DCS
CO₂			
Danner (1970)	1–19 1.5°–150°	$\Delta E = 0.083, 0.163,$ $0.172, 0.255, 0.291,$ $0.339, 0.422, 0.459,$ $0.506, 0.591, 0.669,$ $0.754, 0.832, 1.006,$ $1.160, 1.314$ eV	Beam–beam DCS, σ_v
Register et al. (1980a)	4–50 5°–140°	$0,1^1,0$ $0,2^0,0$ $1,0^0,0$ $0,0^0,1$ $0,3^1,0$ $1,1^1,0$ $0,4^0,0$ $1,2^0,0$ $2,0^0,0$	Beam–beam DCS, σ_v

SO_2			
Andrić et al. 1981; (1982, private communication)	3.4	$n, 0, 0$ $n = 1$–20	Beam–beam DCS
CH_4			
Rohr (1979b)	Threshold to 4 eV 60°	$(v_2; v_4)$	Beam–beam DCS
Rohr (1980)	1, 60° 20, 20°–120°	$(v_1; v_3)$ $(v_2; v_4)$	Beam–beam DCS
Sohn et al. (1983)	Threshold to 4 eV, 60° 1, 45° Threshold to 1 eV 35°, 75°, 90°, 105° 1, 30°–105° 0.3, 30°–105° 0.6, 30°–105°	$(v_1; v_3)$ $(v_2; v_4)$	Beam–beam DCS
Kubo et al. (1981)	3–20 30°–140°	$(v_1; v_3)$ $(v_2; v_4)$	Beam–beam DCS
C_2H_4 and C_2D_4			
Walker et al. (1978)	1–11 90° 1.95 40°–110° 1–3.5 90° 7.5 30°–110°	v_1, v_2, v_3 v_2, v_3, v_7 $v_2, v_3, v_7, 2v_3$ v_1, v_3, v_7	Beam–beam DCS

TABLE I (*cont.*)

Reference	Energy (eV) and angular range	Excitation	Method, type of cross section
CCl$_3$F, CCl$_2$F$_2$			
Rohr (1979a)	1–4 20°–60°	ν_1, ν_4, ν_2 $2\nu_1, {}^2\nu_4, {}^3\nu_3$	Beam–beam DCS

[a] For earlier measurements and numerical data see Massey *et al.* (1969), Trajmar *et al.* (1983), Kieffer (1969a,b, 1971, 1973), Schulz (1976).
[b] For identifying the vibrational excitation, we use the standard notation [G. Herzberg, *Infrared and Raman Spectra* (Van Nostrand, Princeton, NJ, 1945)]. Vibrational state designations in perentheses refer to unresolved excitations. In cases where the rotational transition is specified, that structure was either resolved or deconvoluted.
[c] When the energy loss ΔE is given in the Excitation column, the measurement was carried out at that particular energy loss point and may represent the overlap of more than one excitation process.

No simple characteristics can be identified for symmetry-forbidden excitations. The integral cross sections for these processes are usually smaller then for optically allowed excitations, reach their peak value at lower impact energies, and their DCSs show a large variety. There is, however, a very unique character associated with parity-unfavored transitions ($\Sigma^+ \leftrightarrow \Sigma^-$) as discussed above, and the DCSs for these excitations goes to zero at 0° and 180° scattering angles.

In practically all of the measurements carried out so far, the initial states for electron scattering processes have been the ground electronic and vibrational state of the molecule. Only for excitation from the $O_2(a^1\Delta_g)$ metastable state are some fragmentary cross section data available (Hall and Trajmar, 1975). The major difficulty here is the generation of a sufficient number of excited species for collision studies; it is hoped that with the application of lasers this situation will improve.

In Table II, the cross section measurements for electron-impact excitation of discrete molecular states have been summarized. As can be seen from this table, with the exception of N_2 only fragmentary data exist. A detailed compilation of the cross section data has been made by Trajmar et al. (1983).

D. Total Electron Scattering Cross Sections

At low impact energies, elastic scattering is the major contributor to the total electron scattering cross section. Although the rotational inelastic channels are open at impact energies of a few milli-electron-volts and above, their contribution is significant only in the resonance-dominated regions. At intermediate and high impact energies the electronic excitation and ionization channels represent the major contributions to total electron scattering cross sections. One notable exception is the highly polar molecules, for which the $\Delta J = \pm 1$ rotational excitation dominates all other cross sections.

As discussed in Section IV, total electron scattering cross sections have been measured by several highly accurate methods and a summary of these measurements is given in Table III.

VI. Generation of Optical Data by Electron-Impact Techniques

It was pointed out in Section III that the generalized oscillator strength obtained from electron collision measurements becomes equivalent to the optical f value in the limit of zero momentum transfer. This equivalence can

TABLE II

Summary of Available Experimental Cross Section Data for Electron-Impact Excitation of Molecular Electronic States[a,b,c]

Excitation	Energy (eV) and angular range	Method	Type of cross section	Reference
H_2				
$b\,^3\Sigma_u^+$	25–60 10°–80°	Beam–beam	DCS, σ	Trajmar et al. (1968)
	10–16 10–120°	Beam–beam	DCS, σ	Weingartshofer et al. (1970)
$B\,^1\Sigma_u^+$	11–13 40°	Beam–beam	DCS, σ	Weingartshofer et al. (1970)
	11–14 8.5°	Beam–beam	DCS	Weingartshofer et al. (1975)
	15–60 10°–135°	Beam–beam	DCS σ, σ^M	Srivastava and Jensen (1977)
$c\,^3\Pi_u$	Threshold, 5°–132° 12.0–13.5 5°–132°	Beam–beam	DCS, σ	Weingartshofer et al. (1975)
$C\,^1\Pi_u$	50–6000	Optical[d]	σ	de Heer and Carriere (1971)
	Threshold to 300 eV	Optical	σ	Stone and Zipf (1972b)
$d\,^3\Pi_u$	Threshold to 100 eV	Optical	σ	Baltayan and Nedelec (1976)
	Threshold to 100 eV	Optical	σ	Mohlmann and de Heer (1976)

State	Energy range	Method	Measurement	Reference
$G\,^1\Sigma_g^+$	Threshold to 500 eV	Optical	σ	Anderson et al. (1977)
$E\,^1\Sigma_g^+, F\,^1\Sigma_g^+, H\,^1\Sigma_g^+$	Threshold to 300 eV	Optical	σ	Watson and Anderson (1977)
$G\,^1\Sigma_g^+, I\,^1\Pi_g$ $K\,^1\Sigma_g^+, M\,^1\Sigma_g^+$ $N\,^1\Sigma_g^+$	50	Optical	σ	Day et al. (1979)
$B\,^1\Sigma_u^+, C\,^1\Pi_u$	Threshold to 300 eV	Optical	σ	Malcolm et al. (1979)
$d\,^3\Pi_u$	Threshold to 100 eV	Optical	σ	Bogdanova et al. (1981)
$B\,^1\Sigma_u^+, C\,^1\Pi_u$	Threshold to 400 eV	Optical	σ	Ajello et al. (1982)
N_2				
$a\,^1\Pi_g$	10–50 5°–138°	Beam–beam	DCS, σ	Cartwright et al. (1977a,b)
	500–2000 4.75–133.5°	Beam–beam	σ	Oda and Osawa (1981)
$a'\,^1\Sigma_u^-, w\,^1\Delta_u,$ $a''\,^1\Sigma_g^+, A\,^3\Sigma_u^+,$ $B\,^3\Pi_g, W\,^3\Delta_u,$ $B'\,^3\Sigma_u^-, C\,^3\Pi_u,$ $E\,^3\Sigma_g^+$	10–50 5°–138°	Beam–beam	DCS, σ	Cartwright et al. (1977a,b)

TABLE II (cont.)

Excitation	Energy (eV) and angular range	Method	Type of cross section	Reference
$b\,^1\Pi_u, c\,^1\Pi_u,$ $c'\,^1\Sigma_u^+, 0\,^1\Pi_u,$ $b'\,^1\Sigma_u^+, F\,^3\Pi_u,$ $G\,^3\Pi_u$	10–50 eV 5°–138°	Beam–beam	DCS, σ	Chutjian et al. (1977)
$a\,^1\Pi_g$	10–208	Optical	σ	Ajello (1970)
$a\,^1\Pi_g, D\,^3\Sigma_u^+$	Threshold to 39 eV	Optical	σ	Freund (1971a)
$a\,^1\Pi_g, p\,^1\Sigma_g^+$	50–5000	Optical	σ	Aarts and de Heer (1971)
$B\,^3\Pi_g, C\,^3\Pi_u$	Threshold to 30 eV	Optical	σ	Shemansky and Broadfoot (1971)
$E\,^3\Sigma_g^+$	11.87–13.0	Optical	σ	Borst et al. (1972)
$C\,^3\Pi_u$	Threshold to 1000 eV	Optical	σ	Imami and Borst (1974)
$c\,^3\Pi_u$	Threshold to 1000 eV	Optical	σ	Kosoruchkina and Trekhov (1978)
$b\,^1\Pi_u$	Threshold to 500 eV	Optical	σ	Zipf and Gorman (1980)
CO				
$A\,^1\Pi, B\,^1\Sigma^+,$ $C\,^1\Sigma^+$	100–5000 eV	Optical	σ	Aarts and de Heer (1970)
$A\,^1\Pi, a\,^3\Pi$	Threshold to 300 eV	Optical	σ	Ajello (1971)

$A^1\Pi$	Threshold to 350 eV	Optical	σ	Mumma et al. (1971)
$b^3\Sigma^+$	Threshold to 40 eV	Optical	σ	Wells et al. (1973)
NO				
$B'^2\Delta$	Threshold to 300 eV	Optical	σ	Stone and Zipf (1972a)
$A^2\Sigma^+$	300 eV	Optical	σ	Van Sprang et al. (1979)
	Threshold to 300 eV	Optical	σ	Povch et al. (1972)
	Threshold to 300 eV	Optical	σ	Imami and Borst (1975)
O_2				
$a^1\Delta_g, b^1\Sigma_g^+$	Threshold to 4 eV	Beam–beam	σ	Linder and Schmidt (1971b)
$a^1\Delta_g, b^1\Sigma_g^+$	4–45 eV 10°–90°	Beam–beam	DCS, σ	Trajmar et al. (1971)
$a^1\Delta_g, b^1\Sigma_g^+$, $\Delta E = 6.1$ eV; $(c^1\Sigma_u^-, C^3\Delta_u, A^3\Sigma_u^+)$; 8.3 eV $(B^3\Sigma^-)$; 9.97, 10.29 eV	20–45 eV 10°–90°	Beam–beam	DCS, σ	Trajmar et al. (1972)
$B^3\Sigma_u^-$ $\Delta E = 9.7, 12.1$ eV	20–500 5°–130°	Static gas	DCS, σ	Wakiya (1978a)
$a^1\Delta_g, b^1\Sigma_g^+$ $\Delta E = 6.1$ eV $(c^1\Sigma_u^-, C^3\Delta_u, A^3\Sigma_u^+)$	20–500 10°–130°	Static gas	DCS, σ	Wakiya (1978b)

TABLE II (*cont.*)

Excitation	Energy (eV) and angular range	Method	Type of cross section	Reference
SO$_2$				
$\Delta E = $ 3.1—3.9	20, 50, 200	Beam–beam	DCS, σ	Vuškovič and Trajmar (1982)
3.9—5.5	2°–127°			
5.5—7.1				
7.1—9.6				
9.6—10.3				
10.3—11.3 eV				
CH$_4$				
$\Delta E = $ 7.5 to 9.0	20, 30, 200	Beam–beam	DCS, σ	Vuškovič and Trajmar (1983)
9.0 to 10.5	8°–130°			
10.5 to 12.0				
= 12.0 to 13.5				
13.5 to 15.0 eV				

[a] Dissociation, ionization, and dissociative ionization cross sections are excluded.
[b] The standard spectroscopic notation of Herzberg [G. Herzberg, *Spectra of Diatomic Molecules* (Van Nostrand, Princeton, NJ, 1950)] is used.
[c] When the energy loss ΔE is given in the Excitation column, the measurement was carried out at that particular energy loss or energy loss range and it may represent the overlap of more than one excitation process.
[d] Optical excitation function.

TABLE III
Summary of Recent Total Electron Scattering Cross Section Measurements for Molecules

Reference	Energy range (eV)	Method	Error limits (%)
H_2			
Ferch et al. (1980, 1981)	0.02–2.0	Transmission with time of flight	±2.5
Dalba et al. (1980b)	0.02–100	Transmission	1.7 to ±5
van Wingerden et al. (1980)	25–750	Transmission	3.1 to ±5.4
van Wingerden et al. (1980)	20–2000	Semiempirical	±5
Jones and Bonham (1981, private communication)	0.6–50	Transmission with time of flight	±3
Hoffman et al. (1982)	2–500	Transmission	±2

Other Molecules	Molecule	Energy range (eV)	Method
Miller and Kasdan (1973)	Na_2, K_2	0.5–50	Recoil
Slater et al. (1974a)	CsF	0–15	Recoil

TABLE III (cont.)

Reference	Molecule	Energy range (eV)	Method
Becker et al. (1974)	CsCl	0–15	Recoil
Slater et al. (1974b)	KI	0–15	Recoil
Mathur and Hasted (1977)	N_2	1–4	Transmission
Szmytkovski and Zubek (1978)	CO, CO_2, OCS	1–8	Transmission
Guškov et al. (1978)	CO	0.5–6.0	Time of flight
Hasted et al. (1979)	$CO, N_2, CO_2, CH_4, C_2H_4$	0.3–5	Transmission
Kennerly et al. (1979)	SF_6	0.5–100	Time of flight
Baldwin (1974)	N_2	0–2	Time of flight
Barbarito et al. (1979)	CH_4	0.05–6.0	Time of flight
Blaauw et al. (1980)	N_2	15–750	Transmission
Kennerly (1980)	N_2	0.5–50	Time of flight
Dalba et al. (1980a)	N_2, O_2, NO	100–1600	Transmission
Bonham et al. (1980, private communication)	$CO_2, CH_4, CCl_4, CCl_3F, CCl_2F_2, CClF_3, CF_4$	0.5–100	Time of flight

Zubek et al. (1981)	H_2S, CO_2, N_2O, SO_2	0.5–10	Transmission
Sokolov and Sokolava (1981)	H_2O, H_2S, SO_2	0–10	Indirect
Ferch et al. (1982)	SF_6	0.036–1.0	Time of flight
Hoffman et al. (1982)	N_2, CO_2	2–700	Transmission
Miller et al. (1982)	Li_2	0.5–10	Recoil
Ferch et al. (1981)	CO_2	0.07–2.0	Time of flight

be exploited to generate optical f values (or, equivalently, photoabsorption cross sections). In practice, one measures electron-impact cross sections as a function of scattering angle at high impact energies and converts them into generalized oscillator strengths (or apparent generalized oscillator strengths) as a function of momentum transfer. An extrapolation of these quantities to zero momentum transfer then yields the optical f value. It should be noted, however, that the extrapolation faces several problems which could seriously influence the accuracy of the data. These problems equally effect the inverse procedure of normalizing electron-impact data to optical f values; this matter was discussed in Section IV.C.

The pioneering work in developing the technique and instrumentation for this type of research was carried out by Lassettre and co-workers (Lassettre, 1964) and Bromberg (1974). Recently Brion and co-workers and Van der Wiel and co-workers have applied elegant coincidence techniques to obtain quantitative photoionization, photofragmentation, and photofluorescence cross sections by high-energy electron-impact methods. Normalization of the experimental data to the absolute scale was (in some cases) achieved by application of the Thomas–Kuhn–Reiche sum rule. A comprehensive review of these techniques and results has been given by Brion and Hamnett (1980). So far this technique has found only limited applications although it may be made useful in the extreme UV and x-ray regions where optical techniques are very difficult to apply.

VII. Recent Developments

A. Study of Superexcited States (Inner-Shell and Multielectron Excitations)

States with excitation energy greater than the lowest ionization potential are called superexcited states. Such states are generated in electron-impact processes by exciting inner-shell electrons or more than one (valence and/or inner-shell) electron at the same time to higher molecular orbitals. Since these states overlap with at least one ionization continuum, the spectral feature corresponding to the excitation shows the typical interference (Beutler–Fano) profile. The decay of the superexcited state can proceed either by photon emission, by electron ejection (autoionization), or by dissociation.

There has been a great deal of interest recently in inner-shell excitation processes. (Brion, 1975; van der Wiel, 1980; Read and King, 1979, Read, 1978, 1980; Shaw et al., 1982). From the point of view of resolution, the electron-impact technique becomes very competitive with optical techniques at high energy losses corresponding to inner-shell excitation (Read, 1978). In addition, it is a much more convenient and inexpensive approach than the use of available UV and x-ray sources or synchrotron radiation. Energy loss spectroscopy has demonstrated that a considerable fraction of the total oscillator strength is associated with such transitions. The investigations so far have mainly been concerned with spectroscopic information and no absolute cross section data exist for inner-shell or multielectron excitation processes.

B. Coincidence Measurements

Electron–photon coincidence measurements have seen extensive applications in atomic collision phenomena. (Blum and Kleinpoppen, 1979; McConkey, 1979; Kleinpoppen and Williams, 1979). These experiments study the fundamental relationships between the magnetic sublevel populations in electron-impact excitation of a particular atomic energy level and can provide scattering amplitude and phase information for magnetic sublevel excitations by electron impact. The basic theory of angular correlations (Macek and Jacks, 1971; Fano and Macek, 1973; Blum and Kleinpoppen, 1979) has recently been extended to molecules (Blum and Jakubowicz, 1978). For the molecular case, however, the analysis of experimental data is more complicated than for atoms because of the customary averaging over rotational (and sometimes even vibrational) levels. Preliminary work on H_2 electron–photon coincidence measurements has been reported by McConkey 1979 and Malcolm and McConkey (1979).

One little-used capability of this technique is in the resolution of two very close-lying, or even degenerate, energy levels in an atom or molecule. When the electron energy channel contains unresolved scattering, one may rely on coincident photon emissions that are separated in wavelength, or on photon coincidence associated with one of the energetically degenerate levels, to measure cross sections. Applications of the first method have been made to He (Pochat et al., 1973; Chutjian, et al., 1980) and of the second technique to the hydrogen atom (Frost and Weigold, 1980).

Electron–ion coincidence techniques are useful for determining partial ionization and molecular fragmentation cross sections. A description of this technique and some results are given by Brion and Hamnett (1980), van der Wiel and Wiebes (1971), and Backx et al. (1976).

C. Electron Collisions in Laser Fields

The availability of high-intensity narrow-frequency-band lasers has opened up new possibilities for electron scattering studies. These studies can yield information on the interaction of molecules, electrons, and photons which could not be obtained previously and which are of practical importance in the field of laser–plasma interactions. The general treatment of the problem of electron scattering in electromagnetic fields and the interpretation of experimental observations is quite complex. The usual problems of treating the interaction of an electron with an N-electron molecule are made more difficult when the laser field is introduced: (1) the energy states of the molecule may be modified (dressed states), (2) multiphoton excitation and ionization of the molecule may take place, (3) multiphoton free–free electron transitions (bremsstrahlung, inverse bremsstrahlung) can occur, and (4) the close dynamics of the scattering process may be affected by n-photon transitions in the temporary $(N + 1)$-electron molecular ion. Both the theoretical and experimental efforts in this area have been limited mainly to atoms and to special situations in which one particular interaction dominates and the others can be neglected or treated as perturbations. The experimental investigations carried out so far fall into two categories: (1) electron scattering by laser-excited species, and (2) free–free radiative transitions.

In the first category low-power tunable dye lasers are utilized to excite the target species to specific states and the electron scattering by these excited species is studied. In this case, the resonant excitation of the target is the dominating process for the photon field and the interaction of the free electron with the field and the distortion of the energy-level scheme of the target can be neglected. Total cross section measurements on laser-excited Na have been reported by Bhaskar *et al.* (1977) and by Jaduszliver *et al.* (1980), and DCSs for laser-excited Ba by Register *et al.* (1978). The technique, however, has not yet been applied to molecular species, mainly because of the difficulty of producing tunable lasers in the UV or IR regions. When the laser is operated in the single-frequency mode, individual isotopes and hyperfine levels can be pumped and, with the selection of proper polarization of the laser light, the population of the magnetic sublevels can also be controlled. Thus a polarized target (aligned and/or oriented) can be prepared for electron scattering measurements. In this case, not only cross sections for individual magnetic sublevel excitations but also scattering amplitudes and phase relationships can be obtained from measuring superelastic scattering intensities. The excitation with polarized light and subsequent measurement of the superelastic electron scattering intensities is equivalent to the time-inverse electron–photon coincidence measurement (Hertel and Stoll, 1977).

The second category of experiment has been concerned with free–free radiative transitions in strong electromagnetic fields. Here the photon energy is chosen to be far from resonance (even for multiphoton processes) and the dominant interaction for the photon field is with the free electron. The current status of this research area has been summarized by Gavrila and van der Wiel (1978). So far only free–free transition associated with elastic scattering have been observed. In the approximate theoretical treatment of the problem, the molecule acts only as a third body (so that the energy and momentum conservation can be satisfied) and its identity plays no role in the process. This assumption needs to be experimentally verified. An interesting way of studying resonances in the free–free transition channels has been pointed out by Jung and Kruger (1978) and Jung and Taylor (1981). In this scheme, the nonresonant background scattering is eliminated by the proper choice of scattering geometry and laser light polarization, and the signal originates only from resonance scattering. Thus, instead of the interference (Beutler–Fano) profile, one observes regular peaks at resonances.

D. Electron Scattering with Spin-Selected Electrons

Scattering studies which utilize spin-polarized electrons represent a step closer to the ideal experiment where averaging over indistinguishable quantum states is avoided. Such studies yield a deeper understanding of the scattering process and are especially important, from the electron-impact excitation point of view, for the study of electron exchange processes. Measurements have been conducted with spin-polarized electrons (mainly elastic scattering on atomic species) but these measurement have been rather difficult owing to the low intensity of the polarized electron beams that have been available. The physics of spin-polarized free electrons, their generation, detection, and utilization in scattering experiments are described by Kessler (1976).

Significant progress has been made in recent years in the production of spin-polarized electron beams by utilizing solid-state sources. Photoemission from GaAs, when irradiated with circularly polarized light, can generate electron beams with a high degree of polarization ($\sim 45\%$), narrow energy spread (~ 130 meV, FWHM), and high beam intensity (~ 10 μA) (Pierce et al., 1975; Pierce et al., 1980). This development offers exciting new possibilities and one can expect increasing utilization of spin-polarized electron beams in future scattering experiments including inelastic scattering involving molecular species (Pierce and Celotta, 1980; Kessler, 1982).

VIII. Theory of Electron–Molecule Collisions

During the past ten years, interest in electron scattering by atoms and molecules has increased substantially and a number of excellent reviews dealing with the relevant theory have been published (e.g., Bransden and McDowell, 1978; Rescigno et al., 1979; Lane, 1980; Amusia and Cherepkov, 1975; Blum and Kleinpoppen, 1979; Callaway, 1980; Norcross and Collins,

1982; and Buckley *et al.*, 1983) to augment the classic review by Massey *et al.* (1969). As a consequence, this chapter will only supplement these reviews by focusing on the most recent developments in the theory for nonresonant vibrational and electronic excitation of molecules in the low- and medium-energy region. In discussing the theoretical techniques, the emphasis will be placed on comparisons between theory and experiment and hence on the ability of theory to reliably predict the relevant cross sections needed for particular applications. Finally, the high-energy region ($E_0 \gtrsim 200$ eV) will not be discussed in this chapter because it has been well covered in a recent review by Inokuti (1971).

In order to evaluate the ability of a particular theory to predict or explain inelastic electron-impact cross sections for a specific process of interest, the differential cross sections, as well as the integral cross sections, need to be examined. This is because in the study of a wide variety of phenomena of current interest, the motion and distribution of electrons in the medium are properties of importance for characterizing the medium. Since the differential cross sections (DSCs) for elastic and inelastic electron scattering by atoms and molecules are highly anisotropic, the scattering process has a pronounced effect on the velocity and spatial distribution of electrons in the medium. The effect of anisotropic scattering has been shown to be important in the interpretation of ionospheric and auroral electron processes (Altshuler, 1963), and also in gas discharge processes (Ferrari, 1977; Lin *et al.*, 1979). Consequently, for any theory of electron scattering by atoms and molecules to be useful for a wide range of applications, it must predict reliable DSCs, as well as reliable integral cross sections, for complex target systems.

For all electronic excitation processes involving incident electron energies less than a few hundred electron volts, it has been known (Bransden and McDowell, 1978; Truhlar *et al.*, 1970; Rice *et al.*, 1972) for some time that the impact parameter, Born, Ochkur–Rudge, and Glauber and related "lowest-order" theories fail for atomic targets for scattering angles greater than about 50°, even for optically allowed transitions. This behavior is characteristic of all theories that are based on some type of lowest-order expansion of the electron–target interaction potential which does not explicitly include distortion of the incident- and scattered-electron wave function. More recent comparisons between experimental and theoretical DCSs for electron scattering by Ar and N_2 show (Padial *et al.*, 1981; Cartwright *et al.*, 1977a; Fliflet *et al.*, 1979) that qualitatively correct results are predicted only by theories that are at the distorted-wave level of sophistication or better, and this will be substantiated by further comparisons to be made below.

Finally, it should be emphasized that reliable theories to treat inelastic electron–molecule scattering are needed, not only to explain and interpret experimental data, but also because there are a number of interesting molecules that are difficult to study experimentally for which the scattering information is important. Examples are the chemically reactive molecules (e.g., CN, CH, OH) that are of considerable interest in astrophysics; transitions from the excited states of the excimer molecules such as XeCl, HgBr, etc.; and metastable species such as $N_2(A\,^3\Sigma_u^+)$ and $O_2(a\,^1\Delta_g, b\,^1\Sigma_g^+)$.

A. Vibrational Excitation

The earliest calculations of nonresonant vibrational excitation of molecules appear to have been performed by Morse (1929), by Massey (1935), and subsequently by Wu (1947) and Carson (1954) in attempts to explain the observed large fractional energy loss of electron swarms in molecular hydrogen. These authors all used variations of the Born approximation but none of the models could explain the large fractional energy loss observed in the swarm measurements for any value of the mean electron energy. Gerjuoy and Stein (1955) subsequently showed that including the interaction of the incident electron with the quadrupole moment of homonuclear molecules (i.e., rotational excitation) produced theoretical results which then agreed with experimental data for low values ($E \lesssim 1.5$ eV) of the mean electron energy of electron swarms in H_2 and N_2. Haas (1957) made the important observation that the large vibrational-excitation probability observed in N_2 for characteristic mean energies near 2 eV could be explained by a "compound-state" model for the target plus electron. This characteristic has indeed turned out to be generally correct and the highest probability for vibrational excitation of many molecules is via the resonance process. Since the resonant process in molecules is the subject of another chapter in this review, attention here will be directed only to what is currently known about *nonresonant* vibrational excitation.

Nonresonant vibrational excitation of molecules also occurs because of the coupling between the electronic and nuclear degrees of freedom of the target molecule. The theory for this process has been well reviewed by Takayanagi (1967) by Schulz (1976) and for polar molecules by Norcross and Collins (1982), so only the recent developments will be discussed in this section. The theory is based on expressing the interaction potential between the incident electron and the target molecule as a multipole expansion which includes the electrostatic terms associated with the unperturbed molecule (e.g., dipole, quadrupole) and the dynamic terms arising from the polarization of the target molecule by the incident electron. In practice, not all the interaction terms can be included in a "full" calculation for all incident electron energies because of the computational difficulties (Norcross and Collins, 1982; Buckley *et al.*, 1983) associated with the coupling of all the energetically open channels. With the availability of the first measured differential cross sections for vibrational excitation of H_2 in 1969, it became possible to examine the accuracy of the various scattering approximations. The first published calculation of a differential cross section for vibrational excitation of a molecule containing comparison to experiment appears to have been by Trajmar et al. (1970), who employed a plane-wave approximation and included lowest-order exchange and polarization effects. Henry and Chang (1972) reported differential cross sections obtained using a static plus exchange model with an effective polarization term containing as adjustable parameter. Klonover and Kaldor (1979) have reported the first results obtained using a completely *ab initio* theory including direct, exchange, and polarization terms in an optical potential. The comparison of their results with experiment shows clearly that target polarization effects need to be included in order to obtain even semiquantitative agreement with experiment. A summary of the previous calculations reported to date for predicting nonresonant vibrational excitation in N_2 is contained in recent papers (Onda and Truhlar, 1980; Jung et al., 1982) in which the authors reported calculations employing the vibrational "sudden" approximation, rotational close coupling, and various effective potentials; and by Choi et al. (1979), who applied a two-potential model to treat the scattering. The differential cross sections that both sets of authors report are in reasonably good agreement with the measurements in N_2 at 50 eV incident electron energy but still show substantial disagreement at small and large scattering angles. The same series of models has been applied (Thirumalai et al., 1981) to treat excitation of the asymmetric stretch in CO_2 at 10 eV and the results are in better agreement with the available data than in the case of N_2. However, these models need to be extended to treat the other vibrational modes of CO_2 and to a wide range of incident electron energies before the general applicability of these models can be established. Comparisons of all the theoretical results with the available experimental data on H_2 and N_2 suggest that considerable theoretical development is still needed before theory will be able to reliably predict (*a priori*) differential, integral, and momentum transfer cross sections for vibrational excitation of nonpolar molecules.

Although there has also been considerable interest in the past few years in electron scattering by polar molecules, most of the experimental and theoretical work has been directed to understanding elastic scattering and rotational excitation rather than vibrational or electronic excitation. Since a thorough review of electron scattering by polar molecules has just been completed by Norcross and Collins (1982), only a few summary comments on the status of theory for treating vibrational excitation of polar molecules will be given here. From theoretical and computational points of view, the effect of a permanent dipole moment in the target molecule dominates both the inelastic and elastic scattering. At this time, there are still relatively few experimental data that can be clearly interpreted as pure vibrational excitation of a polar molecule. Furthermore, for these few data that are available, it has not yet been possible to clearly separate the resonant and nonresonant contributions to the vibrational excitation. It appears that significant progress in our understanding of vibrational excitation of polar molecules must await more experimental data on a variety of polar molecules, over a wider range of incident electron energies, from which the relative contributions of resonant and nonresonant scattering can be determined.

B. Electronic Excitation

Theoretical models to reliably predict differential and integral cross sections for electronic excitation of molecules in the medium- and low-energy regions continue to elude theorists although considerable progress has been made in the past ten years. The origin of most of the difficulty in electron molecule scattering can be traced to the computational problems associated with accurately representing the many-electron, nonspherical, electronic, and nuclear force fields experienced by the incident and scattered electrons. That is, the theoretical techniques that are being employed for treating the scattering physics of electronic excitations of molecules are all direct applications of techniques originally derived to treat excitation of atoms (Bransden and McDowell, 1978; Callaway, 1980), but the nonspherical nature of the molecular force field generally requires that more severe mathematical approximations be employed to evaluate numerically the T-matrix than in the atomic case.

Before going into detailed comparisons between theory and experiment, a few comments are made about the strengths and weaknesses of the various theories (sometimes called models) now used to treat electron-impact excitation of atoms and molecules.

1. Theories Lowest-Order in the Interaction Potential

The Born, Born–Oppenheimer, Ochkur–Rudge, and impact-parameter models are all variations of the same philosophy based on retaining just the leading term in a power-series expansion of the scattering T-matrix with respect to the incident electron energy. As expected, these approximations work best at high incident energy ($E_0 \gtrsim 1000$ eV) but the energy at which they begin to break down (as the incident energy decreases), appears (Moiseiwitsch and Smith, 1968) to depend on the specific target and also on the symmetry of the electronic states involved. Furthermore, all "lowest-order"-type theories predict qualitatively, as well as quantitatively, incorrect DCSs. As shown in Fig. 6 this failure of lowest-order theories to predict the correct DCSs holds both for atomic targets such as argon and for molecular targets such as N_2.

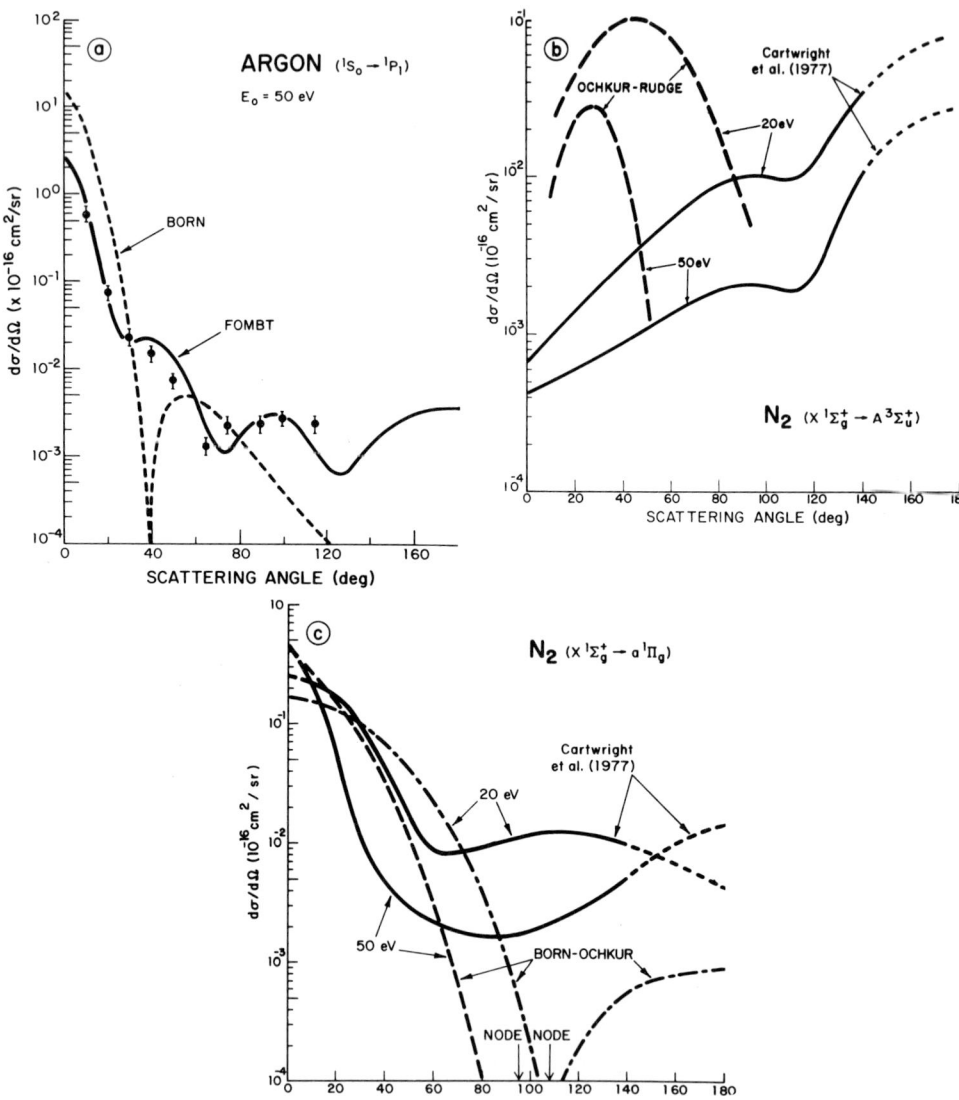

Fig. 6. Differential cross sections for excitation of (a) the $^1P_1$ state in atomic argon, and (b) the $A\ ^3\Sigma_u^+$ and (c) a $^1\Pi_g$ states in molecular nitrogen, at the incident electron energies indicated in the figure. The results from Born, Ochkur–Rudge, and Born–Ochkur scattering models are compared with experimental data in panels (a), (b), and (c), respectively, and these comparisons show the inability of Born-type theories to predict the correct differential cross sections. The argon results are from Padial *et al.* (1981) and the N_2 results from Cartwright *et al.* (1977a).

This characteristic is a direct result of the fact that lowest-order theories contain *only* the collision physics appropriate at high electron energies ($E \approx 1$ keV), and therefore cannot be expected to contain the correct physics at low and medium electron energies. In the case of an optically allowed transition in argon, Fig. 6 shows that the DCS predicted by the Born theory overestimates the small-angle scattering, has a shape which only crudely resembles the measured shape, and is in error by orders of magnitude at intermediate and large scattering angles. For N_2, the similar inadequacy in lowest-order theories is found for both symmetry-forbidden (a $^1\Pi_g$) and pure exchange (A $^3\Sigma_u^+$) transitions by electron impact from the ground electronic state (X $^1\Sigma_g^+$).

However, results continue to appear on electronic excitation of atoms and molecules obtained by use of lowest-order theories because

(i) such theories are substantially easier to apply computationally, and

(ii) even though the DCSs are known to be qualitatively incorrect, the integral cross sections will occasionally be only a factor of 2–5 larger than the true cross section, even in the medium-energy range.

The reason why scattering models which predict qualitatively incorrect DCSs can sometimes predict integral cross sections within a factor of 2–5 of experiment is explained as follows. The integral cross section σ is obtained from the DCS ($d\sigma/d\Omega$) by quadrature according to Eq. (3), which, because of the $\sin\theta$ factor in the integration element, emphasizes the region of $d\sigma/d\Omega$ near $\theta = 90°$ more than those near $0°$ or $180°$. The relationship in Eq. (3) shows that the integral cross section contains no physics that is not included in the DCSs; i.e., $d\sigma/d\Omega$ contains all the physics of the scattering event. One also observes that any quantitative agreement between integral cross sections obtained from lowest-order theory and experiment is likely to be strictly fortuitous because, as shown in Fig. 6 and discussed below, lowest-order theories predict quantitatively incorrect DCSs. The fortuitous cases of agreement between lowest-order integral cross sections and experiment are generally unpredictable and are due to a combination of the particular shape of the DCS and the nonlinear $\sin\theta$ weighting factor in Eq. (3). In general, lowest-order models for electron-impact excitation of optically allowed transitions substantially overestimate the small-angle DCS, underestimate at all other angles, and occasionally (but only for certain energies) the $\sin\theta$ weighting factor compensates just right for these two large errors of opposite effect to produce a correct integral cross section.

a. Impact-Parameter Model

The impact-parameter model (IPM), a semiclassical approach to the electron–molecule scattering, contains somewhat less physics for the electron scattering process than the Born approximation because straight-line electron trajectories are usually assumed to simplify the calculations and an arbitrary cutoff parameter has to be introduced into the T-matrix in order to avoid divergent results (Hazi, 1981). A method of renormalizing the integral cross sections obtained from the IPM to optical oscillator strengths at high incident electron energies has recently been reported (Hazi, 1981) which results in the IPM giving integral cross sections to about the same accuracy as the Born approximation, that is, a factor of 2–5 larger than experiment. However, for the reasons mentioned above, DCSs obtained by the IPM are qualitatively as well as quantitatively incorrect and the integral cross sections are usually as large or larger than Born results.

b. Born and Born-Type Models

The Born approximation, and its many variations to treat electron exchange (e.g., Born–Oppenheimer, Ochkur–Rudge, first-order exchange, etc.), has applied to electron-impact excitation of molecules far more than any other scattering model. Although originally derived as a high-energy approximation (Moiseiwitsch and Smith, 1968), it is still routinely used to obtain integral cross sections at low and medium energies. This is probably due to the facts that these approximations are relatively easy to apply computationally and, prior to 1977, there were no extensive experimental studies of the DCSs for excitation of molecules against which the accuracy of the theory could be calibrated (Cartwright *et al.*, 1977a). In fact, before 1977 the only published comparison between theoretical and experimental DCSs for molecular excitation was reported (Trajmar *et al.*, 1968) for a single transition ($X\ ^1\Sigma_g^+ \to b\ ^3\Sigma_u^+$) in H_2 which actually suggested that the Ochkur–Rudge approximation was suprisingly better than one had reason to expect. In 1977, a relatively thorough study of the differential and integral cross sections for excitation of 19 electronic states of N_2 was reported in a series of three papers (Cartwright *et al.*, 1977a, b; Chutjian *et al.*, 1977) which contains detailed comparisons between experiment and the results of lowest-order theory. These comparisons strongly suggest (e.g., Fig. 6) that lowest-order theories cannot reliably predict DCSs in the low- and medium-energy regions. This conclusion also holds for pure exchange transitions even though the Ochkur–Rudge approximation, for reasons which are not clear at this moment (see below), gave results in good agreement with experiment for the $X\ ^1\Sigma_g^+ \to b\ ^3\Sigma_u^+$ transition in H_2. Extensive comparisons between experimental data acquired in the past five years and results from Born and Born-related approximations show conclusively that these models do not predict reliable reliable DCSs, and generally overestimate integral cross sections, except at high incident electron energies.

c. *Eikonal (e.g., Glauber) Approximations*

The eikonal approximations, which are generalizations of the Glauber approximation (Glauber, 1959) first used in nuclear physics, represent an improvement over the first-order approximations in that they can be extended to lower incident electron energy but, although this model has not been used much to study excitation of molecules, some insight as to its usefulness can be gained from its application to excitation of atoms. For incident electron energies three or four times threshold, the predicted differential cross sections for excitation of atoms are in better agreement (Bransden and McDowell, 1978) with experiment for scattering angles less than about 50° than those obtained from the Born approximation. However, for excitation of helium, the eikonal methods do not predict the observed deep minima (at 50°) in the 2^1S excitation cross section at 40 eV, nor do they predict the observed increase in backward scattering. A serious defficiency in the original Glauber theory was its inability to treat electron exchange, and some efforts to correct this have been made (Bransden and McDowell, 1978). Consequently, although the eikonal approximations do represent an improvement over first-order theories (Burke, 1972), they do not contain the essential physics necessary to reliably predict integral and differential electron scattering cross sections in the medium-energy region for atoms and should not be expected to do any better for molecular targets.

2. *Distorted-Wave Approximation*

The distored-wave (DW) approximations are well known in both nuclear and atomic physics, having been employed (Mott and Massey, 1965) in one form or another since the late 1930s. In one particular form, namely, for the incident and scattered electron moving in the field of the ground state of the target, it is equivalent to the first-order many-body theory, which has been shown to predict reasonably accurate differential and integral cross sections for excitation of helium, (Thomas *et al.*, 1974) neon, and argon (Padial *et al.*, 1981). This form of the DW approximation has more recently been applied to treat electronic excitation of H_2, N_2, CO, F_2, and CO_2 (see Table IV) and, where comparisons can be made, the agreement is encouraging. This form of the DW approximation appears to contain the minimum physics required to predict DCSs in qualitative agreement with experiment.

3. *Close-Coupling and Related Methods*

The close-coupling (CC) method for treating electron–atom scattering has been highly successful in predicting the resonant structure frequently found in atomic targets and is just beginning to be applied to electronic excitation of molecules. Two-state CC calculations have been reported by Chung and Lin (1978) and by Weatherford (1980) for selected transitions in H_2 and by Holley et al. (1981) for excitation of the a $^1\Pi_g$ state in N_2. The agreement between the results from close-coupling and distorted-wave calculations is not particularly good (see below for details) for the one transition that has been considered by both (b $^3\Sigma_u^+$ in H_2). Unfortunately, the CC results (Holley et al., 1981) reported for N_2 did not include DCSs, so the accuracy of the two-state CC approximation for molecular excitation cannot be very well assessed.

A great deal of work has been devoted to obtaining satisfactory solutions to the CC equations for the case of atomic excitation and this effort has spawned a number of formal and computational approximations. Above the ionization thresholds, the CC approximation converges slowly, or inadequately represents physically important terms such as polarization, because of the infinite number of open channels that are present. Awesome computational problems (Burke, 1971–1972) associated with "psuedostates," and the large number of excited states that need to be coupled in some cases, have resulted in certain modifications of the equations in order to get results. The problems associated with solving the CC equations for molecular targets are substantially more severe (Buckley et al., 1983) than for atomic targets. This is the primary reason that no calculation more sophisticated than the DW and two-state CC calculations have been reported for electron-impact excitation of molecules. However, there is currently considerably more effort being devoted (Norcross and Collins, 1982; Buckley et al., 1983) to obtaining more complete solutions of the CC equations than in previous years and consequently the prospect of going beyond the two-state CC approximation for electronic excitation of molecules in the next five years is quite good.

IX. Comparison of Theory and Experiment

In the following paragraphs, comparisons will be made between theoretical and experimental differential and integral cross sections for vibrational and electronic excitation in order to assess the reliability of the theoretical cross sections.

A. Vibrational Excitation

Differential cross sections for vibrational excitation for a wide range of incident electron energies have been directly measured (i.e., as opposed to inferred from other data) for very few molecules (see Table I) and attention here will be directed to the ability of theory to predict the DCSs for vibrational excitation.

2 Excitation of molecules by Electron Impact

1. *Molecular Hydrogen*

Since H_2 has the largest vibrational spacing and the smallest number of electrons among molecules, it is not surprising that considerable experimental and theoretical work has been done on it. As stated in a previous section, significant progress in understanding the most important physical interactions (and hence developing good theories) for nonresonant vibrational excitation of molecules was not made until after 1970, when good experimental differential cross sections were reported. Differential cross sections for vibrational excitation of H_2 were reported first by Ehrhardt *et al.* (1968), by Trajmar *et al.* (1970), and by Linder and Schmidt (1971a).

The distinction between resonant and nonresonant vibrational excitation of H_2 is not at all clear at the moment and, because of the very short lifetime of the H_2^- state, such a distinction may not be possible. In fact, theory has been fairly successful in explaining the data without a resonance component (Schultz, 1976).

Figure 7 illustrate DCSs for excitation of the first vibrational level in molecular hydrogen for incident electron energies of 3.5 and 10.0 eV. The most

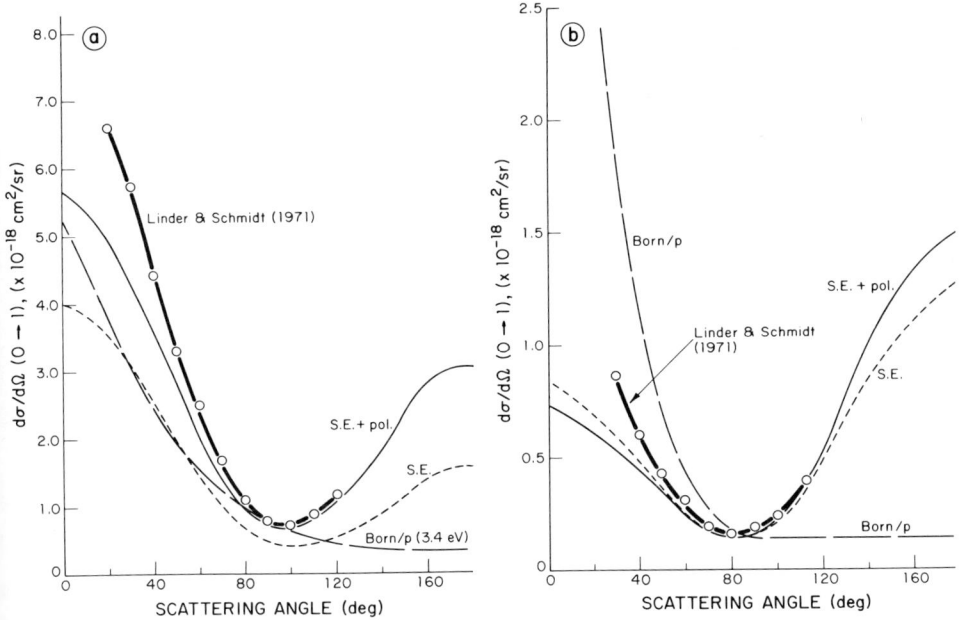

Fig. 7. DCSs for excitation of the first vibrational level in H_2 for incident electron energies of 3.5 (a) and 10 eV (b). The experimental data are from Linder and Schmidt (1971) and theoretical results from Born plus polarization (Born/p) (Trajmar *et al.*, 1970), static exchange (S.E.) (Klonover and Kaldor, 1979), static exchange plus polarization (S.E. + pol.) (Klonover and Kaldor, 1979) models are shown for comparison. See text for details.

complete *ab initio* theoretical method applied to date to treat vibrational excitation has been by Klonover and Kaldor (1979), and their results agree well with experiment at immediate scattering angles ($50° \lesssim \theta \lesssim 120°$). For scattering angles less than about 50°, their results generally underestimate the DCS at all incident electron energies. For scattering angles greater than 120°, the reliability of theory cannot yet be established because there are no data for comparison in that angular range. Comparing the results of other theories to experimental data (and to the results of Klonover and Kaldor) shows that certain physical effects are essential in a theory for nonresonant vibrational excitation if its results are to agree with experiment. First, comparing the results obtained with a static exchange (SE) model to experiment shows that polarization effects are essential to obtain semiquantitative agreement with experiment at all scattering angles except near 90°. Second, "lowest-order" approximations such as the Born model, even with a polarization potential, are in qualitative agreement only at small scattering angles.

The general conclusions stated above also hold for the integral cross sections as illustrated in Fig. 8. Since the SE model and *ab initio* calculations

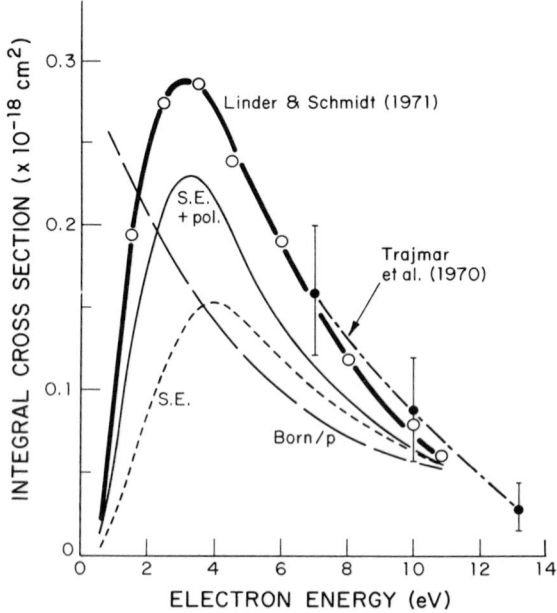

Fig. 8. Integral cross section for excitation of the first vibrational level in H_2. The experimental data are from Linder and Schmidt (1971a) and Trajmar *et al.* (1970). The theoretical results are from the same models as in Fig. 7 and are discussed in the text.

both underestimate the forward scattering (and the backward scattering for the SE model) and are generally smaller than experiment at all angles, their corresponding integral cross sections are also smaller than experiment. The Born/p integral cross section is overall in relatively poor agreement with the experimental integral cross section.

It should be noted that vibrational excitation of H_2 has a strong dependence on the rotational quantum number for which the relative importance appears to be correctly accounted for by the electric-quadrupole rigid-rotor model of Gerjuoy and Stein (1955). For treating nonresonant vibrational excitation, this rotational quantum number dependence is probably important only for H_2 because of the relatively large energy spacing of the rotational levels.

2. *Molecular Nitrogen*

To date, no calculation has been reported for nonresonant vibrational excitation of N_2 employing fully *ab initio* static plus exchange plus polarization potential as was done by Klonover and Kaldor for H_2. This is probably due to the fact that most of the recent theoretical works have been devoted to developing models to explain the resonant path for vibrational excitation, which is the dominant mechanism. Onda and Truhlar (1980) have reported results from calculations using a variety of effective polarization potentials combined with rotational CC and the vibrational sudden approximation and compared them to the reanalyzed experimental data of Srivastava *et al.* (1976) and the work of Chandra and Temkin (1976). Figure 9 contains a comparison between theory, the reanalyzed data of Srivastava *et al.* (1976), and the newer results of Tanaka *et al.* (1981) and shows good agreement at 10 eV incident electron energy but relatively poor agreement at 50 eV for the model potential results of Onda and Truhlar (1980). Choi *et al.* (1979) have recently reported N_2 vibrational excitation results from an interesting two-potential model for the scattering process which appears to work as well in predicting the DCS for the $0 \to 1$ excitation in N_2 at 50 eV as the model of Onda and Truhlar (1980). Additional detailed model calculations and some using *ab initio* static plus exchange plus polarization potentials (analogous to the one done by Klonover and Kaldor) are needed to determine the dominant physics at the higher impact energy.

3. *Other Molecules*

Relatively little theoretical work has been done on nonresonant vibrational excitation of other molecules. The effective polarization potential method has been recently applied (Thirumalai *et al.*, 1981) and Rumble *et al.* (1981) to treat excitation of the asymmetric stretch in CO_2 at 10 eV and the results were found to be in reasonably good agreement with the recent data of Register

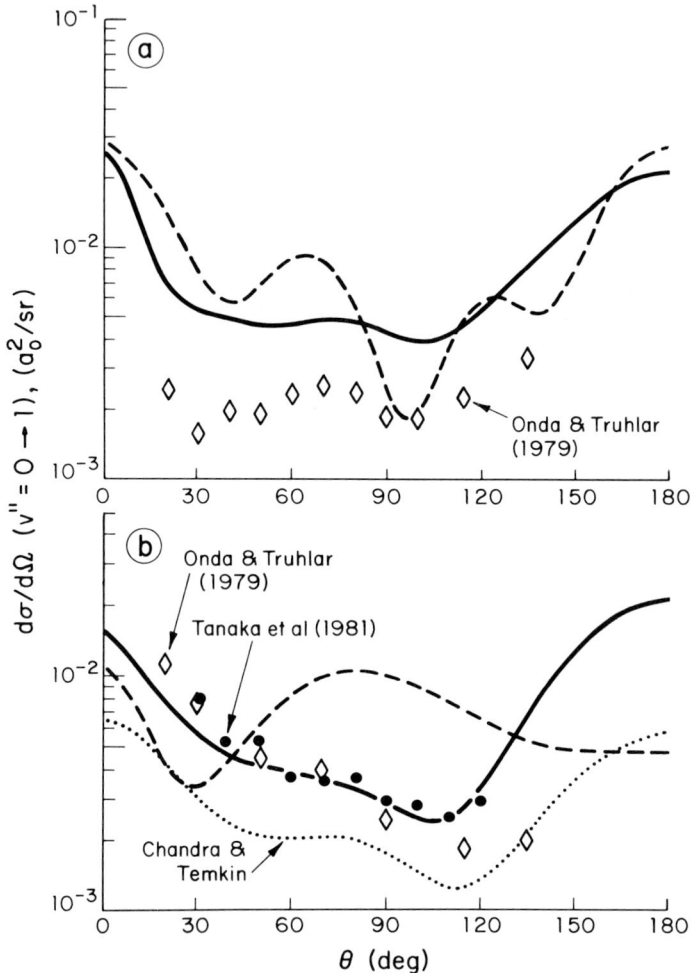

Fig. 9. DCSs for excitation of the first vibrational level in N_2 for 50 (a) and 10 eV (b). The experimental data are taken from Tanaka et al. (1981) and from Onda and Truhlar (1980), who renormalized the original relative N_2 vibrational DCSs of Srivastava et al. (1976) to newer absolute He elastic scattering DCSs. The theoretical curves are model potential results from Onda and Truhlar (1979, 1980) (solid and dashed lines) and from Chandra and Temkin (1976) (dotted line). See text for details.

et al. (1980a). However, although experimental data were also reported (Register et al., 1980a) for excitation of asymmetric stretch and other vibrational modes of CO_2 at higher incident electron energies, corresponding theoretical results have not yet been reported so the general usefulness of the effective polarization potential method cannot yet be established.

For vibrational excitation of polar molecules, the recent review by Norcross and Collins (1982) contains a good summary of both the theoretical and experimental work that has been done through 1981.

B. Electronic Excitation

Table IV contains a summary of the theoretical work that has been reported on electronic excitation of molecules in the low- to medium-energy region ($E \lesssim 200$ eV). For the high-energy region, the Born theory (and its variations) works well and has been the subject of a number of thorough reviews (Inokuti, 1971) and therefore will not be discussed here. We will make detailed comparisons between theory and experiment for excitation of H_2 and N_2 in order to evaluate the reliability of theory because by far the most theoretical and experimental work has been performed on these two molecules. The Franck–Condon approximation has been shown to be accurate for treating electronic excitation of molecules such as H_2 and N_2 and is assumed valid in the following discussions.

1. Molecular Hydrogen

More theoretical work has been done on electronic excitation of H_2 than on any other single molecule because of its simple electronic structure. Experimental and theoretical DCSs for the $X\,^1\Sigma_g^+ \to b\,^3\Sigma_u^+$ excitation in the medium- and low-energy region were first reported by Trajmar et al. (1968) for five incident electron energies from 25 to 60 eV. Although these DCSs are relative, the shapes were found to agree surprisingly well with those predicted by the lowest-order Ochkur–Rudge approximation. Eight years elapsed before the next calculations of the DCSs for this transition, the distorted-wave–RPA model of Rescigno et al. (1976), were reported. Although these authors did not report calculations that can be directly compared to the previously published experimental and theoretical data, inspection of their results at 20 eV suggests that there will be only qualitative agreement with the shape of the measured DCS at 25 eV. An improved DW calculation of the DCS for the $X\,^1\Sigma_g^+ \to b\,^3\Sigma_u^+$, removing some of the deficiencies of the distorted-wave–RPA model, was reported by Fliflet and McKoy (1980) as well as the first two-state CC calculation of the DCSs for this transition by Weatherford (1980). Figure 10 contains a comparison of these theoretical DCSs with the experimental results (Trajmar et al., 1968) for the $X\,^1\Sigma_g^+ \to b\,^3\Sigma_u^+$ excitation at 25 eV. Note that in this particular case, the relative experimental DCSs were least-squares normalized to the Ochkur–Rudge DCSs because there was no other absolute scale available at that time. However, comparison of the theoretical integral cross section with the

TABLE IV

Theoretical Results

Theory	Reference	Excitation from ground state to	Quantities reported	Comments
H_2^+				
Born	Kerner (1953)	$2p\sigma_u$	$\sigma(E)$	$\Psi =$ LCAO[a]
Born	Ivash (1958)	$2p\sigma_u$	$\sigma(E)$	$R = R_e$, LCAO
Born	Peek (1964)	$2p\sigma_u, 2p\pi_u, 2s\sigma_g$	$\sigma(E)$	Exact $\Psi(R)$
Born	Peek (1965)	$2p\sigma_u$	$\sigma(E)$	Exact $\Psi(R)$
H_2				
Born	Massey and Mohr (1932)	$B\ ^1\Sigma_u^+$	$\sigma(E)$	$\Psi =$ LACO
Born	Roscoe (1941)	$B\ ^1\Sigma_u^+, C\ ^1\Pi_u, D\ ^1\Pi_u, X\ ^1\Sigma_g^+$	$\dfrac{d\sigma}{d\Omega}$ (400 eV)	$\Psi =$ LCAO
First-order exchange BO-O	Khare and Moiseiwitsch (1966)	$b\ ^3\Sigma_u^+$	$\sigma(E)$	separated-atom approximation
Born–Ochkur	Khare (1966a)	$B\ ^1\Sigma_u^+, C\ ^1\Pi_u$	GOS, $\sigma(E)$	One-center Ψ
Born–Ochkur	Khare (1966b)	$D\ ^1\Pi_u$	GOS, $\sigma(E)$	One-center Ψ
Ochkur	Khare (1967)	$a\ ^3\Sigma_g^+, b\ ^3\Sigma_u^+, C\ ^3\Pi_u$	$\sigma(E)$	One-center Ψ
Born	Rozsnyai (1967)	$C\ ^1\Pi_u$	$\sigma(E)$	Small-k limit, STO[b]
Ochkur–Rudge	Cartwright and Kuppermann (1967)	$b\ ^3\Sigma_u^+, a\ ^3\Sigma_g^+$	$\sigma(E)$	Two-center Ψ, STO

Method	Reference	States	Quantity	Notes
Born	Miller and Krauss (1967)	$C^1\Pi_u, D^1\Pi_u, D'^1\Pi_u, B^1\Sigma_u^+$	GOS (300 eV)	Gaussian Ψ
Ochkur–Rudge	Trajmar et al. (1968)	$b^3\Sigma_u^+, a^3\Sigma_g^+$	$\frac{d\sigma}{d\Omega}$ (25, 35, 40, 60, eV)	Two-center Ψ, STO
Born/Ochkur–Rudge	Chung et al. (1975)	$b^3\Sigma_u^+, a^3\Sigma_g^+, c^3\Pi_u, d^3\Pi_u, e^3\Sigma_u^+$	$\sigma(E)$	Gaussian Ψ
Distorted wave	Rescigno et al. (1976)	$b^3\Sigma_u^+, a^3\Sigma_g^+$	$\frac{d\sigma}{d\Omega}, \sigma(E)$	RPA[c] transition density
Born	Domenicucci and Miller (1977)	$n\sigma_u, n\Pi_u$	GOS, $\sigma(E)$	Independent-particle Ψ
Two-state close coupling	Chung and Lin (1978)	$b^3\Sigma_u^+, a^3\Sigma_g^+, e^3\Sigma_u^+, C^3\Pi_u, B^1\Sigma_u^+$	$\sigma(E)$	Gaussian Ψ
Distorted wave	Fliflet and McKoy (1980)	$b^3\Sigma_u^+, B^1\Sigma_u^+$	$\frac{d\sigma}{d\Omega}, \sigma(E)$	Single-center expansion
Two-state close coupling	Weatherford (1980)	$b^3\Sigma_u^+$	$\frac{d\sigma}{d\Omega}, \sigma(E)$	Semiclassical exchange
Impact parameter	Hazi (1981)	$B^1\Sigma_u^+$	$\sigma(E)$	renormalized to GOS[d]
Distorted wave	Mu-Tao et al. (1982)	$C^1\Pi_u, c^3\Pi_u, B'^1\Sigma_u^+, E(F)^1\Sigma_g^+$	$\frac{d\sigma}{d\Omega}, \sigma(E)$	Single-center expansion
N_2				
Born	Rozsnyai (1967)	$a^1\Pi_g, b^1\Pi_u$	$\sigma(E)$	Small-k limit
Ochkur–Rudge	Cartwright (1970, 1972)	$A^3\Sigma_u^+, B^3\Pi_g, W^3\Delta_u, B'^3\Sigma_u^-, C^3\Pi_u, E^3\Sigma_g^+, D^3\Sigma_u^+$	$\sigma(E)$	Two-center Ψ
Born	Chung and Lin (1971)	$B^3\Pi_g$	$\sigma(E)$	See next entry

TABLE IV (cont.)

Theory	Reference	Excitation from ground state to	Quantities reported	Comments
Born/Ochkur–Rudge	Chung and Lin (1972)	$a^1\Pi_g, c'^1\Sigma_u^+, a''^1\Sigma_g^+,$ $w^1\Delta_u, b'^1\Sigma_u^+, b^1\Pi_u,$ $A^3\Sigma_u^+, B^3\Pi_g, C^3\Pi_u,$ $W^3\Delta_u, D^3\Sigma_u^+, E^3\Sigma_g^+$	$\sigma(E)$	Gaussian Ψ
Born	Domenicucci and Miller (1977)	$n\pi_g, n\pi_u, n\sigma_g, n\sigma_u$	$GOS, \sigma(E)$	Inadequate Ψ
Ochkur–Rudge	Cartwright et al. (1977a, b) Chung and Lin (1972)	$a^1\Pi_g, A^3\Sigma_u^+, W^3\Delta_u,$ $C^3\Pi_u, E^3\Sigma_g^+$	$\dfrac{d\sigma}{d\Omega}, \sigma$	Gaussian Ψ
Distorted wave	Fliflet et al. (1979)	$B^3\Pi_g, C^3\Pi_u, E^3\Sigma_g^+$	$\dfrac{d\sigma}{d\Omega}, \sigma$	Single-center expansions
Impact parameter	Hazi (1981)	$b'^1\Sigma_u^+, c'^1\Sigma_u^+$	σ	Renormalized to GOS
Two-state close coupling	Holley et al. (1981)	$a^1\Pi_g$	σ	No polarization
Distorted wave	Mu-Tao and McKoy (1983a)	$w^1\Delta_u, W^3\Delta_u, A^3\Sigma_u^+,$	$\dfrac{d\sigma}{d\Omega}, \sigma(E)$	Single-center expansions
CO				
Born/Ochkur–Rudge	Chung and Lin (1974)	$A^1\Pi, E^1\Pi, B^1\Sigma^+, C^1\Sigma^+,$ $D^1\Delta, a^3\Pi, c^3\Pi, b^3\Sigma^+,$ $j^3\Sigma^+, a'^3\Sigma^+, d^3\Delta$	$GOS, \sigma(E)$	Gaussian Ψ
Distorted wave	Mu-Tao and McKoy (1982)	$A^1\Pi, a^3\Pi, a'^3\Sigma^+, D^1\Delta, d^3\Delta$	$\dfrac{d\sigma}{d\Omega}, \sigma(E)$	Single-center expansions

O_2				
Born/Ochkur–Rudge	Chung and Lin (1980)	$B\,^3\Sigma_u^-$	GOS, $\sigma(E)$	Gaussian Ψ
F_2				
Distorted wave	Fliflet et al. (1980)	$^3\Pi_u$	$\dfrac{d\sigma}{d\Omega}$, $\sigma(E)$	Single-center expansions
CO_2				
Distorted wave	Mu-Tao and McKoy (1983b)	$^{1,3}\Sigma_u^+$, $^{1,3}\Pi_g$, $^{1,3}\Pi_u$, $^{1,3}\Delta_u$	$\dfrac{d\sigma}{d\Omega}$, $\sigma(E)$	Single-center expansions

[a] Linear combination of atomic orbitals.
[b] Slater-type orbitals.
[c] Random phase approximation.
[d] Generalized oscillator strength (GOS).

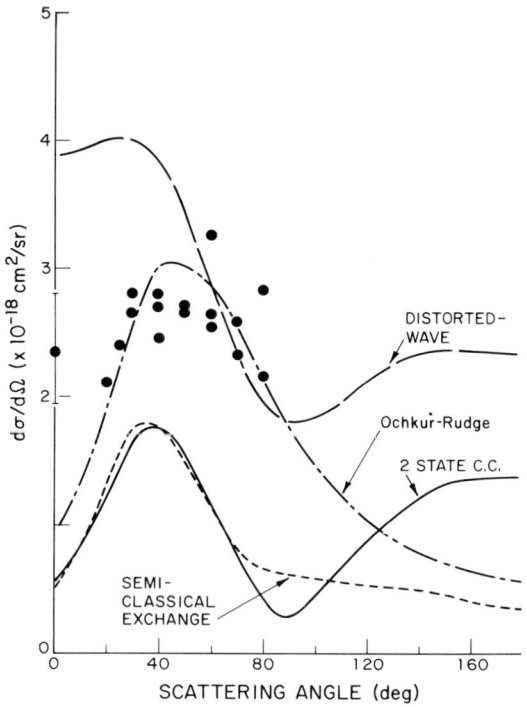

Fig. 10. DCS for excitation of the b $^3\Sigma_u^+$ electronic state in H_2, for an incident electron energy of 25 eV. The experimental data ( ) are from Trajmar et al. (1968) normalized to the Ochkur–Rudge results in a least-squares sense. The theoretical results shown are from Fliflet and McKoy (1980) (distorted wave) and from Weatherford (1980) (two-state C.C. and semiclassical exchange). See text for details.

independently measured H_2 dissociation cross section shows that this normalization procedure is reasonably accurate for this particular transition in H_2. From Fig. 10, it is seen that the DW theory overestimates the DCSs for scattering angles out to about 60°. On the other hand, the DCSs predicted by both the two-state CC and the Ochkur–Rudge model have similar shapes and are in better agreement with the data. For scattering angles greater than 90°, the shape of the DCSs predicted by the DW and two-state CC methods agree well with each other. The monotonically decreasing DCS predicted by the Ochkur–Rudge theory for scattering angles greater than 90° is a general characteristic of "lowest-order" models and is not correct for most transitions. The two-state CC results obtained using a semiclassical exchange potential are included in Fig. 10 because comparison with the results from the other models (and experimental data to be discussed below on N_2) strongly suggests that exchange must be treated at the DW level or higher in order to obtain the correct shape for the DCS for scattering angles greater than 90°. The fairly large differences between the DCSs predicted by the DW and two-state CC models is somewhat puzzling because a two-state CC approximation is only one version of the DW model. As is well known, however, there is no single DW approximation because a variety of approximations can be used to represent the distortion of the incident and scattered electrons, all of them equally valid. Based on recent results (Thomas et al., 1974; Padial et al., 1981) obtained by applying the first-order many-body theory (FOMBT), Fliflet and McKoy have used an algorithm in which both the incident- and scattered-electron wave functions are distorted by the field of the ground-state target. That is, final-state distortion effects are incorporated in the many-body formalism at the level of second order and there is no ambiguity as to how to treat distortion. The FOMBT has been very successful in predicting quantitatively accurate DCSs for excitation of He (Thomas et al., 1974), Ne, and Ar (Padial et al., 1981), but, as will be shown below, it does not appear to work quite as well for transitions in H_2 and N_2. It is likely that the various additional computational approximations and series truncations associated with application of the theory to treat the non-spherical molecular target (as opposed to atomic targets) are responsible for the somewhat poorer agreement with experiment.

Figure 11 contains comparisons of the integral cross sections for the excitation $X\ ^1\Sigma_g^+ \to b\ ^3\Sigma_u^+$ as obtained by the same models discussed above. Also shown in this figure is another integral cross section obtained by a two-state CC model reported by Chung and Lin (1978). Unfortunately Chung and Lin did not report DCSs, which are the more fundamental quantities for evaluating scattering theories, and their integral cross section is about 50% larger and has a different energy dependence than the results of Weatherford (1980). This differencce must manifest itself in the DCSs and a comparison with the results of the other theories would have been interesting. The experimental cross section shown in this figure was obtained from the analysis of Cartwright and Kuppermann (1967) and includes a small contribution from excited atom fragments as well as ground-state fragments. It has been shown (Chung and Lin, 1978) that cascade from a $^3\Sigma_g^+$ and c $^3\Pi_u$ electronic states into the b $^3\Sigma_u^+$ are major contributors to the dissociation of H_2 into ground-state fragments and, when these integral cross sections obtained by the two-state CC approximation of Chung and Lin (1978) are added together, reasonable agreement with the experimental dissociation cross section (Fig. 11) is obtained. This result is further indication that the DW theory as applied by Fliflet and McKoy (1980) overestimates the cross sections for excitation of the triplet states of H_2.

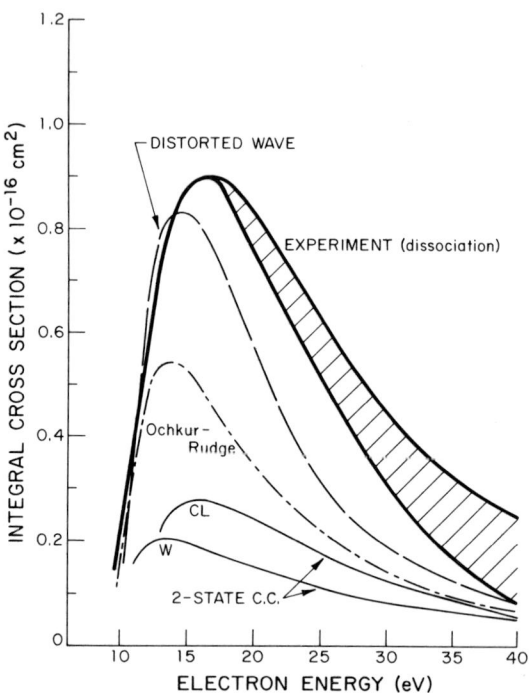

Fig. 11. Integral cross section for electron-impact excitation of the b $^3\Sigma_u^+$ state in H_2. The experimental curve shown is the total for dissociation into neutral fragments and was taken from (Cartwright and Kuppermann, 1967) and serves as an upper limit to the b $^3\Sigma_u^+$ cross section. The shaded region denotes the range of uncertainty in the dissociation cross section. The theoretical results shown are from Fliflet and McKoy (1980) (distorted wave), Cartwright and Kuppermann (1967) (Ochkur–Rudge), Weatherford (1980) (W, two-state C.C.), and Chung and Lin (1978) (CL, two-state C.C.). See text for details.

The only DCSs for excitation of the singlet states of H_2 that have been reported are the early Born approximation results by Roscoe (1941) and the recent DW results by Fliflet and McKoy (1980) on excitation of the B $^1\Sigma_u^+$ state and the calculations of Mu-Tao *et al.* (1982) for excitation of the C $^1\Pi_u$, B' $^1\Sigma_u^+$, and E(F) $^1\Sigma_g^+$ states. Chung and Lin (1978) have reported integral cross sections for excitation of the B $^1\Sigma_u^+$ state obtained by the two-state CC method but unfortunately did not report any DCSs so that the most important comparison between these two models cannot be made at this time. Figure 12 contains a comparison between the experimental DCSs of Srivastava and Jensen (1977) and the theoretical results of Fliflet and McKoy (1980) for the X $^1\Sigma_g^+ \to$ B $^1\Sigma_u^+$ ($v' = 2$) excitation in H_2, at six incident electron energies. The DCSs predicted by the Born-type models are seen to be in qualitative and

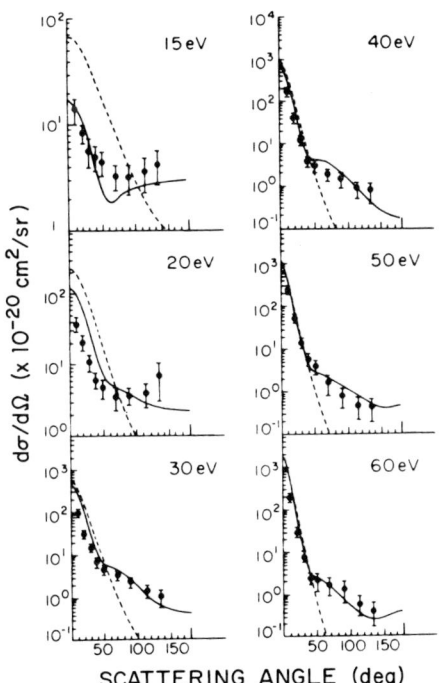

Fig. 12. DCSs for excitation of the B $^1\Sigma_u^+$ ($v' = 2$) electronic sate of H_2 for six incident electron energies, as taken directly from Fliflet and McKoy (1980). The experimental data are from Srivastava and Jensen (1977), the dashed lines are from the Born approximation, and the solid lines are from the distorted-wave model. Born-type approximations do their best for optically allowed transitions (such as this one), for scattering angles less than about 50°, and for the higher incident electron energies.

quantitative disagreement with the experimental results for scattering angles greater than about 50°, for all incident electron energies. The DCSs predicted by the DW model are generally in good qualitative agreement with the experimental data at all incident electron energies and in quantitative agreement for energies of 50 eV and greater. The DW model generally overestimates (by as much as a factor of 2) the forward scattering part of the DCS and underestimate the backward scattering for incident electron energies less than 50 eV. In general, however, as was previously found to be the case of excitation of the inert gases, a DW model which incorporates the distortion and orthogonality effects that follow (Padial et al., 1981) automatically from the FOMBT predicts DCSs which agree reasonably well with experiment for optically allowed transitions as shown in Fig. 12.

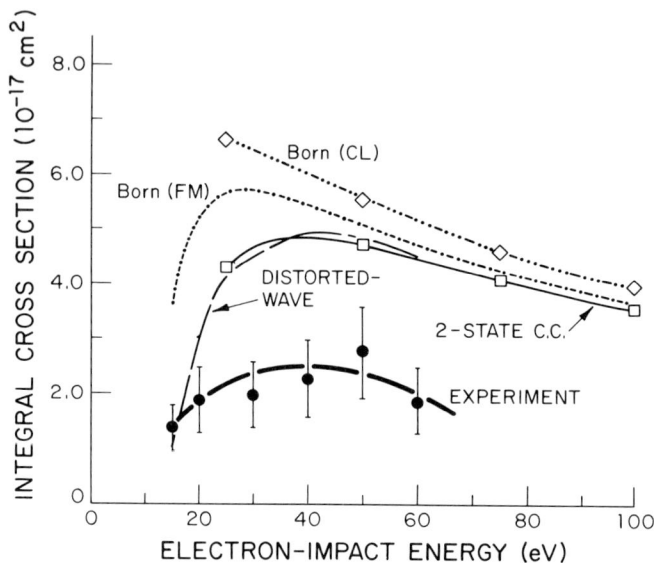

Fig. 13. Integral cross sections for excitation of the B $^1\Sigma_u^+$ state of H_2. The experimental data are from Srivastava and Jensen (1977) and the results from various theoretical models are by Fliflet and McKoy (1980) [Born (FM) and distorted wave] and by Chung and Lin (1978) [Born (CL) and two-state C.C.]. The differences between the two sets of Born results are due, at least in part, to differences in the bound state.

Figure 13 contains a comparison of the measured and various theoretical integral cross sections for excitation of the B $^1\Sigma_u^+$ state of H_2. As seen from the figure, the results from the DW and two-state CC models agree very well, which is probably a result of the fact that the optically allowed character of the transition (i.e., the direct part of the potential is long range) so dominates the scattering process that other differences in the two calculations have a small effect. (This leads one to speculate that the DCSs predicted by the two-state CC model would also agree very well with those shown in Fig. 12 from the DW model.) The fact that the integral cross sections for both models are more than a factor of 2 larger than experiment is due to the fact that both models overestimate the forward scattering by about that amount. That is, for optically allowed transitions, the integral cross section is primarily determined by the strong forward peak in the DCS. This deficiency in DW-type models (e.g., two-state CC) is largely corrected when dynamic polarization is included in the formulation. This physics is automatically included as a second-order effect in the many-body formulation of electron scattering (Csanak et al., 1970) and is also contained in CC calculations that include more than two states. Born integral cross sections, as calculated by Fliflet and

McKoy (1980) and by Chung and Lin (1978), are also shown in Fig. 13 and are both larger than the DW and CC results, particularly at incident electron energies less than 50 eV as expected. Although not shown in Fig. 13, the integral cross section obtained in the modified impact-parameter model of Hazi (1981) lies close to the Born (FM) results and similarly overestimates the integral cross section by a factor of about 2.5. The integral cross sections obtained for this excitation by Domenicucci and Miller (1977) using the Born approximation and independent-particle wave functions are about a factor of 2 larger than the other Born results and hence are not shown in Fig. 13.

For molecular hydrogen, the following statements summarize the status of theory for treating electronic excitation. The "lowest-order" theories give qualitatively correct DCSs for scattering angles less than about 50° and surprisingly accurate DCSs for the one pure exchange transition (b $^3\Sigma_u^+$) for which sufficient data exist for comparison. Perhaps more surprising is the qualitative (for $\theta \lesssim 40°$) and quantitative (all angles) differences between the results from the DW and the two-state CC models, which both contain the same major physical interactions. Quantitatively, the DW results are more than a factor of 4 larger than the results from the two-state CC model and surprisingly larger than the results from the lowest-order Ochkur–Rudge model. Integral cross sections from two different two-state CC calculations also differ considerably. These comparisons indicate that substantially more work remains to be done computationally so that differences in calculations containing different scattering physics can be assigned to the different scattering physics rather than computational differences.

2. *Molecular Nitrogen*

There are more experimental data on the electronic excitation of molecular nitrogen by electron impact than any other atom or molecule. This work was motivated by the very important role played by (e, N_2) processes in a variety of atmospheric phenomena and was made possible by the fact that it is easy to handle experimentally. The most recent data include differential and integral cross sections for excitation from the ground state to

(a) nineteen different singlet and triplet electronic states including both valence and Rydberg character;
(b) final electronic states for which the excitation process is
  (i) pure electron exchange,
  (ii) dipole allowed, and
  (iii) symmetry forbidden;
(c) for incident electron energies from just above the electronic thresholds to 50 eV.

Because of the existence of these data for a wide variety of final-electronic-state symmetries, bonding character, and incident electron energies, this molecule is one of the best for thoroughly testing various scattering models.

a. Differential Cross Sections

Figure 14 shows comparisons between the experimental DCSs and those from the Ochkur–Rudge model based on the GOS results of Chung and Lin (1972) (both from Cartwright *et al.*, 1977a) and DW models (Fliflet *et al.*, 1979; Mu-Tao and McKay, 1983a) for excitation of the A $^3\Sigma_u^+$ and C $^3\Pi_u$ electronic states of N_2, at 20 and 50 eV. Figure 15 shows the corresponding comparisons for excitation of the E $^3\Sigma_g^+$ and c' $^1\Sigma_u^+$ electronic states. Since the spin-orbit coupling in N_2 is very small, the triplet excitation processes occur by electron exchange. As seen in Figs. 14 and 15 and discussed by Cartwright *et al.*, (1977a), the Ochkur–Rudge approximation does poorly in predicting the DCSs for excitation of the A $^3\Sigma_u^+$ and E $^3\Sigma_g^+$ states and only slightly better for excitation of the C $^3\Pi_u$ state. The DW model, however, does considerably better in predicting the correct shape for the various DCSs but is somewhat too large for most scattering angles, as was also the case for an analogous transition in H_2. The comparisons shown in Figs. 10 and 14 for H_2 and N_2 are important because they show that the minimum scattering physics necessary to obtain at least qualitatively correct DCSs for exchange transitions must be at the DW level. Unfortunately, the lowest-order theories do not appear to predict qualitatively correct DCSs in the case of N_2. This fact then raises the question as to why the Ochkur–Rudge model does much better in the case of H_2 (Fig. 10) than for N_2? A plausible (but not necessarily correct) explanation is that the distortion from plane waves caused by N_2, which is substantially less spherical than H_2, is very important for obtaining the correct DCSs in N_2.

The theoretical results that are shown in Figs. 15c and 15d for excitation of the E state are from the Ochkur–Rudge model (Chung and Lin, 1972 and Cartwright *et al.*, 1977a) and from the recent DW calculations of Fliflet *et al.* (1979). For this transition, to a final electronic state of definite Rydberg character, neither of the models do very well in predicting either the shape or the magnitudes of the measured DCSs (Cartwright *et al.*, 1977a). The results from the lowest-order Ochkur–Rudge model are in the poorest agreement of all the transitions studied in N_2 but for reasons that are not obvious at this moment, although it is tempting to assign some of the difficulty to the fact that the E state is a Rydberg state. The distorted-wave theory does better in a qualitative sense but underestimates the small-angle scattering, and overestimates the large-angle scattering, by as much as an order of magnitude. As was suggested earlier (Cartwright *et al.*, 1977a), the general forward peaking of the DCSs for excitation of the E state may be associated with the accepted fact that the Rydberg electron in the E state is loosely coupled to the molec-

2 Excitation of Molecules by Electron Impact

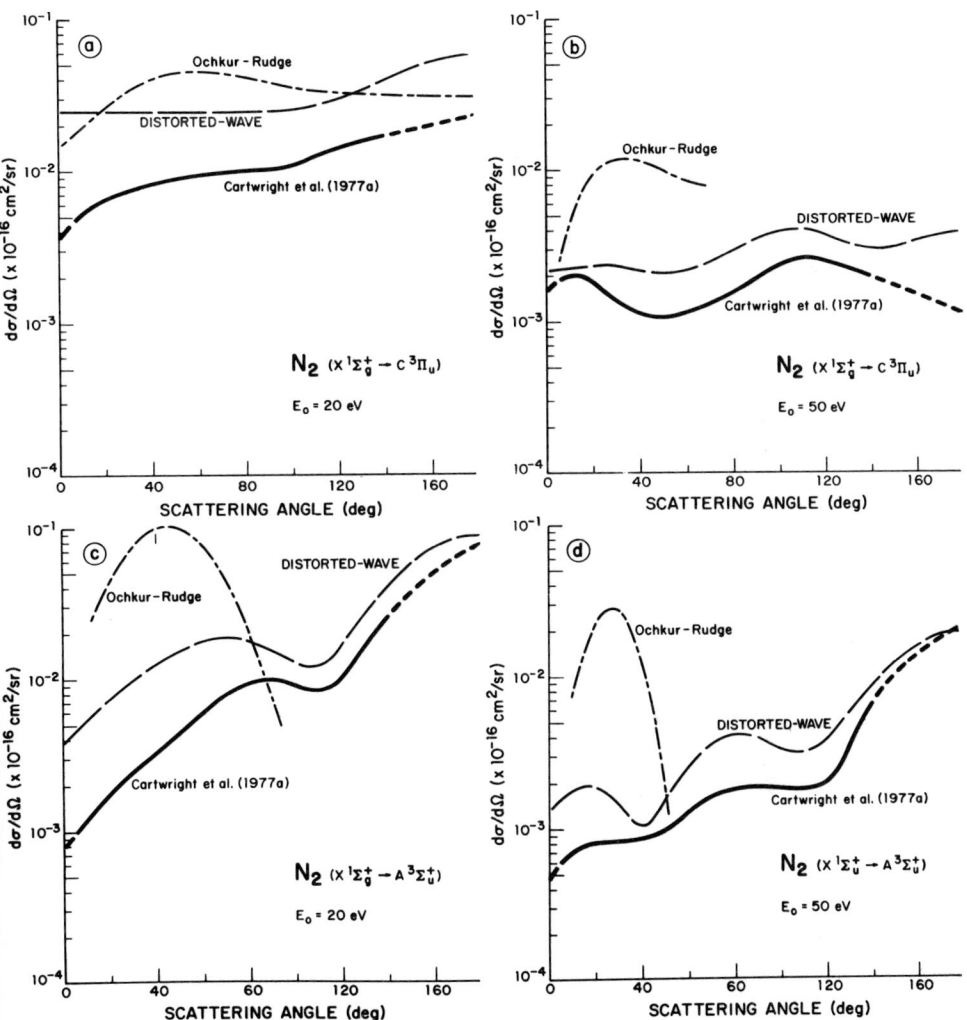

Fig. 14. DCSs for excitation of the $C\ ^3\Pi_u$ electronic state at 20 (a) and 50 eV (b), and for excitation of the $A\ ^3\Sigma_u^+$ state at 20 (c) and 50 eV (d). The heavy solid line denotes the experimental results of Cartwright *et al.* (1977a). The theoretical results are from Cartwright *et al.* (1977a) (Ochkur–Rudge) and the distorted-wave results are from Fliflet *et al.* (1979) (C state) and Mu-Tao and McKoy (1983a) (A state).

ular core. This concept was in fact suggested by the observation (Cartwright *et al.*, 1977a) that the DCSs for excitation of the $a''\ ^1\Sigma_g^+$ and $E\ ^3\Sigma_g^+$ states (both $3\sigma_g^{-1}\ 3s\sigma_g$) have very similar shapes and incident electron energy dependence.

The DW results shown in Figs. 15a and 15b for excitation of the $c'\ ^1\Sigma_u^+$

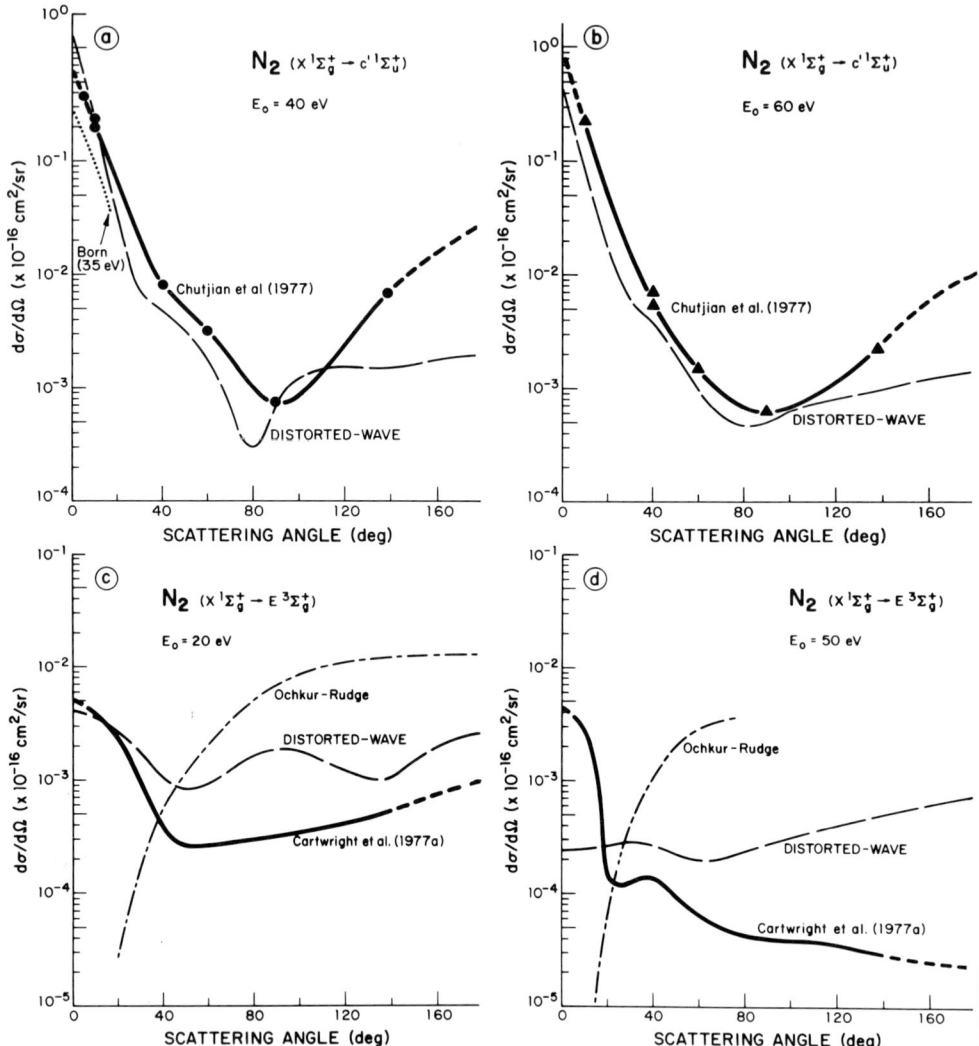

Fig. 15. DCSs for excitation of the c' $^1\Sigma_u^+$ electronic state at 40 (a) and 60 eV (b), and for excitation of the E $^3\Sigma_g^+$ state at 20 (c) and 50 eV (d). For the c' $^1\Sigma_u^+$ excitation, the experimental data are from Chutjian *et al.* (1977) and for the E $^3\Sigma_g^+$ excitation they are from Cartwright *et al.* (1977a). The distorted-wave results shown for excitation of the c' state are from Mu-Tao and McKoy (1983a) while those for excitation of the E state are from Fliflet *et al.* (1979). The curves labeled Ochkur–Rudge are from Cartwright *et al.* (1977a).

electronic state are the configuration–interaction (CI) results of Mu-Tao and McKoy (1983a). As expected, the CI-DW results agree reasonably well in shape for this particular transition for scattering angles out to about 80° but are in poorer agreement for larger scattering angles. The good agreement is expected because *all* scattering theories for electron-impact excitation do best in predicting the forward scattering for optically allowed transitions. This is a result of the fact that the scattering physics responsible for the forward scattering in optically allowed transitions is dominated by potential scattering for which large impact parameters contribute significantly. This characteristic, which has been known for over 40 years and is the essential physics behind the Born (and related) approximations (Moiseiwitsch and Smith, 1968; Inokuti, 1971), is reasonably well incorporated in the lowest-order and DW approximations. The comparison between theory and experiment for larger scattering angles, however, suggests that the scattering physics that determines the shape of the DCSs at these angles is not included at the DW level and will require a more developed theory incorporating higher-order many-body and nonlocal effects. One such physical effect known to be important is the dynamical polarization of the target (and free-electron wave function), which is not included in the customary DW model.

Figure 6c shows the DCSs for excitation of the a $^1\Pi_g$ stage, for incident electron energies of 20 and 50 eV, in order to illustrate an excitation process for which the direct and exchange contributions are both important. The results of the Born–Ochkur model (Chung and Lin, 1972 and Cartwright *et al.*, 1977a) are in fair agreement for scattering angles less than about 40° but is in poor agreement for all greater scattering angles. Although two-state CC calculations of integral cross sections for this excitation process have been reported by Holley *et al.* (1981), these authors did not report DCSs. This is unfortunate since the $X\ ^1\Sigma_g^+ \to a\ ^1\Pi_g$ transition serves as a good test of theory because the direct and exchange components of the excitation process are both important for this transition in the medium energy region.

b. Integral Cross Sections

Figures 16a and 16b show integral cross sections for excitation of the B $^3\Pi_g$ and C $^3\Pi_u$ electronic states of N_2, respectively, and Figs. 16c and 16d contain the same for excitation of the E $^3\Sigma_g^+$ and A $^3\Sigma_u^+$ states. In addition to the experimental data of Cartwright *et al.* (1977b), the earlier experimental results of Brinkmann and Trajmar (1970), Stanton and St. John (1969), Borst (1972), Borst *et al.* (1972), and of Imami and Borst (1974) are included for comparison. A discussion of the experimental results and their comparison with the results from the Ochkur–Rudge model is contained in Cartwright *et al.* (1977b) and will not be repeated in detail here. The DW results are new, however, and the

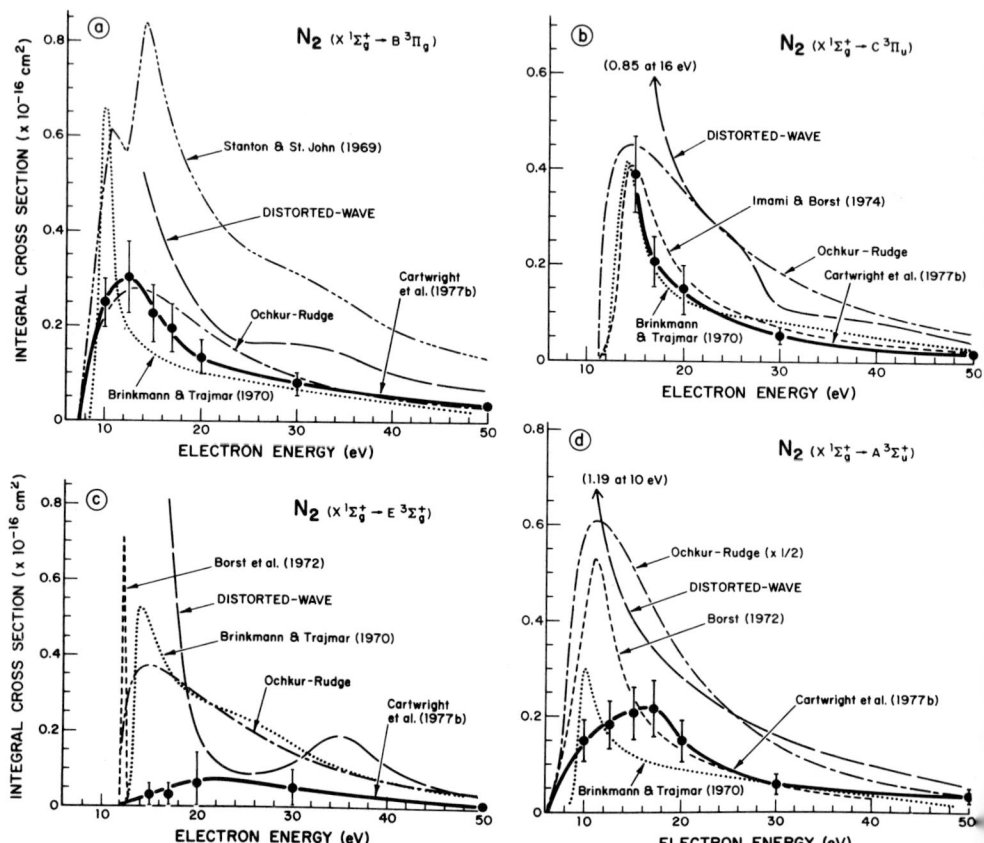

Fig. 16. Integral cross sections for excitation of the $B\ ^3\Pi_g$, $C\ ^3\Pi_u$, $E\ ^3\Sigma_g^+$, and $A\ ^3\Sigma_u^+$ electronic states of N_2, in panels (a), (b), (c), and (d), respectively. All the data shown in these four panels *except* the distorted-wave results have been discussed by Cartwright *et al.* (1977b). Note that the Ochkur–Rudge [Chung and Lin (1972)] results (d) have been reduced by a factor of 2 before plotting. The distorted-wave results shown for excitation of the B, C, and E states are from Fliflet *et al.* (1979), while those for excitation of the A state are from Mu-Tao and McKoy (1983a). The overestimation of the cross sections from this distorted-wave model is due to a shape resonance in the elastic scattering channels (Mu-Tao and McKoy 1983a). See also text.

comparison of these new results with the earlier work is worth making. For excitation of all the triplet states of N_2, the integral cross sections obtained in the DW approximation are larger than experiment and, for excitation of the C, B, and E states, even larger than the results from the Ochkur–Rudge model. It needs to be emphasized that this fact does not imply that the Ochkur–Rudge model is better than the DW model for describing pure exchange transitions in atoms and molecules because, as was shown in the section comparing the DCSs obtained by various theoretical models with experiment, the Ochkur–Rudge model predicts DCSs that are qualitatively and quantitatively incorrect for N_2. The occasional good agreement between experiment and results from the Ochkur–Rudge model for integral cross sections is therefore a manifestation of the $\sin \theta$ weighting factor discussed above. The reason why the DCSs, and hence integral cross sections, predicted by the DW theory for pure exchange transitions in N_2 (and H_2 as noted earlier) are generally too large is not clear at the moment. Since the analogous FOMBT results (Padial et al., 1981) for argon are in better quantitative agreement with experiment than the DW results are for N_2 and H_2, it is tempting to assign the relatively poorer agreement for molecular targets to the additional approximations necessary for computing the relevant quantities for molecular targets. Because of the importance of developing improved theoretical models for electron scattering processes, additional DW calculations should be performed on molecular targets to determine if the shortcomings of the DW theory identified above are intrinsic to the theory or the result of various approximations in its application to specific molecular, as opposed to atomic, targets.

Figure 17 contains a comparison of experimental and theoretical integral cross sections for excitation of the $a\,^1\Pi_g$ and $c'\,^1\Sigma_u^+$ singlet states of N_2. For excitation of the $a\,^1\Pi_g$ state, experimental data of Finn and Doering (1975) are also shown and have been discussed by Cartwright (1977b). The integral cross section for excitation of the $a\,^1\Pi_g$ state is the only electronic transition in N_2 for which CC results have been reported (Holley et al., 1981) but no DW results have been reported for this transition. As seen from Figs. 5a and 15b, the results from both the two-state CC and the Born–Ochkur models are in good agreement with experiment. The good agreement for the Born–Ochkur integral cross section is fortuitous because, as discussed above (Fig. 6c), the corresponding DCSs are in poor agreement with experiment. Unfortunately, no DCSs from the two-state CC model have been reported, so the accuracy of this model cannot be thoroughly evaluated.

For the case of excitation of the $c'\,^1\Sigma_u^+$ state, the results obtained from the CI-DW and CI-impact parameter model (Hazi, 1981) (IPM) are also shown. In the case of both theoretical calculations, the authors found it necessary to include CI with the neighboring $b'\,^1\Sigma_u^+$ in order to obtain better agreement with the experimental data. Although no DCSs obtained from the IPM have been reported, they will certainly be no better than those obtained from the Born approximation. As a consequence, DCSs obtained from the IPM will be valid only for a limited scattering angle region in the forward direction. The IPM also generally overestimates integral cross sections in much the same fashion as with the Born approximation.

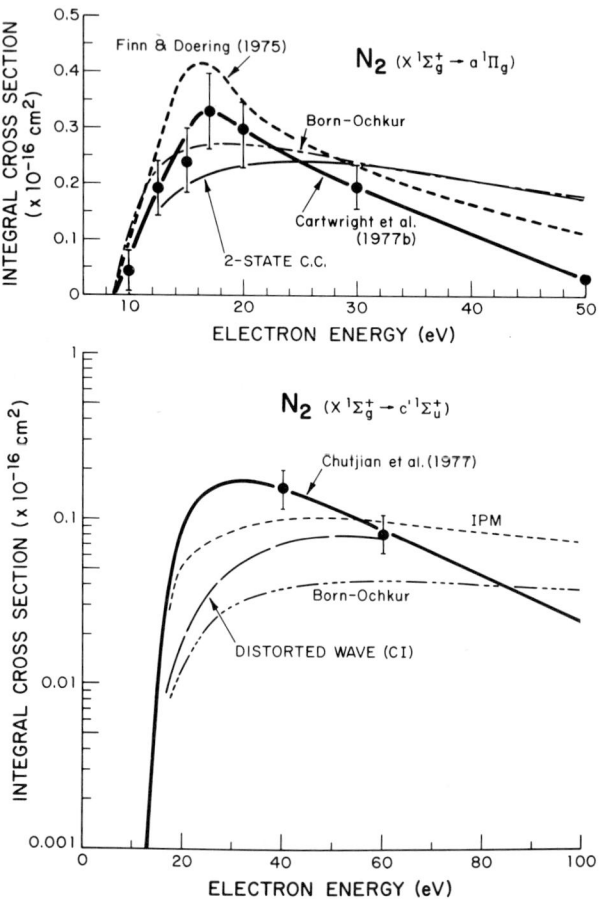

Fig. 17. Integral cross sections for excitation of the a $^1\Pi_g$ and c′ $^1\Sigma_u^+$ electronic states of N_2. The experimental data are taken from Cartwright et al. (1977b) and Chutjian et al. (1977) for the a and c′ states, respectively, and are discussed in those references. For excitation of the a state, the Born–Ochkur results are taken from Chung and Lin (1972) and the two-state C.C. results from Holley et al. (1981). For excitation of the c′ state, the distorted-wave (CI) results are from Mu-Tao and McKoy (1983a), the Born–Ochkur results are from Chung and Lin (1972), and the impact-parameter model (IPM) results are from Hazi (1981).

3. Other Molecules

Relatively little theoretical work has been reported for the electronic excitation of molecules other than H_2 and N_2 and that work will be summarized here.

a. Carbon Monoxide

Chung and Lin (1980) have reported generalized oscillator strengths and integral cross sections obtained using the Born/Ochkur–Rudge approximation for a large number of transitions (see Table IV) in CO and compared with the relatively meager experimental data available. Mu-Tao and McKoy (1982) have recently reported DW calculations of differential and integral cross sections for excitation of a number of states in CO analogous to those reported for N_2 and discussed above. These authors have compared their DW results to the integral cross sections predicted by the BOR model and with unpublished data for selected transitions by Cartwright and Trajmar. These comparisons lead to the same conclusions as were drawn from the corresponding comparisons in N_2 discussed in the previous section. That is, the DW model overestimates both the differential and integral cross sections for electronic excitation for CO in much the same way as it does in N_2 but gives qualitatively correct shapes for the DCSs. Mu-Tao and McKoy (1983a) attribute this overestimation of the cross sections for both CO and N_2 to the overestimation of the effects of a shape resonance in calculating the distorted waves. As was the case in N_2, the BOR results for the integral cross sections are usually somewhat smaller than those obtained by the DW model, and will predict DCSs that are in relatively poor agreement with experiment. This latter conclusion can be drawn from the fact that the DCSs for excitation of the electronic states of CO appear to be very similar in shape and magnitude to those for the corresponding states of N_2 for which detailed comparison have been made (Cartwright et al., 1977a). Since CO and N_2 are isoelectronic, and CO has a small permanent dipole moment, the similarity in their electronic excitation cross section is not surprising. A comprehensive experimental study of the electronic excitation of CO, similar to that done on N_2, is needed before a more detailed comparison with the theoretical results can be made.

b. Molecular Oxygen

Relatively little theoretical work has been reported on the electron-impact excitation of electronic states of O_2. Watson et al. (1967) used a modified Born approximation and generalized oscillator strength data to synthesize a set semiempirical integral cross section to be used for modeling atmospheric processes. Chung and Lin (1980) have recently reported Born integral cross sections for excitation of the $B\,^3\Sigma_u^-$ state of O_2 but reported no DCSs nor considered excitation to any other electronic states of O_2. Because of the very strong perturbations in the excited electronic states of O_2, no thorough and systematic experimental or theoretical study of electronic excitation of O_2 has been reported. More experimental and theoretical work on the electronic excitation of O_2 should be done because of the importance of these processes in understanding the composition of planetary atmospheres.

c. Molecular Fluorine

The only published results for the electronic excitation of F_2 appear to be the recent work by Fliflet *et al.* (1980) reporting results obtained with their DW model for excitation of the $^3\Pi_u$ state. There are, unfortunately, no experimental data which can be used for comparison but, comparisons discussed above for H_2 and N_2 suggest that the theoretical integral cross sections are probably a factor of 2–4 too large and the DCSs are probably in error at small scattering angles. This molecule also warrants experimental and theoretical study because of the central role played by (e, F_2) processes in rare-gas fluoride laser systems.

X. Applications of Electron Collision Data

A. Atmospheric Processes

Electron-impact processes play a major role in two naturally occurring atmospheric phenomena. In the earth's ionosphere, processes associated with production and loss of free electrons, and the related ion chemistry, determine the characteristics of the secondary-electron distribution and the ionospheric emissions. Auroral emissions are produced as a result of energy deposition by charged particles, primarily electrons, entering the earth's polar atmospheres along magnetic field lines. These two atmospheric phenomena involve electrons of somewhat different energies, but require approximately the same cross section information.

1. The Earth's Ionosphere

The general characteristics of the ionosphere are determined by the production, transport, slowing down, and disappearance of electrons. Free electrons in the ionosphere are produced by photoionization of atmospheric constituents and the subsequent cascade ionizations of these "primary" electrons as they move through the atmosphere. These electrons lose their energy by ionization; electronic, vibrational, and rotational excitation of the atmospheric constituents; and collisions with other electrons, until they thermalize and ultimately recombine with the positive ions. An excellent review of the elementary collision processes involving electrons in the earth's ionosphere was published in 1970 by Takayanagi and Itikawa and, although there is a considerable amount of new data, their identification and discussion of important physical and chemical processes is still valid.

The full range of electron-impact processes, elastic scattering through ionization, is possible because the flux of normal daytime photoelectrons, illustrated in Fig. 18, ranges from zero to greater than 100 eV. The photo-

2   Excitation of Molecules by Electron Impact   227

electron data shown in Fig. 18 are representative of the ionospheric spectrum throughout most of a typical day, for altitudes from 150 to about 300 km, as measured (Lee *et al.*, 1980) by photoelectron spectrometers on the Atmospheric Explorer satellites. Since N_2 is the major atmospheric constituent below 200 km, integral electron-impact cross sections for two important processes in N_2 are also shown in Fig. 18. The lower of the two cross sections, that for excitation of 337.1-nm radiation, characterizes the well-known (Imami and Borst, 1975) N_2 second-positive (0, 0) band recently used (Hernandez *et al.*, 1983) to assess the absolute calibration of the Atmospheric Explorer photoelectron spectrometers. The upper cross section, that for

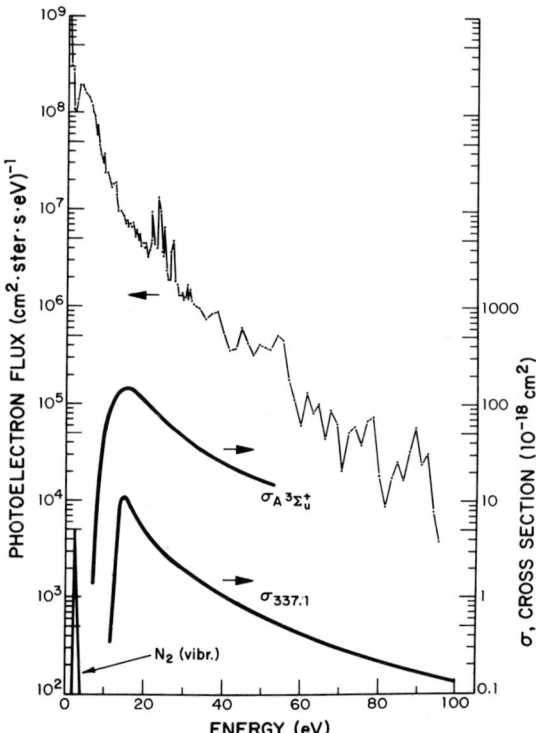

Fig. 18. Normal daytime photoelectron flux (upper curve, left ordinate) as a function of the electron energy, as measured by the Atmospheric Explorer satellite at 195 km (Lee *et al.*, 1980). The two curves in the lower portion of the figure (right-hand ordinate) denote the integral cross section for electron-impact excitation of 337.1-nm radiation (Imami and Borst, 1975) in N_2 and the total (direct plus cascade) integral cross section for excitation of the $N_2(A)$ state and have been included to illustrate those portions of the photoelectric spectrum that excite these states. The vibrational excitation cross section at the 2.3-eV resonance energy region is also indicated, and it is responsible for the dip in the electron energy distribution curve at that energy.

production of the $A\,^3\Sigma_u^+$ electronic state, represents an estimated total excitation cross section and includes direct excitation as well as cascade population from higher N_2 triplet states. This estimate was obtained by simply adding together the direct excitation cross sections for the A state and those (Cartwright et al., 1977b) for excitation of the $B\,^3\Pi_g$, $W\,^3\Delta_u$, $B'\,^3\Sigma_u^-$, $C\,^3\Pi_u$, and $E\,^3\Sigma_g^+$ states. A composite (rather than direct only) integral cross section for excitation of the A state is shown because the relatively large metastable population produced in the A state in this electron environment (Cartwright, 1978) is believed to play a role in energy pooling processes (Nadler et al., 1982) and energy transfer processes (Piper, 1982) of importance in the atmosphere. Comparison of the shapes of these integral cross sections, which are representative of those for excitation of most atmospheric molecules, with the shape of the photoelectron spectrum shows that a significant fraction of the total photoelectron flux (which is dominated by low-energy electrons) is involved in electronic excitation of molecules. Specifically, the recent laboratory measurements by Piper (1982) show that quenching of $N_2(A\,^3\Sigma_u^+)$ by $O(^3P)$ has a large cross section and that about 75% of all quenching events of $N_2(A)$ by O lead to $O(^1S)$ excitation. This is an important (and controversial) result and is one example of why knowledge of the relevant electron-impact cross sections is essential for understanding the earth's ionosphere.

Molecular nitrogen, however, is the only atmospheric constituent for which extensive results have been reported for the differential and integral cross sections for electronic excitation and, even for N_2, there remain a number of atomspherically important issues for which the available data are in major disagreement. The leading issue involving electronic excitation has to do with the absolute cross sections for dissociation of N_2 into $N(^4S)$, $N(^2D)$, and $N(^2P)$ fragments. Zipf and McLaughlin (1978) have reported integral "apparent" cross sections for excitation of the dipole-allowed states that are a factor of two larger than those determined (Chutjian et al., 1977) from electron energy loss spectra and the origin of this difference has not yet been explained. This difference needs to be resolved because nitrogen atoms, in their ground and excited states, play key roles in the formation of NO and similar constituents in the ionosphere and dissociation of N_2 by electron impact appears to be the major mechanism for producing N atoms.

The data situation for O_2, also a major molecular constituent in the ionosphere, is currently worse than for N_2 because there are relatively few absolute differential and integral cross sections for excitation of specific states in O_2. One of the major reasons for this deficiency is that the very stong perturbations present in the excited states of O_2 so complicate (see Fig. 4) the electron energy loss spectra in O_2 that a straightforward analysis of the data is not possible. Typical O_2 energy loss spectra (Trajmar et al., 1976) are shown

2 Excitation of Molecules by Electron Impact

in Fig. 19 and illustrate the complicated mixture of broad and narrow peaks associated with transitions to both bound and dissociating electronic states. In the atmospheric sciences, the primary interest in electron-impact excitation of O_2 is as a source of excited-state O atoms which are both chemically reactive and sources of the well-known 630.0- and 557.7-nm atmospheric emissions.

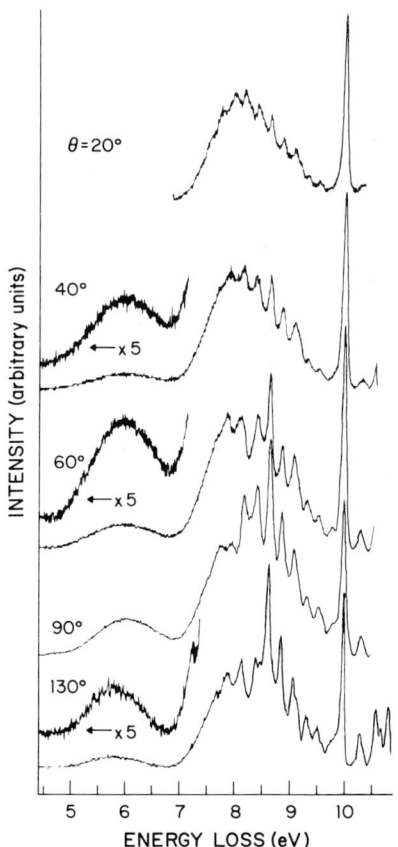

Fig. 19. Electron energy loss spectra in O_2 (Trajmar et al., 1976) for scattering angles of 20°, 40°, 60°, 90°, and 130°, for energy losses in the range 4.5–11.0 eV. In all of these spectra, the (residual) energy of the scattered electron was held fixed at $E_R = 1.0$ eV, which required that the incident electron energy and energy lost by the electron vary simultaneously according to $E_R = 1.0 = E_0 - \Delta E$ as ΔE varied over the range 4.5–11.0 eV. The rapidly changing shape of these spectra with changing scattering angle indicates that the energy loss region 7–11 eV in O_2 is composed of a relatively large number of bound electronic states in addition to the known repulsive states. The irregular spacings of the peaks are a result of the strong perturbations among the excited states.

That is, electron-impact excitation of essentially every electronic state above the b $^1\Sigma_g^+$ state (at 1.6 eV) results in dissociation into O-atom fragments. Two important (and interrelated) sets of data are needed for O_2: the magnitude and energy dependence for direct electron-impact excitation of the electronic states of O_2 (higher than 4 eV in the molecule), and the excited-state and kinetic-energy distribution of the atomic dissociation fragments. Progress has been made recently in understanding the photodissociation of O_2 via the well-known and atmospherically important Schumann–Runge continuum (Stone *et al.*, 1976; Lee *et al.*, 1977) the electron-impact dissociation of higher-lying molecular Rydberg states (Freund, 1971b; Mumma and Zipf, 1971) and the magnitudes of the electron-impact cross sections for excitation of composite (i.e., more than one) electronic state in the energy loss spectra (Wakiya, 1978a,b). However, differential and integral cross sections for excitation of individual electronic states, and the correlation to excited-state fragments, is still not available. Providing this information will require a particularly close collaboration between theorists and experimentalists because the very strong perturbations in the electronic states of O_2 preclude application of existing theoretical techniques on the one hand, and make straightforward interpretation of the energy loss data impossible on the other.

As a final comment on molecules for which electron-impact excitation data are useful in the atmospheric sciences, we mention the molecule NO. This molecule, although not a dominant neutral species in the earth's atmosphere, is of particular interest because of the key role (Takayanagi and Itikawa, 1970) it plays in the ion chemistry in the ionosphere and the aurora (Vallance-Jones, 1974). The excited electronic states of this molecule are also strongly perturbed (Miescher and Alberti, 1976; Miescher and Huber, 1978) and, again, very few data are currently available on the direct electronic excitation and/or the excited-state and kinetic-energy distribution of the dissociation atomic fragments. As in the case of O_2, determination of the cross sections for electronic excitation of NO is a fertile area for collaborative theoretical and experimental efforts for which the results would be of considerable interest to researchers in the atmospheric sciences.

2. The Aurora

Auroral processes that are driven by electron impact are very similar to those that occur in the ionosphere except that the electron spectrum is "harder" (i.e., more electrons with $E \gtrsim 50$ eV; see Fig. 18) and the resulting photoemissions are correspondingly more intense than in the ionosphere. An excellent review of the morphology, optical emissions, coupling to the ionosphere, and detailed excitation mechanisms has been provided by Vallance-Jones (1974). Since the primary energy source for auroral processes is energetic (1–10 keV) electrons from outside the earth's atmosphere, electronic excitation, and ionization of atmospheric species is more extensive in aurora than in the ionosphere. However, from the standpoint of the basic electron-impact excitation cross section information, the study of auroral and ionospheric processes have the same basic needs. Consequently, all the comments made in the previous section on ionospheric processes also apply for the aurora. It is worth pointing out here that since the "normal" aurora penetrates through most of the normal ionosphere to about 105 km, the coupling between the aurora and earth's ionosphere is of considerable interest. This coupling is accomplished mainly by transport of both neutral and charged particles. For a more complete understanding of the electron transport, differential (as well as integral) cross sections are needed (Buckley et al., 1983) in order to correctly incorporate the effects of elastic and inelastic scattering.

3. Stimulated Atmospheric Effects

The delicate balance that exists in the earth's atmosphere between the production and loss of charged particles was made apparent by two recent experiments. In the first (Whalen et al., 1981), the electron deposition in an aurora, and the related ionospheric plasma, was modified by the rocket release of an efficient electron scavenger. The results of this experiment suggest that an intimate relationship exists between the auroral acceleration mechanism and the thermal ionospheric plasma, but the mechanism for altering the local electron precipitation pattern is not yet understood. In the second (Vas'kov et al., 1981), an artificial increase in the F-layer ionospheric electron density was produced by enhancing the ionizing effectiveness of the local photoelectrons by use of intense radio waves. The mechanism for this effect is similarly not yet well understood.

Electron-impact processes involving N_2, O_2, and NO, as well as their ions and atomic components, are intimately involved in both the naturally occurring and stimulated atmospheric phenomena and our quantitative understanding of these phenomena will be improved as more fundamental electron collision data become available.

B. Electron-Beam and Discharge Processes

The past 20 years have seen considerable research activity directed toward the discovery and development of efficient methods for extracting energy (for specific applications) from the energy contained in a partially (or fully) ionized plasma. The most intense activity has been associated with the various magnetic-confinement fusion programs, but renewed interest in gas discharge processes has developed in the past 10 years along with a new interest in relativistic electron beams. The role of electron collision processes in the magnetic-confinement fusion programs has been well documented and therefore will not be discussed here. However, the role of electron–molecule collisions in the areas of new, and renewed, interest will be outlined as examples of where electron-impact cross sections are needed. This is not meant to be a comprehensive review of these fields of research, but rather just a sampling to provide perspective on the role of electron–molecule collisions in various processes.

1. Electron-Beam Transport

There is now considerable interest in the transport properties of, and the extraction of energy from, relativistic electron beams passing through atomic and molecular gases. Depending on the composition of the gas medium, its pressure, and the electron-beam energy, a variety of phenomena occur which are currently incompletely understood. In the case of H_2 as the gas medium, it has been determined (de Haan et al., 1981a) experimentally that there is a region around an H_2 pressure of 1 Torr for which the electron-beam energy loss is relatively small and the propagation is stable. A quantitative understanding of the reasons for this behavior is difficult because the energy coupling to the medium comes not only from the electron beam, but also from plasma-wave instabilities produced in the plasma. Under conditions of partial ionization, nearly all possible atomic and molecular species (e.g., H_3^+ as well as H_2, H, and H^+) are present (de Haan et al., 1981b). Such phenomena are not in thermal equilibrium and the detailed response of the gas medium for a given incident beam energy (e.g., dissipation of plasma waves, buildup of plasma instabilities, power balance, plasma current, and plasma decay), has to depend on the fundamental characteristics of the medium. For example, N_2 is very stable against electron-impact dissociation relative to H_2 and O_2, which dissociate readily. In addition, the vibrational-excitation cross sections in H_2 are substantially different from those for N_2 and O_2 and vibrational excitation of H_2 has recently been shown (Mozumder, 1982) to be important in the thermalization of electrons in H_2. The detailed characteristics of the electron-beam transport in N_2 can be expected, therefore, to be substantially different than in H_2 and O_2. Electron-impact cross sections serve as the foundation for all studies involving plasma properties and, as pointed out in the preceding pages, many of the electronic excitation cross sections for H_2 (for example) have yet to be determined. Accurate characterization of the plasma in molecular gases requires all the relevant cross sections, so a considerable amount of basic research remains to be done before the plasma properties in molecular gases can be well characterized.

2. Gas Discharges

The initial motivation for studying electron-impact processes in atoms and molecules came from efforts to interpret gas-discharge and electron swarm experiments performed over 50 years ago, so recognition of their role in such phenomena is certainly not new. However, there is considerable renewed interest in the general properties of gas discharges, primarily because they serve as efficient means to drive gas lasers. The next section contains comments about gas lasers, and in this section we will sketch the importance of vibrational and electronic excitation of molecules in other gas discharge processes.

The electron transport properties in media for which the characteristic electron energy is well below the first inelastic vibrational or electronic excitation threshold are reasonably well understood. Specifically, the low-energy electron transport properties in the inert gases are generally well described by the stationary solutions of the Boltzmann equation with the appropriate elastic momentum transfer cross sections. The substantial amount of experimental and theoretical work that has been done over the past 50 years on low-energy electron transport properties in gases is well documented in a number of excellent references (Huxley and Crompton, 1974; Gilardini, 1972; McDaniel, 1964) on the subject.

However, gas breakdown and the transport of moderate-energy electrons through gas media are not yet very well understood and remain the subject of considerable research. The origin of the current difficulties with modeling these phenomena can be traced to a combination of (1) the mathematical and computational problems in dealing with a non-LTE plasma, and (2) the general absence of the cross section data necessary to model the charged-particle interactions in the medium.

a. Electron Transport Properties

Although the details have been the subject of some controversy, it is generally agreed (Ferrari, 1977; Braglia *et al.*, 1977; Kleban and Davis, 1977; Lin *et al.*, 1979) that inelastic channels play an important role in determining the transport of energetic electrons through gases and that the customary two-term solution of the Boltzmann equation is not adequate. The role of inelastic effects has been examined specifically for the cases of N_2 (Pitchford *et al.*, 1981) and for CH_4 (Foreman *et al.*, 1981; Kleban *et al.*, 1981), for which it has been shown that higher-order solutions to the Boltzmann equation are necessary in order to explain the measured transport properties. It also needs to be emphasized here that, in order to answer detailed questions involving the motion and energy distribution of electrons in a partially ionized plasma, the

DCSs are as necessary (Ferrar, 1977; Braglia et al., 1977; Kleban and Davis, 1977; Lin et al., 1979; Pitchford et al., 1981; Foreman et al., 1981; Kleban et al., 1981) as the integral cross sections.

b. Gas Breakdown

For plasma processes involving gas breakdown and the properties of multicomponent plasmas, the depth of understanding is, as expected, less than for stable discharges and properties of single-component systems. The production (ionization) and loss (recombination and attachment) of electrons obviously are key processes that determine the formation, stability, and decay of the plasma, and the development of a self-consistent theory to describe the plasma is a topic of current research interest (Taniguchi et al., 1977; Edgley and von Engel, 1980; Rusanov et al., 1981). In this section we give two examples of plasmas involving molecules to illustrate how vibrational and electronic excitation of the molecules play a role in the overall properties of a plasma.

In the case of plasmas involving rare-gas atoms, molecules, and alkali-metal vapors, a number of different studies have been performed to determine the electron energy balance and plasma stability, and the results of these studies have been reviewed by Mnatsakanyan (1974). In these particular studies the plasma is characterized by

$$[\text{rare gas}] \sim 10^{18} \text{ cm}^{-3},$$
$$[\text{molecule}] \sim 10^{15}-10^{17} \text{ cm}^{-3},$$
$$[\text{alkali}] \sim 10^{15} \text{ cm}^{-3},$$
$$[e^-] \sim 10^{12}-10^{14} \text{ cm}^{-3},$$
$$T_e \sim 0.2-3.0 \text{ eV}.$$

The electrons in such plasmas are generally not in thermodynamic equilibrium and a single temperature cannot be used (Rusanov et al., 1981) to characterize the kinetic temperatures and the populations of the quantum states. Furthermore, energy transfer between the excited states of the molecule and the alkali atoms has been found to have a very important effect (Mnatsakanyan, 1974) on both the plasma parameters and distribution of excited quantum states. Figure 20, constructed from the data compiled by Mnatsakanyan (1974), illustrates the dramatic effect that a small amount of Cs has on the electron energy balance. The energy loss fraction δ is defined here for each elementary process (rotational, vibrational, electronic excitation, etc.) as the ratio of the rate of electron energy lost in the particular excitation process to that lost by elastic electron scattering with heavy particles. In Fig. 20 for the $N_2 + Ar$ mixture, the rapid rise in the "inelastic loss factor" near $T_e = 10,000$ K is due to vibrational excitation and some electronic excitation,

2 Excitation of Molecules by Electron Impact

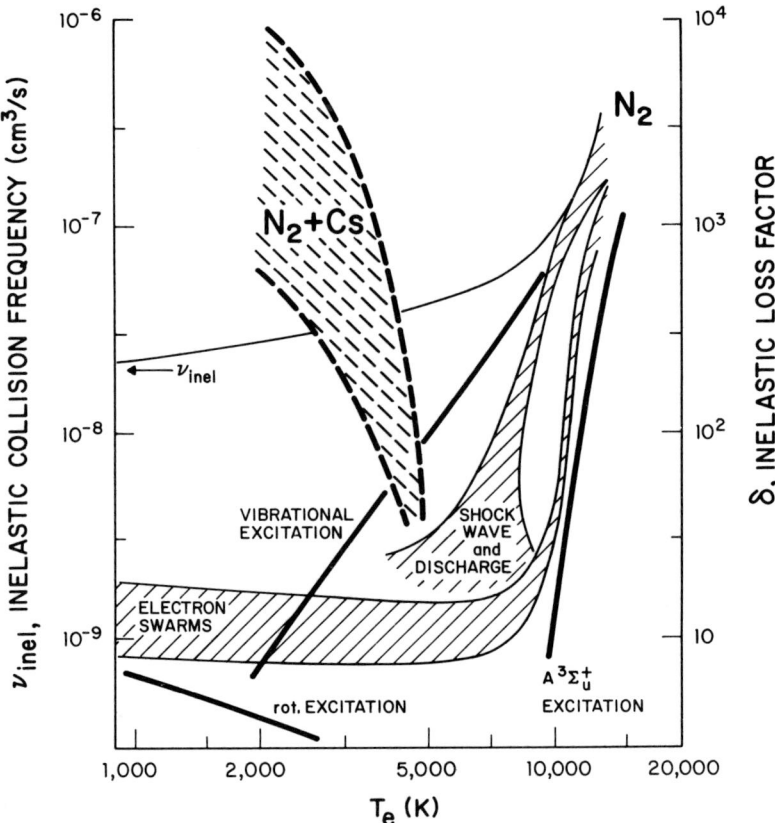

Fig. 20. Inelastic collision frequency (left ordinate) and inelastic loss factor (right ordinate) as a function of the electron temperature (K) adapted from the review by Mnatsakanyan (1974). The *solid* lines enclosing hatched regions denote data for a $N_2 + Ar$ plasma from a variety of experiments including electron swarms, arc and pulse discharges, and shock heating. The heavy solid lines denote theoretical results (Rusanov *et al.*, 1981) for the indicated process. The *dashed* lines enclosing the hatched regions denote the analogous data obtained, under similar conditions, when cesium has been added to the $N_2 + Ar$ plasma. The inelastic loss factor is defined as the ratio of the rate of electron energy exchange for a particular excitation process, to that lost by the electrons in elastic collisions. (Mnatsakanyan, 1974).

the latter produced by the "tail" of the electron energy distribution. However, when Cs is added to the $N_2 + Ar$ plasma, a change in the ionization and recombination occurs that produces a significantly different functional dependence of the inelastic loss factor on the electron temperature, as shown in Fig. 20. There are clearly complicated processes associated with the addition of the molecular species, possibly involving vibrational and electronic states as well as energy transfer between the molecular species and the alkali, that are still not understood. A considerable amount of work remains to be done on these particular types of plasmas.

3. Gas Lasers

The first gaseous optical laser was developed in 1961 by Javan *et al.* (1961) some seven years after the first maser was demonstrated, and has been followed by a tremendous amount of research and development into their physics and chemistry as well as their application to a variety of research (Shimoda, 1976; Hinkley, 1976; Ben-Shaul *et al.*, 1981) and industrial (Crone, 1982) processes. There are a number of excellent textbooks (e.g., Willett, 1974) on the mechanisms for achieving, and maintaining, population inversions in a wide variety of lasing media, and a number of articles reviewing the new developments (Grasyuk *et al.*, 1980) appear in the literature each year. A general review of the role of electron-impact excitation of molecules in gas laser technology is completely beyond the scope of this chapter and, instead, a brief discussion will be given of the role of electron–molecule collisions in excimer lasers. Excimers are the class of exciplexes formed from identical atoms or atomic groups (e.g., rare-gas excimers). This general class of lasers, which are generally of "short" wavelength ($\lambda \lesssim 600$ nm), has been recently developed (first demonstrated in 1975) and is currently the subject of intense research activity because of the tremendous wealth of applications for laser photons at wavelengths less than 450 nm (Crone, 1982). Exciplex molecules are defined as those which are bound in their excited electronic states but dissociate readily in their ground electronic state. The general characteristics of, and detailed processes occurring in, exciplex lasers have been discussed in a number of excellent review articles (e.g., Rhodes, 1979; *IEEE J. Quantum Electron.*, 1979; Lakoba and Yakovlenko, 1980; Huestis, 1982), and attention here will be directed to the role of electron–molecule processes in the production and loss of the excited exciplex molecules. Because of their importance in a wide variety of applications, the rare-gas excimer systems (and in particular KrF) will be used as the example to illustrate the mechanisms although much of what will be said applies generally to gaseous exciplex laser systems.

Figure 21 is a table of the rare-gas monohalide molecules indicating which ones are known (Brau, 1979) to lase and/or fluoresce. A complete explanation as to why certain of the rare-gas monohalide molecules do not lase or fluoresce is apparently not yet available but is believed (Brau, 1979) to be a result of predissociation (e.g., ArBr and KrI) or the location of the ionic states (e.g., HeCl). Figure 22 illustrates (Brau, 1979) the potential-energy curves for those rare-gas monohalide molecules that are known to lase. The upper lasing level is generally a low-lying ionic electronic state that is dipole connected to the

	He	Ne	Ar	Kr	Xe
F		FLUORESCENCE	LASING	LASING	LASING
Cl			FLUORESCENCE	LASING	LASING
Br				FLUORESCENCE	LASING
I					FLUORESCENCE

Fig. 21. Emission properties of the rare-gas monohalides excited under conditions that can produce stimulated emission. Six of the species have been observed to lase and four have shown fluorescence. No spectral properties have been reported for the other ten potential monohalide molecules. [From Brau (1979).]

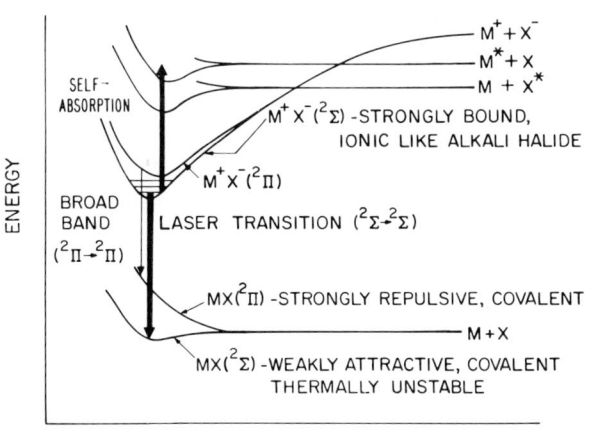

Fig. 22. General electronic structure characteristics of the lasing rare-gas monohalide molecules. The lowest electronic state is repulsive, or only slightly bound, and therefore the ground-state molecule is thermally unstable against dissociation into atomic fragments. The upper laser level is an ionic electronic state that dissociates into the indicated ionic fragments. [From Brau (1979).]

weekly bound or dissociative ground electronic state. The rare-gas dimers, He_2^*, Ar_2^*, Kr_2^*, Xe_2^* are also known (Lakoba and Yakovlenko, 1980) to lase and, as will be discussed below, the electron-pumped excited states of these dimers play an important role in the performance characteristics of the rare-gas monohalide lasers.

The KrF excimer laser is usually produced from an atmospheric-pressure mixture of Ar-Kr-F_2 having percentage proportions about 94.5-0.050-0.005, respectively. That is, the constituents that form the lasing molecule are diluents in a carrier gas. The exciplex active lasing medium was first produced by using

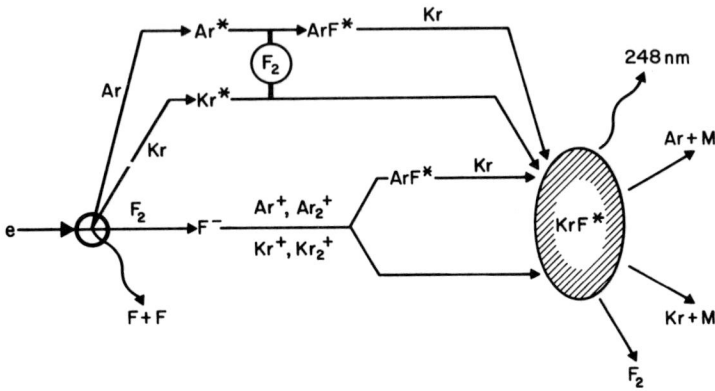

Fig. 23. Schematic diagram of the primary reactions for formation of KrF* by electron-impact excitation of atmospheric-pressure mixtures of Ar-Kr-F_2 [Adapted from Nighan (1978).]

high-current electron beams as the source of energy, and it was the newly acquired electron-beam technology that made this first demonstration of rare-gas halogen lasers possible. A variety of electrical discharge techniques (*IEEE J. Quantum Electron.*, 1979; Huestis, 1982) have recently been developed which are efficient as well as having the greatest potential for scaling to high average power. The production and loss of the active exciplex media involve a large number of electron and heavy-particle collision processes. A thorough analysis of the basic collision processes that take place in an electron-beam stabilized electrical discharge in atmospheric Ar-Kr-F_2 mixtures has been provided by Nighan (1978), and Fig. 23 is an adaptation of his schematic summary of the primary collision processes contributing to KrF* formation and destruction. Electron-impact excitation of the rare-gas atoms and dissociation of F_2 initiate the production of the KrF* molecule which, as shown in Fig. 23, occurs via either ion–molecule reactions or reactions involving metastable rare-gas atoms. Although no single mechanism completely dominates the production of KrF*, Nighan's analysis shows that two- and three-body positive and negative ion recombination is most important for electron-beam excitation (lowest path in Fig. 23), while reactions between F_2 and metastable rare-gas atoms are most important in the case of gas-discharge pumping (upper two paths in Fig. 23).

A key factor in achieving high efficiency for these lasers is the efficient production of rare-gas atoms in metastable states. Figure 24 [from Nighan (1978)] summarizes the electron–neutral power transfer, for the example gas mixture, as a function of the gas-discharge characteristic E (V/cm)/n (cm^{-3}). The quantity E is the electric field strength and n is the total neutral gas density. The results in Fig. 24 show that there is a substantial electron energy

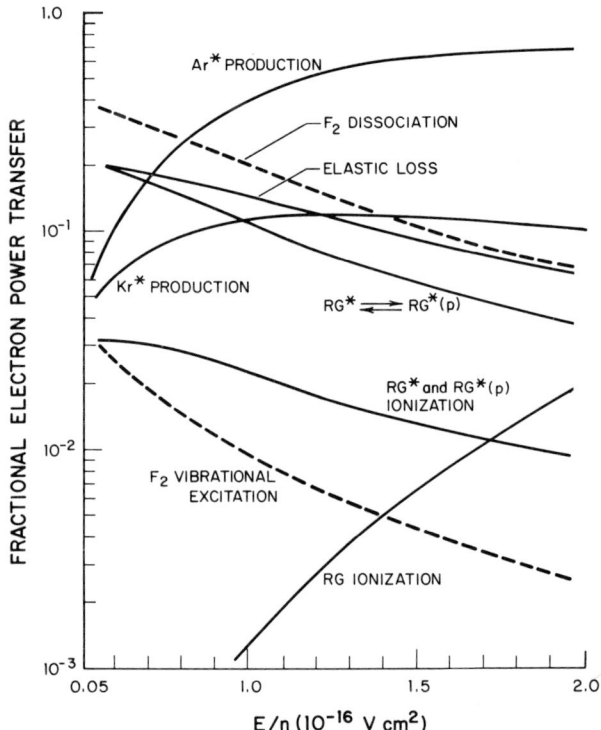

Fig. 24. Fractional contributions to electron–neutral power transfer for the example Ar-Kr-F_2 mixture. The symbol RG denotes the combined effort of Ar and Kr and the dashed curves denote the contributions from vibrational and electronic excitation of the molecules. [Adapted from Nighan (1978).]

loss to electron-impact dissociation of F_2. Vibrational excitation of F_2 is not an important energy loss mechanism (because the energy lost per collision is relatively small) but is very important in the formation of KrF* because of its possible effect (Morgan and Szöke, 1981) on dissociative attachment. Accurate assessments of these loss mechanisms are, unfortunately, not yet possible because there are currently very few data available on the electronic and vibrational excitation of F_2, and these deficiencies hamper the assessment of the role of electron-F_2 processes in the KrF laser.

As shown schematically in Fig. 23, F^- plays a central role in the formation of KrF* but the role of excitation to electronic states of F_2 that dissociate into $F^+ + F^-$ could not be determined because the relevant cross section data are not available. Theoretical work (Hay and Cartwright, 1976) suggests that ionic-character states that dissociate into $F^+ + F^-$ are abundant in F_2 and, consequently, the electron-impact excitation of these states should be studied to determine if this source of F^- is important in the KrF laser. In the short term, theoretical techniques may be the best way to obtain this information.

Electron-impact excitation of rare-gas dimer molecules also plays an important role in determining how efficiently the energy available in the lasing medium can be extracted as laser photons. Specifically, it is the photoabsorption by the rare-gas dimers that reduces the extraction efficiency and tunability of these laser systems. The effects of these transient intermediates on laser performance is a topic of current research interest (Zamir et al., 1979; Sauerbrey et al., 1982) and a schematic diagram of the energy-flow kinetics is provided in Fig. 25. The heavy dashed arrows denote the electron-impact excitation (and deexcitation) processes that are important in the kinetics. Since the electron density is reasonably high, electrons are effective in collisional mixing of the $^1\Sigma^+$ and $^3\Sigma^+$ states, as well as producing excitation to higher electronic states, but with cross sections that are currently unknown. These excitation (and deexcitation) processes play a major role (Zamir et al., 1979) in controlling the competition between excimer formation and metastable atom production. Furthermore, since these transient dimers absorb strongly throughout the visible portion of the spectrum, the desired development of a

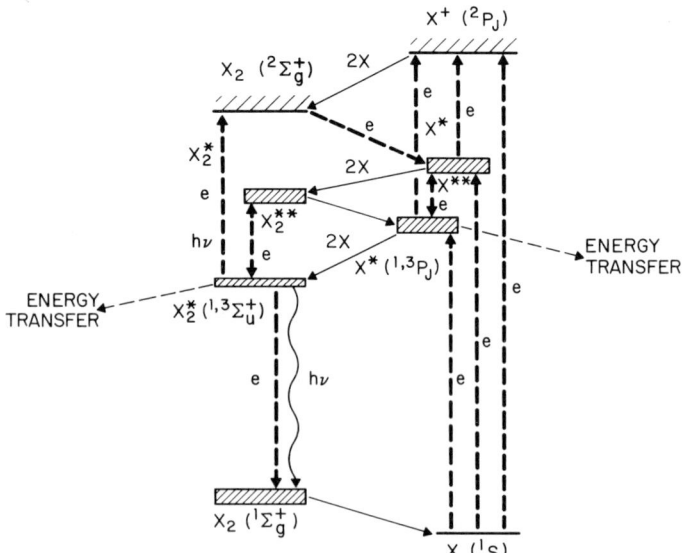

Fig. 25. Schematic energy-level diagram of the energy flow kinetics involving rare-gas atoms and dimers. The electron-impact processes are indicated by heavy dashed arrows and the heavy-particle collision processes by the light solid lines. [Adapted from Zamir et al. (1979).]

high-power visible laser using rare-gas exciplexes requires control of these excited-state dimers, and hence better knowledge of their production and loss mechanisms.

C. Astrophysics

The scientific world's understanding of the role of molecules in a variety of processes of astrophysical interest is rapidly changing as a result of new spectral measurements at infrared, and microwave frequencies, and satellite measurements at UV frequencies. For example, it has just recently been recognized (Andrew, 1980) that most of the mass of the interstellar medium is contained in molecular clouds, which means that molecular processes, in general, play a greater role in stellar formation than previously appreciated. We cite two cases of current research interest for which electron-impact excitation cross section information is needed in order to better understand the astrophysical processes, planetary nebulae, and comets.

Black (1978) has recently shown that significant abundances of simple molecules such as H_2, H_2^+, HeH^+, OH, and CH^+ can exist in the transition zones of ionized nebulae where neutral atoms and free electrons mingle at moderately high temperatures. Dalgarno (1976) has summarized the formation and destruction mechanisms that may be responsible for the observed abundances of CH and CH^+ in the interstellar media, which also apply to planetary nebulae. The transition zone of an ionization-bounded nebula is where the electron temperature drops from a few electron volts to tenths of an electron volt and simple molecules composed of H, He, C, N, and O will be formed by radiative association. Vibrational excitation and excitation of fine-structure levels of the simple diatomic molecules formed will play an important role in the kinetics that determine the abundances of these molecules as well as the cooling of the medium. Except for H_2 and N_2, relatively few data are currently available for these molecules.

Comets have been of considerable interest for many years in astrophysics because the material in comets is primordial and, as such, provides unique information on galactic evolution. Extensive analyses of cometary spectroscopic data obtained from ground-based observatories in the visible region of the spectrum have resulted in identification (Arpigny, 1965a,b) of the molecules C_2, CN, CH, NH, OH, CO, N_2, and H_2O (and their positive ions) as primary constituents in comets. The molecules H_2O, H_2, and OH have long been believed to be major cometary constituents, but only fragmentary experimental evidence was available until the early 1970s, at which time ultraviolet measurements (using new techniques) were made on comets Bennett (Keller and Lillie, 1974) and Kohoutek (Wyckoff and Wehinger,

1976). These measurements revealed that, as expected, the water molecule is a major constituent of cometary nuclei. Although solar photons are the primary energy source for excitation and ionization of the cometary constituents, electron-impact excitation by solar wind electrons in the cometary coma, and by photoelectrons in closer to the comet nucleus, may make significant contributions to the overall kinetics in cometary processes. Of the list of major cometary molecules given above, cross sections for vibrational and electronic excitation are available only for some transitions in CO and N_2. Consistent sets of cross sections for vibrational and electronic excitation of CN, CH, NH, OH, and H_2O would help immensely in the effort to improve our understanding of cometary emissions.

Acknowledgments

We would like to thank Dr. G. Csanak and Professor U. Fano for their constructive comments on this chapter. Special thanks go to Howard Shirley (LANL) who expertly prepared the illustrations for this chapter.

References

Aarts, J. F. M., and de Heer, F. J. (1970). *J. Chem. Phys.* **52**, 5354–5360.
Aarts, J. F. M., and de Heer, F. J. (1971). *Physica* **52**, 45–69.
Ajello, J. M. (1970). *J. Chem. Phys.* **53**, 1156–1165.
Ajello, J. M. (1971). *J. Chem. Phys.* **55**, 3158–3168.
Ajello, J. M., Srivastava, S. K., and Yung, Y. (1982). *Phys. Rev. A* **25**, 2485–2498.
Alder, H. L., and Roessler, E. B. (1968). "Introduction to Probability and Statistics", 4th edition. Freemand, San Francisco.
Altshuler, S. (1963). *J. Geophys. Res.* **68**, 4707.
Amusia, M. Ya., and Cherepkov, N. A. (1975). *Case Stud. At. Phys.* **5**, 47–179.
Anderson, R. J., Watson, J., and Sharpton, F. A. (1977). *J. Opt. Soc. Am.* **67**, 1641–1643.
Andrew, B. H., ed. (1980). Interstellar Molecules. Symposium No. 87, August 1979. Reidel, Dordrecht, Holland, and private communication (1982).
Andric, L., Cadez, K., and Hall, R. I. (1981). *Trend in Physics 5th General Conference*, Instanbul, Turkey, September 7–11.
Arpigny, C. (1965a). *Mem. Soc. R. Sci. Liege* **12**, 155–183 (fasc 5).
Arpigny, C. (1965b). *Annu. Rev. Astron. Astrophys.* **3**, 351–376.
Backx, C., Wright, G. R., and van der Wiel, M. J. (1976). *J. Phys. B At. Mol. Phys.* **9**, 315–331.
Baldwin, G. C. (1974). *Phys. Rev. A* **9**, 1225–1229.
Baltayan, P., and Nedelec, O. (1976). *J. Quant. Spectrosc. Radiat. Transfer* **16**, 207–211.
Barbarito, E., Basta, M., Calicchio, M., and Tessari, G. (1979). *J. Chem. Phys.* **71**, 54–59.
Becker, W. G., Fickes, M. G., Slater, R. C., and Stern, R. C. (1974). *J. Chem. Phys.* **61**, 2283–2289.
Ben-Shaul, A., Haas, Y., Kompa, K. L., and Levine, R. D. (1981). "Lasers and Chemical Change" (V. I. Goldanskii, R. Gomer, F. P. Schafer, and J. P. Toennies, eds.), Vol. 10 (Series in Chemical Physics). Springer-Verlag, Berlin.

Bethe, H. A. (1930). *Ann. Phys.* **5**, 325–400.
Bevington, P. R. (1969). "Data Reduction and Error Analysis for the Physical Sciences." McGraw-Hill, New York.
Bhaskar, N. D., Jaduszliver, B., and Bederson, B. (1977). *Phys. Rev. Letters* **38**, 14–17.
Blaauw, H. J., Wagenaar, R. W., Barends, D. H., and de Heer, F. J. (1980). *J. Phys. B* **13**, 359–376.
Black, J. H. (1978). *Astrophys. J.* **222**, 125–131.
Blum, K., and Jakubowicz, J. (1978). *J. Phys. B* **11**, 909–925.
Blum, K., and Kleinpoppen, H. (1979). *Phys. Rep.* **52**, 204–261.
Bogdanova, I. P., Efremova, G. V., Lavrov, B. P., Ostroviskii, V. N., Ultimor, V. I., and Yakovleva, V. I. (1981). *Opt. Spektrosk. USSR* **50**, 63–65.
Bonham, R. A. (1980). Private communication.
Borst, W. L. (1972). *Phys. Rev. A* **5**, 648–656.
Borst, W. L., Wells, W. C., and Zipf, E. C. (1972). *Phys. Rev. A* **5**, 1744–1747.
Braglia, G. L., Caraffini, G. L., and Iori, M. (1977). *Lett. Nuovo Cimento* **19**, 193–200.
Bransden, B. H., and McDowell, M. R. C. (1978). *Phys. Rep.* **46**, 249–394.
Brau, C. (1979). Rare Gas Halogen Excimers. *In* "Excimer Lasers" (C. K. Rhodes, ed.) Vol. 30, p. 87 (Topics in Applied Physics). Springer-Verlag, Berlin.
Brinkmann, R. T., and Trajmar, S. (1970). *Ann. Geophys.* **26**, 201–207.
Brinkmann, R. T., and Trajmar, S. (1981). *J. Phys. E* **14**, 245–255.
Brion, C. E. (1975). *Radiat. Res.* **64**, 37–52.
Brion, C. E., and Hamnett, A. (1980). *In* "Advances in Chemical Physics", Vol. 45, Part 2. Wiley, New York.
Bromberg, J. P. (1974). *J. Chem. Phys.* **61**, 963–969.
Buckley, B. D., Burke, P. G., and Noble, C. J. (1983). Recent Developments in the Theory of Electron-Molecule Scattering. *In* "Electron Molecule Collisions" (I. Shimamura and K. Takayanagi, eds.). Plenum, New York.
Burke, P. G. (1971–1972). *Comments At. Mol. Phys.* **III**, 31–45.
Burke, P. G. (1972). *In* "Atoms and Molecules in Astrophysics" (T. R. Carson and M. J. Roberts, eds.), p. 1. Academic Press, London.
Callaway, J. (1980). *Adv. Phys.* **29**, 771–867.
Carson, T. R. (1954). *Proc. Phys. Soc. A* **67**, 909–916.
Cartwright, D. C. (1970). *Phys. Rev. A* **2**, 1331–1348.
Cartwright, D. C. (1972). *Phys. Rev. A* **5**, 1974.
Cartwright, D. C. (1978). *J. Geophys. Res.* **83**, 517–531.
Cartwright, D. C., and Kuppermann, A. (1967). *Phys. Rev.* **163**, 86–102.
Cartwright, D. C., Trajmar, S., Williams, W., and Huestis, D. L., (1971). *Phys. Rev. Lett.* **27**, 704–707.
Cartwright, D. C., Chutjian, A., Trajmar, S., and Williams, W. (1977a). *Phys. Rev. A* **16**, 1013–1040.
Cartwright, D. C., Trajmar, S., Chutjian, A., and Williams, W. (1977b). *Phys. Rev. A* **16**, 1041–1051.
Chandra, N., and Temkin, A. (1976). *J. Chem. Phys.* **65**, 4537–4539.
Choi, B. H., Poe, R. T., Sun, J. C., and Shan, Y. (1979). *Phys. Rev. A* **19**, 116–124.
Chung, S., and Lin, C. C. (1971). *Appl. Opt.* **10**, 1790–1794.
Chung, S., and Lin, C. C. (1972). *Phys. Rev. A* **6**, 988–1002.
Chung, S., and Lin, C. C. (1974). *Phys. Rev. A* **9**, 1954–1964.
Chung, S., and Lin, C. C. (1978). *Phys. Rev. A* **17**, 1874–1891.
Chung, S., and Lin, C. C. (1980). *Phys. Rev. A* **21**, 1075–1081.
Chung, S., Lin, C. C., and Lee, E. T. P. (1975). *Phys. Rev. A* **12**, 1340–1357.
Chutjian, A. (1979). *Rev. Sci. Instrum.* **50**, 347–355.
Chutjian, A., and Tanaka, H. (1980). *J. Phys. B B* **13**, 1901–1908.
Chutjian, A., Truhlar, D. G., Williams, W., and Trajmar, S. (1972). *Phys. Rev. Lett.* **29**, 1580–1583; (1974). Erratum **33**, 1524.
Chutjian, A., Cartwright, D. C., and Trajmar, S. (1977). *Phys. Rev. A* **16**, 1052–1060.

Chutjian, A., Hippler, R., McGregor, I., and Kleinpoppen, H. (1980). "XIth ICPEAC" (N. Oda and K. Takayanagi, eds.), p. 172. North Holland, Amsterdam.
Crompton, R. W., Gibson, D. K., and Robertson, A. G. (1970). *Phys. Rev. A* **2**, 1386-1395.
Crone, W. R., ed. (1982). *Proceedings of the IEEE* **70**, No. 6, special issue.
Csanak, Gy., Taylor, H. S., and Yaris, R. (1970). *Adv. At. Mol. Phys.* **7**, 287-361.
Dalba, G., Fornasini, P., Grisenti, R., Ranieri, G., and Zecca, A. (1980a). *J. Phys. B* **13**, 4695-4701.
Dalba, G., Fornasini, P., Lazzizzera, I., Ranieri, G., and Zecca, A. (1980b). *J. Phys. B* **13**, 2839-2848.
Dalgarno, A. (1976). The Interstellar Molecules CH and CH^+. *In* "Atomic Processes and Applications" (P. G. Burke and B. L. Moiseiwitsch, eds.), p. 109. North-Holland, Amsterdam.
Danner, D. (1970). Ph.D. Thesis, Physikalisches Institute de Universitat Freiburg, West Germany.
Day, R. L., Anderson, R. J., and Sharpton, F. A. (1979). *J. Chem. Phys.* **71**, 3683-3688.
de Haan, P. H., Hopman, H. J., Janssen, G. C. A. M., and Yamagiva, K. (1981a). *Proceeding of the 4th International Topical Conference of High-Power Electron and Ion-Beam Research and Technology, July 29* (H. J. Doucent and J. M. Buzzi, eds.). Laboratoire de Physique des Milieux Ionises, Groupe de Recherche du CNRS, Ecole Polytechnique-Palaiseau, France.
de Haan, P. H., Singh, R. N., Hopman, H. J., Janssen, G. C. A. M., Granneman, E. H. A., and Strelkov, P. S. (1981b). *J. Phys. E* **14**, 373-377.
de Heer, F. J., and Carriere, J. D. (1971). *J. Chem. Phys.* **55**, 3829-3835.
Doering, J. P., and McDiarmid, R. (1980). *J. Chem. Phys.* **73**, 3617-3624.
Doering, J. P., and McDiarmid, R. (1982). *J. Chem. Phys.* **76**, 1838-1844.
Domenicucci, A. G., and Miller, K. J. (1977). *J. Chem. Phys.* **66**, 3927-3937.
Dymond, E. G. (1927). *Phys. Rev.* **29**, 433-441.
Edgley, P. D., and von Engle, A. (1980). *Proc. R. Soc. L* **370**, 375-387.
Ehrhardt, H., Langhans, L., Linder, F., and Taylor, H. S. (1968). *Phys. Rev.* **173**, 222-230.
Fano, U., and Macek, J. (1973). *Rev. Mod. Phys.* **45**, 553-573.
Ferch, J., Raith, W., and Schroder, K. (1980). *J. Phys. B* **13**, 1481-1490.
Ferch, J., Masche, C., and Raith, W. (1981). *J. Phys. B* **14**, L97-L100.
Ferch, J., Raith, W., and Schörder, K. (1982). *J. Phys. B* **15**, L175-L178.
Ferrari, L. (1977). *Physica* **85c**, 161-179.
Finn, T. G., and Doering, J. P. (1975). *J. Chem. Phys.* **63**, 4399-4404.
Fliflet, A. W., and McKoy, V. (1980). *Phys. Rev. A* **21**, 1863-1875.
Fliflet, A. W., McKoy, V., and Rescigno, T. N. (1979). *J. Phys. B* **12**, 3281-3293.
Fliflet, A. W., McKoy, V., and Rescigno, T. N. (1980). *Phys. Rev. A* **21**, 788-792.
Foreman, L., Kleban, P., Schmidt, L. D., and Davis, H. T. (1981). *Phys. Rev. A* **23**, 1553-1557.
Franck, J., and Hertz, G. (1914). *Verh. Dtsch. Phys. Ges. (Berlin)* **16**, 457-467.
Freund, R. S. (1971a). *J. Chem. Phys.* **54**, 1407-1409.
Freund, R. S. (1971b). *J. Chem. Phys.* **54**, 3125-3141.
Frost, L., and Weigold, E. (1980). *Phys. Rev. Lett.* **45**, 247-250.
Gavrila, M., and van der Wiel, M. J. (1978). *Comments At. Mol. Phys.* **8**, 1-20.
Gerjuoy, E., and Stein, S. (1955). *Phys. Rev.* **97**, 1671-1679.
Gibson, D. K. (1970). *Aust. J. Phys.* **23**, 683-696.
Gilardini, A. (1972). "Low Energy Electron Collisions in Gases; Swarm and Plasma Methods Applied to Their Study." New York.
Glauber, R. J. (1959). *In* "Lectures in Theoretical Physics" (W. E. Brittin *et al.*, eds.), Vol. 1, p. 315. Wiley (Interscience), New York.
Goddard, W. A., Huestis, D. L., Cartwright, D. C., and Trajmar, S. (1971). *Chem. Phys. Lett.* **11**, 329-333.
Grasyuk, A. Z., Letokhov, V. S., and Lobko, V. V. (1980). *Sov. J. Quantum Electron. Engl. Transl.* **10**, 1317-1338.

Guśkov, Y. K., Savvov, R. V., and Slobodyanyuk, V. A. (1978). *Sov. J. Plasma Phys. Engl. Transl.* **4**, 527–528.
Haas, R. Z. (1957). *Z. Phys.* **148**, 177–191.
Hall, R. I., and Trajmar, S. (1975). *J. Phys. B* **8**, L293–L296.
Hasted, J. B., Kadifachi, S., and Solovyev, T. (1979). *XI Int. Conf. on the Physics of Electronic and Atomic Collisions, Kyoto, Japan, Abstract of Contributed Papers*, pp. 334–335.
Hay, P. J., and Cartwright, D. C. (1976). *Chem. Phys. Lett.* **41**, 80–83.
Hazi, A. U. (1981). *Phys. Rev. A* **23**, 2332–2240.
Henry, R. J. W., and Chang, E. S. (1972). *Phys. Rev. A* **5**, 276–284.
Hernandez, S. P., Doering, J. P., Abreu, V. J., and Victor, G. A. (1983). Planetary and Space Science, *31*, 221–233.
Hertel, I. V., and Stoll, W. (1977). *Adv. At. Mol. Phys.* **13**, 113–228.
Hinkley, E. D. (1976). "Laser Monitoring of the Atmosphere", Springer-Verlag, Berlin.
Hoffman, K. R., Dababneh, M. S., Hsieh, Y. F., Kauppila, W. E., Pol, V., Smart, J. H., and Stein, T. S. (1982). *Phys. Rev. A* **25**, 1393–1403.
Holley, T. K., Chung, S., Lin, C. C., and Lee, E. T. P. (1981). *Phys. Rev. A* **24**, 2946–2952.
Huestis, D. (1982). "Gas Lasers" (E.W. McDaniel and W. L. Nighan, eds.), Chap. 1. Academic Press, New York.
Huetz, A., Cadez, I., Grestau, F., Hall, R. I., Vichon, D., and Mazeau, J. (1980). *Phys. Rev. A* **21**, 622–628.
Huxley, L. G. H., and Crompton, R. W. (1974). *In* "The Diffusion and Drift of Electrons in Gases." New York.
IEEE J. of Quant. Elec. **15**, No. 5 May 1979, Special Issue on Excimer Lasers. C. K. Rhodes editor.
Imami, M., and Borst, W. L. (1974). *J. Chem. Phys.* **61**, 1115–1117.
Imami, M., and Borst, W. L. (1975). *J. Chem. Phys.* **63**, 3602–3605.
Inokuti, M. (1971). *Rev. Mod. Phys.* **43**, 297–347.
Ivash, E. V. (1958). *Phys. Rev.* **112**, 155–158.
Jaduszliver, B., Dang, R., Weiss, P., and Bederson, B. (1980). *In* "Coherence and Correlation in Atomic Collisions" (H. Kleinpoppen and J. F. Williams, eds.), p. 613–623. Plenum, New York.
Javan, A., Bennett, W. R., and Herriott, D. R. (1961). *Phys. Rev. Lett.* **6**, 106–110.
Jones, R. K., and Bonham, R. A. (1981). Private communication.
Jung, C., and Kruger, H. (1978). *Z. Phys. A* **287**, 7–13.
Jung, C., and Taylor, H. S. (1981). *Phys. Rev. A* **23**, 1115–1126.
Jung, K., Antoni, Th. Muller, R., Kochem, K. H., and Ehrhardt, H. (1982). *J. Phys. B* **15**, 3535–3555.
Keller, H. U., and Lillie, C. F. (1974). *Astron. Astrophys.* **34**, 187–196.
Kennerly, R. E. (1980). *Phys. Rev. A* **21**, 1876–1883.
Kennerly, R. E., Bonham, R. A., and McMillan, M. (1979). *J. Chem. Phys.* **70**, 2039–2041.
Kerner, E. H. (1953). *Phys. Rev.* **92**, 1441–1447.
Kessler, J. (1976). "Polarized Electrons." Springer-Verlag, Berlin.
Kessler, J. (1982). *Comments At. Mol. Phys.* **12**, 53–60.
Khare, S. P. (1966a). *Phys. Rev.* **149**, 33–37.
Khare, S. P. (1966b). *Phys. Rev.* **152**, 74–75.
Khare, S. P. (1967). *Phys. Rev.* **157**, 107–112.
Khare, S. P., and Moiseiwitsch, B. L. (1966). *Proc. Phys. Soc.* **88**, 605–610.
Kieffer, L. J. (1969a). *At. Data* **1**, 19–89.
Kieffer, L. J. (1969b). *At. Data* **1**, 121–287.
Kieffer, L. J. (1971). Atomic *Data* **2**, 293–391.
Kieffer, L. J. (1973). *JILA Rep. ICR* **13**, September 30.
King, G. C., McConkey, J. W., and Read, F. H. (1977). *J. Phys. B* **10**, L541–L543.
Kleban, P., and Davis, H. T. (1977). *Phys. Rev. Lett.* **39**, 456–459.

Kleban, P., Foreman, L., and Davis, H. T. (1981). *Phys. Rev. A* **23**, 1546–1552.
Kleinpoppen, H., and Williams, J. F., eds. (1979). "Coherence and Correlation in Atomic Collisions." Plenum, New York.
Klonover, A., and Kaldor, U. (1979). *J. Phys. B* **12**, 323–330.
Kosoruchkina, A. D., and Trekhov, E. S. (1978). *Opt. Spectrosc. USSR* **44**, 359–360.
Kubo, M., Matsunaga, D., and Tanaka, H. (1981). *XIIth Int. Conf. on the Physics of Electronic and Atomic Collisions, Gatlinburg, TN, Abstracts*, pp. 344–345.
Kuppermann, A., Rice, J. K., and Trajmar, S. (1968). *J. Phys. Chem.* **72**, 3894–3903.
Kuppermann, A., Flicker, W. M., and Mosher, O. A. (1979). *Chem. Rev.* **79**, 77–90.
Kuyatt, C. E., and Simpson, J. A. (1967). *Rev. Sci. Instrum.* **38**, 103–111.
Lakoba, I. S., and Yakovlenko, S. I. (1980). *Sov. J. Quantum Electron.* **10**, 389–410.
Land, J. E. (1978). *J. Appl. Phys.* **49**, 5716–5721.
Lane, N. F. (1980). *Rev. Mod. Phys.* **52**, 29–119.
Lassettre, E. N. (1965). *J. Chem. Phys.* **43**, 4479–4486.
Lassettre, E. N., Berman, A. S., Silverman, S. M., and Krasnow, M. E. (1964). *J. Chem. Phys.* **40**, 1232–1242.
Lassettre, E. N., Skerbele, A., and Dillon, M. A. (1969). *J. Chem. Phys.* **50**, 1829–1839.
Lee, L. C., Slanger, T. G., Blck, G., and Sharpless, R. L. (1977). *J. Chem. Phys.* **67**, 5602–5606.
Lee, J. S., Doering, J. P., Potemra, T. A., and Brace, L. H. (1980). *Planet. Space Sci.* **28**, 947–971.
Lenard, P. (1902). *Ann. Phys.* **8**, 149–198.
Lin, S. L., Robson, R. E., and Mason, E. A. (1979). *J. Chem. Phys.* **71**, 3483–3498.
Linder, F., and Schmidt, H. (1971a). *Z. Naturforsch.* **26a**, 1603–1617.
Linder, F., and Schmidt, H. (1971b). *Z. Naturforsch.* **26a**, 1617–1625.
Macek, J., and Jacks, D. J. (1971). *Phys. Rev. A* **4**, 2288–2300.
McConkey, J. W. (1979). *In* "Symposium on Electron-Molecule Collisions, Invited Papers" (I. Shimamura and M. Matsuzawa, eds.), pp. 163–178. University of Tokyo Press, Tokyo.
McDaniel, E. W. (1964). "Collision Phenomena in Ionized Gases." Wiley, New York.
Malcolm, I. C., and McConkey, J. (1979). *J. Phys. B* **12**, L167–L170.
Malcolm, I. C., Dassen, H. W., and McConkey, J. W. (1979). *J. Phys. B* **12**, 1003–1017.
Marmet, P., and Kerwin, L. (1960). *Can. J. Phys.* **38**, 787–796.
Massey, H. S. W. (1935). *Trans. Faraday Soc.* **31**, 556–563.
Massey, H. S. W., and Mohr, C. B. O. (1931). *Proc. R. Soc. A* **132**, 605–630.
Massey, H. S. W., and Mohr, C. B. O. (1932). *Proc. R. Soc. A* **135**, 258–275.
Massey, H. S. W., Burhop, E. H. S., and Gilbody, H. B. (1969). "Electronic and Ionic Impact Phenomena", Vol. I. and Vol. II, 2nd Ed. Clarendon Press, Oxford.
Mathur, D., and Hasted, J. B. (1977). *J. Phys. B* **10**, L265–L267.
Miescher, E., and Alberti, F. (1976). *J. Phys. Chem. Ref. Data* **5**, 309–317.
Miescher, E., and Huber, K. P. (1978). *In* "Physical Chemistry Series Two," Vol. 3 (Spectroscopy). Butterworths, London.
Miller, K. J., and Krauss, M. (1967). *J. Chem. Phys.* **47**, 3754–3762.
Miller, T. M., and Kasdan, A. (1973). *J. Chem. Phys.* **59**, 3913–3919.
Miller, T. M., Kasdan, A., and Bederson, B. (1982). *Phys. Rev. A* **25**, 1777–1778.
Mnatsakanyan, A. Kh. (1974). *High Temp.* **12**, 745–762.
Mohlmann, G. R., and de Heer, F. J. (1976). *Chem. Phys. Lett.* **43**, 240–244.
Mohr, C. B. O., and Nicoll, F. H. (1932). *Proc. R. Soc. London A* **138**, 229–244, 469–478.
Moiseiwitsch, B. L., and Smith, S. J. (1968). *Rev. Mod. Phys.* **40**, 238–353.
Morgan, W. L., and Szöke, A. (1981). *Phys. Rev. A* **23**, 1256–1265.
Morse, P. M. (1929). *Phys. Rev.* **34**, 57–64.
Mott, N. F., and Massey, H. S. W. (1965). "The Theory of Atomic Collisions," 3rd Ed., p. 349. Oxford Press.

Mozumder, A. (1982). *J. Chem. Phys.* **76**, 3277–3284.
Mumma, M. J., and Zipf, E. C. (1971). *J. Chem. Phys.* **55**, 1661–1669.
Mumma, M. J., Stone, E. J., and Zipf, E. C. (1971). *J. Chem. Phys.* **54**, 2627–2634.
Mu-Tao, L., and McKoy, V. (1982). *J. Phys. B* **15**, 3971–3983.
Mu-Tao, L., Lucchese, R. R., and McKoy, V. (1982). *Phys. Rev. A* **16**, 3240–3248.
Mu-Tao, L., and McKoy, V. (1983a). *Phys. Rev. A* **28**, 697–705.
Mu-Tao, L., and McKoy, V. (1983b). *J. Phys. B* **16**, 657–669.
Nadler, I., Rotem, A., and Rosenwaks, S. (1982). *Chem. Phys.* **69**, 375–380.
Nighan, W. L. (1978). *IEEE J. Quantum Electron.* **QE-14**, 714–726.
Norcross, D. W., and Collins, L. A. (1982). Recent Developments in the Theory of Electron Scattering by Highly Polar Molecules. *In* "Advances in Atomic and Molecular Physics 18 (D. R. Bates and B. Bederson, eds.). Academic Press, New York.
Oda, N., and Osawa, T. (1981). *J. Phys. B* **14**, L563–L568.
Onda, K., and Truhlar, D. G. (1979). *J. Chem. Phys.* **71**, 5107–5123.
Onda, K., and Truhlar, D. G. (1980). *J. Chem. Phys.* **72**, 5249–5262.
Oppenheimer, J. R. (1928). *Phys. Rev.* **32**, 361–376.
Padial, N. T., Meneses, G. D., da Paixão, F. J., Csanak, G., and Cartwright, D. C. (1981). *Phys. Rev. A* **23**, 2194–2212.
Pavlovic, Z., Boness, M. J. W., Herzenberg, A., and Schulz, G. J. (1972). *Phys. Rev. A* **6**, 676–685.
Peek, J. M. (1964). *Phys. Rev.* **134**, A877–A883.
Peek, J. M. (1965). *Phys. Rev.* **140**, A11–A18.
Pierce, D. T., and Celotta, R. J. (1980). *Nature* **284**, 308.
Pierce, D. T., Meier, F., and Zurcher, P. (1975). *Phys. Lett. A* **51**, 465–466.
Pierce, D. T., Celotta, R. J., Wamg, G. C., Unerti, W. N., Galeji, A., Kuyatt, C. E., and Mielczarek, S. R. (1980). *Rev. Sci. Instrum.* **51**, 478–499.
Piper, L. G. (1982). *J. Chem. Phys.* **77**, 2372–2377.
Pitchford, L. C., O'Neill, S. V., and Rumble, J. R. (1981). *Phys. Rev. A* **23**, 294–304.
Pochat, A., Rozuel, D., and Peresse, J. (1973). *J. Phys.* **34**, 701–709.
Povch, M. M., Subenich, V. V., and Zapesochnyi, I. P. (1972). *Opt. Spectrosc.* **32**, 565.
Read, F. H. (1978). *J. Phys.* **39**, C182–C186.
Read, F. H. (1980). *In* "The Physics of Ionized Gases" (M. Matec and Boris Kidric, eds.), pp. 33–48. Institute of Nuclear Sciences, Beograd, Yugoslavia.
Read, F. H., and King, G. C. (1979). *In* "Symp. on Electron Molecule Collisions" (I. Shimamura and M. Matsuzawa, eds.), pp. 155–162. University of Tokyo Press, Japan.
Register, D. F., Trajmar, S., Jensen, S. W., and Poe, R. T. (1978). *Phys. Rev. Lett.* **41**, 749–752.
Register, D. F., Nishimura, H., and Trajmar, S. (1980a). *J. Phys. B* **13**, 1651–1662.
Register, D. F., Trajmar, S., and Srivastava, S. K. (1980b). *Phys. Rev. A* **21**, 1134–1151.
Rescigno, T. N., McKoy, V., and Schneioler, B., eds. (1979). "Electron–Molecule and Photon–Molecule Collisions." Plenum, New York.
Rescigno, T. N., McCurdy, C. W., McKoy, V., and Bender, C. F. (1976). *Phys. Rev. A* **13**, 216–228.
Rice, J. K., Truhlar, D. G., Cartwright, D. C., and Trajmar, S. (1972). *Phys. Rev. A* **5**, 762–782.
Rhodes, C. K. ed. (1979). Excimer Lasers Vol. 30. *In* "Topics in Applied Physics." Springer-Verlag, Berlin.
Rohr, K. (1977a). *J. Phys. B* **10**, L399–L402.
Rohr, K. (1977b). *J. Phys. B* **10**, L735–L738.
Rohr, K. (1978a). *J. Phys. B* **11**, 1849–1860.
Rohr, K. (1978b). *J. Phys. B* **11**, 4109–4117.
Rohr, K. (1979a). *XIth Int. Conf. on the Physics of Electronic and Atomic Collisions, Kyoto, Japan, Abstract of Contributed Papers*, pp. 322–323.

Rohr, K. (1979b). *XIth Int. Conf. on the Physics of Electronic and Atomic Collisions, Kyoto, Japan, Abstracts of Contributed Papers*, pp. 324–325.
Rohr, K. (1979c). *In* Symposium on Electron Molecule Collisions Invited Papers, I. Shimamura and M. Mortsuzawa, eds. pp. 67–76. University of Tokyo Press, Japan.
Rohr, K. (1980). *J. Phys. B* **13**, 4897–4905.
Rohr, K., and Linder, F. (1976). *J. Phys. B* **9**, 2521–2537.
Roscoe, R. (1941). *Philos. Mag.* **31**, 349–362.
Rozsnyai, B. F. (1967). *J. Chem. Phys.* **47**, 4102–4112.
Rumble, J. R., Truhlar, D. G., and Morrison, M. A. (1981). *J. Phys. B* **114**, L301–L305.
Rusanov, V. D., Fridman, A. A., and Sholin, G. V. (1981). *Sov. Phys. Usp. Engl. Transl.* **24**, 447–474.
Sauerbrey, R., Tittel, F. K., Wilson, W. L., and Nighan, W. L. (1982). *IEEE J. Quantum Electron* **QE-18**, 1336–1340.
Schulz, G. J. (1958). *Phys. Rev.* **112**, 150–154.
Schulz, G. J. (1976). *In* "Principles of Laser Plasmas" (G. Bekefi, ed.), Chap. 2. Wiley, New York.
Seng, G., and Linder, F. (1974). *J. Phys. B* **7**, L509.
Seng, G., and Linder, F. (1976). *J. Phys. B: Atom. Phys.* **9**, 2539–2551.
Shaw, D. A., King, G. C., Read, F. H., and Cvejanovic, D. (1982). *J. Phys. B* **15**, 1785–1793.
Shemansky, D. E., and Broadfoot, A. L. (1971). *J. Quant. Spectrosc. Radiat. Transfer* **11**, 1401–1437.
Shimoda, K., ed. (1976). High-Resolution Laser Spectroscopy, Vol. 13. *In* "Topics in Applied Physics." Springer-Verlag, Berlin.
Simpson, J. A. (1964). *Rev. Sci. Instrum.* **35**, 1698–1704.
Skerbele, A. M., and Lassettre, E. N. (1964). *J. Chem. Phys.* **40**, 1271–1275.
Slater, R. C., Fickes, M. G., Becker, W. G., and Stern, R. C. (1974a). *J. Chem. Phys.* **60**, 4697–4709.
Slater, R. C., Fickes, M. G., Becker, W. G., and Stern, R. C. (1974b). *J. Chem. Phys.* **61**, 2290–2293.
Sohn, W., Jung, K., and Ehrhardt, H. (1983). *J. Phys. B* **16**, 891–902.
Sokolov, V. J., and Sokolava, Y. A. (1981). *Sov. Tech. Phys. Lett. Engl. Transl.* **7**, 268–269.
Spence, D. (1981). *J. Chem. Phys.* **74**, 3898–3904.
Srivastava, S. K., and Jensen, S. (1977). *J. Phys. B* **10**, 3341–3346.
Srivastava, S. K., Chutjian, A., and Trajmar, S. (1975). *J. Chem. Phys.* **63**, 2659–2665.
Srivastava, S. K., Chutjian, A., and Trajmar, S. (1976). *J. Chem. Phys.* **64**, 1340–1344.
Stanton, P. N., and St. John, R. (1969). *J. Opt. Soc. Am.* **59**, 252–260.
Stone, E. J., and Zipf, E. C. (1972a). *J. Chem. Phys.* **56**, 2870–2874.
Stone, E. J., and Zipf, E. C. (1972b). *J. Chem. Phys.* **56**, 4646–4650.
Stone, E. J., Lawrence, G. M., and Fairchild, C. E. (1976). *J. Chem. Phys.* **65**, 5083–5092.
Szmythovski, C., and Zubek, M. (1978). *Chem. Phys. Lett.* **57**, 105–108.
Takayanagi, K. (1967). *Prog. Theor. Phys. Suppl. Jpn.* **40**, 216–248.
Takayanagi, K., and Itikawa, Y. (1970). *Space Sci. Rev.* **11**, 380–450.
Tanaka, H., Yamamoto, T., and Okada, T. (1981). *J. Phys. B* **14**, 2081–2088.
Taniguchi, T., Tagashira, H., and Sakai, Y. (1977). *J. Phys. D* **10**, 2301–2306.
Thirumalai, D., Onda, K., and Truhlar, D. G. (1981). *J. Chem. Phys.* **74**, 6792–6805.
Thomas, L. D., Csanak, G., Taylor, H. S., and Yarlagadda, B. S. (1974). *J. Phys. B* **7**, 1719–1733.
Trajmar, S. (1980). *Acc. Chem. Res.* **13**, 14–20.
Trajmar, S., and Register, D. F. (1984). Experimental Techniques for Cross Section Measurements. *In* "Electron Molecule Collisions" (I. Shimamura and K. Takayanagi, eds.). Plenum, New York.
Trajmar, S., Cartwright, D. C., Rice, J. K., Brinkmann, R. T., and Kupperman, A. (1968). *J. Chem. Phys.* **49**, 5464–5472.
Trajmar, S., Truhlar, D. G., Rice, J. K., and Kuppermann, A. (1970). *J. Chem. Phys.* **52**, 4516–4533.
Trajmar, S., Cartwright, D. C., and Williams, W. (1971). *Phys. Rev. A* **4**, 1482–1492.
Trajmar, S., Williams, W., and Kuppermann, A. (1972). *J. Chem. Phys.* **56**, 3759–3765.

Trajmar, S., Cartwright, D. C., and Hall, R. I. (1976). *J. Chem. Phys.* **65**, 5275–5279.
Trajmar, S., Register, D. F., and Chutjian, A. (1983). *Phys. Rep.* **97**, 219–356.
Truhlar, D. G., Rice, J. K., Kuppermann, A., Trajmar, S., and Cartwright, D. C. (1970). *Phys. Rev. A* **1**, 778–802.
Truhlar, D. G., Brandt, M. A., Chutjian, A., Srivastava, S. K., and Trajmar, S. (1976). *J. Chem. Phys.* **65**, 2962–2969.
Truhlar, D. G., Brandt, M. A., Srivastava, S. K., Trajmar, S., and Chutjian, A. (1977). *J. Chem. Phys.* **66**, 655–663.
Vallance-Jones, A. (1974). "Aurora." Reidel, Boston.
van der Wiel, M. J. (1980). *In* "Electronic and Atomic Collisions" (N. Oda and K. Takayanagi, eds.), pp. 209–218. North Holland, Amsterdam.
van der Wiel, M. J., and Wiebes G. (1971). *Physica* **53**, 225–255.
van Sprang, H. A., Brongersma, H. H., and de Heer, F. J. (1979). *Chem. Phys. Lett.* **65**, 55–60.
van Wingerden, B., Wagenaar, R. W., and de Heer, F. J. (1980). *J. Phys. B* **13**, 3481–3491.
Vas'kov, V. V., Golyan, S. F., Gruzdev, Yu. V., Gurevich, A. V., Dimant, Ya. S., Kim, V. Yu., Lobachevskii, L. A., Migulin, V. V., Panchenko, V. A., Petrov, M. S., Polimatidi, V. P., Sitnikov, V. I., Shoya, L. D., Shlyuger, I. S., and Yurin, K. I. (1981). *JETP Lett.* **34**, 558–561.
Vušković, L., Srivastava, S. K., and Trajmar, S. (1978). *J. Phys. B* **11**, 1643–1652.
Vušković, L., and Trajmar, S. (1982). *J. Chem. Phys.* **77**, 5436–5440.
Vušković, L., and Trajmar, S. (1983). *J. Chem. Phys.* **78**, 4947–4951.
Wakiya, K. (1978a). *J. Phys. B* **11**, 3913–3930.
Wakiya, K. (1978b). *J. Phys. B* **11**, 3931–3938.
Walker, I. C., Stamatovic, A., and Wong, S. F. (1978). *J. Chem. Phys.* **69**, 5532–5537.
Watson, C. E., Dulock, V. A., Stolarksi, R. S., and Green, A. E. S. (1967). *J. Geophys. Res.* **72**, 3961–3966.
Watson, Jr., J., and Anderson, R. J. (1977). *J. Chem. Phys.* **66**, 4025–4030.
Weatherford, C. A. (1980). *Phys. Rev. A* **22**, 2519–2528.
Weingartshofer, A., Ehrhardt, H., Hermann, V., and Linder, F. (1970). *Phys. Rev. A* **2**, 294–304.
Weingartshofer, A., Clarke, E. M., Holmes, J. K., and McGowan, J. W. (1975). *J. Phys. B* **8**, 1552–1569.
Wells, W. C., Borst, W. L., and Zipf, E. C. (1973). *Phys. Rev. A* **8**, 2463–2468.
Whalen, B. A., Yau, A. W., Crentzberg, F., Gattinger, R. L., Harris, F. R., Vallance-Jones, A., McNamara, A. B., Pongratz, M. B., Smith, G. M., Forsyth, P. A., and Koehlelr, J. A. (1981). *Can. J. Phys.* **59**, 1175–1182.
Whitten, R. C., and Poppoff, I. G. (1971). "Fundamentals of Aeronomy." Wiley, New York.
Wilden, D. C., Hicks, P. G., and Comer, J. (1979). *J. Phys. B* **12**, 1579–1590.
Willett, C. S. (1974). "Introduction to Gas Lasers: Population Inversion Mechanisms." Pergamon, New York.
Wong, S. F. and Schulz, G. J. (1974). *Phys. Rev. Lett.* **32**, 1089–1092.
Wu, T. Y. (1947). *Phys. Rev.* **71**, 111–118.
Wyckoff, S., and Wehinger, P. A. (1976). *Astrophys. J.* **204**, 616–625.
Zamir, E., Huestis, D. L., Nakano, H. H., Hill, R. M., and Lorents, D. C. (1979). *IEEE J. Quantum Electron.* **QE-15**, 281–288.
Zipf, E. C., and Gorman, M. R. (1980). *J. Chem. Phys.* **73**, 813–819.
Zipf, E. C., and McLaughlin, R. W. (1978). *Planet. Space Sci.* **26**, 449–462.
Zubek, M., and Szmytkowski, C. (1979). *Phys. Lett. A* **74**, 60–62.
Zubek, M., Kadifachi, S., and Hasted, J. B. (1981). *European Conf. on Atomic Physics*, April 6–10, Heidelberg, West Germany, Book of Abstracts, p. 763.

3 Ionization of Molecules by Electron Impact

T. D. Märk
Institut für Experimentalphysik
Leopold Franzens Universität
Innsbruck, Austria

I.	Introduction	251
II.	Ionization Mechanisms and Types of Ions Produced	252
	A. Ionization Mechanisms	252
	B. Types of Ions Produced	261
	C. Quasiequilibrium Theory	277
III.	Total and Partial Ionization Cross Sections	279
	A. Theoretical Considerations	279
	B. Experimental Techniques	287
	C. Results	296
IV.	Differential Ionization Cross Sections	306
	A. Electron–Electron Spectroscopy	307
	B. Fragment-Ion Kinematics	310
	References	315

I. Introduction

A review of electron-impact ionization of molecules is a formidable task because the field is so broad (e.g., Märk and Dunn 1984). Our understanding of how molecules become ions has deepened a great deal during the past years. I have tried to limit this review to those aspects which allow the development of a complete picture of the ionization process. This inevitably involves the illustrative (rather than exhaustive) discussion of phenomena with the help of small molecules, which have been studied in more detail and whose properties are better understood than those of large molecules. However, proper reference has been made, where necessary, to polyatomic molecules. Particular emphasis has been put on those phenomena which will help us in our ultimate quest, i.e., the ability to describe the mechanisms of ionization and to predict qualitatively or quantitatively the outcome of the ionization process in terms of differential, partial, and total ionization cross sections.

II. Ionization Mechanisms and Types of Ions Produced

A. Ionization Mechanisms

1. Ionization Processes

Electrons accelerated through an electric potential gradient can collide with molecules of a gas-phase target either elastically or inelastically. Provided that the energy of this electron beam is greater than a critical value, some molecules in the gas-phase target will be ionized. As the energy of the electron beam is increased, the abundance and variety of ionized species will increase. Moreover, in addition to the ionization channels possible for an atom, in the case of a molecular target, some of the ionized molecules may dissociate if the electron energy is sufficiently great.

Hence, the electron-impact ionization of a molecule may occur in several different channels, each of which gives rise to characteristic ionized and neutral products. They are, for the simplest case of a diatomic molecule XY (not including the production of negative ions, e.g., see Dillard 1973, Christophorou 1971, Melton 1970),

$$XY + e^- \longrightarrow XY^+ + e_s^- + e_e^- \quad \text{single ionization} \quad (1)$$

$$\longrightarrow XY^{2+} + e_s^- + 2e_e^- \quad \text{double ionization} \quad (2)$$

$$\longrightarrow XY^{n+} + e_s^- + ne_e^- \quad \text{multiple ionization} \quad (3)$$

$$\longrightarrow X^+ + Y + e_s^- + e_e^- \quad \text{dissociative ionization} \quad (4)$$

$$\longrightarrow X^+ + Y^- + e_s^- \quad \text{ion pair formation} \quad (5)$$

with e_s^- 'scattered' electron and e_e^- 'ejected' electron. Other products (e.g., excited ions) may be obtained, especially when using more complex molecules. However, a discussion of these is deferred.

2. Frank–Condon Principle

Inelastic collisions between electrons and molecules involve transitions between two well-defined (electronic, vibrational, and rotational) molecular states. The energy losses of the scattered electron in the excitation of molecular vibration and rotation are relatively small compared to the energy of the electronic transition, at least assuming that only one vibrational quantum is excited in a single collision. For instance, the greatest separation between a ground and a first excited vibrational state of any molecular ion, that in H_2^+, is 0.27 eV (Massey et al., 1969) compared to the ionization potential of 15.426 eV (Rosenstock et al., 1977). The changes in the vibrational levels from v to v' that result in the ionization process can be described, as is well known, in terms of the *Franck–Condon principle*, whereas rotational excitation depends on the validity (above 800 eV electron energy) or breakdown (below 800 eV) of the electric dipole selection rule (Hernandez et al., 1982).

Qualitatively, the Franck–Condon principle may be summarized as follows: in an electronic transition no (or only negligible) changes occur in the nuclear separation and velocity of relative nuclear motion. Because of the great ratio of nuclear to electronic mass and the short interaction time, the point on the upper potential-energy curve (corresponding to the configuration after the transition) lies directly above (*vertical transition*) the starting point on the initial potential energy curve. This leads to a number of possible electronic transitions which depend on the relative shapes of the potential-energy curves available in a specific system. The various possible Franck–Condon transitions can be discussed with help of Fig. 1 showing the potential-energy curves for electronic states of H_2 and H_2^+ (see also Pichou et al., 1983). The nuclear separation in the ground vibrational state lies between the limits R_1 and R_2 (effective Franck–Condon region; shaded area in Fig. 1) for all possible transitions, and according to the Franck–Condon principle it must still lie between R_1 and R_2 after any of these transitions. From this it follows that several cases are possible:

(1) The final level accessible lies within the region of discrete vibrational states of the upper potential-energy curve [e.g., transition $H_2(X\,^1\Sigma_g^+) \to H_2(B\,^1\Sigma_u^+)$ in Fig. 1]. The probability that the vibrational quantum number will change depends on the relative position of the potential-energy curves, which itself depends on geometric and energetic properties of the respective molecular states. This probability can be calculated for diatomic molecules using Born's first approximation, yielding the Franck–Condon overlap integral

$$p_{l,u} \sim \left(\int \psi_l^v \psi_u^{v'} \, dR \right)^2, \qquad (6)$$

with $p_{l,u}$ the probability that the transition involves the vibrational level v in the lower state l and v' in the upper state u, ψ_l^v and $\psi_u^{v'}$ the corresponding

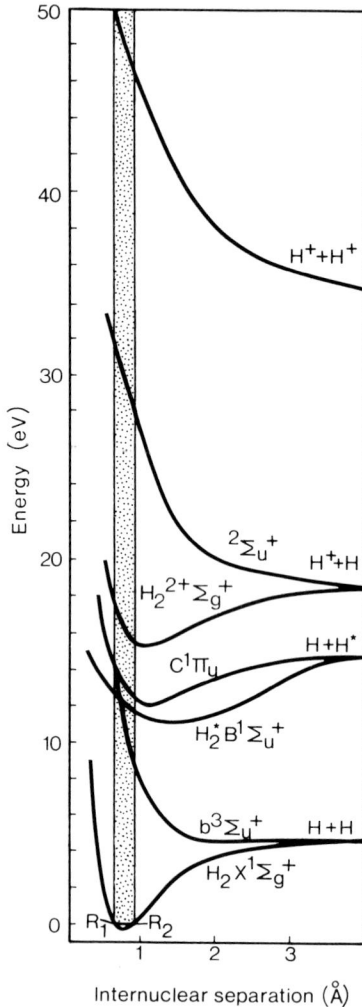

Fig. 1. Schematic potential-energy diagram for ground state of H_2 and some states of H_2^*, H_2^+, and H_2^{2+}. [After Massey et al. (1969).]

vibrational wave functions, and R the internuclear distance (Massey et al., 1969).

(2) The final level accessible lies not only within the region of discrete vibrational states but includes some part of the continuum [e.g., transition $H_2(X\,^1\Sigma_g^+) \rightarrow H_2^+(^2\Sigma_g^+)$]. Hence some of the transitions will lead to dissociation (e.g., Section II.B.2.)

(3) The final level accessible lies within the continuum of a repulsive state and all transitions lead to dissociation [e.g. transition $H_2(X\,^1\Sigma_g^+) \to H_2^+(^2\Sigma_u^+)$].

It will of course be realized that the above considerations are applicable only to diatomic molecules, allowing for this species a quantitative description of the ionization process (appearance potentials, cross section ratios; see, for example, Sections II.B.2 and IV.B.1). In the case of polyatomic molecules (with the exception of small polyatomic molecules of high symmetry, which can be treated theoretically in a similar manner as diatomic molecules by using correlation rules; see Section II.B.2) the potential curves are to be replaced by n-dimensional potential-energy surfaces where n refers to the number of atoms in the molecule. Although electron-impact ionization will proceed in a way that the ion is formed without changes in the nuclear separations (Franck–Condon principle), the resulting vibrationally excited molecular ion may by internal transitions undergo further unimolecular decompositions along energetically available paths. Hence, for large polyatomic molecules it is impossible to interpret fully the ionization process in terms of detailed potential hypersurfaces. Therefore it becomes necessary to apply statistical concepts, and the so-called quasiequilibrium theory (QET) based on statistical dynamics is used for this purpose (Rosenstock *et al.*, 1952). A discussion of its applicability is deferred to Section II.C.

3. Autoionization

The ionization processes discussed so far may be classified as direct ionization processes [e.g., processes (1)–(4)], where the 'ejected' and the 'scattered' electron leave the ion simultaneously or within 10^{-16} s of each other (Berry, 1974; Johnson and Morrison, 1975). There exists, however, a second type of ionization process (competing very strongly with direct ionization), in which the electrons are ejected one after another, a process known as autoionization via a bound metastable state (when using electrons instead of photons there is a relaxation of the selection rules governing the excitation of these states):

$$XY + e \longrightarrow XY^{**} + e_s \qquad (7)$$
$$\hookrightarrow XY^{+*} + e_e$$
$$\hookrightarrow XY^{2+} + 2e_e$$
$$\hookrightarrow X^+ + Y^-,$$

$$XY + e \longrightarrow XY^{+*} + e_s + e_e \qquad (8)$$
$$\hookrightarrow XY^{2+} + e_e.$$

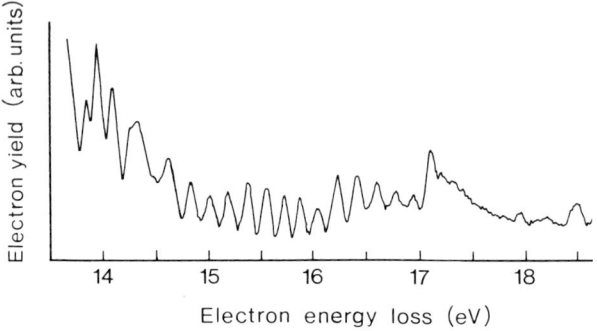

Fig. 2. Electron energy loss spectrum of CO of the region above the ionization potential measured with a time-of-flight spectrometer (Incident electron energy 50 eV, scattering angle 40°, energy resolution 55 meV.) [After Comer (1978).]

The autoionization event can be described in terms of a two-step process. First, a molecule (XY) is raised into a superexcited, bound state (XY**), which exists for some finite time [neutral or ionized (see Section II.B.4) metastable state]. This excited state then couples radiationless to the continuum and ionization occurs. The occurrence of autoionization is tied to the existence of degrees of freedom (electronic excitation, vibration, rotation, spin angular momentum) among which energy may be distributed and from which this energy eventually is redistributed to a mode that leads irreversibly to a continuum state (autoionization or predissociation). The upper autoionization rate is limited by the characteristic energy storage mode frequency. If predissociation of the excited neutral state is faster than autoionization the latter will not occur at an appreciable rate. Autoionization is a resonance process and may complicate the interpretation of ionization cross section functions (see Section III.C.1; see also Märk 1984a).

The two electrons emitted, e_e^- and e_s^-, share the energy difference between the incident electron energy and the final-ion-state potential energy. Energy loss spectra for the scattered electron generally show a wealth of structure above the ionization potential due to the excitation of autoionizing molecular states (see Fig. 2). Additional measurement of the corresponding electron ejection spectra yields information about the respective decay modes of the excited states into final ionic states. In practice, however, the interpretation of such results for molecules is difficult owing to the large number of possible initial and final states. This problem can be overcome by making measurements on the scattered and ejected electrons in coincidence (e, 2e spectroscopy; see also Section IV.A). This allows the identification of the final states [see, e.g., Fig. 3; (Comer, 1978)] and the measurement of binding energies and electronic momentum distributions (Weigold and McCarthy, 1978; Erhardt et al., 1980).

3 Ionization of Molecules by Electron Impact   257

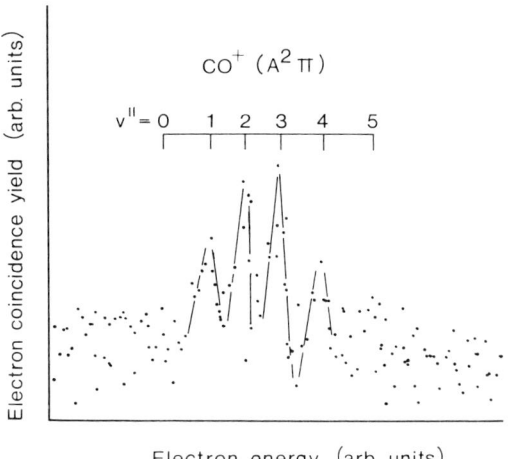

Fig. 3. Ejected-electron coincidence spectrum of CO. This spectrum shows at an electron energy loss of 18.44 eV peaks representing electron emitting transitions to different vibrational levels of the A $^2\pi$ state. The ejected-electron energy increases from right to left. [After Comer (1978).]

4. *Inner-shell Ionization*

Usually, electron-impact ionization refers to the removal of a valence-shell electron, but if the energy of the incoming electron is high enough an inner-shell or core electron can be removed. This latter process is termed inner-shell ionization and can be identified either by the subsequent emission of characteristic x rays and/or Auger electrons or by the analysis of energy loss spectra.

The normal decay path for most of the light-element core holes is Auger decay. Thus, for inner-shell ionized molecules we must expect at least doubly charged final states, except for those cases where the initial excitation of an inner-shell electron proceeds to unoccupied discrete states (yielding a core–hole neutral excited state). However, in general the dominant decay path is dissociative ionization, since the majority of doubly ionized molecules is not stable (see Section II.B.3), e.g., the fraction of inner-shell ionization events leading to dissociation is $\sim 80\%$ for K-shell ionization of CO (Kay et al., 1977) and nearly as much for SF_6 (Hitchcock et al., 1978) comprising a number of ion pairs (as seen from the ion autocorrelation diagram given in Fig. 4).

The question whether a core–hole remains localized at one nucleus of a homonuclear molecule or whether hole hopping occurs during the lifetime against Auger decay has recently been discussed by Van der Wiel (1980) (see also Camilloni et al., 1982). Moreover, Van der Wiel has shown that electron

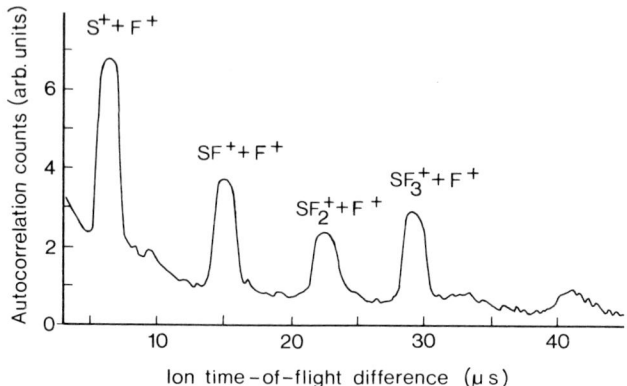

Fig. 4. Schematic autocorrelation spectrum of ion pairs formed by dissociative double ionization [DDI, van der Wiel (1980)] of SF_6 (Incident electron energy 8 keV.) [After Hitchcock et al. (1978).]

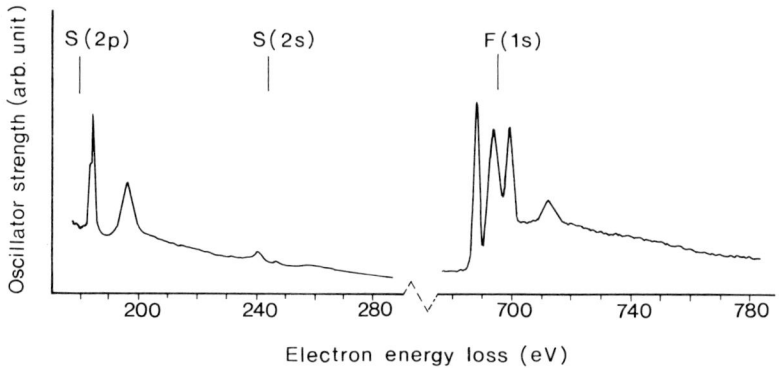

Fig. 5. Electron energy loss spectrum of SF_6 near the F(1s) and S(2p)edges (Incident electron energy 2.5 keV; ordinate calibration see Section IV.A). [After Hitchcock and Brion (1978).]

energy loss spectroscopy (see also Shaw et al., 1979; Oda et al., 1979, Brion and Hitchcock, 1981) has made significant contributions to the present understanding of the strikingly "nonatomic" features of molecular inner-shell absorption spectra around inner-shell edges [see as an example the energy loss spectrum of SF_6, given in Fig. 5 as measured by Hitchcock and Brion (1978)].

5. *Electron–Electron Correlation*

In concluding Section A it is interesting to consider long-range electron–electron correlations present in near-threshold electron-impact ionization

3 Ionization of Molecules by Electron Impact

(Read, 1980; Klar, 1980) in order to gain a full picture of the ionization mechanism and its kinetic aspects. We shall consider the process

$$X + e^- \longrightarrow X^+ + e_s^- + e_e^-, \qquad (9)$$

with the total excess energy of the two free electrons after the ionization process being very small. Using hyperspherical coordinates and assuming the ion to be a point charge we express the potential energy of the system of the scattered and ejected electron and a positive ion of charge e as (in atomic units)

$$V = -C(\alpha, \theta_{1,2})/2R, \qquad (10)$$

where

$$C(\alpha, \theta_{1,2}) = \frac{2}{\cos \alpha} + \frac{2}{\sin \alpha} - \frac{2}{(1 - \sin 2\alpha \cos \theta_{1,2})^{1/2}}, \qquad (11)$$

with $R = (r_1^2 + r_2^2)^{1/2}$, $\alpha = \tan^{-1}(r_2/r_1)$, $\theta_{1,2} = \cos^{-1}(\mathbf{r}_1, \mathbf{r}_2)$, and r_1 and r_2 are the positions of the two electrons with respect to the ion. Figure 6 shows, according to Lin (1974) and Fano and Lin (1975), the functional dependence ("relief plot") given in Eq. (11) for $R = 1$. The two ditches (at $\alpha = 0$ and $\pi/2$, and $0 \leq \theta_{1,2} \leq \pi$) correspond to attractive forces between the electrons and the ion, whereas the spike (at $\alpha = \pi/4$ and $\theta_{1,2} = 0$) corresponds to the

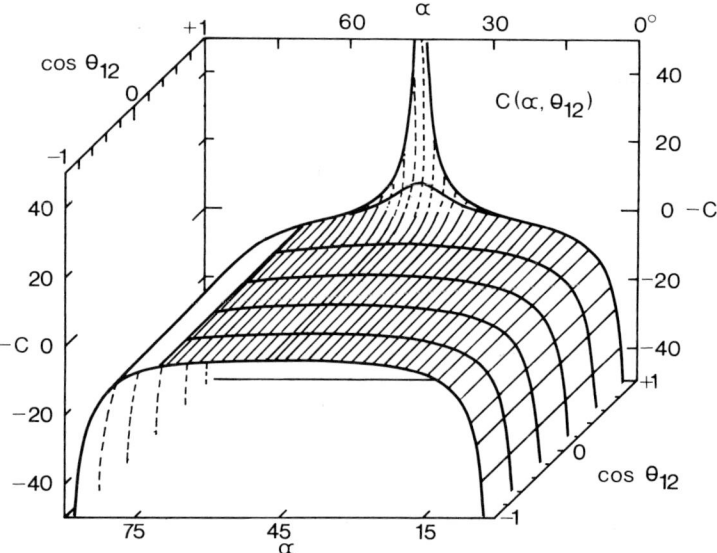

Fig. 6. Functional dependence (relief plot) given in Eq. (11): $C = C(\alpha, \theta_{1,2})$. The ordinates represent a potential surface at $R = 1$ [see Eq. (10)]. [After Lin (1974) and Fano and Lin (1975).]

repulsive force between the two electrons. In the neighborhood of the *Wannier point* (at $\alpha = \pi/4$ and $\theta_{1,2} = \pi$) exists a broad saddle-shaped region. The potential energy $V(\alpha)$ is maximum at $\alpha = \pi/4$.

This relief plot can be used to explain the escape process of the three-particle system after the collision without any knowledge of the ionization process in the reaction zone itself. This was first done by Wannier (1953) following a similar procedure by Wigner (1948). An important parameter in Wannier's treatment is the critical distance R_c at which $|V|$ equals the total excess kinetic energy. It follows that ionization occurs if the three particles start after the collision in the neighborhood of the Wannier point and remain in this region until $R > R_c$. In this case $|V|$ is able to continue to decrease with increasing R with the result that both electrons eventually become free. Conversely, if the particles are far from the Wannier point (in terms of α) when $R = R_c$, one of the electrons becomes bound to the positive ion owing to the dominating action of the attractive forces in the ditches. Hence, for double escape (single ionization) to occur after the reaction the system must proceed within the narrow cone in hyperspace ($\alpha = \pi/4$ and $\theta_{1,2} = \pi$), leading to actual ionization. In other words, actual ionization occurs when the distances r_1 and r_2 of the two electrons remain approximately equal (radial correlation). If one electron lags behind it is decelerated in the attractive field of the ion and may end up in a bound state (van de Water and Heideman, 1981). Moreover, the repulsive Coulomb interaction forces the two electrons to move in nearly opposite directions (angular correlation; see also Section IV.A).

The lower the value of the total excess kinetic energy, the closer the three particles must pass through the Wannier point in order to allow double escape of the electrons. This is due to the fact that for low total excess kinetic energy the system has longer times to move from the unstable Wannier saddle region into one of the ditches. In the absence of correlations between the two electrons the available phase space corresponding to ionization, and thus the cross section, is proportional to the total excess energy (Read, 1980). Owing to the correlation effect close to the threshold, the ionization cross section starts more slowly than linearly. This and other threshold laws (in particular as applied to molecules) will be discussed further in Section III.A.4.

A similar escape model has been invoked to explain the apparent exchange of energy that can occur between two electrons receding from a positive atomic ion, when the incident electron energy is near the energy of an autoionizing state of the target system (Read, 1975). These energy exchanges manifest themselves, e.g., by an increase of the observed threshold for the excited autoionizing state, and this effect is termed postcollision interaction (Heideman *et al.*, 1974; Hicks *et al.*, 1974; Smith *et al.*, 1974; Spence, 1975; King *et al.*, 1975; Morgenstern *et al.*, 1976; Heideman and Van de Water 1981). Work is in progress to obtain evidence for the existence of energy shifts on some simple diatomic molecules (Read, 1975).

B. Types of Ions Produced

1. Parent Molecular Ions

The term molecular ion or parent ion designates a positively charged ion as produced by reaction (1) through removal of one electron from a neutral molecule with a lifetime of $\geq 10^{-5}$ s (see below). Most organic molecules have an even number of electrons. When a single electron is removed, an ion with an odd number of electrons results; this is sometimes indicated by a dot, e.g., in the case of methane

$$CH_4 + e^- \longrightarrow CH_4^{+\cdot} + 2e. \qquad (12)$$

Conversely, if the target molecule is a free radical, the ionization process leads to an even-electron ion, e.g., in the case of nitric oxide,

$$NO^{\cdot} + e^- \longrightarrow NO^+ + 2e. \qquad (13)$$

The production of these molecular ions relative to other (fragment) ions (see below) is dependent on the electron energy and on the nature of the molecule. If a sample is bombarded with electrons of gradually increasing energy it can be found that below the ionization potential (for molecules usually 7–15 eV) the probability of ion formation is zero, but it rises as the ionizing electron energy is increased above the ionization potential E_i. At and just above the ionization potential only singly charged parent ions are produced, but at higher electron energies other ions (fragment ions and multiply charged ions) are observed. Moreover, a molecular ion has, of course, many excited states and it may be produced in any one of the vibrational-electronic levels accessible according to the Franck–Condon principle. This is mainly due to the fact that the range of allowed kinetic energies of the scattered electrons is effectively continuous and hence any amount of excess energy can be removed by the scattered electrons.

The relative importance of the various ions as the electron energy is varied can be illustrated with help of clastograms (see, for example, Fig. 7). In general, a direct relationship exists between the stability (against decomposition into fragment ions) of a parent ion and its relative abundance in the clastogram. For small molecules (diatomics, triatomics, and some tetratomics) the parent ion is the dominant ion at all electron energies, although there are noted exceptions, e.g., CF_4, CCl_4 (Stephan et al., 1983a,b), IF_7 (Schack et al., 1968). On the other hand, for large molecules the parent-ion intensity usually decreases with higher electron energies (Fig. 8) and with increasing molecular weight (Fig. 9) (Brunnee and Voshage, 1964; Spiteller, 1966; Field and Franklin, 1970; Litzow and Spalding, 1973). Although this is valid in increasing chain lengths, many exceptions have been found at higher

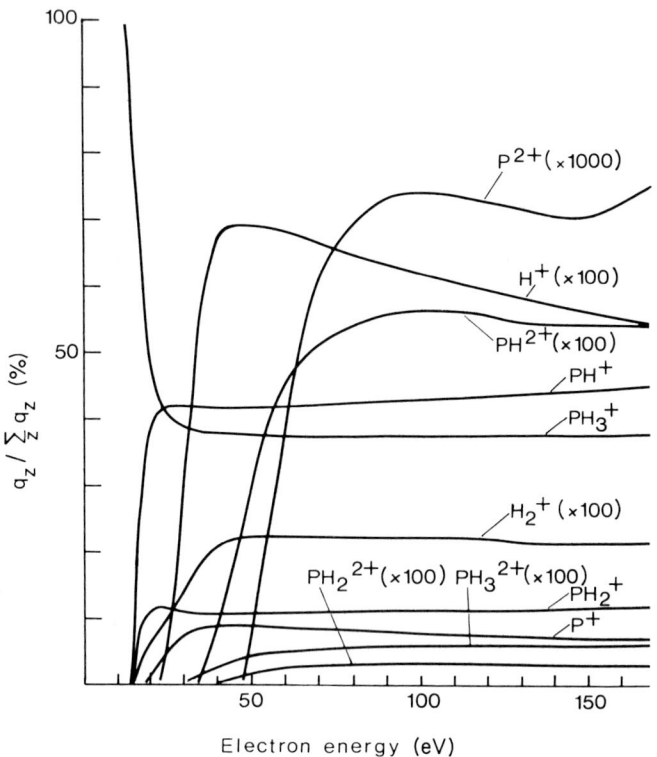

Fig. 7. Clastogram for PH_3 using data of Märk and Egger (1977). Plotted are the partial ionization cross sections divided by the counting ionization cross section as a function of electron energy.

molecular weights (Mc Lafferty, 1966). In some cases, the intensity of the parent ion is too small to be detected. This is, in general, due to the fact that the rate of parent-ion decomposition is so high that all parent ions have decayed in the time between formation and collection ($\sim 10^{-5}$ s in most mass spectrometers).

The choice of electron energies of 50–100 eV for running mass spectra was decided largely on the basis that fragmentation patterns do not vary very much with electron energy in this energy range and that the ionization cross section (and hence detection efficiency) has its maximum in this energy range. Because so much energy may be imparted in this energy range to the molecular ion that several generations of fragment ions (see next section) may be produced, a superfluity of peaks is obtained. Because of this fact electron-impact ionization is considered a "nonsoft" ionization technique as compared to photoionization. However, if similar energies are used for photoionization

3 Ionization of Molecules by Electron Impact

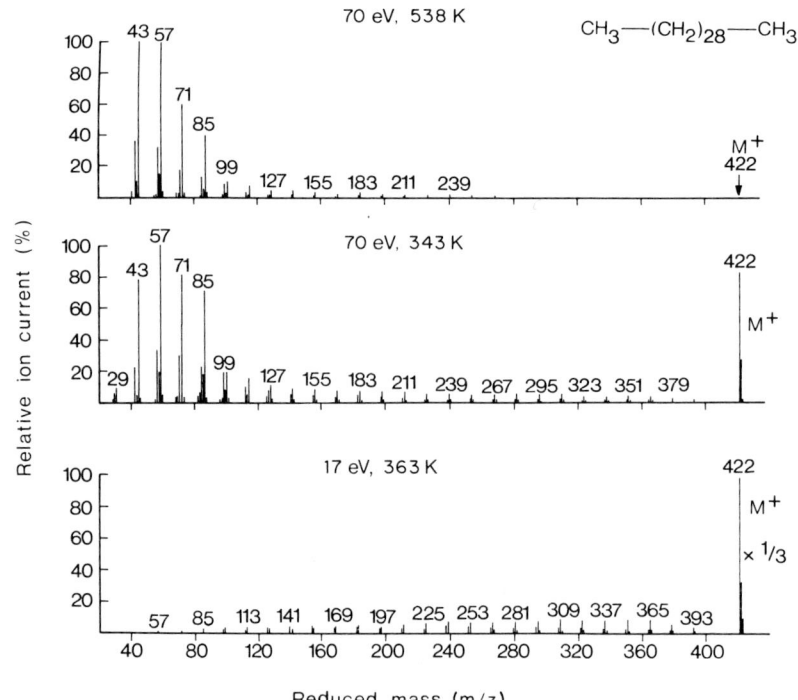

Fig. 8. Schematic mass spectra of triacontane. [After Remberg *et al.* (1968).]

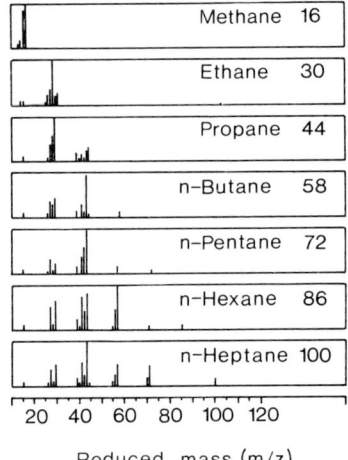

Fig. 9. Schematic mass spectra of some light paraffins (Incident electron energy 70 eV.) [After Brunnee and Voshage (1964).]

and electron-impact ionization (see Fig. 10), then very similar mass spectra are observed; i.e., at low electron energies it is possible to simplify the mass spectrum and to obtain less fragmentation ("soft ionization"). Moreover, in specific cases photoionization appears to lead to more fragmentation than electron impact (see Fig. 11).

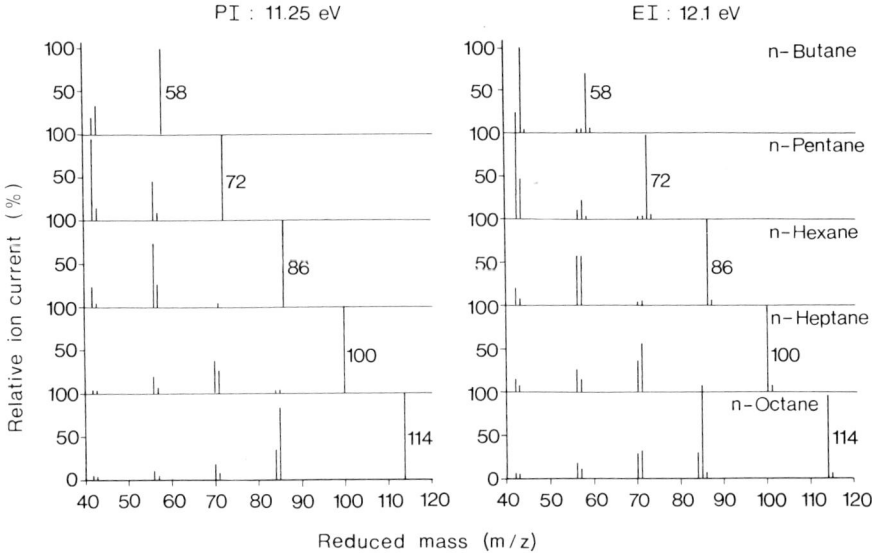

Fig. 10. Schematic mass spectra of the *n*-alkanes obtained by photoionization with photons of 11.25 eV energy [after Steiner *et al.* (1961)] and by electron-impact ionization at 12.1 eV [after Maccoll (1982)].

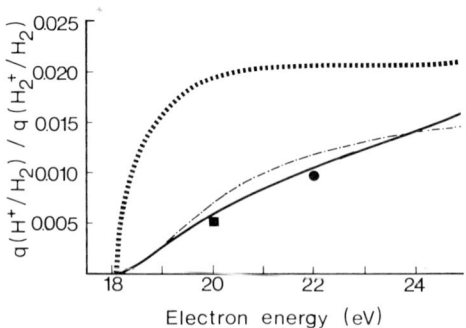

Fig. 11. Partial ionization cross section ratio for the production of H^+ and H_2^+ in hydrogen as a function of incident energy: (■) Adamczyk *et al.* (1966), (●) Hipple (see Crowe and Mc Conkey, 1973a), (—) electron-impact data (Crowe and Mc Conkey, 1973a), (– · –) electron-impact curve predicted by Browning and Fryar (1973), (- - -) photon-impact data (Browning and Fryar, 1973).

It should also be mentioned that the abundance of the parent ion may critically depend on the temperature of the molecular gas target (see Fig. 8); i.e., with higher temperature (1) the specific intensities of all ions decrease (owing to the fact that molecular kinetic energy is a function of temperature) and (2) the relative intensity of the parent ion decreases appreciably [owing to the increase in the vibrational energies of the neutral precursor, which leads to a higher subsequent fragmentation rate of the molecular ion; Erhardt and Osberghaus (1958), Jackson et al. (1974)]. The latter can be explained for polyatomic ions in terms of the QET (see below), where an increase in temperature increases the maximum of the distribution function of the molecular excess energy with the consequence that the fragmentation rate of the various decomposition processes increases (Erhardt and Osberghaus, 1960). It is interesting to note that well-defined relations appear to exist for the temperature dependence of the relative parent-ion intensities, although they are quite different for different systems (Field and Franklin, 1970).

2. Fragment Ions

It has already been mentioned that as the electron energy is increased above the ionization potential of a molecule fragment ions appear as produced by reaction (4). In the case of a polyatomic molecule, however, a wide range of fragment ions may be produced. Dissociation of the parent ion can occur by various channels; e.g., the molecular ion $M^{+\cdot}$ can decay into an even-electron fragment ion $(M^{+\cdot} \rightarrow F_i^+ + N_j^{\cdot})$ or into an odd-electron fragment ion $(M^{+\cdot} \rightarrow F_i^{+\cdot} + N_j)$. A fragment ion F_1^+ produced in such a single transition is termed a primary fragment ion. These primary fragment ions may be produced in excited states and subsequently decay into further fragments, termed secondary, tertiary, or further fragment ions. This secondary fragmentation again may be depicted as yielding either an even-electron ion and a neutral molecule $(F_1^+ \rightarrow F_2^+ + N)$ or a positively charged radical and a neutral radical $(F_1^+ \rightarrow F_2^{+\cdot} + N^{\cdot})$. Both reactions are present in the fragmentation of organic ions, but the even-electron fragments containing only paired electrons tend to be more stable and hence more abundant.

The possible fragmentation paths of a hypothetical triatomic molecule XYZ may be represented in the following way:

$$\begin{aligned}
XYZ + e^- &\longrightarrow XYZ^+ + 2e^- \\
XYZ^+ &\longrightarrow X^+ + YZ \longrightarrow X^+ + Y + Z \\
XYZ^+ &\longrightarrow XY^+ + Z \longrightarrow X^+ + Y + Z \\
& \longrightarrow X + Y^+ + Z \\
XYZ^+ &\longrightarrow X + YZ^+ \longrightarrow X + Y + Z^+ \\
& \longrightarrow X + Y^+ + Z \\
XYZ^+ &\longrightarrow XY + Z^+ \longrightarrow X + Y + Z^+
\end{aligned} \qquad (14)$$

yielding five different fragment ions by these straightforward fragmentations. In addition other ions (multiply charged and rearrangement ions) may appear (see the following sections).

The number of fragment ions produced and their relative abundance as a function of electron energy are characteristic of the corresponding parent molecule. There exist three main categories of fragment ions: (1) fragment ions which contain the functional group of the molecule, (2) ions which are produced by the rupture of the functional group (e.g., rupture of hydrocarbon bonds), and (3) rearrangement ions. In general, the occurrence of certain fragment ions reflects both the relative strengths of the various bonds in the ion and the stability of the neutral fragments. Hence, the study of cracking patterns and fragmentation pathways yields information on the molecular structure [correlation studies; see, e.g., Reed (1962)]. At the time of writing, over 31,000 compounds are tabulated in the *Eight Peak Index of Mass Spectra* (1975) and 55,000 compounds in the Registry of mass spectral data (1982), and new spectra are continuously being added. Such a wealth of data makes it impossible to discuss here the fragmentation patterns of individual compounds. This is done in review books especially dedicated to certain subjects, e.g., organic, inorganic, or organometallic compounds (e.g., Beynon, 1960; Biemann, 1962; Mc Lafferty, 1963, 1966; Budzikiewicz *et al.*, 1964; Beynon *et al.*, 1965, 1968; Spiteller, 1966; Reed, 1966; Kienitz, 1968; Kiser *et al.*, 1968; Benz 1969; von Ardenne *et al.*, 1971; Porter and Baldas 1971; Williams and Howe 1972; Hamming and Foster 1972; Litzow and Spalding, 1973; Charalambous, 1975; Levsen, 1978; Williams, 1978; Budzikiewicz, 1981; Johnstone, 1980). Several attempts exist to find common features and to formulate empirical rules for the decomposition of different types of compounds; some of these will be summarized below.

In the case of diatomic molecules the production of fragment ions can, in principle, be treated quantitatively in terms of the Franck–Condon principle (Dunn, 1966; Browning and Fryar, 1973; Crowe and Mc Conkey, 1973a) and compared with observed ratios (fragment to parent ion) (Stevenson, 1947; Schaeffer and Hastings, 1950; Dibeler *et al.*, 1950; Schaeffer, 1955; Stevenson, 1960; Adamczyk *et al.*, 1966; Crowe and Mc Conkey, 1973a; see also III.C.2); e.g., see Fig. 11. At higher electron energies the formation of fragment ions from repulsive potential-energy curves renders this calculation more difficult, but even so it is possible to adequately explain the mass spectra of diatomic molecules in terms of the Franck–Condon principle.

As mentioned earlier, for the ionization and dissociation of small polyatomic molecules an approach is used in which spectroscopic and quantum-theoretical ideas are utilized to determine the dissociation paths and to predict the electronic states and optical allowed or forbidden (Van Sprang *et al.*, 1979) excitation mechanisms of fragments produced. Basically, in its beginning this approach consisted of extending the correlation rules given by Wigner and Witmer (1928) to small linear polyatomic ions. Moreover, Mulliken (1933) has used group-theoretical methods to classify the electronic states of small nonlinear polyatomic molecules with high degrees of symmetry. This method has been applied successfully to elucidate the dissociation process in a number of small molecular ions, e.g., H_2O (Laidler, 1954); CH_4, NH_3 (Mc Dowell, 1954; Liehr, 1957); CH_3X, H_2, N_2, O_2, HCN (Mc Dowell, 1963a); CH_3OH (Momigny et al., 1980); and more recent references given in Section IV.B. More recently, with the advent of *ab initio* configuration interaction calculations (see, e.g., Pople, 1975) detailed information on the potential-energy surface of the different electronic states and of their crossings is becoming available. This field has recently been reviewed by Lorquet (1980), including a discussion of the dissociation of such ions as H_2O^+, HCN^+, H_2CO^+, and $C_2H_4^+$.

With regard to the fragmentation of large polyatomic molecules, the theoretical considerations applied in the above cases are no longer usable because of the large number of electronic and vibrational states with less and less spacing between the energy levels. To deal with these molecules the quasiequilibrium theory (Rosenstock et al., 1952) provides some means to elucidate the theoretical origin of fragmentation patterns. But because of its complexity and the large number of parameters employed in calculations this theory has not been applied in a widespread fashion to the interpretation of mass spectra. Basic aspects and present applications will be discussed in Section II.C. There exist, in addition, alternative treatments, which take into account the fact that primary ionization of the molecule might occur at preferred locations depending on the binding energy of the respective electrons (bonding or nonbonding). This leads to situations in which the most probable charge centers can be quite remote from the most probable centers of fission (Read, 1962; Spiteller, 1968).

Because of this the interpretation of fragmentation patterns relies heavily on empiricism. In general, the relative abundance of any fragment ion is related to its rate of formation and its rate of decomposition. Hence, a measured mass spectrum is a record in time of the position of the quasiequilibrium of these rates for the respective ions. There are several factors that will influence these rates: (1) ionization potential, (2) electron configuration, (3) bond dissociation energies, (4) rearrangement and elimination channels. With regard to the first factor, a useful generalization was first reported by Stevenson (1951), who noted that for hydrocarbons the appearance potential $AP(X^+/XY)$ for reaction (4) is only given as the sum of the ionization potential of X and the dissociation energy $DE(X-Y)$ if $IP(X) < IP(Y)$. Later this finding was extended by Audier (1969) and is used as Stevenson's rule, stating that the decay path that is favored, in case a molecule

AB^+ can either decay to $A^+ + B$ or to $A + B^+$, is the one that produces the ion whose neutral species has the lower ionization potential. Moreover, in this case the products will be in their ground state, whereas in the other case (production of ion with higher ionization potential) the products will be excited. Although Stevenson's rule is based on observation of paraffin hydrocarbons, it is followed with a few exceptions (Field and Franklin, 1970). Another generalization is that odd-electron ions are much less stable (abundant) than even-electron ions, which contain no unpaired electrons. Exceptions to this rule have, however, been reported (see, e.g., Reed, 1962). There are other generalizations and classifications with respect to factors (2)–(4), and the reader is referred to extensive previous reviews in this matter (Beynon, 1960; Biemann, 1962; Reed, 1962; Mc Lafferty, 1963; Spiteller, 1966; Kienitz, 1968; Field and Franklin, 1970; Litzow and Spalding, 1973; Levsen, 1978; Budzikiewicz, 1981).

3. *Multiply Charged Ions*

Although singly charged parent and fragment ions account for the vast majority of ions occurring by electron-impact ionization, multiply-charged ions may also be produced as shown in reactions (2) and (3). Multiply-charged *atomic* ions were observed very early (Thomson, 1912b); the first *molecular* multiply-charged ions were observed by Thomson (1921), Conrad (1930), and Vaughan (1931). Subsequent observation of numerous doubly charged polyatomic species followed. In addition, triply charged molecular ions have been detected (Märk, 1984b) in low abundance in the mass spectra of a number of aromatic hydrocarbons and some derivatives of such hydrocarbons (Dibeler *et al.*, 1953; Quinn and Mohler, 1959; Dorman and Morrison, 1961; Meyerson and Van der Haar, 1962; Van Brunt and Wacks, 1964), in phosphonitrilic chlorides (Brion and Paddock, 1968), in small molecules such as COS, CS_2, C_2N_2 (Newton, 1964; Fiquet-Fayard *et al.*, 1968), and in field evaporation spectra (Kellogg, 1981, 1982). Conversely, for certain molecules it is not possible to detect any doubly charged ions [e.g., H_2O (Märk and Egger, 1976) and CH_4 (Hanner and Moran, 1981)] or triply charged ions [e.g., HCl, HBr (Johnston and Arnold, 1953; see also Becker and Dietze, 1983 and Märk, 1984b)]. In addition, for some molecules only the parent ion can be observed doubly ionized, e.g., NH_3 (Märk *et al.*, 1977a), whereas in case of other molecules doubly charged fragment ions can also be observed, e.g., PH_3 (Märk and Egger, 1977).

All of the diatomic doubly charged ions XY^{2+} observed (e.g., O_2^{2+}, N_2^{2+}, CO^{2+}, NO^{2+}, Cl_2^{2+}, Br_2^{2+}, I_2^{2+}, ICl^{2+}, IBr^{2+}, HCl^{2+}, HBr^{2+}, HI^{2+}, $SBCl^{2+}$ (Thomson, 1921; Conrad, 1930; Vaughan, 1931; Friedländer *et al.*, 1932; Tate *et al.*, 1935; Nier and Hanson, 1936; Kusch *et al.*, 1937; Johnston and Arnold, 1953; Herron and Dibeler, 1960; Dorman and Morrison, 1961;

Kupriyanov, 1964; Stalherm, et al., 1969; Beynon et al., 1971; Märk, 1973, 1975; Newton and Sciamanna, 1969, 1970b; Appell et al., 1973; Hille and Märk, 1978; Brehm and De Frenes, 1978; Kim et al., 1981; Agee et al., 1981; Nasrallah et al., 1983) satisfy the energetic condition

$$IP(X^+) + IP(Y^+) < IP(X^{2+}) \leq IP(Y^{2+}), \quad (15)$$

where IP denotes the respective single and double ionization potential for the atoms X and Y. XY^{2+} ions characterized by Eq. (15) exist when short-range chemical forces impose a sufficiently strong attractive well in the repulsive Coulomb interaction between X^+ and Y^+ [potential curves; see, e.g., Friedländer et al. (1932), Magee and Gurnee (1952), Hurley and Maslen (1961), Hurley (1962, 1971), Stalherm et al. (1969), Beynon et al. (1971), Thulstrup et al. (1974), Thulstrup and Anderson (1975), Beebe et al. (1976), Moran et al. (1980), Cobb et al. (1980), Sramek et al. (1980), Taylor (1983)]. Figure 12 gives as an example the potential curves for He_2^{2+} as shown by Helm et al. (1981) according to a calculation by Cohen and Bardsley (1978). He_2^{2+} has not yet been observed directly (Johnson and Biondi, 1978), but it can be used to illustrate the above-mentioned properties. The lowest lying doubly charged state $X\ ^1\Sigma_g^+$ exhibits a relative minimum at about $1.4a_0$ due to valence forces similar to those present in the isoelectronic H_2 molecule. There also exists an attractive state $B\ ^1\Sigma_u^+$ arising from the interaction of $He^{2+} + He$. This state is short lived, because the $B \to X$ transition is dipole allowed.

As has recently been pointed out by Helm et al. (1981) and Stephan et al. (1981a), there is a second class of doubly charged diatomic ions satisfying the relation

$$IP(X^+) + IP(Y^+) > IP(X^{2+}). \quad (16)$$

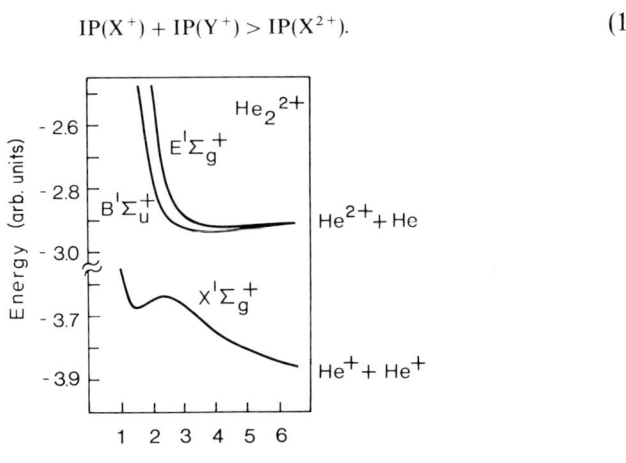

Fig. 12. Schematic potential-energy diagram for He_2^{2+}. [After Cohen and Bardsley (1978) and Helm et al. (1981).]

In this case the repulsive Coulomb state arising from X^+ and Y^+ lies energetically above the weakly bound state arising from $X^{2+} + Y$ or $X + Y^{2+}$. This allows, in principle, the formation of stable XY^{2+} at energies below the dissociation limit $X^+ + Y^+$. Three combinations of rare-gas dimers fall into this group, namely $HeXe^{2+}$, $HeKr^{2+}$, and $NeXe^{2+}$. One of these ions, $NeXe^{2+}$, has recently been observed experimentally (Johnson and Biondi, 1979; Helm et al., 1981), and Helm et al. (1981) reported semiquantitative potential curves derived from isoelectronic neutral molecules. They also reported the existence of long-lived $ArXe^{2+}$ and $NeKr^{2+}$ ions which belong to the first class of doubly charged diatomic ions [e.g., potential-energy curves of $NeKr^{2+}$; see Fig. 13 (Stephan et al., 1981a)]. Measured appearance potentials of these ions are consistent with the predictions from these potential-energy curves.

Information concerning the production of polyatomic doubly ionized species had been compiled up to 1949 by Mohler et al. (1949). More recent studies concerning the molecular structure, fragmentation, fragment-ion kinetic-energy distribution (see also Section IV.B), charge localization and the method of doubly charged ion mass spectrometry have been referenced in the papers of Mathur et al. (1980), Jones et al. (1982), Hanner et al. (1982) and Appling et al. (1983).

The ionization efficiency for the production of doubly charged ions of organic compounds rarely exceeds 1–5% of the base peak production

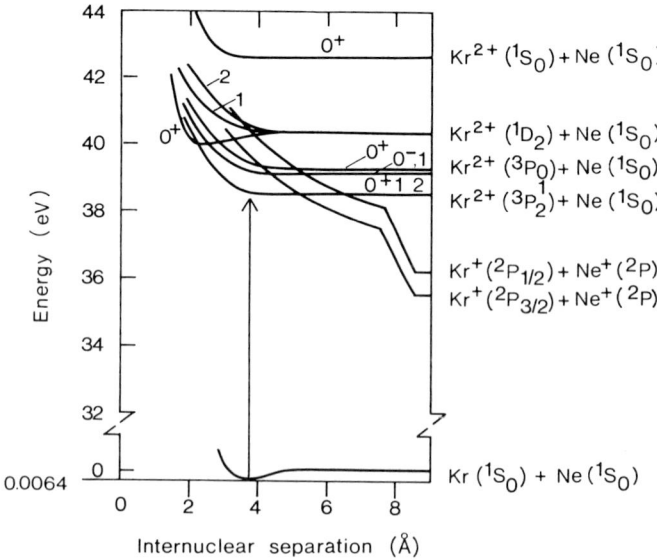

Fig. 13. Potential-energy diagram for NeKr and $NeKr^{2+}$. [After Stephan et al. (1981a).]

(Beynon, 1960; Biemann, 1962). Aromatic or heteroaromatic molecules which do not contain bonds which are easily ruptured have been found to possess an increased ability to sustain two positive charges (Kienitz, 1968). This appears to be also true for certain aminoboranes (Dewar and Rona, 1965), polyfluor compounds (Kienitz, 1968), and organometallic compounds (King, 1969). Moreover, in some cases it was noted that the doubly charged parent molecular ion was more abundant than the corresponding singly charged species (Solomon and Mandelbaum, 1969; Waight, 1969; Hellwinkel, and Wünsche, 1969, Stimpson et al., 1978). Conversely, mass spectra of saturated aliphatic hydrocarbons contain almost no doubly charged ions (Kienitz, 1969). This very low abundance of doubly charged ions in aliphatic hydrocarbons has been partially attributed to the possibility that doubly charged ions may decompose to two singly charged fragments (Mohler et al., 1954). The presence of olefinic bonds and/or heteroatoms was found to enhance the stability (and hence abundance) of doubly charged ions, since in this case the ionization process involves the removal of π (nonbonding) electrons and not σ-bonded electrons as for hydrocarbons (Vouros and Biemann, 1969).

Doubly ionized molecules can decompose to fragment ions in a way similar to their singly ionized analogs. The two charges, however, may either be shared by the two separating fragment ions or else remain with one of the resulting fragments, i.e.,

$$XY^{2+} \longrightarrow X^+ + Y^+, \tag{17}$$

$$XY^{2+} \longrightarrow X^{2+} + Y. \tag{18}$$

According to Brehm and De Frenes (1978, 1980), in general a large fraction of the electron-impact multiple-ionization processes of molecules leads to singly charged fragments [process (17)] rather than to one multiply-charged fragment and a neutral [process (18)]. The product ions X^+ and Y^+ of reaction (17) are identical with those formed in the fragmentation of the singly ionized parent species XY^+, and hence a decomposition of this type is usually established only if the corresponding "metastable peak" is detected. This "charge separation" process (sometimes termed Coulomb explosion, see also Sattler et al., 1981 and Jentsch et al., 1982) has recently been reviewed by Ast (1980) and will be treated also in the next section.

The first evidence of a doubly charged ion decomposing via reaction (18) into a second doubly ionized species and a neutral fragment was reported by Beynon and Williams (1959) for the metastable transition $(p - 2\,CO)^{2+} \rightarrow (p - 3\,CO)^{2+} + CO$, with p = 2-hydroxyanthraquinone. This type of fragmentation has also been reported for loss of H_2, C_2H_2, and C_2H_4 in many spectra involving ions of even mass number (Meyerson and Van der Haar, 1962), for boron hydrides (Fehlner and Koski, 1964), for azulene and naphthalene (Van Brunt and Wacks, 1964), for metal and transition-metal carbonyl (Winters and Kiser, 1966a; Winters and Collins, 1966) parent molecule ions. In case of tungsten hexacarbonyl (Winters and Kiser, 1966a) it was observed that, contrary to previous observations—on the spectra of hydrocarbons (Biemann, 1962) where no direct relationship has been observed between singly and doubly charged ions—a definite relation appears to exist between the separate spectra of singly and doubly charged ions. In fact, the occurrence of triply charged fragment ions (Meyerson and Van der Haar, 1962) implies that also triple ionization may be followed by a loss of a neutral particle with the formation of a new triply charged molecule.

It is interesting to add at the end of this section that electron-impact ionization of van der Waals clusters [such as H_{2n}, N_{2n}, $(CO_2)_n$] can lead, according to reports by Henkes and Isenberg (1970), Gspann and Körting (1973), and Echt et al. (1982), to multiply-charged cluster ions (see also Dole et al., 1968) with up to four and five elementary charges. Stimpson et al. (1978) observed decomposition of doubly charged cluster ions via reaction (18).

4. Metastable Ions

Many of the ions produced by electron impact are formed with sufficient internal energy to dissociate before reaching the collector of the respective detecting system (i.e., in a time of about $< 10^{-5}$ s). These ions are referred to as *metastable ions*. Conversely, a *stable ion* (in terms of detection) is defined as one which is produced within the ion source and travels from this source to the collector without decomposition. Moreover, an *unstable ion* is defined as one which is produced with an amount of internal energy so that it decomposes immediately into fragment ions in the ion source before leaving the ionization regime (i.e., in a time of about $< 10^6$ s). In a more general sense, however, *metastable ions* with lifetimes from 10^{-3} to 10^{-11} s have been observed (Lifshitz, 1978; Brenton et al., 1979).

If the metastable decomposition occurs in flight before the analyzer of a magnetic-type instrument this process gives rise to a metastable peak in the mass spectrum. Hipple and Condon (1945), Hipple et al. (1946), and Hipple (1948) were the first to correctly interpret the occurrence of these peaks, showing that the resultant daughter ion of mass m_2 appears at a position on the mass scale designated as the apparent mass m^* (Beynon et al., 1965),

$$m^* = \frac{(m_2/z_2)^2}{m_1/z_1} \left(1 + \frac{m_1 - m_2}{m_2} \frac{U_A - U}{U_A} \right), \tag{19}$$

where U_A is the accelerating voltage, U the potential difference through which the parent (precursor) ion of mass m_1 falls before decomposition, and z_1 and z_2 the respective charges. It follows from this equation that metastable peaks need not and in general do not occur at an integral m/z value. A considerable part of today's knowledge about metastable ions is based on this technique and has been extensively reviewed (Beynon, 1968; Cooks et al., 1973; Beynon and Cooks, 1976; Brenton et al., 1979).

The existence of metastable ions can be explained by metastable transitions corresponding to different mechanisms. Normally, dissociation occurs during the first oscillation, or predissociation that involves a transition from one potential-energy hypersurface to another, also occurs rapidly; hence the product ions are formed in the ion source (see fragment ions). However, if crossings of potential-energy surfaces are for instance spin forbidden, some ions will nevertheless undergo electronic predissociation (Herzberg, 1967) but will do so at a reduced and delayed rate.

Hence, *electronic (forbidden) predissociation* (Friedländer et al., 1932; Momigny, 1961; Mc Dowell, 1963b) via repulsive states is one origin of metastable ions and its occurrence has been documented for several small molecules; i.e., $H_2S^+ \to S^+ + H_2$ (Dibeler and Rosenstock, 1963b; Fiquet-Fayard and Guyon, 1966; Jones et al., 1972; Terwillinger et al., 1975; Jarrold et al., 1982a), $NO_2^+ \to NO^+ + N$ (Newton and Sciamanna, 1966), $HNCO^+ \to HCO^+ + N$ (Rowland et al., 1968), $CO^+ \to C^+ + O$ and $C + O^+$ (Wankenne and Momigny, 1969), $N_2^+ \to N^+ + N$ (Newton and Sciamanna, 1969, 1970a; Wankenne and Momigny, 1969, 1971; Fournier et al., 1970, 1971, 1972; Locht et al., 1975), $NO^+ \to O^+ + N$ and $O + N^+$ (Newton and Sciamanna, 1969, 1970a; Wankenne and Momigny, 1969; Govers and Schopman, 1971; Pham et al., 1971), $C_2H_3N_3^+ \to C_2H_3N^+ + N_2$ (Shannon, 1970), $CH^+ \to C^+ + H$ (Lorquet et al., 1970, 1971), $C_6H_5CO-CH_2X^+ \to C_6H_5CO^+ + CH_2X$ [X = F, Cl, Br; Cooks et al. (1974a)], $CF_4^+ \to CF_3^+ + F$ (Brehm et al., 1974; Stephan et al., 1983a,b), and $N_2O^+ \to NO^+ + N$ (Märk et al., 1981, and references therein). It is interesting to note that fine structure is seen for the transition $NO^+ \to O^+ + N$ when high-energy resolution is employed (Govers and Schopman, 1971). Furthermore, recently a number of small cluster ions have been observed and their existence explained by electronic and/or rotational predissociation (e.g., Stephan and Märk, 1982a–c, 1983, Stephan et al., 1983c–f).

As an example Fig. 14 shows the potential-energy surface correlation according to Fiquet-Fayard and Guyon (1966) for $H_2S^+ \to S^+ + H_2$. It can be seen that all three bonding states of H_2S^+ ($^2B_1$, $^2A_1$, and $^2B_2$) can undergo predissociation via the repulsive $^4A_2$ state to form the ground-state products $S^+(^4S)$ and $H_2(^1\Sigma_g^+)$ (for details see Jarrold et al., 1982a).

A second possible mechanism giving rise to metastable ions from small molecules is *dissociation by tunneling*. In this case the initial excitation is to bound levels of an excited state of the molecular ion which are above the dissociation limit of this state, but below the top of a weak dissociation

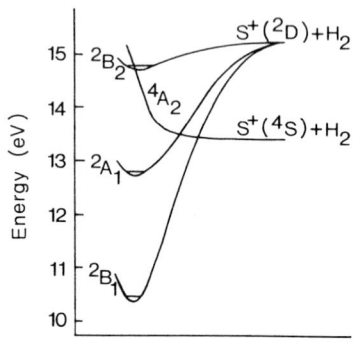

Fig. 14. Potential-energy diagram for H_2S^+ as a function of S^+-H_2 internuclear separation. [After Fiquet-Fayard and Guyon (1966).]

barrier (see e.g., Fig. 12 and references in Section II.B.3). Tunneling through these barriers has first been proposed for the dissociation of doubly charged ions (Friedländer et al., 1932; Kupriyanov, 1964; Newton and Sciamanna, 1969; Dorman and Morrison, 1961; Hurley, 1962; Beynon et al., 1971). Morever, it has been reported for singly charged ions as tunneling through the centrifugal barrier [*rotational predissociation,* (Herzberg, 1967)], i.e., hydrogen loss from metastable methane ions (Dibeler and Rosenstock, 1963a; Rosenstock, 1968; Klots, 1971; Solka et al., 1975; Flamme et al., 1975, 1978; Illies et al., 1982), from HeH^+ (Houver et al., 1970; Schopman et al., 1969, 1970, 1973; Schopman and Los, 1970; Peek, 1974; Fournier et al., 1976; Locht et al., 1976); from NH_3^+ and isotopic analogs (Jarrold et al., 1982b); and for the decay of He_2^+ (Maas et al., 1976; Flamme et al., 1980) and H_2^+ (Maas et al. 1975; Brenton et al., 1983). Moreover, Medved et al. (1976) have reported that the kinetic energy released when H is lost in a metastable decay of the *n*-propanol molecular ion displays a temperature dependence (Solka et al., 1975) and that this temperature dependence is likely to be due to a centrifugally controlled decay process. Charge separation [process (17)] with possible tunneling has been observed for many doubly charged diatomic ions (e.g., Kupriyanov, 1964; Beynon et al., 1971; Newton and Sciamanna, 1969, 1970b; Hirsch et al., 1975; Hille and Märk, 1978), triatomic ions (Newton and Sciamanna, 1964; Fuchs and Taubert, 1965; Newton and Sciamanna, 1970a, c; Cooks et al., 1974b) and other ions (e.g., Rabrenovic et al., 1983).

The existence of large metastable ions can be rationalized in the framework of the quasiequilibrium theory (see Section II.C) and is termed according to Herzberg (1967) *vibrational predissociation.* After electron-impact ionization radiationless transitions yield highly vibrational excited ground-state ions with a distribution of internal energies and hence a distribution of decay rate constants including those which lead to unimolecular decompositions in the metastable time range (see also Section II.B.2). For detailed information about the numerous metastable transitions studied so far the reader is referred to the reviews mentioned at the beginning of this section. Metastable cluster dissociations have been discussed by Märk and Castleman (1984).

A fourth possible metastable fragmentation path arises from the fact that it might be necessary for the parent ion to undergo an internal rearrangement involving actual transfer of atoms or ions between different parts of the molecule before the metastable transition can occur (Cooks et al., 1973). Similarly, Futrell et al. (1981, 1982) and Stephan et al. (1983c) have reported that in the case of ammonia cluster ions the availability of a rearrangement decomposition channel is a relevant factor for metastable dissociation.

The intensity of metastable peaks in a mass spectrum is usually less than 1% of the base peak. For certain molecules, e.g., $CF_4^+ \to CF_3^+$, $CCl_4^+ \to CCl_3^+$ (Brehm et al., 1974; Stephan et al., 1983a,b) it is not possible to detect any precursor ion signal. In addition, it has recently been found that the metastable to precursor ion ratio for certain cluster ions varies between 0.01 and 10 (Märk, 1982a; Stephan and Märk, 1982b,c, 1983) depending on the properties of the neutral precursor molecule. Other properties determined for metastable ions include the kinetic-energy release, the appearance potential, and the half-life of the metastable dissociation. Determination of the two former values will be discussed in Sections III.C.1 and IV.B. Experiments with ions of known half-lives have been discussed in detail by Brenton et al., 1979.

5. Rearrangement Ions

The term *rearrangement ions* is usually applied to the occurrence of ions containing bonds between atoms that were not directly bonded in the precursor molecule, i.e., produced by dissociation reactions of the following nature:

$$WXYZ + e \longrightarrow WXYZ^+ + 2e \longrightarrow XZ^+ + WY + 2e. \tag{20}$$

In order to be observable these rearrangements or transfers must occur in the short time between formation of the presursor molecular ion and either its acceleration from the ion source or its entering the analyzer region. Hence, for rearrangement ions to occur in observable amounts, their unfavorable entropies of activation [low-frequency factors of the right side of reaction (20)] must be balanced by favorable activation energies [$\Delta H_f^\circ(XZ^+) + \Delta H_f^\circ(WY)$ small]. Transfer of hydrogen or fluorine is common [see also rearrangement reactions in cluster ions; e.g., Stephan et al. (1982b); Märk and Castleman (1984)], but rearrangements involving larger groups such as CH_3 or C_6H_5 have also been frequently observed. Mc Lafferty

(1959) has distinguished between specific rearrangement, which leads to a high abundance of the rearranged ion, and a random form of rearrangement. Consecutive rearrangements may lead to a statistical distribution of the transferred system (atom) between different positions in the ion. As a result, an isotopic label (e.g., D) may be statistically scrambled within the ion (Levsen, 1978). For more detailed information on this important subject see reviews by Reed (1962), Mc Lafferty (1966), Spiteller (1968), Field and Franklin (1970), and Levsen (1978).

6. *Ion Pairs*

Fragment ions as discussed in Section II.B.2 can also be produced by the process of ion pair production. In this case, a positively and negatively charged fragment ion (and possibly a neutral) are formed simultaneously by electron-impact ionization of the parent molecule without the intermediate formation of a molecular ion, i.e.,

$$AB + e \longrightarrow A^+ + B^- + e \tag{21}$$

with A and B either atoms or molecules, e.g., $CO + e \rightarrow C^+ + O^- + e$ (Hagstrum and Tate, 1941). For ionization by this mechanism the molecule makes an electronic transition within the Franck–Condon region to a repulsive curve and dissociates into the respective fragments, all of which may be excited (see Fig. 15). The scattered electron can carry away excess energy and thus the process can [in contrast to other

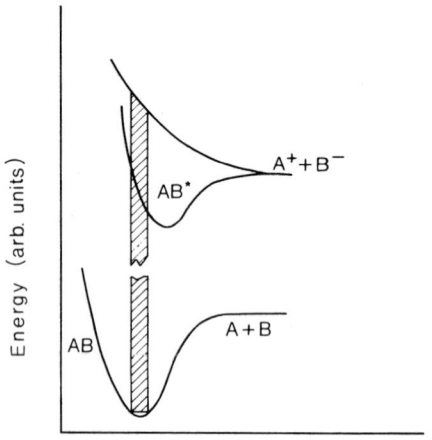

Fig. 15. Schematic potential-energy curves of a diatomic molecule AB, illustrating possible ion pair production by electron impact.

negative-ion production processes (Dillard, 1973; Bowie and Williams, 1975)] proceed over a wide energy range from above the appearance potential. Obviously a pair production process can give rise only to ions that can be formed by fragmentation and is most likely to occur when both ions are particularly stable. This process has been found and studied for a number of systems (e.g., Melton, 1963, and references cited therein; Winters and Kiser, 1966b; Kiser *et al.*, 1968; Collin *et al.*, 1968; Melton, 1970; Locht and Momigny, 1971; Van Brunt and Kieffer, 1974; Köllmann 1975a, Kurepa *et al.*, 1981; Mathur, 1981; Scheunemann *et al.*, 1982). Special experimental setups are necessary for detecting and analyzing the simultaneous formation of a positive and negative ion pair (e.g., Svec and Flesch, 1968; Ahnell and Koski, 1975).

C. Quasiequilibrium Theory

The quasiequilibrium theory [recently reviewed by Rosenstock and Krauss (1963), Rosenstock (1968); Vestal (1968); Wahrhaftig (1972); Cooks *et al.* (1973); Lifshitz (1978)] still provides the most successful theoretical approach to the discussion of dissociative ionization of large molecules, especially when combined with quantum mechanics (Lorquet, 1980, 1981). Levsen (1978) has demonstrated that when an adequate mathematical treatment is performed at least qualitative and in some cases quantitative agreement between QET and experimental observation (breakdown graphs, rate constants, and kinetic-energy release distributions) has been observed, thus supporting the validity of this theory.

The salient features of this theoretical approach are the following. The initial effect of electron-impact ionization of a molecule is to bring about the formation of the molecular ion by a Franck–Condon-type transition. This leads to the fact that the molecular nuclei are initially not necessarily in their most favorable locations, minimizing the potential energy of the ion, but mainly in a configuration corresponding to a minimum potential energy of the neutral molecule before electron impact. As a consequence of vibrational excitation energy the molecular ionic system starts to oscillate. Provided that electronic states differing little in energy exist in this energy region, radiationless energy transfer occurs between closely situated potential-energy surfaces. Consequently, much of the electronic energy deposited initially will be eventually converted into vibrational energy distributed among the numerous vibrational degrees of freedom. According to QET, decomposition of the molecular ionic system occurs after sufficient time has elapsed for the accumulation of energy equal to the dissociation energy in one of the system's degrees of freedom (termed an activated complex). The essential task of QET is to calculate the relative probabilities of excess internal energy among the available degrees of freedom populated in a completely random fashion.

Because QET theory is mathematically very complex (see references given above), only an outline of some of the basic underlying assumptions and the *rate expression* will be given here:

Assumption 1. The initial interaction time leading to the formation of the excited molecular ion ($\sim 10^{-16}$ s) is short compared with the time required for unimolecular dissociation; i.e., the dissociation rate is independent of the ionization mode.

Assumption 2. The rate of dissociation is slow relative to the rate of redistribution of the initial vibrational and electronic excitation energy over all degrees of freedom.

Rate Expression. As a result of these assumptions the rate coefficient $k(E)$ can be calculated by applying the Absolute Rate Theory (Glasstone et al., 1941) modified to apply to isolated systems (Rosenstock et al., 1952), yielding

$$k(E) = \frac{\sigma}{h} \frac{\rho^*(E - E_0)}{\rho(E)}, \qquad (22)$$

with h Planck's constant, σ a symmetry factor (number of identical reactions), $\rho^*(E - E_0)$ the number of states of activated complexes with energy between E and $E - E_0$, E_0 the activation energy, and $\rho(E)$ the density of states of the molecular ion with energies between E and $E + dE$. An equivalent equation has been obtained by Marcus and Rice (1951) for thermal reactions by reformulating the Rice–Ramsberger–Kassel theory, also termed the RRKM theory (Marcus and Rice, 1951; Marcus, 1952; Forst, 1973). To calculate the value of $k(E)$ from Eq. (22) for a specific system the density of states and the number of states must be known. For this, a number of methods and approximations have been developed and reviewed in detail by Forst (1971, 1973).

Since its introduction in 1952, numerous experiments have been performed to test the assumptions and the validity of QET. In recent years experiments have been designed with well-defined state-selected ions, e.g., the determination of dissociation lifetimes of molecular ions produced by charge exchange (Andlauer and Ottinger, 1971, 1972) and the sophisticated method of photoion–photoelectron coincidence spectroscopy (PIPECO) (e.g., Brehm and von Puttkamer, 1967; Danby and Eland, 1972; Stockbauer, 1973; Werner and Baer, 1975; Eland and Schulte, 1975; Baer et al., 1975, 1976; Dannacher and Vogt, 1978; Nenner et al., 1980; Rosenstock et al., 1981). As a result of this, more basic aspects of QET could be tested (Lifshitz, 1978). In addition, the recent advent of studies with the objective of using ion clusters to test

3 Ionization of Molecules by Electron Impact

statistical theories has the important advantage that cluster formation offers the possibility of producing a series of molecular ions having different numbers of degrees of freedom, but similar chemical properties (Stace and Shukla, 1980, 1982; Futrell *et al.*, 1981, 1982; Sunner and Kebarle, 1981; Stephan and Märk, 1982a–c, 1983; Stephan *et al.*, 1983c–f; Märk, 1982a; Märk *et al.*, 1982; Stamatovic and Miletic, 1982; Lau *et al.*, 1982, Stace and Moore 1983; Illies *et al.*, 1983, Märk and Castleman, 1984).

III. Total and Partial Ionization Cross Sections

A. Theoretical Considerations

1. Introduction

A great deal of experimental and theoretical work has been devoted in recent years to the determination of electron-impact ionization cross sections of atoms and molecules (Märk and Dunn, 1984). An important motivation for the quantitative study of these inelastic collision processes is the fact that the interpretation of a wide variety of physical phenomena demand an accurate knowledge of reaction rates for ionization by electron impact.

Inelastic collisions of electrons with molecules can be classified into two groups by comparing the impacting electron velocity with the mean orbital velocity of the molecular electrons in the (sub)shell responsible for the inelastic process under consideration, i.e., *fast* collisions and *slow* collisions. For fast collisions, the influence of the incoming electron upon a molecule can be treated as a sudden and small external perturbation. In this case the cross section formula consists of two different factors, one involving the properties of the incoming electron and the other dealing with the properties of the target molecule, i.e., the generalized oscillator strengths. This picture has been used by Bethe (1930, 1932) to derive a quantum-mechanical theory of ionization cross sections based on the Born approximation [see also review by Inokuti (1971) and Kim (1983)]. Conversely, for slow collisions the combined system of incoming electron plus target molecule has to be considered. Hence, a theory of slow electron-impact ionization of a molecule XYZ has to deal with the dynamics of XYZ^- in its excited continuum states.

For convenience we shall derive the Born–Bethe cross section formula to introduce the appropriate notation and then advance to the discussion of cross section calculations for the more general case of the low- and low-plus-high-energy regime.

2. Born–Bethe Approximation

The Bethe (1930) approximation is a simple application of the first Born approximation (Bell and Kingston, 1974), and its main application is the determination of the shape of the ionization cross section function at high electron energy. If an electron with velocity v (energy $T = \frac{1}{2}mv^2$) and mass m collides with a target i of mass M_i, the (single) differential cross section dq_{in} describes the probability that this electron will be deflected into the solid-angle element $d\Omega$ along the direction with polar angles, ϑ, φ (CM system) and the target be excited to a state n, with excitation energy E_n (n represents a set of quantum numbers which designates a molecular state). According to Inokuti (1971) dq_{in} can be expressed by

$$dq_{in} = \frac{4\pi a_0^2}{T/R} \frac{f_n(K)}{E_n/R} d\ln(Ka_0)^2, \qquad (23)$$

where $a_0 = \hbar^2/me^2 = 0.52918 \times 10^{-8}$ cm (Bohr radius), $R = me^4/2\hbar^2 = 13.606$ eV (Rydberg energy–ionization potential of H), K is the momentum transfer of the electron, and $f_n(K)$ a generalized oscillator strength. Ionization, however, involves excitation into the continuum, hence E_n has to be replaced by a continuous variable E, of target i. In this case the (double) differential cross section for excitation into a continuum state between E and $E + dE$ (at fixed T) is $d^2q_i/dE\,d\Omega$. Hence, using the density of the generalized oscillator strength per unit range of E, i.e., $df(K,E)/dE$, Eq. (23) yields

$$d\left(\frac{dq_i}{dE}\right) = \frac{4\pi a_0^2}{T/R} \frac{df(K,E)/dE}{E/R} d\ln(Ka_0)^2. \qquad (24)$$

The incident electron is in principle indistinguishable from the target electrons. This allows electron exchange effects, which are not included in this treatment. According to Moiseiwitsch and Smith (1968) electron exchange effects are unimportant in fast collisions, but have to be accounted for in slow collisions (see below and Kim, 1983). Integration of $d(dq_i/dE)$ yields an integrated cross section for transitions into the continuum per unit range of excitation energy E, regardless of the angle of scattering of the incoming electron, i.e.,

$$\frac{dq_i}{dE} = \frac{4\pi a_0^2}{T/R} \int_{(Ka_0)^2_{\min}}^{(Ka_0)^2_{\max}} \frac{df(K,E)/dE}{E/R} \frac{d(Ka_0)^2}{(Ka_0)^2}. \qquad (25)$$

Applying the Bethe procedure (Inokuti, 1971), Eq. (25) yields

$$\frac{dq_i}{dE} = \frac{4\pi a_0^2}{T/R} \left[\frac{R}{E}\frac{df}{dE}\ln\frac{4c_E T}{R} + \frac{d\gamma}{dE}\frac{R}{T} + O\left(\frac{E^2}{T^2}\right)\right], \qquad (26)$$

with $df/dE = [df(K,E)/dE]_{K=0}$ the differential optical oscillator strength, $c_E = (\bar{K}a_0)^2(R/E)^2$, $\gamma = (m/2\mu)df/dE - (E/4R)d(df/dE)/d(Ka_0)^2$; μ is the reduced mass of the colliding system, i.e., see Inokuti (1971).

The integration of Eq. (26) over the continuum energy E yields the so-called counting ionization cross section q_i. The possibility of multiple ionization requires the definition of an additional quantity [i.e., the probability $\eta_z(E)$ that, as a result of energy transfer E by a single collision z electrons are ejected] and its use for the above integration, yielding partial ionization cross sections q_z for z-fold ionization. Hence partial ionization cross sections define the cross section describing the electron-impact-induced production of an ion of specific charge ze at an electron energy T via process

$$XY + e^- \longrightarrow XY^{z+} + ze_e^- + e_s^-. \tag{27}$$

A more stringent definition would include separation of initial and final states, but they are usually not resolved.

The counting cross section defined above (de Heer et al., 1979) is the simple sum of the partial cross sections

$$q_i = \sum_z q_z. \tag{28}$$

This counting cross section, however, is only accessible by experiment if the number of ionization events (regardless of the number of electrons ejected in each event) is measured (Mc Clure, 1953; Rieke and Prepejchal, 1969, 1970, 1972). In most experiments, however, not the counting but the gross (or total) ionization cross section q_t is determined by measuring the total current produced by all different ionization mechanisms,

$$q_t = \sum_z z q_z. \tag{29}$$

According to Platzman (1963) an excitation to a molecular state above the first ionization potential, however, does not always give rise to ionization (as assumed in the above derivation), but sometimes leads to other processes such as decomposition into neutral fragments. Again one has to introduce an additional quantity $\eta(E)$, which describes the probability that the molecule ionizes when it has received an amount of energy E, with

$$\eta(E) = \sum_z \eta_z(E). \tag{30}$$

A useful way of representing ionization cross sections at high energies is, according to Fano (1954), to plot $Tq/4\pi a_0^2 R$ vs. $\ln(T/R)$ in the nonrelativistic energy region (Fano plot), because the ionization cross section q_i [integration of Eq. (26)] is given asymptotically by the simple expression (Miller and Platzman, 1957)

$$q_i = \frac{4\pi a_0^2 R}{T} M_i^2 \ln(4c_i T/R), \tag{31}$$

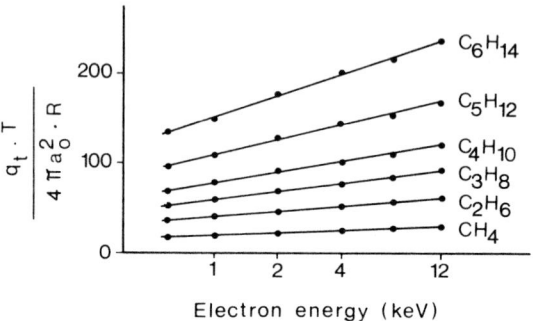

Fig. 16. Fano plot of total ionization cross sections for various hydrocarbons. [After Schram et al. (1966c).]

where M_i^2 and c_i are constants. The quantity M_i^2 is related to the differential oscillator strength by

$$M_i^2 = \int_{E_i}^{\infty} \frac{df}{dE} \frac{R}{E} \, dE, \quad (32)$$

and c_i is defined by

$$M_i^2 \ln c_i = \int_{E_i}^{\infty} \frac{df}{dE} \frac{R}{E} \ln c_E \, dE. \quad (33)$$

Hence, a Fano plot should, at sufficiently high energies, become a straight line whose slope and intercept relate to the integrals given in Eqs. (32) and (33). This has been verified (see, e.g., Fig. 16 and Kim, 1983) for total ionization cross sections of diatomics, H_2O, and hydrocarbons (Hooper et al., 1962; Schram et al., 1965, 1966c,d; Schutten et al., 1966). See also presentation by Rieke and Prepejchal (1972). Studies by Monahan and Stanton (1962) and by Fiquet-Fayard et al. (1968) have shown that Eq. (31) holds also for single ionization of various molecules (such as NO, N_2, CO, COS, CS_2, H_2S, NH_3, and several hydrocarbons), but direct double ionization at high energies is only proportional to $1/T$.

3. Other Approximations and Empirical Formulas

The basic formulation (quantum mechanical and classical) of the problem of electron-impact ionization cross sections has received a great deal of attention as described in review by Bates et al. (1950), Massey (1956), Craggs and Massey (1959), Veldre (1965), Burgess and Percival (1968), Rudge (1968), Vriens (1969), Inokuti (1971), Vainstein et al. (1973), Bell and Kingston (1974), Peterkop (1977), Kim (1983), and Märk and Dunn (1984). Quantum-

3 Ionization of Molecules by Electron Impact

mechanical calculations even in their simplest approximations are lengthy and not yet as accurate as could be wished, except for the simplest cases, i.e., He, Ne, Ar (Märk, 1982b,c). Alternative treatments have therefore been pursued with the aim of obtaining reasonably accurate cross section functions in a simple fashion. Such treatments are possible by using classical or semiclassical theory and by devising semiempirical formulas. Hence, in general three different approaches have been used to calculate ionization cross sections of atoms and molecules (initially in the ground state), i.e.,

(1) *Empirical and semiempirical rules and formulas* (e.g., Morgulis, 1934; Fano, 1946; de la Ripelle, 1949; Elwert, 1952, Mc Clure, 1953; Otvos and Stevenson, 1956; Lampe et al., 1957; Lorquet, 1960; Drawin, 1961; Green and Barth, 1965, Green and Dutta, 1967; Mann, 1967; Stolarski et al., 1967; Watson et al., 1967; Lotz, 1967, 1968; Peterson et al., 1969; Beran and Kevan, 1969; Krinberg, 1970; Mann, 1970; Green and McNeal, 1971; Rieke and Prepejchal, 1972; Vainstein et al., 1973; Jain and Khare 1975, 1976; Franco and Daltabuit, 1978; Rao et al., 1979; Djuric et al., 1981; Franco, 1981; Younger and Märk, 1984).

(2) *Classical and semiclassical collision theories* (e.g., Thomson, 1912a; Thomas, 1927; Webster et al., 1933; Gryzinski, 1959, 1963, 1965; Ochkur and Petrunkin, 1963; Prasad and Prasad, 1963; Burgess, 1963; Kingston, 1964; Stabler, 1964; Vriens, 1964, 1966, 1969; Bauer and Bartky, 1965; Tiwari et al., 1969; Flannery, 1971; Tripathi and Rai, 1972; Roy et al., 1972; Roy and Rai, 1973; Ton-That and Flannery, 1977; Ton-That et al., 1977; Kumar and Roy, 1978; Mc Cann et al., 1979; Hyman, 1979; Kunc, 1980; Flannery et al., 1981; Younger and Märk, 1984).

(3) *Quantum-mechanical approximations* [mostly for atomic targets e.g., see extensive reviews by Rudge (1968), Inokuti (1971), Bell and Kingston (1974), and Peterkop (1977); in addition see e.g. Peach (1971), Tweed (1973), Wallace et al. (1973), Mc Guire (1977), Kim (1983), Younger (1984)].

Some of these methods have been applied to calculate cross sections for ionization of excited molecules, i.e., N_2^* (Tannen, 1973), N_2^*, CO^*, Ne_2^*, and Ar_2^* (Ton-That and Flannery, 1977; Mc Cann et al., 1979; Flannery et al., 1981), N_2^* (Kukulin et al., 1979). Some of the most successfully and widely used empirical, classical, and semiclassical formulas are given below:

Drawin (1961):

$$q_{zn} = 2.66\pi a_0^2 \zeta_n (R/E_{in})^2 f_1 [(U-1)/U^2] \ln(1.25 f_2 U); \quad \text{var: } f_1, f_{2i}(\approx 1) \quad (34)$$

Lotz I (1967):

$$q_{zn} = a_n \zeta_n [\ln(U)/E_{in}T] \{1 - b_n \exp[-c_n(U-1)]\};$$
$$\text{var: } a_n, b_n, c_n (n = 1, \ldots, N). \quad (35)$$

Lotz II (1967):

$$q_{zn} = a\zeta_n[\ln(U)/TE_{in}]\{1 - b\exp[-c(U - 1)]\}; \qquad \text{var: a,b,c.} \quad (36)$$

Green and Mc Neal (1971):

$$q_z = (Za)^\Omega(T - E_i)/(j^{\Omega+v} + T^{\Omega+v}); \qquad \text{var: } v, \Omega, a, j(\Omega = 0.75). \quad (37)$$

Vainstein et al. (1973):

$$q_{zn} = \pi a_0^2 \frac{\zeta_n}{2l+1}\left(\frac{R}{E_{in}}\right)^2\left(\frac{U-1}{U}\right)^{3/2}\frac{C}{U-1+\Phi}; \qquad \text{var: } C, \Phi. \quad (38)$$

Gryzinski (1959)–Stabler (1964) (see Rudge, 1968):

$$q_{zn} = 4\pi a_0^2 \zeta_n (R/E_{in})^2 [U/(U+1)]^{3/2} g(U),$$

$$g(U) = (5/U - 2/U^2), \qquad U \geq 2, \quad (39)$$

$$g(U) = (4\sqrt{2}/3U)(1 - 1/U)^{3/2}, \qquad 1 \leq U \leq 2.$$

Gryzinski (1965) II:

$$q_{zn} = 4\pi a_0^2 \zeta_n \left(\frac{R}{E_{in}}\right)^2 \frac{1}{U}\left(\frac{U-1}{U+1}\right)^{3/2} \cdot \left[1 + \frac{2}{3}\left(1 - \frac{1}{2U}\right)\ln(2.7 + (U-1)^{1/2})\right]. \quad (40)$$

Vriens (1966)–Burgess (1963):

$$q_{zn} = 4\pi a_0^2 \zeta_n \frac{(R)^2}{T + E_{in} + E_2}\left[\left(\frac{1}{E_{in}} - \frac{1}{T}\right) + \frac{2}{3}E_2\left(\frac{1}{E_{in}^2} - \frac{1}{T^2}\right) - \frac{\Phi}{T + E_{in}}\ln U\right];$$

$$\Phi = 1 \ (E_{in} > R), \Phi = 0 \ (E_{in} < R), \quad (41)$$

where q_{zn} is the partial cross section for electron-impact ionization taking only electrons into account from nth (sub)shells, i.e., $q_z = \sum_{n=1}^{N} q_{zn}$; N numbers of subshells considered; ζ_n is the number of equivalent electrons in the nth shell; E_{in} is the binding energy of electrons in the nth shell; $U = T/E_{in}$; Z is the total number of electrons in the target; and E_2 is the kinetic energy of the target electron ($E_2 = E_i$). In addition, Lorquet (1960) has given an empirical relation for the partial cross section into a specific final state of the produced ion.

Recently, Jain and Khare (1975, 1976) have given a successful semiempirical formula for the calculation of total ionization cross sections of molecules, using a combination of the Born–Bethe approximation (describing high-incident-electron-energy collisions with small amount of energy transfer: soft collisions) and of the Mott–Möller formula (describing collisions with fast secondary electrons: hard collisions), i.e., in differential form (energy loss cross section $dq(T, E)/dE$):

$$\frac{dq}{dE} = \frac{4\pi a_0^2 R^2}{T}\left[f_1\left(\frac{1}{E}\frac{df}{dE}\ln\frac{4c_E T}{R}\right) + f_2\left(\frac{1}{\varepsilon^2} - \frac{1}{\varepsilon(T-\varepsilon)} + \frac{1}{(T-\varepsilon)^2}\right)s\right], \quad (42)$$

with ε the energy of an ejected secondary electron ($\varepsilon = E - E_i$), s the number of electrons which can participate in hard collisions,

$$f_1 = \frac{1}{1 + E_i/T}\left(1 - \frac{\varepsilon}{T - E_i}\frac{\ln[1 + (4c_E/R)(T - E_i)]}{\ln(4c_E T/R)}\right),$$

$$f_2 = \varepsilon^3 \frac{(1 - E_i/T)}{\varepsilon^3 + \varepsilon_0^3},$$

and ε_0 is a parameter. The empirical functions f_1 and f_2 extrapolate the Born–Bethe and Mott–Möller cross sections to low incident electron energies and control the mixing of the soft and hard collisions.

Another way to obtain quantitative information on relative atomic and molecular ionization cross section is based (Fano, 1946) on a result by Bethe (1930) which states that the ionization of an atomic electron with quantum number (n, l) is approximately proportional to the mean-square radius of the electron shell (n, l) (Otvos and Stevenson, 1956; Batabyal et al., 1965; Mann, 1967, 1970; Tiwari et al., 1969). This leads to proposed correlations of (in some cases maximum) electron-impact ionization cross sections with polarizability, diamagnetic susceptibility, number of electrons, and additivity of atomic ionization cross sections (Otvos and Stevenson, 1956; Lampe et al., 1957; Stevenson and Schissler, 1961; Batabyal et al., 1965; Schram et al., 1966c; Harrison et al., 1966; Pottie, 1966; Drowart and Goldfinger, 1967; Stafford, 1968, 1971; Beran and Kevan, 1969; Grosse and Bothe, 1970; Flaim and Ownby, 1971; Center and Mandl, 1972; Blackburn and Danielson, 1972; Alberti et al., 1974; Arai and Hotta, 1975; Rao et al., 1979; Bartmess and Georgiadis, 1983; Fitch and Sauter, 1983). None of these correlations, however, seems to be generally valid, but each is valid for certain molecular classes. For instance, the concept of additivity applies to hydrocarbons (Otvos and Stevenson, 1956), there is a single linear correlation with diamagnetic susceptibility for all non-fluorine-substituted compounds (Beran and Kevan, 1969), and there is an excellent correlation with polarizability for monofunctional compounds (Bartmess and Georgiadis, 1983). In addition, Franco and Daltabuit (1978) (see also Franco, 1981) pointed out a simple empirical relation between the maximum ionization cross section (its position) and the ionization potential as already given in Thomson's classical formula (1912b).

4. Threshold Laws

The energy dependence of the partial ionization cross section in the threshold region has been extensively investigated (mostly for atomic targets) from a theoretical point of view, but difficulties due to the three-body problem involved have given contradictory results. The threshold law has usually been expressed in the form

$$q_z \sim (T - E_i)^{nz}. \tag{43}$$

Various theoretical analyses have led to values of n in the range of 1–1.5. Some of these calculations also predict angular and energy distributions of the outgoing electrons.

In the renowned work of Wannier (1953) [later extended by Vinkalns and Gailitis (1967)] it is assumed that the electrons move classically when outside some inner reaction zone (see Section II.A.5), and that the trajectories from the inner zone fill the available phase space with a smooth probability. The prediction of this classical approach is that $n = 1.127$ (for the production of atomic ions with $z = 1$). Extending this law to multiple ionization, in 1955 Wannier (1955) obtained, on the basis of simple geometric considerations, the approximate threshold law that bears his name, with $n = 1$. This dependence would be exact in the absence of the Coulomb interaction between the electrons. Wannier reasons by analogy to single ionization that the true exponent is probably slightly larger than z.

The early quantum-mechanical treatments of the threshold process (using a Coulomb modified Born approximation) yielded $n = 1$ (for $z = 1$ and hydrogenic systems) (Geltman, 1956; Rudge and Seaton, 1964, 1965). Generalization of this dependence regarding multiple ionization is found to yield again $n = 1$ (Geltman, 1956). Later quantum-mechanical calculations for $z = 1$ have yielded values of n ranging from 1 to 1.5 (Temkin, 1966; Temkin et al., 1968; Kang and Kerch, 1970). In more recent quantum-mechanical calculations Rau (1971) has obtained the classical value of $n = 1.127$ for $z = 1$ and Temkin and Hahn (1974) and Temkin (1974) obtained a cross section proportional to $(T-E_i)\{1-C\sin[A\ln(T-E_i)+B]\}$, pointing out that, for a suitable set of constants A, B, and C, this law would not appreciably differ from that of Wannier. In a study of the multiple escape of electrons from a Coulomb field Klar and Schlecht (1976) found that the dominant contribution close to threshold comes from the doublet state [similar to the situation for single ionization where the dominant state is the singlet state; see also Klar (1980) and Greene and Rau (1982)] with an exponent for q_2 of $nz = 2.270$. This threshold behavior has been recently confirmed by Grujic (1983) using classical dynamics generalizing the method of Vilkalns and Gailitis (1967) (see Grujic 1982).

Recently, Bottcher (1981) has used finite-element integration of the Schrödinger equation for the time evolution of a wave packet and was able to reproduce the quantitative features of Wannier's theory not only at threshold, but at much higher energies. A similar result was obtained by Klar (1981a) including screening effects. This theory has recently been extended by Klar (1980, 1982) to general fragmentation reactions producing three or more charged particles of arbitrary masses and charges. It is shown that the nuclear mass in electron-impact ionization processes has only little influence on the threshold behavior. In addition, with increasing ion charge Z the electron correlation becomes less and less important. Conversely, the threshold behavior for positron-impact ionization is quite different; i.e., $n = 2.650$ for $z = 1$ (Klar, 1981b) (see also Grujic, 1982; Temkin, 1982a; Geltman, 1983).

Other approaches have also been used, one semiclassical approximation again yielding (for $z = 1$) $n = 1.127$ (Peterkop and Liepinsh, 1969, 1981; Peterkop, 1971) and one in which classical trajectories are integrated numerically, with (for $z = 1$) $n \sim 1.127$ (Banks et al., 1969; Peterkop and Tsukerman, 1969, 1970; Grujic, 1972) and (for $z = 2$) $nz = 4.069$ (Dimitrijevic and Grujic, 1981; see also Cvejanovic and Grujic, 1974 and Dimitrijevic and Grujic, 1979). Furthermore, Temkin (1982a,b) has recently derived a threshold law for electron- and positron-atom ionization on the basis of the Coulomb dipole theory (the quantum mechanical implications have been discussed by Temkin, 1982c) yielding a modulated quasilinear law (see also experimental results by Donahue et al., 1982). For more information on this subject see also results presented at a special session of the XIII ICPEAC (Berlin, 1983) and the review by Temkin (1982c).

B. Experimental Techniques

Experimental determination of the ionization cross section usually involves an experiment of a simple conceptual design (Fig. 17). Consider a parallel, monoenergetic, and homogeneous beam of electrons impinging on a semi-infinite medium containing N target molecules per cubic centimeter at rest. Of

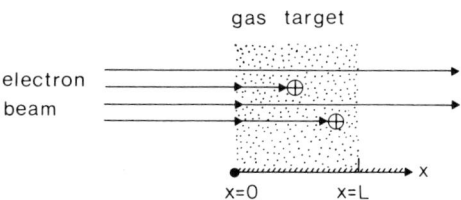

Fig. 17. Schematic of electron-impact ionization experiment (see text).

those electrons that penetrate to depth x in the target without ionizing, a fraction equal to $Nq\,dx$ will ionize in passing from x to $x + dx$. If $n(0)_e$ represents the initial intensity of the incident electrons per square centimeter–second and $n(x)_e$ is the intensity of the electron beam at depth x, then

$$dn(x)_e = -n(x)_e Nq\,dx \qquad (44)$$

and

$$n(x)_e = n(0)_e \exp(-Nqx). \qquad (45)$$

Assuming single collision condition (i.e., $Nqx \ll 1$), the number of ions per second generated along the collision path length L (over which the ions are collected) is

$$n^*(L)_i = n(0)_e Nq_i L, \qquad (46)$$

whereas the ion current $i_i = n^*(L)_i ze$ is

$$i_i = i_e NL \sum_z q_z z = i_e NLq_t = i_e Ls, \qquad (47)$$

with i_e the total electron current bombarding the target and s (in cm^{-1}) the macroscopic cross section representing the total effective cross-sectional area for ionization of all target molecules in 1 cm^3 of the target medium. This quantity is numerically equal to the ionization efficiency, usually defined as the number of charges (or ion pairs) produced per centimeter of path at 1 Torr and 0°C (von Engel, 1965).

Kieffer and Dunn (1966) have summarized the most important experimental conditions necessary for obtaining accurate measurements of the quantities in Eq. (47) needed to obtain the ionization cross section. In addition, the electron energy must be defined accurately if the ionization cross section function is to be determined. Depending on the type of cross section to be measured and the molecular system to be studied various types of apparatus have been employed and in the following subsections the most advanced ones will be discussed [for more detail see also reviews by Smyth (1931), Massey (1956), Craggs and Massey (1959), Kieffer and Dunn (1966), Field and Franklin (1970), Märk (1982b,c, 1984a), Märk and Dunn (1984)].

3. Total Ionization Apparatus

In the beginning several types of apparatus were used (Lenard, 1902, 1903, 1904; Kossel 1912; Mayer 1914; Hughes and Klein, 1924; Compton and Van Voorhis, 1925, 1926; Jesse, 1925; Jones, 1927; von Hippel, 1928; Bleakney, 1929; Funk, 1930; Smith, 1930; Lozier, 1930; Tate and Lozier, 1932; Liska 1934) but the type used today yielding the most satisfactory results is that of Jones (1927) and Smith (1930). Some measurements have, however, been made with Lozier-type setups (Craggs et al., 1957; Schulz, 1962; Asundi et al., 1963) or by the summation method (e.g. Märk and Egger, 1977; Märk et al., 1977a). The total ionization condensor tube (Jones, 1927; Smith, 1930) has also been used successfully for the measurement of dissociative ionization cross sections (Bleakney, 1930a; Rapp et al., 1965). In this case discrimination effects due to the procedure of drawing the fragment ions out of the ion source of a mass-spectrometric system (see next section) can be avoided, because fragment ions with excess kinetic energy are collected on the condensor plates by using a reversed (retarding) collecting field across the plates (see below).

A schematic drawing of this total condensor tube [as used by Rapp and Englander-Golden (1965)] is shown in Fig. 18. An electron beam from a differentially pumped cathode (A) is collimated and accelerated by apertures B,C,D and confined by a magnetic field (typically 500 G). This beam passes through an ionization chamber (containing the gas under study at about 10^{-5} Torr) containing two guard plates E and the ion collector plate F. The electrons then pass an electron collection shield G before being collected in the electron trap H. An electric field perpendicular to the electron beam accelerates ions out of the ionization region to the ion collector F.

Consistency checks necessary for accurate cross section measurements are discussed in detail by Rapp and Englander-Golden (1965), Kieffer and Dunn (1966), van der Wiel (1973), Cowling and Fletcher (1973), and Kurepa et al. (1974). One must demonstrate that the electron and ion currents are completely measured, and that path-length corrections (Taylor et al., 1974) and space-charge effects are negligible. The perfect gas law is usually used to determine the gas density, which necessitates the absolute measurement of the gas pressure. Gas pressures are typically obtained by using a McLeod gauge, although it has been shown that large errors can occur (Ishii and Nakayama, 1961; Meinke and Reich, 1962, 1963; Rothe 1964). Recently, dynamic techniques (Rapp and Englander-Golden, 1965; Kurepa et al., 1974; Märk, 1975) and/or capacitance manometers (Märk and Egger, 1977; Kurepa et al., 1981) have been used.

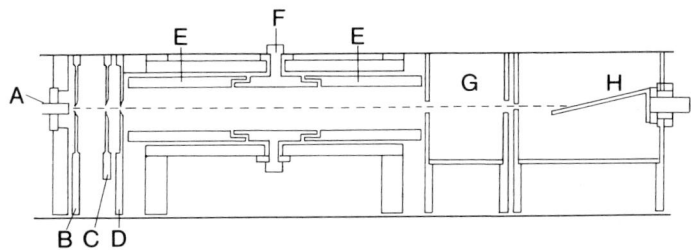

Fig. 18. Schematic drawing of total ionization condensor tube as used by Rapp et al. (1965): (A) electron-emitting cathode; (B,C,D) collimating and accelerating electrodes; (E) guard plates; (F) ion collector plate; (G) shield; (H) electron trap.

2. Mass Spectrometry

It is not possible here to give an elaborate discussion of the principles of mass spectrometry. Standard operation techniques and apparatus descriptions are extensively discussed in excellent reviews of this subject (see, e.g., some recent books: Beynon, 1960; Mc Lafferty, 1963; Mc Dowell, 1963b; Brunnee and Voshage, 1964; Kiser, 1965; Blauth, 1965; Mc Lafferty, 1966; Jayaram 1966; Hill 1966; Spiteller, 1966; Kienitz, 1968; Roboz, 1968; White, 1965; Field and Franklin, 1970; Melton, 1970; Milne 1971; Shrader, 1971; Budzikiewicz, 1972; Johnstone, 1972, 1981; Litzow and Spalding 1973; Maccoll, 1975; Dawson, 1976; Millard, 1978; Beynon and Morgan, 1978; Merritt and Mc Ewen, 1979; Rose and Johnstone, 1982).

The first mass spectrometric studies of ionization cross section functions date back to the 1930s: Bleakney (1930a,b,c), Smyth (1931), Tate and Smith (1934), Tate et al. (1935), Bleakney and Smith (1936), Nier (1938), Nottingham (1939), Hagstrum and Tate (1941), Stevenson and Hipple (1942). These early studies have been repeated only in a few cases, however, the new measurements being in general (exceptions see below) no conceptual extension or improvement. Even for the simplest gases large differences existed for these studies in magnitude and shape of experimentally determined partial ionization cross section functions (e.g., see Kieffer and Dunn, 1966; Märk, 1982b,c, 1984a). The crucial condition to be met for measuring accurate partial ionization cross section functions is a constant and/or complete ion-source mass spectrometer collection efficiency independent of the mass to charge ratio, the electron energy, and the initial kinetic energy. According to Kieffer and Dunn (1966) the fulfillment of this condition has not been demonstrated in these earlier measurements. Moreover, one problem which has never been solved satisfactorily is the absolute calibration of these functions. Closely related to this problem is the fact that discrimination may occur at the detector, especially if an electron multiplier is used as detector, since the multiplier response depends quite strongly on the mass to charge ratio (e.g., Higatsberger *et al.*, 1954; Klein, 1965; Dietz, 1965; Schram *et al.*, 1966b; Collins, 1969; Van Gorkom and Glick, 1970; Van Gorkom *et al.*, 1970; Tuithof and Boerboom, 1974; Märk, 1977; Nagy *et al.*, 1980).

It has been pointed out by a large number of authors that discrimination effects also occur at the ion-source exist slits and the mass spectrometer slits (Bleakney, 1936; Hagstrum and Tate, 1941; Jordan and Coggeshall, 1942; Coggeshall, 1944, 1946, 1962; Washburn and Berry, 1946; Bainbridge, 1947; Berry, 1950; Bertein, 1950; Careri and Nencini, 1950; Vauthier, 1950, 1955; Barnard, 1953, 1956; Schaeffer, 1954; Brubaker, 1955; Taubert, 1959; Fiquet-Fayard, 1962; Fiquet-Fayard and Lahmani, 1962; Naidu and Westphal, 1966; Kieffer and Dunn, 1966; Adamczyk *et al.*, 1966, 1973; Schram *et al.*, 1966a; Schulz *et al.*, 1968; Fock, 1969; Loveless and Russell, 1969; Chantreu and

Vauthier, 1970; Wallington, 1970, 1971; Rüdenauer, 1970; Märk, 1973, 1975; Werner et al., 1972; Werner and Linssen, 1974; Werner, 1974a,b; Drewitz, 1976; Drewitz and Taubert, 1976; Märk et al., 1977b; Egger and Märk, 1978; Helm et al., 1979, 1981; Märk and Castleman, 1980; Stephan et al., 1980a,b; Nagy et al., 1980; Märk, 1982b,c). In several of these theoretical and experimental studies it has been noted that errors due to discrimination occur in isotope ratio measurements. A possible influence of discrimination effects on the shape and magnitude of measured partial cross section curves has recently been studied for the three-electrode Nier-type ion source (Nier, 1940) by Märk and co-workers (Märk et al., 1977b; Märk and Castleman, 1980; Stephan et al., 1980a,b, 1983a,b; Märk, 1982b,c). On the basis of these findings a new approach to the mass spectrometric measurement of partial ionization cross section functions (see below) with particular regard to the low-electron energy regime (<200 eV) has been developed and applied to a number of atomic and molecular gases. This method will be discussed in detail in the following paragraph, because the Nier-type ion source is used in most present-day sector-type mass spectrometers (Elliot, 1963; Brunnee and Voshage, 1964).

The extraction of ions from the ionization region in the Nier-type ion source (Fig. 19) depends under usual experimental conditions on various parameters, such as the initial energy of the ions, the mass to charge ratio, the guiding magnetic field, the electron-beam space charge, and the applied extraction field. Usually, ions are extracted from the ionization region (in which there is a crossed electric and magnetic field) by a weak electric field applied between the collision-chamber exit slit and an electrode opposite the exit slit (pusher). This extraction, however, is not complete and results in discrimination for ions with different m/ze.

In an alternative approach a penetrating field from external electrodes may be used; i.e., all electrodes confining the collision chamber (P, C, and L_1 in Fig. 19) are kept at the same potential (e.g., 3 kV) and ions are drawn out of the collision chamber through the slit in L_1 under the action of an electric potential applied to L_2. It has been shown that this *penetrating field extraction* assures saturation of the ion current; i.e., complete ion collection is achieved [saturation condition L_2 at ≤ 2.8 kV (Stephan et al., 1980a,b, 1983a,b)].

Ions extracted in this manner are then centered by elements L_3 and L_4, while D is a defining aperture and L_5 (earth slit) is the end of the accelerating region (0V). Stephan et al. (1980a,b) additionally introduced deflection plates $L_{6,7}$ and $L_{8,9}$, which serve to sweep the ion beam across the mass spectrometer entrance slit S_1 in the y direction (perpendicular to S_1) and z direction (parallel to S_1). This allows the recording and/or integration of the ion beam profile, and hence discrimination at S_1 can be avoided. In this conjunction it is interesting to note that some authors (Hagstrum, 1951, 1953; Drewitz and Taubert, 1976; Schmidt et al., 1976; Nagy et al., 1980) attempt to focus by

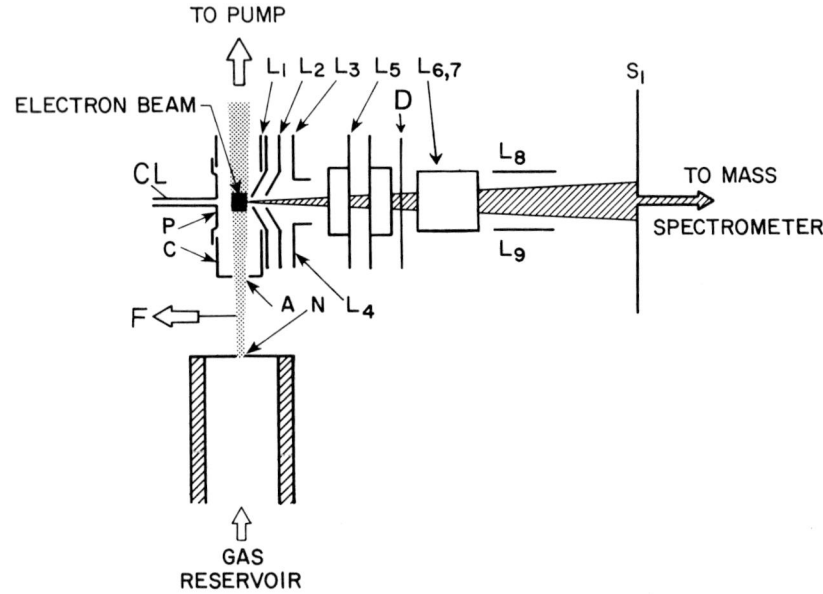

Fig. 19. Schematic view of three-electrode-type ion source as used by Märk and co-workers (see text): (P) pusher, (C) collision chamber, (CL) capillary leak, (A) aperture, (N) nozzle for molecular beam, (L_1) collision-chamber exit-slit electrodes, (L_2–L_4) ion-beam focusing and accelerating electrodes, (L_5) earth electrode, (D) defining aperture, ($L_{6,7}$, $L_{8,9}$) y and z deflection plates, (S_1) mass spectrometer entrance slit.

means of ion optics the entire ion beam into S_1 in order to avoid discrimination at S_1, whereas other authors eliminate discrimination (in the z direction) at S_2 by monitoring the z profile of the ion beam with deflection plates mounted between S_1 and S_2 (Rees and Hipple, 1949; Berry, 1950). This latter deflection method has been used by several authors to measure initial energy distribution functions of fragment ions (see Section IV.B.1 and also Durup and Durup, 1967). Moreover, the above described y and z *deflection method* is a useful technique for distinguishing ions produced in the static target gas (capillary leak gas inlet) and ions produced in the neutral beam (molecular-beam gas inlet through nozzle expansion; see Fig. 19) (Helm et al., 1979, 1981; Stephan et al., 1981a,b, 1982a,b, 1983c–f; Stephan and Märk, 1982a–c, 1983; Futrell et al., 1981, 1982).

In addition, there exist some other recent experimental approaches to overcome the difficulties described above (see also Märk, 1982c, 1984a): (*i*) Cook and Peterson (1962), Peterson (1964), and, more recently, Brook et al. (1978) (see also Dixon and Harrison, 1971; Dixon et al., 1975, 1976; Märk and de Heer, 1979; Harrison et al., 1979) used the fast-atom-beam technique. This method employs a crossed-beam arrangement, where the target beam is obtained from fast-ion-beam charge transfer. Absolute ionization cross sections are obtained directly without measurement of gas density. (*ii*) To avoid mass and kinetic-energy discrimination effects Schram *et al.* (1966a), Adamczyk *et al.* (1966), Schutten *et al.* (1966), and Adamczyk (1970) developed a special cycloidal mass spectrometer with a high ion extraction efficiency, using an open ion source without any slits. A number of studies concerning partial cross sections of parent and fragment ions have recently been made with this type of apparatus (see Section III.C.2). (*iii*) Crowe *et al.* (1972) and Mc Conkey *et al.* (1972) have recently carried out electron-impact ionization studies taking great care to fulfill the requirements outlined by Kieffer and Dunn (1966). Their apparatus consisted of a modified Pierce-type electron gun, capable of firing a beam of electrons with an extremely small beam divergence. Ions produced in a field-free interaction region are allowed to drift through apertures defining an angular resolution into an ion lens, where they are accelerated and focused into a quadrupole mass spectrometer and detected using a channeltron multiplier in pulse counting mode. The ion lens and the quadrupole were tuned to ensure minimum discrimination with regard to ions of different mass and energy.

3. Threshold Experiments

The importance of low-energy electron-impact ionization studies (threshold studies) has been recognized for many years with emphasis on the verification of the threshold laws, studies of excited states (fine structure), and measurements of appearance potentials (Beynon *et al.*, 1975; Rosenstock, 1976). In the early work a practical limitation was set by the lack of resolution arising from the quasi-Maxwellian energy spread of electrons produced by a thermionic filament, and by the fact that the actual energy maximum of the beam can differ from the nominal energy expected from the applied potentials. Metal oxide cathodes had been employed initially, producing improved energy spreads of $\lesssim 0.20$ eV (Smith, 1937; Morrison and Nicholson, 1959; Dorman *et al.*, 1960). Lawrence (1926), Bell (1939), and Nottingham (1939) made an early attempt to achieve energy selection using a magnetic analyzer. A very useful electron monochromator for use in an axial magnetic field has been developed recently by Stamatovic and Schulz (1968, 1970). With this trochoidal monochromator, where electrons are injected parallel to the magnetic field and an electric field is applied in a perpendicular direction, an energy FWHM of 0.020 eV with a transmitted current of 10^{-9} A can be obtained.

A completely different approach to overcoming the energy spread is the RPD (retarding potential difference) technique developed by Fox and Hickam (1953, 1954). In this method, which sometimes produces anomalous results (Johnson *et al.*, 1978), the electron gun using a thermionic filament is designed with a (retarding) potential distribution such that the low-energy portion of the energy distribution is not transmitted into the ion source. This cutoff potential is varied between two slightly differing values and the corresponding difference in ion signal is recorded. This ion signal corresponds in principle to ions produced by electrons having an energy distribution defined by the cutoff potentials. Since its development this method has been widely used. Anomalous effects and improvements have been discussed by Wei and Kuppermann (1969) and by Cloutier and Schiff (1959), Brongersma and Oosterhoff (1967), Lifshitz et al. (1973), and Hubin-Franskin et al. (1974), respectively. The energy resolution is limited to about 60 meV, in accordance with mathematical analyses of Anderson et al. (1967) and Wiesemann (1981).

Another quasimonoenergetic method is the EDD (energy distribution difference) technique developed by Winters et al. (1966). In this method (see also Winters and Collins, 1968; Collins et al., 1968; Giessner and Meisels, 1971; Vogt and Pascual, 1972) the measured ion current at an electron energy E is "corrected" by subtracting from it a constant fraction of the ion current measured at an electron energy $E + \Delta E$. Winters et al. argue that this difference current corresponds to an ion current due to a much narrower electron energy distribution. Several other approaches (e.g., determining the first and/or second derivative) to observe details in the ionization cross section curve have been proposed by Morrison (1953, 1954, 1963). Bolduc et al. (1971) have developed the "straightening through smoothing" technique, which allows one to observe small variations in the cross section curve. Recently, Hashizume and Wasada (1979, 1980) have used the inverse convolution method proposed by Vogt and Pascual (1972) and Johnstone and Mc Master (1974). For all of these methods, however, a calibration of the energy scale with a gas of known ionization threshold is necessary, giving rise to a number of errors.

Most recent research, however, has been devoted to the construction of electrostatic velocity selectors, such as parallel-plate, 127° radial field cylindrical, 180° spherical condensor, and cylindrical mirror system analyzers. Maeda et al. (1968), Kerwin et al. (1969), Read et al. (1974), and Johnson et al. (1978) have recently given detailed reviews on the design criteria and the energy spreads obtained with these analyzers (see also Märk and Dunn, 1984). Simpson (1964), using a double hemispherical analyzer, reports a remarkable energy spread FWHM of 0.005 eV at 3.35 eV primary energy, but only at currents of $\sim 10^{-14}$ A. Recently, Cvejanovic and Read (1974) have developed a new technique for obtaining high-resolution (40 meV) threshold ionization cross sections for low-energy inelastically scattered electrons. This is achieved by using a very weak field penetrating into the target chamber, which produces a saddle point in the potential distribution. This has the effect of focusing and enhancing the extraction efficiency of very low-energy electrons.

4. Others

Some molecular target species are not stable under normal experimental conditions, and in these cases special techniques (see, e.g., Kieffer and Dunn, 1966) must be used which vary significantly in detail from the aforementioned setups. In addition, it is of interest to know the final states of the ions observed in ionization cross section measurements. One method to determine the final state is to measure photons emitted from excited molecular ions. Such a fluorescence technique involves the tedious task of absolute radiometry, if absolute cross section functions are to be measured. For instance, a number of studies exist about emission cross sections of excited N_2^+ (see, e.g., Stewart, 1956; Hayakawa and Nishimura, 1964; Mc Conkey and Latimer, 1965; Mc Conkey et al., 1967; Aarts et al., 1968; Nishimura, 1968; Srivastava and Mirza, 1968; Stanton and John, 1969; Simpson and Mc Conkey, 1969; Borst and Zipf, 1970; Koppe et al., 1971; Holland and Maier, 1972; Haasz and de Leeuw, 1976; Mandelbaum and Feldman, 1976). Similar problems are encountered in the limited number of inner-shell excitation/ionization cross section measurements, a subject which is not discussed in the present review. The interested reader is referred to recent reviews by Mohr (1968), Powell (1976), Morita (1976), Tawara (1978), Burhop (1979), Hitchcock (1982), and Casnati et al. (1982) (see also Märk and Dunn, 1984).

Most studies of electron-impact ionization have been carried out, both experimentally and theoretically, for atoms and molecules in the ground state. Little work has been done using excited species as the target, largely because of the difficulties associated with producing sufficiently intense sources of excited states, despite the fact that information on such cross sections would be useful to characterize stepwise ionization processes. Armentrout et al. (1981) have recently reported a first study on the electron-impact ionization of the metastable $N_2(A\ ^3\Sigma_u^+)$ state from threshold up to 240 eV using a charge transfer neutralization technique (as discussed in Section III.B.2). This method involves neutralizing an ion beam ($\sim$ keV) in a collision chamber filled with the neutral charge transfer gas (Fig. 20). This reactant gas has to be selected such that production of a neutral in the excited state is near resonance, while production of the ground state is not. Ions which have not reacted are extracted by an electric field (ion beam dump in Fig. 20). The remaining beam is then crossed by the electron beam. The resulting ion signal can be related to ionization of the excited state. In addition, Evans et al. (1980) have recently reported a first study of vibrationally excited O_2, using a shock-heated molecular beam, and De Koven et al. (1981) a study of rotational excitation of supercooled N_2 (see also Hernandez et al., 1982).

Another interesting new technique is based on the fact that positive ions may be trapped in the negative potential well produced by an electron space

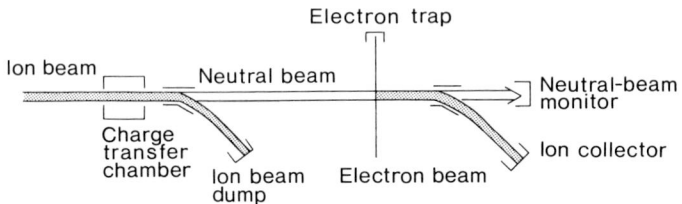

Fig. 20. Schematic view of the charge transfer neutralization technique used by Armentrout et al. (1981) to measure the electron-impact ionization of $N_2(A\ ^3\Sigma_u^+)$.

charge (see Baker and Hasted, 1968). Recently, Lifshitz and Gefen (1980) have used this trapped-ion mass spectrometry to measure ionization cross sections for parent and fragment ions of 1.5 hexadiyne as a function of time after the electron-impact ionization event took place (see also Section III.C.1 and Lifshitz and Weiss, 1980; Lifshitz and Eaton, 1983; and Lifshitz et al., 1983).

C. Results

1. Threshold Behavior and Appearance Potentials

The use of monoenergetic electrons to study electron-impact ionization close to threshold and results obtained have been recently reviewed by Kerwin et al. (1969) and Rosenstock (1976). Above the ground-state ion threshold the ionization cross section rises according to the corresponding threshold law until the appearance of excited ion states adds new contributions to the cross section. The situation may be different if the cross section is influenced by competitive channels (Lifshitz and Long, 1964; Rapp, 1971; Lifshitz, 1971; Stephan et al., 1980b). In addition, autoionization is possible and can contribute to the cross section shape.

The available experimental results indicate that close to threshold the cross section energy dependence is, at least for H and He, close to the theoretical 1.127 power law (Brion and Thomas, 1968; Mc Gowan and Clarke, 1968; Marchand et al. 1969; Cvejanovic and Read, 1974). In addition, Schmitt (1978) confirmed this threshold law for K-shell ionization of Ne up to about 100 eV above threshold. For the heavier rare gases, which all have an excited state above threshold, there is evidence of more complicated behavior due to autoionization and other levels (Morrison, 1964; Winters et al., 1966; Collins et al., 1968; Maeda et al., 1968; Marchand et al., 1969; Marmet et al., 1972; Johnson et al., 1978; Miletic et al., 1980; Hashizume and Wasada, 1980), and different studies yield different results.

In earlier studies of the threshold behavior of the electron-impact ionization of molecules it was expected that a series of straight line segments

3 Ionization of Molecules by Electron Impact

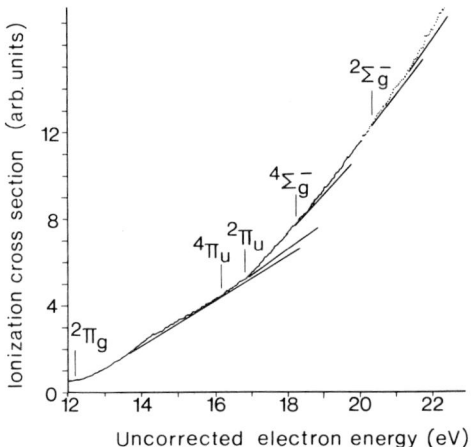

Fig. 21. Ionization cross section function for O_2 close to threshold (measured with 127° electron analyzer). [After Brion and Thomas (1968).]

would be observed in the threshold region, with the onset of each new segment representing the occurrence of a new excited level. Hence, numerous studies have employed a data treatment based on this hypothesis, and in some instances plausible results are obtained (e.g., Frost and McDowell, 1960 and references therein; Sjögren, 1964; Fehlner and Callen, 1968; Kerwin et al., 1969; Lossing, 1972; Maeda et al., 1974; Suzuki and Maeda, 1974, 1977a,b, 1979; Ravishankara and Hanrahan, 1975; Hsieh and Hanrahan, 1977; Miletic et al., 1980; Boesten and Heideman, 1980; Pichou et al., 1983). For instance, the ionization cross section curve of O_2 has been measured by several workers (see Kerwin et al., 1969). Figure 21 shows a curve recently obtained by Brion and Thomas (1968), who used a 127° electron analyzer. Good signal-to-noise is very difficult to obtain and the exact position and contribution of each new excited state is difficult to evaluate, especially for states with small relative Franck–Condon factors. Conversely, one generally (Rosenstock, 1976) obtains, owing to a combination of direct ionization and autoionization, smooth threshold curves of more or less similar shape, no matter what molecule is studied [see, e.g., N_2^+ in Fig. 22, or Figs. 1–5 in Lefaivre and Marmet (1978)]. Using EED techniques, smoothening, straightening, differential plots, or monoenergetic electron beams, a wealth of fine structure due to autoionization of vibrational levels of excited states or to different dissociative channels can be resolved (e.g., Carboneau and Marmet, 1972; Morrison and Traeger, 1973; Johnson et al., 1978; Lefaivre and Marmet, 1978; Locht et al., 1979; Selim, 1980; Hubin-Franskin et al., 1980; Mathur, 1980, 1981; Rabbih

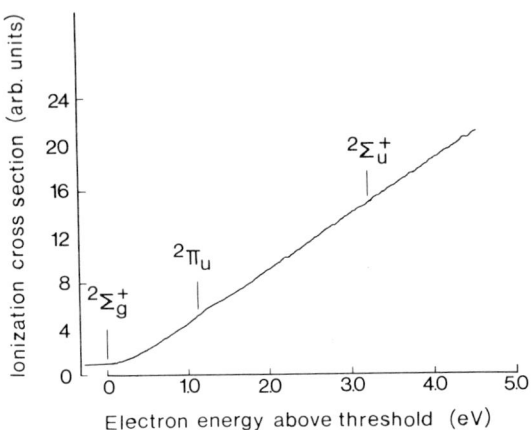

Fig. 22. Ionization cross section function for N_2 close to threshold (measured with $127°$ electron analyzer). [After Brion and Thomas (1968).]

et al., 1981). See also threshold behavior of cluster ions reported by Stephan et al. (1982a,b, 1983c) and Märk (1982a).

Investigations of the threshold laws for multiple ionization have been made in a number of laboratories. Most of these studies employed, however, at best only quasimonoenergetic electrons. For He (where there are no low-lying excited doubly ionized states), double ionization has been recently found by Brion and Thomas (1968) using monoenergetic electrons to follow a quadratic threshold law up to 20 eV above threshold. Determinations of the threshold law of the other rare gases are in conflict [for earlier references see Kerwin et al. (1969) and Rosenstock (1976)]. However, several recent experimental results have confirmed that the ionization cross section of the doubly charged rare gases roughly follows the $nz = 2$ law predicted (Johnson and Morrison, 1975; Marmet et al., 1975; Dutil and Marmet, 1980; Stephan et al., 1980c; Kim et al., 1981).

Following the pioneering work of Dorman and Morrison (1961) a number of workers have interpreted the threshold behavior of molecules in terms of a nzth power law (e.g., Dorman and Morrison, 1963; Newton and Sciamanna, 1964, 1970b; Fischler and Halman, 1964; Daly and Powell, 1966, 1967; Märk, 1975; Märk and Egger, 1977; Märk et al., 1977a, 1978; Märk and Hille, 1978; Hille and Märk, 1978; Brehm and De Frenes, 1978; Stephan et al., 1980c, 1981a, 1983a,b; Kim et al., 1981; Helm et al., 1981). Figure 23 shows as an example the threshold cross section region of NO_2^{2+} and Ar^{2+} (Stephan et al., 1980c). Appearance potentials determined by nzth-root extrapolation are in good agreement with theoretical results (see Section II.B.3 and, e.g., Asbrink et al., 1979).

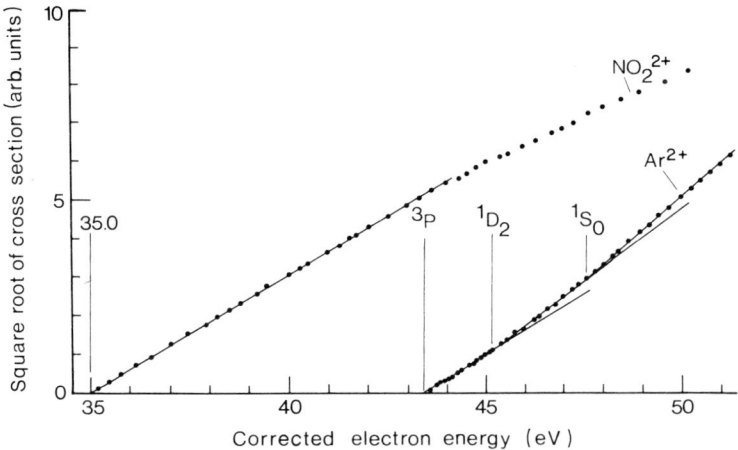

Fig. 23. Square root of the ionization cross section function close to threshold for the processes $NO_2 + e \rightarrow NO_2^{2+}$ and $Ar + e \rightarrow Ar^{2+}$. [After Stephan et al. (1980c).]

Obviously, the study of the ionization cross section function close to threshold invites the determination of the threshold itself, i.e., the appearance potential. The importance, techniques, and problems involved in determining ionization and appearance potentials have recently been discussed in detail (Giessner and Meisels, 1971; Occolowitz et al., 1974; Beynon et al., 1975; Rosenstock, 1976). In addition, there exist recent extensive compilations on the energetics of gaseous ions, including appearance potentials measured by electron-impact ionization and thermochemical information (e.g., Holmes and Lossing, 1980; Maccoll, 1982) deduced from those measurements (Franklin et al., 1969; Field and Franklin, 1970; Rosenstock et al., 1977; Levin and Lias, 1982). Problems inherent in obtaining thermochemical information from studies of ion appearance potentials were first recognized by Chupka (1959) and recently summarized by Klots (1973). Apart from experimental difficulties discussed above, there are two basic reasons why the measured appearance potential is only an upper limit to a true thermochemical value. First, the reverse of an observed decomposition reaction (dissociative ionization) may require activation energy (Cooks et al., 1973), and second, the possible occurrence of a kinetic shift (Chupka, 1959; Cooks et al., 1973; Levsen, 1978). Vestal (1968) has made calculations of the excess energy required to obtain fragmentation rates in the range accessible by mass spectrometry, and Gross (1972), Gordon and Reid (1975), Lifshitz et al. (1974), and Lifshitz and Gefen (1980) have made ion-trapping experimental studies to detect the kinetic-energy shift. Moreover, the experimental determination is further complicated by the occurrence of a competitive shift (Lifshitz and Long, 1964) and by the thermal shift (Chupka, 1971).

2. Ionization Cross Section Functions

There are considerable differences in the ionization cross section data reported in the literature and in addition there are large gaps in the data available (Kieffer and Dunn, 1966). The literature was searched for data through 1968 and comprehensive compilations of low-energy electron ionization cross sections have been issued by the JILA Information Centre [Kieffer and Dunn (1966); Kieffer (1968, 1969)]. It was concluded by Kieffer (1968, 1969) that virtually all the measurements made so far and reported in these reviews were defective; i.e., that known sources of systematic error were not taken into account. Hence, there is a great uncertainty in the absolute magnitude as well as the shape of the data reported. According to Kieffer the range of uncertainties is probably best illustrated by the difference between independent measurements, i.e., varying from a minimum of about 10% to a maximum of factors of 2 and more (this is particularly true with molecules). This situation has been improved in the meantime for several molecules, but data have not been collected in a new compilation [except for specific purposes; e.g., Drowart and Goldfinger (1967), Kieffer (1973), Mc Daniel et al. (1977), de Heer (1981)], but are referenced in several bibliographies (Chamberlain and Kieffer, 1970; Kieffer, 1976; Gallagher et al., 1979; Gallagher and Beaty, 1980; Katsonis, 1981).

In the present work absolute cross section function data and/or reference will be listed below only for studies reported in the meantime of those molecules which are important in aeronomy, gaseous electronics, and plasma physics. The same criteria for selection of data as applied by JILA have been used. For more information see Märk and Dunn (1984).

Diatomics.

H_2: $q_1(H^+/H_2)$ and $q_1(H_2^+/H_2)$ (threshold up to 1000 eV); Adamczyk et al. (1966).

q_t (threshold up to 500 eV); Cowling and Fletcher (1973).

$q_1(H^+/H_2)/q_1(H_2^+/H_2)$ (threshold up to 25 eV, see Fig. 11), $q_1(H_2^+/H_2)$, $q_1(H^+/H_2)$ from $H_2^+(^2\Sigma_g^+)$ and $H_2^+(^2\Sigma_u^+)$ close to threshold; Crowe and Mc Conkey (1973a).

q_i (threshold up to 50 eV, theoretically); Kukulin et al. (1979).

D_2: q_t (threshold up to 500 eV); Cowling and Fletcher (1973).

In addition, there exists a recent review by de Heer (1981) considering excitation, dissociation, and ionization processes for H_2, D_2, T_2, and simple hydrocarbons.

N_2: q_i (threshold up to 120 eV, theoretically); Bauer and Bartky (1965).

q_t (threshold up to 80 eV); Srinivasan et al. (1967).

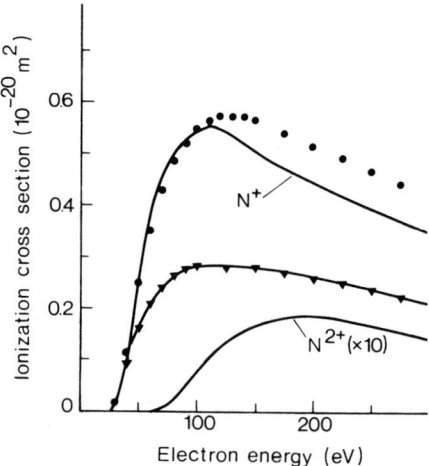

Fig. 24. Absolute partial ionization cross section functions for the dissociative process $N_2 + e \rightarrow N^+$ and $N_2 + e \rightarrow N^{2+}$: (●) data for $N^+ + 2N^{2+}$ (Rapp and Englander-Golden, 1965; Rapp et al., 1965); (▼) N^+ data (Halas and Adamczyk, 1972); (—) (Crowe and Mc Conkey, 1973b).

$q_1(N_2^+/N_2)$, $q_1(N^+/N_2)$, $q_2(N_2^{2+}/N_2)$ (threshold up to 600 eV); Halas and Adamczyk (1972).

$q_1(N_2^+/N_2)$, $q_1(N^+/N_2)$, $q_2(N^{2+}/N_2)$ (threshold up to 300 eV, see as an example Fig. 24); Crowe and Mc Conkey (1973b).

$q_1(N_2^+/N_2)$, $q_2(N_2^{2+}/N_2)$ (threshold up to 170 eV, see as an example Figs. 25 and 26); Märk (1975).

$q_1(N_2^+(A\ ^2\Pi_u)/N_2(A\ ^3\Sigma_u^+))$, $q_1(N_2^+(A\ ^2\Pi_u)/N_2(a'\ ^1\Sigma_u^-))$ (threshold up to 200 eV, theoretically); Ton-That and Flannery (1977).

$q_i(\text{ions}/N_2(A\ ^3\Sigma_u^+))$ (threshold up to 50 eV, theoretically); Kukulin et al. (1979).

$q_1(N_2^+/N_2(A\ ^3\Sigma_u^+))$ (threshold up to 240 eV); Armentrout et al. (1981).

Ionization to excited ionic states see also Section III.B.4.

O_2: $q_1(O_2^+/O_2)$, $q_2(O_2^{2+}/O_2)$ (threshold up to 170 eV); Märk (1975).

$q_1(O_2^+/O_2(v))$ (threshold up to 400 eV); Evans et al. (1980).

NO: $q_1(NO^+/NO)$, $q_2(NO^{2+}/NO)$ (threshold up to 180 eV); Kim et al. (1981).

CO: q_t (threshold up to 100 eV); Srinivasan and Rees (1967).

q_t (threshold up to 10^4 eV, theoretically); Jain and Khare (1976).

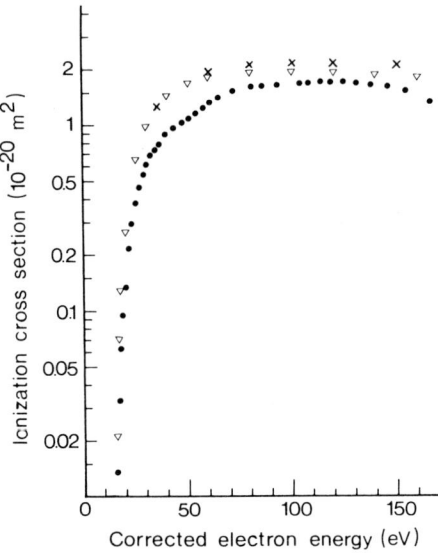

Fig. 25. Absolute partial ionization cross section function for the process $N_2 + e \rightarrow N_2^+$: ($\nabla$) curve derived from measurements of Rapp and Englander-Golden (1965) and Rapp et al. (1965) for $q(N_2^+ + 2\,N_2^{2+})$, ($\times$) Halas and Adamczyk (1972), ($\bullet$) Märk (1975).

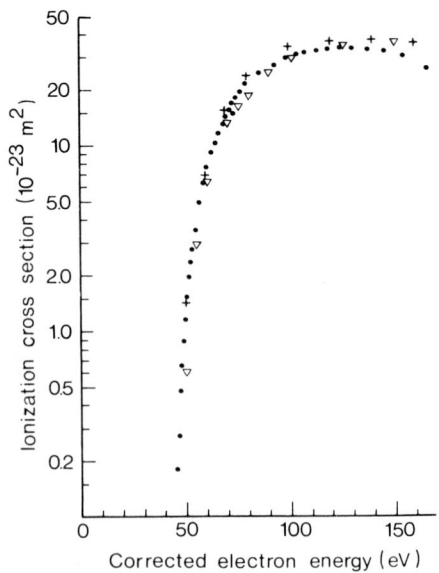

Fig. 26. Absolute partial ionization cross section function for the process $N_2 + e \rightarrow N_2^{2+}$: ($+$) Daly and Powell (1966), (∇) Halas and Adamczyk (1972), ($\bullet$) Märk (1975).

$q_1(CO^+(X\,^2\Sigma^+)/CO(a\,^3\Pi))$ (threshold up to 200 eV, theoretically); Ton-That and Flannery (1977).

$q_1(CO^+/CO)$, $q_2(CO^{2+}/CO)$, $q_2(CO^{2+m}/CO)$ (threshold up to 180 eV); Hille and Märk (1978).

F_2: q_t (threshold up to 100 eV); Center and Mandl (1972).
q_t (threshold up to 100 eV); Stevie and Vasile (1981).

Cl_2: q_t (threshold up to 100 eV); Center and Mandl (1972).
q_t (threshold up to 100 eV); Kurepa and Belic (1978).
q_t (threshold up to 100 eV); Stevie and Vasile (1981).

Br_2: q_t (threshold up to 100 eV); Kurepa *et al.* (1981).

Triatomics.

O_3: $q_1(O_3^+/O_3)$, $q_1(O_2^+/O_3)$, $q_1(O^+/O_3)$ (threshold up to 1000 eV); Siegel (1982).

H_2O: $q_1(H_2O^+/H_2O)$, $q_1(OH^+/H_2O)$, $q_1(O^+/H_2O)$, $q_1(H_2^+/H_2O)$, $q_1(H^+/H_2O)$, $q_2(O^{2+}/H_2O)$, q_t (threshold up to 2 keV); Schutten *et al.* (1966).

$q_1(H_2O^+/H_2O)$, $q_1(D_2O^+/D_2O)$ (threshold up to 170 eV); Märk and Egger (1976).

q_t (threshold up to 10^4 eV, theoretically); Jain and Khare (1976).

CO_2: $q_1((CO_2^+/CO_2))$, $q_1(CO^+/CO_2)$, $q_1(O^+/CO_2)$, $q_1(C^+/CO_2)$, $q_1(O_2^+/CO_2)$ (threshold up to 600 eV); Adamczyk *et al.* (1972).

$q_1(CO_2^+/CO_2)$, $q_1(CO^+/CO_2)$, $q_1(O^+/CO_2)$, $q_1(C^+/CO_2)$ (threshold up to 300 eV); Crowe and Mc Conkey (1974).

q_t (threshold up to 10^4 eV, theoretically); Jain and Khare (1976).

$q_1(CO_2^+/CO_2)$, $q_2(CO_2^{2+}/CO_2)$ (threshold up to 170 eV); Märk and Hille (1978).

NO_2: $q_1(NO_2^+/NO_2)$, $q_2(NO_2^{2+}/NO_2)$ (threshold up to 180 eV); Stephan *et al.* (1980c).

N_2O: $q_1(N_2O^+\,(A\,^2\Sigma^+)/N_2O)$ (threshold up to 2000 eV); Van Sprang *et al.* (1978).

$q_1(N_2O^+/N_2O)$, $q_1(N_2O^{+m}/N_2O)$ (threshold up to 180 eV); Märk *et al.* (1981).

SO_2: q_t (threshold up to 230 eV); Cadez *et al.* (1972).
q_t, $q_1(SO_2^+/SO_2)$, $q_1(SO^+/SO_2)$ (threshold up to 40 eV); Smith and Stevenson (1981).

HgBr$_2$: q_1(HgBr$_2^+$/HgBr$_2$), q_1(HgBr$^+$/HgBr$_2$), q_1(Br$^+$/HgBr$_2$), q_t (threshold up to 70 eV); Wiegand and Boedeker (1982).

Polyatomics.

BF$_3$: q_t (threshold up to 250 eV); Kurepa et al. (1976).

NH$_3$: q_t (threshold up to 10^4 eV, theoretically); Jain and Khare (1976).

q_1(NH$_3^+$/NH$_3$), q_1(NH$_2^+$/NH$_3$), q_1(NH$^+$/NH$_3$), q_1(N$^+$/NH$_3$), q_1(H$_2^+$/NH$_3$), q_1(H$^+$/NH$_3$), q_2(NH$_3^{2+}$/NH$_3$), q_t (threshold up to 180 eV); Märk et al. (1977a).

q_1(NH$_3^+$/NH$_3$), q_1(NH$_2^+$/NH$_3$), q_1(NH$^+$/NH$_3$), q_t (threshold up to 300 eV); Crowe and Mc Conkey (1977).

q_1(NH$_3^+$/NH$_3$), q_1(NH$_2^+$/NH$_3$), q_1(NH$^+$/NH$_3$), q_1(N$^+$/NH$_3$), q_1(H$_2^+$/NH$_3$), q_1(H$^+$/NH$_3$), q_2(NH$_3^{2+}$/NH$_3$), q_t (threshold up to 1000 eV); Bederski et al. (1980).

q_t (threshold up to 200 eV, experimentally and theoretically); Djuric et al. (1981).

PH$_3$: q_1(PH$_3^+$/PH$_3$), q_1(PH$_2^+$/PH$_3$), q_1(PH$^+$/PH$_3$), q_1(P$^+$/PH$_3$), q_1(H$_2^+$/PH$_3$), q_1(H$^+$/PH$_3$), q_2(PH$_3^{2+}$/PH$_3$), q_2(PH$_2^{2+}$/PH$_3$), q_2(PH^{2+}/PH$_3$), q_2(P^{2+}/PH$_3$), q_t (threshold up to 180 eV); Märk and Egger (1977).

SO$_3$: q_1(SO$_3^+$/SO$_3$), q_1(SO$_2^+$/SO$_3$), q_1(SO$^+$/SO$_3$), q_t (threshold up to 30 eV); Smith and Stevenson (1981).

CH$_4$: q_1(CH$_4^+$/CH$_4$), q_1(CH$_3^+$/CH$_4$), q_1(CH$_2^+$/CH$_4$), q_1(CH$^+$/CH$_4$), q_1(C$^+$/CH$_4$), q_1(H$_2^+$/CH$_4$), q_1(H$^+$/CH$_4$) (threshold up to 2000 eV); Adamczyk et al. (1966).

q_t (threshold up to 10^4 eV, theoretically); Jain and Khare (1976).

q_t (threshold up to 1000 eV, theoretically); Deutsch (1982, private communication) (see also Fig. 27).

Cl and F methanes:

q_t(CCl$_2$F$_2$) (threshold up to 250 eV); Pejcev et al. (1979).

q_t(CCl$_4$, CHCl$_3$, CF$_4$) (threshold up to 1000 eV, theoretically); Deutsch (1982, private communication) (see Fig. 27).

q_1(CCl$_3^+$/CCl$_4$), q_1(CCl$_2^+$/CCl$_4$), q_1(CCl$^+$/CCl$_4$), q_1(C$^+$/CCl$_4$), q_1(Cl$_2^+$/CCl$_4$), q_1(Cl$^+$/CCl$_4$), q_2(CCl$_3^{2+}$/CCl$_4$), q_2(CCl$_2^{2+}$/CCl$_4$), (threshold up to 180 eV); Stephan et al. (1983a,b).

q_1(CF$_3^+$/CF$_4$), q_1(CF$_2^+$/CF$_4$), q_1(CF$^+$/CF$_4$), q_1(C$^+$/CF$_4$), q_1(F$_2^+$/CF$_4$), q_1(F$^+$/CF$_4$), q_2(CF$_3^{2+}$/CF$_4$), q_2(CF$_2^{2+}$/CF$_4$) (threshold up to 180 eV); Stephan et al. (1983a,b).

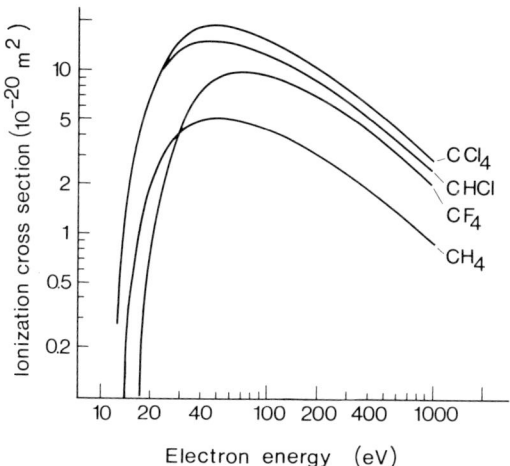

Fig. 27. Absolute total (counting) ionization cross function for CCl_4, $CHCl_3$, CF_4, and CH_4 calculated by Deutsch (1982, private communication) using Gryzinski's (1965) formula [Eq. (40)].

SiH$_4$: $q_1(SiH^+(A\ ^1\Pi)/SiH_4)$, $q_1(Si^+(4p\ ^2P_j^0)/SiH_4)$ (threshold up to 65 eV); Perrin and Schmitt (1982). See also Turban et al. (1980).

UF$_6$: q_t (threshold up to 1000 eV); Compton (1977).

SF$_6$: $q_1(SF_5^+/SF_6)$, $q_1(SF_4^+/SF_6)$, $q_1(SF_3^+/SF_6)$, $q_1(SF_2^+/SF_6)$, $q_1(SF^+/SF_6)$, $q_1(S^+/SF_6)$, $q_1(F^+/SF_6)$, $q_2(SF_4^{2+}/SF_6)$, $q_2(SF_2^{2+}/SF_6)$ (threshold up to 600 eV); Stanski and Adamczyk (1983).

Clusters.

Recently, there has been a growing interest in studies of the formation, properties, and reactions of neutral and ionic clusters comprised of atoms and/or molecules (Knof, 1974; Kebarle, 1977; Castleman, 1979, 1982; Märk, 1982 a,b; Märk et al., 1982; Märk and Castleman, 1984). Neutral van der Waals clusters are produced in free jet nozzle expansion yielding beams of vibrationally and rotationally cooled species (Andres, 1969; Anderson, 1974; Hagena, 1974). Most experiments use electron-impact ionization in combination with mass spectrometry for the detection of these weakly bound molecules. However, very little information is yet available on absolute cross section functions or on fragmentation yields in electron-impact ionization of clusters. This includes fragmentation yields of Ar dimers (Lee and Fenn, 1978), CO_2 dimers (Gough and Miller, 1982), SF_6 dimers and trimers (Geraedts et al., 1982); partial ionization cross sections in Ar_2, $(CO_2)_2$ (Leckenby and Robbins, 1966), Ar_2, Kr_2, Xe_2, ArKr, KrXe (Helm *et al.*, 1979; Märk, 1982b), H_2O clusters (Milne *et al.*, 1970; Castleman *et al.*, 1981); and ionization cross sections as a function of electron energy and mean value of the number of monomers per cluster in Ar, CO_2 (Hagena and Henkes, 1965; Falter *et al.*, 1970), H_2 (Tay *et al.*, 1970; Henkes and Mikosch, 1974), H_2, N_2, and CO_2 (Bottiglioni *et al.*, 1972). For more information see Märk, 1984a and Märk and Castleman, 1984.

IV. Differential Ionization Cross Sections

Partial and total ionization cross sections are of great importance for practical applications (Märk and Dunn, 1984), but they give only limited insight into the ionization process itself. Conversely, a great deal of information may be obtained from the study of differential cross sections, i.e.,

Single differential cross section: $dq(T, W)/dW$. (48)

Double differential cross section: $d^2q(T, W, \vartheta)/dW\, d\Omega$. (49)

Triple differential cross section: $d^3q(T, W_1, \vartheta_1, \vartheta_2, \varphi_2)/dW\, d\Omega_1\, d\Omega_2$. (50)

where W is the energy and ϑ and φ the polar angles of detection of the electrons after the collision (see Fig. 28). Triple differential cross sections contain all kinematic information [even recoil effects (Mc Conkey *et al.*, 1972)], but owing to experimental difficulties were not determined until 1969 (Amaldi *et al.*, 1969; Erhardt *et al.*, 1969). In addition to these electron–electron spectroscopy studies, ion kinetic-energy spectroscopy [called IKES (Beynon and Cooks, 1975)] and the measurement of angular distributions of fragment ions is necessary for a full picture of the ionization process.

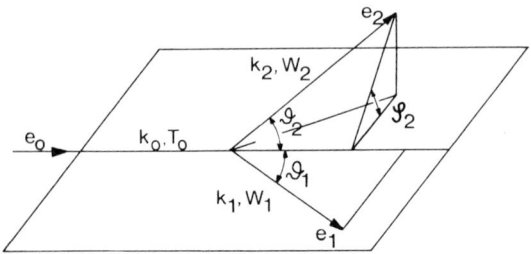

Fig. 28. Schematic view of the kinematics of an electron-impact ionization process: e_0 incident electron with kinetic energy T_0 and momentum k_0; e_1 and e_2 scattered and ejected electron with kinetic energy W_1 and W_2 and momentum k_1 and k_2, respectively.

A. Electron–Electron Spectroscopy

It is outside the scope of this chapter to give a detailed review of this subject. However, there exist several excellent up-to-date reviews in the literature (e.g., Erhardt *et al.* 1972, 1980; Opal *et al.*, 1972; van der Wiel, 1973, 1976, 1980; Oda, 1973; Weigold and Mc Carthy, 1978; Weigold, 1980; Giardini-Guidoni *et al.*, 1978, 1980, 1981; Mc Carthy, 1980; Brion and Hamnett, 1980; Erhardt, 1983). Hence this section will be illustrative rather than exhaustive.

1. Single Differential Cross Section

The single differential cross section [see Eq. (48)] measured at the collision energy T_0 represents the energy distribution (integrated over all angles) of the two outgoing electrons after the ionization event. The energy distribution is in principle symmetrical (see Fig. 29) with respect to $\frac{1}{2}(W_1 + W_2) = \frac{1}{2}(T_0 - E_i)$, since for each fast electron a corresponding slow electron of complementary energy is ejected. If the incident energy T_0 is large enough, the left-hand part of the distribution is mainly due to 'ejected' electrons and the right-hand part mainly due to 'scattered' electrons (hence this part can also be interpreted in terms of an energy loss spectrum). Although in principle the two outgoing electrons are indistinguishable, this interpretation is supported by the different angular behavior of the scattered (fast) and ejected (slow) electrons. From theoretical and experimental studies (e.g., Opal *et al.*, 1971, 1972; Oda, 1973; Jain and Khare, 1976) it is known that the single differential

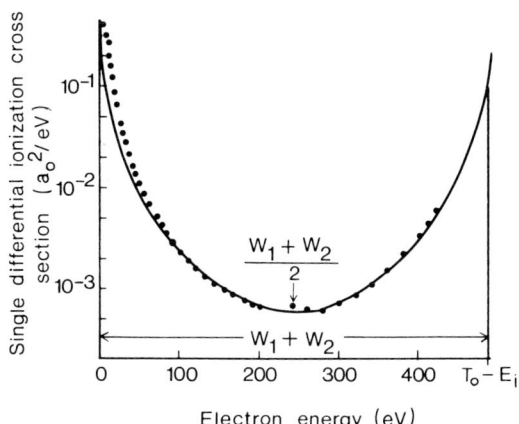

Fig. 29. Single differential ionization cross section function for CH_4 (incident electron energy 500 eV): (●) Opal *et al.* (1971, 1972) and Oda and Nishimura (1975), (—) Mott cross section corrected for binding energies of orbital electrons. [After Oda and Nishimura (1975).]

cross section function has maxima close to $W = 0$ and $W = T_0 - E_i$ (see Fig. 29). Conversely, at incident energies close to the ionization threshold this function becomes at least for atoms independent of W (Cvejanovic and Read, 1974; Pichou *et al.*, 1978).

2. Double Differential Cross Section

The double differential cross section is higher by one order with respect to the number of kinetic parameters, i.e., the energy W and the angle ϑ of the outgoing electrons have to be measured. Moreover, for a meaningful interpretation of the results it is useful to know the initial state of the target system before the ionizing collision. In practice, one of the three experimental parameters T, W, or ϑ is varied, with the other two remaining constant. This allows the determination of angular distributions (T and W constant), and of ejected-electron and/or of energy loss spectra, i.e., of the first kind (T and ϑ constant; see, e.g., Fig. 2) and of the second kind (W and ϑ constant). In general, the higher the energy of the scattered electrons, the more their intensity is peaked forward, whereas the angular distribution of very low-energy electrons is nearly isotropic, especially if autoionization of molecules is involved (see, e.g., Erhardt and Erbse, 1963; Erhardt *et al.*, 1965; Erhardt and Linder, 1967; Opal *et al.*, 1972). Energy loss studies in the range of validity of the first Born approximation (high electron energies) allow quantitative determination of the differential generalized oscillator strengths (Bethe surface) and dipole oscillator strengths using only forward-scattered electrons. Double differential cross sections obtained for various molecules are given for instance by Erhardt (1968), Tisone (1972), Opal *et al.* (1972), Erhardt *et al.* (1972), van der Wiel (1973, 1976, 1980), Oda (1973), Lee *et al.* (1975), Brion (1975), Dubois and Rudd (1975), Hamnett *et al.* (1976), Oda and Nishimura (1975), Oda *et al.* (1979), van Wingerden *et al.* (1979), and Brion and Hamnett (1980). In addition, energy distributions of ejected electrons from water clusters have been reported recently (Mathis and Vroom, 1975).

3. Triple Differential Cross Section

The triple differential cross section gives the most detailed information available on the ionization process. It has to be measured in coincidence ($e, 2e$ experiments), because the energies and angles of both outgoing electrons have to be measured. Since it is a function of five parameters [Eq. (50)] a full graphical representation is impossible. Various groups have measured different sets of data, depending on the type of information to be obtained about the ionization process of molecules:

(1) At low and intermediate incident electron energies with asymmetric kinematics (i.e., ϑ_1 small, $T_0 \lesssim 200$ eV, $W_1 > W_2$) the wave functions of the

slow electrons are strongly influenced by the presence of the other collision partners. These studies provide stringent tests for theoretical methods (three-body process with long-range forces). The first study of this kind (Erhardt et al., 1969, 1972, 1980) of the angular correlation of the outgoing electrons in helium and also more complex targets has laid the basis for the present understanding of the ionization process. Very briefly, for intermediate incident energy and a given scattering angle and energy loss of the projectile, one finds—in coincidence—two intensity lobes of ejected electrons. One points more or less in the direction of the momentum transfer (binary peak), the other in the opposite direction (recoil peak). At low incident energy rapid variations of the angular distributions with changing scattering angle indicate interference between the amplitude of direct scattering and exchange scattering. Further recent work includes studies by Cvejanovic and Read (1974) (near threshold), Beaty et al. (1977) (out of plane), and Weigold et al. (1979).

(2) This technique can obviously also be applied to the study of autoionizing transitions (Balashov et al., 1972, Weigold et al., 1975) and has been applied to a number of molecular systems by Wilden et al. (1977) and Comer (1978) (see Section II.A.3).

(3) At high incident electron energy one may obtain information on the momentum distribution of target electrons in the molecule and on the structure of the neutral target molecule and the ion being produced in the ionization event. Symmetric and asymmetric (forward scattering) kinematics have been employed:

Symmetric Kinematics. This involves measurements where the two indistinguishable electrons in the exit channel are treated equally, i.e., $W_1 = W_2$ and $\vartheta_1 = \vartheta_2$, with either $\varphi = 0°$ [coplanar symmetry (Giardini-Guidoni et al., 1978, 1980; Fantoni et al. (1982)] or φ variable [noncoplanar symmetry (Weigold and Mc Carthy 1978; Weigold, 1980)]. For very high momentum transfer one approaches the situation of binary scattering.

Asymmetric Kinematics. In this case high-energy electrons are scattered through a very small angle ($\vartheta_1 \cong 0$) with small energy loss and approximately zero momentum transfer. The ionization process involves therefore distant collisions and takes place under conditions corresponding to the optical limit (dipole excitation), thus allowing the determination of partial dipole oscillator strengths. Hamnett et al. (1976) have shown that this (e, 2e) reaction becomes equivalent to photoelectron spectroscopy and that an angle $\vartheta_2 = 54.7°$ corresponds to a magic angle at which the coincidence intensity is independent of the angular anisotropy. Extending this concept (e.g., to electron–ion coincidence experiments), a number of "photon simulation" experiments have been carried out recently, including "photoionization, photoabsorption, photofluorescence, and photoelectron spectroscopy" (van der Wiel, 1973, 1976; Hamnett et al., 1976; Brion and Hamnett, 1980).

B. Fragment-Ion Kinematics

Partial dissociative ionization cross section measurements are especially susceptible to error. Not only can the absolute magnitude of the cross section measured be in error owing to discrimination effects (see Section III.B.2), but the shape of the cross section function may be distorted since the anisotropy in the fragment-ion angular distribution and kinetic-energy distribution is, in general, a function of the incident electron energy.

Quite impressive work on the kinetic-energy distribution of fragment ions was done in the early thirties and forties (Hagstrum, 1951, and references cited therein). Conversely, real interest in the fragment-ion angular distribution was only aroused by Dunn (1962), who used basic symmetry arguments to demonstrate that fragment ions should be emitted anisotropically [see also earlier work by Sasaki and Nakao (1935, 1941) and Kerner (1953)]. In order to completely specify the dissociative ionization process these kinematical properties must be known.

1. *Ion Kinetic-Energy Spectrometry*

Ion kinetic-energy spectrometry (IKES) has its roots in two areas of mass spectrometry; first, studies of metastable ions, in particular, relating metastable peak width to translational energy release (MIKE); and second, measurement of the kinetic-energy distribution of fragment ions formed by direct ionization within the ion source. As comprehensive reviews on the subject of metastable and collision-induced dissociation kinetic-energy studies have been published recently (Beynon and Cooks, 1975, 1976; Cooks, 1978; Los and Govers, 1978; Beynon et al., 1982), the present work will attempt to complement this by discussing the measurement of fragment-ion kinetic distributions.

The translational energy with which fragment ions are formed in the ion source and the dependence of the appearance potential on the kinetic energy have been measured with a number of different methods, i.e., *retarding potential analysis* [Hagstrum (1951) (H_2, N_2, O_2, CO, NO), and references cited therein; Kandel (1953, 1954) (N_2, C_2H_6, C_3H_8, C_2N_2, $C_6H_5CH_3$, nitromethane); Locht and Schopman (1974) (O_2); Brehm and De Frenes (1978) (H_2, N_2, O_2, CO, NO); Locht and Momigny (1978) (N_2, O_2); Locht et al.

(1979) (CH_4), Locht and Momigny (1980) (CH_4); Johnson and Franklin (1980) (H_2); Olivier et al. (1982) (N_2O)]; *magnetic field and peak shape analysis with sector field mass spectrometers* [Hagstrum and Tate (1941) (H_2, N_2, O_2, CO, NO); Mohler et al. (1954) (methyl ions from hydrocarbons); Lagergren (1954); Taubert (1959) (chloroethane); Tsuchiya (1962) (propane); Dunn and Kieffer (1963) (H_2); Olmsted et al. (1964) (methyl ions from organic compounds); Kieffer and Van Brunt (1967) (N_2); Kieffer and Dunn (1967) (H_2, D_2); Van Brunt and Kieffer (1970) (H_2); Van Brunt (1977) (H_2)]; *time-of-flight analysis* (Franklin et al. (1967); Hierl and Franklin (1967); Haney and Franklin (1968) (average and maximum translational energies); Stockdale et al. (1975); Deleanu and Stockdale (1975) (N_2); Weiner and Ossinger (1976) (apparatus); Ossinger and Weiner (1976) (hydrocarbons, organic halides); Stockdale (1977) (N_2); Burrows et al. (1980) (H_2, D_2, HD); Armenante et al. (1980) (N_2); Khakoo and Srivastava (1983) (H_2); see also pulsed ion source method by Platzner et al. (1971)]; *electrostatic energy analysis* (Morrison and Stanton (1958) (CH_4 and methylbromide); Stanton (1959) (propane, butane); Monahan and Stanton (1962) (hydrocarbons); Stanton and Monahan (1964) (CO, H_2O, NH_3, CH_4, C_6H_6); Mc Culloh and Rosenstock (1968) (H_2); Crowe and Mc Conkey (1973c) (H_2); Crowe and Mc Conkey, (1975) (N_2); Köllmann (1975a,b) (H_2, N_2, CO); Köllmann (1978) (H_2); Pichou et al. (1983) (H_2, D_2); Khakoo and Srivastava (1983) (H_2)]; *deflection method* [Rees and Hipple (1949) (*n*-butane); Berry (1950) (CO, O_2, N_2, NO, CH_4, *n*-butane); Osberghaus and Taubert (1951) (H_2, N_2); Dibeler et al. (1956) (trifluoromethylhalides); Taubert (1959) (hydrocarbons); Bracher et al. (1963) (hydrocarbons); Durup and Heitz (1964) (H_2, N_2, NO, C_2H_5OD, C_3H_8); Appell et al. (1964) (N_2, NO, CO_2, C_3H_8); Taubert (1964) (paraffins); Erhardt and Tekaat (1964) (propane); Fuchs and Taubert (1965) (CO_2); Erhardt and Kresling (1967) (N_2, O_2, H_2O, CO_2, ethane); Rowland et al. (1969) (HNCO); Rowland (1971) (anthracene, phenanthrene, diphenylacetylene); Appell and Durup (1972) (H_2O); Sen Sharma and Franklin (1974) (propane and others)].

When a diatomic molecule is ionized into an excited state above one of its decomposition asymptotes, this ion dissociates into two fragments [reactions, (4), (5), (17), (18), (21)] whose translational energy sum will be equal to the difference between the excitation energy and the ground-state heat of reaction. The kinetic energy will depend on the effective Franck–Condon region, and one important application of the Franck–Condon principle is the prediction of kinetic energies of fragment ions. Detailed measurements and calculations have more frequently been made on the hydrogen molecule than on any other and hence this system will be discussed below as an illustrative example. However, even in the simplest case a completely satisfactory picture of the processes involved has yet to emerge. Although some early studies were in disagreement (Stevenson, 1960), Dunn and Kieffer (1963) concluded from their studies that the measured proton kinetic-energy distributions above ~ 1.5 eV are in agreement with distributions calculated using the Franck–Condon principle assuming excitation to the $^2\Sigma_u^+$ state. Subsequent results (Kieffer and Dunn, 1967; Van Brunt and Kieffer, 1970) showed that other processes were also present. Crowe and Mc Conkey (1973a) reported several resolved peaks in the proton energy distributions between 1 and 8 eV. These well-resolved peaks have not been observed in other studies, save for some structure seen by Stockdale et al. (1975) and van Brunt (1977). Additional evidence of the complex nature of this process (see Fig. 30) is provided by the recent work of (1) Köllmann (1975b, 1978) on H^+ from H_2 (autoionizing states) using incident electron energies below 29 eV; (2) Johnson and Franklin (1980) indicating strong contributions from other mechanisms; and (3) Burrows et al. (1980). The last authors observed significant isotope effects for dissociative ionization in the H_2-D_2-HD molecular family, consistent with a large dissociative autoionization contribution. For more information see recent studies by Pichou et al. (1983) and Khakoo and Srivastava (1983). Kinetic-energy distributions of other diatomic molecules have also been studied in detail recently (see references given above).

Only a few studies are concerned with the kinematics of doubly ionized species. Kieffer and Van Brunt (1967) measured the appearance potential and

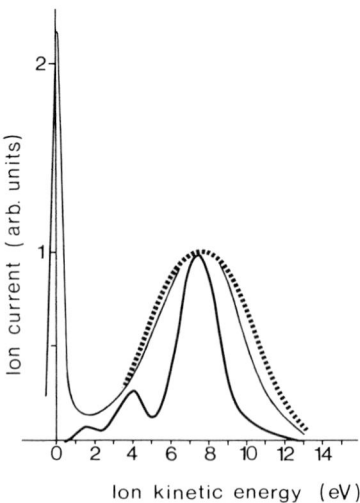

Fig. 30. Ion kinetic-energy distribution for H^+ produced by electron-impact ionization of H_2: (- - -) Van Brunt and Kieffer (1970) for $T_0 = 160$ eV, $\vartheta = 23°$; (—) Crowe and Mc Conkey (1973a) for $T_0 = 100$ eV, $\vartheta = 27°$; (—) Köllmann (1975) for $T_0 = 150$ eV, $\vartheta = 35°$.

the kinetic-energy distribution of N^{2+} produced by electron-impact ionization of N_2. Crowe and Mc Conkey (1975) and Deleanu and Stockdale (1975) measured the energy and angular dependence of the N^{2+} distribution. However, a large discrepancy exists for $AP(N^{2+})$, as determined by these authors, and no thermal N^{2+} have been observed. Conversely, Locht and Momigny (1979) have confirmed the existence of thermal N^{2+} ions and O^{2+} ions (from O_2) and have eliminated the existing discrepancy for $AP(N^{2+})$.

In addition, in several studies (Mohler *et al.*, 1954; Tsuchiya, 1962; Bracher *et al.*, 1963; Olmsted *et al.*, 1964; Stanton and Monahan, 1964) the energies of the fragment ions were found to be consistent with that expected from the fragmentation of a doubly charged molecular ion into a pair of singly charged products [see reaction (17)]. Recent experiments in which both fragments produced in the dissociations of a diatomic molecule are detected in coincidence can provide definitive information about the charge state of the parent ion. Stockdale (1977) reported a coincidence study in which N^+ ions from the process $e + N_2 \rightarrow N_2^{2+} + 3e \rightarrow N^+ + N^+ + 3e$ are observed for incident electron energies of 100, 150, and 300 eV, indicating the presence of N^+ ions with kinetic energies between 5 and 14 eV (see also Edwards and Wood, 1982). Brehm and De Frenes (1978) used another technique to observe the charge separation process [reaction (17)] in H_2, N_2, O_2, CO, and NO. Results in H_2 agree well with earlier coincidence experiments and Franck–Condon calculations by Mc Culloh and Rosenstock (1968). Similarly, Köllman (1975a) reported the kinetic-energy distributions of ion pair production in H_2 and CO.

For polyatomic systems (see also references given above), there is no such simple energy relationship (Franck–Condon principle), because part, or all, of the excess energy can survive in rotational and/or vibrational modes of the products. The fraction depends upon the corresponding potential-energy surface of the activated complex, which are in general unknown. Average translational energies have then to be calculated in the framework of QET theory (e.g., Haney and Franklin, 1968). One means of investigating the dissociative ionization process in polyatomics and to identify the molecular states involved is the measurement of the appearance potentials of the ionic fragments as a function of their kinetic energy (see e.g., Olivier *et al.*, 1982, Appell and Durup, 1972, and references cited therein). Appell and Durup (1972) were able to observe six different formation processes for H^+ from H_2O by this method, using a tentative model of the molecular states established from theoretical and experimental data in the literature. Ossinger and Weiner (1976) report the observation of three different groups of methyl ions produced by dissociation of propane, 1-butene, and methyl bromide, i.e., including quasithermal ions (see, e.g., Taubert, 1964) and intermediate- and high-energy ions (satellites).

2. Angular Distribution of Fragment Ions

In some cases (H_2, D_2, N_2, CO, CO_2) translational kinetic-energy distribution measurements have been supplemented with data on angular distributions measured relative to the incident electron direction (Sasaki and Nakao, 1935, 1941; Dunn and Kieffer, 1963; Kieffer and Van Brunt, 1967; Van Brunt and Kieffer, 1970; Crowe and Mc Conkey, 1973a,b, 1974, 1975; Deleanu and Stockdale, 1975; Köllmann 1975b, 1978; Van Brunt, 1977; Johnson and Franklin, 1980). As expected from theoretical considerations of symmetry and effects of momentum transfer (Sasaki and Nakao, 1935; Kerner, 1953; Dunn, 1962; Zare, 1967; Van Brunt and Kieffer, 1970; Van Brunt, 1974) angular distributions in the dissociative ionization of diatomic molecules were found to be quite anisotropic [except N^{2+} from N_2 (Crowe and Mc Conkey, 1975)]. The degree of anisotropy is greatest near threshold and depends both on the incident electron energy and the fragment-ion kinetic energy (see Fig. 31). Moreover, Crowe and Mc Conkey (1974) demonstrated for CO_2 that anisotropies can also exist in the angular distributions of dissociative ions produced from a polyatomic molecule. Knowledge of both the kinetic energy and angular distribution of dissociative ions reveals information about the shape and position of ion and molecular potential-energy curves and the ionization process involved.

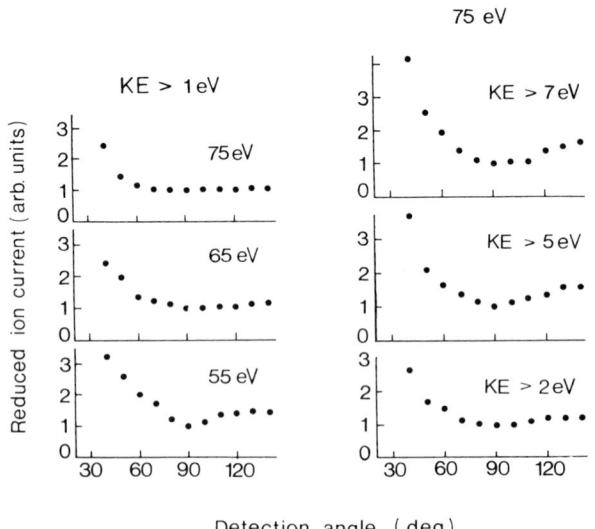

Fig. 31. Angular distributions of H^+ produced by electron-impact ionization of H_2. Parameter: incident electron energy and ion kinetic energy (KE) after ionization [After Johnson and Franklin (1980).]

Acknowledgments

This work was partially supported by the Österreichischer Fonds zur Förderung der wissenchaftlichen Forschung.

References

Aarts, J. F. M., de Heer, F. J., and Vroom, D. A. (1968). *Physica* **40**, 197–206.
Adamczyk, B. (1970). *Ann. Univ. Mariae Curie-Sklodowska* **24**, 141–156.
Adamczyk, B., Boerboom, A. J. H., Schram, B. L., and Kistemaker, J. (1966). *J. Chem. Phys.* **44**, 1–2.
Adamczyk, B., Boerboom, A. J. H., and Lukasiewicz, M. (1972). *Int. J. Mass Spectrom. Ion Phys.* **9**, 407–412.
Adamczyk, B., Boerboom, A. J. H., and De Vries, A. E. (1973). *Int. J. Mass Spectrom. Ion Phys.* **12**, 314–315.
Agee, J. H., Wilcox, J. B., Abbey, L. E., and Moran, T. F. (1981). *Chem. Phys.* **61**, 171–179.
Ahnell, J. E., and Koski, W. S. (1975). *J. Chem. Phys.* **62**, 4474–4476.
Alberti, R., Genoni, M. M., Pascual, C., and Vogt, J. (1974). *Int. J. Mass Spectrom. Ion Phys.* **14**, 89–98.
Amaldi, U., Egidi, A., Marconero, R., and Pizzella, G. (1969). *Rev. Sci. Instrum.* **40**, 1001–1004.
Anderson, J. B. (1974). *In* "Molecular Beams and Low Density Gas Dynamics" (P.P. Wegener, ed.), pp. 1–92. Dekker, New York.
Anderson, N., Eggleton, P. P., and Keesing, R. G. W. (1967). *Rev. Sci. Instrum.* **38**, 924–927.
Andlauer, B., and Ottinger, C. (1971). *J. Chem. Phys.* **55**, 1471–1472.
Andlauer, B., and Ottinger, C. (1972). *Z. Naturforsch.* **27a**, 293–309.
Andres, R. P. (1969). *In* "Nucleation" (A. C. Zettlemoyer, ed.), pp. 69–108. Dekker, New York.
Appell, J., and Durup, J. (1972). *Int. J. Mass Spectrom. Ion Phys.* **10**, 247–265.
Appell, J., Durup, J., and Heitz, F. (1964). *Ad. Mass Spectrom.* **3**, 457–469.
Appell, J., Durup, J., Fehsenfeld, F. C., and Fournier, P. (1973). *J. Phys. B.* **6**, 197–205.
Appling, J. R., Jones, B. E., Abbey, L. E., Bostwick, D. E., and Moran, T. F. (1983). *Org. Mass Spectrom.* **18**, 282–294.
Arai, H., and Hotta, H. (1975). *Radiat. Res.* **64**, 407–415.
Armenante, M., Brancaccio, A., Burattini, E., Santoro, V., Spinelli, N., and Vanoli, F. (1980). *Int. J. Mass Spectrom. Ion Phys.* **36**, 213–230.
Armentrout, P. B., Tarr, S. M., Dori, A., and Freund, R. S. (1981). *J. Chem. Phys.* **75**, 2786–2794.
Asbrink, L., Fridh, C., and Lindholm, E. (1979). *Int. J. Mass Spectrom. Ion Phys.* **32**, 93–96.
Ast, T. (1980). *Adv. Mass Spectrom.* **8A**, 555–576.
Asundi, R. K., Craggs, J. D., and Kurepa, M. V. (1963). *Proc. Phys. Soc. London* **82**, 967–978.
Audier, H. E. (1969). *Org. Mass Spectrom.* **2**, 283–298.
Baer, T., Werner, A. S., and Tsai, B. P. (1975). *J. Chem. Phys.* **62**, 2497–2499.
Baer, T., Tsai, B. P., Smith, D., and Murray, P. T. (1976). *J. Chem. Phys.* **64**, 2460–2465.
Bainbridge, K. T. (1947). Solvay Report (R. Stoops, ed.), pp. 55–100. Brussels.
Baker, F. A., and Hasted, J. B. (1968). *Adv. Mass Spectrom.* **4**, 727–734.
Balashov, V. V., Lipovestsky, S. S., and Senaskenko, V. S. (1972). *Phys. Lett. A* **39**, 103–104.
Banks, D., Percival, I. C., and Valentine, N. A. (1969). "Proc. 6th ICPEAC," pp. 215–216. MIT Press, Cambridge.
Barnard, G. P. (1953). "Modern Mass Spectrometry." The Institute of Physics, London.
Barnard, G. P. (1956). "Mass Spectrometer Researches." Her Majesty Stationary Office, London.

Bartmess, J. E., and Georgiadis, R. M. (1983). *Vacuum TAIP* **33**, 149–153.
Batabyal, A. K., Barna, A. K., and Srivastava, B. N. (1965). *Indian J. Phys.* **39**, 219–226.
Bates, D. R., Fundaminsky, A., Leech, J. W., and Massey, H. S. W. (1950). *Philos. Trans. R. Soc. London A* **243**, 93–143.
Bauer, E., and Bartky, C. D. (1965). *J. Chem. Phys.* **43**, 2466–2476.
Beaty, E. C., Hesselbacher, K. H., Hong, S. P., and Moore, J. H. (1977). *J. Phys. B* **10**, 611–620.
Bederski, K., Wojcik, L., and Adamczyk, B. (1980). *Int. J. Mass Spectrom. Ion Phys.* **35**, 171–178.
Beebe, N. H. F., Thulstrup, E. W., and Anderson, A. (1976). *J. Chem. Phys.* **64**, 2080–2093.
Becker, S., and Dietze, H. J. (1983). *Int. J. Mass Spectrom. Ion Phys.* **51**, 325–345.
Bell, K. L., and Kingston, A. E. (1974). "Adv. Atom. Molec. Phys.," (D. R. Bates, and B. Bederson, ed.), Vol. 10, pp. 53–130 Academic Press, New York.
Bell, M. (1939). *Phys. Rev.* **55**, 201–202.
Benz, W. (1969). "Massenspektrometrie organischer Verbindungen" (unter Mitwirkung von D. Henneberg) Akademische Verl. Ges., Frankfurt.
Beran, J. A., and Kevan, L. (1969). *J. Phys. Chem.* **73**, 3866–3876.
Berry, C. E. (1950). *Phys. Rev.* **78**, 597–605.
Berry, C. E. (1974). *Adv. Mass Spectrom.* **6**, 1–13.
Bertein, F. (1950). *C. R. Acad. Sci. Paris* **231**, 1134 1135, 1448–1449.
Bethe, H. (1930). *Ann Phys.* **5**, 325–400.
Bethe, H. (1932). *Z. Phys.* **76**, 293–299.
Beynon, J. H. (1960). "Mass Spectrometry and its Applications to Organic Chemistry." Elsevier, Amsterdam.
Beynon, J. H. (1968). *Adv. Spectrom.* **4**, 123–138.
Beynon, J. H., and Cooks, R. G. (1975). *In* "Mass Spectrometry" (A. Maccoll, ed.), pp. 159–205 (Physical Chemistry, Series Two). Butterworths, London.
Benyon, J. H., and Cooks R. G. (1976). *Int. J. Mass Spectrom. Ion Phys.* **19**, 107–137.
Beynon, J. H., and Morgan, R. P. (1978). *Int. J. Mass Spectrom. Ion Phys.* **27**, 1–30.
Beynon, J. H., and Williams, A. E. (1959). *Appl. Spectrosc.* **13**, 101–105; **14**, 27–27.
Beynon, J. H., Saunders, R. A., and Williams, A. E. (1965). "Table of Metastable Transitions for use in Mass Spectrometry." Elsevier, Amsterdam.
Beynon, J. H., Saunders, R. A., and Williams, A. E. (1968). "The mass spectra of organic molecules." Elsevier, New York.
Beynon, J. H., Caprioli, R. M., and Richardson, J. W. (1971). *J. Am. Chem. Soc.* **93**, 1852–1857.
Beynon, J. H., Cooks, R. G., Jennings, K. R., and Ferrer-Correia, A. J. (1975). *Int. J. Mass Spectrom. Ion Phys.* **18**, 87–99.
Beynon, J. H., Harris, F. M., Green, B. N., and Bateman, R. H. (1982). *Org. Mass Spectrom.* **17**, 55–66.
Biemann, K. (1962). "Mass Spectrometry, Organic Chemical Application." McGraw-Hill, New York.
Blackburn, P. E., and Danielson, P.M. (1972). *J. Chem. Phys.* **56**, 6156–6164.
Blauth, E. W. (1965). "Dynamische Massenspektrometer." Vieweg, Braunschweig.
Bleakney, W. (1929). *Phys. Rev.* **34**, 157–160.
Bleakney, W. (1930a). *Phys. Rev.* **35**, 1180–1186.
Bleakney, W. (1930b). *Phys. Rev.* **35**, 139–148.
Bleakney, W. (1930c). *Phys. Rev.* **36**, 1303–1308.
Bleakney, W. (1936). *Am. Phys. T.* **4**, 12–23.
Bleakney, W., and Smith, L. G. (1936). *Phys. Rev.* **49**, 402–402.
Boesten, L. G. J., and Heideman, H. G. M. (1980). *Physica C* **98**, 242–245.
Bolduc, E., Quemener, J. J., and Marmet, P. (1971). *Can. J. Phys.* **49**, 3095–3098.
Borst, W. L., and Zipf, E. C. (1970). *Phys. Rev. A* **1**, 834–840.

Bottcher, C. (1981). *J. Phys. B* **14**, L349–L355.
Bottiglioni, F., Contant, J., and Fois, M. (1972). *Phys. Rev. A* **6**, 1830–1843.
Bowie, J. H., and Williams, B. D. (1975). *In* "Mass Spectrometry" (A. Maccoll, ed.), Vol. 5, pp. 89–128. Butterworths, London.
Bracher, J., Erhardt, H., Fuchs, R., Osberghaus, O., and Taubert, R. (1963). *Adv. Mass Spectrom.* **2**, 285–295.
Brehm, B., and De Frenes, G. (1978). *Int. J. Mass Spectrom. Ion Phys.* **26**, 251–266.
Brehm, B., and De Frenes, G. (1980). *Adv. Mass Spectrom.* **8A**, 138–143.
Brehm, B., and von Puttkamer, E. (1967). *Z. Naturforschg.* **22a**, 8–10.
Brehm, B., Frey, R., Küstler, A., and Eland J. H. D. (1974). *Int. J. Mass Spectrom. Ion Phys.* **13**, 251–260.
Brenton, A. G., Morgan, R. P., and Beynon, J. H. (1979). *Annu. Rev. Phys. Chem.* **30**, 51–78.
Brenton, A. G., Beynon, J. H., Richard, E. G., and Fournier, P. G. (1983). *J. Chem. Phys.* **79**, 1834–1845.
Brion, C. E. (1975). *Radiat. Res.* **64**, 37–52.
Brion, C. E., and Hamnett, A. (1980). *Adv. Chem. Phys.* **48**, 1–80.
Brion, C. E., and Hitchcock, A. P. (1981). *Proc. XIIth ICPEAC, Tennessee*, **1**, 355–356.
Brion, C. E., and Paddock, N. L. (1968). *J. Chem. Soc. A* 388–392.
Brion, C. E., and Thomas, G. E. (1968). *Int. J. Mass Spectrom. Ion Phys.* **1**, 25–39.
Brongersma, H. H., and Oosterhoff, L. J. (1967). *Chem. Phys. Lett.* **1**, 169–172.
Brook, E., Harrison, M. F. A., and Smith, A. C. H. (1978). *J. Phys. B* **11**, 3115–3132.
Browning, R., and Fryar, J. (1973). *J. Phys. B* **6**, 364–371.
Brubaker, W. M. (1955). *J. Appl. Phys.* **26**, 1007–1012.
Brunnee, C., and Voshage, H. (1964). "Massenspektrometrie." Thiemig, München.
Budzikiewicz, H. (1972). "Massenspektrometrie." Verlag Chemie, Weinheim.
Budzikiewicz, H. (1981). "Interpretation of Mass Spectra of Organic Compounds." Univ. Microfilms International, Ann Arbor.
Budzikiewicz, H., Djerassi, C., and Williams, D. H. (1964). "Structure elucidation of natural products by mass spectrometry" Holden-Day, San Francisco.
Burgess, A. (1963). "Proc. 3rd ICPEAC," pp. 237–242. North-Holland, Amsterdam.
Burgess, A., and Percival, I. C. (1968). "Adv. Atomic Molecular Physics" (D. R. Bates and I. Estermann, ed.), Vol. 4, pp. 109–141. Academic Press, New York.
Burhop, E. H. S. (1979). *In* "Adv. Atomic and Molecular Physics" (D. R. Bates and B. Bedersen, eds.), Vol. 15, pp. 329–380. Academic Press, New York.
Burrows, M. D., Mc Intyre, L. C., Ryan, S. R., and Lamb, Jr., W. E. (1980). *Phys. Rev. A* **21**, 1841–1847.
Cadez, I., Pejcev, V., and Kurepa, M. V. (1972). *Proc. 6th Symp. Physics Ionized Gases, Miljevac*, pp. 21–24.
Camilloni, R., Fainelli, E., Petrocelli, G., and Stefani, G. (1982). *Book of Abstract, MOLEC IV, Nijmegen.*
Carbonneau, R., and Marmet, P. (1972). *Int. J. Mass Spectrom. Ion Phys.* **10**, 143–155.
Careri, G., and Nencini, G. (1950). *J. Chem. Phys.* **18**, 897–898; *Nuovo Cimento* **7**, 64–68.
Casnati, E., Tartari, A., and Baraldi, C. (1982). *J. Phys. B* **15**, 155–167.
Castleman, A. W., Jr., (1979). *Adv. Colloid Interface Sci.* **10**, 73–128.
Castleman, A. W., Jr., (1982). "Contr. Symp. Atomic Surface Phys." (W. Lindinger *et al.*, eds.), Vol. 3, pp. 1–6.
Castleman, A. W., Jr., Kay, B. D., Hermann, V., Holland, P. M., and Märk, T. D. (1981). *Surface Sci.* **106**, 179–187.
Center, R. E., and Mandl, A. (1972). *J. Chem. Phys.* **57**, 4104–4106.
Chamberlain, G. E., and Kieffer, L. J. (1970). JILA Information Center Report No. 10.

Chantreau, J., and Vauthier, R. (1970). "Recent Developments in Mass Spectroscopy" (K. Ogata and T. Hayakawa, ed.), pp. 198–204. University Park Press, Baltimore.
Charalambous, J. (1975). (Ed.) "Mass spectrometry of metal compounds" Butterworths, London.
Christophorou, L. G. (1971). "Atomic and Molecular Radiation Physics" Wiley, New York.
Chupka, W. A. (1959). *J. Chem. Phys.* **30**, 191–211.
Chupka, W. A. (1971). *J. Chem. Phys.* **54**, 1936–1947.
Cloutier, G. G., and Schiff, H. I. (1959). *Adv. Mass Spectrom.* **1**, 473–488.
Cobb, M., Moran, T. F., Borkman, R. F., and Childs, R. (1980). *J. Chem. Phys.* **72**, 4463–4470.
Coggeshall, N. D. (1944). *J. Chem. Phys.* **12**, 19–23.
Coggeshall, N. D. (1946). *Phys. Rev.* **70**, 270–280.
Coggeshall, N. D. (1962). *J. Chem. Phys.* **36**, 1640–1647.
Cohen, J. S., and Bardsley, J. N. (1978). *Phys. Rev. A* **18**, 1004–1008.
Collin, J. E., Hubin-Franskin, M. J., and D'Or, L. (1968). *Adv. Mass Spectrom.* **4**, 713–726.
Collins, J. H., Winters, R. E., and Engerholm G. G. (1968). *J. Chem. Phys.* **49**, 2469–2472.
Collins, R. D. (1969). *Vacuum* **19**, 105–111.
Comer, J. (1978). *In* "Electronic and Atomic Collisions" (G. Watel, ed.), pp. 43–50. North Holland, Amsterdam.
Compton, K. T., and Van Voorhis, C. C. (1925). *Phys. Rev.* **26**, 724–731.
Compton, K. T., and Van Voorhis, C. C. (1926). *Phys. Rev.* **27**, 436–453.
Compton, R. N. (1977). *J. Chem. Phys.* **66**, 4478–4485.
Conrad, R. (1930). *Phys. Z.* **31**, 888–892.
Cook, C. J., and Peterson, J. R. (1962). *Phys. Rev. Lett.* **9**, 164–166.
Cooks, R. G. (1978). *In* "Collision Spectroscopy" (R. G., Cooks, ed.), pp. 357–450. Plenum, New York.
Cooks, R. G., Beynon, J. H., Caprioli, R. M., and Lester G. R. (1973). "Metastable Ions." Elsevier, Amsterdam.
Cooks, R. G., Kim, K. C., and Beynon, J. H. (1974a). *Chem. Phys. Lett.* **26**, 131–133.
Cooks, R. G., Terwilliger, D. T., and Beynon J. H. (1974b). *J. Chem. Phys.* **61**, 1208–1213.
Cowling, I. R., and Fletcher, J. (1973). *J. Phys. B* **6**, 665–674.
Craggs, J. D., and Massey, H. S. W. (1959). *In* "Handbuch d. Physik" (S. Flügge, ed.), Vol. 37, pp. 314–415. Springer Verlag, Berlin.
Craggs, J. D., Thorburn, R., and Tozer, B. A. (1957). *Proc. R. Soc. London A* **240**, 473–483.
Crowe, A., and Mc Conkey, J. W. (1973a). *J. Phys. B* **6**, 2088–2107.
Crowe, A., and Mc Conkey, J. W. (1973b). *J. Phys. B* **6**, 2108–2117.
Crowe, A., and Mc Conkey, J. W. (1973c). *Phys. Rev. Lett.* **31**, 192–196.
Crowe, A., and Mc Conkey, J. W. (1974). *J. Phys. B* **7**, 349–361.
Crowe, A., and Mc Conkey, J. W. (1975). *J. Phys. B* **8**, 1765–1769.
Crowe, A., and Mc Conkey, J. W. (1977). *Int. J. Mass Spectrom. Ion Phys.* **24**, 181–189.
Crowe, A., Preston, J. A., and Mc Conkey, J. W. (1972). *J. Chem. Phys.* **57**, 1620–1625.
Cvejanovic, S., and Grujic, P. (1974). *Proc. SPIG, Rovinj* 65–68.
Cvejanovic, S., and Read, F. H. (1974). *J. Phys. B* **7**, 1180–1193, 1841–1852.
Daly, R. N., and Powell, R. E. (1966). *Proc. Phys. Soc.* **89**, 273–280.
Daly, R. N., and Powell, R. E. (1967). *Proc. Phys. Soc.* **90**, 629–635.
Danby, C. J., and Eland, J. H. D. (1972). *Int. J. Mass Spectrom. Ion Phys.* **8**, 153–161.
Dannacher, J., and Vogt, J. (1978). *Helv. Chim. Acta* **61**, 361–372.
Dawson, P. H. (1976). "Quadrupole mass spectrometry and its applications." Elsevier, Amsterdam.
de Heer, F. J. (1981). *Phys. Scr.* **23**, 170–178.
de Heer, F. J., Jansen, R. H. J., and van der Kaay, W. (1979). *J. Phys. B* **12**, 979–1001.
De Koven, B. M., Levy, D. H., Harris, H. H., Zegarski, B. R., and Miller, T. A. (1981). *J. Chem. Phys.* **74**, 5659–5668.

de la Ripelle, F. (1949). *J. Phys. Radium* **10**, 318–329.
Deleanu, L., and Stockdale, J. A. D. (1975). *J. Chem. Phys.* **63**, 3898–3906.
Dewar, M. J. S., and Rona P. (1965). *J. Am. Chem. Soc.* **87**, 5510–5510.
Dibeler, V. H., and Rosenstock, H. M. (1963a). *J. Chem. Phys.* **39**, 1326–1329.
Dibeler, V. H., and Rosenstock, H. M. (1963b). *J. Chem. Phys.* **39**, 3106–3111.
Dibeler, V. H., Wells, Jr., E. J., and Reese, R. M. (1950). *Phys. Rev.* **79**, 223–223.
Dibeler, V. H., Mohler, F. L., and Reese, R. M. (1953). *J. Chem. Phys.* **21**, 180–181.
Dibeler, V. H., Reese, R. M., and Mohler F. L. (1956). *J. Res. Natl. Bur. Stand.* **57**, 113–118.
Dietz, L. A. (1965). *Rev. Sci. Instrum.* **36**, 1763–1770.
Dillard, J. G. (1973). *Chem. Rev.* **73**, 589–641.
Dimitrijevic, M. S., and Grujic, P. (1979). *J. Phys. B* **12**, 1873–1880.
Dimitrijevic, M. S., and Grujic, P. (1981). *J. Phys. B* **14**, 1663–1674.
Dixon, A. J., and Harrison, M. F. A. (1971). *Proc. 7th ICPEAC Amsterdam, North Holland*, pp. 892–894.
Dixon, A. J., von Engel, A. and Harrison, M. F. A. (1975). *Proc. R. Soc. A* **343**, 333–349.
Dixon, A. J., Harrison, M. F. A., and Smith, A. C. H. (1976). *J. Phys. B* **9**, 2617–2631.
Djuric, N., Belic, D., Kurepa, M., Mack, J. U., Rothleitner, J., and Märk T. D. (1981). *Proc. 12th ICPEAC, Gatlinburg*, pp. 384–385.
Dole, M., Mack, L. L., Hines, R. L., Mobley, R. C., Ferguson, L. D., and Alice, M. B. (1968). *J. Chem. Phys.* **49**, 2240–2249.
Donahue, J. B., Gram, P. A. M., Hynes, M. V., Hamm, R. W., Frost, C. A., Bryant, H. C., Butterfield, K. B., Clark, D. A., and Smith, W. W. (1982). *Phys. Rev. Lett.* **48**, 1538–1541.
Dorman, F. H., and Morrison, J. D. (1961). *J. Chem. Phys.* **35**, 575–581.
Dorman, F. H., and Morrison, J. D. (1963). *J. Chem. Phys.* **39**, 1906–1907.
Dorman, F. H., Morrison, J. D., and Nicholson, J. C. (1960). *J. Chem. Phys.* **32**, 378–384.
Drawin, H. W. (1961). *Z. Phys.* **164**, 513–521.
Drewitz, H. J. (1976). *Int. J. Mass Spectrom. Ion Phys.* **19**, 313–325, **21**, 212–212.
Drewitz, H. J., and Taubert, R. (1976). *Int. J. Mass Spectrom. Ion Phys.* **19**, 293–312.
Drowart, J., and Goldfinger, P. (1967). *Angew. Chemie*, **6**, 581–648.
Dubois, R. D., and Rudd, M. E. (1975). *Proc. 9th ICPEAC, Seattle*, Vol. 1, pp. 479–480.
Dunn, G. H. (1962). *Phys. Rev. Lett.* **8**, 62–64.
Dunn, G. H. (1966). *J. Chem. Phys.* **44**, 2592–2594.
Dunn, G. H., and Kieffer, L. J. (1963). *Phys Rev.* **132**, 2109–2117.
Durup, J., and Durup, M. (1967). *J. Chim. Phys. Phys. Chim. Biol.* **64**, 386–394.
Durup, J., and Heitz, F. (1964). *J. Chim. Phys. Phys. Chim. Biol.* **61**, 470–479.
Dutil, R., and Marmet, P. (1980). *Int. J. Mass Spectrom. Ion Phys.* **35**, 371–379.
Echt, O., Sattler, K., and Recknagel, R. (1982). *Phys. Lett.* **90A**, 185–189.
Edwards, A. K., and Wood, R. M. (1982). *J. Chem. Phys.* **76**, 2938–2942.
Egger, F., and Märk, T. D. (1978). *Z. Naturforsch.* **33a**, 1111–1113.
The Eight Peak Index of Mass Spectra (1975). MSDC, The Royal Society of Chemistry, Nottingham.
Eland, J. H. D., and Schulte, H. (1975). *J. Chem. Phys.* **62**, 3835–3836.
Elliot, R. M. (1963). In "Mass Spectrometry" (C. A. Mc Dowell, ed.), pp. 69–103. McGraw-Hill, New York.
Elwert, G. (1952). *Z. Naturforsch.* **7a**, 432–439.
Erhardt, H. (1968). In "Survey of Phenomena in Ionized Gases," pp. 61–74. IAEA, Wien.
Erhardt, H. (1983). *Proc. 13th ICPEAC*, Berlin, and *Com. Atom. Molec. Phys.* **13**, 115–125.
Erhardt, H., and Erbse, U. (1963). *Z. Phys.* **172**, 210–218.
Erhardt, H., and Kresling, A. (1967). *Z. Naturforsch.* **22a**, 2036–2043.
Erhardt, H., and Linder, F. (1967). *Z. Naturforsch.* **22a**, 444–453.
Erhardt, H., and Osberghaus, O. (1958). *Z. Naturforsch.* **13a**, 16–21.

Erhardt, H., and Osberghaus, O. (1960). *Z. Naturforsch.* **15a**, 575–584.
Erhardt, H., and Tekaat, T. (1964). *Z. Naturforsch.* **19a**, 1382–1388.
Erhardt, H., Linder, F., and Meister, G. (1965). *Z. Naturforsch.* **20a**, 989–997.
Erhardt, H., Schulz, M., Tekaat, T., and Willmann, K. (1969). *Phys. Rev. Lett.* **22**, 89–92.
Erhardt, H., Hesselbacher, K. H., Jung, K., and Willmann, K. (1972). *In* "Case Studies Atomic Collisions Physics II" (E. W. Mc Daniel, and M. R. C. Mc Dowell, eds.), pp. 160–208. North Holland, Amsterdam.
Erhardt, H., Jung, K., and Schubert, E. (1980). *In* "Coherence and Correlation in Atomic Collisions" (H. Kleinpoppen and J. F. Williams, eds.), pp. 41–58. Plenum, New York.
Evans, B., Hobson, R. M., and Chang, J. (1980). *Eur. Conf. Abstr.* **4D**, 49–49.
Falter, H., Hagena, O. F., Henkes, W., and Wedel, H. (1970). *Int. J. Mass Spectrom. Ion Phys.* **4**, 145–163.
Fano, U. (1946). *Phys. Rev.* **70**, 44–52.
Fano, U. (1954). *Phys. Rev.* **95**, 1198–1200.
Fano, U., and Lin, C. D. (1975). *In* "Atomic Physics, 4" (G. Putlitz, E. W. Weber, and A. Winnacker, eds.), pp. 47–70. Plenum, New York.
Fantoni, R., Giardoni-Guidoni, A., and Tiribelli, R. (1982). *J. Electron Spectrosc. Relat. Phenom.* **26**, 99–105.
Fehlner, T. P., and Callen, R. B. (1968). *Adv. Chem. Ser.* **72**, 181–190.
Fehlner, T. P., and Koski, W. S. (1964). *J. Am. Chem. Soc.* **86**, 581–586.
Field, F. H., and Franklin J. L. (1970). "Electron Impact Phenomena." Academic Press, New York.
Fischler, J., and Halman, M. (1964). *J. Chem. Soc.* 31–36.
Fiquet-Fayard, F. (1962). *J. Chim. Phys.* **59**, 439–441.
Fiquet-Fayard, F., and Guyon, P. M. (1966). *Mol. Phys.* **11**, 17–30.
Fiquet-Fayard, F., and Lahmani, M. (1962). *J. Chim. Phys.* **59**, 1050–1055.
Fiquet-Fayard, F., Chiari, J., Muller, F., and Ziesel, J. P. (1968). *J. Chem. Phys.* **48**, 478–482.
Fitch, W. L., and Sauter, A. D. (1983). *Anal. Chem.* **55**, 832–835.
Flaim, T. A., and Ownby, P. D. (1971). *J. Vac. Sci. Techn.*, **8**, 661–662 and references therein.
Flamme, J. P., Momigny, J., and Wankenne, H. (1975). *J. Am. Chem. Soc.* **98**, 1045–1047.
Flamme, J. P., Wankenne, H., Locht, R., Momigny, J., Nowak, P. J. C. M., and Los, J. (1978). *Chem. Phys.* **27**, 45–49.
Flamme, J. P., Märk, T. D., and Los, J. (1980). *Chem. Phys. Lett.* **75**, 419–422.
Flannery, M. R. (1971). *J. Phys. B* **4**, 892–895.
Flannery, M. R., Mc Cann, K. J., and Winter, N. W. (1981). *J. Phys. B* **14**, 3789–3796.
Fock, W. (1969). *Int. J. Mass Spectrom. Ion Phys.* **3**, 285–291.
Forst, W. (1971). *Chem. Rev.* **71**, 339–356.
Forst, W. (1973). "Theory of Unimolecular Reactions." Academic Press, New York.
Fournier, P., Ozenne, J. B., and Durup, J. (1970). *J. Chem. Phys.* **53**, 4095–4096.
Fournier, P., van de Runstraat, C. A., Govers, T. R., Schopman, J., de Heer, F. J., and Los, J. (1971). *Phys. Lett.* **9**, 426–428.
Fournier, P. G., Govers, T. R., van de Runstraat, C. A., Schopman, J., and Los, J. (1972). *J. Phys.* **33**, 755–759.
Fournier, P. G., Comtet, G., Odom, R. W., Maas, J. G., van Asselt N. P. F. B., and Los J. (1976). *Chem. Phys. Lett.* **40**, 170–174.
Fox, R. E., and Hickam, W. M. (1953). *Phys. Rev.* **89**, 555–558.
Fox, R. E., and Hickam, W. M. (1954). *J. Chem. Phys.* **22**, 2059–2063.
Franco, J. (1981). *Rev. Mex. Fis.* **4**, 475–485.
Franco, J., and Daltabuit, E. (1978). *Rev. Mex. Astron. Astrofis.* **2**, 325–329.
Franklin, J. L., Hierl, P. M., and Whan, D. A. (1967). *J. Chem. Phys.* **47**, 3148–3153.
Franklin, J. L., Dillard, J. G., Rosenstock, H. M., Herron, J. T., Draxl, K., and Field, F. H. (1969). *Natl. Stand. Ref. Data Ser. Natl. Bur. Stand.* **26**, 1–289.

Friedländer, E., Kallmann, H., Lasareff, W., and Rosen, B. (1932). *Z. Phys.* **76**, 60–79.
Frost, D. C., and Mc Dowell, C. A. (1960). *Can. J. Chem.* **38**, 407–420.
Fuchs, R., and Taubert, R. (1965). *Z. Naturforsch.* **20a**, 823–826.
Funk, H. (1930). *Ann. Phys.* **4**, 149–184.
Futrell, J. H., Stephan, K., Castleman, Jr., A. W., and Märk, T. D. (1981). *Proc. 8th Int. Symp. Molecular Beams, Cannes*, pp. 262–265.
Futrell, J. H., Stephan, K., and Märk, T. D. (1982). *J. Chem. Phys.* **76**, 5893–5901.
Gallagher, J. W., and Beaty, E. C. (1980). *JILA Inf. Cent. Rep.* **18**, 1–142.
Gallagher, J. W., Rumble, Jr., J. R., and Beaty, E. C. (1979). *NBS Spec. Publ.* **426**, Suppl. 1, 1–106.
Geltman, S. (1956). *Phys. Rev.* **102**, 171–179.
Geltman, S. (1983). *J. Phys. B* **16**, L525–L528.
Geraedts, J., Stolte, S., and Reuss, J. (1982). *Z. Phys. A* **304**, 167–175.
Giardini-Guidoni, A., Missone, G., Camilloni, R., and Stefani, G. (1978). *Adv. Mass Spectrom.* **7A**, 175–181.
Giardini-Guidoni, A., Camilloni, R., and Stefani, G. (1980). In "Coherence and Correlation in Atomic Collisions" (H. Kleinpoppen and J. F. Williams, eds.), pp. 13–40. Plenum, New York.
Giardini-Guidoni, A., Fantoni, R., Camilloni, R., and Stefani, G. (1981). *Com. Atom. Molec. Phys.* **10**, 107–119.
Giessner, B. G., and Meisels, G. G. (1971). *J. Chem. Phys.* **55**, 2269–2275.
Glasstone, S., Laidler, K. J., and Eyring, H. (1941). "Theory of Rate Processes." McGraw-Hill, New York.
Gordon, S. M., and Reid, N. W. (1975). *Int. J. Mass Spectrom. Ion Phys.* **18**, 379–391.
Gough, T. E., and Miller, R. E. (1982). *Chem. Phys. Lett.* **87**, 280–283.
Govers, T. R., and Schopman, J. (1971). *Chem. Phys. Lett.* **12**, 414–418.
Green, A. E. S., and Barth, C. A. (1965). *J. Geophys. Res.* **70**, 1083–1092.
Green, A. E. S., and Dutta, S. K. (1967). *J. Geophys. Res.* **72**, 3933–3941.
Green, A. E. S., and Mc Neal, R. J. (1971). *J. Geophys. Res.* **76**, 133–144.
Greene, C. H., and Rau, A. R. P. (1982). *Phys. Rev. Lett.* **48**, 533–537.
Gross, M. L. (1972). *Org. Mass Spectrom.* **3**, 827–831.
Grosse, H. J., and Bothe H. K. (1970). *Z. Naturforsch.* **25a**, 1970–1976.
Grujic, P. (1972). *J. Phys. B* **5**, L137–L139.
Grujic, P. (1982). *J. Phys. B* **15**, 1913–1928.
Grujic, P. (1983). *J. Phys. B* **16**, 2567–2576.
Gryzinski, M. (1959). *Phys. Rev.* **115**, 374–382.
Gryzinski, M. (1963). *Proc. 3rd ICPEAC, London*, pp. 226–236.
Gryzinski, M. (1965). *Phys. Rev.* **138**, A305–A358; *Phys. Rev. Lett.* **14**, 1059–1059.
Gspann, J., and Körting, K. (1973). *J. Chem. Phys.* **59**, 4726–4734.
Haasz, A. A., and de Leeuw, J. H. (1976). *J. Geophys. Res.* **81**, 4031–4034.
Hagena, O. F. (1974). In "Molecular Beams and Low Density Gas Dynamics" (P. P. Wegener, ed.), pp. 93–182, Dekker, New York.
Hagena, O. F., and Henkes, W. (1965). *Z. Naturforsch.* **20a**, 1344–1348.
Hagstrum, H. D. (1951). *Rev. Mod. Phys.* **23**, 185–203.
Hagstrum, H. D. (1953). *Rev. Sci. Instrum.* **24**, 1122–1142.
Hagstrum, H. D., and Tate, J. T. (1941). *Phys. Rev.* **59**, 354–370.
Halas, S., and Adamczyk, B. (1972). *Int. J. Mass Spectrom. Ion Phys.* **10**, 157–160.
Hamming, M. C., and Foster, N. G. (1972). "Interpretation of mass spectra of organic compounds." Academic Press, New York.
Hamnett, A., Stoll, W., Branton, G., van der Wiel, M. J., and Brion, C. E. (1976). *J. Phys. B* **9**, 945–957.
Haney, M. A., and Franklin, J. L. (1968). *J. Chem. Phys.* **48**, 4093–4097.
Hanner, A. W., and Moran, T. F. (1981). *Org. Mass Spectrom.* **16**, 512–514.

Hanner, A. W., Abbey, L. E., Bostwick, D. E., Burgess, E. M., and Moran, T. F. (1982). *Org. Mass Spectrom.* **17**, 19–28.
Harrison, A. G., Jones, E. G., Gupta, S. K., and Nagy, G. P. (1966). *Can. J. Chem.* **44**, 1967–1973.
Harrison, M. F. A., Smith, A. C. H., and Brook, E. (1979). *J. Phys. B* **12**, L433–L435.
Hashizume, A., and Wasada, N. (1979). *Jpn. J. Appl. Phys.* **18**, 429–438.
Hashizume, A., and Wasada, N. (1980). *J. Phys. B* **13**, 4865–4875; *Int. J. Mass Spectrom. Ion Phys.* **36**, 291–299.
Hayakawa, S., and Nishimura, N. (1964). *J. Geomagn. Geoelectr. Jpn.* **16**, 72–74.
Heideman, H. G. M., Nienhuis, G., and van Ittersum, T. (1974). *J. Phys. B* **7**, L493–L495.
Heidman, H. G. M., and Van de Water, W. (1981). *Com. Atom. Molec. Phys.* **10**, 87–98.
Hellwinkel, D., and Wünsche, C. (1969). *Chem. Commun.* 1412–1413.
Helm, H., Stephan, K., and Märk, T. D. (1979). *Phys. Rev. A* **19**, 2154–2160.
Helm, H., Stephan, K., Märk T. D., and Huestis D. L. (1981). *J. Chem. Phys.* **74**, 3844–3851.
Henkes, W., and Isenberg, G. (1970). *Int. J. Mass Spectrom. Ion Phys.* **5**, 249–254.
Henkes, W., and Mikosch, F. (1974). *Int. J. Mass Spectrom. Ion Phys.* **13**, 151–161.
Hernandez, S. P., Dagdigian, P. J., and Doering, J. P. (1982). *J. Chem. Phys.* **77**, 6021–6026.
Herron J T, and Dibeler, V. H. (1960). *J. Chem. Phys.* **32**, 1884–1885.
Herzberg, G. (1967). "Molecular Spectra and Molecular Structure, III. Electronic Spectra and Electronic Structure of Polyatomic Molecules," pp. 471–473. Van Nostrand, Princeton.
Hicks, P. J., Cvejanovic, S., Comer, J., Read, F. H., and Sharp, J. M. (1974). *Vacuum* **24**, 573–580.
Hierl, P., and Franklin, J. L. (1967). *J. Chem. Phys.* **47**, 3154–3161.
Higatsberger, M. J., Demorest, H. L., and Nier A. O. (1954). *J. Appl. Phys.* **25**, 883–886.
Hill, H. C. (1966). "Introduction to mass spectrometry." Heyden, London.
Hille, E., and Märk, T. D. (1978). *J. Chem. Phys.* **69**, 4600–4605.
Hipple, J. A. (1948). *J. Phys. Colloid. Chem.* **52**, 456–462.
Hipple, J. A., and Condon E. U. (1945). *Phys. Rev.* **68**, 54–55.
Hipple, J. A., Fox, R. E., and Condon, E. U. (1946). *Phys. Rev.* **69**, 347–356.
Hirsch, R. G., Van Brunt, R. J., and Whitehead, W. D. (1975). *Int. J. Mass Spectrom. Ion Phys.* **17**, 335–449.
Hitchcock, A. P. (1982). *J. Electron Spectrosc. Relat. Phenom.* **25**, 245–275.
Hitchcock, A. P., and Brion, C. E. (1978). *Chem. Phys.* **33**, 55–64.
Hitchcock, A. P., Brion, C. E., and Van der Wiel, M. J. (1978). *J. Phys. B* **11**, 3245–3261.
Holland, R. F., and Maier II, W. B. (1972). *J. Chem. Phys.* **56**, 5229–5246.
Holmes, J. L., and Lossing, F. P. (1980). *J. Am. Chem. Soc.* **102**, 1951–1955, 3732–3735.
Hooper, J. W., Harmer, D. S., Martin, D. W., and Mc Daniel, E. W. (1962). *Phys. Rev.* **125**, 2000–2004.
Houver, J. C., Bandon, J., Abignoli, M., Barat, M., Fournier, P., and Durup, J. (1970). *Int. J. Mass Spectrom. Ion Phys.* **4**, 137–144.
Hsieh, T., and Hanrahan, R. J. (1977). *Int. J. Mass Spectrom. Ion Phys.* **23**, 201–207.
Hubin-Franskin, M. J., Heinisch, J., and Collin, J. E. (1974). *Int. J. Mass Spectrom. Ion Phys.* **13**, 131–137.
Hubin-Franskin, M. J., Marmet, P. and Huard, D. (1980). *Int. J. Mass Spectrom. Ion Phys.* **33**, 311–324.
Hughes, A. L., and Klein, E. (1924). *Phys. Rev.* **23**, 450–463.
Hurley, A. C. (1962). *J. Mol. Spectrosc.* **9**, 18–29.
Hurley, A. C. (1971). *J. Chem. Phys.* **54**, 3656–3657.
Hurley, A. C., and Maslen V. W. (1961). *J. Chem. Phys.* **34**, 1919–1925.
Hyman, H. A. (1979). *Phys. Rev. A* **20**, 855–859.
Illies, A. J., Jarrold, M. F., and Bowers, M. T. (1982). *J. Am. Chem. Soc.* **104**, 3587–3595.
Illies, A. J., Jarrold, M. F., Bass, L. M., and Bowers, M. T. (1983). *J. Am. Chem. Soc.* **105**, 5775–5781.

Inokuti, M. (1971). *Rev. Mod. Phys.* **43**, 297–347
Ishii, H., and Nakayama, K. (1961). *Proc. 8th Natl. Vac. Symp.* Vol. 1, p. 519.
Jackson, W. M., Brackmann, R. T., and Fite, W. L. (1974). *Int. J. Mass Spectrom. Ion Phys.* **13**, 237–250.
Jain, D. K., and Khare, S. P. (1975). *Proc. 9th ICPEAC, Seattle,* pp. 484–485.
Jain, D. K., and Khare, S. P. (1976). *J. Phys.* **B 9**, 1429–1438.
Jarrold, M. F., Illies, A. J., and Bowers, M. T. (1982a). *Chem. Phys.* **65**, 19–28.
Jarrold, M. F., Illies, A. J., and Bowers, M. T. (1982b). *Chem. Phys. Lett.* **92**, 653–658.
Jayaram, R. (1966). "Mass spectrometry; theory and applications." Plenum, New York.
Jentsch, T., Drachsel, W., and Block, J. H. (1982). *Chem. Phys. Lett.* **93**, 144–147.
Jesse, W. P. (1925). *Phys. Rev.* **26**, 208–220.
Johnson, J. P., and Franklin, J. L. (1980). *Int. J. Mass Spectrom. Ion Phys.* **33**, 393–407.
Johnson, L. P., and Morrison, J. D. (1975). *Int. J. Mass Spectrom. Ion Phys.* **18**, 355–366.
Johnson, L. P., Morrison, J. D., and Wahrhaftig, A. L. (1978). *Int. J. Mass Spectrom. Ion Phys.* **26**, 1–21.
Johnson, R., and Biondi, M. A. (1978). *Phys. Rev. A* **18**, 996–1003.
Johnson, R., and Biondi, M. A. (1979). *Phys. Rev. A* **20**, 87–97.
Johnstone, R. A. W. (1972). "Mass spectrometry for organic chemists." Cambridge Univ. Press, London and New York.
Johnstone, R. A. W. (1981). "Mass Spectrometry," Vol. 6. Royal Society of Chemistry, London.
Johnstone, R. A. W., and Mc Master, B. N. (1974). *Adv. Mass Spectrom.* **6**, 451–456.
Johnston, W. H., and Arnold, J. R. (1953). *J. Chem. Phys.* **21**, 1499–1502.
Jones, B. E., Abbey, L. E., Chatham, H. L., Hanner, A. W., Teleshefsky, L. A., Burgess, E. M., and Moran, T. F. (1982). *Org. Mass Spectrom.* **17**, 10–18.
Jones, E. G., Beynon, J. H., and Cooks, R. G. (1972). *J. Chem. Phys.* **57**, 3207–3212.
Jones, T. (1927). *Phys. Rev.* **29**, 822–829.
Jordan, E. B., and Coggeshall, N. D. (1942). *J. Appl. Phys.* **13**, 539–550.
Kandel, R. J. (1953). *Phys. Rev.* **91**, 436–436.
Kandel, R. J. (1954). *J. Chem. Phys.* **22**, 1496–1499.
Kang, I. J., and Kerch, R. L. (1970). *Phys. Lett. A* **31**, 172–173.
Katsonis, K. (1981). Recent References for Atomic Collision Data of Interest to Fusion IAEA-NDS-AM 11, pp. 1–29, Wien.
Kay, R. B., Van der Leeuw, P. E., and Van der Wiel, M. J. (1977). *J. Phys. B* **10**, 2521–2529.
Kebarle, P. (1977). *Annu. Rev. Phys. Chem.* **28**, 445–476.
Kellogg, G. L. (1981). *Phys. Rev. B* **24**, 1848–1851.
Kellogg, G. L. (1982). *Surf. Sci.* **120**, 319–313.
Kerner, E. H. (1953). *Phys. Rev.* **92**, 1441–1447.
Kerwin, L., Marmet, P., and Carette, J. D. (1969). "Case Studies in Atomic Collision Physics" (E. W. Mc Daniel and M. R. C. Mc Dowell, ed.), Vol. 1, pp. 527–581. North Holland, Amsterdam.
Khakoo, M. A., and Srivastava, S. K. (1983). *Proc. 13th ICPEAC*, Berlin, p. 289.
Kieffer, L. J. (1968). *JILA Inf. Cent. Rep.* **6**, 1–95.
Kieffer, L. J. (1969). *At. Data* **1**, 19–89.
Kieffer, L. J. (1973). *JILA Inf. Cent. Rep.* **13**, 1–139.
Kieffer, L. J. (1976). *NBS Spec. Publ.* **426**, 1–212.
Kieffer, L. J., and Dunn, G. H. (1966). *Rev. Mod. Phys.* **38**, 1–35.
Kieffer, L. J., and Dunn, G. H. (1967). *Phys. Rev.* **158**, 61–65.
Kieffer, L. J., and Van Brunt, R. J. (1967). *J. Chem. Phys.* **46**, 2728–2734.
Kienitz, H. (1968). "Massenspektrometrie." Verlag Chemie, Weinheim.
Kim, Y. B., Stephan, K., Märk, E., and Märk, T. D. (1981). *J. Chem. Phys.* **74**, 6771–6776.
Kim, Y. K. (1983). *In* "Physics of Ion-Ion and Electron-Ion Collisions (F. Brouillard and J. W. McGowan, eds.), pp. 101–165. Plenum, New York.

King, G. C., Read, F. H., and Bradford, R. C. (1975). *J. Phys. B.* **8**, 2210–2224.
King, R. B. (1969). *Can. J. Chem.* **47**, 559–568.
Kingston, A. E. (1964). *Phys. Rev. A* **135**, 1537–1539.
Kiser, R. W. (1965). "Introduction to mass spectrometry and its applications." Prentice-Hall, Englewood Cliffs, New Jersey.
Kiser, R. W., Dillard, J. G., and Dugger, D. L. (1968). *Adv. Chem. Ser.* **72**, 153–180.
Klar, H. (1980). *In* "Coherence and Correlation in Atomic Collisions" (H. Kleinpoppen and J. F. Williams, eds.), pp. 67–77. Plenum, New York.
Klar, H. (1981a). *J. Phys. B* **14**, 3255–3265.
Klar, H. (1981b). *J. Phys. B* **14**, 4165–4170.
Klar, H. (1982). *Z. Phys. A* **307**, 75–81.
Klar, H., and Schlecht, W. (1976). *J. Phys. B* **9**, 1699–1711.
Klein, H. J. (1965). *Z. Phys.* **188**, 78–92.
Klots, C. E. (1971). *J. Phys. Chem.* **75**, 1526–1532.
Klots, C. E. (1973). *J. Chem. Phys.* **58**, 5364–5367.
Knof, H. (1974) "Massenspektrometrie von Kondensationskeimen in der Gasphase." Physik-Verlag, Weinheim.
Köllmann, K. (1975a). *J. Chem. Phys.* **63**, 1314–1316.
Köllmann, K. (1975b). *Int. J. Mass Spectrom. Ion Phys.* **17**, 261–285.
Köllmann, K. (1978). *J. Phys. B* **11**, 339–355.
Koppe, V. T., Koval, A. G., Fisgeer, B. M., Fogel, Y. M., and Ivanov, S. I. (1971). *Sov. Phys. JETP* **32**, 1016–1018.
Kossel, W. (1912). *Ann. Phys.* **37**, 393–424.
Krinberg, I. A. (1970). *Sov. Astron. Engl. Transl.* **13**, 780–783.
Kukulin, V. I., Osipov, A. P., Chuvilskii, Y. M. (1979). *Sov. Phys. Tech. Phys.* **24**, 883–885.
Kumar, A., and Roy, B. N., (1978). *Can. J. Phys.* **56**, 1255–1260.
Kunc, J. A. (1980). *J. Phys. B* **13**, 587–602.
Kupriyanov, S. E. (1964). *Zh. Tekh. Fiz.* **34**, 861; *Sov. Phys. Techn. Phys.* **9**, 659–664.
Kurepa, M. V., and Belic, D. S. (1978). *J. Phys. B* **11**, 3719–3729.
Kurepa, M. V., Cadez, I. M., and Pejcev, V. M. (1974). *Fizika* **6**, 185–209.
Kurepa, M. V., Pejcev, V. M., and Cadez, I. M. (1976). *J. Phys. D* **9**, 481–484.
Kurepa, M. V., Babic, D. S., and Belic, D. S. (1981). *J. Phys. B* **14**, 375–384.
Kusch, P., Hustrulid, A., and Tate, J. H. (1937). *Phys. Rev.* **52**, 840–842.
Lagergren, C. R. (1954). *Phys. Rev.* **96**, 823–823.
Laidler, K. J. (1954). *J. Chem. Phys.* **22**, 1740–1745.
Lampe, F. W., Franklin, J. L., and Field, F. H. (1957). *J. Am. Chem. Soc.* **79**, 6129–6132.
Lau, Y. K., Ikuta, S., and Kebarle, P. (1982). *J. Am. Chem. Soc.* **104**, 1462–1469.
Lawrence, E. O. (1926). *Phys. Rev.* **28**, 947–961.
Leckenby, R. E., and Robbins, E. J. (1966). *Proc. R. Soc. London A* **291**, 389–412.
Lee, J. S., Wong, T. C., and Bonham, R. A. (1975). *J. Chem. Phys.* **63**, 1609–1611.
Lee, N., and Fenn, J. B. (1978). *Rev. Sci. Instrum.* **49**, 1269–1272.
Lefaivre, D. and Marmet, P. (1978). *Can. J. Phys.* **56**, 1549–1558.
Lenard, P. (1902). *Ann. Phys.* **8**, 149–199.
Lenard, P. (1903). *Ann. Phys.* **12**, 474–490.
Lenard, P. (1904). *Ann. Phys.* **15**, 485–508.
Levin, R. D., and Lias, S. G. (1982). *Natl. Stand. Ref. Data Ser., Natl. Bur. Stand (US)* **71**, 1–634.
Levsen, K. (1978). "Fundamental Aspects of Organic Mass Spectrometry." Verlag Chemie, Weinheim.
Liehr, A. D. (1957). *J. Chem. Phys.* **27**, 476–477.
Lifshitz, C. (1971). *J. Chem. Phys.* **55**, 4155–4156.

Lifshitz, C. (1978). *Adv. Mass Spectrom.* **7A**, 3–18.
Lifshitz, C., and Gefen, S. (1980). *Int. J. Mass Spectrom. Ion Phys.* **35**, 31–37.
Lifshitz, C., and Long, F. A. (1964). *J. Chem. Phys.* **41**, 2468–2471.
Lifshitz, C., and Eaton, P. E. (1983). *Int. J. Mass Spectrom. Ion Phys.* **49**, 337–345.
Lifshitz, C., and Weiss, M. (1980). *Int. J. Mass Spectrom. Ion Phys.* **34**, 311–315.
Lifshitz, C., Agam, J., Weinberg, A., Kantor, D., Shainok, U., and Peres, M. (1973). *Int. J. Mass Spectrom. Ion Phys.* **11**, 243–253.
Lifshitz, C., MacKenzie Peers, A., Weiss, M., and Weiss, M. J. (1974). *Adv. Mass Spectrom.* **6**, 871–875.
Lifshitz, C., Gotchiguian, P., and Roller, R. (1983). *Chem. Phys. Lett.* **95**, 106–108.
Lin, C. D. (1974). *Phys. Rev. A* **10**, 1986–2001.
Liska, J. W. (1934). *Phys. Rev.* **46**, 169–176.
Litzow, M. R., and Spalding, T. R. (1973). "Mass Spectrometry of Inorganic and Organometallic Compounds." Elsevier, Amsterdam.
Locht, R., and Momigny, J. (1971). *Int. J. Mass Spectrom. Ion Phys.* **7**, 121–144.
Locht, R., and Momigny, J. (1979). *Chem. Phys. Lett.* **66**, 574–577.
Locht, R., and Momigny, J. (1980). *Chem. Phys.* **49**, 173–180.
Locht, R., and Schopman, J. (1974). *Int. J. Mass Spectrom. Ion Phys.* **15**, 361–378.
Locht, R., Schopman, J., Wankenne, H., and Momigny, J. (1975). *Chem. Phys.* **7**, 393–404.
Locht, R., Maas, J. G., van Asselt, N. P. F. B., and Los, J. (1976). *Chem. Phys.* **15**, 179–184.
Locht, R., Olivier, J. L., and Momigny, J. (1979). *Chem. Phys.* **43**, 425–432.
Lorquet, J. C. (1960). *J. Chim. Phys.* **57**, 1078–1084.
Lorquet, J. C. (1980). *Adv. Mass Spectrom.* **8A**, 3–16.
Lorquet, J. C. (1981). *Org. Mass Spectrom.* **16**, 469–482.
Lorquet, J. C., Lorquet, J., Momigny, J., and Wankenne, H. (1970). *J. Chim. Phys.* **67**, 64–68.
Lorquet, J. C., Lorquet, J., Momigny, J., and Wankenne, H. (1971). *J. Chem. Phys.* **55**, 4053–4061.
Los, J., and Govers, T. (1978). *In* "Collision Spectroscopy" (R. G. Cooks, ed.), pp. 289–356. Plenum, New York.
Lossing, F. P. (1972). *Can. J. Chem.* **50**, 3973–3981.
Lotz, W. (1967). *Z. Phys.* **206**, 205–211; *Astrophys. Journal Suppl. 128*, **14**, 207–238.
Lotz, W. (1968). *Z. Phys.* **215**, 241–247.
Loveless, A. J., and Russell, R. D. (1969). *Int. J. Mass Spectrom. Ion Phys.* **3**, 257–266.
Lozier, W. W. (1930). *Phys. Rev.* **36**, 1285–1292.
Maas, J. G., van Asselt, N. P. F. B., and Los, J. (1975). *Chem. Phys.* **8**, 37–45.
Maas, J. G., van Asselt, N. P. F. B., Nowak, P. J. C. M., Los, J., Peyerimhoff, S. D., and Bünker, R. J. (1976). *Chem. Phys.* **17**, 217–225.
Mc Cann, K. J., Flannery, M. R., and Hazi, A. (1979). *Appl. Phys. Lett.* **34**, 543–545.
Mc Carthy, E. (1980). *In* "Coherence and Correlation in Atomic Collisions" (H. Kleinpoppen and J. F. Williams, ed.), pp. 1–11. Plenum, New York.
Maccoll, A. (1975). "Mass Spectrometry" (A. D. Buckingham, ed.), Vol. 5 (Physical Chemistry, Series Two). Butterworths, London.
Maccoll, A. (1982). *Org. Mass Spectrom.* **17**, 1–9.
Mc Clure, G. W. (1953). *Phys. Rev.* **90**, 796–803.
Mc Conkey, J. W. and Latimer, I. D. (1965). *Proc. Phys. Soc.* **86**, 463–466.
Mc Conkey, J. W., Woolsey, J. M., and Burns, D. J. (1967). *Planet. Space Sci.* **15**, 1332–1334.
Mc Conkey, J. W., Crowe, A., and Hender, M. A. (1972). *Phys. Rev. Lett.* **29**, 1–4.
Mc Culloh, K. E., and Rosenstock, H. M. (1968). *J. Chem. Phys.* **48**, 2084–2089.
Mc Daniel, E. W., Flannery, M. R., Ellis, H. W., Eisele, F. L., Pope, W. and Roberts, T. G. (1977). Tech. Rep. H-78-1, Vol. 2, pp. 526–587. (US Army Missile Research, Redstone Arsenal, AL 35809.)

Mc Dowell, C. A. (1954). *Trans. Faraday Soc.* **50**, 423–430.
Mc Dowell, C. A. (1963a). *In* "Mass Spectrometry" (C. A. Mc Dowell, ed.), pp. 506–588. McGraw-Hill, New York.
Mc Dowell, C. A. (1963b). "Mass Spectrometry." McGraw-Hill, New York.
Mc Gowan, J. W., and Clarke, E. M. (1968). *Phys. Rev.* **167**, 43–51.
Mc Guire, E. J. (1977). *Phys. Rev. A* **16**, 62–72.
Mc Lafferty, F. W. (1959). *Anal. Chem.* **31**, 82–87.
Mc Lafferty, F. W. (1963). "Mass Spectrometry of Organic Ions." Academic Press, New York.
Mc Lafferty, F. W. (1966). "Interpretation of mass spectra." Benjamin, New York.
Maeda, K., Semeluk, G. P., and Lossing, F. P. (1968). *Int. J. Mass Spectrom. Ion Phys.* **1**, 395–407.
Maeda, K., Suzuki, I. H., and Koyama, Y. (1974). *Int. J. Mass Spectrom. Ion Phys.* **14**, 273–283.
Magee, J. L., and Gurnee, E. F. (1952). *J. Chem. Phys.* **20**, 894–898.
Mandelbaum, D., and Feldman, P. D. (1976). *J. Chem. Phys.* **65**, 672–677.
Mann, J. B. (1967). *J. Chem. Phys.* **46**, 1646–1651.
Mann, J. B. (1970). *In* "Recent Developments in Mass Spectrometry" (K. Ogata and T. Hayakawa, cd.), pp. 814–819. Univ. Park Press, Baltimore.
Marchand, P., Paquet, C., and Marmet, P. (1969). *Phys. Rev.* **180**, 123–132.
Marcus, R. A. (1952). *J. Chem. Phys.* **20**, 359–364.
Marcus, R. A., and Rice, O. K. (1951). *J. Phys. Colloid Chem.* **55**, 894–908.
Märk, E., Märk, T. D., Kim, Y. B., and Stephan, K. (1981). *J. Chem. Phys.* **75**, 4446–4453.
Märk, T. D. (1973). *Acta Phys. Austriaca* **37**, 381–384.
Märk, T. D. (1975). *J. Chem. Phys.* **63**, 3731–3736.
Märk, T. D. (1977). *Z. Naturforsch.* **32a**, 1559–1560.
Märk, T. D. (1982a). "Properties and reactions of cluster ions, Book of Invited Papers of 4th Symp. Elementary Processes and Chemical Reactions in LTP" (V. Martisovic and P. Lukac, ed.), pp. 55–73. Stara Lesna, Slovakia.
Märk, T. D. (1982b). *Beitr. Plasmaphys.* **22**, 257–294.
Märk, T. D. (1982c). *Int. J. Mass Spectrom. Ion Phys.* **45**, 125–145.
Märk, T. D. (1984a). *In* "Electron impact ionization." (T. D. Märk and G. H. Dunn, eds.) Springer-Verlag. Wien.
Märk, T. D. (1984b). *Int. J. Mass Spectrom. Ion Phys.* **55**, 325–327.
Märk, T. D., and Castleman, Jr., A. W. (1980). *J. Phys. E* **13**, 1121–1124.
Märk, T. D., and Castleman, Jr., A. W. (1984). *Adv. Atom. Mol. Phys.*, **20**.
Märk, T. D., and de Heer, F. J. (1979). *J. Phys. B* **12**, L429–L432.
Märk, T. D., and Dunn, G. H. (1984). "Electron impact ionization." Springer-Verlag, Wien.
Märk, T. D., and Egger F. (1976). *Int. J. Mass Spectrom. Ion Phys.* **20**, 89–99.
Märk, T. D., and Egger, F. (1977). *J. Chem. Phys.* **67**, 2629–2635.
Märk, T. D., and Hille, E. (1978). *J. Chem. Phys.* **69**, 2492–2496.
Märk, T. D., Egger, F., and Cheret, M. (1977a). *J. Chem. Phys.* **67**, 3795–3802.
Märk, T. D., Egger, F., Hille, E., Cheret, M., Störi, H., and Stephan, K. (1977b). *Proc. 10th ICPEAC, Paris*, pp. 1070–1071.
Märk, T. D., Stephan, K. Helm, H., Futrell, J. H., Märk, E. Peterson, K. I., Castleman, Jr., A. W., Djuric, N., and Stamatovic, A. (1982). *Contr. Symp. Atomic Surface Physics, Maria Alm* (W. Lindinger *et al., ed.*), Vol. 3, pp. 7–18.
Marmet, P., Bolduc, E., and Quemener, J. J. (1972). *J. Chem. Phys.* **56**, 3463–3468.
Marmet, P., Bolduc, E., and Quemener, J. J. (1975). *Can. J. Phys.* **53**, 2438–2444.
Massey, H. S. W. (1956). Handbuch d. Physik (S. Flügge, ed.), Vol. 36, pp. 232–408. Springer-Verlag, Berlin.
Massey, H. S. W., Burhop, E. H. S., and Gilbody, H. B. (1969). "Electronic and Ionic Impact Phenomena." Clarendon Press, Oxford.

Mathis, R. F., and Vroom, D. A. (1975). *Proc. 9th ICPEAC, Seattle*, pp. 483.
Mathur, B. P., Abbey, L. E., Burgess, E. M., and Moran T. F. (1980). *Org. Mass Spectrom.* **15**, 312–316.
Mathur, D. (1980). *J. Phys. B* **13**, 4703–4716.
Mathur, D. (1981). *Int. J. Mass Spectrom. Ion Phys.* **40**, 235–239.
Mayer, F. (1914). *Ann. Phys.* **45**, 1–28.
Medved, M., Cooks, R. G., and Beynon, J. H. (1976). *Int. J. Mass Spectrom. Ion Phys.* **19**, 179–183.
Meinke, C., and Reich, G. (1962). *Vak.-Tech.* **11**, 86–88.
Meinke, C., and Reich, G. (1963). *Vak.-Tech.* **12**, 79–82.
Melton, C. E. (1963). *In* "Mass Spectrometry of Organic Ions" (F. W. Mc Lafferty, ed.), pp. 163–206. Academic Press, New York.
Melton, C. E. (1970). "Principles of Mass Spectrometry and Negative Ions." Dekker, New York.
Merritt, C., Jr., and Mc Ewen, C. N. (1979) (Eds.) "Mass Spectrometry" Dekker, New York.
Meyerson, S., and Van der Haar, R. W. (1962). *J. Chem. Phys.* **37**, 2458–2462.
Middleditch, B. S. (1979). (Ed.) "Practical mass spectrometry: a contemporary introduction." Plenum, New York.
Miletic, M., Eres, D., Veljkovic, M., and Zmbov, K. F. (1980). *Int. J. Mass Spectrom. Ion Phys.* **35**, 231–242.
Millard, B. J. (1978). "Quantitative Mass Spectrometry," Heyden, London.
Miller, W. F., and Platzman, R. L. (1957). *Proc. Phys. Soc. London A* **70**, 299–303.
Milne, G. W. (1971). (Ed.) "Mass Spectrometry: techniques and applications," Wiley, New York.
Milne, T. A., Beachey, J. E., and Greene, F. T. (1970). Air Force Cambridge Research Laboratories Rep. AFCR L-70-0341.
Mohler, F. L., Bloom, E. G., Wells, E. J., Lengel, J. H., and Wise, C. E. (1949). *Phys. Rev.* **74**, 1332–1332.
Mohler, F. L., Dibeler, V. H., and Reese, R. M. (1954). *J. Chem. Phys.* **22**, 394–397.
Mohr, C. B. O. (1968). "Adv. Atomic Molecular Physics" (D. R. Bates and I. Estermann, ed.), Vol. 4, pp. 221–236. Academic Press, New York.
Moiseiwitsch, B. L., and Smith S. J. (1968). *Rev. Mod. Phys.* **40**, 238–349.
Momigny, J. (1961). *Bull. Soc. Chim. Belg.* **70**, 291–306.
Momigny, J., Wankenne, H., and Krier, C. (1980). *Int. J. Mass Spectrom. Ion Phys.* **35**, 151–170.
Monahan, J. E., and Stanton, H. E. (1962). *J. Chem. Phys.* **37**, 2654–2661.
Moran, T. F., Cobb, M., and Borkmann, R. F. (1980). *Chem. Phys. Lett.* **70**, 166–169.
Morgenstern, R., Niehaus, A., and Thielmann, U. (1976). *J. Phys. B* **9**, L363–L367.
Morgulis, N. (1934). *Phys. Z. Sowjet Union* **5**, 407–417.
Morita, S. (1976). *Proc. 2nd Int. Conf. Innershell Ionization Phenomena*, pp. 339–350.
Morrison, J. D. (1951). *J. Chem. Phys.* **19**, 1305–1308.
Morrison, J. D. (1953). *J. Chem. Phys.* **21**, 1767–1772.
Morrison, J. D. (1963). *J. Chem. Phys.* **39**, 200–207.
Morrison, J. D. (1964). *J. Chem. Phys.* **40**, 2488–2492.
Morrison, J. D. (1954). *J. Chem. Phys.* **22**, 1219–1223.
Morrison, J. D., and Nicholson, A. J. C. (1959). *J. Chem. Phys.* **31**, 1320–1323.
Morrison, J. D., and Stanton, H. E. (1958). *J. Chem. Phys.* **28**, 9–11.
Morrison, J. D., and Traeger, J. C. (1973). *Int. J. Mass Spectrom. Ion Phys.* **11**, 77–88, 277–288, 289–300.
Mulliken, R. S. (1933). *Phys. Rev.* **43**, 279–302.
Nagy, P., Skutlartz, A., and Schmidt, V. (1980). *J. Phys. B* **13**, 1249–1267.
Naidu, P. S., and Westphal, K. O. (1966). *Br. J. Appl. Phys.* **17**, 645–656.
Nasrallah, H. K., Marmet, P., and Dutil, R. (1983). *Int. J. Mass Spectrom. Ion Phys.* **55**, 1–14.
Nenner, I., Guyon, P. M., Baer, T., and Govers, R. T. (1980). *J. Chem. Phys.* **72**, 6587–6591.
Newton, A. S. (1964). *J. Chem. Phys.* **40**, 607–608.

Newton, A. S., and Sciamanna, A. F. (1964). *J. Chem. Phys.* **40**, 718–723.
Newton, A. S., and Sciamanna, A. F. (1966). *J. Chem. Phys.* **44**, 4327–4332.
Newton, A. S., and Sciamanna, A. F. (1969). *J. Chem. Phys.* **50**, 4868–4877.
Newton, A. S., and Sciamanna, A. F. (1970a). *J. Chem. Phys.* **52**, 327–336.
Newton, A. S., and Sciamanna, A. F. (1970b). *J. Chem. Phys.* **53**, 132–136.
Newton, A. S., and Sciamanna, A. F. (1970c). *In* "Recent Developments in Mass Spectroscopy" (K. Ogata and T. Hayakawa, eds.), pp. 828–829. Univ. Park Press, Baltimore.
Nier, A. O. (1938). *Phys. Rev.* **53**, 282–286.
Nier, A. O. (1940). *Rev. Sci. Instrum.* **11**, 212–216.
Nier, A. O., and Hanson, E. E. (1936). *Phys. Rev.* **50**, 722–726.
Nishimura, H. (1968). *J. Phys. Soc. Jpn.* **24**, 130–143.
Nottingham, W. B. (1939). *Phys. Rev.* **55**, 203–219.
Occolowitz, J. L., Cerimele, B. J., and Brown, P. (1974). *Org. Mass Spectrom.* **8**, 61–76.
Ochkur, V. I., and Petrunkin, A. M. (1963). *Opt. Spectrosc. USSR* **14**, 245–248.
Oda, N. (1973). *In* "Physics of Electronic and Atomic Collisions" (B. C. Cobic and M. V. Kurepa, eds.), pp. 443–463. Institute Physics, Beograd.
Oda, N., and Nishimura, F. (1975). *Proc. 9th ICPEAC, Seattle*, Vol. 1, pp. 481–482.
Oda, N., Nishimura, F., and Osawa, T. (1979). *Proc. 11th ICPEAC, Kyoto*, pp. 408–409.
Olivier, J. L., Locht, R., and Momigny, J. (1982). *Chem. Phys.* **68**, 201–211.
Olmsted, J., Street, K., and Newton, A. S. (1964). *J. Chem. Phys.* **40**, 2114–2122.
Opal, C. B., Peterson, W. K., and Beaty, E. C. (1971). *J. Chem. Phys.* **55**, 4100–4106.
Opal, C. B., Beaty, E. C., and Peterson, W. K. (1972). *At. Data* **4**, 209–253.
Osberghaus, O., and Taubert, R. (1951). *Angew. Chem.* **63**, 287–287.
Ossinger, A. I., and Weiner, E. R. (1976). *J. Chem. Phys.* **65**, 2892–2900.
Otvos, J. W., and Stevenson, D. P. (1956). *J. Am. Chem. Soc.* **78**, 546–551.
Peach, G. (1971). *J. Phys. B* **4**, 1670–1677.
Peek, J. (1974). *Physica* **64**, 93–113.
Pejcev, V. M., Kurepa, M. V., and Cadez, I. M. (1979). *Chem. Phys. Lett.* **63**, 301–304.
Perrin, J., and Schmitt, J. P. M. (1982). *Chem. Phys.* **67**, 167–176.
Peterkop, R. (1971). *J. Phys. B* **4**, 513–521.
Peterkop, R. K. (1977). "Theory of ionization of atoms by electron impact" (D. G. Hummer, transl.). Colorado Ass. Univ. Press, Boulder.
Peterkop, R., and Liepinsh, A. (1969). *Proc. 6th ICPEAC* pp. 212–214.
Peterkop, R., and Liepinsh, A. (1981). *J. Phys. B* **14**, 4125–4135.
Peterkop, R., and Tsukerman, P. B. (1969). *Proc. 6th ICPEAC* pp. 209–211.
Peterkop, R., and Tsukerman, P. B. (1970). *Zh. Eksp. Teor. Fiz.* **58**, 699–705; *JETP* **31**, 374–377.
Peterson, J. R. (1964). *In* "Atomic Collision Processes" (M. R. C. Mc Dowell, ed.), pp. 465–473. North Holland, Amsterdam.
Peterson, L. R., Prasad, S. S., and Green, A. E. S. (1969). *Can J. Chem.* **47**, 1774–1780.
Pham, D., Bizot, M., Durup, J., Fourmann, B., and Ozenne, J. B. (1971). *Proc. 8th ICPEAC*, p. 427.
Pichou, F., Huetz, A., Joyez, G., and Landau, M. (1978). *J. Phys. B* **11**, 3683–3692.
Pichou, F., Hall, R. I., Landau, M., and Schermann, C. (1983). *J. Phys. B* **16**, 2445–2456.
Platzman, R. L. (1963). *J. Chem. Phys.* **38**, 2775–2776.
Platzner, I., Levin, G., and Klein, F. S. (1971). *J. Chem. Phys.* **55**, 2276–2278.
Pople, J. A. (1975). *Int. J. Mass Spectrom. Ion Phys.* **19**, 89–106.
Porter, Q. N., and Baldas, J. (1971). "Mass spectrometry of heterocyclic compounds", Wiley, New York.
Pottie, R. F. (1966). *J. Chem. Phys.* **44**, 916–922.
Powell, C. J. (1976). *Rev. Mod. Phys.* **48**, 33–47.
Prasad, S. S., and Prasad, K. (1963). *Proc. Phys. Soc.* **82**, 655–658.

Quinn, E. I., and Mohler, F. L. (1959). *J. Res. Natl. Bur. Stand.* **62**, 39–42.
Rabbih, M. A., Selim, E. T. M., and Fahmey, M.A. (1981). *Ind. J. Pure Appl. Phys.* **19**, 962–966.
Rabrenovic, M., Brenton, A. G., and Beynon, J. H. (1983). *Int. J. Mass Spectrom. Ion Phys.* **52**, 175–182.
Rao, B. P., Murthy, V. R., Sabbaiah, D. V., and Naidu, S. V. (1979). *Acta Cien. Indica* **5**, 118–123.
Rapp, D. (1971). *J. Chem. Phys.* **55**, 4154–4155.
Rapp, D. and Englander-Golden, P. (1965). *J. Chem. Phys.* **43**, 1464–1479.
Rapp, D. Englander-Golden, P., and Briglia, D. D. (1965). *J. Chem. Phys.* **42**, 4081–4085.
Rau, A. R. P. (1971). *Phys. Rev. A* **4**, 207–220.
Ravishankara, A. R., and Hanrahan, R. J. (1975). *J. Phys. Chem.* **79**, 876–881.
Read, F. H. (1975). *Radiat. Res.* **64**, 23–36.
Read, F. H. (1980). In "The Physics of Ionized gases" (M. Malic, ed.), pp. 11–32. Boris Kidric Institute, Beograd.
Read, F. H., Comer, J., Imhof, R. E., Brunt, J. N.H., and Harting, E. (1974). *J. Electron Spectrosc. Relat. Phenom.* **4**, 293–312.
Reed, R. I. (1962). "Ion Production by Electron Impact." Academic Press, London.
Reed, R. I. (1966). "Applications of mass spectrometry to organic chemistry." Academic Press, New York.
Reese, R. M., and Hipple, J. A. (1949). *Phys. Rev.* **75**, 1332–1332.
Registry of mass spectral data. (1982). Wiley, New York.
Remberg, G., Remberg, E., Spiteller-Friedman, M., and Spiteller, G. (1968). *Org. Mass Spectrom.* **1**, 87–113.
Rieke, F. F., and Prepejchal, W. (1969). *Proc. 6th Int. Conf. Physics Electronic Atomic Collisions*, pp. 623–625.
Rieke, F. F., and Prepejchal, W. (1970). Argonne National Laboratory Report ANL-7760-I, p. 87.
Rieke, F. F., and Prepejchal, W. (1972). *Phys. Rev. A* **6**, 1507–1519.
Roboz, J. (1968). "Introduction to mass spectrometry; instrumentation and techniques." Wiley, New York.
Rose, M. E., and Johnstone, R. A. W. (1982). "Mass spectrometry for chemists and biochemists." Cambridge University Press, London and New York.
Rosenstock, H. M. (1968). *Adv. Mass Spectrom.* **4**, 523–546.
Rosenstock, H. M. (1976). *Int. J. Mass Spectrom. Ion Phys.* **20**, 139–190.
Rosenstock, H. M. and Krauss, M. (1963). *Adv. Mass Spectrom.* **2**, 251–284.
Rosenstock, H. M., Wallenstein, M. B., Wahrhaftig, A. L. and Eyring, H. (1952). *Proc. Natl. Acad. Sci. U S.* **38**, 667–678.
Rosenstock, H. M., Draxl, K., Steiner, B. W., and Herron, J. T. (1977). *J. Phys. Chem. Ref. Data* **6**, Suppl. 1, 1–783.
Rosenstock, H. M., Stockbauer, R., and Parr, A. C. (1981). *Int. J. Mass Spectrom. Ion Phys.* **38**, 323–331.
Rothe, E. W. (1964). *J. Vac. Sci. Technol.* **1**, 66–68.
Rowland, C. G. (1971). *Int. J. Mass Spectrom. Ion Phys.* **7**, 79–87.
Rowland, C. G., Eland, J. H. D., and Danby, C. J. (1968). *Chem. Commun.* 1535–1536.
Rowland, C. G., Eland, J. H. D., and Danby, C. J. (1969). *Int. J. Mass Spectrom. Ion Phys.* **2**, 457–469.
Roy, B. N., and Rai, D. K. (1973). *J. Phys. B* **6**, 816–822.
Roy, B. N., Tripathi, D. N., Rai, D. K. (1972). *Can. J. Phys.* **50**, 2961–2966.
Rudge, M. R. H. (1968). *Rev. Mod. Phys.* **40**, 564–590.
Rudge, M. R. H., and Seaton, M. J. (1964). *Proc. Phys. Soc.* **83**, 680–682.
Rudge, M. R. H., and Seaton, M. J. (1965). *Proc. R. Soc. A* **283**, 262–290.
Rüdenauer, F. G. (1970). *Rev. Sci. Instrum.* **41**, 1487–1488.

Sasaki, V. N., and Nakao, T. (1935). *Proc. Imp. Acad. Jpn.* **11**, 138–140, 413–415.
Sasaki, V. N., and Nakao, T. (1941). *Proc. Imp. Acad. Jpn.* **17**, 75–77.
Sattler, K., Mühlbach, J., Echt, O., Pfau, D., and Recknagel, E. (1981). *Phys. Rev. Lett.* **47**, 160–163.
Schack, C. J., Pilipovich, D., Cohz, S. N., and Sheehan, D. F. (1968). *J. Phys. Chem.* **72**, 4697–4698.
Schaeffer, O. A. (1954). *Rev. Sci. Instrum.* **25**, 660–662.
Schaeffer, O. A. (1955). *J. Chem. Phys.* **23**, 1309–1313.
Schaeffer, O. A., and Hastings, J. M. (1950). *J. Chem. Phys.* **18**, 1048–1050.
Scheunemann, H. U., Heni, M., Illenberger, E., and Baumgärtl, H. (1982). *Ber. Bunsenges, Phys. Chem.* **86**, 321–326.
Schmidt, V., Sandner, N., and Kuntzemüller, H. (1976). *Phys. Rev. A* **13**, 1743–1747.
Schmitt, H. P. (1978). Diplomarbeit, Universität Würzburg (see Klar 1981a).
Schopman, J., and Los, J. (1970). *Physica* **48**, 190–216.
Schopman, J., Barna, A. K., and Los, J. (1969). *Phys. Lett. A* **29**, 112–133.
Schopman, J., Barna, A. K., and Los, J. (1970). In "Recent Developments in Mass Spectroscopy" (K. Ogata and T. Hayakawa, eds.), pp. 888–896. Univ. Park Press, Baltimore.
Schopman, J., Fournier, P. G., and Los, J. (1973). *Physica* **63**, 518–526.
Schram, B. L., de Heer, F. J., van Wiel, M. J., and Kistemaker, J. (1965). *Physica* **31**, 94–112.
Schram, B. L., Adamczyk, B., and Boerboom, A. J. H. (1966a). *Rev. Sci Instrum.* **43**, 638–640.
Schram, B. L., Boerboom, A. J. H., Kleine, W., and Kistemaker, J. (1966b). *Physica* **32**, 749–761.
Schram, B. L., van der Wiel, M. J., de Heer, F. J., and Moustafa, H. R. (1966c). *J. Chem. Phys.* **44**, 49–54.
Schram, B. L., Moustafa, H. R., Schutten, J., and de Heer, F. J. (1966d). *Physica* **32**, 734–740.
Schulz, G. J. (1962). *Phys. Rev.* **128**, 178–186.
Schulz, W., Drost, H., and Klotz, H. D. (1968). *Exp. Technik. Phys.* **16**, 16–22.
Schutten, J., de Heer, F. J., Moustafa, H. R., Boerboom, A. J. H., and Kistemaker, J. (1966). *J. Chem. Phys.* **44**, 3924–3928.
Selim, E. T. M. (1980). *Ind. J. Pure Appl. Phys.* **18**, 31–35.
Sen Sharma, D. K., and Franklin, J. L. (1974). *Int. J. Mass Spectrom. Ion Phys.* **13**, 139–150.
Shannon, T. W. (1970). *Int. J. Mass Spectrom. Ion Phys.* **3**, App. 12–14.
Shaw, D., King, G. C., and Read, F. H. (1979). *Proc. 11th ICPEAC, Kyoto*, pp. 400–401.
Shrader, S. R. (1971). "Introductory mass spectrometry," Allyn and Bacon, Rockleigh, New Jersey.
Siegel, M. W. (1982). *Int. J. Mass Spectrom. Ion Phys.* **44**, 19–36.
Simpson, F. R., and Mc Conkey, J. W. (1969). *Planet. Space Sci.* **16**, 1941–1948.
Simpson, J. A. (1964). *Rev. Sci. Instrum.* **35**, 1698–1704.
Sjögren, H. (1964). In "Atomic Collision Processes" (M. R. C. Mc Dowell, ed.), pp. 474–477. North Holland, Amsterdam.
Smith, A. J., Hicks, P. J., Read, F. H., Cvejanovic, S., King, G. C. M., Comer, J., and Sharp, J. M. (1974). *J. Phys. B* **7**, L496–L502.
Smith, L. G. (1937). *Phys. Rev.* **51**, 263–275.
Smith, O. I., and Stevenson, J. S. (1981). *J. Chem. Phys.* **74**, 6777–6783.
Smith, P. T. (1930). *Phys. Rev.* **36**, 1293–1302.
Smyth, H. D. (1931). *Rev. Mod. Phys.* **3**, 347–391.
Solka, B. H., Beynon, J. H., and Cooks, R. G. (1975). *J. Phys. Chem.* **79**, 859–862.
Solomon, M., and Mandelbaum, A. (1969). *Chem. Commun.* 890–890.
Spence, D. (1975). *Phys. Rev. A* **12**, 2353–2360.
Spiteller, G. (1966). "Massenspektrometrische Strukturanalyse Organischer Verbindungen." Verlag Chemie, Weinheim.
Spiteller, G. (1968). In "Massenspektrometrie" (H. Kienitz, ed.) pp. 423–578. Verlag Chemie, Weinheim.
Sramek, S. J., Macek, J. H., and Gallup, G. A. (1980). *Phys. Rev. A* **21**, 1361–1363.
Srinivasan, V., and Rees, J. A. (1967). *Br. J. Appl. Phys.* **18**, 59–64.

Srinivasan, V., Rees, J. A., and Craggs, J. D. (1967). *5th ICPEAC, Leningrad, Nauka*, pp. 56–57.
Srivastava, B. N., and Mirza, I. M. (1968). *Phys. Rev.* **168**, 86–88.
Stabler, R. C. (1964). *Phys. Rev.* **133**, A1268–A1273.
Stace, A. J., and Shukla, A. K. (1980). *Int. J. Mass Spectrom. Ion Phys.* **36**, 119–122.
Stace, A. J., and Shukla, A. K. (1982). *Chem. Phys. Lett.* **85**, 157–160.
Stace, A. J., and Moore, C. (1983). *Chem. Phys. Lett.* **96**, 80–84.
Stafford, F. E. (1968). *Adv. Chem. Ser.* **72**, 115–126.
Stafford, F. E. (1971). *High Temp.-High Pressures* **3**, 213–224.
Stalherm, D., Cleff, B., Hillig, H., and Mehlhorn, W. (1969). *Z. Naturforsch.* **24a**, 1728–1733.
Stamatovic, A., and Miletic, M. (1982). "Contr. Symp. Atomic Surface Physics, Maria Alm" (W. Lindinger *et al.*, ed.), Vol. 3, pp. 42–47.
Stamatovic, A., and Schulz, G. J. (1968). *Rev. Sci. Instrum.* **39**, 1752–1753.
Stamatovic, A., and Schulz, G. J. (1970). *Rev. Sci. Instrum.* **41**, 423–427.
Stanski, T., and Adamczyk, B. (1983). *Int. J. Mass Spectrom. Ion Phys.* **46**, 31–34.
Stanton, H. E. (1959). *J. Chem. Phys.* **30**, 1116–1117.
Stanton, H. E., and Monahan, J. E. (1964). *J. Chem. Phys.* **41**, 3694–3702.
Stanton, P. N., and John, R. M. S. (1969). *J. Opt. Soc. Am.* **59**, 252–260.
Steiner, B., Giese, C. F., and Inghram, M. G. (1961). *J. Chem. Phys.* **34**, 189–220.
Stephan, K., and Märk, T. D. (1982a). *Chem. Phys. Lett.* **87**, 226–228.
Stephan, K., and Märk, T. D. (1982b). "Contr. Symp. Atomic Surface Physics, Maria Alm" (W. Lindinger *et al.*, ed.), Vol. 3, pp. 48–53.
Stephan, K., and Märk, T. D. (1982c). *Chem. Phys. Lett.* **90**, 51–54.
Stephan, K., and Märk, T. D. (1983). *Int. J. Mass Spectrom. Ion Phys.* **47**, 195–198.
Stephan, K., Helm, H., and Märk, T. D. (1980a). *Adv. Mass Spectrom.* **8A**, 122–132.
Stephan, K., Helm, H., and Märk, T. D. (1980b). *J. Chem. Phys.* **73**, 3763–3778.
Stephan, K., Helm, H., Kim, Y. B., Seykora, G., Ramler, J., Grössl, M., Märk, E., and Märk, T. D. (1980c). *J. Chem. Phys.* **73**, 303–308.
Stephan, K., Märk, T. D., and Helm, H. (1981a). *Eur. Conf. Abstr.* **5A**, 741–742; (1982) *Phys. Rev. A*, **26**, 2981–2982.
Stephan, K., Futrell, J. H., Peterson, K. I., Castleman, Jr., A. W., Wagner, H. E., Djuric, N., and Märk, T. D. (1981b). *Proc. 8th Intern. Symp. Molec. Beams, Cannes*, pp. 211–215.
Stephan, K., Futrell, J. H., Peterson, K. I., Castleman, Jr., A. W., and Märk, T. D. (1982a). *J. Chem. Phys.* **77**, 2408–2415.
Stephan, K., Futrell, J. H., Peterson, K. I., Castleman, Jr., A. W., Wagner, H. E., Djuric, N., and Märk, T. D. (1982b). *Int. J. Mass Spectrom. Ion Phys.* **44**, 167–181.
Stephan, K., Deutsch, H., and Märk, T. D. (1983a). *Proc 31st Ann. Conf. Mass Spectrom. Allied Topics, Boston*, 734–734.
Stephan, K., Leiter, K., Deutsch, H., and Märk, T. D. (1983b). *Proc. 13th ICPEAC, Berlin*, 276–276.
Stephan, K., Futrell, J. H., Märk, T. D., and Castleman, Jr. (1983c). *Vacuum TAIP*, **33**, 77–85.
Stephan, K., Märk, T. D., Märk, E., Stamatovic, A., Djuric, N., and Castleman, Jr., A. W. (1983d). *Beitr. Plasmaphys.* **23**, 369–372.
Stephan, K., Märk, T. D., and Castleman, Jr., A. W. (1983e). *J. Chem. Phys.* **78**, 2953–2956.
Stephen, K., Stamatovic, A., and Märk, T. D. (1983f). *Phys. Rev. A* **28**, 3105–3108.
Stevenson, D. P. (1947). *J. Chem. Phys.* **15**, 409–411.
Stevenson, D. P. (1951). *Discuss. Faraday Soc.* **10**, 35–45.
Stevenson, D. P. (1960). *J. Am. Chem. Soc.* **82**, 5961–5965.
Stevenson, D. P., and Hipple, J. A. (1942). *Phys. Rev.* **62**, 237–240.
Stevenson, D. P., and Schissler, D. O. (1961). "The Chemical and Biological Action of Radiation" (M. Harssinsky, ed.), Vol. 5 Academic Press, London.
Stevie, F. A., and Vasile, M. J. (1981). *J. Chem. Phys.* **74**, 5106–5110.

Stewart, D. T. (1956). *Proc. Phys. Soc. London, A* **69**, 437–440.
Stimpson, B. P., Simons, D. S., and Evans, Jr., C. A. (1978). *J. Phys. Chem.* **82**, 660–670.
Stockbauer, R. (1973). *J. Chem. Phys.* **58**, 3800–3815.
Stockdale, J. A. D. (1977). *J. Chem. Phys.* **66**, 1792–1794.
Stockdale, J. A. D., Anderson, V. E., Carter, A. E., and Deleanu, L. (1975). *J. Chem. Phys.* **63**, 3886–3897.
Stolarski, R. S., Dulock, V. A., Watson, C. E., and Green, A. E. S. (1967). *J. Geophys. Res.* **72**, 3953–3960.
Sunner, J., and Kebarle, P. (1981). *J. Phys. Chem.* **85** 327–335.
Suzuki, I. H., and Maeda, K. (1974). *Int. J. Mass Spectrom. Ion Phys.* **15**, 281–290.
Suzuki, I. H., and Maeda, K. (1977a). *Mass Spectrosc. Jpn.* **25**, 223–232.
Suzuki, I. H., and Maeda, K. (1977b). *J. Can. Chem.* **55**, 3124–3131.
Suzuki, I. H., and Maeda, K. (1979). *Mass Spectrom. Jpn.* **27**, 31–40,
Svec, H. J., and Flesch, G. D. (1968). *Int. J. Mass Spectrom. Ion Phys.* **1**, 41–52.
Tannen, P. D. (1973). Thesis (Air Force Institute of Technology) Univ. Microfilms, Ann Arbor Michigan, Order No. 74–14940.
Tate, J. T., and Lozier, W. W. (1932). *Phys. Rev.* **39**, 254–277.
Tate, J. T., and Smith, P. T. (1934). *Phys. Rev.* **46**, 773–776.
Tate, J. T., Smith, P. T., and Vaughan, A. L. (1935). *Phys. Rev.* **48**, 525–531.
Taubert, R. (1959). *Adv. Mass Spectrom.* **1**, 489–503.
Taubert, R. (1964). *Z. Naturforsch* **19a**, 484–493, 494–506, 911–925.
Tawara, H. (1978). "Electronic and Atomic Collisions" (G. Watel, ed.), pp. 311–329. North Holland, Amsterdam.
Tay, E. S., Dawson, P. G., and Mosson, G. A. G. (1970). *Proc. 6th Europ. Symp. Fusion Techn.*, Aachen, pp. 51–58.
Taylor, P. O., Dolder, K. T., Kauppila, W. E., and Dunn, G. H. (1974). *Rev. Sci. Instrum.* **45**, 538–544.
Taylor, P. R. (1983). *Mol. Phys.* **49**, 1297–1314.
Temkin, A. (1966). *Phys. Rev. Lett.* **16**, 835–839.
Temkin, A. (1974). *J. Phys. B* **7**, L450–L453.
Temkin, A. (1982a). *Phys. Rev. Lett.* **49**, 365–368.
Temkin, A. (1982b). *J. Phys. B* **15**, L301–L304.
Temkin, A. (1982c). *Com. At. Mol. Phys.* **11**, 287–295.
Temkin, A., and Hahn, Y. (1974). *Phys. Rev. A* **9**, 708–724.
Temkin, A., Bhatia, A. K., and Sullivan, E. (1968). *Phys. Rev.* **176**, 80–89.
Terwillinger, D. T., Cooks, R. G., and Beynon J. H. (1975). *Int. J. Mass Spectrom. Ion Phys.* **18**, 43–56.
Thomas, L. H. (1927). *Proc. Cambridge Philos. Soc.* **23**, 829–831.
Thomson, J. J. (1912a). *Philos. Mag.* **23**, 449–457.
Thomson, J. J. (1912b). *Philos. Mag.* **24**, 668–672.
Thomson, J. J. (1921). see ref. in Vaughan, 1931.
Thulstrup. E. W., and Anderson, A. (1975). *J. Phys. B* **6**, 965–976.
Thulstrup, P. W., Thulstrup, E. W., Anderson, A., and Öhrn, Y. (1974). *J. Chem. Phys.* **60**, 3975–3980.
Tisone, G. C. (1972). *J. Chem. Phys.* **56**, 108–112; **57**, 3686–3689.
Tiwari, P., Rai, D. K., and Rustgi, M. L. (1969). *J. Chem. Phys.* **50**, 3040–3045.
Ton-That, D., and Flannery, M. R. (1977). *Phys. Rev. A* **15**, 517–526.
Ton-That, D., Manson, S. T., and Flannery, M. R. (1977). *J. Phys. B* **10**, 621–635.
Tripathi, D. N., and Rai, D. K. (1972). *Indian J. Pure Appl. Phys.* **10**, 185–186.
Tsuchiya, T. (1962). *J. Chem. Phys.* **36**, 568–569.

Tuithof, H. H., and Boerboom, A. J. H. (1974). *Int. J. Mass Spectrom. Ion Phys.* **15**, 105–109.
Turban, G., Catherine, Y., and Grolleau, B. (1980). *Thin Solid Films*, **67**, 309–320.
Tweed, R. J. (1973). *J. Phys. B* **6**, 270–284.
Vainstein, L. A., Sobelman, I. I., and Yukov, E. A. (1973). "Electron-Excitation Cross Sections of Atoms and Ions" (in Russian). Nauka Moscow.
Van Brunt, R. J. (1974). *J. Chem. Phys.* **60**, 3064–3070.
Van Brunt, R. J. (1977). *Phys. Rev. A* **16**, 1309–1311.
Van Brunt, R. J., and Kieffer, L. J. (1970). *Phys. Rev. A* **2**, 1293–1304.
Van Brunt, R. J., and Kieffer, L. J. (1974). *J. Chem. Phys.* **60**, 3057–3063.
Van Brunt, R. J., and Wacks, M. E. (1964). *J. Chem. Phys.* **41**, 3195–3199.
van der Wiel, M. J. (1973). "Physics of Electronic and Atomic Collisions" (B. C. Cobic and M. V. Kurepa, ed.), pp. 417–442. Institute Physics, Beograd.
van der Wiel, M. J. (1976). *In* "Photoionization and Other Probes of Many Electron Interactions" (F. J. Wuilleumier, ed.), pp. 187–208. Plenum, New York.
van der Wiel, M. J. (1980). "Electronic and Atomic Collisions" (N. Oda and K. Takayanagi, ed.), pp. 209–218. North Holland, Amsterdam.
van de Water, E., and Heideman, H. G. M. (1981). *Proc. 12th ICPEAC, Gatlinburg*, pp. 270–271.
Van Gorkom, M., and Glick, R. E. (1970). *Int. J. Mass Spectrom. Ion Phys.* **4**, 203–218.
Van Gorkom, M., Beggs, D. P., and Glick, R. E. (1970). *Int. J. Mass Spectrom. Ion Phys.* **4**, 441–450.
Van Sprang, H. A., Möhlmann, G. R., and de Heer, F. J. (1978). *Chem. Phys.* **33**, 65–72.
Van Sprang, H. A., Brongersma, H. H., and de Heer, F. J. (1979). *Chem. Phys.* **38**, 277–284.
van Wingerden, B., van der Leeuw, P. E., de Heer, F. J., and van der Wiel, M. J. (1979). *J. Phys. B* **12**, 1559–1576.
Vaughan, A. L. (1931). *Phys. Rev.* **38**, 1687–1695.
Vauthier, R. (1950). *C. R. Acad. Sci. Paris* **231**, 764–765, 1218–1220.
Vauthier, R. (1955). *C. R. Acad. Sci. Paris* **241**, 1033–1036; *Ann. Phys. Paris* **10**, 968–1025.
Veldre, V. (1965). Riga Report, Translation TT-66-1239. John Crerar Library, Chicago.
Vestal, M. L. (1968). Ionic Fragmentation Processes, Chapter 2. *In* "Fund. Proc. Radiation Chemistry" (P. Ausloos, ed.), pp. 59–118. Wiley(Interscience), New York.
Vinkalns, I., and Gailitis, M. (1967). *Proc. 5th ICPEAC Leningrad, Nauka* pp. 648–650.
Vogt, J., and Pascual, C. (1972). *Int. J. Mass Spectrom. Ion Phys.* **9**, 441–448.
von Ardenne, M., Steinfelder, K., and Tümmler, R. (1971). "Elektronenanlagerungs-Massenspektrographie Organischer Substanzen." Springer Verlag, Berlin and New York.
von Engel, A. (1965). "Ionized Gases." Clarendon Press, Oxford.
von Hippel, A. (1928). *Ann. Phys.* **87**, 1035–1087.
Vouros, R., and Biemann, K. (1969). *Org. Mass Spectrom.* **2**, 375–386.
Vriens, L. (1964). *Phys. Lett.* **8**, 260–261; **9**, 295–296; **10**, 170–171.
Vriens, L. (1966). *Proc. Phys. Soc.* **89**, 13–21; *Phys. Rev.* **141**, 88–92.
Vriens, L. (1969). "Case Studies in Atomic Collision Physics I" (E. W. Mc Daniel, and M. R. C. Mc Dowell, ed.), pp. 337–398. North Holland, Amsterdam.
Wahrhaftig, A. L. (1972). *In* "Mass Spectrometry" (E. A. Maccoll, ed.), Vol. 5, (Physical Chemistry, Series 1). pp. 1–24 Butterworths, London.
Waight, E. S. (1969). *Chem. Commun.* 1258–1258.
Wallace, S. J., Berg, R. A., and Green, A. E. S. (1973). *Phys. Rev. A* **7**, 1616–1629.
Wallington, M. J. (1970). *J. Phys. E* **3**, 599–604.
Wallington, M. J. (1971). *J. Phys. E* **4**, 1–8.
Wankenne, H., and Momigny, J. (1969). *Chem. Phys. Lett.* **4**, 132–134.
Wankenne, H., and Momigny, J. (1971). *Int. J. Mass Spectrom. Ion Phys.* **7**, 227–243.
Wannier, G. H. (1953). *Phys. Rev.* **90**, 817–825.
Wannier, G. H. (1955). *Phys. Rev.* **100**, 1180–1180.

Washburn, H. W., and Berry, C. E. (1946). *Phys. Rev.* **70**, 559–559.
Watson, C. E., Dulock, V. A., Stolarski, R. S., and Green, A. E. S. (1967). *J. Geophys. Res.* **72**, 3961–3966.
Webster, D. C., Hansen, W. W., and Duvaneck, F. B. (1933). *Phys. Rev.* **43**, 839–858.
Wei, P. S. D., and Kuppermann, A. (1969). *Rev. Sci. Instrum.* **40**, 783–785.
Weigold, E. (1980). "Electronic and Atomic Collisions" (W. Oda and K. Tagayanagi, eds.), pp. 81–93. North Holland, Amsterdam.
Weigold, E., and Mc Carthy, I. E. (1978). *Adv. At. Mol. Phys.* **14**, 127–179.
Weigold, E., Ugbabe, A., and Teubner, P. J. O. (1975). *Phys. Rev. Lett.* **35**, 209–212.
Weigold, E., Noble, C. J., Hood, S. T., and Fuss, I. (1979). *J. Phys. B* **12**, 291–313.
Werner, A. S., and Baer, T. (1975). *J. Chem. Phys.* **62**, 2900–2910.
Weiner, E. R., and Ossinger, A. I. (1976). *Rev. Sci. Instrum.* **47**, 84–87.
Werner, H. W. (1974a). *Int. J. Mass Spectrom. Ion Phys.* **14**, 189–204.
Werner, H. W. (1974b). *J. Phys. E* **7**, 115–121.
Werner, H. W., and Linssen, A. J. (1974). *J. Vac. Sci. Technol.* **11**, 843–847.
Werner, H. W., Venema, A., and Linssen, A. J. (1972). *J. Vac. Sci. Technol.* **9**, 216–219.
White, F. A. (1968). "Mass Spectrometry in Science and Technology." Wiley, New York.
Wiegand, W. J., and Boedeker, L. R. (1982). *Appl. Phys. Lett.* **40**, 225–227.
Wiesemann, K. (1981). *J. Phys. E* **14**, 1404–1406.
Wigner, E. P. (1948). *Phys. Rev.* **73**, 1002–1009.
Wigner, E. P., and Witmer, E. E. (1928). *Z. Phys.* **31**, 859–886.
Wilden, D. G., Hicks, D. J., and Comer, J. (1977). *Proc. 10th ICPEAC*, pp. 679, 680.
Williams, D. H. (1978). *Adv. Mass Spectrom.* **7B**, 1157–1175.
Williams, D. H., and Howe, I. (1972). "Principles of organic mass spectrometry." McGraw-Hill, New York.
Winters, R. E., and Collins, J. H. (1966). *J. Phys. Chem.* **70**, 2057–2058.
Winters, R. E., and Collins, J. H. (1968). *J. Am. Chem. Soc.* **90**, 1235–1239.
Winters, R. E., and Kiser, R. W. (1966a). *J. Phys. Chem.* **70**, 1680–1681.
Winters, R. E., and Kiser, R. W. (1966b). *J. Chem. Phys.* **44**, 1964–1966.
Winters, R. E., Collins, J. H., and Courchenne, W. L. (1966). *J Chem. Phys.* **45**, 1931–1937.
Younger, S. M. (1984). *In* "Electron impact ionization." (T. D. Märk and G. H. Dunn, eds.), Springer-Verlag, Wien. Chapter 1.
Younger, S. M., and Märk, T. D. (1984). *In* "Electron impact ionization," (T. D. Märk and G. H. Dunn, eds.), Springer-Verlag, Wien. Chapter 2.
Zare, R. N. (1967). *J. Chem. Phys.* **47**, 204–215.

4 Dissociation of Molecules by Electron Impact

E. C. Zipf
Department of Physics and Astronomy
University of Pittsburgh
Pittsburgh, Pennsylvania

I.	Introduction.	335
II.	Models	338
III.	Experimental Techniques	346
	A. Optical Cross Section Measurements	347
	B. Time-of-Flight Measurements	354
	C. Linewidth and Collisional Quenching Studies	362
IV.	Measurements.	363
	A. Molecular Nitrogen	365
	B. Molecular Oxygen	376
	C. Carbon Monoxide and Dioxide	390
	D. Other Molecules	396
	References	397

I. Introduction

Dissociative processes play a major role in the physics and chemistry of planetary and cometary atmospheres. Because they are a copious source of photons, kinetically energetic atoms and ions, chemically active metastable species, and short-lived excited states, dissociative processes are also quite important in many industrial applications which use gaseous discharges or in laser systems that are pumped by pulsed beams of high-energy electrons. The need to understand the fundamental molecular processes involved in dissociative excitation and to assemble a comprehensive set of accurate cross section values for a wide range of incident electron energies arises, in part, because of the important role of dissociative excitation in a wide range of interdisciplinary problems. Such research, for example, bears directly on practical questions in plasma chemistry, on environmental problems concerning the terrestrial ozone layer, on the effects of major solar-proton events in the middle atmosphere, on the interpretation of the large body of airglow data accumulated after two decades of space exploration, and on energy degradation in the atmosphere following a nuclear explosion. But this work also encompasses some novel topics including the role of dissociative excitation in the synthesis of odd-nitrogen compounds in lightning strokes, as a factor in paleobiology, where it may have been an important source of nutrients for primitive life, or as the crucial element controlling planetary escape and orchestrating the evolution of planetary atmospheres. Thus, in addition to its interest at the fundamental level, the study of dissociative excitation has enjoyed a particular timeliness because it services such a broad constituency.

Three categories of dissociative processes have now been widely studied.

Dissociative recombination:

$$X_2^+ + e \longrightarrow X + X \tag{1}$$

Photodissociative absorption:

$$X_2 + h\nu \longrightarrow X + X \tag{2}$$

$$X_2 + h\nu \longrightarrow X + X^+ + e \tag{3}$$

Dissociative excitation by electron impact:

$$X_2 + e \longrightarrow X + X^* + e \tag{4}$$

$$X_2 + e \longrightarrow X^+ + X^* + 2e \tag{5}$$

where, in each case, the fragment atoms and/or molecules may be electronically, vibrationally, and/or rotationally excited. The release of kinetic energy is also an important aspect of these processes, since it provides an effective method for transforming a significant proportion of the energy possessed by an incident photon or an impacting electron into heat, and it produces kinetically energetic atoms whose chemical reactivity may be substantially enhanced. Such processes are particularly important in auroral substorms and in nuclear explosions where much of the energy deposited in the atmosphere appears briefly in the form of large fluxes of energetic secondary electrons. The surplus kinetic energy released by dissociative excitation also results in the emission of radiation with highly Doppler-broadened linewidths, a factor which complicates the analysis of the role of dissociative processes in optically thick astrophysical and planetary environments as well as in more mundane gaseous discharges (Zipf and Wells, 1980). Unfortunately, the possibility of this complication compromises the use of dissociatively excited emission features as a diagnostic tool unless the observations are made with high-resolution instruments and the opacity of the medium is known. The released kinetic energy also contributes to the isotope separation that has been observed in the Martian atmosphere by the Mariner spacecrafts (Nier and McElroy, 1977) and to the atomic hydrogen corona that surrounds Titan (Shemansky, 1982). Thus, a comprehensive investigation of dissociative excitation asks a wide variety of questions and requires not only absolute cross section measurements and threshold studies, but must also include linewidth and kinetic-energy studies because of their importance in so many applied problems.

Although the study of the three broad categories of dissociative processes developed historically more or less independently, the substantial experimental and theoretical advances of the last two decades make it clear that these investigations are really complementary facets of a broader study of a common set of intermediate, predissociating, and repulsive quantum states and that a comprehensive and unified understanding of how molecular systems dissociate will ultimately emerge from a synthesis of these efforts. Our progress toward that goal has been enhanced strikingly by recent technological advances in all three areas. For example although photodissociation studies have been a timely topic for at least 60 years, it is only recently with the availability of synchroton light sources that such highly bound molecules as CO and N_2 could be effectively studied (Wu et al., 1983). With the emergence of this new capability, energetic repulsive and predissociating states previously inaccessible can now be investigated and high-resolution probing of the potential-energy surfaces leading to the formation of high-Rydberg atoms and other excited states will be feasible. The use of multiphoton techniques to study molecular predissociation also opens exciting new possibilities in photodissociation work, and the recent study on N_2 dissociation using this approach is an excellent example of how photodissociation and electron-impact dissociation studies complement each other (Baravian et al., 1982). In a similar fashion, the advent of tunable dye lasers has spectacularly expanded the scope of dissociative recombination studies because they allow process (1) to be studied as a function of the quantum state of the parent ion for the first time (Zipf, 1980). Perhaps the most notable advances have occurred in the study of the dissociative excitation of molecular gases by electron impact. Here, the development of new types of photomultiplier tubes, the availability of low-scattered-light holographic gratings, and, above all, the development of computer-supported data acquisition techniques have made it possible to measure dissociative excitation cross sections as small as 10^{-24} cm^2 at a spectral resolution of less than 0.05 Å. As the result of these advances, nearly complete cross section sets for the major diatomic and polyatomic molecules of planetary and cometary importance (e.g., H_2, HD, D_2, O_2, N_2, NO, N_2O, CO, CO_2, CH_4, NH_3, and the light hydrocarbons) over the wavelength range 500 Å to 1.3 μm have been obtained. Now that major improvements have also been made in the quality of the absolute calibration standards and in the spectral purity of the laboratory data, the next few years should see the emergence of revised cross section data sets with such small probable errors that these cross sections can then be used themselves as secondary calibration standards. This development will then make the atomic and molecular branching ratio technique a very convenient and accurate method for calibrating optical instruments absolutely.

An indication of how rapidly this field is developing and of the new opportunities opened up by the technological advances made in the last decade is provided by recent studies on inelastic electron scattering by metastable $N_2(A\ ^3\Sigma_u^+)$ molecules,

$$e^- + N_2(A\ ^3\Sigma_u^+) \longrightarrow N^* + N + e, \qquad (6)$$

and the discovery that this process is a copious source of excited nitrogen atoms N^* (Gorman and Zipf, 1981). Preliminary studies on excitation of the $NI[2p^2\,3s\,^4P - 2p^3\,^4S^0]$ and other NI states from which the VUV transitions of nitrogen originate indicate a net cross section for this subset of $\sim 10^{-16}$ cm^2 at 25 eV. This development suggests that the total cross section for process (6) may be unusually large and peaks at a comparatively low energy. If further studies show that this is the case, then an earlier surmise that process (6) is of major importance in the development of shock waves in air will have been demonstrated. These results also suggest that under some circumstances processes like process (6) may contribute significantly to the atomic multiplet radiation observed in dissociative excitation experiments and thus would be a potential source of a systematic error in a cross section assessment. Some of the variability in the literature cross section values for a number of gases may be due to such nonlinear effects.

This chapter is concerned specifically with the production of excited states by electron-impact dissociation of molecular targets. This process has access to the largest manifold of excited states because both electric-dipole-allowed and spin exchange channels are open to it. Thus, the states normally exploited by photodissociation and dissociative recombination are also used by processes (4) and (5) and, in fact, the cross-comparison of the data from the different classes of dissociation processes has been particularly valuable. In the following sections, the current status of this study is reviewed with particular attention to advances in the areas of theory, technique, and measurement.

II. Models

A number of models have been developed which account qualitatively for the principal features observed in dissociative excitation experiments. Some

4 Dissociation of Molecules by Electron Impact

quantitative work has been carried out on H_2 with partial success. Two broad categories of processes are considered in these models. In the simplest case, the impacting electron excites the target molecule to a discrete state which subsequently predissociates, autoionizes, or relaxes radiatively. The outcome of the competition between these channels then determines the effective atomic production rate. Under very simple circumstances, the issue is determined solely by the fundamental physical properties of the excited molecular state. Among the key physical factors that are amenable to measurement are the radiative lifetime of the initial state, the nature of the predissociation process (forbidden or allowed), the location of the pertinent potential surfaces, and the predissociation or autoionization branching ratio. Molecular nitrogen affords a particularly good example of the importance of predissociation which accounts for more than 75% of the total cross section for electron-impact dissociation of N_2. This process is also noteworthy for the discrete kinetic-energy spectrum exhibited by the fragment atoms and the fact that it is forbidden.

When the excitation of the initial, discrete quantum state takes place in an optically thick medium, determining the atomic production rates and the released kinetic-energy spectrum is much more complicated. Under these circumstances, photons normally emitted by the relaxing molecule are trapped in the medium. The competition then shifts to one in which predissociation, cascade emission to a lower state other than the ground state, and collisional quenching of the initial state vie with one another. In many gas discharges and at lower altitudes in planetary atmospheres, collisional quenching becomes the dominant loss channel so that little atomic production occurs, and the upper state behaves as though it were pseudometastable with a chemistry of its own (Zipf, 1982). Little work has been done on these states in dense, optically thick environments, in part because of the lack of suitable single-event detectors for most of the ground-state configuration atoms produced by the discrete process; theoretical work on these problems is virtually nonexistent.

The second broad category of dissociation processes is shown schematically in Fig. 1, which depicts the dissociative excitation of the metastable $O(^5S)$ state $(t_r = 182 \; \mu s)$,

$$e^- + O_2 \longrightarrow O(^5S) + O + e^- \tag{7a}$$

$$O(^5S) \longrightarrow O(^3P) + h\nu \quad (\lambda 1356 \, \text{Å}). \tag{7b}$$

In process A oxygen molecules in the $X\,^3\Sigma_g^-$ ground state are excited by electron impact to a purely repulsive state followed by dissociation. The two vertical lines drawn through the turning points of the $v' = 0$ level of the $X\,^3\Sigma_g^-$ state define the Franck–Condon region. Based on a consideration of the Franck–Condon principle, the onset energy for process A is shown in the figure as $E_{\min}$, which is considerably larger than the dissociation limit E_D. The

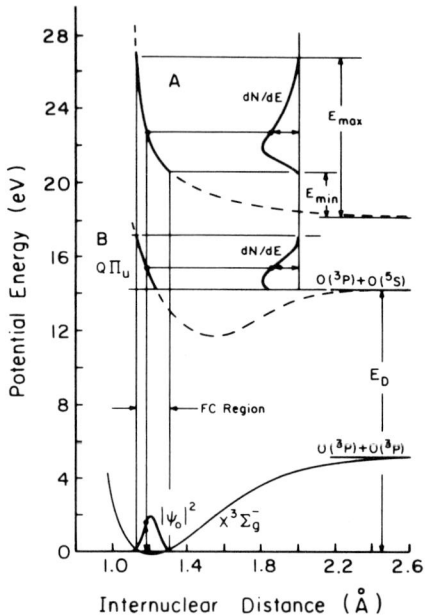

Fig. 1. Semiquantitative illustration of two dissociation processes observed in the excitation of the metastable $O(^5S^0)$ state by electron impact on O_2 and the kinetic-energy distributions that result. In case A, excitation to a purely repulsive potential-energy curve in the Franck–Condon region results in a zero value in the translational-energy spectrum at a finite kinetic energy E_{min}. However, a finite value for dN/dE is observed at zero kinetic energy when the dissociation process involves the repulsive part of a bound curve (case B). [From Borst and Zipf (1971b).]

kinetic-energy distribution dN/dE for the purely repulsive curve is plotted in Fig. 1. It can be seen that the translational energy spectrum has a zero value below E_{min} and that the shape and location of the repulsive-potential curve could be deduced in principle from high-quality time-of-flight data. The amount of kinetic energy released by these processes can be very large, with important practical effects. Figure 2 illustrates this point for the $O(^5S)$ and OI high-lying Rydberg atoms formed in the dissociative excitation of CO. In addition to the enhanced chemical reactivity exhibited by these kinetically hot atoms, the highly Doppler-broadened radiation emitted by the fragment atoms complicates radiation entrapment problems. The potential magnitude of this effect can be appreciated by examining Fig. 3, which shows the linewidth of hydrogen Balmer-alpha emission produced by electron-impact dissociation of H_2. For comparison purposes, keep in mind that H_α radiation emitted with a room-temperature Doppler width has a FWHM of approximately 81 mÅ. It is also noteworthy that a significant fraction of the energy possessed by the incident electron is converted directly into heat (cf. Fig. 2); this is an important

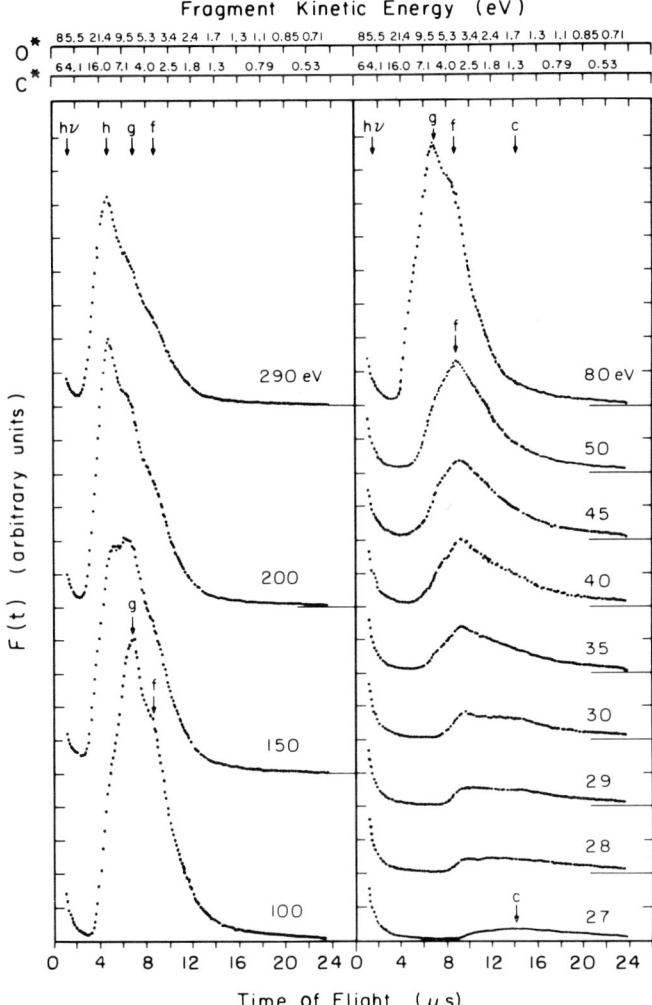

Fig. 2. Time-of-flight spectra of long-lived high-Rydberg carbon and oxygen atoms and $O(^5S^0)$ atoms produced in the dissociative excitation of CO by electron impact. Note the large fragment kinetic energies. The energy of the incident electron is noted at the right of each TOF. [From Wells et al. (1978).]

factor in the thermal economy of a planetary atmosphere—particularly in auroral substorms—and in planetary escape.

In process B (Fig. 1), the excitation involves a transition from the $X\ ^3\Sigma_g^-$ ground state to the repulsive part of a bound potential curve. The translational energy spectrum in this case can have a finite value at zero kinetic energy depending on the relative position of the two potential-energy curves.

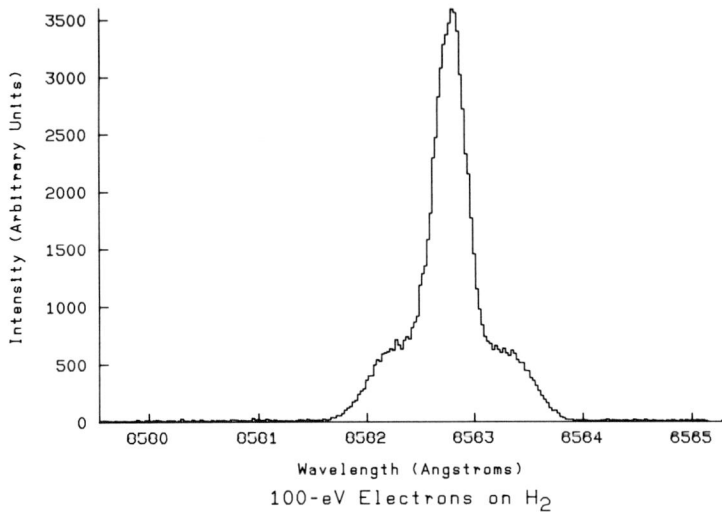

Fig. 3. HI Balmer-alpha line shape excited by 100 eV on H_2. These data were obtained with an instrumental resolution of 0.1 Å (Erdman and Zipf, unpublished results).

It is also worth noting that the threshold energy for process B corresponds to the minimum dissociation limit. The cross section for process B can be quite large ($\sim 5 \times 10^{-18}$ cm^2) when valence states that can be excited by an electric-dipole-allowed transition, are involved. Excitation of the Schumann–Runge system in $O_2(X\,^3\Sigma_g^- \longrightarrow B\,^3\Sigma_u^-)$ is a particularly important example of this process, and it is one of the few cases in which the kinetic-energy spectrum of the metastable $O(^1D)$ atoms produced by dissociative excitation have been measured in the laboratory. Figure 4 shows the remarkable results obtained by Stone *et al.* (1975).

A final model of dissociative excitation that has been extensively studied and has proven quite useful crystallized out of an important observation: namely, that the kinetic-energy spectrum of ions and high-Rydberg states produced dissociatively, e.g.,

$$e^- + N_2 \longrightarrow N^{Ryd} + N + e^- \tag{8}$$

$$e^- + N_2 \longrightarrow N^+ + N + 2e^- \tag{9}$$

bear a striking resemblance to one another. This can be seen readily in Fig. 5, in which the translational kinetic energy of N^+ ion and high-Rydberg fragments are compared (Kieffer and Van Brunt, 1967; Smyth *et al.*, 1973; Deleanu and Stockdale, 1975; Van Brunt and Kieffer, 1975; Wells *et al.*, 1976). The core-ion model, which was first suggested by Kupriyanov (1968), provides a simplified explanation of these similarities.

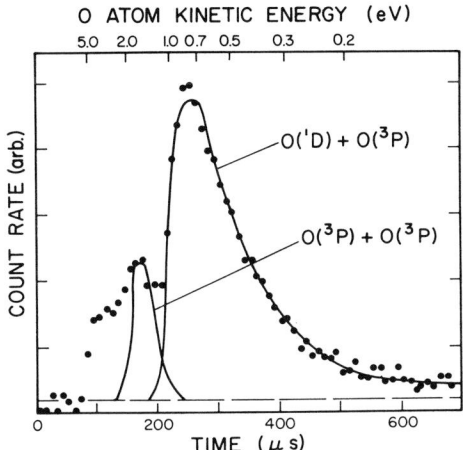

Fig. 4. Time-of-flight spectra showing the oxygen atoms produced by electron-impact dissociation of O_2. The principal peak is associated with excitation of the ground-state molecule to the repulsive portion of the slightly bound $O_2(B\ ^3\Sigma_u^-)$ state. [From Stone et al. (1975).]

In this model, a ground-state electron is excited to a high level in a Rydberg series which converges to one of the ionization limits of the molecule, leaving the core in a repulsive state. The high-lying Rydberg electrons can be treated as a "spectator" orbiting the molecule at such large distances (hundreds of angstroms) that it leaves the core ion relatively unperturbed. Hence, it is the repulsive states of the molecular ion which determine the dissociation processes and products. In one scenario, when the molecular core ion dissociates into a neutral atom and ion, the ion captures the orbiting molecular Rydberg electron into a corresponding high-Rydberg atomic orbital. Thus, energetic Rydberg atoms are formed with a kinetic-energy spectrum determined by the physical properties of the core ion. Alternatively, the newly formed atomic ion may not bind the orbiting Rydberg electron, and we have dissociative ionization.

The core-ion model has a number of distinctive characteristics that are amenable to experimental test and are quite useful in determining the initial state of the parent ions. These observable features arise because the potential-energy curves of the high-Rydberg molecule lie parallel to those of the core ions and less than 0.1 eV lower in energy. Thus, the observed dissociation limits, threshold energies, and translational energy distributions should correspond closely to predictions based upon the potential curves of the parent ions. Hence, by carefully studying the fragment kinetic-energy spectra as a function of incident electron energy, the threshold for dissociation channels can be measured and the correlation limits established. Figure 6 and

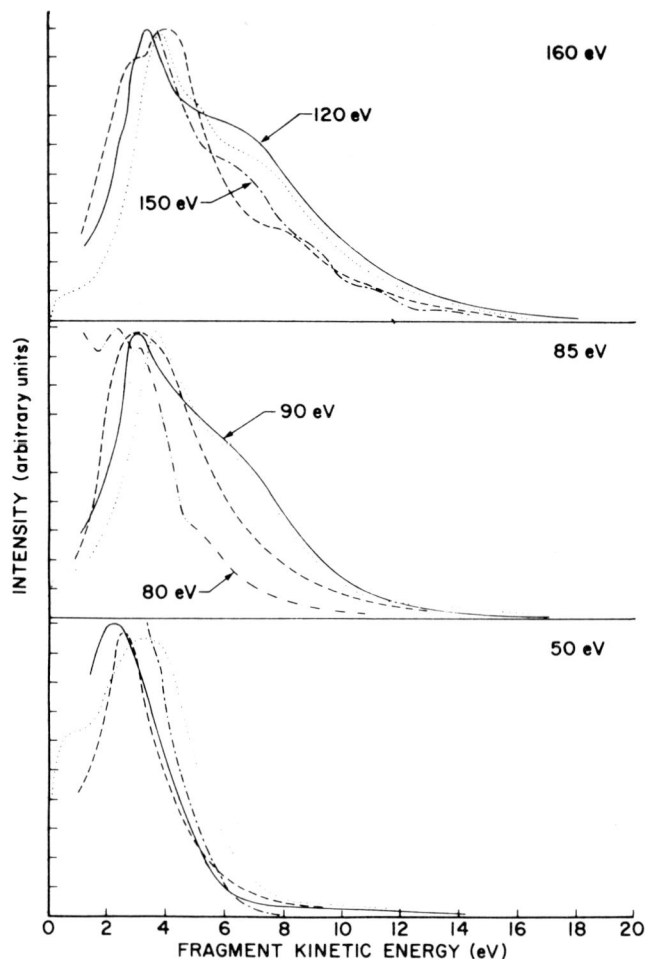

Fig. 5. Comparison of the translational-energy spectra of the high-Rydberg NI atoms [solid line, Wells et al. (1976); dotted lines, Smyth et al. (1973)] and N^+ ion fragments [dash-dot line, Deleanu and Stockdale (1975); dashed line, Kieffer and Van Brunt (1967); dash-double-dot line, Locht et al. (1973)] produced by electron-impact dissociation of N_2. All data were taken at an angle of 90° with the exception of Kieffer and Van Brunt (55°). Electron-impact energies are noted in the upper right-hand corner. The similarity between the N^{Ryd} and N^+ kinetic-energy spectra is qualitatively in accord with the core-ion model of dissociation. [From Wells et al. (1976).]

Table I illustrate the kind of data and analysis that can be deduced in this manner.

Extensive studies have now been carried out on a large number of molecular targets. Detailed information exists for species such as H_2, O_2, CO, CO_2, N_2, NO, NO_2, N_2O, CH_4, and other light hydrocarbons, and for most of these

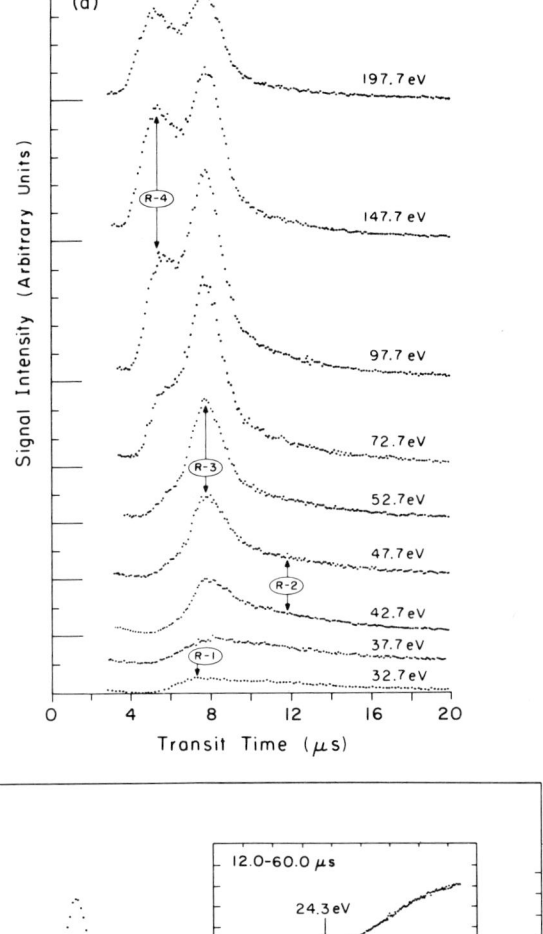
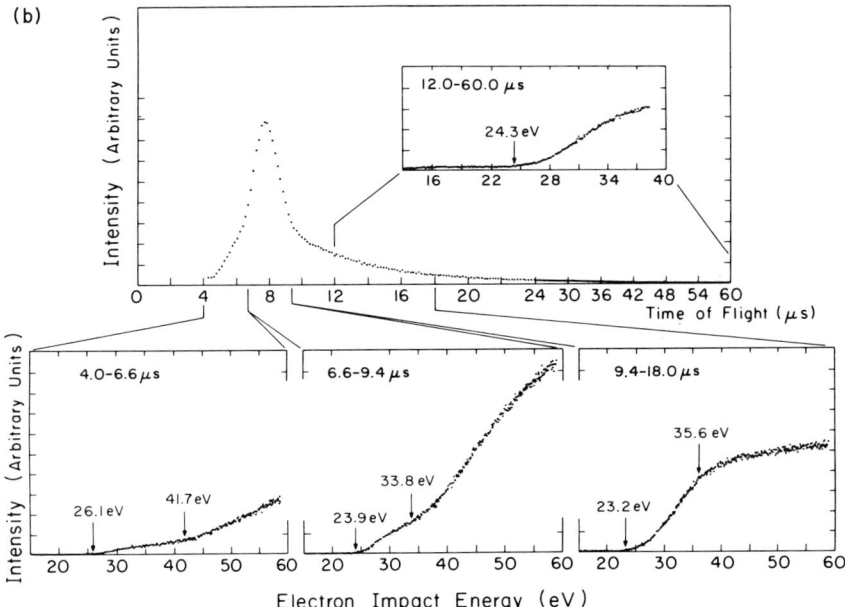

Fig. 6. (a) Time-of-flight distributions for atomic-hydrogen Rydberg atoms produced by electron-impact dissociation of ammonia. The energy of the incident electrons is shown to the far right of each curve. The location of the four fast HI(Ryd) features are indicated by the arrows. [From Carnahan et al. (1981).] (b) Onset excitation functions for several segments of the HI(Ryd) distributions produced from ammonia. The indicated structures in these curves are correlated to the HI(Ryd) TOF features in (a). [From Carnahan et al. (1981).]

TABLE I

The Dissociative Excitation of NH_3: H(2s) and H(Rydberg) Onset Energies, Kinetic-Energy Ranges, and Possible Dissociation Process

H* fragment	Onset energy (eV)	Approximate range of fragment kinetic energy (eV)	Dissociation products
2s-1	17.8 ± 0.8	0.1–2.3	$NH_2(\tilde{X}\,^2B_1) + H(3p)$
2s-2	21.8 ± 0.8	2.0–8.0	$NH_2(\tilde{X}\,^2B_1) + H(2s)$
2s-3	27.9 ± 1.4	1.0–3.8	
2s-4	35.4 ± 1.0	3.4–6.9	
2s-5	39.3 ± 1.5	6.0–19.0	
R-1	23.2 ± 0.7	0.4–10.0	$NH_2(\tilde{X}\,^2B_1) + H(Ryd)$
R-2	31.0 ± 2.0	1.3–4.5	$N(^4S^0) + 2H(1s) + H(Ryd)$
R-3	33.8 ± 0.9	3.5–8.2	
R-4	41.7 ± 1.0	8.0–23.5	

molecules the core-ion model has been reasonably successful. The large data base that exists in support of the core-ion model in contrast to the meager information on dissociation into atomic fragments belonging to the ground-state atomic configuration, which account for the bulk of the total dissociation cross section, is a natural consequence of limitations imposed by detector technology. High-Rydberg atoms and ions have large internal energies and can be readily detected by Auger detectors in the latter case and by field ionization techniques in the former. These detectors have yielded interesting information on the Rydberg atom distribution as a function of principal quantum number n, on the effective lifetimes of these long-lived states, and on their excitation and collision cross sections.

Although the total cross section for the dissociative excitation of high-Rydberg atoms is comparatively small ($<5 \times 10^{-18}$ cm^2), the excitation of the same Rydberg states by direct electron impact on atomic oxygen or nitrogen is believed to have a very large cross section ($\sim 10^{-16}$ cm^2) and may be an important factor in planetary airglows (Porter *et al.*, 1976; Jackman *et al.*, 1977; Zipf, 1982). Because of the difficulty in preparing suitable OI and NI sources for electron scattering work, the only direct knowledge of the actual physical properties of OI and NI high-Rydberg states formed in these processes has been derived from dissociative excitation studies.

III. Experimental Techniques

Although it has been known for at least a century that the dissociation of molecular species in a gaseous discharge produces spectra that are rich in atomic emission features, detailed studies of the underlying fundamental molecular processes are comparatively new (~ 15 years). A major impetus for pursuing this investigation arose from the challenge to interpret the rich spectra of planetary airglows (particularly in the vacuum-ultraviolet wavelength region) that were obtained by sounding rockets, satellites, and interplanetary spacecraft. This field has progressed so rapidly that it is easy to forget that there was no information at all on the VUV airglow spectra (~ 1050–1800 Å) of any planet until 1961 and that analogous EUV data down to 500 Å became available for the first time only in 1975.

Experimental work on dissociative excitation has evolved along several distinct but complementary lines: (1) the measurement of absolute emission cross sections and excitation functions using standard spectroscopic techniques; (2) the study of fragment atoms and molecules in metastable or high-lying Rydberg states using time-of-flight (TOF) techniques which are restricted to comparatively long-lived states ($\tau_r > 10^{-6}$ s) with term energies > 5 eV; (3) TOF experiments and related high-resolution linewidth measurements that concentrate on the kinetic-energy distribution of the fragment particles, on rotational temperature enhancements due to dissociative excitation, on the nonstatistical population of the fine-structure levels of an excited atom, and on the Doppler broadening exhibited by most of the radiation excited dissociatively; (4) studies aimed at using the dissociative excitation process as the basis of a practical method for calibrating optical systems absolutely and measuring lifetime and branching ratio values; (5) investigation of the chemistry of the short-lived excited states ($\tau \sim 10^{-7}$–10^{-9} s) produced dissociatively in optically thick or dense media where collisional deactivation competes effectively with radiation in depopulating the state; and (6) polarization studies. Each of these research areas has developed its own specialized techniques, which we will now briefly summarize, and though these endeavors have been quite prolific, it is still sobering to realize that the sum of the dissociative excitation cross section measured so far for any given species in general accounts for less than 20% of the known total dissociation cross section for that gas. It is only for molecular nitrogen that we have a nearly complete picture but, chiefly, as the result of serendipity and some insights from rocket-borne studies of the aurora borealis.

A. Optical Cross Section Measurements

Figure 7 shows schematically an example of an apparatus developed for absolute emission cross section measurements from 500 Å to ~ 1.3 μm. In this system, the target gas flows from an ultrahigh-vacuum gas-handling manifold into a small collision chamber situated inside a large rapidly pumped vacuum

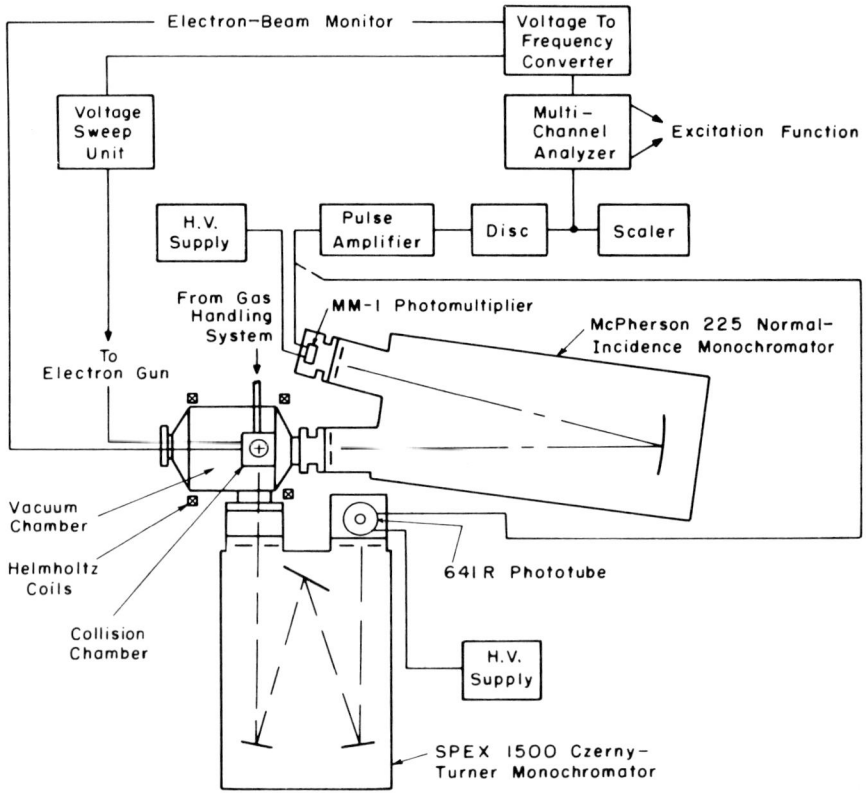

Fig. 7. Schematic diagram of a typical apparatus for excitation cross section measurements from 500 Å to 1.3 μm. [From Zipf and McLaughlin (1978).]

chamber. The all-metal collision chamber is carefully shielded to prevent extraneous electric fields from reaching the collision region, and its interior is coated with a dag-suspension to minimize the creation of electric fields due to charging spots. Achieving this field-free objective is facilitated by the use of high-speed cryogenic UHV pumps which can reduce the base pressures of the system to $\sim 5 \times 10^{-9}$ Torr and are oil free. Minimizing oil contamination is also very important for the protection of the VUV optical instrumentation, whose gratings and mirrors can be seriously compromised by oil films polymerized by ultraviolet radiation.

A collimated electron beam passes through the gas in the collision chamber, producing a variety of atomic and molecular states through dissociative excitation. The prompt radiation which results from the relaxation of these states is observed by a one-meter, normal-incidence vacuum monochromator for studies in the 400–1800 Å wavelength range (McPherson model 225) or by a 3/4-meter Czerny-Turner monochromator in the 1800 Å–1.2 μm region (Spex model 1500); both instruments view the interaction region simultaneously. This arrangement facilitates atomic branching ratio measurement as a study in its own right or as a convenient technique for determining the spectral sensitivity of the apparatus. Extreme ultraviolet (EUV) photons transmitted by the normal-incidence monochromator are detected by a windowless cesium-iodide-coated photocathode mounted behind the exit slit. Photoelectrons ejected from this surface, which has been formed into a polygon for enhanced efficiency, are collected by a Johnston Laboratories MM-1 electron multiplier. This detection system has an exceptionally high quantum efficiency in the EUV range (50–90%) and a minimal background counting rate of only a few counts per minute (Sampson, 1967). When this detector is used in conjunction with osmium-coated diffraction gratings, dissociative excitation cross sections as small as 10^{-24} cm^2 can be measured throughout the vacuum-ultraviolet wavelength region. In the visible and near-infrared ranges [1800–9000 Å] photomultiplier tubes with GaAs photocathodes permit similar cross section measurements when the PMT is suitably cooled. However, in order to achieve such sensitivities considerable attention has to be focused on reducing scattered light in the collision chamber and in the monochromator, and on choosing lenses, windows, and blocking filters that do not fluoresce or exhibit photoluminescence. Working at such low light levels also requires the extensive use of pulse counting techniques and computer assisted data acquisition. The routine use of coherent summing and fast Fourier-transform filtering now permit the detection of emission features that had been observed previously only in astrophysical sources or under unusual conditions in cometary or planetary atmospheres. An example of these advances is given in Fig. 8, which shows the $\lambda_{2143.55}$ and $\lambda_{2139.68}$ doublet emitted by metastable $N^+(^5S)$ ions. The $N^+(^5S)$ ions, which were produced by electron-impact dissociation of N_2,

$$e + N_2 \longrightarrow N^+(^5S) + N + 2e, \qquad (10)$$

are difficult to observe in the laboratory because of their comparatively long radiative lifetime [4 ms; Hibbert and Bates (1981)], their efficient quenching by the N_2 in the collision region (Knight, 1982), and their rapid loss from the field of view due to the large kinetic energy they gain in the dissociative ionization process. The data shown in Fig. 8 required over 700 hours of integration in order to achieve the necessary statistics and were obtained with 0.5-Å resolution (Erdman et al., 1981). The two major features that dominate Fig. 8 are the N_2 Lyman–Birge–Hopfield (5, 14) and (6, 15) bands. Under the conditions of the experiment, the forbidden NII doublet lines have an apparent cross section of $\sim 3 \times 10^{-21}$ cm^2. It is interesting to note that the

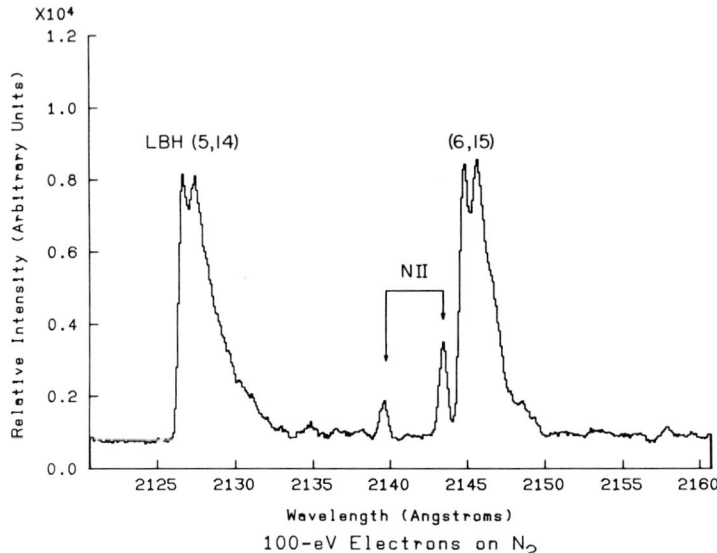

Fig. 8. First laboratory spectrum showing the $\lambda 2143.55$ and 2139.68 Å doublet emitted by metastable $N^+(^5S)$ ions produced by dissociative ionization of N_2 by electron impact. Over 700 h of integration were required to obtain these data. The apparent NII doublet cross section is $\sim 3 \times 10^{-21}$ cm^2. [From Erdman et al. (1981).] The small apparent cross section value is an artifact of the effects of the high translational energy and long lifetime of the $N^+(^5S)$ state. The actual $N^+(^5S)$ cross section appears to have a peak value of $\sim 3.8 \times 10^{-18}$ cm^2 at 175 eV.

$\lambda_{2143.55}/\lambda_{2139.68}$ intensity ratio is not 5/3 as predicted by LS coupling rules (Allen, 1963) but is in good agreement with the ratio of 107/48 predicted by the quantum-mechanical calculation of Hibbert and Bates (1981).

1. Spectral Purity Problems

Although the need to study weak emission features was a major factor in spurring on the instrumentation developments noted above, these welcome gains have now opened up several new areas for investigation. Particular high-spectral-resolution monochromator and interferometer studies (0.01–0.04 Å) on the linewidths and fine-structure populations of dissociatively excited atomic states and on rotational anomalies in molecular band systems in which the transition probabilities vary markedly with the rotation quantum number or which exhibit complex behavior in optically thick media, have benefited dramatically from these technological advances. Early cross section work frequently compromised spectral purity in the search for sensitivity. Thus, it was inevitable, as we progressed technologically, that a number of dissociative excitation emission features that are widely used today as cross section standards would be found, in fact, to be blends of a number of features.

4 Dissociation of Molecules by Electron Impact

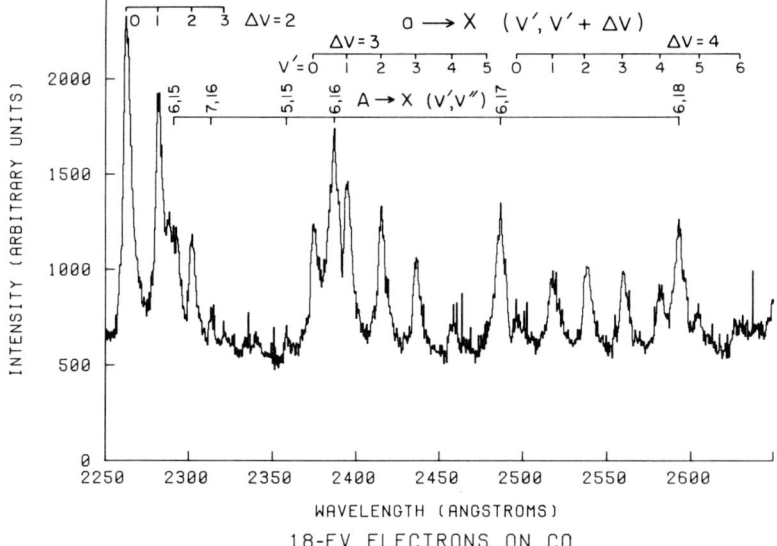

Fig. 9. UV spectrum of CO excited by the impact of 18-eV electrons on carbon monoxide. The data were obtained at a resolution of 2 Å. The complex character of the $\lambda 2389$ Å emission feature, which has been used previously to estimate the magnitude of the total cross section for the excitation of $CO(a\,^3\Pi)$ by electron-impact dissociation of CO_2, can be seen. [From Erdman and Zipf (1983).]

Figure 9 illustrates this problem, which, unfortunately, can lead to order of magnitude errors, In this example, the excitation of the forbidden Cameron electron bands of $CO(a\,^3\Pi \to X\,^1\Sigma)$ by electron impact on CO,

$$e + CO \longrightarrow CO(a\,^3\Pi) + e, \qquad (11)$$

is shown. The medium-resolution spectrum ($\Delta\lambda \sim 2$ Å) shows that the (4, 0) Cameron band, which was used by Ajello (1971a,b) to normalize the total cross section for the entire band system, is actually a complex blend of Cameron band and CO fourth positive $[CO(A\,^1\Pi \to X\,^1\Sigma)]$ features. The higher-spectral-purity data require that the published cross section for process (11) be reduced by a factor of ~ 8 (see Erdman and Zipf, 1983). Since process (11) was used by Ajello in estimating the cross section for the dissociative excitation of the Cameron bands by electron impact on CO_2,

$$e + CO_2 \longrightarrow CO(a\,^3\Pi) + O + e, \qquad (12)$$

the error propagates. Thus, much of the apparent discrepancy between the TOF and optical studies of process (12) (Wells *et al.*, 1972) has been caused by this spectral-purity problem.

Unfortunately, similar problems affect several of the most important calibration standards that have been traditionally used to normalize dissociative excitation cross sections in the vacuum-ultraviolet wavelength region. Perhaps the most important historical example of this problem is evaluation of the cross section for the dissociative excitation of HI Lyman alpha ($\lambda 1215.17$ Å),

$$e^- + H_2 \longrightarrow H(2p) + H + e^- \tag{13}$$

$$H(2p) \longrightarrow H(1s) + h\nu \quad (\lambda 1215.17 \text{ Å}) \tag{14}$$

The earliest work on this process used O_2-filtered photometers as detectors and was compromised by contamination due to the H_2 Werner and Lyman band emission that occurred in the numerous O_2 transmission windows.

Various attempts have been made to correct the countable-ultraviolet (CUV) data. For 100-eV electrons incident on H_2, approximately $80 \pm 10\%$ of the radiation passed by the LiF-O_2 filter consists of Lyman-alpha photons (Mumma and Zipf, 1971a). Unfortunately, even at the resolution used by Mumma and Zipf to separate Lyman alpha from the nearby $\lambda 1217.4$ Å rotational line of the (6, 6) Werner band (c $^1\Pi_u - X\,^1\Sigma_g^+$) the actual spectrum of H_2 in the vicinity of Lyman alpha is considerably more complex than supposed in their model (cf. Fig. 10), with a number of rotational lines

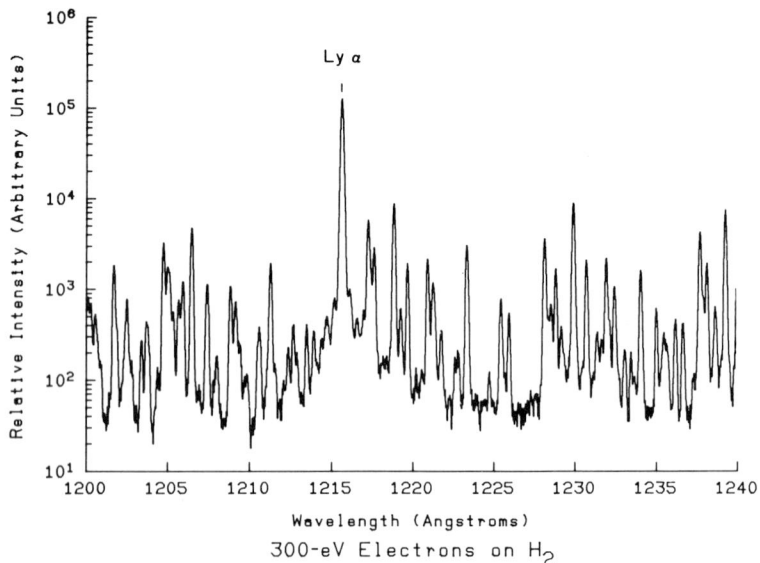

Fig. 10. VUV spectrum of H_2 excited by the impact of 300-eV electrons. The data were obtained at a resolution of 0.1 Å. The HI Lyman-alpha line ($\lambda 1215.17$ Å) is overlapped by numerous rotational lines belonging to the Werner, Lyman, and Rydberg band systems of hydrogen as well as by continuum radiation from the transition c $^1\Pi_u \to X\,^1\Sigma_g^+$ [From Erdman and Zipf (1983).]

4 Dissociation of Molecules by Electron Impact

belonging to the Werner, Lyman, and H_2 Rydberg band systems coincident with Lyman alpha and not resolved even at 0.1 Å resolution. Furthermore, the Werner band system, which has been used widely to determine the relative spectral response of the optical systems in the 1000–1300 Å region, has now been found to exhibit intensity anomalies and is overlapped by other hydrogen band systems that compromise its use for spectral calibration work unless the Werner band data are obtained with comparatively high resolution ($\Delta\lambda \sim 0.1$ Å). Comparative studies on the dissociative excitation of H_2, HD, and D_2 using oxygen filter technology are also subject to significant systematic errors due, in part, to the striking differences in the HD and D_2 rotational distributions in the near vicinity of Lyman alpha. An additional difficulty is now posed as the result of high-resolution ($\Delta\lambda = 0.04$ Å) measurements of the linewidth of $\lambda 1215.7$ Å emission produced dissociatively, which shows that this feature is substantially Doppler broadened (~ 0.3 Å). Since the transmission of the LiF-O_2 filter varies by nearly a factor of 1.6 over this bandwidth, the comparison of the direct versus dissociative excitation of Lyman alpha is considerably more complex than previously imagined. An additional complication is caused by the presence of continuum radiation underneath the Lyman alpha feature due to the $c\,^1\Pi_u \to X\,^1\Sigma_g^+$ transition (Erdman and Zipf, 1983). Thus, further laboratory work is needed on the Lyman-alpha dissociative excitation in order to eliminate some of these sources of error. Fortunately, with the availability of synchrotron radiation sources for VUV and EUV calibration work, highly accurate cross sections ($\pm 5\%$) over the wavelength interval from 500 to 2000 Å should be available in the next few years, and it is likely that methane rather than H_2 will be the most useful traget gas for calibration work in the λ_{900}–λ_{2000} region. These advances will eliminate much of the confusion that presently exists because of a lack of adequate standards.

There is one additional calibration problem that affects the cross section values for the numerous vacuum ultraviolet emission features that can be excited dissociatively by electron impact (see Tables II and V). The absolute values of these cross sections are traceable ultimately to the cross section σ_A for the excitation of Lyman alpha by electron impact on atomic hydrogen:

$$e^- + H \longrightarrow H(2p) + e^-$$
$$H(2p) \longrightarrow H(1s) + h\nu\;(\lambda 1215.17\,\text{Å})$$

This process has been studied theoretically by many workers, and the results of calculations based on the Born approximation have been used to normalize relative excitation functions obtained experimentally.

For many years the laboratory results of Long et al. (1968) have served as the standard. Based on the theoretical work of Moiseiwitsch and Smith (1968), they found that $\sigma_A = 6.0 \times 10^{-17}$ cm^2 ($\pm 3\%$) at 100 eV. Mumma and Zipf

(1971) subsequently used this result and improved Lyman alpha spectral data to deduce a revised cross section for the dissociative excitation of $\lambda 1215.17$ Å of $\sigma_D = 1.2 \times 10^{-17}$ cm^2 ($\pm 11\%$).

More recent experimental (Williams and Willis, 1974; Williams, 1976) and theoretical work (Morgan et al., 1977) indicates that the Born approximation values may underestimate the actual cross section for σ_A by as much as 20%. Should further work verify this conclusion, then at 100 eV, $\sigma_A = 7.2 \times 10^{-17}$ cm^2 and $\sigma_D = 1.44 \times 10^{-17}$ cm^2, and the body of VUV dissociative excitation cross sections provided in Tables III and V would have to be scaled upward in magnitude by a factor of 1.2.

B. Time-of-Flight Measurements

The measurement of the translational energy of the fragment atoms formed by dissociative excitation provides fundamental information on the repulsive and predissociated states of the parent molecule. The kinetic-energy data are also needed in a wide range of applied problems as well as in optical cross section work in order to make field-of-view corrections. Although a large body of kinetic data has been assembled from time-of-flight (TOF) experiments, the results pertain to a fairly limited class of metastable or pseudometastable excited states, with internal energies generally >5 eV and effective lifetimes $>10^{-6}$ s. This narrow focus is due chiefly to detector limitations and transit-time constraints. Little is actually known from direct measurement about the kinetic-energy distribution of the highly excited, short-lived atomic states responsible for most of the vacuum-ultraviolet radiation that dominates the spectrum of gaseous discharges or planetary airglows. Very high-resolution optical studies at these short wavelengths is a research field still in its infancy.

As we have noted in Section II, similar detector problems plague the TOF study of the chemically important metastable states (e.g., ^{1}D, ^{1}S, ^{2}D, ^{2}P) that belong to the ground-state configuration of such atoms as OI, NI, CI, and SI and that are likely to be excited with considerable efficiency in electron-impact dissociation of a wide variety of commonplace molecules.

Fortunately, limited success has been achieved for atomic oxygen fragments by exploiting associative ionization processes, e.g.,

$$S_m + O \longrightarrow S_mO^+ + e^- \tag{15}$$

with mass spectrometric techniques (Stone et al., 1975), and for the metastable nitrogen species [N(^{2}D), N(^{2}P)] that dominate the odd-nitrogen chemistry of the earth's atmosphere, by augmenting laboratory data with rocket-borne studies of these species in the aurora borealis (Zipf et al., 1980). Further advances on these very important species await major improvements in detector technology, with multiphoton techniques appearing quite promising (Baravian et al., 1982).

4   Dissociation of Molecules by Electron Impact

In spite of these limitations, TOF measurements nonetheless provide very useful insights into the dissociative excitation process. A schematic diagram of a representative time-of-flight apparatus is shown in Fig. 11. The basic elements include a periodically pulsed source for metastable production, a metastable detector, and a digital processing system for data acquisition and experiment control. The principle of the TOF experiment is comparatively simple. However, there are numerous practical problems that make absolute cross section work using this technique very difficult. These problems will be discussed in more detail below. In the simplest arrangement, a geometrically

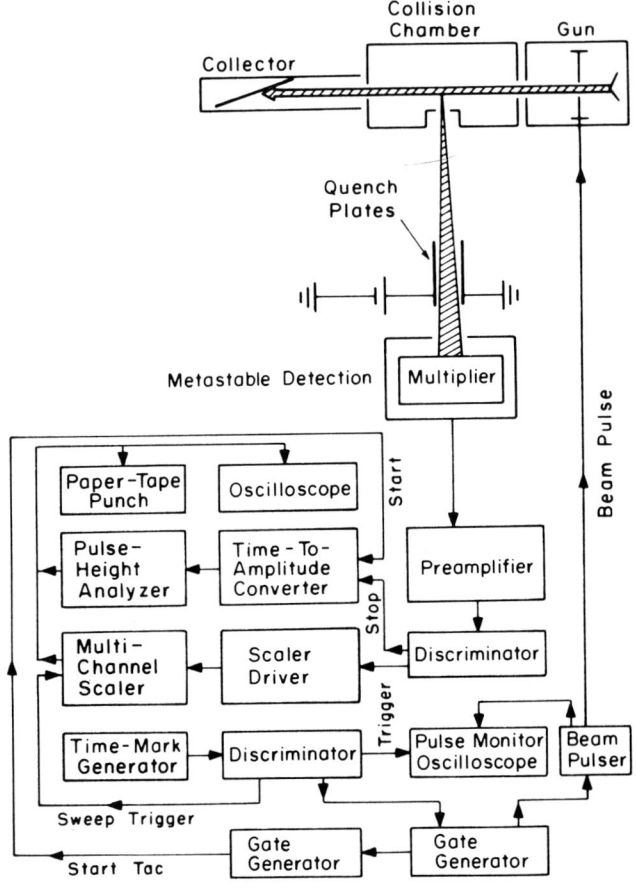

Fig. 11. Schematic diagram of a representative time-of-flight apparatus. [From Borst and Zipf (1971a, b).]

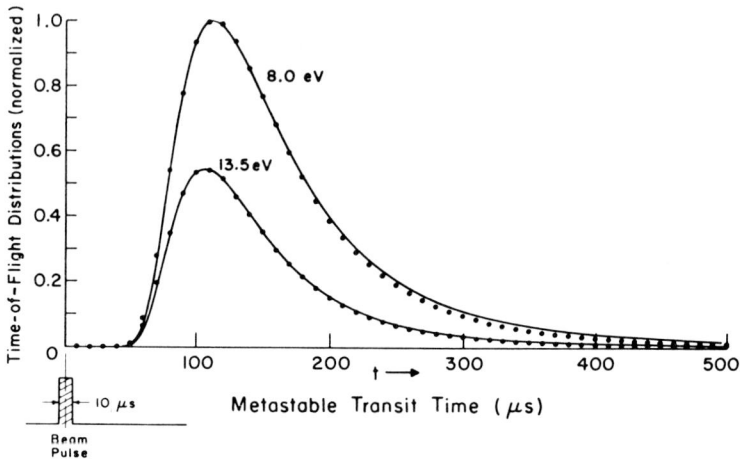

Fig. 12. Normalized time-of-flight distributions of metastable N_2 molecules. The solid dots represent experimental points. The solid lines represent Maxwellian distributions (without decay for the 8-eV data and with decay for the 13.5-eV data) fitted to the data. The 8.0-eV data contain only metastables in the A $^3\Sigma_u^+$ state and thus show no metastable decay for the purposes of this experiment. The 13.5-eV data contain metastables in both the A $^3\Sigma_u^+$ and a $^1\Pi_g$ states. The normalization for the 13.5-eV curve is such that without decay of the a $^1\Pi_g$ state, the maximum of the curve would coincide with that for the 8.0-eV curve. It is seen that the decay of a $^1\Pi_g$ metastables shifts the maximum toward shorter transit times and causes a noticeable reduction in the amplitude of TOF curve. [From Borst and Zipf (1971a).]

narrow, submicrosecond pulse of energetic electrons forms metastable fragments at the center of the diffuse-gas collision chamber. The metastable atoms then diffuse through a small slit in the side of the collision chamber and ultimately are incident on a Cu–Be Auger detector positioned at 90° to the electron beam. The transit time of the excited fragments is measured using digital, coherent summing techniques, and the TOF data are gathered at several collision cell to detector distances. A comparison of the latter results often provides information on radiative lifetimes and on the collisional deactivation of the excited species by the background gas. Figure 12 shows a typical time-of-flight spectrum for metastable N_2 molecules, while Fig. 13 illustrates the technique for extracting the lifetime of the $N_2(a\ ^1\Pi g)$ state from such data (Borst and Zipf, 1971a).

The TOF spectra shown in Fig. 12 are directly related to the Maxwellian distribution at 300 K. Because the excited states formed under these conditions have comparatively small velocities, their transit times can be measured quite accurately. However, when the fragments are produced by dissociative excitation, excited atoms with very high velocities (see Fig. 2) are

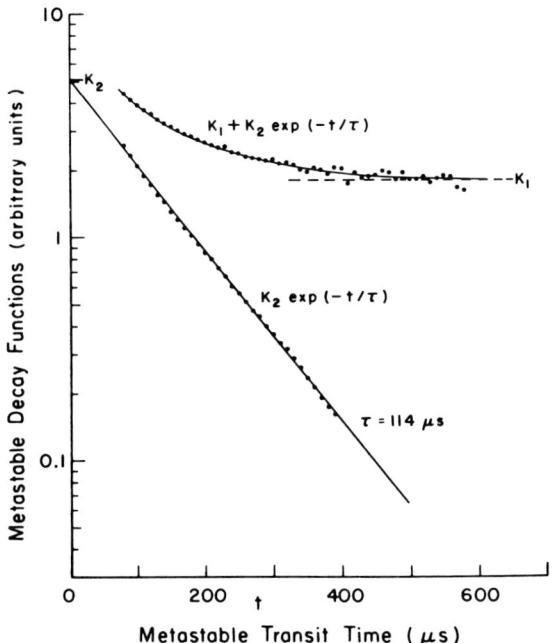

Fig. 13. Metastable N_2 decay function at 13.5 eV. After subtraction of the constant K_1 representing the contribution of the A $^3\Sigma_u^+$ state, one obtains a single metastable decay with $\tau = 114$ μs due to the a $^1\Pi_g$ state. The solid curve through the data was calculated. [From Borst and Zipf (1971a).]

frequently encountered. Extracting the kinetic-energy spectra and potential-energy curve information by the procedures outlined below is at best a primitive art, and the experimental difficulties that confront TOF workers are numerous; these procedures and the most vexing problems are also summarized below.

1. Kinetic-Energy Spectra

An important objective of the time-of-flight experiment is to derive the kinetic-energy spectrum $D(\varepsilon)$ from such data. Under the simplest of circumstances, where the thermal energy of the parent molecule is ignored, the analysis begins by noting that the kinetic energy ε of an excited species A* is inversely proportional to the square of the arrival time t so that

$$|d\varepsilon| \propto t^{-3}|dt|. \tag{16}$$

Then $T(t)\,dt$, the number of excited atoms arriving at the detector between time t and $t + dt$, can be expressed as

$$T_A(t)|dt| = T_A(t)t^3|d\varepsilon|, \tag{17}$$

ignoring a constant factor. Thus the kinetic-energy spectrum is given by

$$D_A(\varepsilon_A) = t^3 T_A(t). \tag{18}$$

From conservation of momentum, the kinetic-energy relationship between fragments A and B with masses m_A and m_B is

$$\varepsilon_B = (m_A/m_B)\varepsilon_A. \tag{19}$$

Thus it follows that the kinetic-energy spectrum for fragment B, $D_B(\varepsilon_B)$, is

$$D_B(\varepsilon_B) = D_A(m_B/m_A)\varepsilon_B, \tag{20}$$

and the total-released-kinetic-energy spectrum $F(E)$ is

$$F(E) = D_A[m_B/(m_A + m_B)]E, \tag{21}$$

where $E = \varepsilon_A + \varepsilon_B$. The kinetic-energy scale shown on the top of Fig. 2 was derived in this manner.

2. Repulsive-Potential-Energy Curves

As can be seen in Fig. 1, the spectrum of the total released kinetic energy $F(E)$ in a dissociation process is determined by the nature of the repulsive energy curve in the Franck–Condon region. Conversely, given the total released kinetic-energy spectrum, it is possible to construct the potential-energy curve in the Franck–Condon region. In such a model, it is assumed that the wave function for the repulsive state can be closely approximated by a delta function and that the Franck–Condon principle holds near threshold. The value of the asymptotic limit E_D of the potential curve is the excitation-function threshold less the minimum kinetic energy E_{min}.

The mathematical prescription for the construction of the repulsive curve from the ground-state probability function $|\psi_0|^2$ and the total-released-kinetic-energy spectra $F(E)$ [both normalized to a maximum height of 1; $|\psi_0(R_0)|^2 = F(E) = 1$] follows.

When the probability function

$$|\psi_0(R_{1,2})|^2 = \exp[-\alpha(R_{1,2} - R_0)^2] \tag{22}$$

is equal to the energy spectra function $F(E_{1,2})$ then

$$V(R_{1,2}) = E_D + E_{1,2}, \tag{23}$$

where R is the internuclear distance, V the potential energy, and E the total kinetic energy released. It is assumed here that the vibrational motion of the molecule can be described as a harmonic oscillator and that the molecule is in its ground vibrational state. Equation (23) is satisfied for two values of R, R_1, and R_2, and it can be inverted to yield

$$R(V) = \begin{cases} R_0 + \left|\left(\dfrac{\ln F(V-E_A)}{-\alpha}\right)^{1/2}\right| & \text{for} \quad V < V(R_0) = E_A + E, \\ R_0 - \left|\left(\dfrac{\ln F(V-E_A)}{-\alpha}\right)^{1/2}\right| & \text{for} \quad V > V(R_0), \end{cases} \quad (24)$$

which is the analytical representation of this numerical process.

The $Q\ ^3\Pi_u$ potential surface shown in Fig. 1 was constructed using these techniques.

3. Practical Techniques

Time-of-flight experiments are vulnerable to a number of uncertainties which arise for the most part from limitations that are encountered in the production and detection of the metastable fragments. In a typical TOF apparatus, metastable atoms are produced dissociatively by electron impact in the energy range from threshold to about 300 eV. Because the signal levels in these experiments are often meager, there is an inevitable compromise between electron-beam quality (spatial and energy resolution) and beam current. Under the most favorable circumstances, magnetically collimated electron beams with an 0.3-eV FWHM and small diameter (1 mm) can be achieved at relatively high beam currents (1 μA). Unfortunately, these experiments are quite sensitive to the effects of secondary processes involving low-energy electrons trapped in the beam by the collimating magnetic field. Numerous optical cross section studies have been marred in the past by this problem, which can induce order of magnitude errors in excitation functions obtained under such circumstances.

The need for computer-assisted data acquisition in these experiments is due in part to the small cross sections involved and to the low duty cycles that are unavoidable if pileup effects are to be avoided. For example, electron-beam pulses 0.5 μs wide with a repetition rate of $\sim 10^3$ pulses/s are typical. The duty cycle is thus about 0.05%, and because of the low signal levels (1 count/s), data runs of 600–700 hours are sometimes required. Thus, considerable effort must be expended to achieve a very low-noise laboratory environment.

A variety of approaches have been used to detect the metastable fragments produced in these experiments. The simplest detector schemes use electron multipliers (e.g., Johnston Laboratories Cu–Be MM-1 mesh multipliers) and rely on the Auger effect for the production of the initial electron(s). The secondary-electron yield γ_m of these surface detectors for different metastable species is a matter of considerable controversy, with assessments of γ_m

differing by as much as a factor of 5 (Borst, 1971; Dunning *et al.*, 1975; Rundel *et al.*, 1973). Thus, the value of these detectors as absolute devices leaves much to be desired. Fortunately, independent optical measurements of the total cross section for the dissociative excitation of the metastable $O(^5S)$ state by electron impact on O_2 now exist with a moderately small absolute probable error ($\pm 15\%$) (see Section II). Thus this process can be used as an absolute standard in TOF work. Cross section values for other metastable species [e.g., $CO(a\ ^3\Pi)$] are also being refined in optical and laser experiments so that the calibration of surface detectors should improve markedly over the next few years.

An important limitation of current TOF detectors is their lack of specificity; that is, they cannot be used to distinguish between excited molecular species in different rotational and vibrational quantum states. Fortunately, with the advent of high-powered tunable dye lasers, multiphoton and laser-induced photofluorescence techniques suggest a promising future for studying low-energy metastable fragments and for coincidence experiments in which a single potential-energy surface can be selected for study.

4. Experimental Uncertainties

The possibility of constructing models of the repulsive potential surfaces for the intermediate states excited in the electron-impact dissociation of a molecule or of distinguishing different dissociation channels based on threshold measurements requires TOF data with high translational energy resolution. Figure 6, which shows TOF data for the hydrogen Rydberg atoms produced in the dissociative excitation of ammonia, illustrates the kinds of threshold results that can be obtained under favorable circumstances from data with adequate signal-to-noise.

The following paragraphs discuss several ways in which this highly desirable resolution, unfortunately, can be degraded.

a. Thermal-Energy Spread

The effect of the thermal motion of the target molecule is to add or subtract a velocity component v_{th} from the velocity of the fragment v_f. In terms of a center-of-mass coordinate transformation, the situation can be viewed with the fragment having a velocity v_f in the center-of-mass system, which itself is moving with velocities between $\pm v_{th}$ with respect to the lab system. The resultant velocity spread in the lab system results in a fragment energy spread given by

$$\Delta E_f = 2(2E_f E_{th})^{1/2}, \tag{25}$$

where E_f is the fragment energy and E_{th} is the thermal energy of the parent molecule. For a diffuse N_2 source at 300 K, we have

$$\Delta E_f = 0.45 E_f^{1/2}. \tag{26}$$

For a fragment kinetic energy of 5 eV, this corresponds to a spread of 1 eV. Thus, the spread in thermal energy is magnified by the fragment kinetic energy and can result in quite poor energy resolution.

b. Recoil Effects

When the electron energy is above threshold, a small fraction of its momentum is imparted to the molecule in the excitation process. The recoil velocity in the worst case is directed either toward or away from the detector and thus leads to a spread in fragment energy as described above. For heavier molecules such as N_2, this is not a serious effect. For example, for a 5-eV fragment and an excess electron energy of 20 eV, we would have a spread in the fragment energy of only 0.1 eV.

c. Beam Pulse Width

The beam pulse width introduces a spread ΔE_f in E_f given by

$$\Delta E_f = 2 E_f \Delta t / t, \tag{27}$$

where Δt is the beam pulse width and t the arrival time at the detector. For a 0.5-μs pulse width and a 5-eV fragment, the spread corresponds to $\Delta E_f \sim 1$ eV. This error manifests itself as a distortion in the high-energy part of the kinetic-energy distribution and as an ambiguity when using narrow time-of-flight windows at short arrival times, e.g., in measuring excitation functions for fast fragments.

d. Timing and Sampling Errors

In the ideal experiment, TOF data would be obtained by using exceedingly narrow electron pulses and high sampling resolution in the data acquisition system. Unfortunately, the low counting rates encountered under typical TOF conditions necessitate compromise choices that introduce possible offset errors in establishing the zero-time base for the transit-time measurement and smearing errors because of the finite channel widths (typically 0.5–2.0 μs) of the memory units used to coherently sum the time-of-flight data. The offset errors can generally be reduced to less than 0.1 μs, which would correspond to a systematic error of 0.2 eV in the example mentioned previously.

C. Linewidth and Collisional Quenching Studies

In many applied problems, dissociative excitation is a major source of the atomic line radiation that can be observed remotely. Measurement of the brightness of these emission features can be used in principle as a diagnostic technique for studying energy deposition in a radiating gas or for deducing information on the electron energy flux distribution. This application of dissociative excitation is widely used in the study of auroral substorms which have now been observed in the atmospheres of Earth, Jupiter, Saturn, and Uranus. One important realization that has emerged from these geophysical studies and has direct impact on future work in the laboratory on dissociative excitation is that most planetary atmospheres are optically thick for the EUV and VUV atomic transitions excited dissociatively and that it is crucial to have detailed information on the actual emission line profiles so that the source function adopted in the radiation entrapment calculation, which are used to analyze the observational data, will be realistic. Although it has been known for some time by the laboratory community that the emission features produced dissociatively will be highly Doppler broadened, this fact has been largely overlooked in a number of entrapment models which assume, for example, that 70% of the nitrogen atoms formed in the $(2p^2\ 3s\ ^4P-2p^3\ ^4S^0,\ \lambda 1200\ \text{Å})$ resonance state by the dissociative excitation of N_2 have thermal energies (Meier et al., 1980; Meier et al., 1982). The early work of Poland and Lawrence (1973), on the $\lambda 1304$ Å resonance line of atomic oxygen when excited by electron-impact dissociation of CO_2 and O_2 made that supposition seem unlikely, but until recently there have been no direct high-resolution studies on the linewidth of the $\lambda 1200$ Å multiplet. Time-of-flight studies on the $\lambda 1356$ Å radiation emitted by $OI(^5S)$ atoms also indicated Doppler widths equivalent to a temperature of 60,000 K [see Fig. 14; Zipf and Wells (1980)], while medium-resolution $\Delta \lambda \sim 0.1$ Å on the hydrogen Balmer-alpha line profile revealed a remarkably broad emission feature (Freund et al., 1976). This can be seen in Fig. 3, which shows that this complex profile consists of a narrow but still broadened (~ 0.34 Å) central peak along with a wide pedestal (~ 1.7 Å), indicating that the molecule is dissociating via a variety of repulsive curves (Freund et al., 1976). Aside from a few transitions, virtually nothing is known about the linewidths of most emission features excited by electron-impact dissociation.

The present need for high-quality linewidth data can ultimately be satisfied from about 2000 Å to 30 µm through the use of Fabry–Perot or Michelson interferometers. The availability of new multidielectric coatings and large-diameter plates with a flatness figure of $\lambda/200$ makes it possible to achieve a working finesse of 30–50, while piezoelectric scanning systems and array detectors permit the full use of digital coherent summing techniques. Such

4 Dissociation of Molecules by Electron Impact

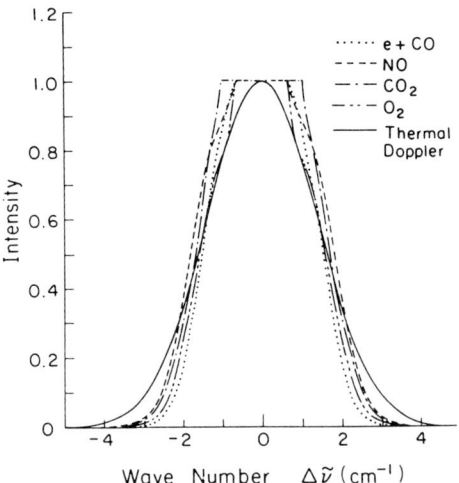

Fig. 14. A comparison of the Doppler line shapes for the OI(^{5}S–^{3}P, λ1356 Å) transition excited by electron-impact dissociation of a variety of target gases with $E = 100$ eV and λ1356 Å. A thermal Doppler line shape is plotted for 60,000 K as a reference. [From Zipf and Wells (1980).]

interferometers can thus achieve low-light-level sensitivities comparable to typical laboratory monochromators but with much higher resolution. These elegant systems are in widespread use for field observations of airglow phenomena but have only recently been pressed into service for dissociative excitation studies in the laboratory.

In addition to the linewidth measurements, high-resolution work is also needed to measure the intensities of individual multiplet line components. Because these components are not necessarily statistically populated and may vary with the energy of the incident electron, this information is needed to correctly characterize the dissociative excitation source function in optically thick media. Figure 15, which shows that the λ1200 Å multiplet components are not statistically populated and do, in fact, vary with bombardment energy, illustrates this point. Perhaps the best known example of this problem is in the dissociative excitation of H_2 where the np sublevels, with the exception of the 2p state, are virtually unpopulated (Mohlmann et al., 1976).

IV. Measurements

A large body of cross section data now exists on the dissociative excitation by electron impact of a variety of aeronomically important diatomic and

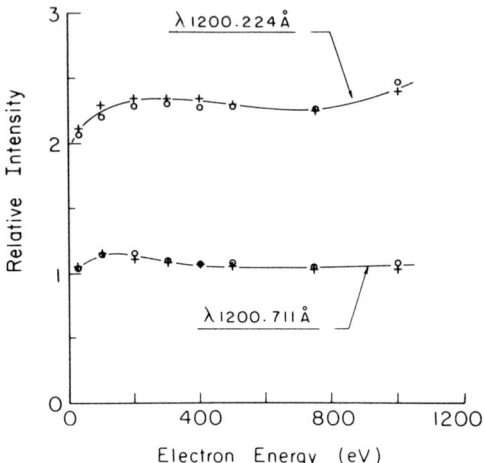

Fig. 15. Relative intensities of the NI $\lambda 1200$ Å triplet components as a function of the electron-impact energy. The target gas is N_2. The $\lambda 1199.55$ Å line intensity has been normalized to a value of 3. The anticipated 3:2:1 intensity ratio based on LS coupling is observed only near threshold (Erdman and Zipf, 1983).

polyatomic molecules. The list of light molecules that have been studied extensively includes H_2, HD, D_2, H_2O, N_2, NO, NO_2, N_2O, NH_3, O_2, CO, CO_2, SO_2, CH_4, and other low-mass hydrocarbons. Very little is known about the excited atomic states that are formed when much heavier molecules are dissociated or when the target gas is a metastable molecular species. For many of the molecules mentioned above, data exist on the absolute emission cross sections for at least the principal spectral features excited dissociatively and representative kinetic-energy spectra are available in some cases for an incident electron energy range from threshold to as high as 6 keV. However, a comprehensive set of cross sections which would permit fairly complete, quantitative modeling of the key dissociation processes exists for only two gases, O_2 and N_2. The contrasts between these two species, which use a variety of dissociation processes but in rather different proportions, are quite illuminating. They are also the only gases for which extensive data exist on the cross sections for electron-impact excitation of atomic nitrogen and oxygen to the same manifold of states. The comparison between the variation of the cross section versus principal quantum number and the differences between the energy dependence of the excitation functions for the dissociative and direct excitation is also interesting and has become the basis for a very sensitive optical mass spectrometer (Zipf et al., 1973). Similar comparisons are possible for some molecular species [e.g., $CO(A\,^1\Pi)$, $CO(a\,^3\Pi)$, $N_2(A\,^3\Sigma_u^+)$] which can be formed in the dissociative excitation of polyatomic targets, and in these cases quite striking differences in the degree of vibrational and rotational excitation are often observed when compared with the direct excitation of these species by electron impact on CO and N_2, respectively.

In this section we focus on gases of major importance in the terrestrial atmosphere, N_2, O_2, CO, and CO_2. Reference to studies of dissociation by electron impact on other molecules is given in Section IV.D.

A. Molecular Nitrogen

Traditionally, chemists have regarded nitrogen as a particulary difficult gas to dissociate. This behavior is attributed, in part, to the large N_2 bond energy (9.76 eV) and to the absence of repulsive potential surfaces that can be reached in thermochemical experiments involving only ground-state N_2 molecules or that can be excited by electron impact at the comparatively low electron temperatures ($T_e \sim 1$ eV) that characterize arcs and gas discharges. This characteristic of nitrogen has impeded the commercial synthesis of odd-nitrogen compounds by plasma chemistry techniques. Thus, in spite of considerable effort to find an alternative, the Haber process, which uses a catalyst in breaking the N_2 bond, is still the dominant method used in fertilizer and munitions production. This same problem is also well known to afterglow workers, who have discovered empirically that as the purity of the N_2 gas is improved, the degree of N_2 dissociation that can be achieved in discharge sources decreases precipitously. Trace amounts of impurities (e.g., O_2) appear to play a major role in enhancing N-atom production in these sources, but little is known about the details of the mechanism. Thus, the cumulative experience of plasma chemists is that molecular nitrogen is difficult to dissociate and that fixed nitrogen, [NO], can be produced by N_2 dissociation with an efficiency of only $\sim 8\%$ under very favorable discharge conditions (Venugopalan, 1971).

A rather different view of the nitrogen dissociation problem has emerged now from recent dissociative excitation experiments involving both electron-impact and EUV photoabsorption techniques (Zipf and McLaughlin, 1978). The results of these studies indicate that N_2 can in fact be dissociated quite easily into chemically important atomic states and can even be ionized by inelastic collision with electrons having energies less than the $N_2{}^+$ ionization potential (15.6 eV). These new insights have emerged from high-resolution inelastic electron scattering studies and complementary optical measurements, which have demonstrated the important role played in the nitrogen dissociation by the family of $^1\Sigma_u^+$ and $^1\Pi_u$ states lying in the energy range 11.5–40 eV. These results make nitrogen one of the few molecules for which there is an essentially complete, quantitative model that accounts for the key dissociation processes in detail. In this connection, it is interesting to note that the odd-nitrogen chemistry encountered in planetary atmospheres makes effective use of the properties of the N_2 singlet manifold which permit N atoms to be formed efficiently at a minimum energy cost (Rusch and Gerard, 1980). These states are also implicated in the formation of NO in coronal discharges and in neutron irradiation of air (Harteck and Dondes, 1956).

Our knowledge of the total cross section for the dissociation of N_2 by electron impact is derived from a complex surface absorption experiment that was carried out by Winters and his collaborators (Winters et al., 1964; Winters, 1966). The magnitude of the total cross section for simple dissociation,

$$e^- + N_2 \longrightarrow N(^4S) + N(^4S) + e^-, \qquad (28)$$

for dissociative excitation,

$$e^- + N_2 \longrightarrow N^* + N + e^-, \qquad (29)$$

and for dissociative ionization,

$$e^- + N_2 \longrightarrow N^+ + N^* + 2e^-, \qquad (30)$$

where N^* is an excited nitrogen atom, was inferred from pumping-speed measurements. The cross section for processes (28), (29), and (30) was found to reach a peak value of 2.0×10^{-16} cm^2 at 90 eV. The dissociation cross section obtained by Winters applies under optically thin conditions (N_2 column densities of $< 10^{14}$ molecules/cm^2). At the pressures found in typical gaseous discharges (several torr), the total cross section for these channels is somewhat larger ($\sim 2.3 \times 10^{-16}$ cm^2) as the result of radiative entrapment processes which will be discussed below (Zipf and McLaughlin, 1978). The cross section for the dissociation of N_2 under optically thick conditions is shown in Fig. 16. The absolute accuracy of Winters's cross section depends critically on the assumption that the probability α that a nitrogen atom will be absorbed by the collection surface was independent of the kinetic energy of the atom; was the same for $N(^4S)$, $N(^2D)$, and $N(^2P)$ atoms as well as for N^+ ions and for nitrogen atoms in Rydberg and high-lying metastable states; and was unity. Although it would seem unlikely that $\alpha = 1$ unilaterally, there is good agreement between the mass spectrometer study of processes (28) and (29) of Niehaus (1967) and Winters's data, when the latter are corrected for the contribution due to dissociative ionization. This result, as well as the good agreement in complementary studies on methane dissociation by a variety of techniques, including the surface absorption method, supports Winters's estimate that the probable error in his results is less than $\pm 20\%$.

The surface absorption experiment does not give any direct information on the details of the dissociation mechanism, the final products, or the translational-energy distribution of the fragments. Strong motivation to pursue such studies was provided by the major expansion of our knowledge of planetary atmospheres since 1960, and interest in the dissociative excitation of

4 Dissociation of Molecules by Electron Impact

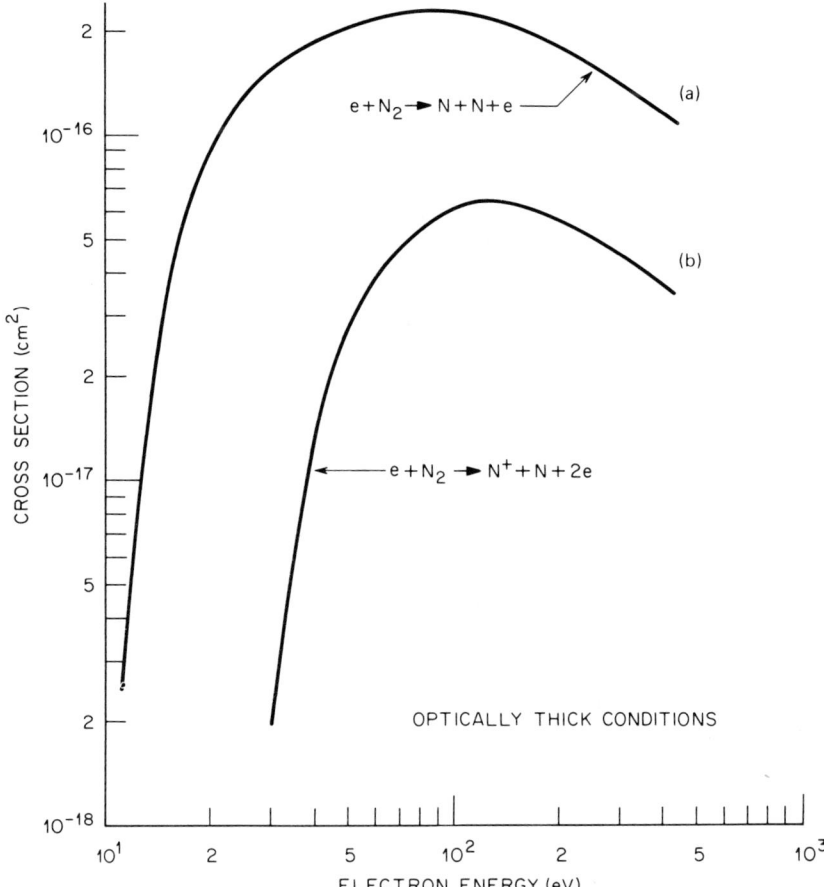

Fig. 16. (a) Total cross section for electron-impact dissociation of N_2 under optically thick conditions and (b) total dissociative ionization cross section. [Data of Winters (1966) and Zipf and McLaughlin (1978).]

N_2, O_2, and H_2, in particular, was heightened when airglow spectra showed numerous emission lines emitted by neutral and ionized O and N atoms. Figure 17 shows a partial term diagram for atomic nitrogen. The principal transitions that characterize its spectrum are noted. Although the most intense NI features are emitted at wavelengths >1100 Å, there are numerous NI multiplets emitted at much shorter wavelengths. However, the cross sections for these transitions are quite small (Morgan and Mentall, 1983). Nonetheless, they are readily observed in the extreme-ultraviolet spectrum of an aurora. In addition to NI features, dissociative excitation of N_2 is also a copious source

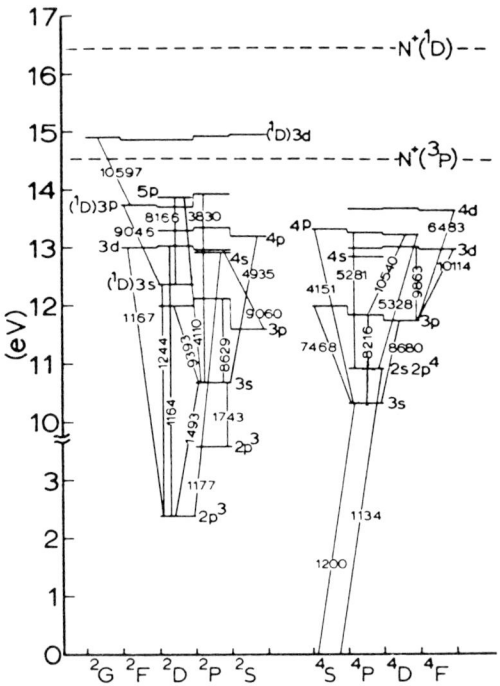

Fig. 17. Atomic-nitrogen term diagram. [From Filippelli et al. (1982).]

of NII multiplets that appear in the 500–1200 Å wavelength range. The branching ratios and cross sections for most of the brighter NI and NII multiplets have now been measured with probable errors of the order of ±20% when comparing the relative cross section values published by different workers. The errors quoted in the absolute cross section values are moderately small for visible transitions (~ ±15%) where adequate calibration standards exist. The situation is considerably poorer in the vacuum-ultraviolet wavelength region. Table II lists cross section values for the brighter of the NI and NII multiplets excited by dissociative excitation. Although the variation in the absolute cross section values given by different workers is sometimes appreciable, the relative values are often in quite good agreement. These differences are due mostly to the use of different normalization standards.

Detailed information on the excitation functions for the emission features given in Table II is also available. Figures 18 and 19 show representative results over the energy range from threshold to ~400 eV. In general, the excitation functions for excited neutral NI atoms exhibit a major broad peak near 85 eV with a structure of varying magnitude near 35 eV. Mumma and Zipf (1971a) suggested that the excitation function may be viewed as the sum

TABLE II

Emission Cross Sections for the Dissociative Excitation of NI and NII Multiplets by Electron Impact on N_2

λ(Å)	Array	Energy (eV)	Cross section	
NII 671	$2p\,3s(^3P^0) \longrightarrow 2p(^3P)$	200		$2.8\ (-19)^{a,b}$
NII 746	$2s\,2p^3(^1P^0) \longrightarrow 2s^2\,2p^2(^1S)$	200	$1.5\ (-19)^c$	$1.7\ (-19)^b$
NII 776	$2s\,2p^3(^1D^0) \longrightarrow 2s^2\,2p^2(^1D)$	200	$1.1\ (-19)^c$	$1.1\ (-19)^b$
NII 916	$2s\,2p^3(^3P^0) \longrightarrow 2s^2\,2p^2(^3P)$	200	$4.4\ (-19)^c$	$4.7\ (-19)^b$
NII 1085	$2s\,2p^3(^3P^0) \longrightarrow 2s^2\,2p^2(^3P)$	200	$2.42\ (-18)^c$	$2.26\ (-18)^b$
NI 1134	$2s\,2p^4(^4P) \longrightarrow 2s^2\,2p^3(^4S^0)$	100	$1.05\ (-18)^c$	$1.24\ (-18)^d$
NI 1164	$2p^2\,3p(^2D) \longrightarrow 2p^3(^2D^0)$	100	$6.2\ (-19)^c$	$1.63\ (-19)^d$
NI 1168	$2p^2\,3p(^2F) \longrightarrow 2p^3(^2D^0)$	100		
NI 1177	$2p^2\,4s(^2P) \longrightarrow 2p^3(^2D^0)$	100	$4.4\ (-19)^c$	$5.16\ (-19)^d$
NI 1200	$2p^2\,3s(^4P) \longrightarrow 2p^3(^4S^0)$	100	$4.72\ (-18)^c$	$6.70\ (-18)^d$
NI 1243	$2p^2\,3s'(^2D) \longrightarrow 2p^3(^2D^0)$	100	$1.52\ (-18)^c$	$1.45\ (-18)^d$
NI 1311	$2p^2\,3p(^2D) \longrightarrow 2p^3(^2P^0)$	100		$3.5\ (-19)^d$
NI 1327		100		$9.0\ (-20)^d$
NI 1411	$2p^2\,3s'(^2D) \longrightarrow 2p^3(^2P^0)$	100		$1.6\ (-19)^d$
NI 1493	$2p^2\,3s(^2P) \longrightarrow 2p^3(^2D^0)$	100	$1.88\ (-18)^c$	$2.62\ (-18)^d$
NI 1743	$2p^2\,3s(^2P) \longrightarrow 2p^3(^2P^0)$	100		$9.1\ (-19)^d$
NI 4107	$2p\,3p'(^1D) \longrightarrow 2p^2\,3s(^3P)$	85		$1.8\ (-20)^e$
NI 4146	$2p^2\,4p(^3P) \longrightarrow 2p^2\,3s(^4P)$	85		$1.3\ (-20)^e$
NI 4928	$2p^2\,4p(^2S^0) \longrightarrow 2p^2\,3s(^2P)$	85		$3.3\ (-20)^e$
NI 5295	$2s^2\,2p^2\,4p(^4P^0) \longrightarrow 2s\,2p^4(^4P)$	85		$9.1\ (-21)^e$
NI 5349	$2s^2\,2p^2\,4p(^4D^0) \longrightarrow 2s\,2p^4(^4P)$	85		$1.6\ (-20)^e$
NII 5680	$2p(^3D) \longrightarrow 3s(^3P)$	160		$2.3\ (-19)^e$
NI 6485	$4d(^4F) \longrightarrow 3p(^4D)$	85		$3.4\ (-20)^e$
NI 7452	$2p\,3p(^4S) \longrightarrow 2p^2\,3s(^4P)$	85		$1.3\ (-19)^e$
NI 8212	$2p\,3p(^4P^0) \longrightarrow 2p^2\,3s(^4P)$	85		$8.2\ (-19)^e$
NI 8618	$2p\,3p(^2P^0) \longrightarrow 2p^2\,3s(^2P)$	85		$4.7\ (-19)^e$
NI 8692	$2p\,3p(^4D^0) \longrightarrow 2p^2\,3s(^4P)$	85		$1.50\ (-18)^e$
NI 9048	$2p^2\,3p'(^2F^0) \longrightarrow 2p^2\,3s'(^2D)$	85		$2.0\ (-19)^e$
NI 9050	$2p^2\,3d(^2P) \longrightarrow 2p^2\,3p(^2S^0)$	85		$7.1\ (-20)^e$
NI 9395	$2p^2\,3p(^2D^0) \longrightarrow 2p^2\,3s(^2P)$	85		$7.0\ (-19)^e$
NI 10,117	$2p^2\,3d(^4F) \longrightarrow 2p^2\,3p(^4D^0)$	85		$5.5\ (-19)^e$
NI 10,525	$2p^2\,3d(^4D) \longrightarrow 2p^2\,3p(^4P^0)$	85		$3.0\ (-19)^e$
NI 10,595	$2p\,3d'(^2G) \longrightarrow 2p^2\,3p'(^2F^0)$	85		$1.8\ (-20)^e$

[a] Read 2.8×10^{-19} cm^2.
[b] Morgan and Mentall (1983).
[c] Aarts and de Heer (1971).
[d] Mumma and Zipf (1971b).
[e] Filippelli et al. (1982).

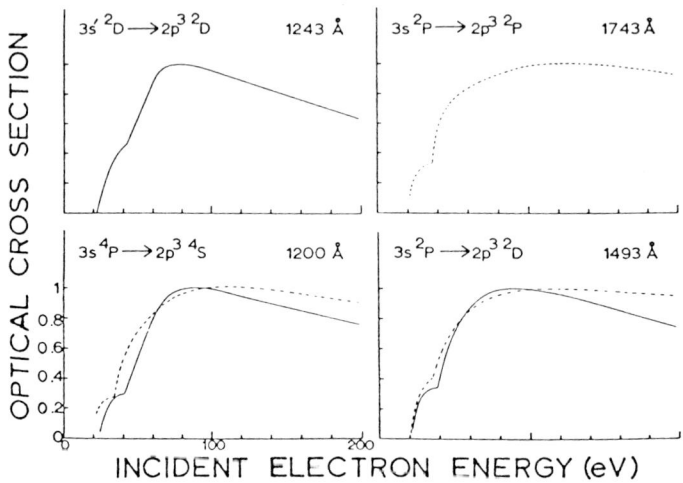

Fig. 18. Optical emission excitation functions for NI multiplet radiation in the vacuum ultraviolet. Solid curves are taken from the data of Mumma and Zipf (1971b), while the dashed curves are data from Ajello (1970). The experimental data for each curve have been normalized to a maximum value of 1.0. All curves shown here are plotted to the same energy scale. The cross section values given for these transitions in Table II can be used to place these curves on a absolute scale. [From Filippelli et al. (1982).]

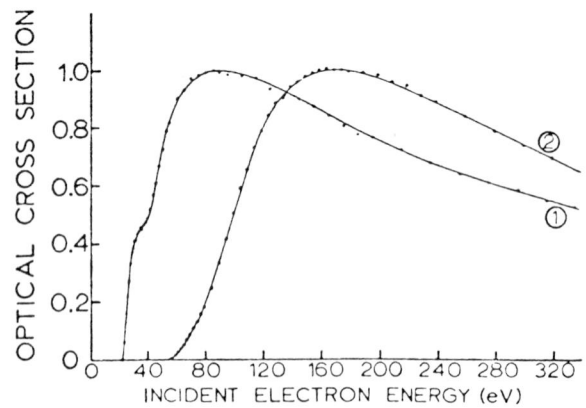

Fig. 19. Comparison of optical emission excitation function shapes for (1) $3p\ ^2D_{5/2} \rightarrow 3s\ ^2P_{3/2}$ ($\lambda 9393$ Å) of NI and (2) $3p\ ^3D_3 \rightarrow 3s\ ^3P_2$ ($\lambda 5680$ Å) of NII produced by electron-impact dissociation of N_2. Peak cross section values are given in Table II. [From Filippelli et al. (1982).]

of two curves, one with a narrow peak at 35 eV and the other with a broad one at 85 eV, corresponding, respectively, to the formation of $N(2p^3) + N(2p^2\,3s)$, and of $N^+(2p^2) + N(2p^2\,3s)$. Thus, the process

$$N_2 + e \longrightarrow N_2^* + e \longrightarrow N(2p^3\,^4S) + N(2p^2\,nl) + e \qquad (nl = 3p, 4p, 3d, \text{etc.}) \qquad (31)$$

dominates below 35 eV whereas the broad peak is a manifestation of

$$N_2 + e^- \longrightarrow (N_2^+)^* + 2e^-$$
$$(N_2^+)^* \longrightarrow N^+(2p^2\,^3P) + N(2p^2\,nl). \qquad (32)$$

Process (32) is an example of the core-ion mechanism of dissociation in which the properties of the excited $(N_2^+)^*$ ions determine the final distribution of excited states and their kinetic-energy spectrum. Large amounts of translational energy are released by this process (cf. Fig. 5), as is easily observed in TOF studies on the N^+ and N(Rydberg) fragments. Recent linewidth studies on the $\lambda 8216$, 8680, and 1200 Å have now demonstrated directly that the $2p\,3p(^4P^o)$, $2p\,3p(^4D^o)$, and $2p^2\,3s(^4P^o)$ states, respectively are kinetically energetic as well; the results provide additional support for the core-ion model (Erdman and Zipf, 1983).

It is also worth noting that because these emission features are substantially Doppler broadened (~ 25 times the mean thermal width at 300 K), their transmission function through an optically thick medium will be complex (Zipf and Wells, 1980) and potentially subject to unusual absorption processes (Zipf et al., 1979).

The dissociation of the core ion may also produce excited NII ions that subsequently radiate,

$$N_2 + e^- \longrightarrow (N_2^+)^* + 2e^-$$
$$(N_2^+)^* \longrightarrow N^+(2p\,nl) + N. \qquad (33)$$

This process, which entails simultaneous ionization and excitation of N_2, is also characterized by an excitation function with a broad maximum (see Figs. 18 and 19). However, the peak occurs at a much higher energy (~ 170 eV), and the shoulder structure due to process (31) is absent. These differences provide a useful tool in the study of the excitation of airglow features in the dayglow, auroral, and cleft regions, where the electron energy distributions differ quite remarkably (Sivjee, 1983).

Although the numerous NI and NII multiplets excited dissociatively by electron impact are conspicuous in the spectra of gas discharges and hollow-cathode tubes, the total cross section leading to the excitation of this subset of states is actually modest and has a value of $\sim 1.4 \times 10^{-17}$ cm^2 at 100 eV. Since the total N_2 dissociation cross section at this energy is approximately 2×10^{-16} cm^2 (Winters, 1966), the excitation of the high-lying NI and NII states only plays a minor role in the overall dissociation process. Dissociative ionization [P(9)] accounts for about one-third of the atomic fragments produced by the impact of 100-eV electrons, but is of negligible importance below 40 eV, where the total dissociative cross section is still quite large ($\sim 1.6 \times 10^{-16}$ cm^2). The first insight into the underlying physical processes responsible for this large cross section emerged from atmospheric studies on the role of metastable N(^{2}D; ^{2}P) atoms produced by photoelectron dissociation of N$_2$ in the odd-nitrogen chemistry of our planet:

$$h\nu(\text{solar}) + N_2 \longrightarrow N_2^+ + e^- + \text{kinetic energy}$$

$$e^- + N_2 \longrightarrow N + N(^2D; {}^2P) \tag{34}$$

$$N(^2D; {}^2P) + O_2 \longrightarrow NO + O$$

Both satellite and sounding rocket studies suggested that process (34) formed the chemically active ^{2}D and ^{2}P states with an overall efficiency of 70% (Rusch and Gerard, 1980), but the field experiments provided no insights into which N$_2$ quantum states were involved, how much kinetic energy was released, or the specific mechanism(s). This information was ultimately provided by laboratory studies on the excitation of molecular band radiation by electron impact on N$_2$, cross section work that was originally stimulated by the sugggestion that as much as 16% of the energy deposited in the lower atmosphere by a magnetospheric substorm is dissipated in exciting the numerous $^1\Sigma_u^+$ and $^1\Pi_u$ states of N$_2$ lying above 11 eV (Rees, 1975).

The physical argument developed by Rees (1975) and Zipf and McLaughlin (1978) proceeded along these lines: inelastic electron scattering studies (Lassettre and Krasnow, 1964; Silverman and Lassettre, 1965; Geiger and Schröder, 1969; Lassettre, 1969, 1974; Williams and Doering 1969a,b; Brinkmann and Trajmar, 1970) as well as extreme-ultraviolet (EUV) absorption measurements (Huffman et al., 1963; Tilford and Wilkinson, 1964a,b, 1965; Cook et al., 1965; Lawrence et al., 1968; Carroll and Collins, 1969; Dressler, 1969; Carroll and Yoshino, 1972) had shown that molecular nitrogen possess a large number of valence and Rydberg states in the energy range 11.5–40 eV. This can be seen readily in Fig. 20. Many of these singlet states relax radiatively in direct dipole-allowed EUV transitions to the ground state (Wilkinson and Houk, 1956; Wallace, 1962; Tilford and Wilkinson, 1964a,b) and through weaker cascade transitions to the a $^1\Pi_g$ state in the 2000–3000 Å wavelength range (Lofthus, 1957; McFarlane, 1965, 1966). Because the total cross section for electron-impact excitation of this manifold of states is comparable in magnitude to the N$_2$ ionization cross section (Porter et al., 1976) and because the radiative lifetimes of these states are generally in the range of 10 ns to 0.1 μs (Hesser and Dressler, 1966), modelers expected that large fluxes of EUV radiation would be excited in the dayglow by energetic photoelectrons and in auroral substorms by precipitating electrons impacting on ambient N$_2$ molecules (Green and Barth, 1967; Stolarski and Green, 1967).

4 Dissociation of Molecules by Electron Impact

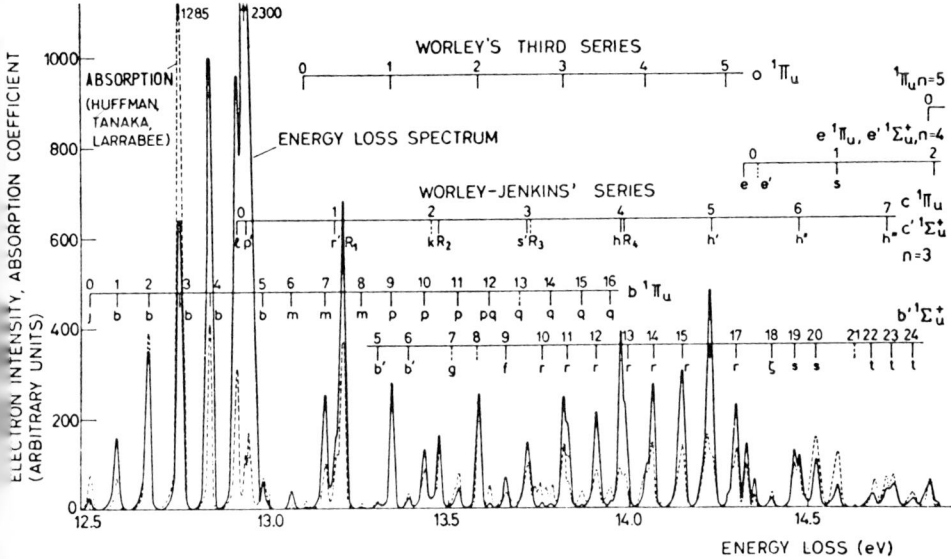

Fig. 20. Inelastic electron scattering spectrum of N_2 obtained by Geiger and Schröder (1969) at an incident energy of 25 keV.

The photochemical implications of an EUV flux of this magnitude were discussed by Zipf (1973a), who suggested that this radiation would produce significant modifications in the ion chemistry of an aurora because of selective photoionization of O_2 and O and that it would also contribute to the excitation of the OI green line ($\lambda = 5577$ Å) by photodissociation of O_2 (Zipf, 1973b; Lawrence and McEwan, 1973). Similar mechanisms, but involving H_2, have been recently proposed by Prasad and Tarafdar (1983) as an important element in the photochemistry of dense interstellar clouds.

Unfortunately, when the auroral EUV spectrum was actually observed (Park et al., 1977), little N_2 band emission was observed. The field observations were thus in complete disagreement with the results anticipated on the basis of our knowledge of the inelastic electron scattering spectrum of N_2 and from auroral energy conservation considerations (Rees, 1975).

The total cross section for directly populating a specific N_2 singlet state at high incident electron energies ($E \geq 200$ eV) can be calculated quite accurately from integrated generalized oscillator strength data using the procedure described by Vriens (1967) and by Vriens et al. (1968), and relative electron scattering spectra such as those given by Geiger and Schröder (1969), and the pertinent excitation functions can be obtained from optical cross section data. Under the simplest circumstances, the total excitation cross section and the

TABLE III

Predissociation Branching Ratios[a]

State	Total cross section[b]	Emission cross section[b]	Predissociation branching ratio
$b\,^1\Pi_u(0-13)$[c]	2.07(-17)[d]	6.5(-19)	0.97
$b'\,^1\Sigma_u^+(0-20)$	1.71(-17)	3.0(-18)	0.83
$c'_4\,^1\Sigma_u^+(0-4)$	2.01(-17)	1.7(-17)	0.15
$c'_4\,^1\Sigma_u^+(5-7)$	3.71(-18)	$<2\,(-19)$	>0.95
$e'\,^1\Sigma_u^+$	5.80(-19)	$<2\,(-20)$	>0.99
$c\,^1\Pi_u(0-7)$	1.20(-17)	$<2\,(-20)$	>0.99
$o\,^1\Pi_u(0-4)$	4.82(-18)	$<2\,(-20)$	>0.99
$e\,^1\Pi_u(0-2)$	2.00(-18)	$<2\,(-20)$	>0.99
$n = 5\,^1\Pi_u$	5.66(-20)	$<2\,(-20)$	>0.6
Total	8.10(-17)	2.2(-17)	0.73

[a] Under optically thin conditions.
[b] At 200 eV.
[c] Means that the vibrational levels $v' = 0-13$ were used in the tabulation.
[d] Read 2.07×10^{-17} cm^2.

emission cross section will be equal in magnitude since radiation is the only depopulating process. If autoionization, predissociation, or quenching are comparable in importance with radiation, then the two cross sections will differ. In nitrogen, the effect is quite dramatic, as Table III shows. The laboratory experiments, in fact, showed that most N_2 molecules excited to the manifold of singlet states observed by Lassettre and others are depopulated by predissociation and not by the emission of extreme-ultraviolet photons as previously assumed, that this process is the principal mechanism by which N_2 is dissociated by solar EUV absorption and by electron impact, and that the total cross section for electron-impact dissociation of N_2 measured by Winters (1966) owes its quantum-mechanical origin to the excitation of these singlet valences and Rydberg states. Thus, these developments provide detailed insights into the competing channels that contribute to the dissociation of N_2 by electron impact and their energy dependence. These results are shown in Fig. 21 under optically thin conditions. In many situations, the singlet radiation will be effectively trapped in the medium, and the EUV photons will be recycled until they are lost via weak cascade transitions or until predissociation converts them into N atoms. Thus, radiation entrapment effects can increase substantially the effective dissociation cross section.

Perhaps equally important is the insight that electron-impact dissociation of N_2 is a major source of metastable $N(^2D)$ and $N(^2P)$ atoms and that the kinetic-energy distribution of these fragments is discrete in character. Both

4 Dissociation of Molecules by Electron Impact

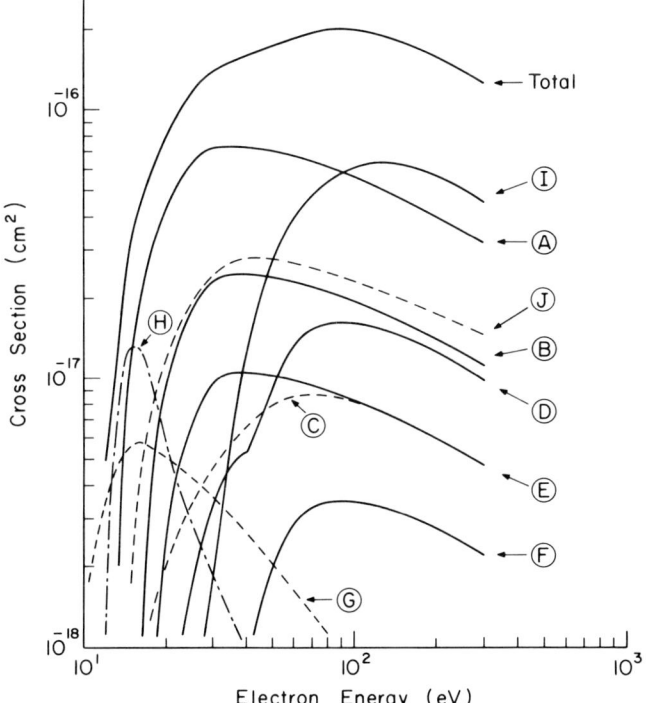

Fig. 21. Total cross section of electron-impact dissociation of N_2 under optically thin conditions. The specific dissociation channels contributing to the total cross section include the excitation of (A) the family of $^1\Pi_g$ states, (B) the states belonging to the 15.8-eV peak in the inelastic scattering data of Lassettre (1974), (C) the $c'_4\ ^1\Sigma_u^+ \to X\ ^1\Sigma_g^+$ system, (D) the states leading to the production of excited N atoms that radiate at VUV wavelengths, (E) the states associated with the 17.3-eV peak of Lassettre, (F) the states contributing to the formation of N Rydberg atoms, (G) the $a\ ^1\Pi_g$ state, (H) the N_2 triplet manifold, (I) the states contributing to dissociative ionization, (J) the $b'\ ^1\Sigma_u^+ \to X\ ^1\Sigma_g^+$ system. [From Zipf and McLaughlin (1978).]

conclusions flow from conservation of energy constraints and the basic nature of the predissociation process (Zipf and McLaughlin, 1978; Zipf et al., 1980). By combining $N(^2D)$ and $N(^2P)$ data obtained in rocket studies of the aurora borealis with the laboratory measurements, it has been possible to evaluate the specific dissociative excitation cross sections for these chemically active N atoms. These results are shown in Fig. 22.

In order to use the NI and NII multiplet radiation emitted by a given source for diagnostic purposes, it is also necessary to have cross section data on a potentially competing excitation channel,

$$e + N \longrightarrow N(2p^2\ nl) + e$$
$$e + N \longrightarrow N^+(2p\ nl) + 2e, \tag{35}$$

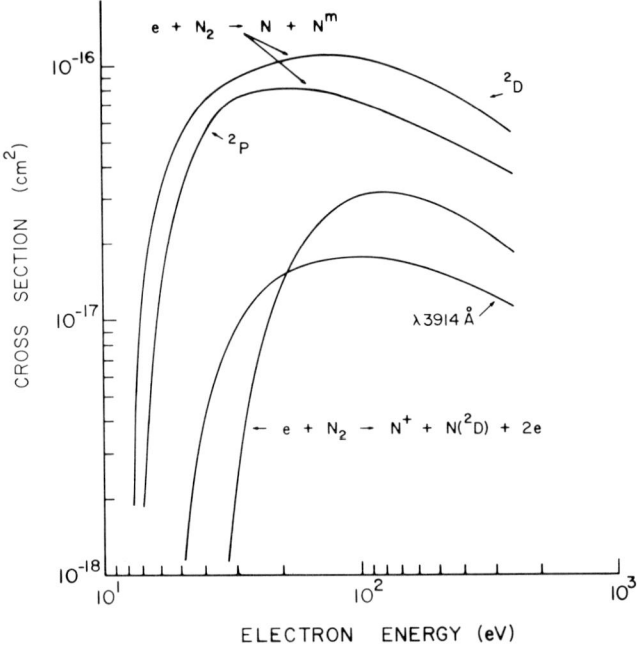

Fig. 22. Cross section for the excitation of metastable N(^{2}D) and N(^{2}P) atoms by electron-impact dissociation of N_2. The cross section for electron-impact excitation of the N_2^+ (0,0) first negative band ($\lambda 3914$ Å) is plotted for comparison. [From Zipf et al., 1980.]

as well as NI and N_2 composition information. Fortunately, in nitrogen, the cross sections for electron-impact excitation of atomic nitrogen have been measured (Stone and Zipf, 1973). The cross sections have been found to be exceptionally large for the VUV transitions ($\sim 2 \times 10^{-16}$ cm^2) and are well developed at low energies. This can be seen in Fig. 23. The remarkable difference between the cross sections for dissociative versus direct atomic excitation as a function of energy has made it possible to construct exceptionally sensitive optical mass spectrometers for NI and N_2 measurements in the upper atmosphere (Zipf et al., 1973).

B. Molecular Oxygen

The dissociative excitation of molecular oxygen is a rich source of emission features which are conspicuous in the spectrum of an aurora and gaseous discharges from 400 Å to 1.3 μm. In addition to a large number of excited OI and OII states that promptly relax radiatively, dissociative excitation is also a

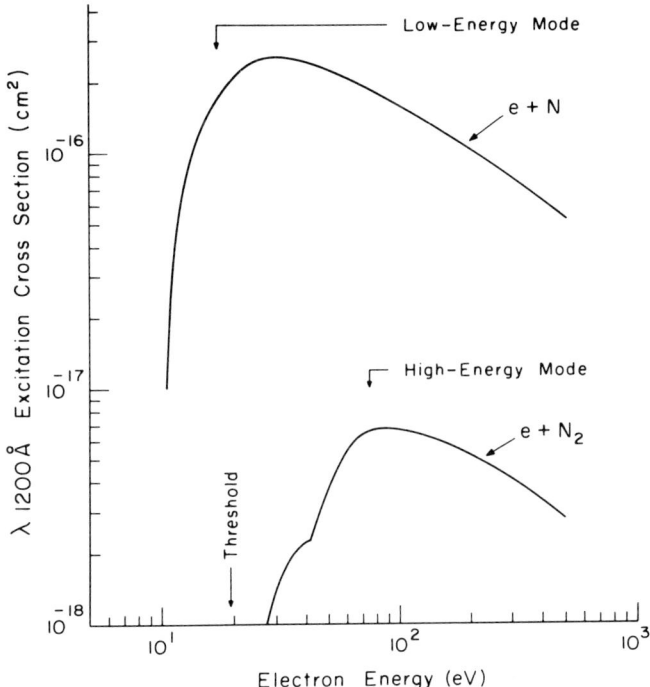

Fig. 23. Excitation cross sections for the NI($3s\ ^4P \to 2p^3\ ^4S$) $\lambda 1200$ Å multiplet excited by electron impact on atomic nitrogen and N_2. By programming the energy of the exciting electron beam NI and N_2 density measurements can be made by observing the intensity of the $\lambda 1200$ Å transition. This technique has been used to measure the atomic nitrogen density in the upper atmosphere. (From Mumma and Zipf, 1971b; Stone and Zipf, 1973.)

major source of the metastable species $O(^1S)$ and $O(^1D)$ which emit the $\lambda 5577$ and 6300 Å lines that dominate the airglow and auroral spectrum and have been particularly useful in the study of high-altitude winds. The study of atomic oxygen emission features in the upper atmosphere has become particularly important because it provides an exceptional opportunity to investigate two- and three-level optical pumping processes under circumstances where there are no walls and where the optical depths vary from thin (<1) to extraordinarily thick ($>10^4$). These investigations have provided a fertile testing ground for complex radiative entrapment codes, and in turn, their need for detailed information on excitation cross sections, absolute transition probabilities, and Doppler linewidth profiles has been an important factor in bringing the laboratory study of the dissociative excitation of O_2 by electron impact to its current mature state.

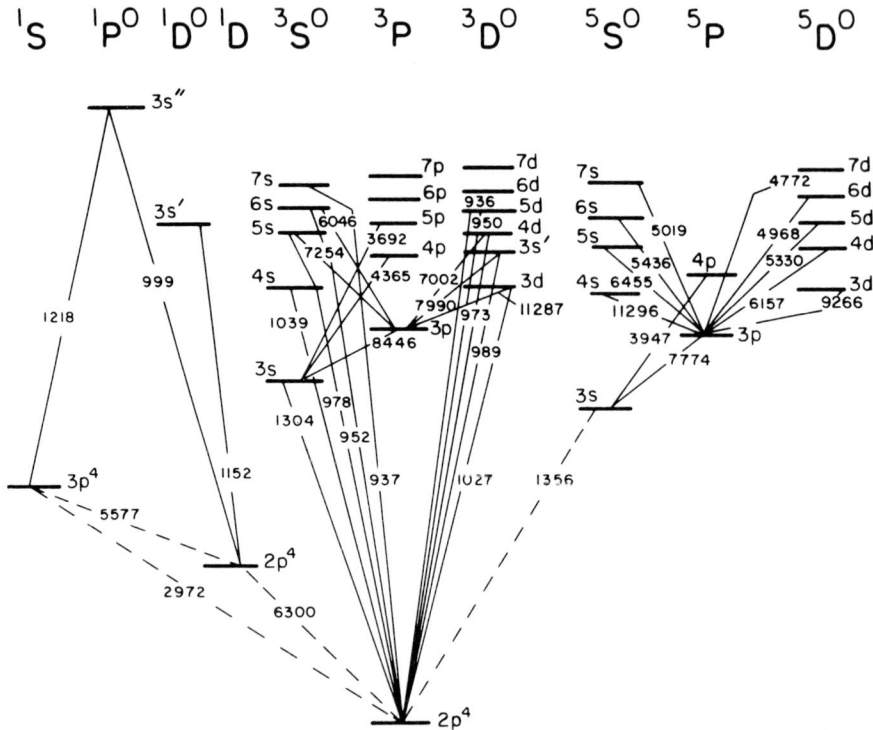

Fig. 24. Term diagram of atomic oxygen showing the numerous emission features that dominate the spectrum of the aurora borealis and the terrestrial airglow. [From Wiese et al. (1966).]

Figure 24 shows the term diagram for atomic oxygen. Absolute cross sections and excitation functions have been measured from threshold to ~400 eV for most of the OI transitions originating from levels with the principal quantum numbers n = 3–8 and l = 0–2. Both the triplet and quintet manifolds have been studied in detail. Some of the brighter transitions, which can be observed readily in gaseous discharges, are indicated in the figure. These measurements have also provided an opportunity to determine the branching ratio of a number of transitions originating from common upper states and to check theoretical oscillator strengths of Pradhan and Saraph (1977). Table IV summarizes these results. The agreement between theory and experiment is reasonably good with several important exceptions that will be discussed below. Table V lists some representative OI dissociative excitation cross sections. Analogous measurements have also been carried out for the excitation of this manifold of OI states by electron impact on atomic oxygen (Zipf et al., 1973).

TABLE IV

Atomic Oxygen Triplet Transition Probabilities and Branching Ratios

Term	n	λ (Å)	A_{calc}[a]	A_{exp}[b]	Experiment[c]	Experiment[d]	Theory
^3P-$^3S^0$	3s	1304	6.3 (8)	6.3 (8)			
	4s	1040	1.7 (8)	4.9 (8)			0.89
	4s	13,164	2.2 (7)				0.11
	5s	977	6.9 (7)	3.2 (7)	0.91		0.86
	5s	7254	6.4 (6)		0.09		0.08
	5s	36,613	4.8 (6)				0.06
	6s	952	3.5 (7)		0.92		0.85
	6s	6046	3.1 (6)		0.08		0.03
	6s	18,230	1.7 (6)				0.04
	6s	77,460	1.2 (6)				0.03
^3P-$^3D^0$	3d	1027	7.7 (7)	1.1 (8)		0.90	0.72
	3d	11,287	3.03 (7)			0.10	0.28
	4d	973	6.3 (7)	5.2 (7)	0.95	0.88	0.93
	4d	7002	2.2 (6)		0.05	0.05	0.03
	4d	32,300	2.8 (6)[d]				0.04
	5d	949	2.9 (7)	3.4 (7)			
	6d	937	1.8 (7)	2.5 (7)			
	3s'	989	2.3 (8)	2.0 (8)	1.00	0.9996	0.9996
	3s'	7990	9.6 (4)		$<5 \times 10^{-4}$		

OII Transition Probabilities and Branching Ratios

Term	n	λ (Å)	A_{cal}[e]	Experiment[f]
$^2D^0$-^2D	3d	482	5.4 (8)	0.56 ± 0.09
	3d	515	1.5 (9)	
^2P-^2P	2p	538	4.34 (9)	1.55 ± 0.3
	2p	581	2.80 (9)	
$^2D^0$-^2D	3s'	555	1.5 (9)	1.85 ± 0.3
	3s'	601	4.3 (8)	
$^2D^0$-^2P	3s	617	3.07 (9)	4.1 ± 0.5
	3s	673	7.48 (8)	
$^2D^0$-^2D	2p	718	1.93 (9)	6.6 ± 0.6
	2p	797	2.92 (8)	

[a] All transition probabilities in this column, with the one exception noted, were calculated from the oscillator strengths given in Pradhan and Saraph (1977). Read 6.3×10^8 s^{-1}. [$A_{ki} = g_i f_{ik}/g_k 1.499 \times 10^{-16} \lambda^2$; k is the upper state and λ is in Å.]
[b] Calculated from the oscillator strengths given in Brooks et al. (1977). Read 6.3×10^8 s^{-1}.
[c] Gorman and Zipf (1983).
[d] Christensen and Cunningham (1978).
[e] Wiese et al. (1966).
[f] Christensen and Cunningham (1978); Morrison et al. (1983).

TABLE V

Emission Cross Sections for the Dissociative Excitation of O_2, CO, and CO_2 by 100-eV Electrons[a]

λ (Å)	Transition array	$\sigma(O_2)$ (cm^2)	$\sigma(CO)$ (cm^2)	$\sigma(CO_2)$ (cm^2)
OI 929	$7d\ ^3D^o \longrightarrow 2p\ ^3P$	1.6 (-20)[b]		
OI 936	$6d\ ^3D^o \longrightarrow 2p\ ^3P$	4.0 (-20)		
OI 940	$7s\ ^3S^o \longrightarrow 2p\ ^3P$	1.6 (-20)		
OI 950	$5d\ ^3D^o \longrightarrow 2p\ ^3P$	1.1 (-20)	3.3 (-20)	
OI 952	$6s\ ^3S^o \longrightarrow 2p\ ^3P$	3.9 (-20)	3.4 (-20)	7.12 (-20)
OI 973	$4d\ ^3D^o \longrightarrow 2p\ ^3P$	3.7 (-19)	1.58 (-19)	1.73 (-19)
OI 978	$5s\ ^3S^o \longrightarrow 2p\ ^3P$	1.0 (-19)	1.56 (-19)	5.4 (-19)
OI 989	$3s'\ ^3D^o \longrightarrow 2p\ ^3P$	1.96 (-18)	8.42 (-19)	5.6 (-19)
OI 999	$3s''\ ^1P^o \longrightarrow 2p^4\ ^1D$			
OI 1027	$3d\ ^3D^o \longrightarrow 2p\ ^3P$	1.46 (-18)	1.64 (-19)	5.54 (-19)
OI 1040	$4s\ ^3S^o \longrightarrow 2p\ ^3P$	3.4 (-19)		
OI 1152	$3s'\ ^1D^o \longrightarrow 2p\ ^1D$			3.47 (-19)
OI 1218	$3s''\ ^1P^o \longrightarrow 2p^4\ ^1S$			
OI 1304	$3s\ ^3S^o \longrightarrow 2p\ ^3P$	3.8 (-18)	1.56 (-18)	1.00 (-18)
OI 3692	$5p\ ^3P \longrightarrow 3s\ ^3S^o$	5 (-21)		
OI 3947	$4p\ ^5P \longrightarrow 3s\ ^5S^o$	4.1 (-20)		
OI 4368	$4p\ ^3P \longrightarrow 3s\ ^3S^o$	5.8 (-20)		
OI 4772	$7d\ ^5D^o \longrightarrow 3p\ ^5P$	7.8 (-21)		
OI 4968	$6d\ ^5D^o \longrightarrow 3p\ ^5P$	2.1 (-20)		
OI 5019	$7s\ ^5S^o \longrightarrow 3p\ ^5P$	4.4 (-21)		
OI 5330	$5d\ ^5D^o \longrightarrow 3p\ ^5P$	4.3 (-20)	8.0 (-21)	1.0 (-20)
OI 5435	$6s\ ^5S^o \longrightarrow 3p\ ^5P$	9.5 (-21)		
OI 6046	$7s\ ^3S^o \longrightarrow 2p\ ^3P$	1.6 (-20)		
OI 6156	$4d\ ^5D^o \longrightarrow 3p\ ^5P$	1.16 (-19)	2.2 (-20)	
OI 6454	$5s\ ^5S^o \longrightarrow 3p\ ^5P$	4.3 (-20)		
OI 7002	$4d\ ^3D^o \longrightarrow 2p\ ^3P$	1.9 (-20)		
OI 7774	$3p\ ^5P \longrightarrow 3s\ ^5S^o$	4.91 (-18)	2.5 (-19)	4.4 (-19)
OI 7990	$3s'\ ^3D^o \longrightarrow 3p\ ^3P$	1 (-21)		
OI 8446	$3p\ ^3P \longrightarrow 3s\ ^3S^o$	2.3 (-18)	5.7 (-19)	3.2 (-19)
CI 1329	$2s\ 2p^3\ ^3P^o \longrightarrow 2s^2\ 2p^2\ ^3P$			2.67 (-19)
CI 1561	$2s\ 2p^3\ ^3D^o \longrightarrow 2s^2\ 2p^2\ ^3P$			7.50 (-19)
CI 1657	$2p\ 3s\ ^3P^o \longrightarrow 2p^2\ ^3P$			1.45 (-18)
CI 5380	$4p\ ^1P \longrightarrow 3s\ ^1P^o$		2.0 (-20)	1.3 (-20)
CI 8335	$3p\ ^1S \longrightarrow 3s\ ^1P^o$		3.7 (-20)	
CII 1324				1.33 (-20)
CII 1335				7.60 (-19)

[a] Compiled from Gorman and Zipf (1984) and Mumma et al. (1972).
[b] Read 1.6×10^{-20} cm^2.

1. Ground-State Configuration

Only indirect information exists on the cross section for exciting the metastable $O(^1D)$ and $O(^1S)$ atoms by electron impact dissociation of O_2. The time-of-flight work of Stone et al. (1975) for the ^1D state is consistent with electron-impact excitation of O_2 to the repulsive portion of the $B\,^3\Sigma_u^-$ potential curve,

$$e + O_2(X\,^3\Sigma_g^-) \longrightarrow O_2(B\,^3\Sigma_u^-) + e \longrightarrow O(^1D) + O(^3P) + e + KE, \qquad (36)$$

followed by rapid dissociation. This process would use the same potential-energy surface exploited by UV photons in the well-known photodissociation of O_2 in the Schumann–Runge continuum. For this process for which the oscillator strength has been accurately measured in photoabsorption and inelastic-electron scattering studies. From these data, phenomenological cross section models have been developed for O_2 by Porter et al. (1976) and Jackman et al. (1977) which have proven to be quite useful in a variety of applied problems. This formalism facilitates numerical calculations by computer since only a few parameters are required to characterize a cross section. For example, the cross section for an allowed dissociative process can be written as

$$\sigma_j(E) = \frac{q_0 F[1 - (W/E)^\alpha]^\beta}{EW} \ln\left(\frac{4EC}{W} + e\right), \qquad (37)$$

where $q_0 = 4\pi a_0^2 R^2 = 6.513 \times 10^{-14}$ eV^2 cm^2, a_0 being the Bohr radius and R the Rydberg energy; W is a parameter that influences the low-energy shape of the cross section and, in most cases, is close to the energy loss W_j associated with excitation of the jth state; F is the optical oscillator strength; C is a factor determined from the high-energy behavior of the cross section; e is the base of the natural logarithm; and α and β are adjustable parameters. Rydberg series can be treated by using the simple rule $F = F^*/(n - \delta)^3$, where F^* is a constant for a given series, and δ is the quantum defect. Extensive tables of these parameters are given by Porter et al. (1976) and Jackman et al. (1977) for both O_2 and N_2, and their analysis shows, for example, that the cross section for process (36) is approximately 2.0×10^{-17} cm^2. Other dissociation channels, which involve the predissociation of a number of discrete states that can be seen readily in the inelastic electron scattering spectrum of O_2, contribute a comparable amount of the total dissociation cross section. Thus, in oxygen, the cross section for producing neutral atomic species by electron impact is slightly smaller than the dissociative ionization cross section at 100 eV, whereas in N_2 dissociative ionization accounts for only 25% of the total dissociation cross section at this energy.

Little is known about the dissociative excitation of the metastable $O(^1S)$ state, which was observed first in the airglow 95 years ago. Lawrence and

McEwan (1973) showed that this species could be excited by O_2 photodissociation and that predissociation appeared to be an important factor. Attempts to excite the same states by electron impact have yielded only negative results so far because of severe interference from O_2^+ first negative bands, which mask the $\lambda 5577$ Å emission and because the $O(^1S)$ are kinetically energetic and leave the field of view before radiating. There have been several conjectures that the cross section for dissociative excitation of the ^1S state might be as large as 10^{-16} cm^2, but the analysis of Porter et al. (1976) seems to rule this possibility out.

Another dissociative process involving excited oxygen atoms belonging to the ground-state configuration, which is important in far-infrared spectroscopy, is the excitation of the fine-structure levels of the $O(^3P_{2,1,0})$ ground state

$$e + O_2 \longrightarrow O(^3P; j = 2,1,0) + O + e. \tag{38}$$

This process gives rise to two emission lines in the remote infrared. They arise from the $j = 1 \rightarrow 2$ and $0 \rightarrow 1$ transitions and have wavelengths of 60 and 114 μm, respectively. The emission of this radiation plays a very important role in the energy balance in the terrestrial thermosphere. Sounding rocket studies of this bright, infrared airglow have been successfully carried out, but no laboratory studies have been reported on these long-lived excited atoms for either direct or for dissociative excitation by electron impact. The direct and dissociative excitation of the OI fine-structure transitions may also be an important cooling mechanism in interstellar gases.

2. *Triplet Manifold*

Major progress has been made in measuring the excitation cross sections for the numerous triplet transitions identified in Fig. 24. Only emission lines in the far infrared ($\lambda > 1.5$ μm) have eluded observation so far, and this is largely due to detector limitations. Part of the motivation for these studies is that many of the OI vacuum-ultraviolet multiplets arise from states that must also emit in the visible and infrared where very accurate cross section measurements can be made. Thus, a set of comparably accurate branching ratio values would provide a convenient means of establishing transfer standards of absolute brightness in the EUV wavelength region, which has lacked such standards for some time; this task has now been partially accomplished (Table IV). However, a number of vexing problems remain. One difficulty deserves particular mention because of the impact on $\lambda 989$ Å entrapment models. The forcus of attention is on the $(2p^3 3s' \, ^3D \rightarrow 3p \, ^3P, \lambda 7990$ Å$)/(2p^3 3s' \, ^3D \rightarrow 2p^4 3P, \lambda 989$ Å$)$ branching ratio and multiplet line strengths under optically thin and thick conditions.

4 Dissociation of Molecules by Electron Impact

The early work of Wiese *et al.* (1966) indicated a branching ratio value of 0.11 for these multiplets. However, a more recent calculation yielded a value of 4×10^{-4} (Pradhan and Saraph, 1977). Subsequent laboratory work (Christensen and Cunningham, 1978; Gorman and Zipf, 1981) placed upper limits on the branching ratio of $< 5 \times 10^{-4}$, and extensive computational work has now been carried out using the lower branching ratio value (Meier, 1982; Christensen *et al.*, 1983). These entrapment models have had some success in predicting the $\lambda 7990$ and 989 Å total intensities and multiplet line strengths observed in an optically thick atmosphere. Figure 25 shows the complex optical pumping effects that take place in such a medium. Unfortunately,

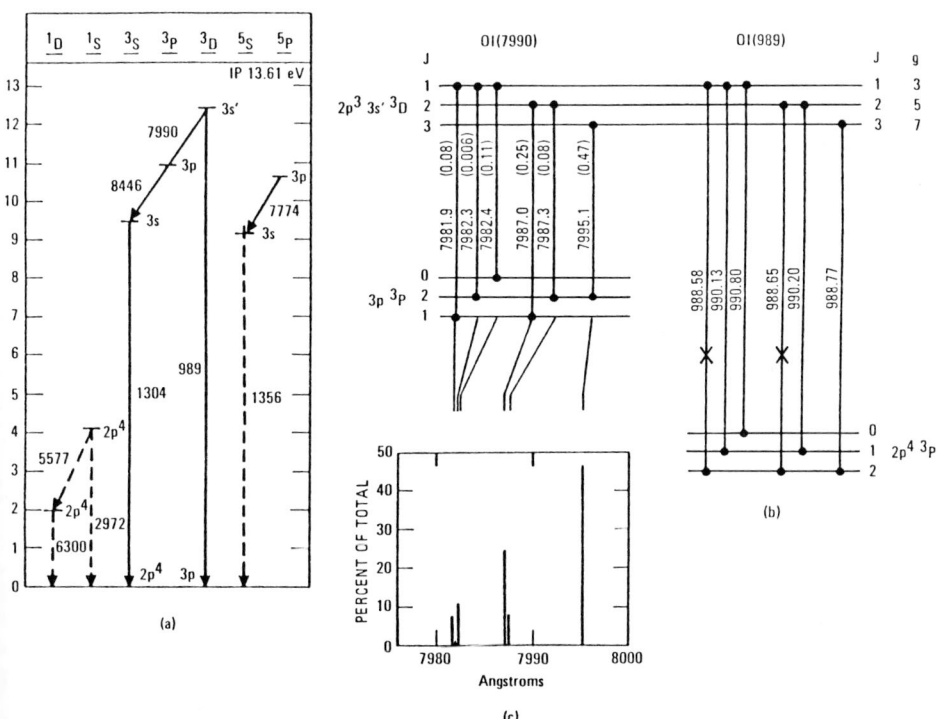

Fig. 25. Partial term diagram for atomic oxygen: (a) terms involved in the entrapment of $\lambda 989$ and 1304 Å, (b) detailed diagram of OI($\lambda 989$ Å) and OI($\lambda 7990$ Å) transitions, and (c) OI($\lambda 7990$ Å) multiplet distribution based on excitation of upper levels in accordance with their statistical weight. Both the $\lambda 989$ and 7990 Å multiplet distributions are highly perturbed in an optically thick medium. [From Christensen *et al.* (1983).] The EUV transitions marked ($\times$) are resonantly absorbed in the upper atmosphere by rotational transitions belonging to the Birge-Hopfield system $X\ ^1\Sigma_g^+ \to b\ ^1\Pi_u$. In an optically thick medium, this process preferentially depletes the 3s' $^3D_1$ and 3s' $^3D_2$ states so that only the $\lambda 7995.1$ (3s' $^3D_3 \to$ 3p $^3P_2$) transition survives.

further laboratory work on this problem has now put a severe upper limit of 5×10^{-5} on the $\lambda_{7990}/\lambda_{989}$ branching ratio, thus underscoring the need for additional work on the theory. The laboratory results have shown that the simple assumptions usually made in current entrapment models with regard to the population rates of the multiplet sublevels in dissociative excitation are too naive and that the velocity distribution of the OI(3s′) state can have a profound effect on the *in situ* observations. These developments point out the need for high-resolution optical studies of the linewidths and individual multiplet line intensities for at least the brighter emission lines excited both by electron impact and photodissociation.

Representative excitation functions for members of the $^3S^0$, ^3P, and ^3D manifolds excited by electron impact on atomic oxygen and O_2 are shown in Fig. 26a and b, respectively. In earlier work Mumma and Zipf (1971a)

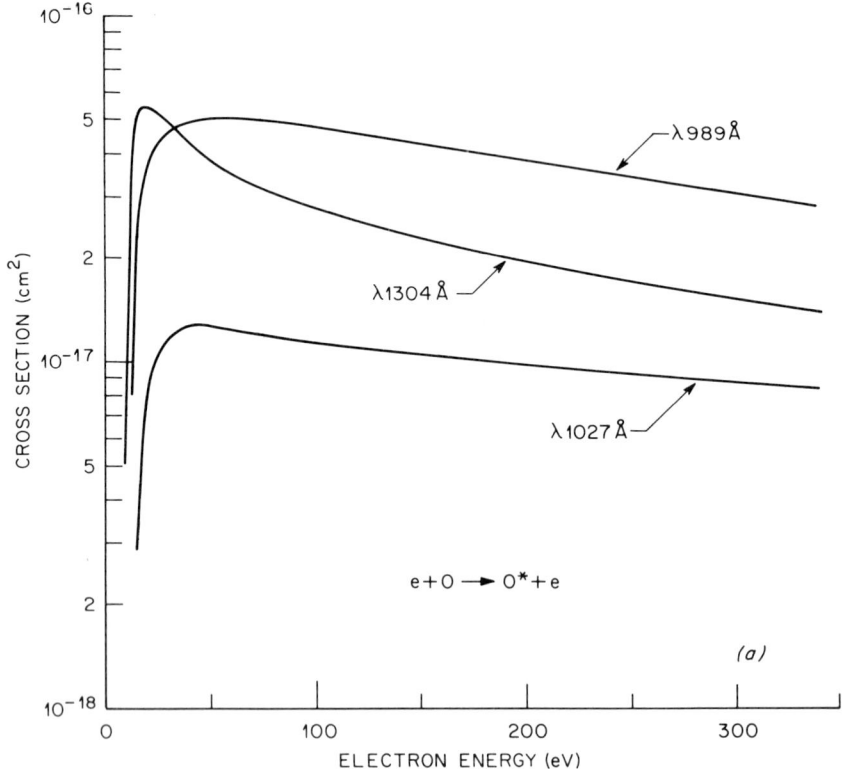

Fig. 26. Excitation cross sections for OI VUV multiplets excited by electron impact on (a) atomic oxygen and (b) O_2: (1) OI(2p ^3P–3s $^3S^0$, $\lambda 1304$ Å), (2) OI(2p ^3P–3s′ $^3D^0$, $\lambda 989$ Å), (3) OI(2p ^3P–3d $^3D^0$, $\lambda 1027$ Å). [Unpublished data of Kao and Zipf (1983) and Zipf *et al.* (1979).]

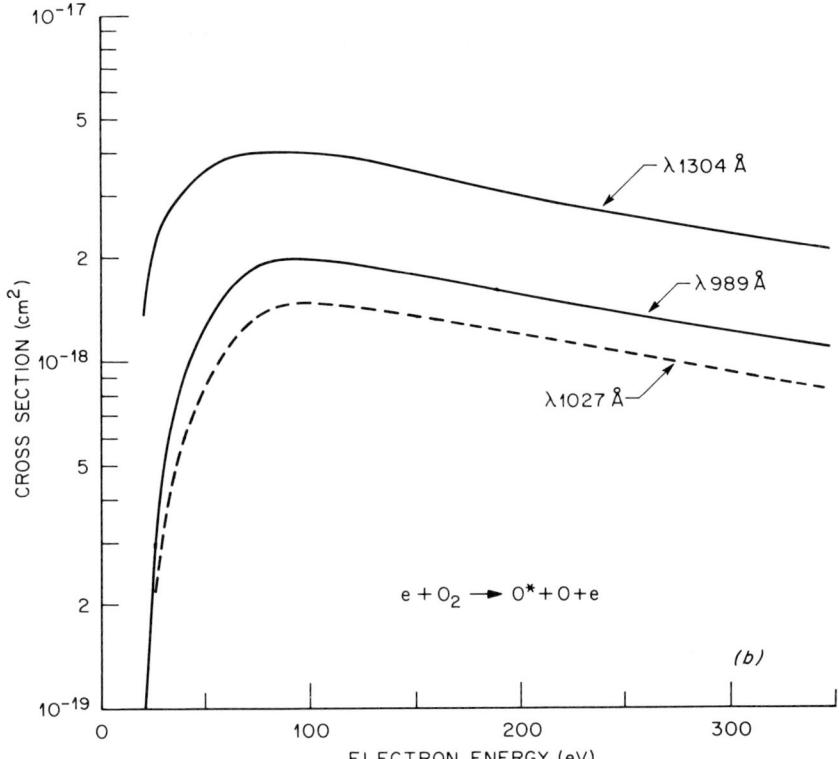

Fig. 26. (*Cont.*)

interpreted the abrupt change in the curvature of the $^3S^0$ excitation function near 40 eV as the onset of the process

$$e + O_2(X\,^3\Sigma_g^-) \longrightarrow O(3s'\,^3S^0) + O^+(2p^3\,^4S) + 2e \tag{39}$$

However, now that an essentially complete set of cross sections exists for the cascade transitions that populate the $^3S^0$ state, a somewhat different picture emerges. Lawrence (1970) has shown that the cascade transition $(3p^1\,^3P-3s^1\,^3S^0, \lambda = 8446.5\text{ Å})$ accounts for $\sim 60\%$ of the total $^3S^0$ excitation cross section. The $3p^1\,^3P$ state, in turn, is strongly populated by the $\lambda = 11{,}287\text{ Å}$ transition that originates from the $3d\,^3D^0$ state. The excitation cross section for the $^3D^0$ manifold is noticeably different from those of transitions originating from ^3P or $^3S^0$ levels. The $^3D^0$ cross section rises linearly from threshold, reaches a maximum near 90 eV, and shows no evidence for a "secondary" process with a 40-eV threshold. When the $\lambda = 11{,}287\text{ Å}$ cascade component is subtracted from the $\lambda = 8446\text{ Å}$ data, a smooth excitation function results. In a similar fashion, when the cascade component is removed from the $\lambda 1304$ Å data, the direct cross sections for exciting the $3s^1\ ^3S^0$ state show little or no evidence of a secondary high-energy process. Thus, in O_2 the structure evident in the cross section shapes is an artifact of the cascade process. The situation in nitrogen is quite different. In this case, the evidence does indicate a significant role for a simultaneous ionization and excitation (Mumma and Zipf, 1971a; Filippelli et al., 1982).

3. Quintet Manifold

The dissociative excitation of the quintet states of atomic oxygen is particularly interesting because it can be studied by both time-of-flight and optical techniques. These studies have yielded line shape, lifetime, cross section, and released kinetic-energy data that have revealed significant differences in the dissociative excitation of the triplet and quintet systems and have provided a convenient technique for calibrating the surface detectors used in TOF work. In contrast to the triplet system, only a single VUV transition at $\lambda 1356$ Å couples the high-lying quintet manifold to the $O(^3P)$ ground state. This doublet is emitted by the metastable $O(^5S^0)$ state, which has a radiative lifetime of 185 μs. A considerable body of translational-energy spectra for the $O(^5S)$ state produced in the dissociation of O_2, NO, CO, and CO_2 exist and from this database the Doppler linewidth of the $\lambda 1356$ Å feature has been deduced (cf. Fig. 14). The kinetic energy gained by the $O(^5S^0)$ atoms when O_2 is dissociated by electron impact is quite large in contrast to the $O(^3S^0)$ state (Fig. 27). Apparently, for electron energies > 20 eV, the

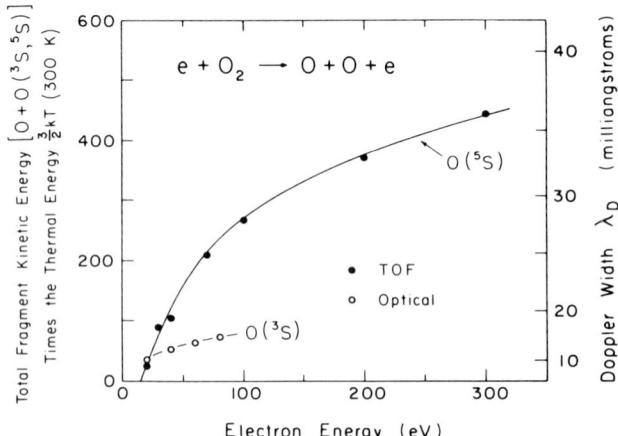

Fig. 27. Total fragment kinetic energy released in the dissociative excitation of the $O(^3S)$ and $O(^5S)$ states by electron impact on O_2 and the Doppler width of the radiation emitted by these species as a function of the energy of the incident electron. [From Zipf and Wells (1980).]

4 Dissociation of Molecules by Electron Impact

dissociation of higher states, which does not lead to excited triplet atoms, rapidly becomes the dominant $O(^5S^0)$ source. Such behavior is not observed in CO_2 where the $O(^3S^0)$ and $O(^5S^0)$ atoms have identical released kinetic-energy spectra (Zipf and Wells, 1980) (Figure 33).

The difference in the triplet–quintet behavior can also be seen quite clearly in a comparison between Figs. 28 and 29, which show the direct cross section for a level plotted versus principal quantum number. Quite remarkably, the cross section for exciting the $n = 2\ ^3S^0$, ^3P, $^3D^0$, $^5S^0$, ^5P, and $^5D^0$ states are about the same (2×10^{-18} cm^2 at 100 eV), but the direct cross section for the quintet states decreases much more rapidly with n. This development implies that the contribution of the quintet states to the total OI Rydberg cross section

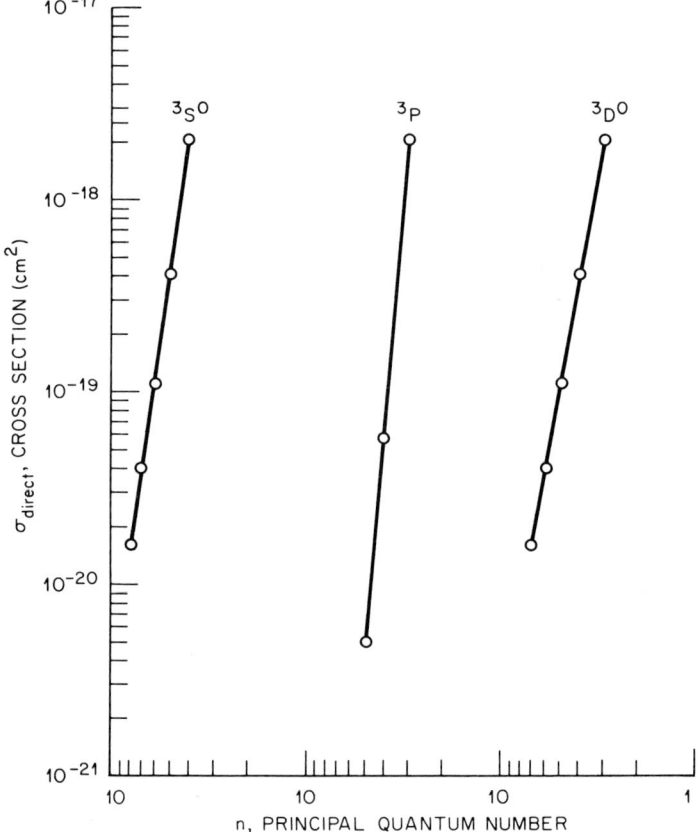

Fig. 28. Cross sections for the direct excitation of OI atoms in specific triplet quantum states as a function of the principal quantum number due to electron-impact dissociation of O_2 at 100 eV (Gorman and Zipf, 1984).

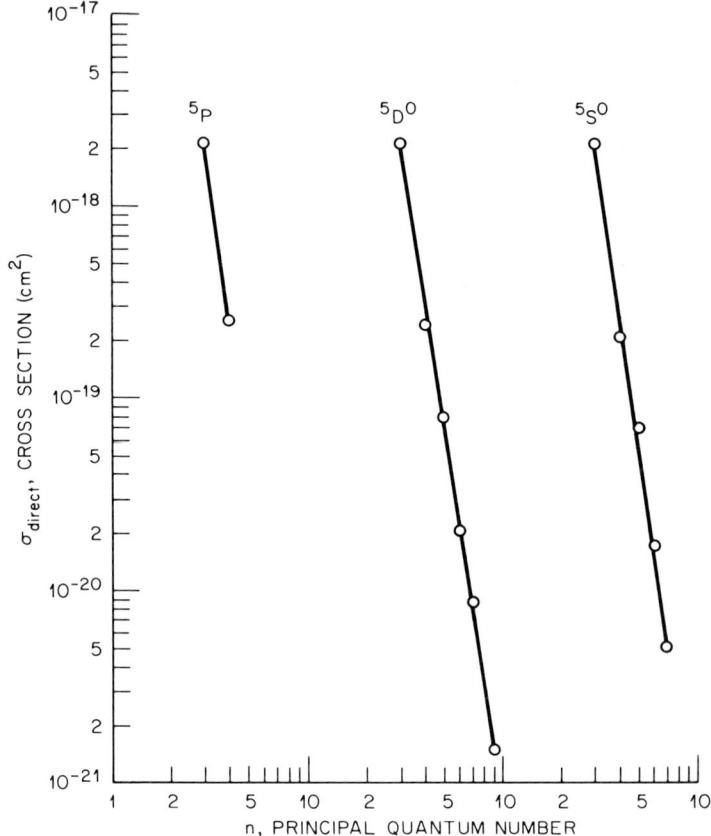

Fig. 29. Cross sections for the direct excitation of OI atoms in specific quintet quantum states as a function of the principal quantum number due to electron-impact dissociation of O_2 at 100 eV (Gorman and Zipf, 1984).

is comparatively small, and Zipf *et al.* (1983) estimate that σ_R(quintets) $\geq$ 1.9 × 10^{-21} cm^2 at 100 eV. Unfortunately, O_2 is the one target gas for which TOF data on high-lying OI Rydberg states are nonexistent, so a direct comparison of the estimate for σ_R based on optical and TOF experiments cannot be made at this time.

The excitation functions for the quintet transitions are quite similar in appearance to those for the triplets. Figure 30 shows such data for the 3p ^{5}P–3s ^{5}S^0 λ7774 Å transition, which accounts for nearly 75% of the O(^{5}S^0) excitation via cascade. Once again, the inflection in the excitation function near 30 eV is mostly a consequence in the shape of the cross section

4 Dissociation of Molecules by Electron Impact

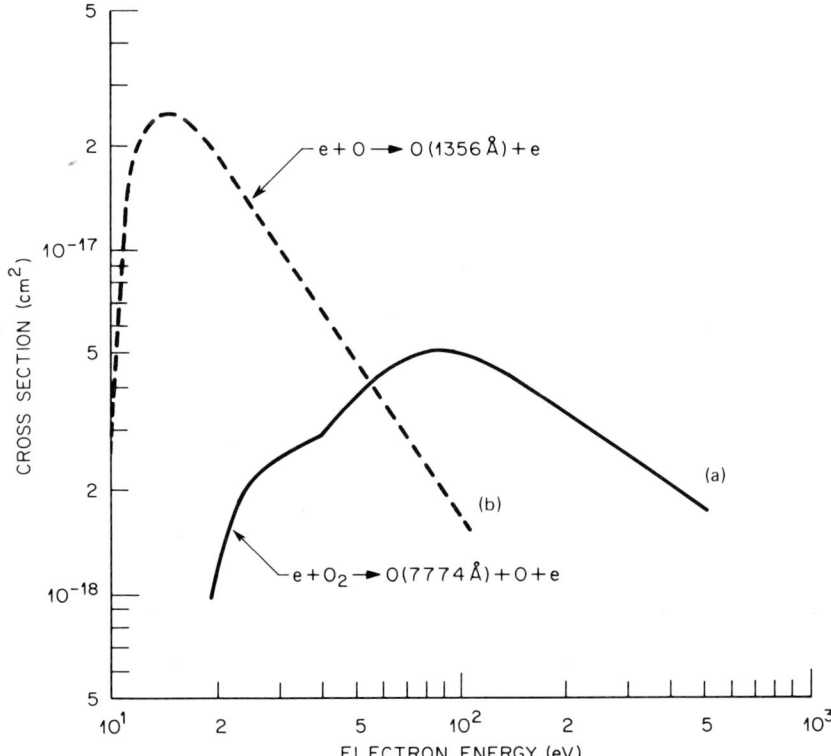

Fig. 30. (a) Excitation cross section for the OI(3s $^5S^0$–3p ^5P) $\lambda 7774$ Å multiplet excited by electron-impact dissociation of O_2. (b) Excitation cross section of the $\lambda 1356$ Å OI(2p ^3P– 3s $^5S^0$ multiplet excited by electron impact on atomic oxygen. Note the striking character of the spin-exchange process for the atomic target (Stone and Zipf, 1974).

for the cascade components rather than of the appearance of a new process with a high energy threshold.

Early attempts to measure the total dissociative excitation cross section for the quintet manifold were compromised by limited data on the velocity distribution of the O($^5S^0$) atoms which are needed in order to make field-of-view corrections and because the measured $^5S^0$ radiative lifetime proved to be a factor of 3 shorter than the theoretical value (Garstang, 1961). The cross section data in Fig. 31, which were derived from an optical study of the quintet manifold, yield a total excitation cross section for the OI quintet system of 6.9×10^{-18} cm^2 ± 15% at 100 eV. This value also applies to the excitation of the metastable O($^5S^0$) state through which all of the excitation energy is ultimately channeled.

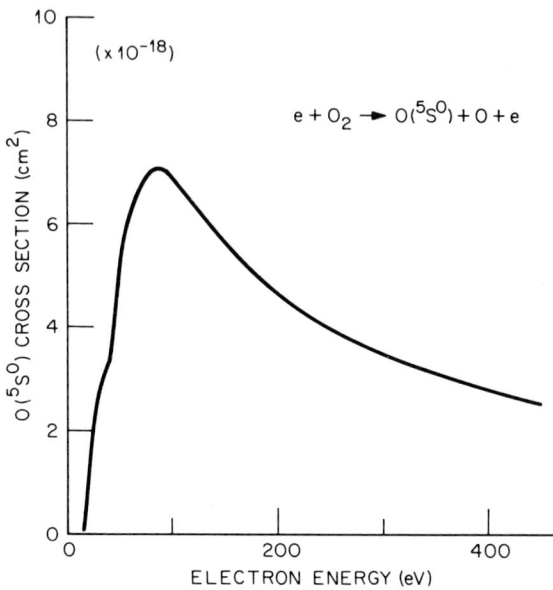

Fig. 31. Total cross section as a function of electron energy for the excitation of the metastable $O(^5S^o)$ state by electron-impact dissociation of O_2 (Gorman and Zipf, 1984).

Finally, it is worth mentioning the TOF study of Kiefl et al. (1980) who have measured the cross section for the process,

$$O(^5S^o) + O_2 + KE \longrightarrow O(^3S^o) + O_2$$
$$O(^3S) \longrightarrow O(^3P) + h\nu \quad (\lambda 1304 \text{ Å}) \tag{40}$$

which attains a peak value of 3×10^{-17} cm^2 for an incident $O(^5S^o)$ kinetic energy of 3 eV. Since the total cross section for exciting the triplet and quintet manifolds in electron-impact dissociation of O_2 are nearly equal, process (40) could more than double the apparent excitation rate for the OI($\lambda 1304$ Å) resonance radiation in moderate-pressure gas discharges and thus complicate the latter's use as a diagnostic tool.

C. Carbon Monoxide and Dioxide

The Earth, Venus, and Mars are planets of comparable size and orbital position. Presumably, they were formed from similar solar material at approximately the same time and thus began the evolution of their planetary atmospheres some 5 billion years ago from essentially the same starting point. Today, these atmospheres are remarkably different. Mars and Venus have atmospheres dominated by carbon dioxide with only trace amounts of N_2 and O_2, and no life. The terrestrial atmosphere, on the other hand, is rich in nitrogen, oxygen, and water vapor. Thus, a major challenge in the study of planetary evolution is to understand the mechanisms that led to such remarkable differences and provided favorable conditions ~ 3.5 billion years ago for the emergence of life on Earth.

The dissociative properties of CO and CO_2 are crucial to this study. In particular, CO_2^+ dissociative recombination,

$$CO_2^+ + e \longrightarrow CO + O \tag{41}$$

and the dissociation of the neutral molecules by solar EUV absorption,

$$h\nu + CO_2 \longrightarrow CO + O \tag{42}$$

and by photoelectron impact,

$$e + CO_2 \longrightarrow CO + O + e \tag{43}$$

$$e + CO \longrightarrow C + O + e \tag{44}$$

figure prominently in the airglows observed by the Mariner and Pioneer spacecraft that have visited Mars and Venus. Furthermore, because these processes release fragments with considerable kinetic energy (cf. Figs. 3 and 5), they have also played an important role in determining the rate of atomic oxygen escape from the exospheres of Mars and Venus.

CO and CO_2 are also major constituents in cometary coma, where photodissociative processes dominate. Since comets were formed in the earliest moments of the solar system and are still thought to have primitive atmospheres largely unaffected by time, the study of the contrasts between the atmospheres of comets and mature planets is likely to be very fruitful, and interest in this research area continues to grow as we approach the first spacecraft rendezvous with comet Halley in 1985.

In this context there is much interest in the variety of excited species produced by the dissociative excitation of CO, CO_2, and H_2O since much of what we know about the physics and chemistry of comets is deduced indirectly from remote optical observations of these species. The cross sections for the excitation of a large number of CI, CII, OI, and OII multiplets in the wavelength interval from 500 Å to 1.3 μm have now been measured (Mumma et al., 1971, 1972; Wells et al., 1972, 1973). Table V gives some representative cross section results and Fig. 32 shows the dissociative excitation cross section for the OI($2p\ ^3P$–$3s\ ^3S^0$, $\lambda 1304$ Å) multiplet excited by electron impact on CO and CO_2.

Once again the use of the carbon and oxygen multiplets for diagnostic purposes requires some care. Because of the ample kinetic energy released in

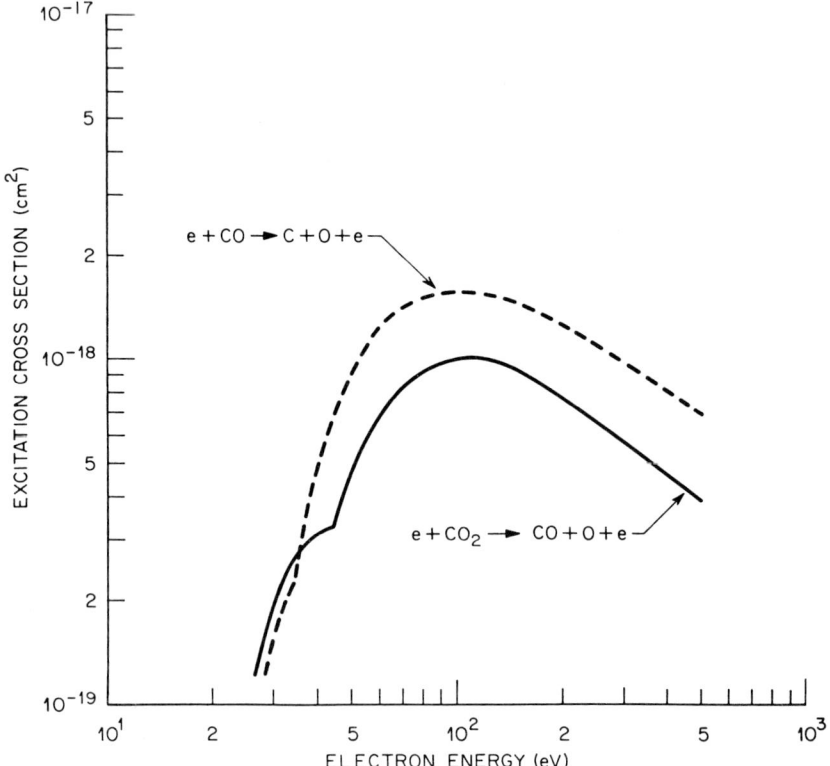

Fig. 32. Cross section for the dissociative excitation of the OI(2p ^{3}P–3s ^{3}S^0, λ1304 Å) multiplet by electron impact on CO and CO_2 (Mumma et al., 1972).

the dissociation process, the radiation emitted by the excited fragments will be highly Doppler broadened. This can be seen in Fig. 14, which shows that the OI(λ1356 Å) transition has a Doppler width equivalent to a kinetic temperature of 60,000 K when excited by electron-impact dissociation of CO and CO_2. The degree of the Doppler broadening varies considerably with the energy of the incident electron (Fig. 33) so that the opacity of a medium optically thick for thermalized atomic radiation will depend on the nature of the electron-impact source for the dissociative component.

In addition to atomic radiation the dissociative excitation of CO_2 is also a copious source of molecular band emission and chemically active metastable states. The excitation of two molecular states [CO(A $^1\Pi$) and CO(a $^3\Pi$)] by process (43) has been at the center of several controversies involving inconsistencies between lifetime, TOF, and inelastic electron scattering measurements on these species. For example, the CO(A $^1\Pi$) state relaxes radiatively

4 Dissociation of Molecules by Electron Impact

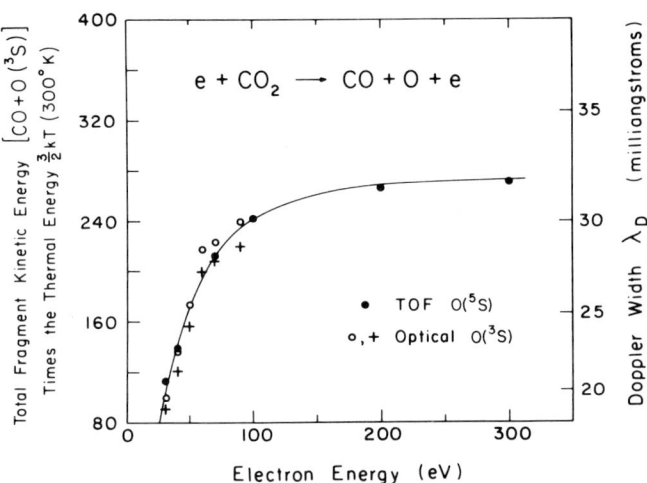

Fig. 33. Total fragment kinetic energy released in the dissociative excitation of the $O(^3S^0)$ and $O(^5S^0)$ states by electron impact on CO_2 and the Doppler width of the radiation emitted by these species as a function of the energy of the incident electron (Zipf and Wells, 1980).

by emitting the $CO(A\,^1\Pi - X\,^1\Sigma^+)$ fourth positive band system and has a lifetime of ~ 10.5 ns. Based on level-crossing spectroscopy, Hesser (1968) found a value of 0.094 for the total absorption oscillator strength for this transition. However, Lassettre and Skerbele (1971) obtained a value $f = 0.195$ from electron energy loss spectra. Mumma et al. (1975) subsequently showed that the source of this discrepancy was due to a nonconstant electronic transition moment R_e. Unfortunately, the variation of R_e with internuclear distance is often large for many molecular band systems and is ignored frequently by applied workers who only use Franck–Condon factors in the analysis of spectroscopic data. Such simplifications can lead to disastrous results when the quantum states involved are perturbed. The manifold of $N_2(^1\Sigma_u^+)$ and $(^1\Pi_u)$ states mentioned earlier provides a good example of this problem. Detailed studies of the N_2 singlet bands (McLaughlin, 1977) have shown that these bands are so highly perturbed that their observed intensities bear little or no resemblance to values calculated using Frank–Condon factors.

In less severely perturbed systems the excitation rate for a particular vibrational level will be proportional to the Frank–Condon factor $q'_{v;0}$ for direct excitation [process (44)]. This can be readily seen in Fig. 34a for the $CO(A\,^1\Pi - X\,^1\Sigma^+)$ band system. However, for dissociative excitation the populations of the vibrational levels of the $A\,^1\Pi$ state will no longer be related to the Franck–Condon factors $q'_{v;0}$ of the $CO(A-X)$ transition but will be a

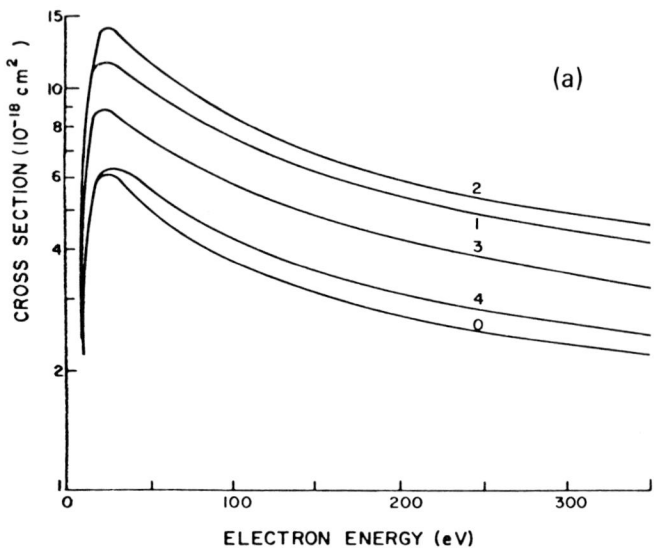

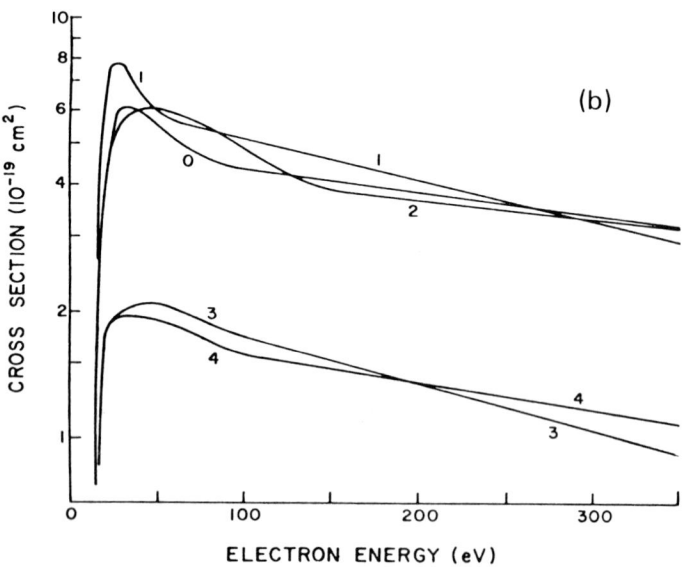

Fig. 34. (a) Absolute cross sections for the excitation of the vibrational levels ($v' = 0 \to 4$) of the CO(A $^1\Pi$) state by electrons impacting on CO. The vibrational population rates are found to be directly proportional to the Franck–Condon factor $q_{v';0}$. (b) Absolute cross sections for the excitation of the ($v' = 0 \to 4$) vibrational levels of the CO(A $^1\Pi$) state by electrons impacting on CO_2. Note the reduced cross section magnitude for $v' \geq 3$ due to competition from autoionization. [From Mumma et al. (1971).]

4 Dissociation of Molecules by Electron Impact

property of the dissociation mechanism of the CO_2 molecule. This can be verified by comparing Figs. 34a and 34b; the vibrational distribution in the $A\,^1\Pi$ state produced dissociatively is quite different and shows a sharp decrease in the excitation cross section for vibrational levels ($v \geq 3'$) of the $A\,^1\Pi$ state. This striking behavior is due to the effects of a competing autoionization channel

$$e + CO_2 \longrightarrow e + CO_2{}^* \longrightarrow CO(A\,^1\Pi, v' \geq 3) + O + e \qquad (45)$$

$$e + CO_2 \longrightarrow e + CO_2{}^* \longrightarrow CO_2{}^+ + 2e \qquad (46)$$

that becomes available when the internal energy of the $CO_2{}^*$ intermediate state is equal to the ionization potential of $CO_2{}^+$.

There are a number of other problems and complications that arise in the dissociative excitation of molecular bands and are worth noting. First, in many molecular systems the radiative lifetime of the upper state varies significantly as a function of the rotational quantum number J. For example, for the very important metastable $N_2(A\,^3\Sigma_u{}^+)$ state the variation is modest with the odd and even fine-structure components for a particular K level, differing by a factor of 2 in radiative lifetime. The $CO(a\,^3\Pi)$ state, on the other hand, represents the opposite extreme. In this instance the radiative lifetime varies profoundly with J with $\tau_r \sim 2$ ms for very large J values but hundreds of milliseconds for the lowest J values. Thus, the actual appearance of the $CO(a\,^3\Pi-X\,^1\Sigma^+)$ Cameron bands in the laboratory depends critically on the geometry of the apparatus and is further complicated in dissociative excitation by the release of kinetic energy, which tends to preferentially remove the longer-lived J states from the field of view. This complication has impeded cross section work on the $CO(a\,^3\Pi)$ state for both direct and dissociative excitation of these bands by electron impact on CO and CO_2, respectively.

In addition to these lifetime effects the apparent rotational temperature of a molecular band can be quite high ($\sim 20{,}000$ K) when the molecular state is excited dissociatively. Mumma et al. (1975) have studied the rotational excitation of the $CO(A\,^1\Pi-X\,^1\Sigma^+)$ fourth positive bands. When this system is excited by direct electron impact on carbon monoxide, the observed rotational temperature is essentially the same as the kinetic temperature of the target gas, and only levels with comparatively small J values are effectively excited. However, when the fourth positive system is excited by process (43), the newly formed $CO(A\,^1\Pi)$ is left rotationally excited and the band structure is well developed to quite large J values (~ 100). The bands are observed to be quite broad in wavelength extent and because of band overlap the spectrum acquires a nearly continuous character (Mumma et al., 1975) which makes absolute intensity measurements and identification very difficult.

In addition to this complication there is a more general problem with the body of dissociative excitation cross sections that have been amassed over the past decade, and that is spectral purity. In Section II.A.1 and Fig. 9, we discussed briefly the $CO(a\ ^3\Pi-X\ ^1\Sigma^+)$ Cameron band cross section measurements as an illustration of how order of magnitude errors are possible when monochromators of limited resolution (5–20 Å) are used in these studies; the situation is even worse when filtered photometers are used to study weak emission features. Future work in this field will require the use of high-resolution, low-noise instruments and better calibration standards. Fortunately, such instrumentation is currently available so the next few years should see the emergence of a second-generation set of highly accurate dissociative excitation cross sections, linewidth measurements, and translational-energy spectra.

D. Other Molecules

Studies of the dissociation by electron impact of a number of other molecules have been made. These include H_2 (D_2, HD) (Corrigan, 1965; Kupriyanov, 1965, 1968; Leventhal et al., 1967; Vroom and de Heer, 1969; Mumma and Zipf, 1971a; Misakian and Zorn, 1972; Stone and Zipf, 1972; Schiavone et al., 1975, 1979; Möhlmann et al., 1977; Boesten and Heideman, 1980; see also review by de Heer, 1981); HCl, HBr (Möhlmann and de Heer, 1979); NO (Lawrence, 1970; Mentall and Morgan, 1972); H_2O (Kupriyanov, 1965, 1968; Lawrence, 1970; Kouchi et al., 1979; Möhlmann and de Heer, 1979); H_2S (Möhlmann and de Heer, 1979); OCS (Van Brunt and Mumma, 1975); SO_2 (Becker et al., 1983); NH_3 (Carnahan et al., 1981; Kouchi et al., 1982); CH_4 (CD_4) (Finn et al., 1975; Winters, 1975; Donohue et al., 1977; Schiavone et al., 1977, 1979; McLaughlin and Zipf, 1978; Kouchi et al., 1982); C_2H_4 (Kupriyanov, 1965, 1968; Donohue et al., 1977; Schiavone et al., 1977); C_2H_6 (Schiavone et al., 1977; Winters, 1979; Schiavone et al., 1979); CH_3OH (Donohue et al., 1977); propane, n-hexane, ethene, allene, cyclopropane, and 1,3-butadiene (Schiavone et al., 1979); SiH_4, Si_2H_6 (Perrin et al., 1982); halogenated methanes (Danilevskii, 1977); CF_4, CF_3H (van Sprang et al., 1978; Winters and Inokuti, 1982); CF_3Cl, CF_2Cl_2, $CFCl_3$ (van Sprang et al., 1978); C_2F_6, C_3F_8 (Winters and Inokuti, 1982).

The work on hydrogen and its isotopes has been quite extensive, with particular emphasis on the dissociative excitation of the Lyman and Balmer transitions of atomic hydrogen (Vroom and de Heer, 1969; de Heer and Carriere, 1971; Carriere and de Heer, 1972; Khayrallah, 1976; Freund et al., 1976; Möhlmann et al., 1976, 1978; Hartfuss et al., 1976; Hartfuss and Roll, 1977; Karolis and Harting, 1978), of the metastable 2S state and high-Rydberg atoms (Vroom and de Heer, 1969; Misakian and Zorn, 1971, 1972; Levanthal *et al.*, 1967; Czuchlewski and Ryan, 1973; Schiavone *et al.*, 1975; Carnahan and Zipf, 1977), and of dissociative ionization (Kieffer and Dunn, 1967; Van Brunt and Kieffer, 1970; Crowe and Mc Conkey, 1973a,b; Stockdale *et al.*, 1975; Burrows *et al.*, 1980). These data have been interpreted with some success in the case of H_2 in terms of the core-ion model of dissociation (Schiavone *et al.*, 1975) and have served as a test of the selection rules governing the angular distribution of the fragment atoms first stated by Dunn (1962). Similar studies on HD have revealed a number of isotope anomalies which neither the core-ion model nor semiclassical models of dissociative autoionization can accommodate (Carnahan and Zipf, 1977; Burrows *et al.*, 1980). There are also continuing controversies concerning the rates for populating the ns, np, and nd sublevels of atomic hydrogen produced in the dissociative excitation of H_2 and a variety of light hydrocarbon molecules (Möhlmann *et al.*, 1976; McLaughlin and Zipf, 1978), and, as yet, there have been no measurements of the total cross section for electron-impact dissociation of H_2. Thus, ample work remains for experimentalists.

References

Aarts, J. F. M., and de Heer, F. J. (1971). *Physica* **52**, 45–73.
Ajello, J. M. (1970). *J. Chem. Phys.* **53**, 1156–1165.
Ajello, J. M. (1971a). *J. Chem. Phys.* **55**, 3158–3168.
Ajello, J. M. (1971b). *J. Chem. Phys.* **55**, 3169–3177.
Allen, C. W. (1963). "Astrophysical Quantities." Athlone, London.
Baravian, G., Godart, J., and Sultan, G. (1982). *Phys. Rev. A* **25**, 1483–1495.
Becker, K., Van Wijngaarden, W., and McConkey, J. W. (1983). *Planet. Space Sci.* **31**, 197–206.
Boesten, L. G. J., and Heideman, H. G. M. (1980). *Physica* **98C**, 242–245.
Borst, W. L. (1971). *Rev. Sci. Instrum.* **42**, 1543–1544.
Borst, W. L., and Zipf, E. C. (1971a). *Phys. Rev. A* **3**, 979–989.
Borst, W. L., and Zipf, E. C. (1971b). *Phys. Rev. A* **4**, 153–161.
Brinkmann, R. T., and Trajmar, S. (1970). *Ann. Geophys.* **26**, 201–207.
Brooks, N. H., Rohrlich, D., and Smith, W. H. (1977). *Astrophys. J.* **214**, 328–330.
Burrows, M. D., McIntyre, L. C., Ryan, S. R., and Lamb, W. E., Jr. (1980). *Phys. Rev. A* **21**, 1841–1847.
Carnahan, B. L., and Zipf, E. C. (1977). *Phys. Rev. A* **16**, 991–1002.
Carnahan, B. L., Kao, W.-W., and Zipf, E. C. (1981). *J. Chem. Phys.* **74**, 5149–5161.
Carriere, J. D., and de Heer, F. J. (1972). *J. Chem. Phys.* **56**, 2993–2996.
Carroll, P. K., and Collins, C. P. (1969). *Can. J. Phys.* **47**, 563–589.
Carroll, P. K., and Yoshino, K. (1972). *J. Phys. B* **5**, 1614–1633.
Christensen, A. B., and Cunningham, A. J. (1978). *J. Geophys. Res.* **83**, 4393–4396.
Christensen, A. B., Sivjee, G. G., and Hecht, J. H. (1983). *J. Geophys. Res.* **88**, 4911–4917.
Cook, G. R., Metzger, P. H., Ogawa, M., Becker, R. A., and Ching, B. K. (1965). Aerospace Corporation Research No. TDR-469(9260-01)-4.
Corrigan, S. J. B. (1965). *J. Chem. Phys.* **43**, 4381–4386.
Crowe, A., and Mc Conkey, J. W. (1973a). *J. Phys. B* **6**, 2088–2107.

Crowe, A., and Mc Conkey, J. W. (1973b). *Phys. Rev. Lett.* **31**, 192–196.
Czuchlewski, S. J. (1973). Ph.D. Thesis, Yale University.
Czuchlewski, S. J., and Ryan, S. R. (1973). *Bull. Am. Phys. Soc.* **18**, 688.
Danilevskii, N. P., Koppe, V. T., Koval, A. G., and Khovanskii, N. A. (1977). *Zh. Eksp. Teor. Fiz.* **73**, 450–453 (*Sov. Phys. JETP* **46**, 236–237, 1977.)
de Heer, F. J. (1981). *Physica Scripta* **23**, 170–178.
de Heer, F. J., and Carriere, J. D. (1971). *J. Chem. Phys.* **55**, 3829–3835.
Deleanu, L., and Stockdale, J. A. D. (1975). *J. Chem. Phys.* **63**, 3898–3906.
Donohue, D. E., Schiavone, J. A., and Freund, R. S. (1977). *J. Chem. Phys.* **67**, 769–780.
Dressler, K. (1969). *Can. J. Phys.* **47**, 547–561.
Dunn, G. H. (1962). *Phys. Rev. Lett.* **8**, 62–64.
Dunn, G. H., and Kieffer, L. J. (1963). *Phys. Rev.* **132**, 2109–2117.
Dunning, F. B., Rundel, R. D., and Stebbings, R. F. (1975). *Rev. Sci. Instrum.* **46**, 697–701.
Erdman, P. W., and Zipf, E. C. (1983). *Planet. Space Sci.* **31**, 317–321.
Erdman, P. W., Espy, P. J., and Zipf, E. C. (1981). *Geophys. Res. Lett.* **8**, 1163–1166.
Filippelli, A. R., Sharpton, F. A., Lin, C. C., and Murphy, R. E. (1982). *J. Chem. Phys.* **76**, 3597–3606.
Finn, T. G., Carnaham, B. L., Wells, W. C., and Zipf, E. C. (1975). *J. Chem. Phys.* **63**, 1596–1604.
Freund, R. S., Schiavone, J. A., and Brader, D. F. (1976). *J. Chem. Phys.* **64**, 1122–1127.
Garstang, R. H. (1961). *Proc. Cambridge Philos. Soc.* **57**, 115–120.
Geiger, J., and Schröder, B. (1969). *J. Chem. Phys.* **50**, 7–11.
Gorman, M. R., and Zipf, E. C. (1981). *Bull. Am. Phys. Soc.* **27**, 100.
Gorman, M. R., and Zipf, E. C. (1984). *J. Chem. Phys.* (submitted)
Green, A. E. S. (1976). *J. Chem. Phys.* **65**, 154–167.
Green, A. E. S., and Barth, C. A. (1967). *J. Geophys. Res.* **72**, 3975–3986.
Harteck, P., and Dondes, S. (1956). *J. Chem. Phys.* **24**, 619.
Hartfuss, H. J., and Roll, H. (1977). *Z. Naturforsch.* **32a**, 720–723.
Hartfuss, H. J., Neumann, J., and Schneider, H. D. (1976). *Z. Naturforsch.* **31a**, 1292–1297.
Hesser, J. E. (1968). *J. Chem. Phys.* **48**, 2518–2535.
Hesser, J. E., and Dressler, K. (1966). *J. Chem. Phys.* **45**, 3149–3150.
Hibbert, A., and Bates, D. R. (1981). *Planet. Space Sci.* **29**, 263–268.
Huffman, R. E., Tanaka, Y., and Larrabee, J. C. (1963). *J. Chem. Phys.* **39**, 910–925.
Jackman, C. H., Garvey, R. H., and Green, A. E. S. (1977). *J. Geophys. Res.* **82**, 5081–5090.
Karolis, C., and Harting, E. (1978). *J. Phys. B* **11**, 357–369.
Khayrallah, G. A. (1976). *Phys. Rev. A* **13**, 1989–2003.
Kieffer, L. J., and Dunn, G. H. (1967). *Phys. Rev.* **158**, 61–65.
Kieffer, L. J., and Van Brunt, R. J. (1967). *J. Chem. Phys.* **46**, 2728–2734.
Kiefl, H. U., Borst, W. L., and Fricke, J. (1980). *Phys. Rev.* **21A**, 518–524.
Knight, R. D. (1982). *Phys. Rev. Lett.* **48**, 792–795.
Kouchi, N., Ito, K., Hatano, Y., Oda, N., and Tsuboi, T. (1979). *Chem. Phys.* **36**, 239–245.
Kouchi, N., Ohno, M., Ito, K., Oda, N., and Hatano, Y. (1982). *Chem. Phys.* **67**, 287–294.
Kupriyanov, S. E. (1965). *Zh. Eksp. Teor. Fiz.* **48**, 467–475. (*Sov. Phys. JETP* **21**, 311–317, 1965.)
Kupriyanov, S. E. (1968). *Zh. Eksp. Teor. Fiz.* **55**, 460–468. (*Sov. Phys. JETP* **28**, 240–244, 1969.)
Lassettre, E. N. (1969). *Can. J. Chem.* **47**, 1733–1773.
Lassettre, E. N. (1974). *In* "Methods of experimental physics" (D. Williams, ed.), Vol. 3, Part B, p. 868. Academic Press, New York.
Lassettre, E. N., and Krasnow, M. E. (1964). *J. Chem. Phys.* **40**, 1248–1255.
Lassettre, E. N., and Skerbele, A. (1971). *J. Chem. Phys.* **54**, 1597–1607.
Lawrence, G. M. (1970). *Phys. Rev. A* **2**, 397–407.
Lawrence, G. M., and McEwan, M. J. (1973). *J. Geophys. Res.* **78**, 8314–8319.

Lawrence, G. M., Mickey, D. L., and Dressler, K. (1968). *J. Chem. Phys.* **48**, 1989–1994.
Leventhal, M., Robiscoe, R. T., and Lea, K. R. (1967). *J. Chem. Phys.* **158**, 49–56.
Locht, R., Schopman, J., Wankenne, H., and Momigny, J. (1975). *Chem. Phys.* **7**, 393–404.
Lofthus, A. (1957). *Can. J. Phys.* **35**, 216–234.
Long, R. L., Cox, D. M., and Smith, S. J. (1968). *J. Res. Natl. Bur. Std.* **72A**, 521.
McFarlane, R. A. (1965). *Phys. Rev.* **140**, 1070–1071.
McFarlane, R. A. (1966). *Phys. Rev.* **146**, 37–39.
McLaughlin, R. W. (1977). "Vacuum Ultraviolet and Visible Radiation from Electron Impact Excitation of Nitrogen, Hydrogen, Oxygen, the Light Hydrocarbons, and the Rare Gases," Ph.D. Thesis, University of Pittsburgh.
McLaughlin, R. W., and Zipf, E. C. (1978). *Chem. Phys. Lett.* **55**, 62–66.
Meier, R. R. (1982). *J. Geophys. Res.* **87**, 6307–6316.
Meier, R. R., Strickland, D. J., Feldman, P. D., and Gentieu, E. P. (1980). *J. Geophys. Res.* **85**, 2177–2184.
Meier, R. R., Conway, R. R., Feldman, P. D., Strickland, D. J., and Gentieu, E. P. (1982). *J. Geophys. Res.* **87**, 2444–2452.
Mentall, J. E., and Morgan, H. D. (1972). *J. Chem. Phys.* **56**, 2271–2277.
Misakian, M., and Zorn, J. C. (1971). *Phys. Rev. Lett.* **27**, 174–177.
Misakian, M., and Zorn, J. C. (1972). *Phys. Rev. A* **6**, 2180–2196.
Misakian, M., Mumma, M. J., and Faris, J. F. (1975). *J. Chem. Phys.* **62**, 3442–3453.
Möhlmann, G. R., and de Heer, F. J. (1979). *Chem. Phys.* **40**, 157–162.
Möhlmann, G. R., Tsurbuchi, S., and de Heer, F. J. (1976). *Chem. Phys.* **18**, 145–154.
Möhlmann, G. R., de Heer, F. J., and Los, J. (1977). *Chem. Phys.* **25**, 103–116.
Möhlmann, G. R., Shima, K. H., and de Heer, F. J. (1978). *Chem. Phys.* **28**, 331–341.
Moiseiwitsch, B. L., and Smith, S. J. (1968). *Rev. Mod. Phys.* **40**, 238–353.
Morgan, H. D., and Mentall, J. E. (1983). *J. Chem. Phys.* **78**, 1747–1757.
Morgan, L. A., Callaway, J., and McDowell, M. R. C. (1977). *J. Phys.* **B10**, 3297–3305.
Morrison, M. D., Cunningham, A. J., and Christensen, A. B. (1983). *J. Quant. Spectrosc. Radiat. Transfer* **29**, 137–143.
Mumma, M. J., and Zipf, E. C. (1971a). *J. Chem. Phys.* **55**, 1661–1669.
Mumma, M. J., and Zipf, E. C. (1971b). *J. Chem. Phys.* **55**, 5582–5588.
Mumma, M. J., Stone, E. J., and Zipf, E. C. (1971). *J. Chem. Phys.* **54**, 2627–2634.
Mumma, M. J., Stone, E. J., Borst, W. L., and Zipf, E. C. (1972). *J. Chem. Phys.* **57**, 68–75.
Mumma, M. J., Stone, E. J., and Zipf, E. C. (1975). *J. Geophys. Res.* **80**, 161–167.
Niehaus, A. (1967). *Z. Naturforsch.* **22a**, 690–700.
Nier, A. O., and McElroy, M. B. (1977). *J. Geophys. Res.* **82**, 4341–4349.
Park, H., Feldman, P. D., and Fastie, W. G. (1977). *Geophys. Res. Lett.* **4**, 41–44.
Perrin, J., Schmitt, J. P. M., De Rosny, G., Drevillon, B., Huc, J., and Lloret, A. (1982). *Chem. Phys.* **73**, 383–394.
Poland, H. M., and Lawrence, G. M. (1973). *J. Chem. Phys.* **58**, 1425–1429.
Porter, H. S., Jackman, C. H., and Green, A. E. S. (1976). *J. Chem. Phys.* **65**, 154–167.
Pradhan, A. K., and Saraph, H. E. (1977). *J. Phys. B* **10**, 3365–3376.
Prasad, S. S., and Tarafdar, S. P. (1983). *Astrophys. J.* **267**, 603–609.
Rees, M. H. (1975). *Planet. Space Sci.* **23**, 1589–1596.
Rundel, R. D., Dunning, F. B., Howard, J. S., Riola, J. P., and Stebbings, R. F. (1973). *Rev. Sci. Instrum.* **44**, 60–63.
Rusch, D. W., and Gerard, J.-C. (1980). *J. Geophys. Res.* **85**, 1285–1290.
Rusch, D. W., Gerard, J.-C., and Sharp, W. E. (1978). *Geophys. Res. Lett.* **5**, 1043–1046.
Sampson, J. A. R. (1967). "Techniques of Vacuum Ultraviolet Spectroscopy," p. 226. Wiley, New York.

Schiavone, J. A., Smyth, K. C., and Freund, R. S. (1975). *J. Chem. Phys.* **63**, 1043–1051.
Schiavone, J. A., Donohue, D. E., and Freund, R. S. (1977). *J. Chem. Phys.* **67**, 759–768.
Schiavone, J. A., Tarr, S. M., and Freund, R. S. (1979). *J. Chem. Phys.* **70**, 4468–4473.
Shemansky, D. E. (1982). *EOS* **63**, 1019.
Silverman, S. M., and Lassettre, E. N. (1965). *J. Chem. Phys.* **42**, 3420–3429.
Sivjee, G. G. (1983). *Geophys. Res. Lett.* **10**, 349–352.
Smyth, K. C., Schiavone, J. A., and Freund, R. S. (1973). *J. Chem. Phys.* **59**, 5225–5241.
Stockdale, J. A. D., Anderson, V. E., Carter, A. E., and Deleanu, L. (1975). *J. Chem. Phys.* **63**, 3886–3897.
Stolarski, R. S., and Green, A. E. S. (1967). *J. Geophys. Res.* **72**, 3967–3974.
Stone, E. J., and Zipf, E. C. (1972). *J. Chem. Phys.* **56**, 4646–4650.
Stone, E. J., and Zipf, E. C. (1973). *J. Chem. Phys.* **58**, 4278–4284.
Stone, E. J., and Zipf, E. C. (1974). *J. Chem. Phys.* **60**, 4237–4243.
Stone, E. J., Lawrence, G. M., and Seitel, S. C. (1975). "Abstracts of the IXth International Conference on the Physics of Electronic and Atomic Collisions," p. 816. University of Washington Press, Seattle.
Tilford, S. G., and Wilkinson, P. G. (1964a). *J. Mol. Spectrosc.* **12**, 231–288.
Tilford, S. G., and Wilkinson, P. G. (1964b). *J. Mol. Spectrosc.* **12**, 347–359.
Tilford, S. G., and Wilkinson, P. G. (1965). *Astrophys. J.* **141**, 427–443.
Van Brunt, R. J., and Kieffer, L. J. (1970). *Phys. Rev. A* **2**, 1293–1304.
Van Brunt, R. J., and Kieffer, L. J. (1975). *J. Chem. Phys.* **63**, 3216–3221.
Van Brunt, R. J., and Mumma, M. J. (1975). *J. Chem. Phys.* **63**, 3210–3215.
van Sprang, H. A., Brongersma, H. H., and de Heer, F. J. (1978). *Chem. Phys.* **35**, 51–61.
Venugopalan, M., (ed. 1971). "Reactions under Plasma Conditions," Vol. 1. Wiley (Interscience), New York.
Vriens, L. (1967). *Phys. Rev.* **160**, 100–108.
Vriens, L., Simpson, J. A., and Mielczarek, S. R. (1968). *Phys. Rev.* **165**, 7–15.
Vroom, D. A., and de Heer, F. J. (1969). *J. Chem. Phys.* **50**, 580–590.
Wallace, L. (1962). *Astrophys. J. Suppl. Ser.* **6**, 445–479.
Wells, W. C., Borst, W. L., and Zipf, E. C. (1972). *J. Geophys. Res.* **77**, 69–75.
Wells, W. C., Borst, W. L., and Zipf, E. C. (1973). *Phys. Rev. A* **8**, 2463–2468.
Wells, W. C., Borst, W. L., and Zipf, E. C. (1976). *Phys. Rev. A* **14**, 695–706.
Wells, W. C., Borst, W. L., and Zipf, E. C. (1978). *Phys. Rev.* **17**, 1357–1365.
Wiese, W. L., Smith, M. W., and Glennon, B. M. (1966). "Atomic Transition Probabilities." National Standard References Services, U. S. Government Printing Office, Washington, D.C.
Wilkinson, P. G., and Houk, N. B. (1956). *J. Chem. Phys.* **24**, 528–534.
Williams, A. J., and Doering, J. P. (1969a). *Planet. Space Sci.* **17**, 1527–1537.
Williams, A. J., and Doering, J. P. (1969b). *J. Chem. Phys.* **51**, 2859–2865.
Williams, J. F. (1976). *In* "Electron and Photon Interactions With Atoms," (H. Kleinpoppen, and M. R. C. McDowell, eds.), p. 309. Plenum, New York.
Williams, J. F., and Willis, B. A. (1974). *J. Phys.* **7B**, L61–L65.
Winters, H. F. (1966). *J. Chem. Phys.* **44**, 1472–1476.
Winters, H. F. (1975). *J. Chem. Phys.* **63**, 3462–3466.
Winters, H. F. (1979). *Chem. Phys.* **36**, 353–364.
Winters, H. F., and Inokuti, M. (1982). *Phys. Rev. A* **25**, 1420–1430.
Winters, H. F., Horne, D. E., and Donaldson, E. E. (1964). *J. Chem. Phys.* **41**, 2766–2772.
Wu, C. Y. R., Chen, J. K., and Judge, D. L. (1983). *J. Geophys. Res.* **88**, 2163–2169.
Zipf, E. C. (1973a). *EOS* **54**, 403.
Zipf, E. C. (1973b). *EOS* **54**, 1155.
Zipf, E. C. (1980). *Geophys. Res. Lett.* **7**, 645–648.

Zipf, E. C. (1982). *EOS* **63**, 1050.
Zipf, E. C., and Gorman, M. R. (1980). *J. Chem. Phys.* **73**, 813–819.
Zipf, E. C., and McLaughlin, R. W. (1978). *Planet. Space Sci.* **26**, 449–462.
Zipf, E. C., and Wells, W. C. (1980). *Planet. Space Sci.* **28**, 859–866.
Zipf, E. C., Erdman, P. W., and Finn, T. G. (1973). *Trans. Am. Geophys. Union* **54**, 1151.
Zipf, E. C., McLaughlin, R. W., and Gorman, M. R. (1979). *Planet. Space Sci.* **27**, 719–732.
Zipf, E. C., Espy, P. J., and Boyle, C. F. (1980). *J. Geophys. Res.* **85**, 687–694.
Zipf, E. C., Kao, W. A., and Erdman, P. W. (1983). *J. Chem. Phys.* (submitted).

5 Electron–Molecule Resonances

J. B. Hasted
Department of Physics
Birkbeck College
University of London
London, England

and

D. Mathur
Tata Institute of Fundamental Research
Bombay, India

I.	Introduction	403
II.	Theory of Resonances	405
	A. Resonance Elastic Scattering of Electrons by Atoms	405
	B. Resonance Scattering of Electrons by Molecules	408
	C. Threshold Resonance Behavior in Polar and Other Molecules	416
	D. Dissociative Channel Decay	420
	E. Resonance States above the First Ionization Limit	421
	F. Resonances and Theoretical Chemistry	422
III.	Experimental Techniques for the Study of Resonances	424
	A. Introduction	424
	B. Spectrometer Configurations	425
IV.	Experimental Results	432
	A. Diatomic Molecules	432
	B. Triatomic Molecules	448
	C. Polyatomic Molecules	453
V.	Conclusions	471
	References	471

I. Introduction

Resonances in electron–molecule scattering have been known since the 1960s and extensively reviewed (Schulz, 1973; Massey, 1976). Electron excitation of molecules, in particular vibrational excitation of nonpolar species, usually proceeds most effectively via resonances. An understanding of the decay of resonance states is necessary in various fields of applied physics: in aeronomy, in the pumping of gas lasers, in high-voltage engineering, and in radiation chemistry and physics. However, we shall not in this chapter discuss the possible role of resonances in heavy-particle collisions.

Resonances are discrete states of molecule plus electron that may be formed during a scattering process; they are embedded in and interact with a continuum of scattering states, so they decay into a state of the molecule together with the free scattered electron; their existence is inferred from structure in the appropriate scattering cross section function, occurring at impact energies around the energy of the resonance state. Similar states of molecular ion plus electron, the two electron excited states of molecules, play a role in autoionization and in dielectronic recombination.

An electron approaching a positive ion experiences a $V \propto r^{-1}$ attractive Coulomb potential at large distances r, and within the potential well a range of discrete bound states are possible. But an electron approaching a neutral atom experiences only a weak attractive field, depending upon a high inverse power of r, or varying exponentially with distance. It is not always the case that even a single stable state can be maintained by such a field; usually one such is possible, but almost never more than one. Thus an atomic negative ion may be stable in its ground state, but in general does not possess excited levels of energy below that of the atom plus free electron.

This state of affairs applies equally well to excited species of neutral atoms, which can often maintain one energy level with a bound electron, but since such a negative-ion energy level is in the continuum of scattering states of the ground-state atom plus free electron, it can readily decay to the ground state, or indeed to any neutral atom state which lies at a lower energy, releasing the electron as it does so. Owing to such decay, structure is observed in the appropriate electron–atom scattering cross section function; decay of the temporary bound state into an excited level of the neutral atom sometimes represents a dominant electron excitation mechanism, and is therefore of great importance in electron–atom scattering. The temporary electron–atom bound state contains essentially *two* electrons which are excited, just as in a neutral atom autoionizing state; the decay process is analogous to that of an autoionizing state, and for that reason has sometimes been termed "autodetachment." The resonance state formed from a neutral atom excited state is termed a Feshbach resonance (Feshbach, 1958, 1962).

All of these considerations apply equally well to molecules. However, the additional energy modes available to molecules may complicate the situation. For example, a stable molecular negative ion often possesses vibrational and rotational energy levels above that of the molecule plus free electron; decay can take place from such levels, which therefore can be said to have resonance properties.

The original excited state of the neutral atom or molecule that gives rise to a Feshbach resonance is termed the "parent" of the resonance state. The parent itself possesses a parent state, of the positive ion; this is termed the "grandparent" of the resonance state.

The parenthood and grandparenthood of a molecular resonance state are to be determined from a study of vibrational spacing and equilibrium nuclear separation. It is usually found that these are similar in the resonance state, in the parent, and in the grandparent (Sanche and Schulz, 1972a,b; Spence, 1974a,b). The parent is a Rydberg rather than a valence state, it is rare for a Feshbach resonance to be associated with the latter.

Molecular negative ions with two excited electrons of identical principal quantum number are more stable than those with other configurations, owing to the minimization of the mutual screening. Where the two principal quantum numbers are different there is considerable screening of one electron by the other. These two electrons, of identical principal quantum number, contribute little to the nuclear motion of the ion core, so that the vibrational spacing and equilibrium nuclear separation are similar to those of the parent and the grandparent. Rydberg electrons are essentially nonbonding, rather than bonding or antibonding.

Of special importance is a further type of resonance state, which does not fall under the heading of Feshbach resonance: the "shape resonance." When the electron approaches the atom or molecule with large collisional angular momentum, that is, far from a head-on collision, classical mechanics shows that there is a repulsive centrifugal potential $L^2/2\mu r^2$, where L is the angular momentum and μ the electron–atom reduced mass; when this repulsive potential is taken in conjunction with the static attractive electron–atom interaction, a barrier can be formed. This potential barrier is of such a shape as to allow the existence of a resonance state whose energy is *above* the limit of free atom and free electron. This resonance state decays into available "channels," the ground state and perhaps vibrationally excited states of the molecule, so that structure is observed in the elastic and perhaps other scattering cross section functions.

It is also possible for excited states of atoms and molecules to exhibit shape resonances.

II. Theory Of Resonances

A. Resonance Elastic Scattering of Electrons by Atoms

The cross section σ for the scattering of an electron of mass m and energy E by an atom is related to the phase shift η_l produced by the central potential by the well-known partial-wave formula

$$\sigma = \frac{4\pi}{k^2} \sum_l (2l+1) \sin^2 \eta_l, \tag{1}$$

where $k^2 = 2mE/h^2$, l is the angular momentum quantum number of the scattered electron, and h is the Planck constant.

Fano (1961), following Feshbach (1958, 1962), obtained an expression for the elastic resonance scattering of electrons of zero angular momentum:

$$\sigma(0) = \frac{4\pi}{k^2} \left| \frac{e^{2i\eta} - 1}{2i} - \frac{\tfrac{1}{2} i \Gamma e^{2i\eta}}{E - E_r + \tfrac{1}{2} i \Gamma} \right|^2, \tag{2}$$

where E_r is the energy of the resonance state whose width is Γ. In the limit of zero Γ this expression reduces to Eq. (1), with η determined by the mean field of the undisturbed atom in its ground state. Γ/h is the rate of breakup of the resonance state. Writing

$$\tfrac{1}{2}\Gamma(E_r - E) = \tan \delta. \tag{3}$$

we can readily show that

$$\sigma(0) = (4\pi/k^2) \sin^2(\eta + \delta), \tag{4}$$

so that a resonance phase shift δ is added to the nonresonant phase shift η; this leads to an expression for elastic scattering:

$$\frac{1}{h} \frac{d\sigma}{dE} = \frac{\Gamma/2h}{(E - E_r)^2 + \Gamma^2/4h^2}. \tag{5}$$

When E is close to the resonance energy E_r, the cross section is approximately

$$\sigma(0) \simeq \frac{\pi}{k^2} \frac{\Gamma^2}{(E - E_r)^2 + \tfrac{1}{4}\Gamma^2}. \tag{6}$$

When all possible collisional angular momenta are taken into account, the cross section can be separated into two parts, resonant and nonresonant:

$$\sigma = \sigma_r + \sigma_b, \tag{7}$$

with

$$\sigma_b = (4\pi/k^2) \sum_{l \neq L} (2l+1) \sin^2 \eta_L, \tag{8}$$

$$\sigma_r = (4\pi/k^2)(2L+1)\sin^2(\eta_L + \delta_L)$$
$$= \sigma_a \sin^2(\eta_L + \delta_L)/\sin^2 \eta_L; \tag{9}$$

5 Electron–Molecule Resonances

σ_a is the partial cross section for angular momentum quantum number L in the absence of the resonance state. These equations refer to a situation when there is a resonance state with total angular momentum quantum number L.

A line profile parameter q_L can be introduced, with the cross section written as

$$\sigma = \sigma_a[(q_L + \varepsilon)^2/(1 + \varepsilon^2)] + \sigma_b \tag{10}$$

or

$$\sigma = \sigma_a Q + \sigma_b,$$

where

$$q_L = -\cot \eta_L \tag{11}$$

and

$$\varepsilon = (E - E_r)/\tfrac{1}{2}\Gamma. \tag{12}$$

The parameter q_L is sometimes known as the "Beutler–Fano parameter," and the variation of Q with ε is illustrated in Fig. 1. Beutler–Fano curves govern the shape of the elastic cross section function for electron scattering in the energy region of the resonance. They also govern the shape of a photoionization or photoabsorption function is the region of an autoionization state, as was demonstrated in Beutler's spectra. For many years the brightening of spectral absorption in the region of these states has been a well-known feature. Electron elastic and total scattering cross section functions for

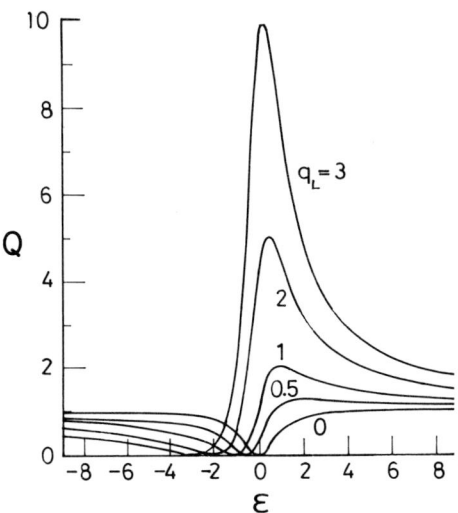

Fig. 1. Variation of Beutler–Fano line shape Q with off-resonance energy ε, for selected values of parameter q_L.

atoms with resonance states demonstrate this shape, although it is often observed in molecular data by the vibrational fine structure. We should note that q_L can take either positive or negative values.

When $q_L = 0$, the cross section function shows no maximum, but only a dip, or a window, as is seen from Fig. 1. Such resonances in atoms have sometimes been referred to as "window resonances."

B. Resonance Scattering of Electrons by Molecules

Many of the additional features of resonance scattering exhibited by molecules are associated with the nuclear motions of the atoms in the molecule itself, and will for simplicity be introduced for the case of diatomic molecules, for which only the nuclear separation R varies as the molecules vibrate. The width Γ of the resonant state is itself a function of R.

Since the electronic motion in a molecule is much faster than the nuclear motion, it is usual to express the molecule wave function as a product $\psi(r, R)\chi(R)$ combining both electronic coordinate r and nuclear coordinate R. For a bound electronic level in a vibrational state with quantum number v, $\chi(R)$ is concentrated between the two sides of the molecular anharmonic oscillator potential well, and can be represented by well-known functions containing quantities which can be deduced from the spectroscopic constants. Transitions between two such electronic levels A and B, which take place much faster than the nuclear motion of vibration, can be represented on a potential-energy diagram, such as that in Fig. 2b, by a vertical coupling, the Franck–Condon overlap factors $\langle A|B \rangle$ being calculated for the vibrational-state distribution as overlap integrals in a single variable R.

For repulsive states which exist in the energy continuum, $\chi(R)$ is negligible inside the classical closest distance of approach, where it has a finite value; it dies away in an oscillatory fashion as R increases.

Both of these types of wave function are found for negative-ion states, the bound ones giving rise to resonance scattering behavior, where this is energetically possible, and the unbound ones giving rise to dissociative attachment and detachment processes, such as are discussed in Chapters 6 and 7.

Consider the impact-energy dependence of an upward electron attachment between molecular ground state A and negative-ion antibonding state B proceeding via such a vertical transition. In the historical graphical procedure illustrated in Fig. 2a the wave function of the lower level A is "reflected" into the potential-energy function of the antibonding level B; an energy dependence shown to the left of the vertical axis is obtained for the collision process. This procedure has been of proven effectiveness for interpreting cross section functions of processes, and is expressed mathematically as the integral

5 Electron–Molecule Resonances

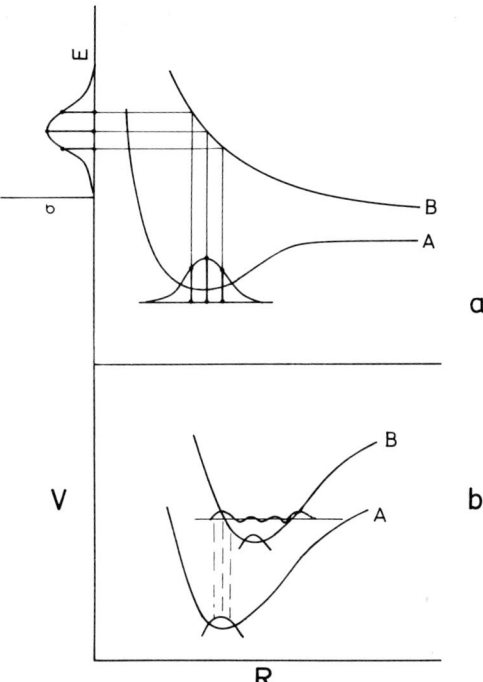

Fig. 2. Schematic potential-energy diagrams for diatomic molecules, showing (a) graphical "reflection" procedure for calculating attachment process $\langle A|V|B\rangle$ where B is negligible except at a single value of R. (b) Franck–Condon overlap $\langle A|B\rangle$ in a vertical transition between two molecular energy levels.

$$\int \chi_A(R) V_B(R) \chi_B \, dR, \quad \text{or} \quad \langle A|V|B\rangle.$$

But in this graphical procedure it is assumed that the wave function of level B is entirely centered at the classical point on the $V_B(R)$ curve. We have seen that in actual fact χ_B has a decaying but oscillatory form which is only approximated by a finite value of V at a single value of R. For a more accurate treatment of a resonant attachment process the actual expresssion $\langle A|V|B\rangle$ must be calculated.

The same expression would be applied to the upward transition to a resonance state.

A resonance state B, since it lies in an electron–molecule continuum, must decay to a lower level of the neutral molecule or decay channel C. When this is a bound level, vibrational excitation is possible, and the usual Franck–Condon factor $\langle B|C\rangle$ applies. A whole range of vibrational states not only of the resonance state but also of the decay channel state must be considered, and each is weighted by the product $\langle A|V|B\rangle \langle B|C\rangle$.

Although early attempts to interpret scattering data were made using the Breit–Wigner formula with this weighting (Herzenberg, 1968; Hasted and Awan, 1969) it became apparent that some unknown factor had not been taken into account. This factor is the variation with R of the width Γ of the resonance state. The width will be zero at the value of R at which both resonance state and decay channel state have the same energy; as R decreases and the two states become further apart in energy, so the width increases. The expression

$$\Gamma(R) = 2\gamma k(R)\rho v_l\{k(R)\} \tag{13}$$

for the variation was used by Blatt and Weisskopf (1952). The reduced width γ and resonance-state "radius" ρ are treated as adjustable parameters. v_l is the penetration factor for the angular momentum barrier for an electron of partial-wave quantum number l, and k the wave number of the electron emitted in the decay.

Birtwistle and Herzenberg (1971) postulated a model in which the electronic energy of the resonance was represented by a local complex potential, the imaginary part of which accounted for coupling to all the open channels. In this model

$$\{\zeta(R)\}^2 = \frac{\hbar^2}{8\pi^2 m} \frac{\Gamma(R)}{k(R)}, \tag{14}$$

with

$$(\hbar^2/2m)k^2(R) = V_r(R) - V_f(R);$$

$\zeta(R)$ is the electronic matrix element between the resonance state of energy V_r and the decay channel, or final state f of energy V_f, plus free electron (i.e., $\langle B|V|C\rangle$). It follows that the integral resonant cross section of the molecule ground state is

$$\sigma_{0\to f} = \frac{64\alpha\pi^5 m^2}{\hbar^4} \frac{|k_f|}{|k_0|} \left| \int \chi_f(R)\zeta(R)\xi(R)dR \right|^2, \tag{15}$$

where $\zeta(R)$ is the resonance-state wave function and k_0 the wave number of the incident electron. $\zeta(R)$ is calculated from Eqs. (13) and (14).

This formula is found to give good agreement for the vibrational excitation of N_2 by electrons, proceeding solely via the $N_2^-(^2\Pi_g)$ resonance. The experimental and calculated vibrational-excitation functions are shown in Fig. 3a, and the potential-energy curves and the chosen $\Gamma(R)$ function in Fig. 3b. Similar calculations have been carried out on the core-excited $^2\Pi_u$ shape resonance associated with $N_2(A\ ^3\Sigma_u^+)$ (Huetz et al., 1980a,b.)

This treatment has been termed the "boomerang" or "compound-state" model. Equation (14) is derived on the assumption that the outgoing spherical

5 Electron–Molecule Resonances

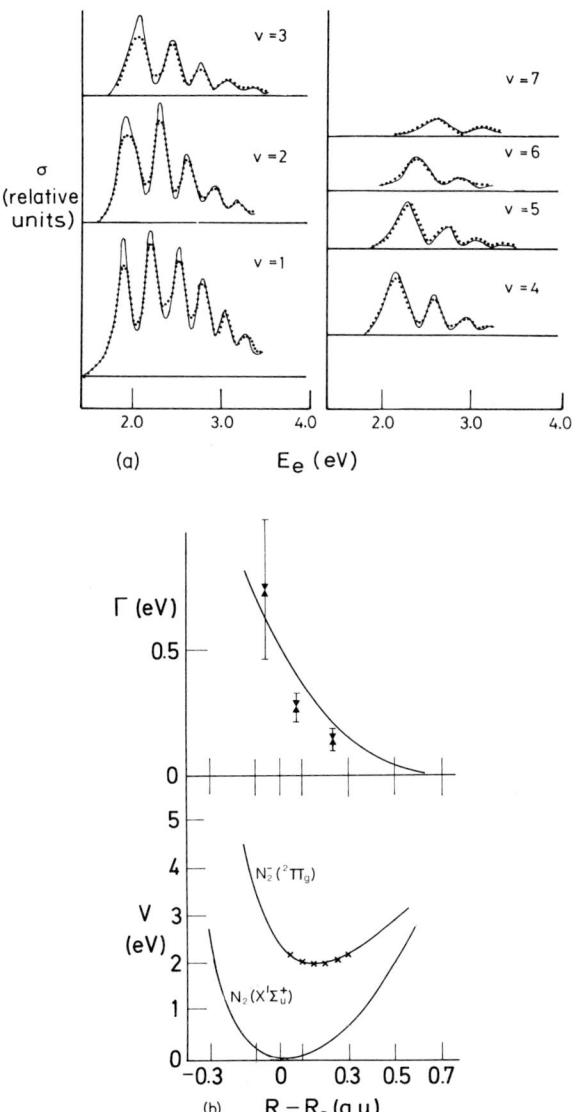

Fig. 3. (a) $N_2(X\,^1\Sigma_u^+)$ vibrational-excitation functions by electrons: full lines, experimental data; dotted lines, calculations using Eq. (15). (b) Chosen resonance-state potential-energy curve $V(R)$ and width variation $\Gamma(R)$ for the calculations shown in (a).

wave, which follows the incoming plane wave, is reflected essentially once by the resonance barrier. There are subsequent reflections, but these are considered sufficiently small to be neglected. This is because of the increasing width (decreasing lifetime) of the resonance state as R is decreased. In atomic resonances, on the contrary, one must consider as many as T^{-2} reflected waves, where $T (\ll 1)$ is the amplitude transmission coefficient of the barrier. At the resonance energy, where the reflected waves are in phase, the electron amplitude builds up to a value T^{-1}. Slightly off-resonance, the higher reflected components get out of phase, giving rise to the Breit–Wigner factor $(E - E_r + \frac{1}{2}i\Gamma)^{-1}$ in Eq. (2); for atomic resonances the width Γ is proportional to T^2.

The angular distributions of electrons scattered via resonance states are governed by the same Legendre polynomials $P_l(\cos\theta)$ of order l that govern the nonresonant scattering of electrons. For an atomic resonance decay

$$A^-(n', L', s') \longrightarrow A(n, L, s) + c(k, l, \tfrac{1}{2}), \tag{16}$$

with $L' = L \pm l$,

$$\frac{d\sigma}{d\Omega}(\theta) = \frac{1}{4k^2}\left|\sum_l (2l+1)(e^{2i\delta_l} - 1)P_l(\cos\theta)\right|^2, \tag{17}$$

The proportionality of the differential cross section to $[P_l(\cos\theta)]^2$ implies the following general features:

	$l = 0$	1	2
$\theta = 0°$	1	1	1
54°	1	$<\tfrac{1}{2}$	0
90°	1	0	$\tfrac{1}{2}$

With the aid of differential scattering data it is possible to identify the angular momentum quantum number L of the resonance state.

For molecules the picture is similar, but complicated by the nonspherical symmetry. Read (1968, 1971) has shown that the method of angular correlation theory employed in nuclear spectroscopy (Hamilton, 1940) can be married to partial-wave theory. The notation (Griffith, 1962) aα, bβ, etc, for molecular states is adopted; thus, homonuclear diatomic molecules belonging to the group $D_{\infty h}$ are labeled both by the component a of the angular momentum along the axis of symmetry (Σ, Π, Δ, etc.) and by the symmetry with respect to inversion at the center. Apart from Σ states, the label α has more than one possible value, so that orbitals pσ, pπ are possible for $l = 1$.

Resonant states of symmetries $C_{3v}E$, $C_{4v}E$, $D_{2d}E$, D_3E, $D_{3h}E$, $D_{3d}E$, $D_{4h}E$, $D_{6h}E$, $C_{\infty v}$, $D_{\infty h}$ (except $\Sigma^\pm$), fall into one class, and in the second class are those of symmetries $C_{\infty v}\Sigma^\pm$; $D_{\infty h}\Sigma^\pm$; $C_{1h}A$; $C_{2h}A,B$; $C_{2v}A,B$; $C_{3v}A$; $C_{4v}A,B$;

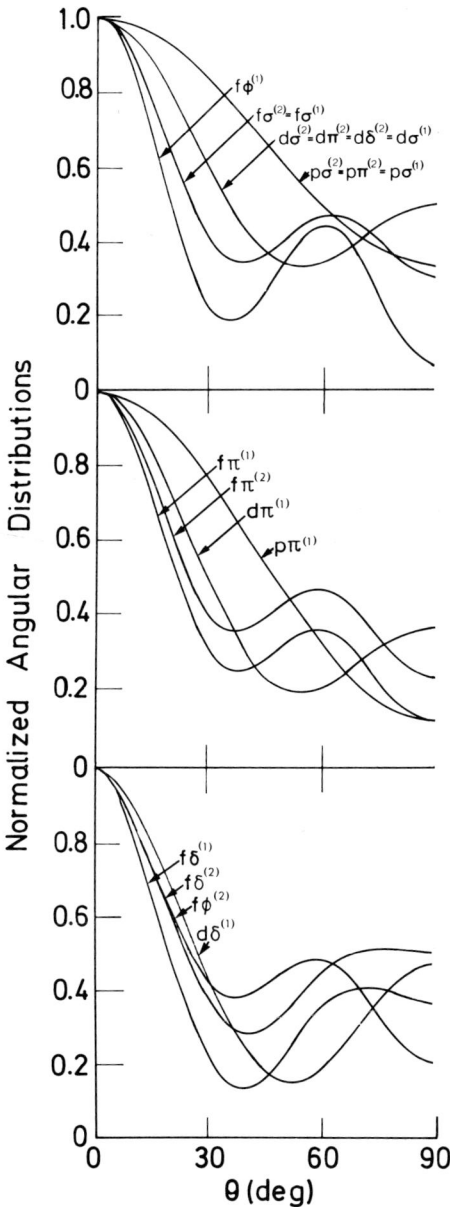

Fig. 4. Calculated scattered-electron angular distributions for some commonly encountered resonance orbitals.

TABLE I

Angular Distribution Coefficients $a_n^{(1)}(l, m)$ and $b_n^{(1)}(l, l+1, m)$

Class 1 symmetry classification

lm	$a_n^{(1)}(l,m)$				$b_n^{(1)}(l, l+1, m)$				
	n = 0	2	4	6	n = 0	1	2	3	4	5
sσ	1.500				1.000	2.000				
pσ	0.100	0.200			0.143	−0.143	0.171			
dσ	0.064	−0.129	0.193		0.084	0.273	−0.101	−0.930	0.175	0.974
fσ	0.022	0.124	−0.386	0.312						
pπ	0.033	0.233								
dπ	0.043	−0.129	0.200		0.086	−0.286	0.143	0.743		
fπ	0.008	0.149	−0.427	0.336	0.047	0.253	−0.039	−1.010	0.117	1.006
dδ	0.054	−0.193	0.254							
fδ	0.022	0.052	−0.295	0.281	0.075	0.175	−0.175	−0.818	0.205	0.851
fφ	0.005	0.230	−0.650	0.481						

Class 2 symmetry classifications

lm	$a_n^{(2)}(l,m)$				$b_n^{(2)}(l,l+1,m)$				
	n = 0	2	4	6	n = 0	1	2	3	4	5
pπ	0.067	0.133			0.114	−0.114	0.057	0.457		
dπ	0.043	−0.086	−0.129							
fπ	0.012	0.104	−0.293	0.226	0.049	0.201	−0.008	−0.679	0.052	0.666
dδ	0.043	−0.086	0.129							
fδ	0.022	0.015	−0.142	0.150	0.065	0.078	−0.078	−0.364	0.091	0.442
fφ	0.010	0.121	−0.322	0.241						

$D_{2h}A,b$; $D_{2d}A$; D_3A; $D_{3h}A$; $D_{3d}A$; $D_{4h}A,B$; $D_{6h}A,B$. Calculated angular distributions (Read, 1968) for resonance decays of some commonly encountered angular momentum states are shown in Fig. 4 (Read, 1968).

It is possible that mixed waves may be responsible for observed angular distributions. The mixing proportions may be represented by two parameters ρ and $\cos^2 \gamma$ and the angular distributions by the formula

$$W^{(i)}(\theta) \propto \sum_{n=0}^{n=m} \cos^n\theta [a_n^{(i)}(l,m) + \rho^2 a_n^{(i)}(l+1,m)$$
$$+ \rho b_n^{(i)}(l, l+1, m) + \rho \cos^2 \gamma b_n^{(i)}(l, l+1, m)] \quad (18)$$

which includes angular distribution coefficients $a_n^{(i)}$ and b_n derived from the appropriate matrix elements and incorporates the Legendre polynomial contribution $P_l(\cos \theta)$ from the partial-wave expansion. Values of these coefficients are given in Table I for the two classes of resonance, $i = 1, 2$.

C. Threshold Resonance Behavior in Polar and Other Molecules

A molecule possessing a permanent dipole moment behaves rather differently from a nonpolar molecule as far as resonance scattering of electrons is concerned.

One of the important features is the strong long-range charge–dipole attractive field. Quite apart from other possible types of interaction this field should be sufficiently strong to maintain one, or even more, stable bound states, provided that the dipole moment D exceeds a critical value $D_{crit} = 1.625D$. This Born–Oppenheimer approximation criterion is satisfied in such cases as hydrogen fluoride HF($D = 1.82D$) and H$_2$O($D = 1.84D$), but not for hydrogen chloride HCl($D = 1.11D$).

However, an even more important feature is the difference that arises from the alignment of the dipole axis and the approaching electron. On this axis the approaching electron sees a potential which is long-range attractive or repulsive according to direction, as appears from the electrostatic calculation shown in Fig. 5. For electron energies below one electron volt this interaction is important even outside the molecule.

We have seen that for a shape resonance to be displayed, there is a centrifugal barrier which arises from a symmetry which does not permit the wave function to match to an s wave outside the molecule. Thus a Σ^+ resonance in a Σ^+ polar molecule would be difficult to maintain, since the s wave would be able to leak through the centrifugal barriers provided by the higher angular momentum components. However, very large vibrational-excitation components of several polar molecules, such as HCl, have been

5 Electron–Molecule Resonances

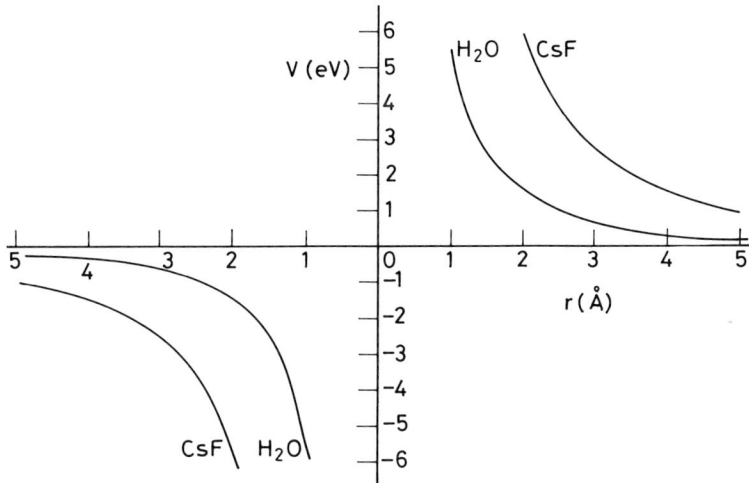

Fig. 5. Calculated electron interactions with strongly polar molecules (Herzenberg, 1979).

reported (Linder, 1979), with isotropic angular distributions appropriate to s wave scattering.

This suggests that the wave function is "piled up" not inside the centrifugal barrier, as is the case for a shape resonance, but actually outside the molecule. The repulsive negative end of the molecule is inaccessible to the electron, which interacts predominantly with the positive side. The dipole potential acts as a channel which ensures that the electron approaches only the attractive end of the molecule. The electron wave must eventually be reflected as it approaches the center of the molecule, and its amplitude is very sensitive to the phase of the reflection coefficient, which depends upon the logarithmic radial derivative at the molecular surface.

There are threshold peaks in the vibrational-excitation functions of the HCl molecule, shown in Fig. 6. These are followed at rather higher energies by broad maxima. The scattering is isotropic, indicating s wave dominance, although, as has been seen, for a conventional shape resonance this is not possible. However, a new treatment (Domcke and Cederbaum, 1981) has succeeded in reproducing the data in terms of a broad s wave resonance state, even without the inclusion of the long-range dipole force. The passage of a broad resonance through the opening of a new decay channel (i.e., through threshold) was first shown to be a nontrivial phenomenon for polar molecules. But it now appears that similar behavior can be expected for nonpolar molecules as well, particularly in new data collected for CH_4 and SF_6. Thus a threshold peak in the spectrum need not neccessarily imply a large dipole moment, nor even a long-range potential.

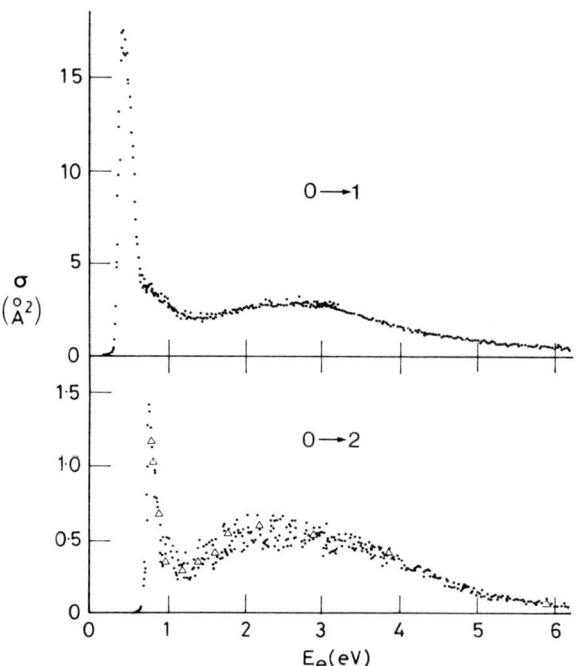

Fig. 6. Threshold peak in electron vibrational-excitation functions (0 → 1 and 0 → 2) for HCl (Rohr and Linder, 1976).

The vibrational-excitation cross sections have been calculated (Domcke and Cedarbaum, 1980, 1981) using the formula

$$\sigma_v(E) = (\pi/k^2)\,\Gamma(E)\,\Gamma(E - v\omega)|R_{v,0}(E)|, \qquad (19)$$

where E is the electron energy above the channel threshold,

$$R_{v,0}(E) = \langle v|(E - \mathcal{H})^{-1}|0\rangle, \qquad (20)$$

and $\mathcal{H}$ is an energy-dependent nonunitary operator describing the vibrational motion in the resonance state.

The long-range scattering potential alters the threshold behavior of $\Gamma(E)$, and for s wave scattering from a molecule with subcritical dipole moment, the authors took the form to be

$$\Gamma(E) = \begin{cases} A(E/B)^{1/2}(2 - E/B)^{1/2}, & 0 < E < 2B, \\ 0, & \text{elsewhere,} \end{cases} \qquad (21)$$

for $0 < E < 2B$, and $\Gamma = 0$ elsewhere.

The form of the cross section function is highly sensitive to the phase shift δ, which varies rapidly with energy near threshold, its derivative being given by

$$d\delta/dE' = -\frac{\pi}{E'[\ln(E'/2B)]^2}. \tag{22}$$

Figure 7a shows the energy variation of the fixed-nuclei phase shift, for different values of nuclear separation $\tilde{R} = R - R_0$. The resulting cross

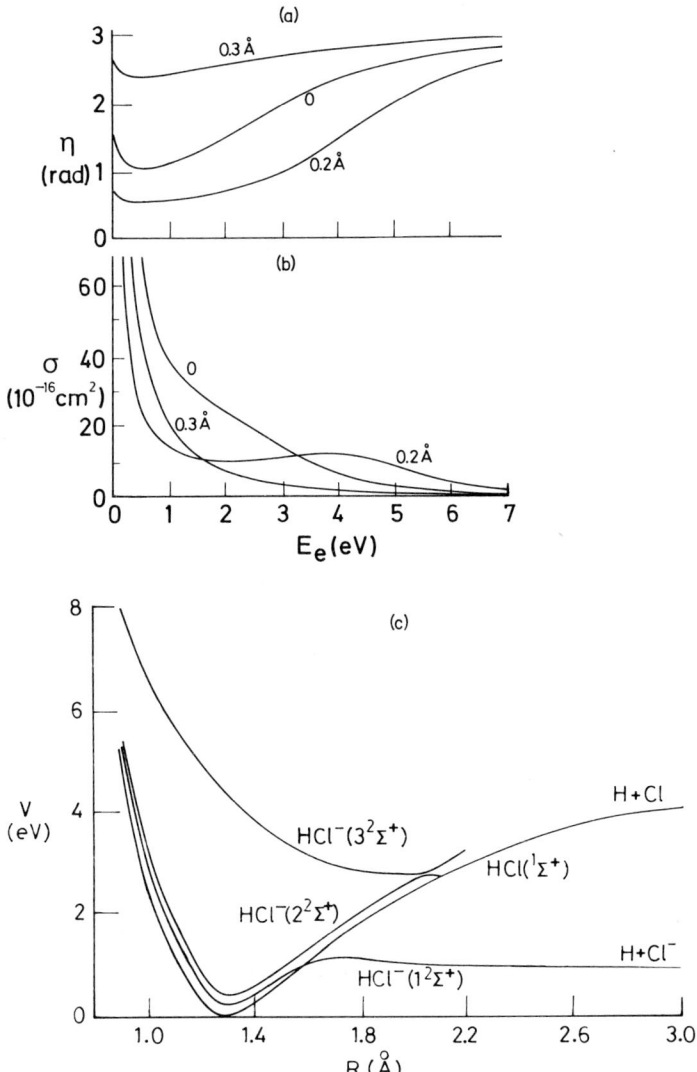

Fig. 7. Vibrational excitation calculations for HCl: (a) fixed-nuclei phase shifts η; (b) vibrational-excitation cross sections, calculated for chosen values of $\tilde{R}$; (c) potential-energy curves of HCl and HCl$^-$.

sections are also shown in this figure; it is seen that for $\tilde{R} = 0.2\text{Å}$ a broad maximum appears at an energy higher than the initial threshold peak. The calculated functions, in Fig. 7b, are seen to shown this feature, although the precise form of the experimental functions is not reproduced.

The potential-energy curves of the resonance state from which these calculations derive lie almost coincident with those of the ground state of the neutral molecule, as is shown in Figure 7c.

D. Dissociative Channel Decay

We have seen that resonance states can decay into the ground state of the neutral atom or molecule, or into vibrationally or electronically excited states. There remains the possibility that one of these states is unstable to dissociation into a negative-ion fragment and a netural fragment. If this negative-ion fragment is stable (an atomic negative ion if a diatomic molecule is under consideration), then the collision process

$$e + XY \longrightarrow X^- + Y \tag{23}$$

falls into the class of processes termed "dissociative attachment" and is more appropriately considered in Chapter 6.

If, on the other hand, the fragment or atomic negative ion is itself unstable toward detachment of the electron, then a neutral fragment or atom will result:

$$e + XY \longrightarrow XY^- \longrightarrow X + Y^- \longrightarrow X + Y + e \tag{24}$$

It could be that the XY^- state might detach to an antibonding neutral state XY' plus free electron, which dissociates to states of $X + Y$; but the overall process is neither attachment nor detachment, but rather dissociation, assisted by the resonance scattering. The term "resonant dissociation by electron impact" has been proposed (Spence and Burrow, 1979).

The process is detected as structure in the appropriate cross section function, namely, that for the production of fragment molecules or atoms. It is not easy to measure a dissociation cross section with highly momentum-resolved electrons; but there is the possibility that structure in the slow-electron production cross section might also indicate the occurrence of this process, and this has proved to be the case with nitrogen (Spence and Burrow, 1979; Mazeau et al., 1978).

Slow-electron production cross sections were first measured by means of the trapped-electron technique (Schulz, 1958, 1959), in which near-thermal-energy electrons are temporarily trapped in an electric and magnetic field, and are then detected. It is possible to vary the energy depth of the trap, thus varying the energy limit above which no electrons will be detected. A modification (Knoop and Brongersma, 1970) enables not just an upper limit to be set, but the selection of an actual near-thermal energy ($0.01 < E < 0.2$ eV) of the scattered electrons. Experiments in which the impacting electron energy is scanned and the yield of near-thermal-energy electrons measured have been termed "threshold spectroscopy experiments."

For example, the threshold spectrum of N_2 shows an important variation with selection of near-thermal energy: a broad feature appears when the selected energy is close to 0.09 eV. This feature arises from resonance-assisted dissociation, ground-state nitrogen atoms being produced by a process which can be understood with the aid of the appropriate potential-energy diagram. Taking account of the Franck–Condon overlap, the upper N_2 state is produced in higher vibrational levels and also in the continuum; temporary N^- ions are therefore produced with energies no more than 1 eV above thermal and the N^4S resulting from them reflect these energies, the emitted electron absorbing the excess (0.09 eV).

This part of the threshold spectrum therefore represents a dissociation cross section function, and can be normalized to other measurements (Cartwright *et al.*, 1977) It maximizes at about 2×10^{-18} cm^2.

E. Resonance States above the First Ionization Limit

Since many two-electron excited states lie above the first ionization limit, one might expect resonance states to be associated with them. The excited states can be of three types:

(i) autoionizing doubly excited valence states;
(ii) autoionizing states in which an electron has been transferred from a shell or subshell immediately below the valence shell;
(iii) states in which an electron has been excited from deeper inner shells, such as the K shell.

Since these resonance states lie above the first, and sometimes above other ionization limits, a positive ion is liberated when they decay, together with an extra electron. They are, therefore, suitably detected by searching for fine structure in an otherwise smooth ionization cross section function. In electron-impact ionization, as has been known for a long time, there is sometimes structure due to the autoionization process itself: autoionization levels show a Beutler–Fano profile in the photoionization function. The resonance-state structure, on the other hand, does not show up in the photoionization function; there are insufficient electrons for the state to be formed.

Near-threshold measurements of electron-impact ionization, using monoenergetic electrons, reveal structure in the cross section function;

TABLE II

Inner-Shell Resonances of Some Diatomic and Triatomic Molecules

Molecule	Onset energy (eV)	Assignment	Parent state and energy (eV)		
NO	397.6	$(1sN)^{-1}(2p\pi)^3$	$(1sN)^{-1}(2p\pi)^2$	$^2\Sigma^-$	399.44
				$^2\Delta$	399.73
				$^2\Sigma^+$	400.35
N_2	400.4	$(1sN)^{-1}(2p\pi)^2$	$(1sN)^{-1}(2p\pi)$	$^1\Pi$	400.86
	(400.8)				
CO	286.1	$(1sC)^{-1}(2p\pi)^2$	$(1sC)^{-1}(2p\pi)$	$^1\Pi$	287.40
	(286.3)				
N_2O	401.2	$(1sN_T)^{-1}(3\pi)^2$	$(1sN_T)^{-1}(3\pi)$		400.4
	403.2	$(1sN_C)^{-1}(3\pi)^2$	$(1sN_C)^{-1}(3\pi)$		403.8
CO_2	289.4	$(1sC)^{-1}(1\pi_u)^2$	$(1sC)^{-1}(1\pi_u)$		289.8

however, it is not possible to distinguish resonance states from autoionizing states simply on the basis of the line shapes (see Mathur and Frost, 1981; Mathur, 1980).

The molecular excited states with which inner-shell resonances are associated contain an atomic inner-shell electron that has been promoted to an unfilled valence or Rydberg orbital of the molecule. These states may be identified using photoabsorption or electron energy loss spectroscopy, and the resonance states give rise to structure in the electron-impact ionization functions (King et al., 1980) Table II summarizes the molecular inner-shell resonances reported. More recently, five positive-ion decay channels for the $(1sC)^{-1}(2p\pi)^2$ state have been studied by Ziesel et al. (1979) and branching ratios have been determined. The decay of the resonance occurs mainly by dissociation of CO^+ by double Auger transitions in which an extra electron is ejected by the shakeoff process (see Kay et al., 1977).

F. Resonances and Theoretical Chemistry

The importance of resonances to theoretical chemistry is becoming increasingly clear, and is likely to stimulate continued research. The tasks of theoretical chemistry include not only the achievement of the ability to predict the stability of molecules by quantitative understanding of the nature and the energies of all molecular orbitals, but also the ability to predict reactivities in a wide variety of situations. One of the factors which influences reactivity is the "electrophilicity" of a molecule—that is, its ability to attract electrons.

The energies of the occupied orbitals in a molecule are determined experimentally by the methods of photoelectron spectroscopy. The first ionization potential provides the energy of the highest occupied molecular orbital (HOMO), and the excitation energies of the first positive ion are those of deeper occupied orbitals.

The first electron affinity represents the energy of the lowest unoccupied molecular orbital (LUMO), and the temporary states of the negative ion correspond to the energies of the higher unoccupied orbitals. In cases in which the negative ion is stable and the electron affinity is positive, resonance spectroscopy cannot determine its value; methods discussed in Chapter 6 of Volume 2 are, however, appropriate. Even in such a case, however, the other temporary ion states, or resonances, provide data about higher unoccupied orbitals.

Experimental data for LUMOs, obtained by electron transmission spectroscopy, are particularly sensitive tests of the computer codes used by theoretical chemists for the calculation of molecular orbital energies.

For many organic molecules the HOMO energy is that of a π orbital, and the LUMO energy is that of a π^* orbital. According to the Hückel and Pariser–Parr–Pople theories, the sum of ionization potential and corresponding electron affinity associated with each π, π^* pair in an alternant hydrocarbon (e.g., ethylene, styrene, benzene) should be constant.

Low-lying excitation energies in related molecules may also be correlated to the quantity $E_i - E_a$. Using perturbation theory it has been shown that the singlet excitation energy E_s should be equal to this quantity plus electron–electron interaction and correlation terms. Measured values of resonance energies have enabled quantitative comparisons to be made; it is found that the two-electron terms are themselves approximately linearly related to $E_i - E_a$. A graph of E_s plotted against $E_i - E_a$ for substituted benzenes is approximately a sloping line. This is also the form of the graph plotted for the first triplet excitation energy, a surprising fact, since triplet transitions involve mixing between the excitation from the HOMO to the second unoccupied orbital and the excitation from the second occupied π orbital to the LUMO.

Organic molecules without carbon–carbon double bonds, such as alkane hydrocarbons, do not possess π^* orbitals, and it will be seen in Section IV that these have no low-lying resonances, or indeed any positive electron affinity. Alkenes, alkynes, and aromatic molecules, however, possess π^* orbitals, and display low-lying resonances. However, the presence of atoms of the second and higher rows of the periodic table changes the situation, since low-lying resonances are found, for example, in silanes. Moreover, fluorine substitution into alkanes produces no such resonances, but chlorine substitution does. This implies that the orbitals concerned do not have pure σ^* character.

III. Experimental Techniques for the Study of Resonances

A. Introduction

The interaction of low-energy electrons with atoms and molecules is studied by two basic types of experimental technique: the electron swarm and the electron beam. In the former, electrons undergo many collisions and drift through a gaseous medium of atomic number density n under the influence of an electric field E. Their energies ε display a considerable spread, usually expressed as a function $f(\varepsilon, E/n)$, where E/n is known as the "pressure-reduced" or simply the "reduced" electric field. Although the energy range covered by this technique is from thermal to around 15 eV, all the measured quantities are averaged over this broad energy distribution and hence the method cannot be used for the *direct* detection of resonant structure (see Chapter 3 of Volume 2).

Beam experiments, on the other hand, employ monochromated electrons with narrow energy distributions; these are either directed at low-pressure gas targets, or are crossed with molecular beams; the maintenance of single-collision conditions makes these experiments ideal for the detection of resonances in atoms and molecules. Experimentally, molecular resonant states are usually detected by observing

(a) the energy dependence of the attenuation of a monoenergetic beam passing through a gas-filled collision chamber;

(b) the energy dependence of the elastic and inelastic electron–molecule cross section functions in either the forward direction or at a fixed scattering angle;

(c) the energy dependence of the total cross section for the dissociative attachment process (see Chapter 6).

In the first two techniques, resonances are identified by changes in the cross section function which cannot be attributed to orthodox quantum interference phenomena such as the Ramsauer effect. The choice between the two alternatives, as to whether the cross section at the resonance energy increases or decreases, is a function of two scattering phases. The first is the phase of the outgoing electron wave of the direct collision process with the molecule AB:

$$e + AB \longrightarrow AB^*(j) + e \qquad (25)$$

and the second is the phase of the partial wave that results from the autodetachment process:

$$e + AB \longrightarrow AB^* \longrightarrow AB^*(j) + e \qquad (26)$$

In different channels j the same resonance may appear differently, since the line shape (the interference structure) is dependent on the direct phase shift and on the scattering angle. The measured interference may be analysed to obtain phase shifts for the direct process.

The experimental study of resonances was made possible by advances in the techniques of producing sufficiently intense monoenergetic electron beams, together with developments in the technologies associated with ultrahigh-vacuum, fast differential pumping, atomic- and molecular-beam sources, electron optics, particle detection, low-noise electronics, and on-line computerized control and data handling. Detailed discussions of these techniques may be found in books by Massey (1976, 1979), Hasted (1970), Massey and Burhop (1969), McDaniel (1964), and Bederson and Fite (1968).

The basic hardware required for resonance spectroscopy comprises a source of energy-selected electrons, an optical system capable of accelerating or retarding the electron beam into a collision zone through a potential which can be varied between wide limits without appreciably changing the beam current; a postcollision energy analyzer; an electron multiplier detection system; a clean bakeable vacuum system capable of background pressures in the region of 10^{-7} Torr or better; and an ambient environment essentially free of external electric and magnetic fields. The pressure in the collision zone (typically $\lesssim 5 \times 10^{-4}$ Torr) should be such that the experiment is sufficiently close to single-collision conditions for the measurements to be dominated by the true cross section function.

B. Spectrometer Configurations

The most widely used spectrometer configurations are shown in Figs. 8–13. The apparatus employing the trochoidal monochromator developed by Stamatovic and Schulz (1970) has been in widespread use to measure the *derivative* of the transmitted (unscattered) electron current as a function of energy, with typical resolution of $\simeq 30$–50 meV. Figure 8 illustrates schematically the components of the monochromator. Electrons from a cathode F are injected into a region R of crossed electric and magnetic fields, where they are dispersed according to their axial velocity. A portion of the electron beam passes through an orifice D and is accelerated either into a collision chamber CC or across a molecular beam. Those electrons which reach the exit of the collision chamber are decelerated to about 0.2 eV by a retarding plate; this ensures that only the unscattered portion of the electron beam can reach the detector E, since all scattered electrons have their axial velocity vector reoriented and do not possess sufficient axial momentum to overcome the retarding potential barrier. Modulation of the electron beam is accomplished by placing within the collision chamber an insulated cylinder CL to which an

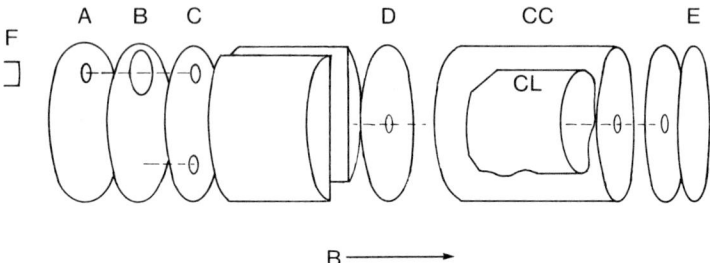

Fig. 8. Trochoidal monochromator configuration, key in text.

ac signal is applied. The synchronously detected unscattered current is then proportional to $d\sigma_t/dE$, where σ_t is the total cross section at impact energy E.

The principal advantage of the derivative technique is the enhancement of resonant structure in comparison with slow variations in cross section function, attributable to direct scattering processes. Also, by applying the modulating voltage only inside the collision chamber, the spurious effects potentials can cause near the entrance and exit slits of the collision chamber are eliminated; surface effects and fringing field effects become unimportant. There are disadvantages to the derivative technique, the major one probably being its inability to detect resonance features much broader than 1 eV. A further problem relates to defining the center of a resonance in the energy differential of the transmitted current (see Spence, 1974a, b). All recent measurements seem to have been reported on the basis that the resonance energy corresponds to the point of steepest slope in the derivative function.

Since the trochoidal monochromator does not select electrons with respect to the energy associated with the radial velocity component, the initial energy spread has to be kept as low as possible; low-temperature coated cathodes are generally used. Verhaart and Brongersma (1980) have suggested an additional slit in the monochromator which physically intercepts electrons with large radial velocities. This modification enables the use of thoriated-tungsten and rhenium filaments and, consequently, the possibility to work at higher gas pressures and with reactive gases.

A new configuration (Fig. 9) employing two trochoidal analyzers (for precollision monochromation and postcollision analysis) has now been developed (Tam and Wong, 1979). A novel feature is the introduction of a set of four equipotential apertures and a soft iron torus (marked T in Fig. 9), which acts as a transverse energy filter by discriminating against those scattered electrons which have a large Larmor radius. This apparatus provides a wide energy range (0.1–30 eV) and the analyzer transmission function can be characterized down to a very low energy.

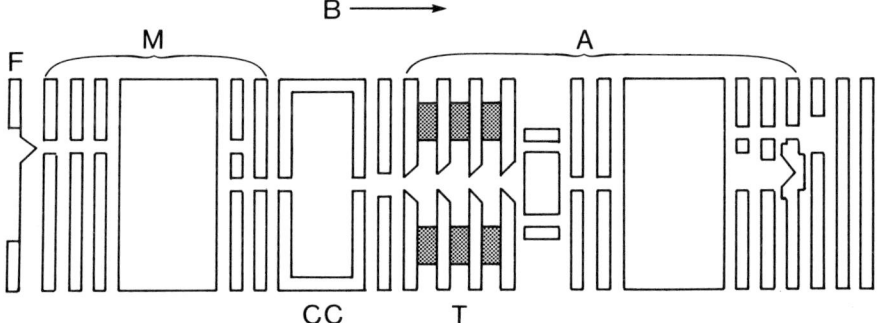

Fig. 9. Electron spectrometer employing trochoidal monochromator M, collision chamber CC, and analyzer A.

An alternative to the trochoidal configuration makes use of purely electrostatic fields to achieve energy monochromation. A typical arrangement is shown in Fig. 10. This type of instrument has been used to detect resonances in the differential elastic or inelastic cross section, with energy resolution of between 10 and 50 meV (Linder, 1979). The experimenter is able to vary the angle between the postcollision analyzer and the monochromator so that the angular distribution of the scattered electrons can be measured, thus providing information not only on the configuration of the resonant state but on the precise resonance width (Weingartshofer et al., 1974) and on the angular momentum of the state.

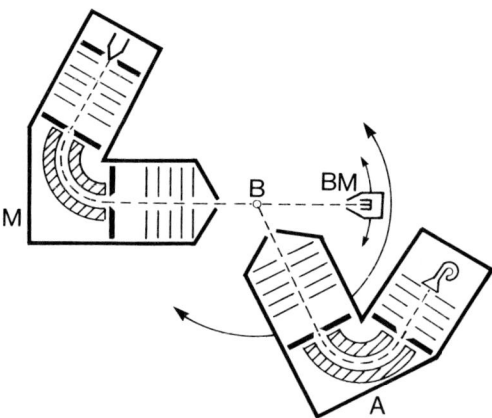

Fig. 10. Electron spectrometer with 127° cylindrical monochromator M, collision beam B, analyzer A, and beam monitor BM.

Referring to Fig. 10, electrons emitted from the filament are focused by electrostatic lenses onto the input orifice of the monochromator. The monochromated beam is focused onto a crossed molecular beam by a lens assembly; the scattered electrons, after passing through further lenses and through the analyzer, are finally detected on a multiplier. The use of a well-collimated molecular-beam target greatly reduces Doppler broadening of resonances. The electron optics are designed to maintain constant transmission characteristics. Shields are necessary to prevent stray electrons from reaching the detector. Hemispherical analyzers have also been used in this type of spectrometer.

A much simpler arrangement can be used for detecting resonances, by means of the electron transmission function (Fig. 11). The current I_t transmitted through the gas-filled collision chamber is related to the incident current I_0 by the relation $I_t = I_0 e^{(-nl\sigma_t)}$, where n is the gas number density, l the effective length of the collision chamber, and σ_t the total cross section. Small excursions in σ_t, which indicate the existence of resonance, are exhibited by structure in the transmitted current, usually under single-collision conditions,

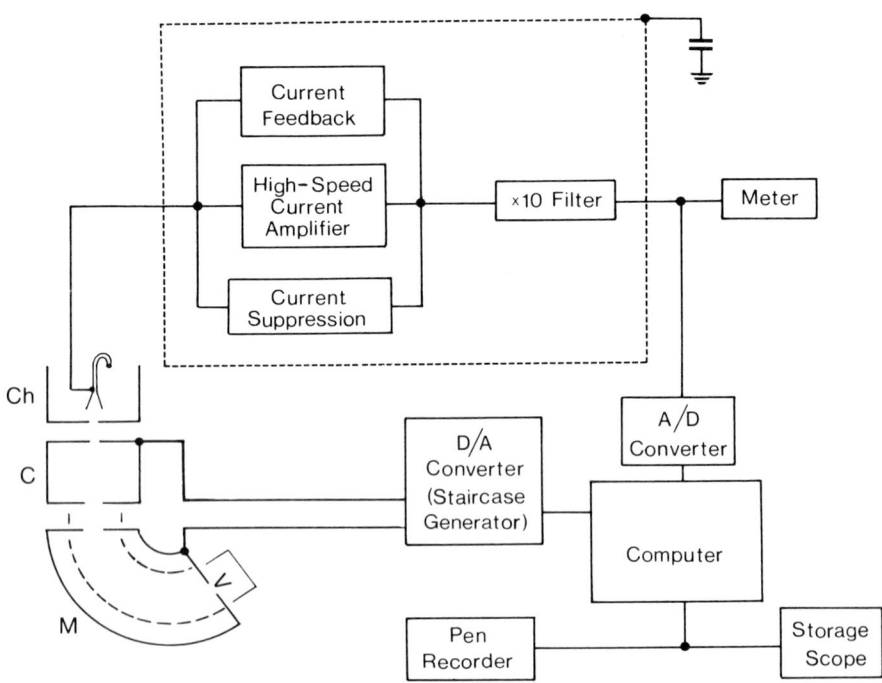

Fig. 11. Simple electron transmission function spectrometer, with 127° cylindrical monochromator M, gas collision chamber C, and channel multiplier detector Ch.

$n l \sigma_t \ll 1$, when only the first approximation term in the series expansion of the exponential need be considered. However, if single-collision conditions are not rigorously maintained, some amplification of the structure takes place. This makes the transmission technique particularly suitable for sensitive measurements of broad, shallow resonant features superimposed on a slowly varying background.

Referring to Fig. 11, we see that a beam of near-thermal-energy electrons, after injection into a cylindrical monochromator, is accelerated into a static gas collision chamber. The unscattered electron current is measured by means of a current amplifier with a background subtraction unit which uses current feedback techniques to obtain low-noise and high-speed performance. The use of an on-line microcomputer not only facilitates fast energy scansion, current detection, and variable control of the incident electron energy, but also provides noise reduction; the multiple scanning technique increases the signal-to-noise ratio by effectively reducing that part of the noise with time periods between the short time taken to digitize the incoming signal and the relatively long period of slow drift of experimental conditions; the random current intensities are averaged over a number of scans and the systematic ones are summed, so that real features emerge from a noisy background. Typical energy resolution of between 20 and 50 meV is obtainable. Resonances as broad as 5 eV have been detected using such an apparatus.

The measurement of inelastic cross sections near the threshold of excitation has also been used for detecting resonances. The principle of these "threshold excitation" experiments is shown in Fig. 12; the scattered electrons, which have lost essentially all their incident energy during collision, are trapped by either a small potential well or a suitable scavenger gas (usually SF_6) which surrounds the collision region.

The trapped-electron method was originally developed by Schulz (1958, 1959). A monoenergetic electron beam is accelerated, preferably with monochromation, into a collision chamber consisting of two end plates and a grid G. The entrance and exit slits and the grid G are held at the same potential, while a cylindrical electrode M surrounding the collision chamber is kept at a positive potential V_M with respect to G. The penetration of V_M into the collision region forms a potential well of depth W and traps any electron making an inelastic collision just above the threshold for an inelastic process. As the trapped electron spirals back and forth in the magnetic field, it eventually undergoes a sufficient number of elastic collisions to diffuse to the trapped-electron collector.

If the incident electron energy exceeds the threshold of an inelastic process by an amount W, the electrons have sufficient energy to escape over the potential barrier (Fig. 12), and the trapped electron current will be zero. As the incident energy increases, the trapped-electron current grows to a peak which

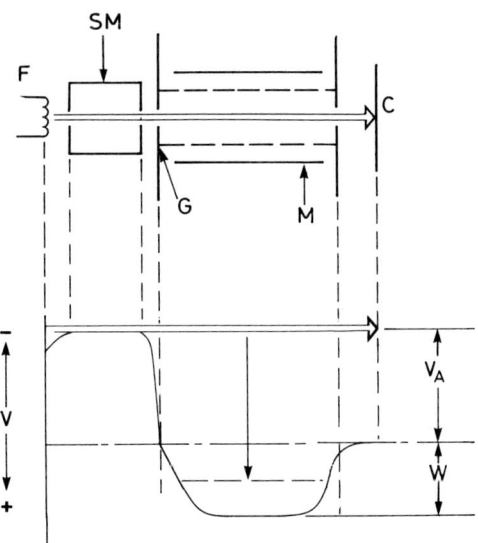

Fig. 12. Threshold excitation spectrometer using trapped-electron technique; filament F, monochromator (shown schematically) SM, cylindrical grid G connected to entrance and exit orifices of collision chamber, cylindrical electrode M, collector C. Axial potential variation shown below.

is proportional to the cross section at an energy that exceeds the threshold by W.

As resonant states often occur in the energy region just below the threshold for electronic excitation of molecules, the trapped-electron method has a high sensitivity for detecting resonances. However, the method offers poor resolution when compared to electrostatic or trochoidal spectrometers, except in some recent experiments in which the conventional collision chamber has been replaced by a molecular beam (Michejda et al., 1981). Complications can arise if negative ions or elastically scattered electrons trapped in the potential barrier contribute to the detected current. These complications can in principle be avoided by admixing a small amount of a scavenger gas (usually SF_6) and mass spectroscopically detecting the SF_6^- current rather than the trapped electrons. The negative-ion current is due to the large attachment cross section that SF_6 possesses for nearly zero-energy electrons. In practice, the scavenger technique is far less sensitive than the trapped-electron method; it is also unreliable for locating resonances below about 0.8 eV, owing to the close proximity of the primary SF_6^- attachment resonance. The trapped-electron method itself is not ideal for determining the exact positions of low-energy resonances, because of a steeply rising background current which develops at energies below about 1 eV.

For low-energy resonance scattering, perhaps the most reliable momentum selection technique will prove to be the time-of-flight (TOF) method (Raith, 1976). A schematic diagram of a spectrometer based on the TOF technique is shown in Fig. 13. An electron beam is accelerated to an energy of about 150 eV and injected into a radio-frequency beam chopper LA, LR. The pulsed electrons are abruptly decelerated at the entrance of an equipotential flight

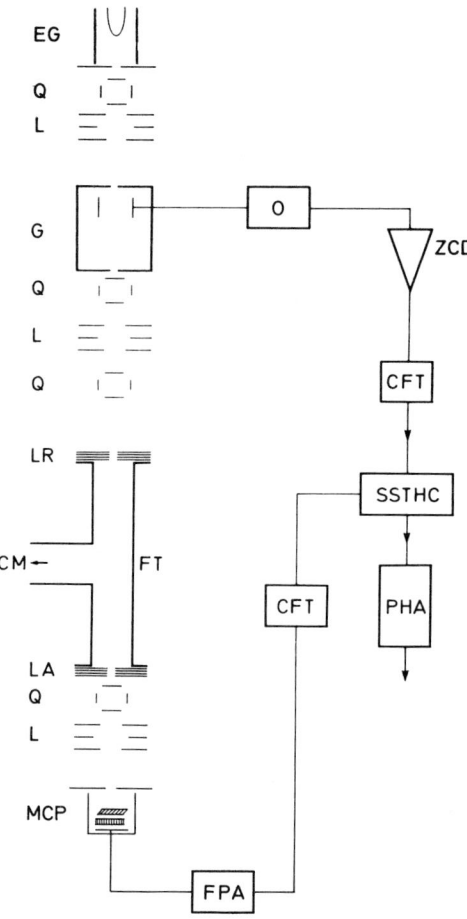

Fig. 13. Time-of-flight monochromator: (EG) electron gun, (Q) quadrupole deflectors, (L) lenses, (G) beam pulsing gate, (LR, LA) retarding and accelerating optics, (FT) flight tube, (CM) capacitance manometer, (MCP) dual-microchannel plate assembly, (CFT) constant-fraction timing discriminator, (O) oscillator, (ZCT) zero-crossing detector, (FPA) fast preamplifier, (SSTHC) stop–start time height convertor, (PHA) pulse height analyzer.

tube FT containing gas. The flight tube acts both as a collision chamber and as an energy-dispersing device. Electrons traversing the flight tube spread out spatially according to their velocities; the flight time is obtained from the measured time difference between the electron pulse and the detection of an electron. It is possible to measure microsecond flight times to nanosecond accuracy. Hence a direct and accurate energy determination is possible in terms of the physical quantities of length and time. The other advantages of the TOF technique are the absence of pressure-dependent effects outside the scattering region, and the availability of a continuous electron energy distribution and continuous data accumulation over the chosen range of energies. The excellent discrimination against small-angle scattering makes it a suitable method for determination of absolute cross section, provided that the gas pressure inside the flight tube is measured by a suitably connected capacitance manometer.

The energy resolution is determined principally by the quality of the fast electronics employed. Typically, for the apparatus shown in Fig. 13, it is of the order of 10 meV FWHM at an electron energy of about 0.1 eV, and 100 meV FWHM in the vicinity of 2 eV (Rajgara and Mathur, 1982).

IV. Experimental Results

A. Diatomic Molecules

1. Hydrogen and Its Isotopes

No fewer than five resonance states of H_2^- been reported as vibrational series in the 11–13.5 eV region (Kuyatt et al., 1964; Weingartshofer et al., 1970; Comer and Read, 1971). They are most effectively studied in the vibrational channels of the ground state $H_2(X\ ^1\Sigma_g^+)$ between $v = 1$ and 13.

Energies of the resonance peaks are represented as spectral lines in Figure 14a; potential-energy curves for H_2 and H_2^- resonant series a, b, and c are shown in Fig. 14b. Only two series appear clearly in this figure, designated a and b. In the highest vibrational exit channels, two separate series appear clearly. The range of contributing resonance vibrational states varies markedly with decay-channel vibrational quantum number; this is expected when there is appreciable difference between the nuclear separations of resonance state and decay channel (which is the same electronic state as that which is originally impacted by the electron).

Analysis was made using a fixed-width formula

$$\sigma(i \to f) \propto |\langle i|r\rangle|^2 |\langle r|f\rangle|^2 \tag{27}$$

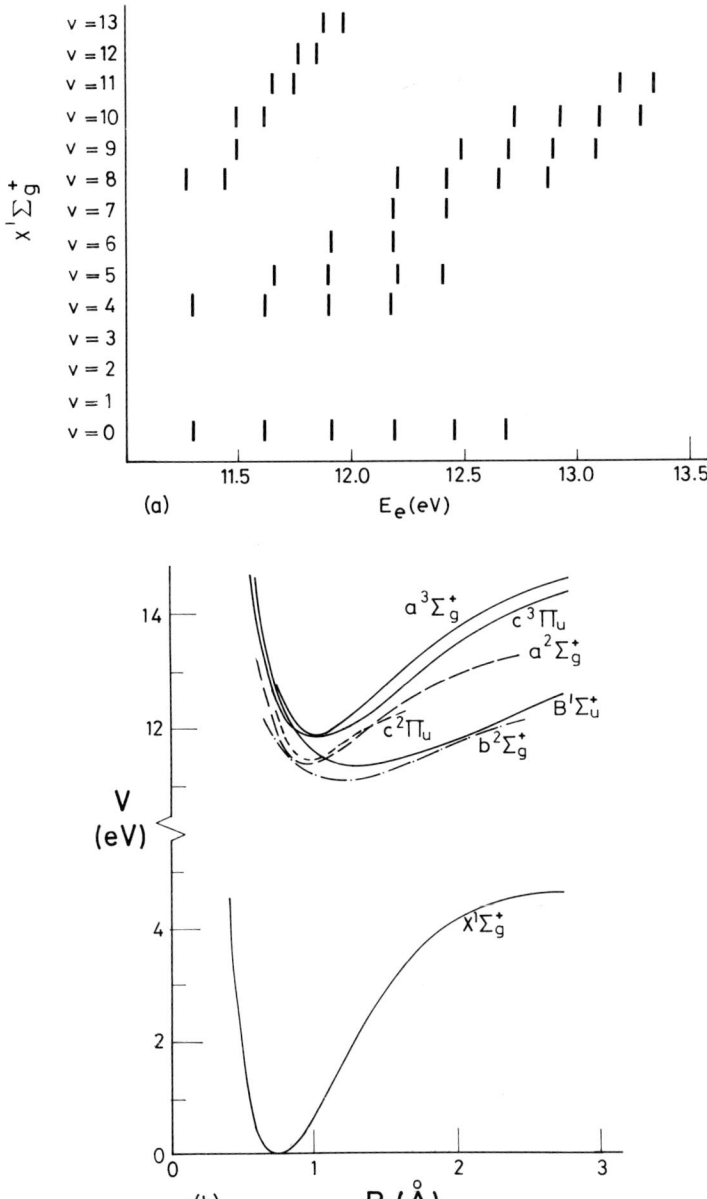

Fig. 14 (a) Spectral line representation of H_2 resonant series a and b. decaying into the $H_2(X\ ^1\Sigma_g^+)$ channels $v = 0, 4-13$. (b) Potential-energy curves of H_2 (full lines) and H_2^- resonant series a, b, and c (broken lines).

for excitation of a final state f from initial state i via resonance state r. A fitting procedure, in which Morse function parameters enable Franck–Condon factors to be calculated for this formula, is used to obtain the most suitable parameters of the resonance states a and b. They are tabulated in Table III together with the available information about series c and d. The potential-energy curves are illustrated in Figure 14b.

Series c is found only in electron scattering from the D_2 molecule, and its parameters are uncertain because there appears to be coherent interference between series a and series c.

The angular distribution of series a has been reported to be isotropic, which implies that it has Σ_g^+ symmetry. However, a further series (d) has been reported (Heideman et al., 1966; Weingartshofer et al., 1970) in the $B\ ^1\Sigma_u^+$ exit channel. The energies of the two series are identical, but the angular distribution of series d is not isotropic but is reasonably close to

$$W(\theta) \propto 3 + 5\cos^2\theta, \tag{28}$$

which is appropriate to Π_g or Π_u. The assignments of the different resonance series have been made on the basis of angular distributions and are given in Table III.

There is also a further resonance series known as II seen in the B exit channel (Heidemann et al., 1966; Weingartshofer et al., 1970). Although the situation is not at all clear, it is possible that this might be designated $(\sigma_g\ 1s)(\pi_u\ 2p)^2\ \Delta_g$.

There is also a repulsive $H_2^-\ (^2\Sigma_g^+)$ state which is found in dissociative attachment data at about 10 eV.

Hydrogen is particularly suitable for the investigation of isotope effects which occur in all electron scattering processes, not simply where resonances are involved. Vibrational excitation from $v = 0$ to $v = v$ behaves as $\mu^{-v/2}$, where μ is the reduced mass (Chang and Wong, 1977).

TABLE III

Hydrogen Resonances

$E_r\ (v=0)$ (eV)	Designation	E_{vib} (eV)	Anharmonicity (eV)	r_e (Å)	Parent	Exit channels
11.40	$(1s\sigma_g)(2p\pi_u)^2\ a\ ^2\Sigma_g^+$	0.345	0.0135	0.97	$c\ ^3\Pi_u$	$X\ ^1\Sigma_g^+$ $b\ ^3\Sigma_u^+$ (?)
11.11	$(1s\sigma_g)(1s\sigma_u)^2\ b\ ^2\Sigma_g^+$	0.19	0.005	1.175	$B\ ^1\Sigma_u^+$	$X\ ^1\Sigma_g^+$
11.45	$(1s\sigma_g)(2s\sigma_g)(2p\pi_u)c\ ^2\Pi_u$	0.28	0.015	0.97	$c\ ^3\Pi_u$ $a\ ^3\Sigma_g^+$	$X\ ^1\Sigma_g^+\ (D_2)$
11.40	$(1s\sigma_g)(1s\sigma_u)(2p\pi_u)$	0.345	0.0135	0.97	$c\ ^3\Pi_u$	$B\ ^1\Sigma_u^+$ $b\ ^3\Sigma_u^+$ (?)

2. Nitrogen and Carbon Monoxide

The nitrogen molecule represents the most interesting study of resonance states, for a variety of reasons:

(i) The 1.9-eV resonance state represents the first discovery in this field, and well illustrates the features discussed in the early sections of this chapter.

(ii) Since the molecule is homonuclear the rotational excitation is particularly suitable for study in the 1.9-eV resonance.

(iii) There exist a whole series of Feshbach resonance states which are especially suitable for interpretation in terms of coupling between the two excited electrons in the field of the core of the N_2^+ grandparent state.

We have seen (Fig. 3) how the electron-impact energy vibrational-excitation functions of nitrogen can be interpreted with the aid of a variable-width compound-state model. There is negligible contribution from nonresonance excitation mechanisms, but for elastic scattering there is a considerable contribution from nonresonant scattering. The total cross section function, which is illustrated in Fig. 15a, is not only a sensitive test of electron scattering theory (Chandra and Temkin, 1976), but is also a convenient, almost a standard, energy calibration for electron spectrometers. Table IV compares the magnitudes and energies of the vibrational peaks published by various workers. Structure reported below 1.8 eV (Golden, 1966) has not been confirmed by other workers.

In vibrational-excitation calculations no account had been taken until recently of the rotational excitation that occurs in the energy loss spectroscopy experiments. Although such excitation must take place, the levels cannot be resolved. The spacing of adjacent rotational lines is 1 meV for $\Delta j = 2$, but adjacent rotational branches are more widely separated: at 300 K, 8.5 meV between $\Delta j = 0$ and $\Delta j = 2$. For a good electron spectrometer of reliable resolution $\simeq 20$ meV, adjacent rotational branches can therefore be detected as a broadening of a narrow resonance convoluted with the apparatus function under conditions when Doppler broadening can be neglected. Broadening in the $v = 0$ to $v = 1$ excitation measured at $E_e = 2.27$ eV for 60° scattering (Wong and Dubé, 1978) is shown in Fig. 15b.

Rotational-excitation cross sections may be assumed to be independent of initial quantum number j, and channels other than $\Delta j = 0, \pm 2$, and ± 4 are neglected. Initial thermal equilibrium is assumed:

$$N(j) \propto g_1(2j + 1)\exp[-B_e j(j+1)/kT], \tag{29}$$

where the nuclear spin degeneracy $g_1 = 6$ for even j. The peak intensity, for a vibrational resonance peak, and for a given Δj branch, is

$$I(\Delta j) = \sum_{j=0}^{\infty} N(j)\sigma(j, \Delta j)\delta[E - B_e(2j + 1 + \Delta j)\Delta j] \tag{30}$$

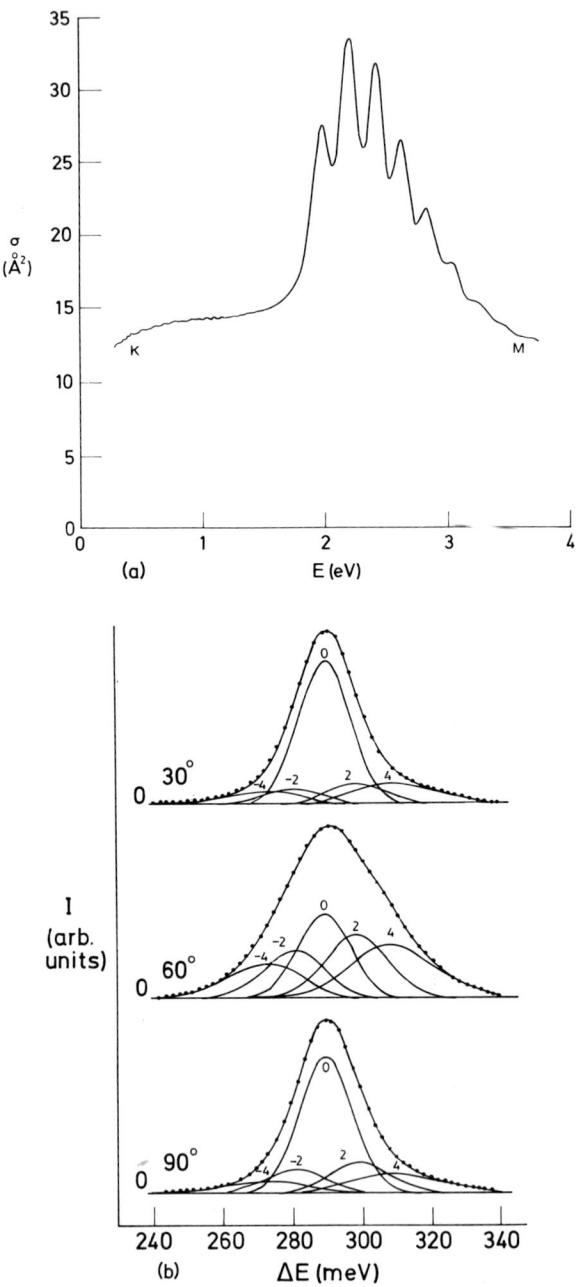

Fig. 15. (a) Total cross section function for N_2 in the region of the 1.89-eV $^2\Pi_g$ shape resonance (Mathur, 1976; Kadifachi, 1982). (b) Broadening of the $v = 0$ to $v = 1$ $N_2(X\,^1\Sigma_u^+)$ vibrational excitation at $E = 2.27$ eV; $\theta = 30°$, $60°$, $90°$. Fitting of $\Delta j = 0, \pm 2, \pm 4$ rotational-excitation functions. (c) Calculated angular variations of N_2 2.27-eV rotational-excitation $\Delta j = 0, 2, 4$ branches (full lines) compared with single values used for adjustment of rotational line broadening to the experimental data.

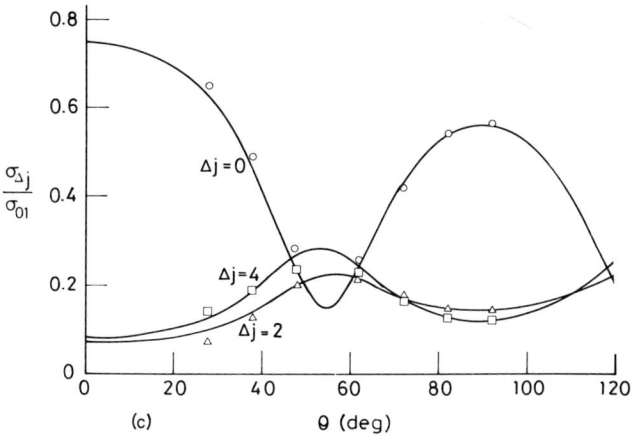

Fig. 15. (c) (cont.)

Calculations are assisted by the use of a continuous Boltzmann distribution (Read, 1972). The experimental line shape can only be reproduced by adjusting the relative magnitudes of the $\Delta j = 0, \pm 2$, and ± 4 cross sections.

This adjustment is carried out on differential measurements of the line shape at various scattering angles, so that the angular variations of the three branches can be determined. These are shown in Fig. 15c. There is a dramatic difference between the branches, which is well predicted by impulse approximation calculations for excitation via the $^2\Pi_g$ resonance (see also Read, 1972; Comer and Harrison, 1973). The expressions are as follows:

Rotational transitions	$\dfrac{d\sigma}{d\Omega}(\theta, \Delta j)$	
$\Delta j = -4$	$0.091 \times \tfrac{45}{56}(1 + \tfrac{2}{3}\cos^2\theta + \tfrac{1}{9}\cos^4\theta)$,	(31)
-2	$0.106 \times \tfrac{15}{14}(1 - \cos^2\theta + \tfrac{4}{3}\cos^4\theta)$,	(32)
0	$0.500 \times \tfrac{135}{112}(1 - \tfrac{46}{9}\cos^2\theta + \tfrac{23}{3}\cos^4\theta)$,	(33)
$+2$	$0.144 \times \tfrac{15}{14}(1 - \cos^2\theta + \tfrac{4}{3}\cos^4\theta)$,	(34)
$+4$	$0.159 \times \tfrac{45}{56}(1 + \tfrac{2}{3}\cos^2\theta + \tfrac{1}{9}\cos^4\theta)$.	(35)

The Feshbach resonances of N_2 have been studied in a great variety of decay channels: in vibrational channels of the ground state $X\,^1\Sigma_g^+$, of $A\,^3\Sigma_u^+$ and $B\,^3\Pi_g$ (Huetz et al., 1980a, b), also in the "total metastable" function (Brunt et al., 1977), which includes participation of $N_2(A\,^3\Sigma_u^+)$, $(a\,^1\Pi_g)$, and $(E\,^3\Sigma_g^+)$, all of which have some metastable character; also in the energy loss modes

TABLE IV

Energies (eV) and Magnitudes (Å²) of Vibrational Peaks in $N_2(^2\Pi_g)$ Resonance

		Energies		
Boness et al. (1968)	Hasted and Awan (1969)	Ehrhardt and Willman (1967)	Mathur (1976)	Kennerly (1980)
1.92	1.90	1.89	1.89	1.98
2.14	2.15	2.15	2.15	2.21
2.38	2.39	2.40	2.39	2.44
2.61	2.62	2.65	2.63	2.67
2.83	2.85	2.89	2.86	2.90
3.04	3.08	3.13	3.10	3.12
	3.31	3.36	3.34	3.35
			3.56	
			3 78	

	Magnitudes		
Chandra and Temkin (1976) (Hybrid theory)	Golden (1966) (Modified Ramsauer)	Mathur (1976) (Transmission)	Kennerly (1980) (Transmission time of flight)
34.0	24.5	27.7	26.3
39.7	32.3	33.7	33.8
32.4	30.8	32.3	33.4
24.0	26.3	26.8	28.0
22.0	22.0	22.0	23.5
18.5	18.6	18.1	
15.7		15.7	
		14.3	
		13.3	

involving these levels; also in the cross section function for the creation of ultraviolet photons, which may contain contributions from $N_2(b\ ^1\Sigma_u^+)$, $(c^1\ ^1\Sigma_u^+)$, and $(b\ ^1\Pi_u)$.

We stress that there is a wealth of information in electron energy loss spectra, in that every single energy loss peak can itself be studied with variation of electron-impact energy; in these variations the resonance vibrational series show up, and they are different in each channel. If there are v_1 vibrational channels, and v_2 resonance peaks, then the energies and peak intensities of $v_1 v_2$ spectral bands must be interpreted over the range of scattering angles. Some examples of the appearance of the $^2\Pi_u$ resonance vibrations in various channels are given in Fig. 16.

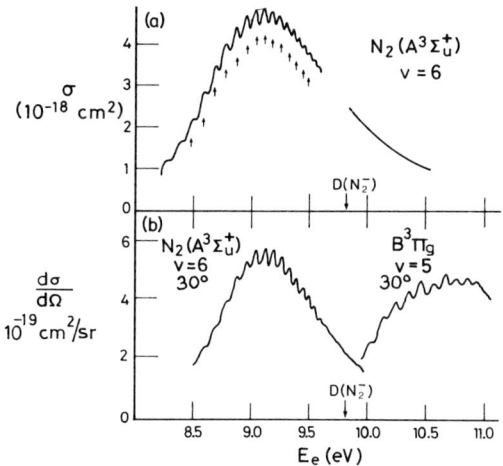

Fig. 16. Appearance of the N_2^- ($^2\Pi_u$) resonance vibration in the channels shown. (a) calculated total cross sections; arrows indicate positions of measured peaks. (b) Observed differential cross section.

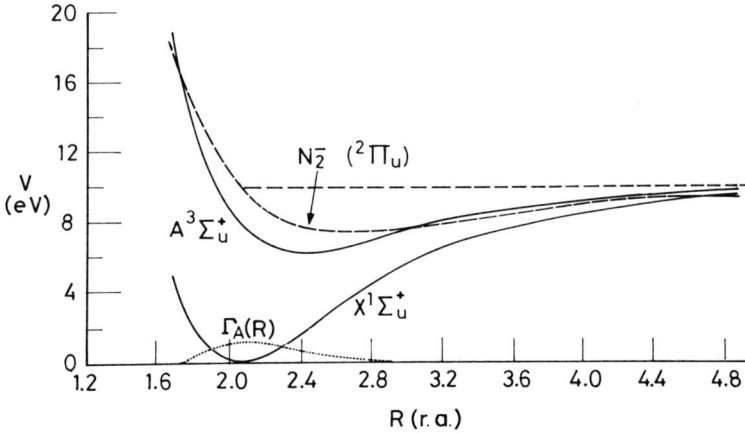

Fig. 17. Potential-energy curves for N_2^- ($^2\Pi_u$) resonance and other N_2 states.

With the aid of variable-width equations the spectroscopic constants of the resonance states of nitrogen have been calculated. Potential-energy curves are shown in Figs. 3b and 17, and data are tabulated in Table IVa.

In all experiments of this type the target is a beam or gas of nitrogen molecules at laboratory temperature ($\simeq 300$ K). On the basis of Boltzmann statistics it may be assumed that most are in the ground vibrational state $v = 0$.

TABLE IVa

N_2 Resonances

$E_r(v=0)$ (eV)	Designation	E_{vib} (eV)	Grandparent and E_{vib} (eV)	Rydberg electron configuration	Screening constant S	References
1.89	Shape $^2\Pi_g$	0.235				
8.0	$^2\Pi_u$					
11.48	$^2\Sigma_g^+$	0.28	$N_2^+(X\,^2\Sigma_g^+)$ 0.270	$(3s\sigma_g)^2\,^1\Sigma_g^+$	0.24	a
11.87	B $^2\Sigma_u^+$	0.26	$N_2^+(X\,^2\Sigma_g^+)$ 0.270	$(3s\sigma_g)(3p\sigma_u)\,^3\Sigma_u^+$	0.23	a
12.14	$^2\Pi_u$	0.26	$N_2^+(X\,^2\Sigma_g^+)$ 0.270	$(3s\sigma_g)(3p\pi_u)\,^3\Pi_u$	0.25	a
12.535	a $^2\Pi_u$	0.245	$N_2^+(A\,^2\Pi_u)$ 0.232	$(3s\sigma_g)^2\,^1\Sigma_g^+$	0.23	a
12.992	b $^2\Pi_g$	0.239	$N_2^+(A\,^2\Pi_u)$ 0.232	$(3s\sigma_g)(3p\sigma_u)\,^3\Sigma_u^+$	0.23	b
14.134	d $^2\Pi_u$	0.235	$N_2^+(A\,^2\Pi_u)$ 0.232	$(3p\sigma_u)^2\,^1\Sigma_g^+$	0.31	b

[a] Comer and Read (1971).
[b] Lefebre-Brion (1973).

5 Electron–Molecule Resonances

However, it is perfectly possible to arrange that the gas target is vibrationally (and rotationally) excited. This can be done by striking an electric discharge upstream in the tube down which the gas passes; a microwave cavity is conventionally used for this purpose.

The interest that is developed in experimenting with vibrationally excited target molecules is the internal consistency check that can be made on the fixed-width or variable-width equations that are used to compute the scattering functions. The double Franck–Condon overlaps are of course reasonably sensitive to the initial vibrational populations of the target molecule.

However, the success of such studies depends upon the accuracy with which these populations can be assessed during the experiment. This is possible by the methods of electron spectroscopy, but using nonresonant excitation, for which a vertical transition ensures a single Franck–Condon overlap. For this purpose direct excitation of the $B\,^3\Sigma_g$ state was used (Huetz et al., 1980a, b), and for the particular microwave discharge used, the population distribution of $N_2(X\,^1\Sigma_g^+)$ was as follows:

$$v = 0, 58.8\%; \quad v = 1, 21.0\%; \quad v = 2, 9.1\%; \quad v = 3, 4.8\%;$$
$$v = 4, 3.6\%; \quad v = 5, 1.8\%; \quad v = 6, 0.9\%.$$

There is an enhancement of higher levels above a Boltzmann distribution.

Experiments with vibrationally excited targets have shown that the overall effect of the vibrational energy is to increase the survival probability of the $A\,^2\Pi_u$ negative-ion resonance state to large internuclear distances. As a consequence, the oscillatory structure which this resonance produces in the excitation cross sections of the $N_2(A\,^3\Sigma_u^+)$ vibrational levels becomes increasingly pronounced.

Carbon monoxide is isoelectronic with nitrogen, and is a weakly polar molecule. There is a low-energy shape resonance of $^2\Pi$ configuration, appearing both in elastic and vibrational-excitation channels, and essentially unchanged since the earliest measurements (Schulz, 1964; Boness and Hasted, 1966; Ehrhardt et al., 1968). It is possible to fit the vibrational excitation data with a single value of Γ, 0.80 ± 0.03 eV (Zubek and Szmytkowski, 1977) with vibrational separation 0.235 ± 0.007 eV, and Franck–Condon separation $\Delta R = 0.092 \pm 0.002$ Å. Figure 18 illustrates the fit. It is surprising how similar the potential parameters are to those of N_2^-, although the width is greater, as a result of the broader potential barrier for $l = 1$.

The Feshbach resonances between 10 and 15.6 eV have been studied in transmission (Sanche and Schulz, 1972a, b), in elastic and inelastic scattering (Comer and Read, 1971; Mazeau et al., 1972; Swanson et al., 1975); also optical and metastable excitation function measurements have contributed. The assignments of these resonances (Brunt et al., 1977), together with their

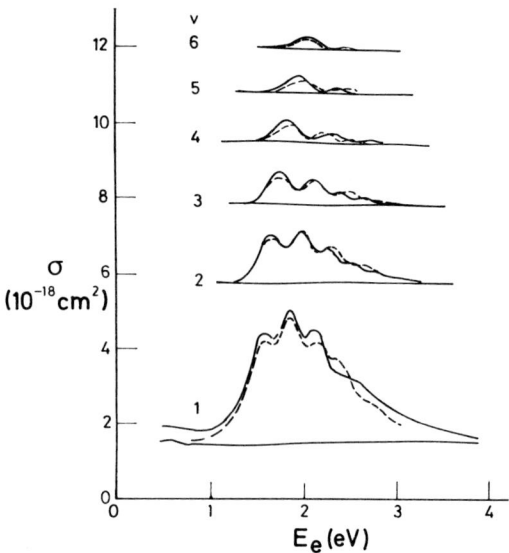

Fig. 18. Single width fit (broken line) of CO vibrational-excitation data (full line) using $^2\Pi$ resonant-state parameters listed in text.

grandparent states, are shown in Table V. However, there are also suggestions of further structures in the 13- and 15-eV regions.

A very broad shape resonance has been reported (Tronc *et al.*, 1980) around 19.5 eV, and the angular distributions indicate Σ character.

3. Nitric Oxide and Oxygen

These molecules are similar in two respects: they both possess negative-ion ground states which are stable in $v = 0$; and they both possess Feshbach resonances which, together with those of N_2 and CO, are particularly suitable for classification using the "grandparent model" proposed by Schulz (1973).

The higher vibrational levels of the negative-ion ground states of both molecules lie at appropriate energies to capture temporarily an electron which impacts the neutral molecule, usually in its ground vibrational state, at a suitable energy. The bound system then decays either into the elastic $v = 0$ channel or into a higher vibrational level of the neutral-molecule ground state, provided that this is energetically possible.

The resonance states of O_2 and NO are tabulated in Tables VI and VII, respectively.

The most stable state of O_2^- has the configuration (Mulliken, 1957)

$$O_2^- (1s\,\sigma_g)^2 (1s\,\sigma_u)^2 (2s\,\sigma_g)^2 (2s\,\sigma_u)^2 (2p\,\sigma_g)^2 (2p\,\pi_u)^3 (2p\,\pi_g)^4 \; ^2\Pi_u$$

TABLE V

CO Resonances

$E_r(v=0)$ (eV)	Designation	E_{vib} (eV)	Grandparent and E_{vib} (eV)	Rydberg electron configuration	S	References
1.52	Shape $^2\Pi$	0.235 ± 0.007	—	—		a
10.044	A $^2\Sigma^+$	0.26	$CO^+(X\,^2\Sigma^+)$ 0.271	$(3s\,\sigma)^2\,^1\Sigma^+$	0.25	b
10.7	$^2\Pi$	—	$CO^+(X\,^2\Sigma^+)$ 0.271	$(3s\,\sigma)(3p\,\pi)\,^3\Pi$	0.26	b
12.514	E $^2\Pi$	—	$CO^+(A\,^2\Pi)$ 0.19	$(3s\,\sigma)^2\,^1\Sigma^+$	0.24	c
13.974	a $^2\Pi$	0.227	$CO^+(A\,^2\Pi)$ 0.19	$(3p\,\sigma)^2\,^1\Sigma^+$	0.31	c
15.60	G $^2\Sigma^+$	—	$CO^+(B\,^2\Sigma^+)$ 0.208	$(3s\,\sigma)^2\,^1\Sigma^+$	0.24	c
$\simeq 19.5$	Shape Σ	—	—	—		d

[a] Zubek and Szmytkowski (1977).
[b] Lefebre-Brion (1973).
[c] Brunt et al. (1976, 1977, 1978).
[d] Tronc et al. (1977, 1980).

TABLE VI
O$_2$ Resonances

$E_r(v=0)$ (eV)	Designation	E_{vib} (eV)	Grandparent and E_{vib} (eV)	Rydberg electron configuration	S	References
−0.4	nuclear excited X $^2\Pi_g$	0.135				a
8.04	$^2\Pi_g$	0.215	O$_2^+$(X $^2\Pi_g$) 0.232	$(3s\,\sigma_g)^2\,^1\Sigma_g^+$	0.27	a
11.69	a $^4\Pi_u$	0.126	O$_2^+$(a $^4\Pi_u$) 0.126	$(3s\,\sigma_g)^2\,^1\Sigma_g^+$	0.23	a
14.27	b $^4\Sigma_g^-$	0.16	O$_2^+$(b $^4\Sigma_g^-$) 0.144	$(3s\,\sigma_g)^2\,^1\Sigma_g^+$	0.28	a

[a] Brunt et al. (1976, 1977, 1978).

TABLE VII

NO Resonances

$E_r(v=0)$ (eV)	Designation	E_{vib} (eV)	Grandparent and E_{vib} (eV)	Rydberg electron configuration	S	References
−0.03	nuclear excited X $^3\Sigma^-$	0.165				
+0.65	$^1\Delta$	0.182				
5.04	a $^1\Sigma^+$	0.286	NO$^+$(X $^1\Sigma^+$) 0.280	$(3s\,\sigma)^2\ ^1\Sigma^+$	0.25	a
5.41	b $^3\Pi$	0.290	NO$^+$(X $^1\Sigma^+$) 0.280	$(3s\,\sigma)(3p\,\pi)\ ^3\Pi$	0.23	b
5.46	c $^3\Sigma^+$	0.292	NO$^+$(X $^1\Sigma^+$) 0.280	$(3s\,\sigma)(3p\,\sigma)\ ^3\Sigma^+$	0.23	b
6.45	d $^3\Pi$	0.282	NO$^+$(X $^1\Sigma^+$) 0.280	$(3p\,\pi)^2\ ^3\Sigma^-$	0.29	c
6.41	d' $^3\Sigma$	0.282	NO$^+$(X $^1\Sigma^+$) 0.280	$(3p\,\pi)^2\ ^3\Sigma^-$		d
12.36	$^3\Pi$	0.21	NO$^+$(b $^3\Pi$) 0.210	$(3s\,\sigma)^2\ ^1\Sigma^+$	0.26	a
14.19	$^1\Pi$	—	NO$^+$(A $^1\Pi$) 0.194	$(3s\,\sigma)^2\ ^1\Sigma^+$	0.26	a
17.51	$^1\Pi$	—	NO$^+$(B $^1\Pi$) —	$(3s\,\sigma)^2\ ^1\Sigma^+$	0.25	a

[a] Brunt et al. (1976, 1977, 1978).
[b] Lefebre-Brion (1973).
[c] Comer and Read (1971).
[d] Gresteau et al. (1979).

The molecular negative ion is stable and its adiabatic electron affinity 0.44 eV. Thus the total electron scattering functions show structure at even the lowest energies, and this was found to be the case down to 0.2 eV in the earliest experiments (Boness and Hasted, 1966). When the structures below 0.3 eV are discarded, because of the rapid variation of electron current with energy in that region, the vibrational constant can be derived (Boness and Schulz, 1970; Linder and Schmidt, 1971; Burrow, 1974) (Table VI) and the anharmonicity estimated as 1–1.5 meV.

Since the NO^- molecular negative ion is basically isoelectronic with O_2, one would expect the lowest states to be designated $^3\Sigma^-$, $^1\Delta_g$, and $^1\Sigma^+$; the lowest state, $^3\Sigma^-$, provides a negative ion which can be stable, but whose vibrational progression extends above the $v = 0$ level of $NO(X\,^2\Pi_r)$ and therefore produces resonance scattering, observed in the early researches (Larkin and Hasted, 1972).

The higher vibrational-level structure due to the $^3\Sigma^-$ state shows irregularity which has been interpreted (Burrow, 1974) as arising from vibrational levels of the $^1\Delta_g$ state; the vibrational constants and energies of these two resonant states are listed in Table VII.

There are also a number of Feshbach resonances which have been observed in inelastic scattering measurements. These narrow resonances have been designated as series a, b, c, d, and d', and are listed in Table VII. They are observed both in vibrationally excited ground-state channels up to $v = 18$ and in the $NO(A\,^2\Sigma^+)$, $C(^2\Pi)$, and $D(^2\Sigma^+)$ Rydberg, and the $B\,^2\Pi$ valence channels.

Calculations (Lefebre-Brion, 1973) predict that the series should be designated as follows:

$$
\begin{array}{lll}
a & (n s \sigma) & ^1\Sigma^+ \\
b & (n s \sigma)(R p \pi) & ^3\Pi \\
c & (n s \sigma)(Rp \sigma) & 2^3\Sigma^+ \\
d & (n p \pi)(R p \sigma) & ^3\Pi \\
d' & (n p \pi) & ^3\Sigma^-
\end{array}
$$

The Franck–Condon factors for the resonance scattering processes cannot be made to fit the data unless the width of each resonance state is assumed to be a continuously varying function of electron energy. The best agreement is obtained when the nuclear separation r_e of the resonance states is within $r_e(NO^+) + 0.01$ Å and -0.005 Å. Thus the parent states of these resonance states are of Rydberg character and the grandparent is $NO^+(X\,^1\Sigma^+)$.

The coupling scheme in the grandparent model is that in which two electrons in Rydberg orbitals $(n_1 l_1 \lambda_1)$ and $(n_2 l_2 \lambda_2)$ couple to form a total spin S and orbital angular momentum component Λ; this state couples to the ionic core of spin s and angular momentum component λ to give a molecular term of total spin S_t and angular momentum component Λ_t. This coupling scheme can be described as $(\lambda s)(n_1 l_1 \lambda_1 n_2 l_2 \lambda_2, \Lambda S)\Lambda_t S_t$; or briefly $\lambda \Lambda S$, by analogy with JLS used for atomic resonances.

It might be expected that the two electrons in the resonance state partially screen one another from the positive charge of the ion core; this would reduce the unit charge to a value $1 - S$, where S is the screening constant. Thus the total binding energy of the two electrons becomes

$$\varepsilon(n_1 l_1 \lambda_1, n_2 l_2 \lambda_2) = \frac{R(1-S)^2}{[n - \delta(n_1 l_1 \lambda_1)]^2} - \frac{R(1-S)^2}{[n - \delta(n_2 l_2 \lambda_2)]^2}$$
$$= (1 - S)^2 [\varepsilon(n_1 l_1 \lambda_1) + \varepsilon(n_2 l_2 \lambda_2)] \tag{36}$$

If the grandparent model is valid the screening constants calculated from the data should have approximately the same values as those obtained for analogous levels of atoms. Tables IVa–VII demonstrate that the screening constants for these diatomic molecules are almost constant, so that the grandparent model is established on firm ground. The angular distributions of scattered electrons are used to verify the overall symmetries of the resonances, and detailed discussions are given by Brunt et al. (1977).

Inner-shell excited states of NO and O_2 are also known (King et al., 1980). They are observed as structure in the yield of positive ions resulting from electron-impact excitation of the molecules. The excited states with which these resonances are associated contain an atomic inner-shell electron which is promoted to an unfilled valence or Rydberg orbital of the molecule; photoabsorption shows up these states. Table II summarizes inner-shell resonances in these molecules.

4. Hydrogen Halides and Halogens

There are several features of interest in electron resonance scattering from hydrogen halides: low-energy dipole resonances, higher-energy Feshbach resonances, and structure from these appearing in the dissociative attachment channel.

The low-energy vibrational-excitation cross sections, with sharp threshold peak and subsequent broad maximum, have been discussed in Section II C.4 above, and one for HCl is illustrated in Fig. 6, with potential-energy curves in Fig. 7c.

The Feshbach resonance structure in hydrogen fluoride (Spence, 1981) commences at 12.82 eV and has vibrational spacings (successively 0.350, 0.340, 0.320 eV) which correspond closely to those of $HF^+(X\,^2\Pi)$ (0.361, 0.339, 0.317 eV). Thus the grandparent model is applicable.

The HF gas is extremely difficult to handle, since because of its great chemical reactivity it does not always remain pure when presented in a metal apparatus as an electron collision target; some incorrect data have been presented (Mathur and Hasted, 1978) and subsequent unpublished mass spectrometric analyses demonstrated that the target was not in fact HF but a mixture of high-molecular-weight fluorinated species (69–150 amu). There seems to be no doubt about the correct Feshbach resonances energies, since they have been confirmed (Abouaf and Teillet-Billy, 1980) both in transmission and in dissociative channels, producing F^-.

The Feshbach resonances in HCl have also been reported in the transmission (Spence and Noguchi, 1975) and dissociative (Azria et al., 1980a, b) channels. However, there is more than one vibrational series; they have been labeled as follows:

$$\begin{array}{ll} 1-1' & 0.20 \text{ or } 9.125 \text{ eV} \\ 2 & 9.60 \text{ or } 9.55 \text{ eV} \\ 3 & 10.05 \text{ eV} \\ 3' & 10.16 \text{ eV} \\ 4' & 10.88 \text{ or } 10.925 \text{ eV} \end{array}$$

Halogen Feshbach resonances can also be interpreted using a grandparent model (Spence, 1974a, b). For F_2, two vibrational series, a and b, are seen, commencing respectively at 11.25 and 11.95 eV. Both have as their grandparent $F_2^+(X\ ^2\Pi)$, for which the spin-orbit splitting is too small to be observed in the resonance scattering.

Only a single vibrational resonance series is seen in Cl_2; it commences at 7.46 eV. However, two series are observed in Br_2, commencing at 6.72 and 7.065 eV. The $^2\Pi_{3/2} - {}^2\Pi_{1/2}$ spin-orbit splitting in Br_2^+ is 0.350 eV, so that there is clear support for the grandparent model. In Cl_2, the positive-ion spin-orbit splitting is approximately the same as the vibrational spacing, so that presumably the second resonance series is obscured.

For I_2 the vibrational spacing of the resonances is too small to be resolved, but, again, the spin-orbit separation between the two resonances at 5.78 and 6.38 eV is similar to that of the positive-ion $^2\Pi_{3/2} - {}^2\Pi_{1/2}$ states (0.650 eV). There are also two smaller resonances at 6.850 and 7.150 eV.

Quantum defect considerations lead to the probability that the principal molecular halogen resonance configurations are $[X\ ^2\Pi_g] (ns\sigma)^2\ ^2\Pi_g$.

B. Triatomic Molecules

The triatomic molecules most widely studied are the isoelectronic CO_2 and N_2O together with the OCS molecule, which is similar but involves higher principle quantum numbers. Measurements have also been made on the NO_2 and SO_2 molecules.

1. CO_2

A fairly broad $^2\Pi\,CO_2^-$ resonance centered at 3.8 eV and exhibiting vibrational structure has been observed in the total and differential cross sections as well as in the excitation function of the symmetric stretch and bending modes of the ground-state neutral molecule (Boness and Hasted, 1966; Andrick et al., 1969; Sanche and Schulz, 1973). This resonance has also been observed in absolute total cross section measurements; coupled-channel calculations (Morrison et al., 1977; Morrison and Lane, 1977) have been carried out, using an interaction potential comprising an *ab initio* electrostatic Hartree potential, a semiempirical polarization, and approximate local exchange potential. The calculations and the absolute measurements, shown in Fig. 19, are in good agreement in the 0.07–2 eV region, and show a steeply rising cross section at energies below 0.5 eV, where the dipole moment caused by bending affects the direct scattering process. In the vicinity of the 3.8-eV resonance, however, the calculated cross section, which does not account for vibrational channels, deviates considerably from the experimental data.

The energy dependence of selected vibrational cross sections in the resonance region has been measured (Boness and Schulz, 1974); the observed vibrational structure shifts toward higher energies as the energy loss increases, in much the same fashion as in resonant vibrational excitation of N_2 and CO. An interpretation in terms of the two-dimensional boomerang model has been presented by Schulz (1976). Very broad shape resonances centered at 10 and 30 eV have also been detected in the vibrational-excitation cross section (Tronc and Azria, 1979).

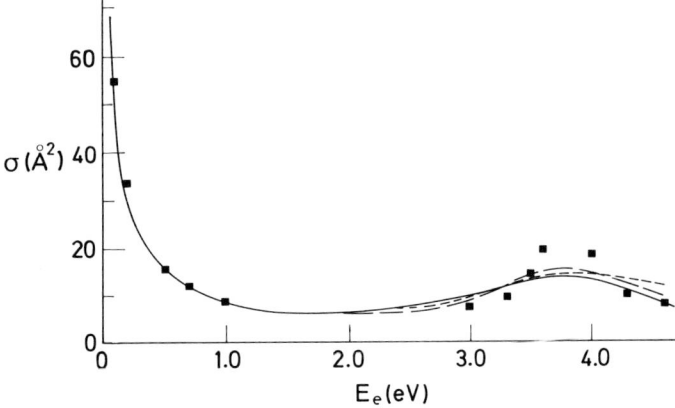

Fig. 19. Total electron–CO_2 scattering cross section. Experimental data of Ferch et al. (1981), full line; Brüche (1927), broken line; Szmytkowski and Zubek (1978), dot–dash line. Calculated values of Morrison et al. (1977) are shown as squares.

A high-energy resonant state assigned as $(1s\,C)^{-1}(1\pi_u)^2$ has been observed at 291.5 eV (King et al., 1980) in the positive ion channel. The parent state is $(1s\,C)^{-1}(1\pi_u)$ at 290.8 eV. The promotion of a carbon K electron results in increased equilibrium internuclear separations in a bent molecular geometry. Hence the resonant state possesses a large width (1.9 eV).

2. N_2O

N_2O possesses a broad, featureless shape resonance (FWHM = 1.1 eV) in the region of 2.3 eV (Boness et al., 1968) which is characterized as the $^2\Sigma^+$ state of N_2O^-. This is not the lowest shape resonance, which, according to dissociative attachment indications, should lie at energies below 0.5 eV (Chantry, 1969; Zecca et al., 1974).

Energy loss spectra obtained at a scattering angle of 40° (Azria et al., 1975) at incident electron energies 0.7, 2.30, and 4 eV show the contribution of the $^2\Sigma^+$ resonance to vibrational excitation. At resonance the vibrational cross sections are approximately one order of magnitude larger than off-resonance, except for the fundamental modes which show a significant nonresonant contribution.

Inner-shell excited resonances have also been observed (King et al., 1980) associated with K-shell excitation from each of the two nitrogen atoms. The peak resonance energies are 402.5 and 404.3 eV, corresponding to two $(1s\,N)^{-1}(3\pi)^2\,N_2O^-$ states. As in the case of CO_2, promotion of a nitrogen K electron probably causes the normally linear N_2O to adopt a bent equilibrium geometry, with increased equilibrium internuclear distances and hence the large widths ($\simeq 1.5$ eV) observed in the two resonances.

3. OCS

A large maximum at 1.23 eV in the total cross section function of OCS has been related to the formation of a $^2\Pi$ resonance state (Szmytkowski and Zubek, 1978). A 2-eV-wide structure, centered at 3.8 eV, was detected in the vibrational-excitation measurements of Tronc and Azria (1979); angular distribution data reveal it to be a Σ shape resonance with the d-wave ($l = 2$) as the dominant term.

4. NO_2

There is a long resonance vibrational series apparent both in the total transmission and v_3 vibrational channels (Larkin and Hasted, 1972). These are shown in Fig. 20. One interpretation of these data is that there is interference between different vibrational modes of the resonance state. Fourier analysis (taking no account of anharmonicity) leads to the following spacings:

5 Electron–Molecule Resonances

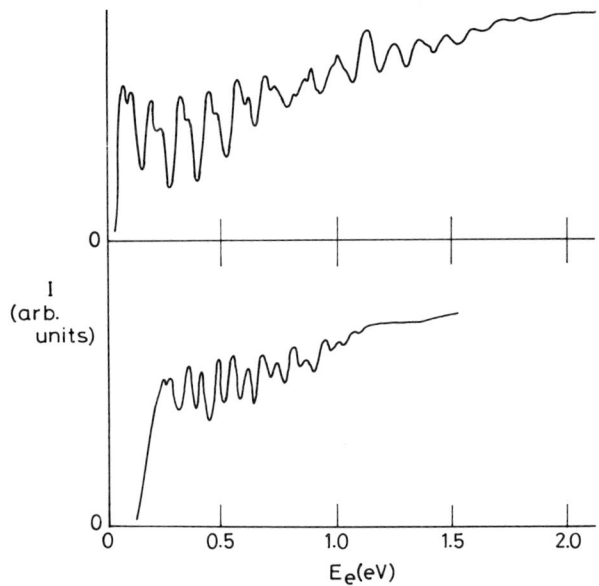

Fig. 20. Transmission and v_3 vibrational-excitation spectra of NO_2.

$v_1 = 0.125 \pm 0.101$ eV, with small contributions from $v_2 = 0.036 \pm 0.003$ eV and $v_3 = 0.060 \pm 0.005$ eV.

The spacing (162 meV) between the first two vibrational levels of the NO_2^- state determined by transmission techniques (Sanche and Schulz, 1973) are in almost exact agreement with the values obtained from an infrared absorption spectrum of NO_2^- isolated in a crystal lattice (Kato and Rolfe, 1967).

5. SO_2

SO_2 has a stable negative ion, and therefore a positive electron affinity (1.12 eV). As in the case of NO_2 one might expect resonant structure near zero energy. Instead, the derivative technique measurements of Sanche and Schulz (1973) displayed sharp vibrational levels only in the 3–6 eV region. The absence of very low-energy resonant structure has been explained by postulating an escape barrier which, because a high-angular-momentum partial wave may be responsible for the resonance, is very large. Therefore the resonant structure at very low energies would be too sharp to be detected. At energies near 3 eV, the lifetime of the SO_2^- state decreases (the width increases) and the parameters become favorable enough for the resonant fine structure to be observed experimentally.

6. Water and its Alkyl Homologs

Ab initio calculations (Weiss and Krauss, 1970) as well as recent dissociative attachment experiments (Belkić *et al.*, 1981) have revealed three negative-ion states, which have been designated as $^2B_1$ (6.5 eV), $^2A_1$ (8.6 eV), and $^2B_2$ (11.8 eV). All three resonant states as well as their parent triplet states of neutral H_2O ($^3B_1$, $^3A_1$, and $^3B_2$, respectively) have potential-energy surfaces which are repulsive in the Franck–Condon region; the autoionization lifetimes of these resonances have been calculated by Jungen *et al.* (1979) to be 20, 60, and 8 fs for the $^2B_1$, $^2A_1$, and $^2B_2$ states, respectively.

Potential-energy curves of the lowest several H_2O^- states have also been calculated by Claydon *et al.* (1971) with the INDO approximation, using the unrestricted Hartree–Fock method. Though the quantitative aspects of these calculations have been questioned (Winter *et al.*, 1973), one may expect a $^2A_1$ H_2O^- state to intersect the Franck–Condon region in the 6–8.5 eV range.

The transmission spectrum of Mathur and Hasted (1975) revealed an extremely broad, shallow dip centered at 7.96 eV which was attributed to the $^2A_1$ state. It was suggested that as the calculated potential surface of this state lay lower than the dissociation limits $OH^-(^1\Sigma^+) + H(^2S)$ and $O^-(^2P) + H_2(X\,^1\Sigma_g^-)$, the resonance state is bound relative to all dissociative coordinates and decay occurs only to ground-state H_2O. This being the case, the experimentally determined width would correspond to the total autoionization width.

The measured full width at half maximum is 3.0 ± 0.2 eV, which yields a lifetime on the order of 1.0 fs. Evidence for such a short-lived H_2O^- state also comes from associative detachment measurements (Howard *et al.*, 1974; Comer and Schulz, 1974). A broad enhancement of the vibrational-excitation cross section around 6 eV has also been attributed (Seng and Linder, 1976) to the $^2A_1$ H_2O^- state.

Alcohols ($C_nH_{2n+1}OH$) may be regarded as alkyl derivatives of H—OH and hence their electron transmission spectrum may be expected to resemble that of water. Table VIII compares resonance energies and indicates a progressive decrease in the negative alcohol ion lifetime as the number of carbon atoms increases.

A broad, featureless structure has also been detected in H_2S at 2.3 eV which correlates well with the results of dissociative attachment experiments (Fiquet-Fayard *et al.*, 1972; Azria *et al.*, 1977). The resonance is formed by the incoming electron occupying the 6a, first unoccupied orbital, giving rise to a $^2A_1$ state. The assignment has recently been verified in the angular measurements of Rohr (1978), who has also found that the total integral cross section for vibrational excitation exceeds 10^{-15} cm^2, nearly three orders of mag-

TABLE VIII

Resonances in Water and Alkyl Homologs[a,b,c]

Molecule	Resonance energy (eV)	Full width at half maximum (eV)
Water	7.96 ± 0.20	3.0
Methanol	7.88 ± 0.10	3.8
Ethanol	7.49 ± 0.10	4.1
Propanol	7.18 ± 0.10	4.2

[a] Transmission method; Mathur and Hasted (1975).

[b] Compare these with the series H_2CO, CH_3CHO, $(CH_3)_2CO$ where resonances appear at increasing energies as one proceeds along the series. This may indicate that the alkyl substitution to the C atom of the carbonyl chromophore destabilises the π^* molecular orbital.

[c] H_2O^- repulsive states not accessible through transmission spectra are as follows: $^2B_1$ (6.5 eV), $^2A_1$ (8.6 eV), $^2B_2$ (11.8 eV) (Trajmar and Hall, 1974).

nitude larger than the cross section for HS^- production (Fiquet-Fayard et al., 1972). This indicates that the $H_2S^-(^2A_1)$ state preferentially decays to the ground state of the neutral molecule.

Dissociative attachment measurements (Azria et al., 1977) also indicate a $^2B_1$ resonance, although this has not been detected in any transmission or differential cross section measurement, probably owing to its large width.

C. Polyatomic Molecules

1. Sulfur Hexafluoride

SF_6 is the only polyatomic molecule for which both theoretical and experimental resonance data exist over a wide energy range.

A resonance in the vicinity of 12 eV was first observed in the elastic and vibrationally inelastic channels by Trajmar and Chutjian (1977). Later measurements of the absolute total cross section function up to 100 eV (Kennerly et al., 1979) revealed three distinct shape resonances which were characterized as a_{1g} (2.5 eV), t_{1u} (7.0 eV), and t_{2g} (11.9 eV). A shallow enhancement of the cross section in the 25–35 eV region was also observed. Total cross section measurements below 1 eV (Ferch et al., 1982) show a smoothly rising function as the energy decreases, in agreement with electron attachment data (Chutjian, 1981).

The total cross section function has been calculated within the framework of multiple-scattering theory (Dehmer et al., 1978), and four resonances have been identified at 2.1, 7.2, 12.7, and 27.0 eV. The highest-energy resonant state has been characterized as e, the other configurations being the same as those proposed by Kennerly et al. (1979). Calculations using the Slater exchange approximation have also been performed in the energy range 10–60 eV (Benedict and Gyemant, 1978) which reveal resonances at 11, 17, and 24 eV. The differences in energy are probably attributable to the lack of convergence obtained in these calculations and to the use of a non-SCF charge distribution for SF_6 (Dehmer et al., 1978).

The agreement between experiment and multiple-scattering theory, though not exact, is good enough to provide strong support for the Dehmer and Dill (1979) model, which compares shape resonances in molecular photoionization and electron scattering and envisages the electron impacting as a molecule becoming temporarily trapped in the same orbital as a low-energy electron ejected from inner-shell x-ray ionization.

2. Aliphatic Hydrocarbons

Collisions of slow electrons with methane CH_4 continue to attract attention [see Mathur (1980) for a compilation of recent data]. Some early transmission experiments detected low-energy features in the total cross section function which are now known to have been due to gas impurity and instrumental effects. There is no low-lying resonance in methane. More recent measurements have established very short-lifetime resonant states in the region of 8 and 18.5 eV (Mathur, 1980) as well as some well-defined structures in the vicinity of the Ramsauer minimum (Barbarito et al., 1979), which have been correlated with the thresholds of a vibrational progression of the v_3, v_4, and v_2 modes and their intercombinations. It is of interest to note that in the vicinity of the Ramsauer minimum the total cross section function for CH_4 is remarkably similar to that for argon, and not to that of neon.

A typical electron transmission function covering the energy range 5–19 eV is shown in Fig. 21a. The broad resonance in the region of 8 eV has been attributed to a T_2 wave function with a strong $l = 2$ contribution in the close-coupling calculations of Gianturco and Thompson (1976). The differential cross section measurement of Rohr (1980) has verified the d-wave character of the resonance (Fig. 21b).

The weak feature centered at 18.5 eV has been interpreted as a resonant state with electronic configuration $(a_1)^2 (2a_1) (t_2)^6 (3sa_1)^2$. The existence of such a state was first predicted by Marmet and Binette (1978), who had determined the energy of the parent $3sa_1$ Rydberg state to be 19.5 eV in high-sensitivity ionization efficiency measurements. The energy separation between the parent state and the resonance yields a rather large value for the electron affinity of the excited state ($\simeq 1$ eV); the resonance can be aptly described as comprising two $3sa_1$ electrons trapped in the field of a positive ion core which constitutes the grandparent state. The energy difference between the

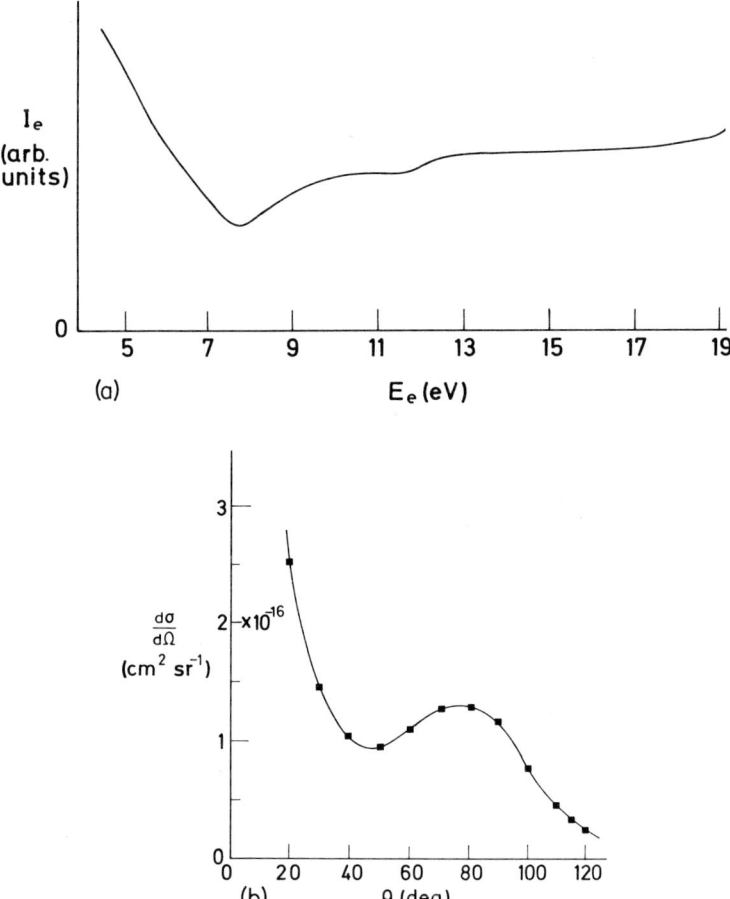

Fig. 21. (a) Transmission spectrum of CH_4. (b) Angular distribution of 5-eV electrons scattered from CH_4.

$^2A_1$ CH_4^+ grandparent and the resonance yields a binding energy of approximately 4.4 eV, in conformity with the rule that the binding energy of a pair of electrons in Rydberg orbits to a positive ion core generally lies between 4.0 and 4.5 eV (Schulz, 1973).

In addition to the CH_4 Rydberg state in the region of 19.5 eV three other such states have been reported in the 20–22 eV region (Mathur, 1981a,b); it is therefore possible that the broad 18.5-eV feature (Fig. 21a) comprises more than one overlapping resonance state.

Higher alkane hydrocarbons such as ethane (C_2H_6), propane (C_3H_8), and butane (C_4H_{10}) also exhibit Ramsauer minima (Massey and Burhop, 1969), though no relationship has been established between such minima and resonances.

3. Alkene Hydrocarbons

The transmission spectra of aliphatic hydrocarbons containing one double bond, such as ethylene, propene and butene (see Table IX) have been extensively investigated by the derivative technique and the trapped-electron method.

In ethylene the highest normally occupied orbital is $1b_{3u}$. The lowest shape resonance is formed by the addition of an electron to the $1b_{2g}$ orbital, yielding a $^2B_{2g}$ temporary negative-ion state. This was first observed as a broad feature in the transmission measurements of Boness et al. (1967) at an energy of 1.8 eV. Some extremely weak fine structure has since been discerned. Burrow and Jordan (1975) have speculated that the small amplitude of such structure (compared, for instance, to the amplitude of structure in benzene) may be indicative of a resonance lifetime which is marginally shorter than the period of a single vibrational mode but is long enough to allow a portion of the nuclear wave packet to cross the potential-energy well along the C—C stretch normal coordinate and be reflected back in "boomerang" fashion.

Prominent vibrational structure is visible in the $^2B_{3u}$ core-excited resonance in the region of 6.6 eV (Verhaart and Brongersma, 1980). Van Veen's (1976b) trapped-electron data had earlier indicated some structure in the region of 5–7 eV which was attributed to the $^2B_{3u}$ excited resonance state. A similar feature has been observed in measurements employing the derivative technique (Sanche and Schulz, 1973).

It is interesting to compare the first electron affinities of methane (-8 eV) and ethylene (-1.55 eV) and note that the ethylenic double bond stabilizes the temporary negative-ion state. Table IX shows that the presence of an additional double bond, as in the case of 1,3-butadiene (electron affinity -0.62 eV), lowers the energy of the first resonant state even further. We have seen that studying the changes in the lowest resonance energies provides a sensible method of investigating unoccupied molecular orbitals.

The data for propene in Table IX indicate that substitution of a methyl group for an H atom raises the lowest resonance energy. The lowering of symmetry and the superposition of lower l values lead to a shorter resonance lifetime; hence there is no indication of vibrational structure in this resonance.

Two resonances have been detected in 1,3-butadiene (Jordan et al., 1976b). The lowest resonance (0.62 eV) occurs when an electron is captured into the $2a_u$ orbital. Vibrational structure corresponding to the C—C symmetric stretch mode is observed. A broad, featureless resonance at 2.8 eV has been characterized as a $C_4H_6^-$ ($^2B_{2g}$) state in which the incident electron is trapped by an $l = 2$ centrifugal barrier.

TABLE IX

Resonances in Aliphatic Molecules

Molecule	Lowest resonances (eV)	Other resonances (eV)	References[a]
Methane	? 0.215 (f.s.)	8	T, Barbarito et al. (1979)
		7.8, 18.5	T, Mathur (1980)
Ethane	? 2.3		S, Pisanias et al. 1972a
Ethylene	1.78 (f.s.)		DT, Burrow and Jordan (1975)
	1.76	6.60	DT, Sanche and Schulz (1973)
		5.7	TE, Van Veen (1976a)
Propene	1.99		DT, Jordon et al. (1976a)
cis-butene	2.22		
trans-butene	2.10		
iso-butene	2.19		DT, Jordan and Burrow (1978)
Trimethylethylene	2.24		DT, Jordan and Burrow (1978)
Tetramethylethylene	2.27		
1,3-butadiene	0.62 (f.s.)	2.80	DT, Jordan et al. (1976a)
Cyclohexene	2.07		
1,3-cyclohexadiene	0.8 (f.s.)	3.43	
1,4-cyclohexadiene	1.75 (f.s.)	2.67	DT, Jordan et al. (1976a,b)
1,5-cyclooctadiene	1.83	2.33	
Norbornadiene	1.04	2.56	
Formaldehyde	0.66 (f.s.)	6.92 (f.s.)	
Acetaldehyde	0.97 (f.s.)	6.60 (f.s.)	TE, Van Veen et al. (1976b)
Acetone	1.41		
Acetylene	2.6		
Acetonitrile	2.84		DT, Jordan and Burrow (1978)
Acrylonitrile	0.21	2.74	
Hexatriene	1.8	3.54	
Cyclooctatetraene	1.91	3.60	DT, Jordan and Burrow (1978)
Acrolein	2.47	—	
2-butyne	3.43	—	
di-t-butylacetylene	3.10	—	
Cyclooctyne	2.18	3.28	DT, Ng et al. (1982)
3,3,6,6-tetramethyl-1-thiacycloheptyne	1.16	2.82	
3,3,6,6-tetramethyl-1-thiacycloheptane	2.8	—	

[a] T, transmission; DT, differential transmission; S, scavenger; T, trapped electron.

Jordan et al. (1976b) have investigated the energetic positions of low-lying resonances as a function of the distances and orientations of the two C—C double bonds in cyclic dienes and shown that as in the series ethane-ethylene-butadiene, two double bonds lower the energy of the lowest resonance state more than does a single double bond (see Table IX). Moreover, as the transmission functions of 1,4-cyclohexadiene, 1,5-cyclohexadiene, 1,5-cyclooctadiene, and norbornadiene show, the splitting between the two lowest π-resonances is also a function of the relative positions of the double bonds within the cyclic skeleton. Discussion in terms of localized bond interactions directly through space or indirectly through adjacent bonds has been presented (Jordan and Burrow, 1978).

Styrene (C_6H_5CH=CH_2) is an interesting alternant hydrocarbon which acts as the link between the ethylenic and aromatic molecules, and has been studied by means of the derivative technique (Burrow et al., 1976). Four resonances are detected, at 0.25, 1.05, 2.48, and 4.67 eV (Table X), corresponding to the ground and successively higher temporary $C_8H_8^-$ states. The lifetimes of the two lowest states are sufficiently long to enable observation of a vibrational progression of the symmetric breathing mode.

Being an alternant hydrocarbon, the π and π^* orbitals ought to obey the pairing theorem whereby the sum of the ionization energy (IE) and electron affinity (EA) associated with each pair of π, π^* orbitals is a constant. Burrow et al. (1976) show that for styrene the sum IE + EA is 8.18, 8.13, 7.97, and 7.63 eV for the first four π, π^* pairs. The constancy is even more remarkable (1%) if one uses the values of IE and EA (8.10, 8.11, 8.14, and 8.18 eV) semiempirically calculated by Younkin et al. (1976).

4. *Halogen Substitution*

In chloroethylenes and chloroethanes, five resonant states have been detected below 2 eV in the attachment studies of Johnson et al. (1977); these are associated with orbitals dominated by the p orbitals of the chlorine atoms. Two such resonances have also been observed in the 0.5–3 eV energy range in recent derivative technique measurements (Burrow et al., 1981) on the chloroethylenes. These measurements have also shown that chlorine substitution appears to destabilize the lower resonant state while at the same time stabilizing the higher-energy resonance.

Photoelectron spectroscopists have long established the fact that perfluorination of a large variety of planar unsaturated hydrocarbons causes large increases in σ ionization energies while only marginally affecting the π ionization energies (Rabalais, 1977). Brundle et al. (1972) have explained this in terms of resonance and inductive effects of the substitutent. The fluorine atoms withdraw charge from the C—C bond, resulting in a stabilization of the π orbital. At the same time, resonance interaction causes destabilization by almost the same amount. Hence the π-ionization energies are relatively unaffected by fluorine substitution. The corresponding effects on normally unoccupied orbitals of the fluoroethylenes have been studied by the derivative technique (Chiu et al., 1979), and the vertical electron affinities of all the fluorinated molecules are found to be more negative than that of ethylene (i.e., resonances at higher energies). The addition of a second fluorine atom produces shifts in resonance energies which are a function of the position of this atom, the largest shift occurring when two fluorine atoms are on the same carbon.

A correlation diagram of the resonance energies and the vertical ionization energies (Sell et al., 1978) (Fig. 22) shows clearly that in contrast to the highest occupied π orbitals, the energies of the lowest normally unoccupied π* orbitals are dependent on the number and relative position of fluorine substitution. A model can be postulated which relates the energies of the π and π* orbitals to the shortening of the C—C and C—F bond lengths which are known to occur upon fluorination (Carlos et al., 1974); a shortening of the C—C bond is associated with stabilization of the π orbital and destabilization of the π* orbital; a decreasing C—F bond length, on the other hand, leads to destabilization of both the π and π* orbitals, since the p_z orbital of the fluorine atom mixes in an antibonding manner with the molecule.

Thus, by using empirical energy shifts of 130 and 40 meV per 0.01-Å change in the C—C and C—F bond lengths, respectively, Chiu et al. (1979) obtain π, π* orbital energies which reflect the trend in the experimental data.

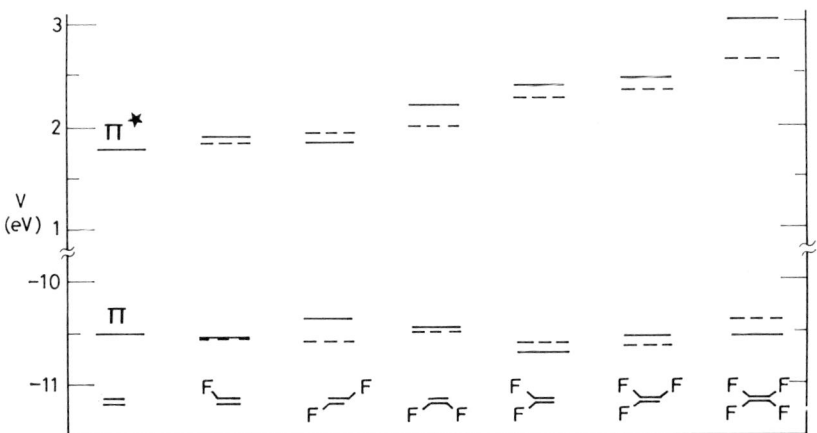

Fig. 22. Resonance energies and vertical ionization energies of fluorocarbons. Full lines, experimental data; dotted lines, model calculations.

5. Perfluorocarbons

Aliphatic perfluorocarbons (C_xF_y) have positive electron affinities, and a large number of resonant states have been detected in electron attachment studies (Chapter 6) in the energy region below $\simeq 2$ eV. The dependence of the energies of these low-lying resonances on molecular structure has been investigated in a large number of singly bonded as well as double- and multiply-bonded molecules (Pai et al., 1979). The findings indicate that for singly bonded molecules such as c-C_4F_8, c-C_6F_{12}, or C_8F_{16}, the size or complexity of the molecule affects neither the energies of the resonance states nor the magnitude of the attachment cross section. In the case of double- and multiply-bonded molecules, c-C_4F_6, 2-C_4F_6, 1,3-C_4F_6, or C_7F_8, though the attachment cross section increases with increasing molecular size, there is no clear-cut evidence regarding any change in the energetic positions of the resonant states.

Despite the importance of these perfluorocarbons (see Chapter 5 of Volume 2), no beam experiments have been reported to date.

6. Aldehydes and Ketones

The simplest molecule containing the highly polar carbonyl group is formaldehyde H_2CO, and replacement of the hydrogen atoms by one or two methyl groups yields acetaldehyde and acetone, respectively. Low-lying resonances in these molecules have been located by means of the derivative technique (Burrow and Michejda, 1976; Van Veen et al., 1976a). Temporary capture of an electron in the normally unoccupied $2b_1$ orbital results in a shape resonance in the region of 0.66 eV in formaldehyde, and at higher energies in acetaldehyde (0.97 eV) and acetone (1.41 eV), the upward shift in resonance energies being a measure of the destabilization of the π^* orbital on alkyl substitution to the C atom of the carbonyl group. Effects on resonance lifetime are also well illustrated: prominent vibrational structure corresponding to the v_2 mode along the C=O stretch coordinate is observed in formaldehyde; such structure is marginally visible in the acetaldehyde transmission function and is not detected at all in acetone (Fig. 23).

Feshbach resonances have also been observed in the region of 6–9 eV in all three molecules. As in the case of the shape resonances, vibrational structure is observed clearly in formaldehyde, less so in acetaldehyde, and not at all in acetone.

7. Aromatic Hydrocarbons

The benzene molecule contains 24 electrons in extended σ orbitals, and 6 electrons in π orbitals shared by the carbon atoms. The ground state configuration of the latter is $C_6H_6[\tilde{X}\,(a_{2u})^2\,(e_{1g})^4\,{}^1A_{1g}]$. The ground state of

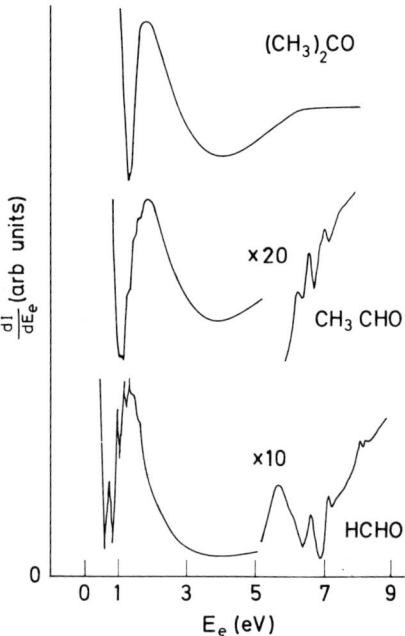

Fig. 23. Derivatives of the transmission functions of acetone, acetaldehyde, and formaldehyde.

the $C_6H_6^-$ ion is formed by adding a π electron with $m = 2$ and $l = 3$ as the first term in its spherical-harmonic expansion; its configuration is $C_6H_6^-[\tilde{X}(a_{2u})^2(e_{1g})^4(e_{2u})^2 E_{2u}$. This short-lived state which decays by electron emission was first detected as a resonance in the vicinity of 1 eV in the inelastic electron scattering measurements of Compton et al. (1966). The transmission measurements of Boness et al. (1967) revealed fine structure in this resonance which could be interpreted as a vibrational progression of the ground-state temporary negative ion. By more refined measurements (Sanche and Schulz, 1973; Mathur and Hasted, 1976a) it was determined that the vibrational spacings in the resonant state corresponded to those obtained on exciting the totally symmetric breathing mode of the A_{1g} ground state of the neutral molecule; just as photoelectron spectroscopy had shown that removal of a π electron does not alter the vibrational frequencies of polyatomic molecule energy levels (Potts et al., 1972), so the addition of an electron does not significantly perturb the vibrational modes in a molecule such as benzene.

Vibrational excitation of C_6H_6 and C_6D_6 via the lowest resonance was studied by Wong and Schulz (1975), who measured the energy loss spectrum at an incident electron energy corresponding to the ground vibrational level of the $^2E_{2u}$ resonance; it was found that though the C_6H_6 molecule possesses 30 normal modes of vibration, only the v_{21}, v_{16}, and v_{20} modes were excited. Boness et al. (1967) and Larkin and Hasted (1972) also reported the excitation of v_3, v_4, v_1, v_5 modes. This selectivity of vibrational modes excited via a resonance points to the existence of selection rules based on the relation between the symmetry of the resonant state and the symmetries of the possible vibrational modes.

The angular distribution of scattered electrons in the region of the lowest benzene resonance was also measured by Wong and Schulz (1975); it confirmed the prediction made by Bardsley and Read (1968) that the ^2E_2u resonance occurs predominantly in the $f\delta$ partial wave.

Scansion of the electron energy over a wider range enables detection of at least two other resonant states (Fig. 24). The broad structure near 5 eV was observed by Sanche and Schulz (1973) and Mathur and Hasted (1976a), though it escaped detection in the earlier measurements of Compton et al. (1968). This short-lived state arises out of the temporary binding of the incident electron in the highest orbital with $m = 3$ and is characterized as having $^2B_{2g}$ symmetry. Smythe et al. (1974) have shown that this resonance

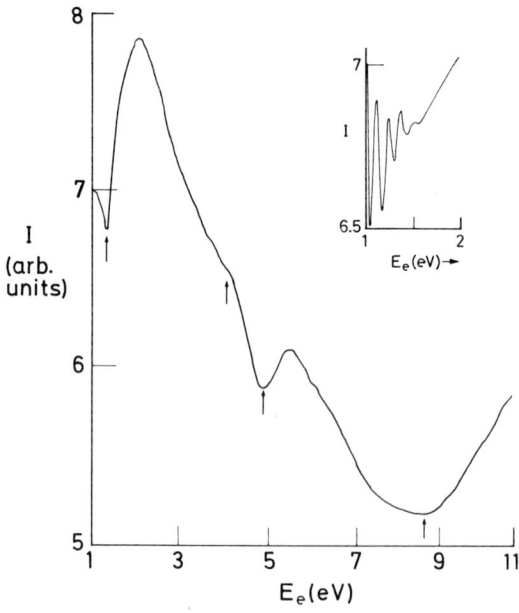

Fig. 24. Transmission spectra of benzene; the 1–2 eV region spectrum is shown expanded in the inset.

decays not only into the ground state of C_6H_6 but also into the three lowest excited states, $^3B_{1 4}$ (3.66 eV), $^3E_{1 4}$ (4.51 eV), and $^1B_{2u}$ (4.72 eV); this indicates that the resonance is likely to be represented by a complex mixture of wave functions and a number of parent states.

Transmission experiments (Mathur and Hasted, 1976a) also reveal a very broad resonance in the 7–10 eV range. The center of this resonance is located at 8.85 eV and its full width at half minimum is over 3 eV, indicating a very short lifetime ($\simeq 10^{-16}$ s). The absence of fine structure in either of the two broad resonances leads one to postulate that the potential-energy surfaces of the two short-lived states are either repulsive in the Franck–Condon region or very short-lived. Several resonances in this energy range have recently been reported by Burrow et al. (1982a).

The lowest unoccupied π orbitals of naphthalene, the simplest of the double-ring compounds, have been studied by transmission technique and by scavenger technique as well as by the trapped-electron method. The energy of the lowest resonance is of considerable interest in that it relates to the first electron affinity of the molecule, which has been the subject of a large number of studies, both theoretical and experimental (see Mathur and Hasted, 1977).

Although the lowest resonance has been placed between zero and 2 eV in two different measurements (see Table X), theoretical studies by both Younkin et al. (1976) and Asbrink et al. (1977a,b, 1979) have indicated that the first electron affinity should be almost zero; it has been suggested that the value of -0.2 eV from transmission experiments relates to the formation of a vibrationally excited ion. At the same time, there are conflicting reports as to whether or not naphthalene negative ions are stable (Pisanias et al., 1972a; Asbrink et al., 1977a,b, 1979). One of the factors responsible for the continuing ambiguity in experimental data may be the fact that low-vapor-pressure substances such as naphthalene present poor targets for low-energy electrons; the vapor produces surface effects detrimental to electron-beam stability, intensity, and resolution; a heated collision chamber is necessary.

Higher-energy resonances might be attributed to occupation of successively higher normally unfilled π^* orbitals, bearing in mind the electron spectroscopic interpretation that the first five ionization energies of naphthalene relate to the five occupied π orbitals of the molecule (Turner et al., 1970).

Early threshold measurements (Compton et al., 1968) had revealed a number of distinct peaks in the energy region between 2.5 and 8 eV, which were assigned as low-lying triplet and singlet states of the neutral molecule. However, one peak, occurring at 5.4 eV, could not be identified, as it did not correspond to any known excited state of napthalene. Birks et al. (1968) postulated that this structure might correspond to the $^3B_{2u}{}^+$ state. The transmission spectrum shown in Fig. 25 reveals a prominent resonance at 5.29 ± 0.1 eV and it appears likely that it was this resonant state that was

TABLE X

Resonances in Aromatic Molecules

Molecule	Lowest resonances (eV)	Other resonances (eV)	Reference[a]
Benzene	1.0 (f.s.)	4.93, 8.85	T, Mathur and Hasted (1976a)
	1.15 (f.s.)	4.85	DT, Sanche and Schulz (1973)
Phenol	1.03, 1.66	4.78	T, Mathur and Hasted (1976a)
	0.61, 1.67		TE, Christophorou et al. (1974)
	1.01, 1.73	4.92	DT, Jordan et al. (1976a,b)
Fluorobenzene	0.89	4.77	DT, Jordan et al. (1976a,b)
	1.27, 1.74		TE, Christophorou et al. (1974)
	0.91, 1.40	4.66	T, Mathur and Hasted (1976a)
p-difluorobenzene	0.62, 1.41	4.51	
1,3,5-trifluorobenzene	0.77	4.48	
2,3,5,6-tetraflurobenzene	0.50, 1.29	4.51	TE, Christophorou (1978)
Pentafluorobenzene	0.36, 1.19	4.53	
Hexafluorobenzene	0.42	4.50	
Chlorobenzene	0.75	4.50	DT, Jordan et al. (1976a,b)
	0.90, 1.74	4.68, 8.22	T, Mathur and Hasted (1976a)
Bromobenzene	0.70	4.42	DT, Jordan et al. (1976a,b)
Toluene	1.27	4.9, 8.37	T, Mathur and Hasted (1976a)
	1.11	4.88	DT, Jordan et al. (1976a,b)
Ethylbenzene	1.17	—	DT, Nenner (1975)
Isopropylbenzene	1.08	4.69	DT, Jordan et al. (1976a,b)
Tert-butylbenzene	1.06	4.67	DT, Jordan et al. (1976a,b)
Cyclopropylbenzene	1.06	4.59	DT, Jordan and Burrow (1978)
o-xylene	1.12	4.90	DT, Jordan et al. (1976a, b)
m-xylene	1.06	4.81	DT, Jordan et al. (1976a,b)
p-xylene	1.07	4.89	DT, Jordan et al. (1976a,b)
1,3,5-trimethylbenzene	1.03	4.78	DT, Jordan et al. (1976a,b)
1,2,4-trimethylbenzene	1.07	4.83	DT, Jordan et al. (1976a,b)
Styrene	0.25, 1.05	2.48, 4.67	DT, Burrow (1976)
Naphthalene	0.19, 0.90	1.67, 3.37, 4.72	DT. Jordan and Burrow (1978)
	0.20, 0.76	1.35, 5.29, 7.77	T, Mathur and Hasted (1977)
	0.8	5.4	SF, Compton et al. (1966)
	0.75	1.3	TE, Pisanias et al. (1972a,b)
cis-stilbene	0.92, 1.68	3.4, 4.67	DT, Jordan and Burrow (1978)
Nitrobenzene	0.62	4.62, 8.18	T, Mathur and Hasted (1976a)
Aniline	1.13, 1.85	5.07	DT, Jordan et al. (1976a,b)
	0.55, 1.88		TE, Christophorou et al. (1974)
N-methylaniline	1.19		DT, Nenner (1975)
N-dimethylaniline	1.24		DT, Nenner (1975)
Anisole	1.90		TE, Christophorou et al. (1974)
	1.09, 1.72	4.92	DT, Jordan et al. (1976a,b)
Thiophenol	0.66, 1.10		TE, Christophorus et al. (1974)
Benzoic acid	0.63, 1.33		TE, Christophorou et al. (1974)
Benzaldehyde	0.71, 1.12		TE, Christophorou et al. (1974)
	0.76, 2.21	4.61	DT, Jordan and Burrow (1978)
Benzonitrile	0.54, 2.49	3.20, 4.90	DT, Jordan and Burrow, (1978)
	0.60, 2.51	4.85	T, Mathur (1981a)

[a] T, transmission; DT, differential transmission; S, scavenger; TE, trapped electron.

5 Electron–Molecule Resonances

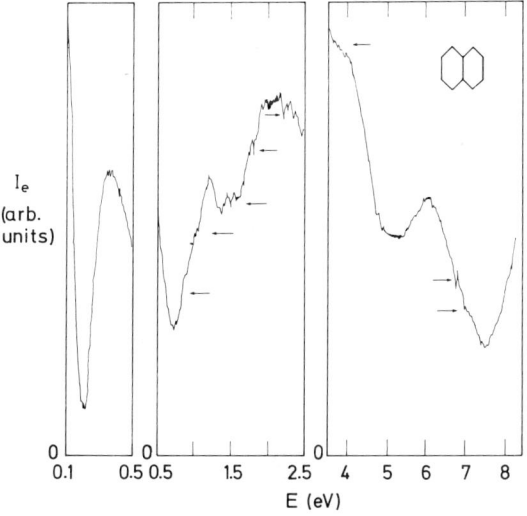

Fig. 25. Transmission spectra of naphthalene covering the energy region 0.1–8 eV.

detected by Compton *et al.* A derivative spectrum has recently been reported by Jordan and Burrow (1981).

The half-widths of the naphthalene resonances appear to be comparable with those of the benzene resonances, though it is not possible to give values for the lifetimes of these states because of the indeterminacy of the resonance natural width due to several underlying continua.

8. Substitued Benzenes

One of the features which have attracted both experimentalists and theoreticians to the benzene molecule is the fact that both the highest occupied and the lowest unoccupied molecular orbitals (e_{1g} and e_{2u}, respectively) are degenerate.

Monosubstitution in the benzene ring splits the degenerate e_{2u} benzene orbital into a_2 and b_1 orbitals, the former being more stable than the latter (see Table X). Transmission studies (Mathur and Hasted, 1976a; Jordan and Burrow, 1978) have indicated that, in general, the magnitude of the splitting of the two lowest resonances depends on the electronegativity of the substituent. In the case of alkyl-substituted benzenes the energy difference between the A_2 and B_1 resonance states is not very different from the zero-point energy of nuclear motion. Hence the ground states of these temporary negative ions may be regarded as pseudo-Jahn–Teller systems whose $v = 0$ states are mixtures of different angular momenta (Alper and Silbey, 1970). This results in much weaker vibrational structure than in the case of benzene because of significant reduction of the resonance lifetime.

Some typical electron transmission spectra of substituted benzenes are shown in Fig. 26a. To appreciate the effect of monosubstitution on both highest occupied and lowest unoccupied molecular orbitals, Fig. 27 compares electron affinities measured mainly by the transmission technique with ionization energies, determined by photoelectron spectroscopy.

The displacements of the normally unoccupied as well as of the highest occupied orbitals appear to be strongly dependent on the nature of the substituent, in particular on its inductive and resonance effects. In the present context, the latter term is a measure of the extent to which electron migration occurs from the substituent to the benzene skeleton, whereas the former term quantifies the opposite effect, namely, electron withdrawal toward the substituent. The reader is referred to Ehrenson et al. (1973) for a fuller discussion of inductive and resonance effects.

In general, substituents with lone-pair electrons, such as OH, OCH_3, NH_2 and SH possess small positive inductive effects and large negative resonance effects. These properties ensure that the lowest resonant states in phenol, anisole, aniline, and thiophenol are slightly more stable than the benzene E_{2u} state. On the other hand, the first excited states of such temporary negative ions—the A $^2B_1$ states—lie well above the benzene ground state (see Table X).

In the case of halogen substitution, the large negative resonance effect is counteracted to a greater extent by a more positive inductive effect. The B_1 state would therefore be expected to occur at somewhat lower energies. The experimental data for the halobenzenes is still not entirely clear. In the case of fluorobenzene the derivative technique data of Jordan et al. (1976a) revealed some structure due to both B_1 states; a later derivative-technique measurement employing almost the same energy resolution (Frazier et al., 1978) revealed both resonances, though at significantly lower energies than trapped-electron data reported by the same group somewhat earlier (Christophorou et al., 1974). In the case of chlorobenzene, Jordan et al. again observed the split resonance detected in the transmission measurements of Mathur and Hasted (1976a). The resonance energies measured in the latter experiment are in accord with dissociative attachment and threshold excitation data (Christophorou and Schweinler, 1979). It is of interest to note that double resonances have also been observed in the fluorobenzene series $C_6H_nF_{6-n}$ for n = 1–6 (Frazier et al., 1978). One of the resonant states lies above and the other below the E_{2u} benzene resonance (see Table X). The splitting is about 0.8 eV except in the highly symmetrical 1,3,5-$C_6H_3F_3$ and C_6F_6, where the two lowest resonant states again become degenerate.

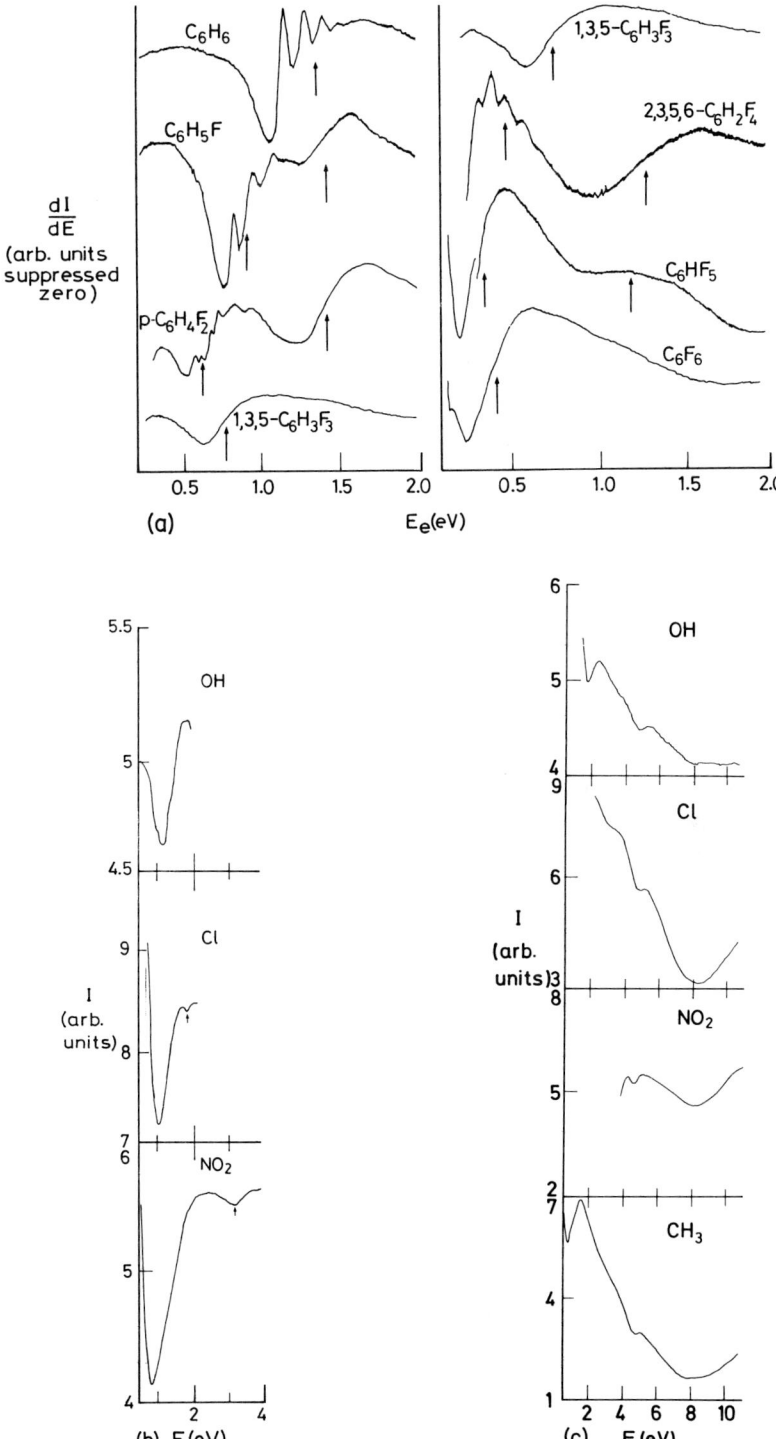

Fig. 26. (a) Derivatives of the transmission functions of some substituted benzenes showing the two lowest resonances. Note the degeneracy of the two resonances in the symmetrical cases. (b, c) Transmission functions of some substituted benzenes.

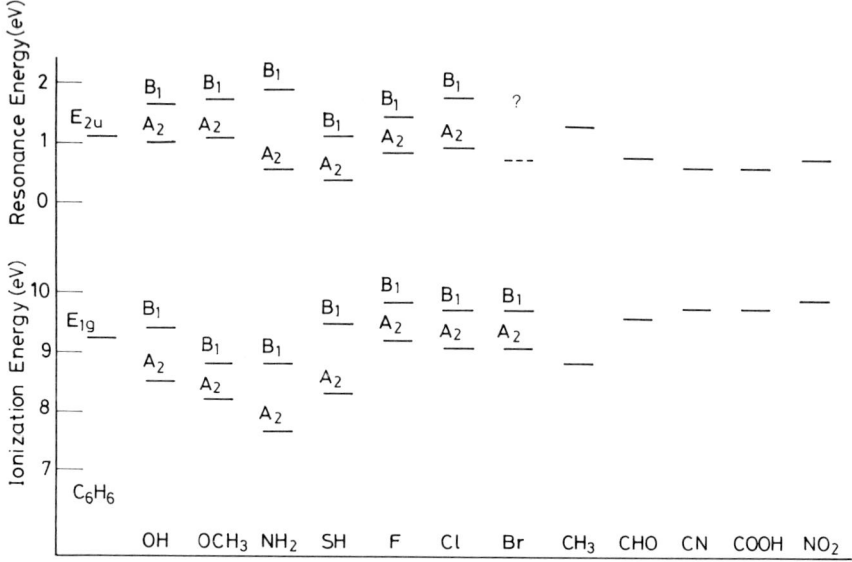

Fig. 27. Ionization energies and resonance energies of substituted benzenes.

Substituents such as CHO, NO_2, CN, and COOH possess large positive inductive effects. Hence molecules such as benzaldehyde, nitrobenzene, benzonitrile, and benzoic acid possess positive electron affinities (i.e., the lowest resonance energy is below 0 eV). The long-lived ground-state negative ions formed can have autodetachment lifetimes of the order of tens of microseconds. In such cases the effect of monosubstitution can be studied by following the changes in the lifetime of the negative ions in swarm experiments (see Christophorou, 1978).

It is worthy of mention that in all the monsubstituted benzenes the energies of the third and fourth resonances are relatively insensitive to the nature of the substituent. This indicates that the effect of the substituent on the carbon p orbital is minimal. These higher-energy resonant states also have shorter lifetimes (10^{-15}–10^{-16} s) and correspondingly larger widths (0.5–4 eV).

9. Heterocyclic Compounds

The replacement of a carbon atom in the benzene skeleton by a heteroatom such as nitrogen or oxygen destroys the ring symmetry, so that the e_{24} orbital is split into two. This splitting has been demonstrated in a number of measurements on pyridine and the diazines (Pisanias *et al.*, 1973; Nenner and Schulz, 1975; Mathur and Hasted, 1976b), just as has been the case in monsubstituted benzenes (Christophorou *et al.*, 1974; Mathur and Hasted,

1976a). Heterocyclic-molecule resonant-state data are summarized in Table XI.

In pyridine the lowest resonance (0.79 eV) occurs at a lower energy than that in benzene; but for the second resonances (1.15 eV) the energies are identical. These energies compare well with calculated values of -0.70 and -1.04 eV for the first two electron affinities of pyridine (Younkin et al., 1976). Pairs of resonances have also been observed in the diazines, pyridazine, pyrimidine, and pyrazine. The energies agree reasonably well with calculated values for the first two electron affinities. An interesting feature revealed by the diazine transmission spectra is that the relative change in the energies of the two lowest unoccupied orbitals is not only a function of the number of nitrogen atoms in the carbon ring but also their relative positions around the ring. Significant discrepancies exist in the published diazine and s-triazine transmission data. Using the derivative technique, Nenner and Schulz (1975) observed in the transmission function fine structure which appeared to extend to energies below 0 eV in the case of the diazines and s-triazine. On this basis they postulated a positive electron affinity for these molecules; the fine

TABLE XI

Resonances in Heterocyclic Molecules

Molecule	Lowest resonances (eV)	Other resonances (eV)	References[a]
Pyridine	0.79, 1.15	4.71, 7.27, 7.86	T; Mathur and Hasted (1976a)
	0.62, 1.20	4.58	DT; Nenner and Schulz (1975)
Pyridazine	0.32, 0.73	4.05	DT; Nenner and Schulz (1975)
	0.49, 0.89	4.10	T; Mathur and Hasted (1976a)
Pyrimidine	0.25, 0.77	4.24	DT; Nenner and Schulz (1975)
	0.33, 0.82	4.23	T; Mathur and Hasted (1976a)
Pyrazine	0.07, 0.87	4.10	DT; Nenner and Schulz (1975)
	0.08, 0.92	4.15	T; Mathur and Hasted (1976a)
s-triazine	0.12	4.0	DT; Nenner and Schulz (1975)
Phosphabenzene	0.64	4.21	DT, Burrow et al. (1982)
Arsabenzene	0.62	3.84	DT, Burrow et al. (1982)
Stilbabenzene	0.60	3.47	DT, Burrow et al. (1982)
Quinoline	.60 (f.s.?)	1.30	TE; Pisanias et al. (1972a,b)
Isoquinoline	0.42 (f.s.)	1.25 (f.s.)	
Pyrrole	2.38	3.44	
Furan	1.76	3.14	
Thiophene	1.17	2.67, 6.77	

[a] T, transmission; DT, differential transmission; TE, trapped electron.

structure was attributed to vibrational levels with $v > 0$; the lowest vibrational level of the negative ion would lie below the ground state of the neutral molecule and the ion in its ground vibrational level would not autodetach. On the other hand, the data of Mathur and Hasted (1976b) and Pisanias et al. (1973) failed to reveal any fine structure in the pyridine and diazine resonances, even though the former experiment was conducted at almost the same energy resolution as that of Nenner and Schulz. Recent derivative measurements (Burrow et al. 1982b) display the vibrational structure, but it does not extend below 0.6 eV.

It is of interest to note that in the case of s-triazine, the molecule belongs to symmetry point group D_{3h} and thus the first two resonant states are once again degenerate.

The extent to which the benzene e_{2u} orbital is split in the positive and the negative ions of pyridine and the diazines is shown in Table XII.

We now briefly examine the double-ring heterocyclic molecules. The condensation of the pyridine nucleus with the benzene ring in the *ortho*position can occur in two ways, giving rise to quinoline and isoquinoline, both of which are π-isoelectronic with naphthalene.

Pisanias et al. (1972a) have carried out threshold experiments on these molecules and ascribed low-energy peaks to resonances, some with detectable fine structure. Stable negative ions of both these molecules have been observed mass spectroscopically and the calculations of Younkin et al. (1976) have confirmed that the first electron affinities of quinoline and isoquinoline are both positive. It is therefore possible that the low-energy trapped-electron current measured by Pisanias et al. may have been due to collection of negative ions.

Resonances have been reported in pyrrole, furan, and thiophene (Van Veen, 1976a), which are five-membered rings in which six π electrons are spread over five centers. There are three occupied orbitals and two normally unoccupied orbitals. The observed double resonances are attributed to capture of

TABLE XII

Splitting of the e_{2u} Orbital in the Positive and Negative Ions of Pyridine and the Diazines (in eV)

	Pyridine	Pyridazine	Pyrimidine	Pyrazine
Positive ion[a]	1.10	0.27	0.29	0.50
Resonance	0.36	0.40	0.49	0.84

[a] Rosentock et al. (1977)

electrons into the latter. In the case of thiophene (see Table IX) a third resonance is detected at 6.77 eV and is characterized as a core-excited shape resonance. Decay of these resonances into low-lying triplet states of the neutral molecule has been verified in threshold excitation spectra.

It is noteworthy that the resonant states become more stable with increasing electron affinity of the heteroatom; i.e., the resonances in pyrrole, furan, and thiophene occur at successively lower energies; the electron affinity increases in the order N (-0.6 eV), O (1.5 eV), and S (2 eV). The measured electron affinities compare reasonably well with semiempirically calculated values (Compton et al., 1980).

V. Conclusions

In this chapter we have described the experimental evidence for resonance scattering of electrons by molecules. Structure in the cross section energy functions provides evidence for the existence of temporary bound states of the electron and molecule. These states rapidly decay into the energetically available channels; states of the neutral molecule, or even the neutral atoms and the free electron. Thus there are many types of experiment by which the resonance structure can be detected. The quantum numbers of the resonance states can be assigned on the basis of measurements of angular distributions of scattered electrons, and by comparing the vibrational energies of the resonance state with its parent and grandparent states.

Much further experimental and theoretical work remains to be done on resonances. Nevertheless there is now a sizable body of data covering almost a hundred molecules, each with usually more than a single resonance state. We look forward to a further period of accumulation of experimental data, it is hoped with improved technique.

Meanwhile the relevance of these studies to various fields of physics and chemistry still attracts the attention of those who will continue to support and carry out this accumulation of data.

References

Abouaf, R., and Teillet-Billy, D. (1980). *J. Phys. B* **13**, L275.
Alper, J. S., and Silbey, R. J. (1970). *J. Chem. Phys.* **52**, 569.
Andrick, D., Danner, D., and Ehrhardt, H. (1969). *Phys. Rev. Lett. A* **29**, 346.
Asbrink, L., Fridh, C., and Lindholm, E. (1977a). *Chem. Phys. Lett.* **52**, 63.
Asbrink, L., Fridh, C., and Lindholm, E. (1977b). *Chem. Phys. Lett.* **52**, 72.
Asbrink, L., Fridh, C., Lindholm, E., and de Bruijn, S. (1979). *Chem. Phys. Lett.* **66**, 411.
Azria, R., Wong, S.-F., and Schulz, G. J. (1975). *Phys. Rev. A* **11**, 1309.

Azria, R., Le Coat, Y., and Lefevre, G. (1977). *Proc. 10th ICPEAC, Paris*, 1977, p. 844.
Azria, R., Le Coat, Y., and Guillotin, J. P. (1980a). *J. Phys. B* **13**, L505.
Azria, R., Le Coat, Y., Simon, D., and Tronc, M. (1980b). *J. Phys. B* **13**, 1909.
Barbarito, E., Basta, M., Callicchio, M., and Tessari, G. (1979). *J. Chem. Phys.* **71**, 54–59.
Bardsley, J. N., and Read, F. H. (1968). *Chem. Phys. Lett.* **2**, 333.
Bederson, B., and Fite, W. L. (1968). "Methods of Experimental Physics," Vol. 7A, p. 67. Academic Press, New York.
Belkić, D. S., Landau, M., and Hall, R. I. (1981). *J. Phys. B* **14**, 175.
Birks, J. B., Christophorou, L. G., and Huebner, R. H. (1968). *Nature* **217**, 809.
Birtwistle, D. T., and Herzenberg, A. (1971). *J. Phys. B* **4**, 53.
Blatt, J. M., and Weisskopf, V. F. (1952). "Theoretical Nuclear Physics." Wiley, New York.
Boness, M. J. W., and Hasted, J. B. (1966). *Phys. Lett.* **21**, 526.
Boness, M. J. W., and Schulz, G. J. (1970). *Phys. Rev. A* **2**, 2182.
Boness, M. J. W., and Schulz, G. J. (1974). *Phys. Rev. A* **9**, 1969.
Boness, M. J. W., Larkin, I. W., Hasted, J. B., and Moore, L. (1967). *Chem. Phys. Lett*, **1**, 292.
Boness, M. J. W., Hasted, J. B., and Larkin, I. W. (1968). *Proc. R. Soc. A* **305**, 493.
Brüche, E. (1927). *Ann. Phys. Lpz.* **82**, 25.
Brundle, C. R., Robin, M. B., Kuebler, N. A., and Basch, H. (1972). *J. Am. Chem. Soc.* **94**, 1451.
Brunt, J. N. H., King, G. C., and Read, F. H. (1976). *J. Phys. B* **9**, 2195.
Brunt, J. N. H., King, G. C., and Read, F. H. (1977). *J. Phys. B* **10**, 433, 3781, 1289.
Brunt, J. N. H., King, G. C., and Read, F. H. (1978). *J. Phys. B* **11**, 173.
Burrow, P. D. (1974). *Chem. Phys. Lett.* **26**, 265.
Burrow, P. D., and Jordan, K. D. (1975). *Chem. Phys. Lett.* **36**, 594.
Burrow, P. D., and Michejda, J. A. (1976). *Chem. Phys. Lett.* **42**, 233.
Burrow, P. D., Michejda, J. A., and Jordan, K. D. (1976). *J. Am. Chem. Soc.* **98**, 6392.
Burrow, P. D., Modelli, A., Chin, N. S., and Jordan, K. D. (1981). *Chem. Phys. Lett.* **82**, 270.
Burrow, P. D., Ashe, A. J., III, Bellville, D. J. Jordan, K. D. (1982b). *J. Am. Chem. Soc.* **104**, 425.
Carlos, J. L. Jr., Karl, R. R. Jr., and Bauer, S. H. (1974). *J. Chem. Soc. Faraday Trans.* **2**, 177.
Cartwright, H. M., Bossomaier, T. R. J., and Grinter, R. (1977). *Theor. Chim. Acta* **44**, 265.
Chandra, N., and Temkin, A. (1976). *Phys. Rev. A* **13**, 188.
Chang, E. S., and Wong, S.-F. (1977). *Phys. Rev. Lett.* **38**, 1327.
Chantry, P. J. (1969). *J. Chem. Phys.* **51**, 3369.
Chiu, N. S., Burrow, P. D., and Jordan, K. D. (1979). *Chem. Phys. Lett.* **68**, 121.
Christophorou, L. G. (1978). *In* "Advances in Electronics and Electron Physics" (L. Marton, ed.), Vol. 46, p. 55. Academic Press, New York.
Christophorou, L. G., and Schweinler, H. C. (1979). *J. Chem. Phys.* **71**, 5385.
Christophorou, L. G., McCorkle, D. L., and Carter, J. G. (1974). *J. Chem. Phys.* **60**, 3779.
Chutjian, A. (1981). *Phys. Rev. Lett.* **46**, 1511.
Claydon, C. R., Segal, G. A., and Taylor, H. S. (1971). *J. Chem. Phys.* **54**, 3799.
Comer, J., and Harrison, M. (1973). *J. Phys. B* **6**, L70.
Comer, J., and Read, F. H. (1971). *J. Phys. B* **4**, 368, 1055, 1678.
Comer, J., and Schulz, G. J. (1974). *J. Phys. B* **7**, L249.
Compton, R. N., Christophorou, L. G., and Huebner, R. N. (1966). *Phys. Lett.* **23**, 656.
Compton, R. N., Huebner, R. H., Reinhardt, P. W., and Christophorou, L. G. (1968). *J. Chem. Phys.* **48**, 901.
Compton, R. N., Yoshioka, Y., and Jordan, K. D. (1980). *Theor. Chim, Acta* **54**, 259.
Dehmer, J. L., and Dill, D. (1979). *In* "Symposium on Electron-Molecule Collisions" (I. Shimamura and M. Matsuza, eds.), p. 95. University of Tokyo, Tokyo.
Dehmer, J. L., Siegel, J., and Dill, D. (1978). *J. Chem. Phys.* **69**, 5205.
Domcke, W., and Cedarbaum, L. S. (1980). *J. Phys. B* **13**, 2829.
Domcke, W., and Cedarbaum, L. S. (1981). *J. Phys. B* **14**, 149.

Ehrenson, S., Brownlee, R. T. C., and Taft, R. W. (1973). *Prog. Phys. Org. Chem.* **10**, 1.
Ehrhardt, H., and Willman, K. (1967). *Z. Phys.* **204**, 462.
Ehrhardt, H., Langhans, L., Linder, F., and Taylor, H. S. (1968). *Phys. Rev.* **173**, 222.
Fano, U., (1961). *Phys. Rev.* **124**, 1866.
Ferch, J., Masche, C., and Raith, W. (1981). *J. Phys. B* **14**, L97.
Ferch, J., Raith, W., and Schroder, K. (1982). *J. Phys. B* **15**, L175.
Feshbach, H. (1958). *Ann. Phys. N.Y.* **5**, 357.
Feshbach, H. (1962). *Ann. Pnys. N.Y.* **19**, 287.
Fiquet-Fayard, F., Ziesel, J. P., Azria, R., Tronc, M., and Chiari, J. (1972). *J. Chem. Phys.* **56**, 2540.
Frazier, J. R., Christophorou, L. G., Carter, J. G., and Schweinler, H. C. (1978). *J. Chem. Phys.* **69**, 3807.
Gianturco, F. A., and Thompson, D. G. (1976). *J. Phys. B* **11**, 727.
Golden, D. E. (1966). *Phys. Rev. Lett.* **17**, 847.
Gresteau, F., Hall, R. I., Huetz, A., Vichon, D., and Mazeau, J. (1979). *J. Phys. B* **12**, 2925, 2937.
Griffith, J. S. (1962). "The Irreducible Tensor Method for Molecular Symmetry Groups." Prentice-Hall, Englewood Cliffs, N.J.
Hamilton, D. R. (1940). *Phys. Rev.* **58**, 122.
Hasted, J. B. (1970). "Physics of Atomic Collisions." 2nd edition. Butterworths, London.
Hasted, J. B., and Awan, A. M. (1969). *J. Phys. B* **2**, 367.
Heideman, H. G. M., Kuyatt, C. E. and Chamberlain, G. E. (1966). *J. Chem. Phys.* **44**, 355.
Herzenberg, A. (1968). *J. Phys. B* **1**, 548.
Herzenberg, A. (1979). *In* "Symposium on Electron-Molecule Collisions" (I. Shimamura and M. Matsuzama, eds.), p. 77. University of Tokyo, Tokyo.
Howard, C. J., Fehsenfeld, F. C., and McFarland, M. (1974). *J. Chem. Phys.* **60**, 5086.
Huetz, A., Gresteau, F., and Mazeau, J. (1980a). *J. Phys. B* **13**, 3275.
Huetz, A., Cădež, I., Gresteau, F., Hall, R. I., Vichon, D., and Mazeau, J. (1980b). *Phys. Rev. A* **21**, 622.
Johnson, J. P., Christophorou, L. G., and Carter, J. G. (1977). *J. Chem. Phys.* **67**, 2196.
Jordan, K. D., and Burrow, P. D. (1978). *Acc. Chem. Res.* **11**, 341–348.
Jordan, K. D., and Burrow, P. D. (1981). *Am. Chem. Soc. Symposium Series*, (D. W. Dwright, T. J. Fabish and H. R. Thomas, ed.) **162**, 1.
Jordan, K. D., Michejda, J. A., and Burrow, P. D. (1976a). *J. Am. Chem. Soc.* **98**, 7189.
Jordan, K. D., Michejda, J. A., and Burrow, P. D. (1976b). *Chem. Phys. Lett.* **42**, 227.
Jungen, M., Jogt, J., and Staemmler, V. (1979). *Chem. Phys.* **37**, 499.
Kadifachi, S. (1983). Unpublished data.
Kato, R., and Rolfe, J. (1967). *J. Chem. Phys.* **47**, 1901.
Kay, R. B., van der Leeuw, Ph. I., and van der Wiel, M. J. (1977). *J. Phys. B* **10**, 2513.
Kennerly, R. E. (1980). *Phys. Rev.* **21**, 1876.
Kennerly, R. E., Bonham, R. A., and McMillan, M. (1979). *J. Phys.* **70**, 2039.
King, G. C., McConkey, J. W., Read, F. H., and Dobson, B. (1980). *J. Phys. B* **13**, 4315.
Knoop, F. W. E., and Brongersma, H. H. (1970). *Chem. Phys. Lett.* **5**, 450. *Chem. Phys. Lett.* **13**, 20.
Kuyatt, C. E., Mielczarek, S. N., and Simpson, J. A. (1964). *Phys. Rev. Lett.* **12**, 293.
Larkin, I. W., and Hasted, J. B. (1972). *J. Phys. B* **5**, 95.
Lefebre-Brion, H. (1973). *Chem. Phys. Lett.* **19**, 456.
Linder, F. (1979). *In* "Electron-Molecule Scattering" (S. C. Brown, ed.), p. 107. Wiley, New York.
Linder, F., and Schmidt, H. (1971). *Z. Naturforsch.* **26a**, 1603.
McDaniel, E. W. (1964). "Collision Phenomena in Ionized Gases." Wiley, New York.
Marmet, P., and Binette, L. (1978). *J. Phys. B* **11**, 3707.
Massey, H. S. W. (1976). "Negative Ions" 3rd edition. Cambridge University Press, Cambridge, England.
Massey, H. S. W. (1979). "Atomic and Molecular Collisions." Taylor and Francis, London.

Massey, H. S. W., and Burhop, E. H. S. (1969). "Electronic and Ionic Impact Phenomena," Vols. 1, 2. Clarendon Press, Oxford.
Mathur, D. (1976). Ph.D. Thesis, University of London.
Mathur, D. (1980). *J. Phys. B* **13**, 4703–4716.
Mathur, D. (1981a). *Chem. Phys. Lett.* **81**, 115.
Mathur, D. (1981b). Unpublished results.
Mathur, D., and Frost, D. C. (1981). *J. Chem. Phys.* **75**, 5381.
Mathur, D., and Hasted, J. B. (1975). *Chem. Phys. Lett.* **34**, 90.
Mathur, D., and Hasted, J. B. (1976a). *J. Phys. B* **9**, L31–37.
Mathur, D., and Hasted, J. B. (1976b). *Chem. Phys.* **16**, 347–352.
Mathur, D., and Hasted, J. B. (1977). *Chem. Phys. Lett.* **48**, 50–54.
Mathur, D., and Hasted, J. B. (1978). *Chem. Phys.* **34**, 29.
Mazeau, J., Gresteau, F., Joyez, G., Reinhardt, J., and Hall, R. I. (1972). *J. Phys. B* **5**, 1890.
Mazeau, J., Gresteau, F., Hall, R. I., and Huetz, A. (1978). *J. Phys. B* **11**, L557.
Michejda, J. A., Dubé, L. J., and Burrow, P. D. (1981). *J. Appl. Phys.* **52**, 3121.
Morrison, M. A., and Lane, N. F. (1977). *Phys. Rev. A* **16**, 975.
Morrison, M. A., Lane, N. F., and Collins, L. A. (1977). *Phys. Rev. A* **15**, 2186.
Mulliken, R. S. (1957). *In* "The Threshold of Space" (M. Zelikoff, ed.). Pergamon, New York.
Nenner, I. (1975). Ph. D. Thesis, L'Université de Paris-Sud.
Nenner, I., and Schulz, G. J. (1975). *J. Chem. Phys.* **62**, 1747.
Ng, L., Jordan, K. D., Krebs, A. and Ruger, W. (1982). *J. Am. Chem. Soc.*, (in press).
Pai, R. Y., Christophorou, L. G., and Christodoulides, A. A. (1979). *J. Chem. Phys.* **70**, 1169.
Pisanias, M. N., Christophorou, L. G., and Carter, J. G. (1972a). *Chem. Phys. Lett.* **13**, 433.
Pisanias, M. N., Christophorou, L. G., Carter, J. G., and McCorkle, B. I. (1973). *J. Chem. Phys.* **58**, 2110.
Potts, A. W., Price, W. C., Streets, D. G., and Williams, T. A. (1972). *Chem. Soc. Faraday Disc.* **54**, 168.
Rabalais, J. W. (1977). "Principles of Ultraviolet Photoelectron Spectroscopy." Wiley, New York.
Raith, W. (1976). *Adv. At. Mol. Phys.* **12**, 281.
Rajgara, F. A., and Mathur, D. (1982). *Proc. 4th Natl. Workshop on At. Mol. Phys.*, Calcutta.
Read, F. H. (1968). *J. Phys. B* **1**, 893.
Read, F. H. (1971). *J. Phys. B* **4**, 911.
Read, F. H. (1972). *J. Phys. B* **5**, 255.
Rohr, K., and Linder, F. (1976). *J. Phys. B* **9**, 2521.
Rosentock, H. M., Draxl, K., Striner, B. W., and Herron, J. T. (1977). *J. Phys. Chem. Ref. Data* **6**, Suppl. 1.
Sanche, L., and Schulz, G. J. (1972a). *Phys. Rev. A* **5**, 1672.
Sanche, L., and Schulz, G. J. (1972b). *Phys. Rev. A* **6**, 69.
Sanche, L., and Schulz, G. J. (1973). *J. Chem. Phys.* **58**, 479.
Schulz, G. J. (1958). *Phys. Rev.* **112**, 150.
Schulz, G. J. (1959). *Phys. Rev.* **116**, 1141.
Schulz, G. J. (1964). *Phys. Rev. A* **135**, 988.
Schulz, G. J. (1973). *Rev. Mod. Phys.* **45**, 378.
Schulz, G. J. (1976). *In* "Principles of Laser Plasmas" (G. Behegi, ed.), p. 34. Wiley, New York.
Sell, J. A., Mintz, D. M., and Kuppermann, A. (1978). *Chem. Phys. Lett.* **58**, 601.
Seng, G., and Linder, F. (1976). *J. Phys. B* **9**, 2539.
Smythe, K. C., Schiavone, J. A., and Freund, R. S. (1974). *J. Chem. Phys.* **61**, 4747.
Spence, D. (1974a). *Phys. Rev.* **10**, 1045.
Spence, D. (1974b). *J. Phys. B* **7**, L87.
Spence, D. (1981). *J. Phys. B* **14**, L107.

Spence, D., and Burrow, P. D. (1979). *J. Phys.* B **12**, L179.
Spence, D., and Noguchi, T. (1975). *J. Chem. Phys.* **63**, 505.
Stamatovic, A., and Schulz, G. J. (1970). *Rev. Sci. Instrum.* **41**, 423.
Swanson, N., Celotta, R. J., Kuyatt, C. E., and Cooper, J. W. (1975). *J. Chem. Phys.* **62**, 4880.
Szmytkowski, C., and Zubek, M. (1978). *Chem. Phys. Lett.* **57**, 105.
Tam, W.-C., and Wong, S.-F. (1979). *Rev. Sci. Instrum.* **50**, 302.
Trajmar, S., and Chutjian, A. (1977). *J. Phys.* B **10**, 2943.
Trajmar, S., and Hall, R. I. (1974). *J. Phys.* B **7**, L458.
Tronc, M., and Azria, R. (1979). *In* "Symposium on Electron-Molecule Collisions" (I. Shimamura and M. Matsuzawa, eds.), p. 105. University of Tokyo, Tokyo.
Tronc, M., Fiquet-Fayard, F., Schermann, C., and Hall, R. I. (1977). *J. Phys.* B **10**, 305.
Tronc, M., Azria, R., and Le Coat, Y. (1980). *J. Phys.* B **13**, 2327.
Turner, D. W., Baker, C., Baker, A. D., and Brundle, C. R. (1970). "Molecular Photoelectron Spectroscopy," p. 332. Wiley, New York.
Van Veen, E. H. (1976a). *Chem. Phys. Lett.* **41**, 535–539.
Van Veen, E. H. (1976b). *Chem. Phys. Lett.* **41**, 540.
Verhaart, G. J., and Brongersma, H. H. (1980). *Chem. Phys.* **52**, 431.
Weingartshofer, A., Ehrhardt, H., Hermann, V., and Linder, F. (1970). *Phys. Rev.* A **2**, 294.
Weingartshofer, A., Willmann, K., and Clarke, E. M. (1974). *J. Phys.* B **7**, 79.
Winter, N. W., Goddard, W. A., and Bobrowicz, F. W. (1973). *J. Chem. Phys.* **62**, 4325.
Wong, S.-F., and Dubé, L. (1978). *Phys. Rev.* **17**, 570.
Wong, S.-F., and Schulz, G. J. (1975). *Phys. Rev. Lett.* **35**, 1429.
Younkin, J. M., Smith, L. J., and Compton, R. N. (1976). *Theor. Chim. Acta. Berl.* **41**, 157.
Zecca, A., Lazzizera, I., Krauss, M., and Kuyatt, C. E. (1974). *J. Chem. Phys.* **61**, 4560.
Ziesel, J.-P., Teillet-Billy, D., Bouby, L., and Paineau, R. (1979). *Chem. Phys. Lett.* **63**, 47.
Zubek, M., and Szmytkowski, C. (1977). *J. Phys.* **10**, L27.

6 Electron Attachment Processes

L. G. Christophorou, D. L. McCorkle, and A. A. Christodoulides*

Oak Ridge National Laboratory
Oak Ridge, Tennessee

and

Department of Physics
The University of Tennessee
Knoxville, Tennessee

I.	Introduction	478
II.	Modes of Production of Negative Ions	479
	A. Molecular Negative-Ion Resonant States	479
	B. The Binding of Attached Electrons to Molecules ("Electron Affinity")	484
	C. The Lifetimes of Metastable Negative Ions	486
	D. Basic Electron Attachment Processes	486
III.	Techniques for the Study of Electron Attachment Processes	495
	A. Techniques for the Study of Negative-Ion States and Transient Anions	496
	B. Measurement of the Total Electron Attachment Cross Section and Rate Constant	496
	C. Techniques for Identification of Negative Ions Following Electron Capture	500
	D. High-Resolution Studies of Dissociative Attachment; Energy and Angular Distribution of Dissociative Attachment Fragment Anions	501
	E. Techniques for the Measurement of Autodetachment Lifetimes and Electron Affinities, and the Study of Bound-Electron Capture Processes	503
IV.	Dissociative Electron Attachment to Ground-State Molecules	504
	A. General Considerations	504
	B. Isotope Effects in Dissociative Attachement	505
	C. Attachment Rate Constants, Cross Sections, and Negative-Ion States	509

---

* Permanent address: Department of Physics, University of Ioannina, Ioannina, Greece.

Electron–Molecule Interactions
and Their Applications, Volume 1.

Copyright © 1984 by Academic Press, Inc.
All rights of reproduction in any form reserved.
ISBN 0-12-174401-9

    D. Dependence of the Magnitude of the Dissociative
       Attachment Cross Section on the Resonance Energy . . 555
 V. Dissociative Electron Attachment to "Hot" Molecules
    (Effects of Temperature on Dissociative Electron
    Attachment) 556
    A. Diatomic Molecules 560
    B. Polyatomic Molecules 565
 VI. Dissociative Electron Attachment to Electronically
     Excited Molecules 568
VII. Molecular Parent Negative Ions 569
     A. Long-Lived Parent Negative Ions 569
     B. Moderately Short-Lived Parent Negative Ions . . 577
     C. Effects of Temperature on Nondissociative Electron
        Attachment 588
     D. Electron Attachment to Molecular Clusters . . . 592
     E. Radiative Attachment to Molecules 592
VIII. Negative Ions Formed by Ion Pair Processes and by
      Collisions of Molecules with Ground-State and
      Rydberg Atoms 593
      A. Ion Pair Formation 593
      B. Capture of Bound Electrons by Molecules in
         Collision with Ground-State Neutral Atoms . . . 595
      C. Capture of Bound Electrons by Molecules in
         Collision with Rydberg Atoms 601
  IX. Doubly Charged Negative Ions 605
      References 606

I. Introduction

There has been an immense interest in electron attachment processes both for fundamental and practical reasons over the last two decades. New studies have constantly enriched our knowledge, increased our understanding, and opened up new possibilities for applications. From the experimental side, impressive progress has been made in the study of electron attachment processes at room and higher temperatures both under isolated conditions with highly monoenergetic electron beams of improved accuracy, and in high-pressure media with a variety of electron swarm methods. From the theoretical side, the advances in our understanding using the resonance model have been most impressive.

Many reviews of the subject, or aspects of it, have appeared over the last decade (e.g., Christophorou, 1971, 1976, 1978a, 1980a; Dillard, 1973; Franklin and Harland, 1974; Dutton, 1975; Caledonia, 1975; Massey, 1976; Berry and Leach, 1981; Smirnov, 1982). To these, to other chapters in this work, and to many original sources we refer the reader frequently. For although we have attempted to treat electron attachment processes *comprehensively*, we have

II. Modes of Production of Negative Ions

A. Molecular Negative-Ion Resonant States

A most significant mode of electron capture by molecules is via negative-ion resonant states (NIRSs). Resonance electron attachment processes occur at low energies ($\lesssim 20$ eV) and are generally discussed within the formalism of resonance scattering theory and the formation of transient negative ions. Molecular NIRSs are the subject of Chapter 5; in this section we comment briefly on these states in connection with electron attachment processes.

A bound electron in a molecule is characterized by a wave function ψ_n which can be expressed as

$$\psi_n \propto \exp[-iE_n(t/\hbar)], \tag{1}$$

where $h = 2\pi\hbar$ is Planck's constant and E_n is the energy eigenvalue. The probability density $|\psi_n|^2$ is time independent and the state is *stationary*. Molecules in an excited stationary state can decay by photon (not electron) emission. A NIRS is a *nonstationary* state of the electron–molecule system. Its energy, $E_n - \tfrac{1}{2}i\Gamma_n$, is complex; the probability density

$$|\psi_n|^2 \propto \exp[-\Gamma_n(t/\hbar)] \tag{2}$$

is time dependent, and the state can decay by electron emission with a characteristic autodetachment lifetime.

$$\tau_a = \hbar/\Gamma_n, \tag{3}$$

where Γ_n is the autodetachment resonance width, which is a function of the internuclear separation(s). Thus, all negative ions produced via NIRSs in their initial step of formation are in a superexcited state; they possess sufficient internal energy to revert back to the neutral molecule and a free electron [i.e., to autoionize (autodetach)].

Four general mechanisms leading to the formation of a transient negative ion can be distinguished (see Chapter 5; Taylor *et al.*, 1966; Bardsley and Mandl, 1968; Schulz, 1973): *shape or single-particle resonances, core-excited shape resonances* (type II), *nuclear-excited Feshbach resonances*, and *electron-excited Feshbach* (core-excited type I) *resonances*. The former two lie energetically above and the latter two lie energetically below the parent state (ground or excited) from which they derive.

1. Shape or Single-Particle Resonances

In this type of resonances, illustrated schematically in Fig. 1a, the incident electron is trapped in a potential well which arises from the interaction between the incident electron and the neutral molecule in its electronic ground state. The potential barrier is the combined effect ($V_{\text{effective}}$ in Fig. 1a) of the attractive polarization potential between the neutral molecule and the incident electron ($V_{\text{attractive}}$ in Fig. 1a) and the repulsive ($V_{\text{repulsive}}$ in Fig. 1a) centrifugal potential which arises from the relative motion of the two bodies; $V_{\text{repulsive}}$ varies as $l(l+1)/2\rho$, where l is the angular momentum quantum number and ρ is the electron–molecule separation. Since the negative-ion potential-energy curve/surface for a shape resonance lies above that of the neutral molecule, the transient anion is subject to autodetachment, decaying back to the neutral molecule in its ground electronic state plus a free electron. The neutral molecule is left with or without vibrational and/or rotational energy. The decay—being in effect electron penetration through the potential barrier—is analogous to the α-particle decay of radioactive nuclei. The lifetime τ_a for this autodetachment process is a function of both the size (height and thickness) of the potential barrier and the internal energy of the anion. The former is strongly dependent on the l value of the state occupied by the captured electron, and the latter is largely dependent on the attractive portion of the potential $V_{\text{attractive}}$ (Fig. 1a). In order for the repulsive term to be nonzero, the l value must be equal to or greater than one. Consequently, as a rule, s waves will not lead to a shape resonance, as do partial p, d, f waves with higher l values.

Shape resonances usually lie at low energies (~ 0–4 eV) and above the potential-energy curve/surface of the neutral molecule [i.e., the molecular electron affinity (EA) is negative (<0 eV)]; they decay by autodetachment preferentially to their parent state with lifetimes ranging from $\sim 10^{-15}$ to $\gtrsim 10^{-10}$ s or, if energetically possible, by dissociative attachment. Shape NIRSs are abundant for diatomic (Chapters 1 and 5; Bardsley and Mandl, 1968; Schulz, 1973), triatomic (Chapters 1 and 5; Sanche and Schulz, 1973); and polyatomic (Chapter 5; Nenner and Schulz, 1975; Christophorou et al., 1977a; Jordan and Burrow, 1978; Christophorou, 1981) molecules.

2. Core-Excited Shape Resonances (Type II)

The mode of formation of these NIRSs is the same as for the Case 1 just discussed, except that now the effective potential arises from the attractive interaction between the incident electron and an excited electronic state of the

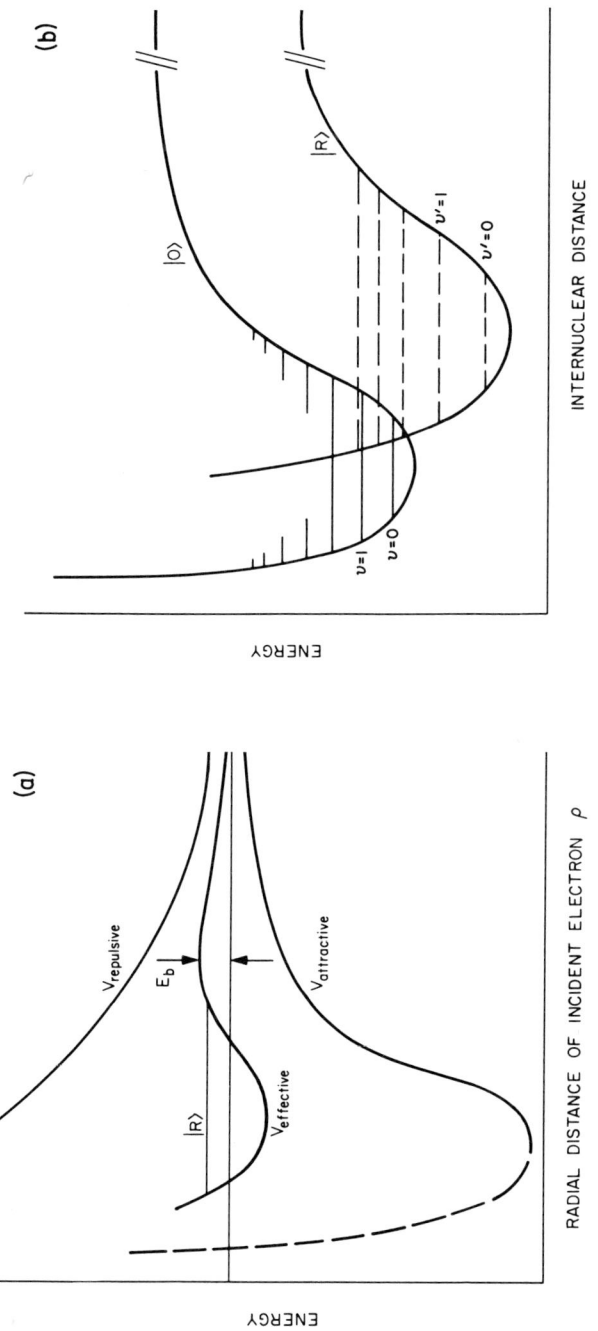

Fig. 1. Schematic illustration of (a) shape and (b) nuclear-excited Feshbach resonances. The symbols $|0\rangle$ and $|R\rangle$ designate, respectively, the electronic ground state of the neutral molecule and the NIRS. [From Christophorou et al. (1977a). Reprinted by permission of John Wiley & Sons, Inc.]

target molecule rather than the ground state. Since the potential barrier is strongly dependent on the l value of the occupied excited-state orbital, one expects p-, d-, and f-wave resonances, but not, as a rule, s-wave resonances of this kind. The NIRS lies *above* the energy of the neutral excited state and decays by autodetachment or, if energetically possible, by dissociative attachment. An example of such NIRS is the H_2^- state at 8–12 eV (Taylor *et al.*, 1966; Schulz, 1973).

3. Nuclear-Excited Feshbach Resonances

This type of resonances involves the coupling of the kinetic energy of the captured electron to molecular vibration (no electronic excitation of the target molecule is involved). They are illustrated schematically in Fig. 1b. The negative-ion state lies energetically below the parent ground state, exhibiting a positive (>0 eV) electron affinity; it thus can be stabilized, leading to stable parent negative ions. In the initial step of their formation (i.e., prior to stabilization) these NIRSs are in vibrational levels v' at or above the lowest vibrational level $v = 0$ of the parent neutral state and can decay into the parent state by autoionization (they can also decay by autodissociation if energetically possible). For many polyatomic molecules the lifetimes of these NIRSs toward autodetachment (and at times toward autodissociation) are long, often $> 10^{-6}$ s. Their cross sections—especially for complex polyatomic molecules—are large (often $> 10^{-14}$ cm^2) and as a rule reach their maximum value at thermal electron energies (Section VII; Christophorou, 1978a). It should be noted that any description of this kind of NIRS that utilizes potential-energy surfaces is artificial, since this type of resonances involves the coupling of the electronic and the nuclear motion resulting from the breakdown of the Born–Oppenheimer approximation. In this respect, their representation by a potential diagram as in Fig. 1b is a schematic illustration and not a physical description of the resonance.

4. Electron-Excited Feshbach Resonances (Core-Excited Type I)

In a nuclear-excited Feshbach resonance electron capture can involve an electronically excited—rather than a ground-state—neutral molecule and in this case the resonance is called *core-excited type I*. The incoming electron excites electronically the target molecule and is concomitantly captured in the field of the excited electronic state(s) of the target. The captured electron must reabsorb energy from the target in order to regain sufficient energy for reemission. In the excited state the nuclei of the molecule are less well screened and the incident electron sees a slightly positive charge, becoming temporarily bound to the molecule. These NIRSs lie below the corresponding excited neutral state(s) (i.e., they have a positive EA)—but above the ground neutral state—and in a way similar to the shape resonances they can decay by autodetachment and/or by dissociative attachment. When the transient anion is in a vibrational level v' that lies energetically below the $v = 0$ of the corresponding excited parent neutral state, decay into the parent state is not energetically possible. In such cases, τ_a is long, since the decay into nonparent states involves a change in configuration. Examples of this type of NIRSs are the H_2^- state at 10–13 eV (Taylor et al., 1966; Kuyatt et al., 1966; Bardsley and Mandl, 1968) and the N_2^- state at 11.48 eV (Heideman et al., 1966; Schulz, 1973). The first experimental evidence for a long-lived ($\tau_a > 10^{-5}$ s) negative-ion state formed by this mechanism was reported for p-benzoquinone by Christophorou et al. (1969a). A classification of Feshbach resonances in electron–molecule scattering was given by Spence (1975).

Molecular negative-ion states are abundant (Chapters 1, 2, and 5; Schulz, 1973; Christophorou, 1978a, 1980a, 1981). Their energies, cross sections, and lifetimes are strongly affected by the details of the molecular structure as well as the medium that surrounds the metastable anions. The cross sections and lifetimes of the NIRSs are functions of the electron energy ε. Often the NIRSs are described—their energies approximated and their numbers rationalized—in terms of the unoccupied molecular orbitals of the neutral molecule. At times, also, molecular geometrical changes concomitant with electron impact and strong electrophilic sites in a polyatomic molecule can constitute effective modes of electron trapping.

As was discussed in Chapters 1, 2, and 5, we usually learn of the properties of NIRSs through their decay channels. Thus, knowledge of the number and energies of NIRSs for a given molecule is normally obtained either from the energy dependence of the total electron scattering cross section, or the cross section for elastic scattering, or the cross section for inelastic scattering involving vibrational and/or electronic excitation—at an angle or in the forward direction—via the decay of the NIRS by electron emission; these cross sections are measured conveniently using electron scattering and electron transmission techniques. In the incident electron energy range in which the temporal trapping of the electron occurs, the magnitude of the electron scattering cross section changes profoundly, and the "resonance" signifies the existence and gives the position of the negative-ion state. Knowlegde of the number and energies of NIRSs of molecules can also be acquired from studies of the energy dependence of the cross sections for specific dissociative attachment fragment ions, total dissociative attachment cross sections, or cross sections for parent negative-ion formation; the latter cross sections are usually determined (especially at low energies) by the swarm and the swarm–beam techniques.

Knowledge of the cross sections for formation and decomposition(s) of NIRSs is acquired from electron-beam, mass spectrometric, and swarm techniques, the last method being most appropriate for low energies and for nuclear-excited Feshbach resonances. Our knowledge of the lifetimes of NIRSs comes from electron scattering methods when EA < 0 eV and $\tau_a \lesssim 10^{-13}$ s, from high-pressure electron swarm studies when EA > 0 eV and τ_a is in the range 10^{-7}–10^{-12} s, from time-of-flight mass-spectrometric studies when EA > 0 eV and $\tau_a \gtrsim 10^{-6}$ s, and from ion cyclotron resonance techniques when EA > 0 eV and $\tau_a \gtrsim 10^{-3}$ s (Christophorou, 1978a).

B. The Binding of Attached Electrons to Molecules ("Electron Affinity")

The electron affinity (EA) of a molecule is normally defined as the difference in energy between the neutral molecule plus an electron at rest at infinity and the molecular negative ion when both, neutral molecule and negative ion, are in their ground electronic, vibrational, and rotational states (Christophorou, 1971). The electron affinity can be positive (>0 eV) or negative (<0 eV) (see Fig. 2). Two other quantities, the vertical detachment energy (VDE) and the vertical attachment energy (VAE) are closely related to EA, the former when EA is positive and the latter when EA is negative. The VDE is defined (Christophorou, 1971) as the minimum energy required to eject the electron from the negative ion in its ground electronic and nuclear state without changing the internuclear separation, and the VAE is defined (Christophorou, 1971) as the difference in energy between the neutral molecule in its ground electronic, vibrational, and rotational states plus the electron at rest at infinity, and the molecular negative ion formed by addition of the electron to the neutral molecule without allowing a change in the internuclear separation of the constituent nuclei. The schematic diagrams in Fig. 2 clarify the physical significance of these quantities and their relation to the various modes of electron capture discussed in Section A.

Figures 2a, 2b and 2c refer to an electron captured in the field of the ground electronic state of an atom (Fig. 2a) and a diatomic molecule (Figs. 2b and 2c). In all three cases the EA is positive and the traditional measurements of EA are typical of these cases. In Fig. 2a, EA(A) = |VDE(A$^-$)|; this is always the case for atoms. For molecules, however, EA(AX) = |VDE(AX$^-$)| only if the equilibrium internuclear separations for AX and AX$^-$ are the same (as shown in Fig. 2c). Although for some small molecular species this is the case (e.g., Christophorou, 1971; Hotop et al., 1974), in the majority of molecular species a situation similar to that in Fig. 2b prevails for which |VDE(AX$^-$)| > EA(AX). Cases (b) and (c) are, of course, typical examples of nuclear-excited Feshbach resonances.

6 Electron Attachment Processes

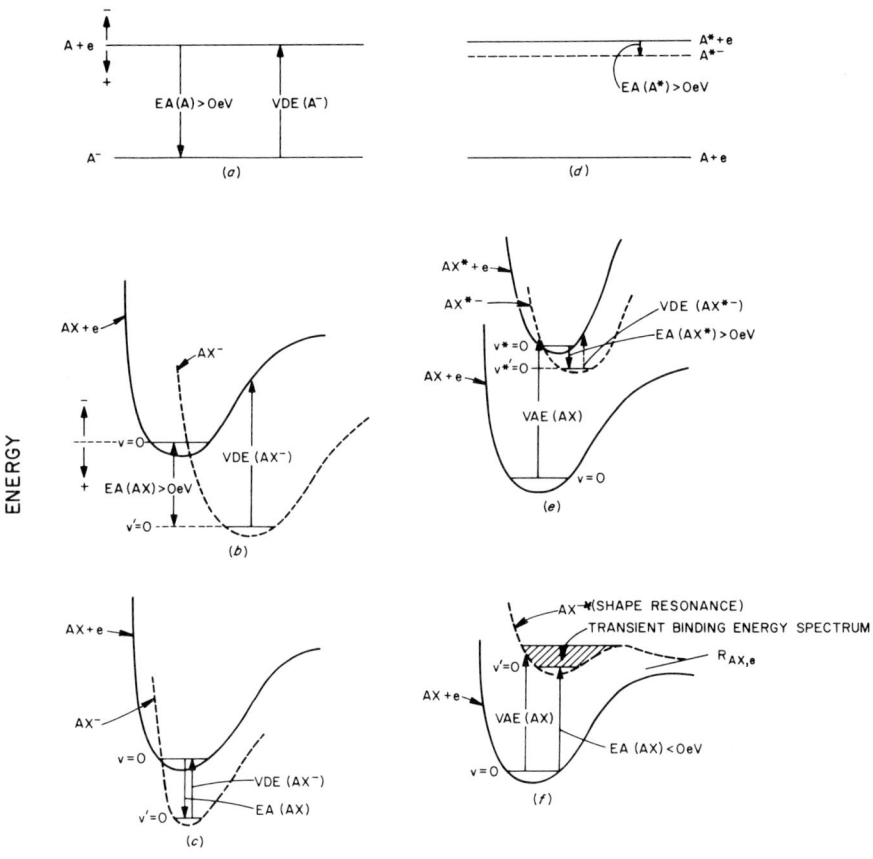

Fig. 2. Schematic diagrams illustrating the positive and negative values of EA and the relation of EA to VAE and VDE and the modes of electron capture (see text). [From Christophorou (1980a).]

Figures 2d and 2e illustrate the case of electron-excited Feshbach resonances (core-excited type I) for atoms and molecules, respectively. Here the electron is captured in the field of an excited atom A^* or an electronically excited molecule AX^*, forming, respectively, A^{*-} and AX^{*-} (note our nomenclature: A^{*-} and AX^{*-} indicate, respectively, a negative ion formed by electron capture by an electronically excited atom or molecule, while A^{-*} and AX^{-*} indicate, respectively, an atomic and molecular negative ion with excess internal energy). The $EA(A^*)$ and $EA(AX^*)$ now refer to the electron affinity of the excited atom A^* and the electronically excited molecule AX^*; they are both positive and can be greater than those of the corresponding ground-state species. Actually, the electron affinity of the ground-state atom or molecule can be negative, while that for the corresponding excited species is positive.

In Fig. 2f a case is depicted which is appropriate for a shape resonance. It should be noted that although in this case $VAE(AX) \leq -EA(AX)$, the electron is temporarily bound to the molecule with a "transient binding-energy spectrum" shown by the shaded area in Fig. 2f. The coordinate system depicting the potential well is indicated in the figure by $R_{AX,e}$, the radial distance of the electron, which is different from that of the internuclear separation.

The electron affinities of atoms and molecules and the methods used for their determination are reviewed in Chapter 6 of Volume 2.

C. The Lifetimes of Metastable Negative Ions

The lifetimes of metastable negative ions vary by more than 13 orders of magnitude, from $\sim 10^{-15}$ to $>10^{-2}$ s; they depend on the total internal energy of the anion. Christophorou (1971, 1978a) classified the metastable atomic and molecular negative ions with regard to their lifetimes broadly as (a) extremely short lived ($10^{-15} \lesssim \tau_a \lesssim 10^{-12}$ s), (b) moderately short lived ($\sim 10^{-12} \lesssim \tau_a \lesssim 10^{-6}$ s), and (c) long lived ($\tau_a > 10^{-6}$ s). The basis of this classification dictates the experimental method employed to determine τ_a: low-energy electron scattering and dissociative attachment techniques for (a), high-pressure electron swarm studies for (b), and time-of-flight mass spectrometry and ion cyclotron resonance techniques for (c).

A comprehensive review and discussion of the lifetimes of metastable molecular (and atomic as well as fragment) negative ions and the theoretical and experimental methods employed for their determination have been given by Christophorou (1978a). The reader is referred to this source and also to Section VII for further discussion.

D. Basic Electron Attachment Processes

1. Resonance Free-Electron Attachment to Molecules

The basic nondissociative and dissociative processes via which parent and fragment negative ions are formed in free electron–molecule collisions at low energies ($\lesssim 20$ eV; in most cases involving parent anions <1 eV) via NIRSs may be classified according to the internal state of excitation of the target molecule as follows (Christophorou, 1980a).

6 Electron Attachment Processes

a. Free-Electron Attachment to Ground-State Molecules Predominantly in their Lowest States (s) of Internal Excitation.

Let us consider the capture of a slow electron of energy ε, $e(\varepsilon)$, by a polyatomic molecule $AB \cdots CD(G; v = 0)$ in its ground electronic state G and predominantly in its lowest ($v = 0$) vibrational state of excitation, forming a transient ion $AB \cdots CD^{-*}$ (G or E) in either the field of the ground (G) or excited (E) electronic state with a capture cross section $\sigma_c(\varepsilon)$. The initial electron capture process and the ensuing decomposition channels of $AB \cdots CD^{-*}$ may be represented as (Christophorou, 1980a).

$$e(\varepsilon) + AB \cdots CD(G; v = 0) \xrightarrow{\sigma_c(\varepsilon)} AB \cdots CD^{-*}(\text{G or E})$$

$$\rightarrow AB \cdots CD + e(\varepsilon) \qquad (4)$$

$$\rightarrow AB \cdots CD^{(*)} + e(\varepsilon') \qquad (5)$$

$$\rightarrow \left.\begin{array}{l} AB^{(*)} + CD^- \\ \{AB \cdots C\}^* + D^- \end{array}\right\} \qquad (6a)$$

or

$$AB^{(*)} + CD^{-*}$$
$$\qquad \rightarrow CD^{(*)} + e$$
$$\qquad \rightarrow C^{(*)} + D^- \qquad (6b)$$

$$\rightarrow AB \cdots CD^- + \text{energy} \qquad (7)$$

where ε' ($< \varepsilon$) is the energy of the scattered electron. The asterisk indicates excess internal energy and the asterisk in parentheses indicates possible increase in the internal energy of the corresponding species. Reactions (4) and (5) are, respectively, indirect elastic and inelastic electron scattering. Reaction (6a) is dissociative attachment producing stable fragment negative ions; for a polyatomic molecule this process can lead to simultaneous multiple fragmentation. Reaction (6b) is dissociative attachment producing metastable negative ion fragment(s) which are subject to autodetachment and/or autodissociation. Finally, reaction (7) represents parent negative-ion formation, which is possible when the electron affinity, $EA(AB \cdots CD)$, of $AB \cdots CD$ is positive (>0 eV) and the excess internal energy is removed, usually by collisions of $AB \cdots CD^{-*}$ with another body. For polyatomic molecules with closely spaced NIRSs the initial state of $AB \cdots CD^{-*}$ may internally convert to a lower-lying negative-ion state which will then decay accordingly. For diatomic as well as polyatomic molecules a bound negative-ion state may predissociate through overlapping with a repulsive negative-ion state (see Section IV).

All four reactions have been studied extensively over the last two decades. At room temperature the $AB \cdots CD$ molecules are predominantly in the $v = 0$

level. However, depending on the molecule, even a small population of higher vibrational and/or rotational levels can affect significantly the cross sections and the onsets, especially for reaction(s) (6) (see Section V).

In molecules, nuclear and electronic motion are separable in the Born–Oppenheimer approximation and application of the Franck–Condon principle permits electron attachment to molecules to proceed by vertical transitions between potential-energy curves (surfaces) from the equilibrium positions of the atoms. Let us then restrict ourselves to a diatomic or a "diatomiclike" molecule AX. Schematic potential-energy curves for AX and AX^{-*} exemplifying the *nondissociative* electron attachment process can be seen in Fig. 1b and Figs. 2b and 2c for $EA(AX) > 0$ eV and in Figs. 2e and 2f for $EA(AX) < 0$ eV. Similar schematic potential-energy curves exemplifying various modes of *dissociative* electron attachment processes are shown in Fig. 3. The energy for the dissociative attachment process is essentially equal to the energy difference between the ground state ((AX + e), the ground-state molecule and the electron at rest at infinite separation) and the NIRS. Thus, we think of the dissociative attachment process as proceeding in two stages and according to the Franck–Condon principle. Electrons within a restricted energy range (between E_1 and E_2; Fig. 3a) are first captured by the neutral molecule without alteration of the position and velocities of the nuclei. The resulting transient ion AX^{-*} dissociates into the final products $A + X^-$ (Fig. 3).

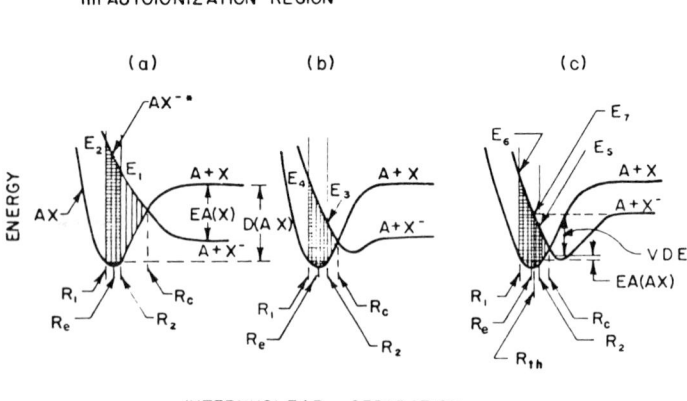

Fig. 3. Schematic potential-energy diagrams for various modes of dissociative electron attachment to diatomic molecules discussed in text. Note that the ground state indicated is for the neutral molecule AX plus the electron at rest at infinity. [From Christophorou (1971). Reprinted by permission of John Wiley & Sons, Inc.]

The position and the shape of the (upper) potential-energy curve of AX^{-*} relative to the (lower) potential-energy curve of AX determines crucially the processes that follow the initial capture process. In Fig. 3a the potential-energy curve for AX^{-*} is purely repulsive. The asymptote of the AX^{-*} curve, $A + X^-$, lies below the asymptote of the AX curve, $A + X$, by an amount equal to the electron affinity EA(X) of X. The difference in energy between the asymptote of the AX curve and the "zero-point energy" (energy of the lowest vibrational level of AX) is equal to the dissociation energy D(AX) of AX. In the region from $R = R_1$ to $R = R_c$, where the potential-energy curve of AX^{-*} lies above that of AX, the molecular negative ion is unstable toward electron ejection. This region will be referred to as the "autoionization region" and is shown in the diagrams of Fig. 3 by vertical lines. In Fig. 3b the AX^{-*} potential-energy curve has an attractive region outside the Franck–Condon and autoionization regions at $R > R_c$, and the minimum electron energy for dissociative attachment is $E_3 [>D(AX) - EA(X)]$. As far as dissociative attachment is concerned, this case is similar to that shown in Fig. 3a. If, however, a situation like that in Fig. 3c exists, where $E_5 < D(AX) - EA(X)$, then both dissociative and nondissociative electron attachment can occur. However, the dissociative attachment cross section $\sigma_{DA}(\varepsilon)$ will exhibit a vertical onset behavior, i.e., the negative ion X^- will appear at E_7 and will show a very sharp rise at or very close to this energy (see Section IV). Because dissociation is energetically possible for only a limited range of interatomic distances, $R_1 \leq R \leq R_{th}$, in the Franck–Condon region (R_{th} is the internuclear distance corresponding to the threshold energy $E_{th} = E_7$), and because the products at the threshold E_7 are formed with essentially zero kinetic energy, $\sigma_{DA}(\varepsilon)$ is smaller (in certain cases very much smaller) than when all R are allowed (see Section IV.B on isotope effects in dissociative attachment processes). On the other hand, some of the final possible states fall at energies less than E_7. Transitions induced by electron attachment in the energy range from E_5 to E_7 (Fig. 3c) will produce vibrationally excited molecular ions AX^{-*}. Since in the case of Fig. 3c EA(AX) < 0, the negative ion is unstable with respect to electron ejection.

In Fig. 4 we further illustrate schematically the pertinent parameters for resonance dissociative and nondissociative electron attachment processes and the energy dependences of the respective cross sections. The widths of the potential-energy curves of AX^{-*} were drawn progressively larger the smaller the value of R to illustrate the increased probability of autodetachment with decreasing R. The magnitude, energy position, and shape of the cross section functions provide most useful information on the states of AX^-. Although resonance dissociative attachment processes leading to X^- occur over a wide energy range (0 to $\sim$20 eV), resonance nondissociative electron attachment processes leading to AX^- occur over a much narrower range of electron

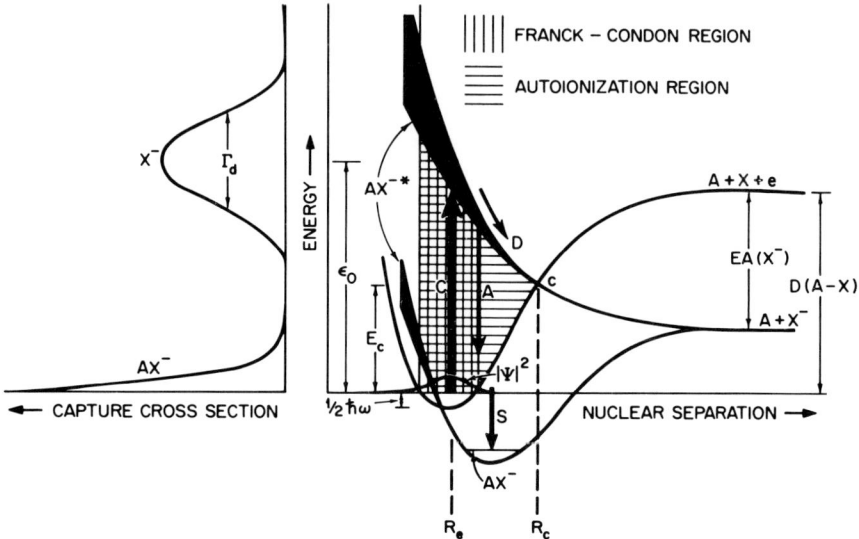

Fig. 4. Schematic diagrams illustrating the pertinent parameters for resonance dissociative and nondissociative electron attachment processes and the energy dependences of the respective cross sections. The arrows identified by C, A, D, and S indicate, respectively, the processes of electron capture, autoionization, dissociation, and stabilization of AX^{-*}. The widths of the AX^{-*} potential-energy curves indicate the variation with the internuclear separation R of the lifetime of AX^{-*}.

energies, usually less than a few electron volts. In fact, in most cases of parent negative-ion formation by electron capture in the field of the molecular ground state, the ions have a maximum probability of formation at ~ 0.0 eV (see Fig. 4 and Section VII). As stated earlier, for AX^- to form, EA(AX) must be positive and the transient anion must be stabilized (see Section VII). Dissociative attachment does not require stabilization. In the latter case considerable insight has been obtained recently (see Section IV) on the dynamics of dissociative attachment, the internal energy of the fragments, and the nature of the AX^- states from measurements of the kinetic (and angular) distributions of the fragment negative ions as a function of the captured-electron energy. The excess energy of the reaction can appear as kinetic energy of separation and/or as electronic excitation of the fragments [for molecules other than diatomic, kinetic-energy analyses are complicated by the internal energy of the molecular fragment(s)]. For dissociative attachment to a diatomic molecule AX, conservation of energy and momentum gives for the most probable kinetic energy E_k^-, of X^-

$$E_k^- = (1 - \beta)[\varepsilon + \text{EA}(X) - D(AX) - E_n], \tag{8}$$

where ε is the captured electron's energy, E_n is the excitation energy of the fragments, and β is the ratio of the mass of X^- to that of AX.

b. *Free-Electron Attachment to "Hot" Molecules*

Electron attachment processes are often affected by the gas temperature T. At elevated T a fraction of the molecules $AB\cdots CD$ in the ground electronic state G are in higher rotational, j, and/or vibrational, v, states; we designate these by $AB\cdots CD^*$ (G; $v > 0$; $j > 0$). The cross section $\sigma_{c,v,j}(\varepsilon)$ for formation of the transient anion $AB\cdots CD^{-*}$ (G or E) and especially the probability of the ensuing decomposition(s) depend on the vibrational and/or rotational quantum states. The electron attachment reaction appropriate to a "hot" molecule can, then, be represented as

$$e(\varepsilon) + AB\cdots CD^*(G; v > 0; j > 0) \xrightarrow{\sigma_{c,v,j}(\varepsilon)} AB\cdots CD^{-*} \text{ (G or E)} \rightleftarrows \text{Decay} \quad (9)$$

Reaction (9) is discussed in Section V.

c. *Free-Electron Attachment to Electronically Excited Molecules*

Under certain conditions (e.g., in plasmas or in situations of intense laser excitation) an appreciable number of gas molecules can be in an excited (metastable) state. Capture of an electron of energy ε, e(ε), by an electronically excited molecule $AB\cdots CD^*(E)$ with a cross section $\sigma_{c,E}(\varepsilon)$ producing $AB\cdots CD^{*-}(E)$ in the field of an excited electronic state may be represented by

$$e(\varepsilon) + AB\cdots CD^*(E) \xrightarrow{\sigma_{c,E}(\varepsilon)} AB\cdots CD^{*-}(E) \rightleftarrows \text{Decay} \quad (10)$$

The very limited experimental data on reaction (10) are discussed in Section VI.

2. *Nonresonance Free-Electron Attachment to Molecules (Ion Pair Formation)*

Ion pair formation by free-electron capture, viz.,

$$e + AX \longrightarrow A^+ + X^- + e \quad (11)$$

is a nonresonance electron attachment process occurring when the electron has adequate kinetic energy to excite the molecule to an unstable state which then dissociates into a positive–negative ion pair. Process (11), referred to by Massey (1976) as polar dissociation, sets in at some definite energy (usually $\gtrsim 10$ eV) and persists to quite high energies (100–200 eV), since the free electron carries away the excess energy. The cross section for (11), as for other inelastic electron collision processes (Chapter 2; Christophorou, 1971), increases (from a zero value at threshold) with energy above threshold approximately linearly close to the threshold, maximizes at some energy about two to four times the threshold energy, and declines monotonically thereafter roughly as ε^{-1} or more slowly. Thus the energy dependence of the cross section for (11) is distinctly different from those of the *resonance* dissociative and nondissociative electron attachment processes discussed earlier in this section; the former is not restricted in energy range as are the latter. Experimental results on process (11) are given in Section VIII.

3. Free-Electron Capture by Dipolar Molecules

Electron–dipolar molecule interactions have been the subject of extensive experimental and theoretical studies over the last 15 years (Chapter 1; Christophorou, 1971; Itikawa, 1978; Lane, 1980). A most interesting finding of these studies is the theoretical result of several authors [see review by Turner (1977)] that the electric field of a *fixed* dipole can bind an electron in an infinite number of bound states if the dipole moment μ is greater than a critical value $\mu_c = 1.625$ D*. The number of bound states for an electron in the dipolar field of a supercritical moment ($\mu > \mu_c$) was found (Garrett, 1970, 1972) to be reduced to a finite number if the dipole is allowed to rotate (*not* fixed). It was also shown by Garrett (1970, 1972) that the minimum dipole moment necessary to sustain at least one bound state for a *finite* dipolar system is increased by ~ 10–30% over that for a *fixed* dipole, depending on the dipole length, the rotational state, and the moments of inertia of the molecule.

These results raised the possibility of formation of "dipolar" negative ions and triggered some thought as to ways to detect them. The difficulty in observing such anions lies in the fact that the binding energy of dipole bound states may be exceedingly small for $\mu \simeq \mu_c$ and the diffuse and weakly bound electrons may thus be stripped away by collisions or by the electric fields normally used in experiments designed to detect them. A number of workers attempted to answer the question "what should the magnitude of the permanent and induced moments be to ensure that dipole bound states are sufficiently stable to endure thermal energy collision processes?" Garrett (1978) (see also Crawford and Garrett, 1977; Jordan et al., 1976; Jordan and Wendoloski, 1977) concluded that polar molecules with $\mu \gtrsim 4$ D will form strong enough anions to be thermally stable. Although the probability of forming such anions—e.g., by charge exchange or by three-body electron attachment—might be small and their observation using mass spectrometers difficult, efforts for their observation and study are encouraged. In this regard, Spence et al. (1982) reported the observation of HI^- using a mass spectrometric technique.

---

* 1.625 D $= 1.625$ debye $= 0.639$ ea_0, where e is the electron charge and a_0 is the Bohr radius.

4. Free-Electron Capture by Molecular Clusters

Electron attachment to molecular clusters and van der Waals dimers has been an interesting recent area of research using sonic nozzle expansion techniques or high-gas-pressure swarm studies (see Section VII.D).

5. Bound-Electron Capture by Molecules

Two broad types of bound-electron capture by molecules can be distinguished depending on the state of excitation of the heavy projectile: capture of bound electrons by molecules from ground-state atoms and capture of electrons from neutral atoms in high-lying Rydberg states. In both cases the atomic projectiles are "hydrogenic." However, in the latter case the size of the electron orbit is very much larger than in the former.

a. Capture of Bound Electrons by Molecules from Ground-State Atoms

Reactions of the form

$$A + XY \longrightarrow A^+ + XY^- \qquad (12a)$$
$$\longrightarrow A^+ + X + Y^- \qquad (12b)$$

or

$$A + XY \cdots Z \longrightarrow \text{Ion pairs} \qquad (13)$$

where A is a thermal-energy or a fast neutral atom of low ionization onset I, and XY and XY$\cdots$Z are, respectively, neutral diatomic and polyatomic target molecules, have received considerable attention recently. They are referred to invariably as chemi-ionization, charge exchange, ion pair formation, or collisional ionization; reactions (12a) and (12b) may be thought of, respectively, as nondissociative ion pair and dissociative ion pair formation processes.[*,†] Such reactions can be exoergic and thus proceed with thermal-energy neutrals (A) if in reaction (12a), for example, the EA(XY) of the target molecule exceeds the ionization threshold energy $I(A)$ of A. Although such situations require large EA(XY), they have been observed for some metal hexafluorides (MF$_6$) whose EA is exceedingly large ($\gtrsim 4.5$ eV; see Chapter 6 of Volume 2). In most cases reactions (12a, b) can proceed with kinetic energies of A in the electron-volt range if use is made of atomic beams consisting of alkali atoms whose I are low (5.390, 5.138, 4.339, and 3.893 eV for Li, Na, K, and Cs, respectively). Alkali-metal atoms are "one-electron"

---

[*] In his review of such processes Baede (1975) referred to processes (12a) and (12b), respectively, as nondissociative and dissociative charge transfer and he listed other processes such as AX$^+$ + Y$^-$ and A + X$^+$ + Y$^-$ which he called, respectively, reactive chemical ionization and collision-induced polar dissociation.

[†] Ion pairs can also be produced in collisions of XY with photons, viz., $h\nu + XY \to X^+ + Y^-$.

systems—and to a first approximation hydrogenic—and therefore more amenable to analysis compared with complex projectiles. For this reason, neutral alkali atom–molecule collisions have been the subject of many recent studies (see Section VIII).

Reactions (12a, b) are thought to proceed via an "electron jump" mechanism through the coupling of the covalent and ionic states at the point where the adiabatic potentials of these states cross. It has been argued (e.g., Zembekov, 1971) that in the case of nondissociative ion formation, the products of the reaction are not limited to those resulting from a single Franck–Condon transition, and that, because of multiple potential-energy surface crossings between the neutral and ionic species, some of the products may be "relaxed" (see Section VIII for further discussion and experimental results).

b. *Capture of Bound Electrons by Molecules from Neutral Atoms in High-Lying Rydberg States*

i. High-Rydberg States. The existence of long-lived high-lying Rydberg states of rare-gas atoms was shown experimentally by Cermac and Herman (1964).* For sufficiently large (>20) principal quantum numbers n, such atoms are nearly hydrogenic—the excited electron moves in a large orbit in a near-Coulombic field of essentially unit charge—and use can be made of the Bohr theory of the H atom to describe some of the characteristics of such highly reactive atoms (see Stebbings, 1976). Thus the radius r_n (in cm), binding energy E_n (in eV), and root-mean-square velocity v_n (in cm s^{-1}) of the electron, and the separation of adjacent n levels $\Delta E_{n,n+1}$ (in eV) of such atoms can be approximated, respectively, by $n^2 a_0$, R_H/n^2, v_1/n, and $2R_H/n^3$, where $a_0 = 5.3 \times 10^{-9}$ cm, $R_H = 13.6$ eV, and $v_1 = 2.2 \times 10^8$ cm s^{-1}. For $n = 30$, $r_{30} \simeq 5 \times 10^{-6}$ cm, $E_{30} \simeq 15$ meV, and $\Delta E_{30,31} \simeq 1$ meV. Thus the electron in high-lying Rydberg atomic states is very weakly bound; it can be ionized easily by electric fields. Field ionization can, thus, be used to detect such states and also to separate Rydberg states with different n values. This, and the fact that radiative lifetimes increase with n (approximately as $n^{4.5}$, averaged over all l substates) and the angular momentum quantum number l, make experiments with Rydberg-excited atoms attractive and feasible.

In collisions of highly excited Rydberg atoms with a target molecule, the target cannot interact simultaneously with both the electron and the core. The electron can be assumed to behave as if it were free with an energy E_n and the highly excited Rydberg atom can, thus, be used as a source of low-energy "free" electrons with a momentum distribution characterized by the particular quantum state of the Rydberg atom.

* See a review of atomic Rydberg states by Feneuille and Jacquinot (1981).

ii. Types of Processes. The general types of processes which have been found to occur in collisions of Rydberg atoms $A^{**}(n, l)$ with molecules XY are (Rundel, 1980)

$$A^{**}(n, l) + XY \longrightarrow A^+ + e + XY \quad \text{Ionization} \quad (14a)$$

$$\longrightarrow AX^+ + e + Y \quad \text{Associative ionization} \quad (14b)$$

$$\longrightarrow A(n', l') + XY \quad \text{State changing} \quad (14c)$$

$$\longrightarrow A^+ + XY^- \quad \text{Electron capture} \quad (14d)$$

$$\longrightarrow A^+ + X^- + Y \quad \text{Dissociative electron capture} \quad (14e)$$

With the exception of reaction (14b), which involves a Rydberg core–target molecule interaction, the rest occur predominantly through a Rydberg electron–target molecule interaction process. Of interest to us are reactions (14d) and (14e).

Reactions (14d) and (14e) were treated theoretically by Matsuzawa (1972a, b, 1975) [see also references to other theoretical work on processes (14a)–(14e) in Matsuzawa (1980)]. Matsuzawa assumed that in the high-Rydberg atom–molecule thermal collision the separation of the excited atom from its ionic core is so large that the target molecule does not interact with both the electron and the core simultaneously and, thus, the excited electron behaves essentially as if it were free, except that it has a momentum distribution characterized by its quantum state. Neglecting the Coulomb attraction of the final channel, Matsuzawa used the "free"-electron model and showed that the rate constants k_{ab} and k_{dab} for bound-electron capture [reaction (14d)] and bound-electron dissociative capture [reaction (14e)] at *thermal* velocities are equal, respectively, to those of the free-*thermal*-electron capture processes

$$e_f + XY \xrightarrow{k_{af}} XY^- \quad (15a)$$

$$e_f + XY \xrightarrow{k_{daf}} X^- + Y \quad (15b)$$

particularly if dissociative attachment occurs in s-wave capture. The n, l dependence of k_{dab} was also investigated by Matsuzawa (1975).

Experimental results on these processes are presented and discussed in Section VIII.

III. Techniques for the Study of Electron Attachment Processes

An impressive variety of methods have been developed and employed to study electron attachment processes over the last two decades. Broadly, these fall into the categories discussed below.

A. Techniques for the Study of Negative-Ion States and Transient Anions

These are generally electron scattering techniques used in various modes of operation depending on the measured quantity. They have been discussed in Chapters 1, 2, and 5 and reviewed by other authors (e.g., Christophorou, 1971; Massey, 1976), and for this reason they will not be discussed here. They include (i) electron transmission arrangements in which the energy loss in the transmitted electron beam is studied or the structure in the derivative of the transmitted current is investigated; (ii) double electrostatic analyzers in which the electron energy loss spectrum at a fixed or a variable angle is measured; (iii) threshold-electron excitation methods in which the "trapped-electron current" due to "zero-energy" scattered electrons is monitored as a function of the incident electron energy; and (iv) the threshold-electron excitation scavenger technique, in which a "zero-energy" electron scavenger (e.g., SF_6) is used as a "detector" of scattered electrons which have lost essentially all of their energy in collisions with the molecules under study and the resultant current (SF_6^-) is monitored as a function of the incident electron energy.

B. Measurement of the Total Electron Attachment Cross Section and Rate Constant

These measurements are made by using both electron-beam and electron swarm techniques.

1. Electron-Beam Methods

In these methods the total attachment cross section is measured under single-collision conditions in a way similar in principle to the measurement of the total ionization cross section. If $\sigma_a(\varepsilon_1)$, $\sigma_i(\varepsilon_2)$, $i_-(\varepsilon_1)$, and $i_+(\varepsilon_2)$ are, respectively, the total attachment cross section at ε_1, the total ionization cross section at ε_2, the saturated negative-ion current at ε_1, and the saturated positive-ion current at ε_2 (both corresponding to the same electron-beam current), $\sigma_a(\varepsilon_1)$ is determined from

$$\sigma_a(\varepsilon_1)/\sigma_i(\varepsilon_2) = i_-(\varepsilon_1)/i_+(\varepsilon_2). \tag{16}$$

This approach, and arrangements similar to the early total-ionization tube of Lozier (Lozier, 1934; Tozer, 1958), have been employed by many workers (e.g., Buchel'nikova, 1959; Asundi et al., 1963; Rapp and Briglia, 1965; Kurepa et al., 1981a,b) to measure total attachment cross sections for many species (see Section IV). Foremost among the total-ionization methods is the trapped-electron method of Schulz (1960). In almost all of these studies the RPD (retarding potential difference) electron source (Fox et al., 1955) was employed which allows electron beams with full widths at half heights usually in the range ~0.2–0.3 eV. A number of such arrangements allowed studies of $\sigma_a(\varepsilon)$ as a function T (e.g., Spence and Schulz, 1969). Spence and Schulz (1972) were also able to increase the gas pressure to ~0.3 Torr in their apparatus and measure three-body attachment coefficients for O_2 using an electron beam with an effective energy width of ~0.1 eV (see Section VII.B).

2. Electron-Swarm Techniques

Most electron swarm studies monitor the rate of removal of electrons in the swarm as a function of the density-reduced electric field E/N (E is the applied uniform electric field and N is the total gas number density). The number density N is such that the electron swarm attains an equilibrium energy distribution $f(\varepsilon, E/N)$ virtually from the beginning of the drift space. The distribution function is broad and is determined by the nature of the gas, E/N and the temperature T.

The most commonly measured quantities in electron swarm experiments are the coefficients* α' (probability of electron attachment per centimeter traveled in the field direction), α (probability of electron attachment per centimeter traveled in the field direction per Torr of attaching gas), and α'/N_a where N_a is the attaching gas number density. Their units are, respectively, cm^{-1}, $Torr^{-1}$ cm^{-1}, and cm^2; they are usually measured as a function of E/N and at times also as a function of N and T. Another commonly measured quantity is the attachment rate constant k_a (in units of cm^3 s^{-1}) at thermal energies, $(k_a)_{th}$, or as a function of E/N, $k_a(E/N)$ (the attachment frequency ν_a is equal to $k_a N_a$). The quantities $\alpha'/N_a(E/N)$, $k_a(E/N)$, and $(k_a)_{th}$ are related to the attachment cross section $\sigma_a(\varepsilon)$ and the normalized $f(\varepsilon, E/N)$ by

$$\frac{\alpha'}{N_a}\left(\frac{E}{N}\right) = \left(\frac{2}{m}\right)^{1/2} w^{-1} \int_0^\infty \sigma_a(\varepsilon)\varepsilon^{1/2} f\left(\varepsilon, \frac{E}{N}\right) d\varepsilon, \tag{17}$$

$$k_a\left(\frac{E}{N}\right) = \left(\frac{2}{m}\right)^{1/2} \int_0^\infty \sigma_a(\varepsilon)\varepsilon^{1/2} f\left(\varepsilon, \frac{E}{N}\right) d\varepsilon, \tag{18}$$

and

$$(k_a)_{th} = \left(\frac{2}{m}\right)^{1/2} \int_0^\infty \sigma_a(\varepsilon)\varepsilon^{1/2} f_M(\varepsilon) d\varepsilon, \tag{19}$$

where m is the electron mass, w is the electron swarm drift velocity, and $f_M(\varepsilon)$ is a Maxwellian distribution function characteristic of T.

---

* The other most commonly used symbol for the attachment coefficient is η.

In most swarm experiments unitary attaching gases were studied and the measurements were thus restricted to $N(=N_a)$ less than a few Torr. Often in such experiments k_a or α'/N_a were plotted as a function of $\langle\varepsilon\rangle_M = 3/2(eD_T/\mu)$ where D_T/μ is the ratio of the transverse electron diffusion coefficient to electron mobility, obtained from independent measurements. These studies, besides being limited to low pressures, suffer from a lack of knowledge of $f(\varepsilon, E/N)$, which is not known for the attaching gases under study.

The aforementioned shortcomings have been overcome by conducting swarm studies in gas mixtures whereby minute amounts of an electron attaching gas are mixed with a buffer, non-electron-attaching, gas for which $f(\varepsilon, E/N)$ is known over a wide range of E/N values convenient for these experiments. In such cases it is important (especially when the buffer gas is a rare gas) to keep N_a much less than N so that its presence does not affect the $f(\varepsilon, E/N)$ or the electron transport properties of the buffer gas. Knowledge of $f(\varepsilon, E/N)$ allows one to transform $k_a(E/N)$ into the more meaningful quantity $k_a(\langle\varepsilon\rangle)$ where $\langle\varepsilon\rangle$ is now the mean electron energy determined at each E/N from the known buffer gas $f(\varepsilon, E/N)$. This most significant advancement has been developed by the Oak Ridge National Laboratory Group (e.g., see Christophorou et al., 1969b; Christophorou, 1971; Goans and Christophorou, 1974; Christodoulides et al., 1979; McCorkle et al., 1980), who used mostly N_2 and Ar as buffer gases.* The distribution functions for N_2 and Ar in the relatively low E/N ranges employed in those experiments can be calculated rather accurately (see Chapter 3 of Volume 2). In Table I are listed values of $\langle\varepsilon\rangle$ for the E/N ranges normally used by the Oak Ridge National Laboratory Group. It is seen that by changing E/N and buffer gas one can scan easily the mean energy range from thermal to ~ 4 eV. Actually, and in spite of the fact that for a given $\langle\varepsilon\rangle$ $f(\varepsilon, E/N)$ for N_2 and Ar may differ in shape, plots of k_a vs. $\langle\varepsilon\rangle$ for many attaching gases studied in both N_2 and Ar show remarkable agreement and smoothly mesh with one another (e.g., see Figs. 24, 25, and 30 in Section IV).

The studies of the Oak Ridge National Laboratory Group can be conducted at any gas density (to $\sim 50,000$ Torr; see Section VII.B). In these high-pressure experiments one should exercise extreme caution to consider possible changes in w with gas density and divergences of the carrier gases from "ideal-gas-law" behavior and also to ensure that, in the ensuing analyses of the high-pressure data, $f(\varepsilon, E/N)$ is not altered from that which is characteristic of the low-pressure gas.

The measurement of $k_a(E/N)$ in buffer gases of known $f(\varepsilon, E/N)$ offers another distinct benefit: it allows through Eq. (18) determination of $\sigma_a(\varepsilon)$ in a

---

* In the experiments of Christophorou and co-workers, α is measured. From an independent measurement of w, $k_a = \alpha' w$ is obtained. At high pressures care is taken to account for the changes in w, especially for N_2.

TABLE I

Mean Electron Energies as a Function of E/N for N_2 and Ar (298 K)

N_2		Ar	
E/N (V cm^2)	$\langle \varepsilon \rangle$ (eV)	E/N (V cm^2)	$\langle \varepsilon \rangle$ (eV)
3.0×10^{-19}	0.040	1.0×10^{-19}	0.305
6.0	0.046	2.0	0.398
9.0	0.054	3.0	0.466
1.2×10^{-18}	0.064	4.5	0.551
1.5	0.074	6.0	0.624
2.0	0.093	9.0	0.751
2.5	0.112	1.2×10^{-18}	0.862
3.0	0.130	1.5	0.960
4.0	0.164	1.8	1.05
5.0	0.195	2.2	1.16
6.0	0.226	2.5	1.23
9.0	0.319	3.0	1.35
1.2×10^{-17}	0.408	4.5	1.64
1.8	0.547	6.0	1.89
2.4	0.638	7.5	2.10
3.0	0.703	9.0	2.29
3.5	0.744	1.2×10^{-17}	2.64
4.0	0.778	1.5	2.95
5.0	0.829	2.2	3.57
6.0	0.866	3.0	4.18

dense medium. Such determinations have been performed in two ways. First, in those cases in which beam studies show that the attachment measured in the swarm study (as, for example, in the case of a number of halogen-containing compounds; see Section IV) is due to a specific product ion whose relative cross section $I(\varepsilon)$ has been measured in beam studies, use can be made of Eq. (18) and the independently measured $k_a(E/N)$ and $I(\varepsilon)$ functions to determine $\sigma_a(\varepsilon)$ through the swarm–beam technique (Christophorou et al., 1965, 1966). Second, since $f(\varepsilon, E/N)$ is known at each E/N value at which k_a is measured, Eq. (18) can be used to determine $\sigma_a(\varepsilon)$ through the swarm unfolding technique (Christophorou et al., 1971c; McCorkle et al., 1980). In this technique the measured rate constant at the mean energy $\langle \varepsilon \rangle_j$ (i.e., at the jth value of E/N), $k_a(\langle \varepsilon \rangle_j)$, is expressed as

$$k_a(\langle \varepsilon \rangle_j) = \int_0^\infty k_a(\varepsilon) f(\varepsilon, \langle \varepsilon \rangle_j) \, d\varepsilon, \qquad (20)$$

where $f(\varepsilon, \langle \varepsilon \rangle_j)$ is the electron energy distribution function that corresponds to $\langle \varepsilon \rangle_j$ and $k_a(\varepsilon)$ is the value of the monoenergetic attachment rate constant at the energy ε (i.e., the value of the attachment rate constant that would be measured had all the electrons in the swarm had the same energy ε). An iterative procedure is then employed (see Christophorou et al., 1971c) which allows $k_a(\varepsilon)$ to be unfolded from the measured functions $k_a(\langle \varepsilon \rangle_j)$ using the corresponding known functions $f(\varepsilon, \langle \varepsilon \rangle_j)$. Once $k_a(\varepsilon)$ is determined, the attachment cross section $\sigma_a(\varepsilon)$ is calculated from

$$\sigma_a(\varepsilon) = k_a(\varepsilon)/[(2/m)^{1/2}\varepsilon^{1/2}]. \tag{21}$$

Examples of such cross section data are shown in Figs. 23 and 25 (Section IV).

In addition, knowledge of $k_a(\varepsilon)$ allows determination of $(k_a)_{th}$ through

$$(k_a)_{th} = \int_0^\infty k_a(\varepsilon) f_M(\varepsilon) \, d\varepsilon. \tag{22}$$

Determining $(k_a)_{th}$ in this way (Christophorou, 1976) may be beneficial in those cases in which measurements with truly thermal electrons are not experimentally feasible.

The measurement of α in the studies of Christophorou and co-workers is made through the pulse-shape technique (Bortner and Hurst, 1958; Christophorou, 1971). A similar technique has been employed by Grünberg (1969) and Nygaard et al. (1978). Four other methods, namely the steady-state diffusion, the steady-state electron filter, the pulsed drift tube, and the avalanche method have been employed for the measurement of $\alpha(E/N)$ or α'/N_a and have been described by Christophorou (1971) and Massey (1976). See Chapter 3 of Volume 2 (Section III.C) for a discussion of recent developments concerning these methods.

A number of other methods have been improved or developed in recent years to measure $(k_a)_{th}$. These include (i) microwave methods (flowing and stationary afterglow techniques) (e.g., see Weller and Biondi, 1968; Puckett et al., 1971; Truby, 1971; Sides et al., 1976); (ii) microwave conductivity techniques (e.g., Warman et al., 1971; Shimamori and Fessenden, 1978); (iii) ion cyclotron resonance techniques (e.g., McMahon and Beauchamp, 1972; Foster and Beauchamp, 1975; Odom et al., 1975; Beauchamp, 1976); (iv) electron cyclotron resonance techniques (e.g., Mothes et al., 1972; Christodoulides et al., 1974); (v) the pulsed sampling technique (e.g., Wentworth and Becker, 1962; Wentworth et al., 1967, 1969); and (vi) the electron-density sampling method (Cavalleri, 1969; Crompton et al., 1980, 1982).

C. Techniques for Identification of Negative Ions Following Electron Capture

These include the various types of mass spectrometers (magnetic, electrostatic, and time-of-flight) used with RPD or electron monochromatic sources.

They identify the stable or long-lived ($\tau_a \gtrsim 10^{-6}$ s) negative ions, monitor their relative abundances as a function of the incident electron energy, and at times measure the kinetic-energy distributions of the dissociative attachment fragments at ambient or higher T [see next section; Chantry (1972, 1976); and references cited in Section IV, V, and VII where experimental results obtained with these techniques are presented and discussed].

D. High-Resolution Studies of Dissociative Attachment; Energy and Angular Distribution of Dissociative Attachment Fragment Anions

Recent advances in electron-beam techniques made possible studies of dissociative attachment processes with monoenergetic electron beams down to a few tenths of an electron volt, the investigation of the role of vibrational and rotational states in dissociative attachment, and detailed measurement of the energy and angular distribution of dissociative attachment fragment anions. Results obtained by these techniques—and references to the original papers—are given in Section IV and V.

Recent electron-impact mass spectrometer arrangements employed a combination of the trochoidal monochromator (Stamatović and Schulz, 1970a), and magnetic mass spectrometers (Abouaf et al., 1976; Abouaf and Teillet-Billy, 1977; Allan and Wong, 1978, 1981; Illenberger et al., 1979; Scheunemann et al., 1980). A diagram of the electron-impact mass spectrometer of Allan and Wong (1981) is shown in Fig. 5; it consists of a trochoidal electron monochromator, a collision chamber located in an iridium oven which can be heated to 1600 K, and an ion analyzer for analyzing the energy and mass of the products which is a magnetic-type spectrometer with

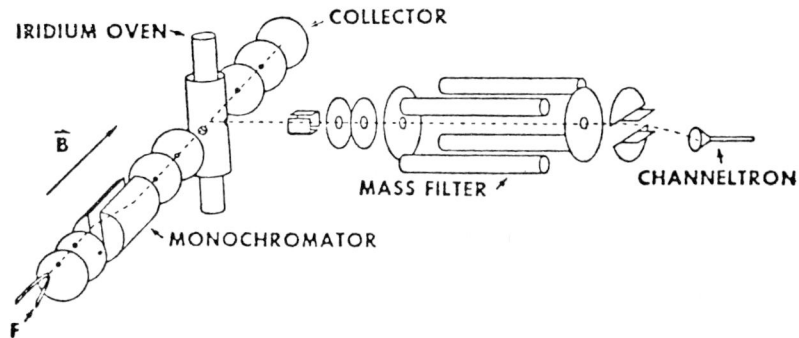

Fig. 5. Layout of the electron-impact mass spectrometer of Allan and Wong (1981).

an observation angle of 90°. The use of a uniform magnetic field B for electron collimation ensured uniform monoenergetic electron beams down to ~ 0.0 eV, and allowed measurement of the total cross sections for ion production based on the scattered current collected at the collision chamber. Focusing and energy selection for the scattered ions prior to the mass analysis was provided by the ion optics between the collision chamber and the quadrupole mass filter.

Measurements of angular distribution of dissociative attachment fragment anions have been made by a number of workers (e.g., Van Brunt and Kieffer, 1970, 1974; Trajmar and Hall, 1974; Hall et al., 1977; see Section IV). A diagram of the electron-impact spectrometer with a magnetic-field momentum filter of Hall and co-workers (Hall et al., 1973, 1977; Schermann et al., 1978) is shown in Fig. 6 (a similar arrangement was employed by Trajmar and co-workers). Electrons produced at the cathode by a hot filament are energy selected by a 127° cylindrical electrostatic analyzer and focused on the gas beam at an angle of 90° in the collision region. Unscattered electrons are collected by a Faraday cup. Scattered electrons and negative ions formed in

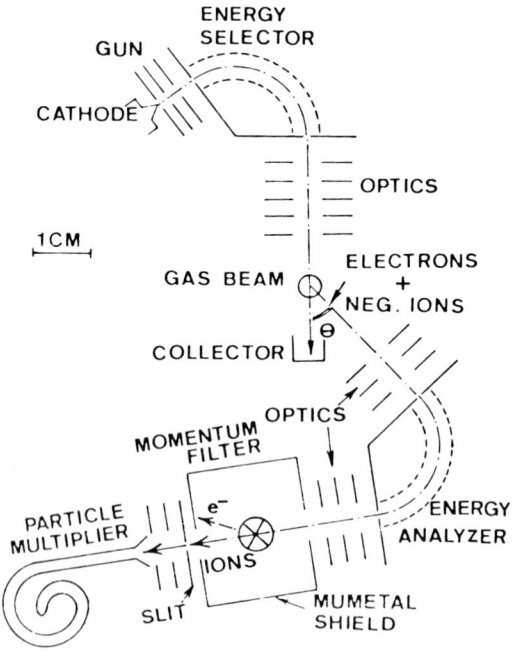

Fig. 6. Schematic diagram of the experimental arrangement of Hall et al. (1977) (see text). (Reproduced with permission from *Physics Review*.)

the collision region are focused on a second *rotatable* 127° electrostatic analyzer. Particles from this energy analyzer then enter the momentum filter, where they are subjected to a magnetic field perpendicular to their velocity. Electrons with less momentum than the negative ions at the same energy are deflected more strongly by the magnetic field and are not detected. Negative ions pass through the momentum filter essentially undeflected and are counted by a particle multiplier.

The spectrometer can be operated in three different modes: (i) In the energy loss mode the incident electron energy ε is fixed and the energy analyzer potential is varied. The ion-energy spectrum yields peaks which correspond to values of the ion residual kinetic energy E_k^- which satisfy Eq. (8). (ii) If E_k^- is kept constant and ε is varied, a constant-ion-energy spectrum is obtained which yields peaks at energies ε which satisfy Eq. (8). If E_k^- is set to zero residual energy, this mode yields the threshold electron excitation spectrum. (iii) An excitation function for a particular dissociation process is obtained by setting ε and E_k^- so that they satisfy Eq. (8) with the constant ratio $1:(1-\beta)$.

Finally, Chutjian (1981) (see also Ajello and Chutjian, 1979) described a so-called threshold photoelectron spectrum by electron attachment technique for the study of the energy dependence of the attachment cross section at *very* low energies, 0–0.2 eV. In this technique a gas A (= Xe) is mixed with the electron attaching gas under study (SF_6 or $CFCl_3$) and the mixture is irradiated by narrow-band radiation from a scanning monochromator which produces *in situ* narrow bands of low energy (0 eV at the photoionization threshold of A to 0.2 eV) electrons by photoionization of A. These electrons attach to the admixed attaching gas (SF_6 or $CFCl_3$) and the resultant anions (SF_6^- or Cl^-) are drawn out of the collision region, mass analyzed, and detected as a function of the photon energy hv, viz.,

$$A + hv \longrightarrow A^+ + e\,(0\text{–}0.2\text{ eV})$$

$$e\,(0\text{–}0.2\text{ eV}) + SF_6\,(\text{or } CFCl_3) \longrightarrow SF_6^-\,(\text{or } CFCl_2 + Cl^-)$$

The relative cross sections are then normalized to absolute cross section data obtained by other means.

E. Techniques for the Measurement of Autodetachment Lifetimes and Electron Affinities, and the Study of Bound-Electron Capture Processes

A discussion of the methods used to measure τ_a can be found in Christophorou (1978a) and EA in Chapter 6 of volume 2. References pertaining to the study of bound-electron capture processes can be found in Section VIII.

IV. Dissociative Electron Attachment to Ground-State Molecules

A. General Considerations

As indicated in Section II, resonance dissociative electron attachment is visualized to proceed through a negative-ion intermediate*—formed by capture of an electron in a restricted energy range defined by a Franck–Condon transition between the initial (neutral molecule and electron at infinite separation) and the final state representing the transient negative ion—which dissociates into (often multiple) neutral and negative-ion fragments. The cross section $\sigma_{DA}(\varepsilon)$ for dissociative attachment as a function of electron energy ε is equal to the product of the cross section $\sigma_c(\varepsilon)$ for capture of the electron by the molecule to form the negative-ion intermediate and the probability $p(\varepsilon)$ that the transient negative ion will decay via dissociative attachment, viz.,

$$\sigma_{DA}(\varepsilon) = \sigma_c(\varepsilon) p(\varepsilon). \tag{23}$$

The quantity $\sigma_c(\varepsilon)$ depends on the symmetry properties of the molecular and resonant states and their Franck–Condon overlap, while $p(\varepsilon)$ depends on the mean lifetime of the resonance and the separation time of the fragments (see below).

Expressions for $\sigma_c(\varepsilon)$ and $p(\varepsilon)$ have been given for diatomic molecules within the resonance scattering theory and the Born–Oppenheimer approximation. Following O'Malley (1966) we write for $\sigma_{DA}(\varepsilon)$ for a diatomic molecule AX initially in its lowest vibrational level ($v = 0$) and with a purely repulsive negative-ion state AX^{-*} (upper AX^{-*} curve in Fig. 4)

$$\sigma_{DA}(\varepsilon)_{v=0} = \underbrace{\frac{4\pi^{3/2}\bar{g}}{(2m/\hbar^2)\varepsilon} \frac{\Gamma_{\bar{a}}}{\Gamma_d} \exp\left(\frac{\Gamma_a^2 - 4(\bar{\varepsilon}_0 - \varepsilon)^2}{\Gamma_d^2}\right)}_{\sigma_c(\varepsilon)} \underbrace{\exp[-\rho(\varepsilon)]}_{p(\varepsilon)}. \tag{24}$$

In Eq. (24) m is the electron mass, $\hbar$ is Planck's constant divided by 2π, $\bar{g}$ is a statistical factor, Γ_a is the total autodetachment width, $\Gamma_{\bar{a}}$ is the partial autodetachment width, Γ_d is the dissociative attachment resonance width, $\bar{\varepsilon}_0 = \varepsilon_0 + \frac{1}{2}\hbar\omega$ (see Fig. 4), ε_0 is the true electron energy at the peak of $\sigma_{DA}(\varepsilon)$, and $\frac{1}{2}\hbar\omega$ is the zero-point energy. The quantity $\exp[-\rho(\varepsilon)]$, defined explicitly (Bardsley et al., 1964, 1966; Chen, 1966; O'Malley, 1966) by

$$\exp[-\rho(\varepsilon)] = \exp\left(-\int_{R_\varepsilon}^{R_c} \frac{\Gamma_a(R)}{\hbar v(R)} dR\right) \simeq \exp\left(-\frac{\tau_s}{\tau_a}\right), \tag{25}$$

---

* Holstein (1951) first discussed the formation of an intermediate in dissociative electron attachment.

is the survival probability of AX^{-*}; that is, the probability that AX^{-*} will not autoionize while separating from the point of formation by capture of an electron with energy ε at an internuclear separation R_ε to the point c (Fig. 4) at an internuclear separation R_c where the potential-energy curves of AX and AX^{-*} cross. For $R > R_c$, AX^{-*} will dissociate with unit efficiency [$p(\varepsilon) = 1$]. The time τ_s needed for A-X$^-$ to separate from R_ε to R_c is

$$\tau_s = \int_{R_\varepsilon}^{R_c} \frac{dR}{v(R)}, \tag{26}$$

where $v(R)$ is the relative velocity of separation of A and X$^-$ and $\bar{\tau}_a(=\hbar/\bar{\Gamma}_a)$ is the mean autoionization lifetime of AX^{-*} ($\bar{\Gamma}_a$ is the average autoionization width).

The energy $\bar{\varepsilon}_0$ at the peak of $\sigma_{DA}(\varepsilon)$ is not always the same as the experimentally measured energy $\varepsilon_{\max}$ at the peak of the cross section, owing to the other energy-dependent factors in Eq. (24). When $\rho(\varepsilon)$ is not varying too rapidly, $\bar{\varepsilon}_0$ and $\varepsilon_{\max}$ are related by the approximate expression (O'Malley, 1966)

$$\bar{\varepsilon}_0 = \varepsilon_{\max} + \frac{1}{8}\Gamma_d^2\left(\frac{1}{\varepsilon_{\max}} + \frac{d\rho}{d\varepsilon}\right), \tag{27}$$

where $d\rho/d\varepsilon$ is always positive. When $d\rho/d\varepsilon$ and Γ_d are small, $\varepsilon_{\max} = \bar{\varepsilon}_0$ (O'Malley, 1966; Christophorou and Stockdale, 1968).

Expression (24) is limited, strictly speaking, to situations in which the potential-energy curve of AX^{-*} in the Franck–Condon region is purely repulsive. Negative-ion potential-energy curves/surfaces have a variety of shapes, and $\sigma_{DA}(\varepsilon)$ is affected by these and the competitive decay channels. This limits the applicability of Eq. (24), especially for polyatomic molecules. However, in certain cases of dissociative attachment to polyatomic molecules such as n-$C_N H_{2N+1}$Br, where the formation of Br$^-$ involves a direct fast cleavage of Br$^-$ along the C-Br coordinate, and the complex molecule may be visualized, as far as this process is concerned, as a diatomiclike R(n-$C_N H_{2N+1}$)-Br system dissociating along a purely repulsive state, Eq. (24) may describe the process satisfactorily (Christophorou et al., 1971a); it relates in a general way σ_{DA} to σ_c, ε, τ_s, τ_a, and the reduced mass M_r of the A-X$^-$ system.

B. Isotope Effects in Dissociative Attachment

It can be seen from Eqs. (24) and (25) that $\sigma_{DA}(\varepsilon)$ depends on the reduced mass M_r of the separating fragments A, X$^-$ in two ways. First, $\sigma_{DA}(\varepsilon)$ depends on M_r through $\rho(\varepsilon)$. Since isotopic species experience the same forces, the relative velocities of separation are proportional to $M_r^{-1/2}$ and hence the

TABLE II

Data on the Isotopic Dependences of σ_{DA} and Γ_d^a

Negative ion/Molecule	Hydrogenated			Deuterated				$\sigma_{DA}(H)/\sigma_{DA}(D)$ at ε'_{max}	Type of isotope effect[b]
	ε_{max} (eV)	$\sigma_{DA}(\varepsilon_{max})$ (cm²)	Γ_d (eV)	ε_{max} (eV)	$\sigma_{DA}(\varepsilon_{max})$ (cm²)	Γ_d (eV)	ε'_{max} (eV)		
[H⁻/H₂], [D⁻/D₂]	10.2	1.27×10^{-20}	~3.6[c]	10.5	0.26×10^{-20}		~10.2	4.7	SD[c]
	~13.9	2.08×10^{-20}	~1.2[c]	~14.0	0.98×10^{-20}		13.9	2.1	SD[e]
	3.75	1.62×10^{-21}	~0.4[c]	3.75	8×10^{-24}		3.75	200	LD[f]
[Cl⁻/HCl], [Cl⁻/DCl]	0.81	1.97×10^{-17}	0.3[c]	0.81	1.4×10^{-17}	0.28	0.81	1.39	SD[g]
	0.84	$(8.9 \pm 0.7) \times 10^{-18}$	~0.3[d]	0.84	1.8×10^{-18}		0.84	5 ± 1.1	SD[h]
[H⁻/HCl], [D⁻/DCl]	6.9	$(5.2 \pm 0.4) \times 10^{-19}$	~1.8[d]	6.9	2.9×10^{-19}		6.9	1.8 ± 0.1	SD[h]
	9.2	$(2.8 \pm 0.2) \times 10^{-19}$	~2.5[d]		≤ 2.1×10^{-19}		9.2	≥1.4	SD[h]
[Br⁻/HBr], [Br⁻/DBr]	0.28	2.70×10^{-16}	0.24	0.28	1.87×10^{-16}	0.23	0.28	1.44	SD[g]
[I⁻/HI], [I⁻/DI]	~0	2.3×10^{-14}		~0	1.38×10^{-14}		~0	1.67	SD[g]
[H⁻/H₂O], [D⁻/D₂O]	6.5	6.9×10^{-18}						1.25 ± 0.06	SD[i]
	6.5 ± 0.1	6.9×10^{-18}	~1.1	6.5 ± 0.1	5.2×10^{-18}	~0.8	6.5 ± 0.1	1.33	SD[j]
	8.6 ± 0.2	1.3×10^{-18}	2.0	8.6 ± 0.2	0.6×10^{-18}	1.2	8.6 + 0.2	2.17	SD[j]
[H⁻/HDO], [D⁻/HDO]							6.5	3.55 ± 0.29	SD[i]
[H⁻/H₂S], [D⁻/D₂S]	5.35	1.2×10^{-18}						1.33 ± 0.07	SD[i]
[H⁻/HDS], [D⁻/HDS]								2.44 ± 0.19	SD[i]
[HS⁻/H₂S], [DS⁻/D₂S]	2.2 ± 0.1	1.7×10^{-18}	~1.1[d]					25 ± 0.3	LD[k]
[S⁻/H₂S], [S⁻/HDS]							9.5	1.72 ± 0.21	SD[l]
[S⁻/H₂S], [S⁻/D₂S]							9.5	2.70 ± 0.15	SD[l]
[NH₂⁻/NH₃], [ND₂⁻/ND₃]	5.65	1.5×10^{-18}	~1.1[d]	5.65	1.2×10^{-18}	~0.9	5.65	1.25	SD[m]
[H⁻/NH₃], [D⁻/ND₃]	5.65	1.4×10^{-18}		5.65	1.4×10^{-18}		5.65	1.0	None[m]
[H⁻/CH₄], [D⁻/CD₄]	9.2	$~7.8 \times 10^{-20}$		9.5[c]	$~9.4 \times 10^{-20}$		8–13	~0.8	SI[n]
[CH₂⁻/CH₄], [CD₂⁻/CD₄]	10.2	9.6×10^{-20}						>100	LD[n]
[Br⁻/C₆H₅Br], [Br⁻/C₆D₅Br]	0.84	9.6×10^{-17}	0.62	0.80	1.04×10^{-16}	0.64	~0.82	0.92	SI[o]

[a] See also Khvostenko et al. (1973) for small direct isotope effects exhibited by hydrogenated and deuterated pairs of organic molecules.
[b] SD = small direct; LD = large direct; SI = small inverse isotope effect.
[c] Determined from $\sigma_{DA}(\varepsilon)$ graphs.
[d] Determined from negative-ion current versus ε graphs.
[e] Rapp et al. (1965).
[f] Schulz and Asundi (1967).
[g] Christophorou et al. (1968).
[h] Azria et al. (1974).
[i] Tronc et al. (1973).
[j] Compton and Christophorou (1967).
[k] Fiquet-Fayard et al. (1972).
[l] Azria et al. (1972).
[m] Sharp and Dowell (1969).
[n] Sharp and Dowell (1967).
[o] Christophorou et al. (1966).

separation times τ_s [and thus $\rho(\varepsilon)$] vary as $M_r^{1/2}$. This can lead to large isotope effects, $\sigma_{DA}(\varepsilon)$ being smaller for the heavier isotope. Second, $\sigma_{DA}(\varepsilon)$ depends on M_r through σ_c [Eq. (24)]. Here the reduced mass enters in two places through the vibrational amplitude Γ_d. The appearance of the vibrational amplitude in the Gaussian term can lead to an isotope effect in both the width and the magnitude of $\sigma_{DA}(\varepsilon)$, making the latter larger for the heavier isotope. The isotope effects resulting from the Gaussian factor, however, are generally small. The term preceding the Gaussian factor in Eq. (24) introduces an $M_r^{1/4}$ dependence of σ_{DA}.

If we neglect the effect of the Gaussian factor, we may write (Christophorou, 1971)

$$\sigma_{DA} \propto M_r^{1/4} \exp(-\text{const} \times M_r^{1/2}). \tag{28}$$

When the probability of autoionization is large, the survival probability is small (large $\bar{\Gamma}_a$) and the second term in Eq. (28) becomes the important factor. In such cases σ_{DA} is smaller for the heavier species; this is called the *direct isotope effect*: depending on the magnitude of $\rho(\varepsilon)$, the direct isotope effect can be small or large. When autoionization is less important, the survival probability is large (small $\bar{\Gamma}_a$), approaching one, and the $M_r^{1/4}$ term may well account for the isotope effect in σ_{DA}. In this case σ_{DA} is larger for the heavier species. This is called the *inverse isotope effect* and is generally small.

The above considerations are appropriate for diatomic molecules; they, however, can be extended to polyatomic species. Experimentally, three classes of isotope effects in σ_{DA} have been observed (see Table II): *large direct, small direct*, and *small inverse*. The isotope effect is more pronounced when the hydrogen atoms in hydrogen-containing molecules are replaced by deuterium atoms. The largest direct isotope effect observed to date is for the production of $H^-(D^-)$ from $H_2(D_2)$ at 3.75 eV (Schulz and Asundi, 1965, 1967). The dissociative attachment process in this case exhibits a vertical onset behavior, and at the peak of the cross section which occurs at the threshold the $H^-(D^-)$-$H(D)$ products separate very slowly; consequently $p(\varepsilon)$ is very small, resulting in $\sigma_{H^-}(H_2)/\sigma_{D^-}(D_2) \simeq 200$. Another large direct isotope effect was reported by Sharp and Dowell (1967) for the production of $CH_2^-(CD_2^-)$ from $CH_4(CD_4)$. At the peak of the cross section at ~ 10 eV, $\sigma_{CH_2^-}(CH_4)/\sigma_{CD_2^-}(CD_4) > 100$. Other large direct isotope effects have been observed (see Table II).

Small direct isotope effects have been found for a large number of molecules (see Table II and Khvostenko et al., 1973).

Inverse isotope effects are expected to be small. The first observation of this kind of isotope effect was observed by Sharp and Dowell (1967) for $H^-(D^-)$ from $CH_4(CD_4)$ in the range 8–13 eV. Their result, $[\sigma_{H^-}(CH_4)]/[\sigma_{D^-}(CD_4)] \simeq 0.8$, was explained by O'Malley (1967b) as being due to the preexponential factor $(M_r^{1/4})$ in Eq. (28).

From the experimental data on σ_{DA} and Γ_d [the full width at half-height of $\sigma_{DA}(\varepsilon)$] for pairs of hydrogenated and deuterated molecules in Table II can be seen that besides σ_{DA}, the Γ_D also depends on M_r; it is generally smaller for the deuterated species, reflecting the narrower width of the square of the ground-state vibrational wave function for the deuterated compared to the hydrogenated species. The isotope dependence of $\sigma_{DA}(\varepsilon)$ for hydrogenated and deuterated pairs of molecules has been used to estimate the τ_a of the dissociating negative ion state (NIS) (see Christophorou, 1978a).

C. Attachment Rate Constants, Cross Sections, and Negative-Ion States

1. Diatomic Molecules

a. H_2, HD, D_2

Resonant dissociative electron attachment to H_2 (and HD and D_2) has been well studied, discussed, and reviewed and served as a prototype for model calculations. In Fig. 7 are shown the cross section data of various authors prior to 1970 as summarized by Christophorou (1971). In the energy range 3–15 eV three resonance dissociative attachment processes can be distinguished. The first process is associated with electron capture from the ground state $^1\Sigma_g^+$ of H_2 to the attractive ground state $^2\Sigma_u^+$ of H_2^-. This process has a vertical onset of 3.75 eV, an exceedingly small cross section with a large isotope effect (Table II), and a short lifetime [$\sim 10^{-15}$ s; see discussion in Christophorou (1978a)]; the products, H^- and H, are in their ground states and have little kinetic energy. The angular distribution measurements of Tronc et al. (1977a) for H^- at 4 eV showed that the relative intensity of H^- at an angle θ to the H^- intensity at 90° can be represented by $1 + 1.96 \cos^2 \theta$.

The second dissociative attachment process lies between 8 and 13 eV. The H^- ions are produced via the repulsive $^2\Sigma_g^+$ state of H_2^-; they are in their ground state and have considerable kinetic energy. Early work indicated the presence of two overlapping resonances in the energy range 11–13 eV. In particular, Dowell and Sharp (1968) observed structure in the cross section between 11.2 and 12.5 eV which they attributed to the vibrational spacings of the attractive $^2\Sigma_g^+$ state of H_2^-: the incident electron is first captured into the vibrational levels of the attractive $^2\Sigma_g^+$ state, which subsequently predissociates via overlapping with the repulsive $^2\Sigma_g^+$ state in this energy range. A more detailed study by Tronc et al. (1977a) showed that $\sigma_{DA}(\varepsilon)$ in the 9–13 eV region exhibits a broad background peak onto which a series of regularly spaced structure is superimposed between 11.30 and 13.30 eV. The broad peak was interpreted as dissociative attachment via the repulsive $^2\Sigma_g^+$ state and the

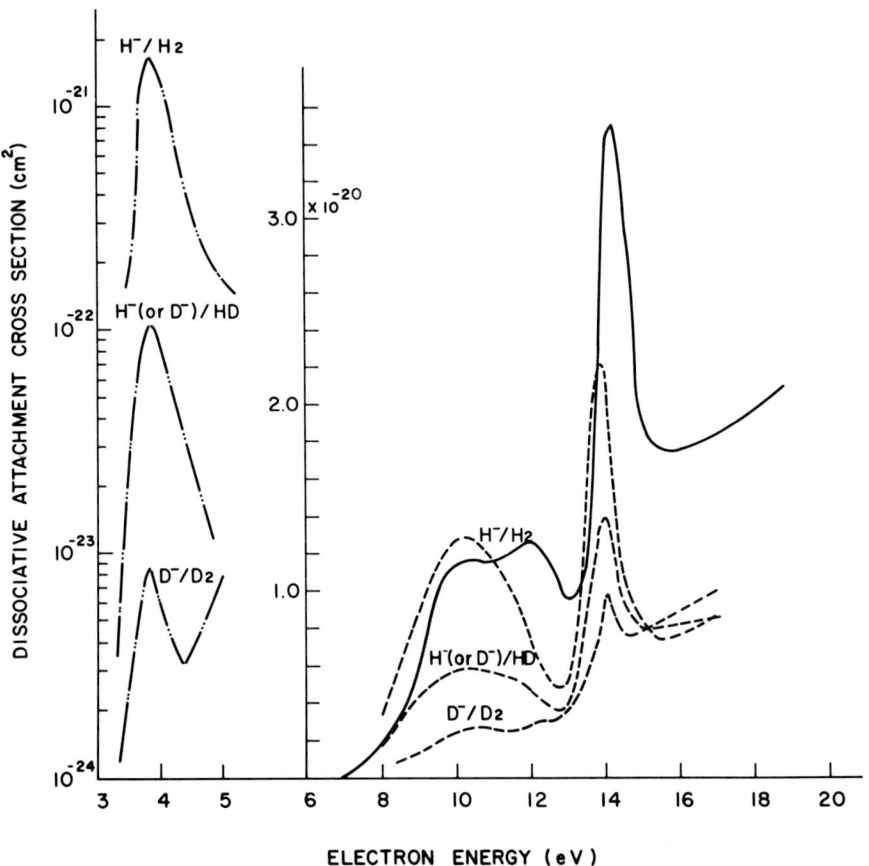

Fig. 7. Dissociative attachment cross sections for the formation of H^- from H_2, H^- (or D^-) from HD and D^- from D_2: (—) Schulz (1959), (---) Rapp *et al.* (1965), (—·—) Schulz and Asundi (1967). [From Christophorou (1971). Reprinted by permission of John Wiley & Sons, Inc.]

vibrational structure as resulting from predissociation of the a resonance bands by the $H_2^-(^2\Sigma_g^+)$ repulsive state [see Tronc *et al.* (1977a), for further discussion of the differential cross sections and angular distributions for dissociative attachment via this process; see also O'Malley and Taylor (1968) for a theoretical treatment of differential cross sections for negative-ion formation].

The third peak between 14.0 and 15.5 eV was associated by Rapp *et al.* (1965) with the repulsive part of a bound $^2\Sigma_g^+$ excited state of H_2^-. Here the H^- has low kinetic energy and the H fragment is in the excited ($n = 2$) state.

Tronc et al. (1979) observed sharp maxima on the high-energy side of the principal maximum in $\sigma_{DA}(\varepsilon)$. These they attributed to predissociation of a bound negative-ion state ($^2\Delta_g$?) which correlates with the $H^- + H$ ($n = 3$) limit and which undergoes rotational coupling with a repulsive $^2\Pi_g$ state, itself dissociating into $H^- + H$ ($n = 2$).

Interestingly, Aberth et al. (1975) and Schnitzer and Anbar (1976) reported mass spectrometric observation of "stable" excited-state HD^{-*}, D_2^{-*} (and H_3^{-*}, H_2D^{-*}, HD_2^{-*}, and D_3^{-*}) with lifetimes $> 10^{-5}$ s using a hollow-cathode duoplasmatron negative-ion source operating between 0.1 and 2 Torr. They suggested that these may be quartet electronic states analogous to He^{-*} [(1s2s2p)^4P_j]. Schnitzer and Anbar (1976) observed isotope effects in the formation of partially or totally deuterated H_2^- and H_3^- ions, on the basis of which they suggested a hydrogen abstraction reaction for the formation of H_2^{-*} and a hydrogen transfer reaction involving excited H_2^* in the formation of H_3^{-*}.

b. *HF, DF; HCl, DCl; HBr, DBr; HI, DI*

Electron attachment to the hydrogen and deuterated-halide molecules H(D)X with X = Cl, Br, and I prior to 1970 has been reviewed by Christophorou (1971). The results of Christophorou et al. (1968) and Azria et al. (1974) on the isotope dependences of $\sigma_{DA}(\varepsilon)$ for these molecules are given in Table II. Dissociative attachment to "hot" HCl, DCl, and HF are discussed in Section V and the negative ion states (NISs) studied in electron scattering experiments have been covered in Chapters 1, 2, and 5. Here we elaborate principally on the structure observed in the $\sigma_{DA}(\varepsilon)$ of the H(D)X molecules.

The most extensively studied member of this group of molecules is HCl. Dissociative attachment to this "prototype" HX molecule proceeds through two distinct reactions, one in the energy range ~ 0.5 to ~ 1.5 eV as

$$e + HCl(^1\Sigma^+) \longrightarrow HCl^{-*}(^2\Sigma^+) \longrightarrow Cl^-(^1S) + H(^2S) \tag{29}$$

and another in the energy range ~ 6 to 11 eV (Azria et al., 1974, 1980) as

$$e + HCl(^1\Sigma^+) \longrightarrow HCl^{-*}(^2\Sigma^+; ^2\Pi) \longrightarrow H^-(^1S) + Cl(^2P) \tag{30}$$

The $\sigma_{DA}(\varepsilon)$ for (29) exhibits (Christophorou et al., 1968; Fig. 8a) a sharp vertical onset and presents, beyond its maximum at 0.81 eV, superimposed on the high-energy tail sharp decreases with spacings corresponding to the vibrational levels of $HCl(^1\Sigma^+)$ (Ziesel et al., 1975a; Abouaf and Teillet-Billy, 1977). The sharp decreases in $\sigma_{DA}(\varepsilon)$ were interpreted by Fiquet-Fayard (1974) as resulting from the fact that as the incident electron energy is increased, σ_{DA} falls to a lower value when a new vibrational level of the neutral molecule is reached which provides a new autodetachment channel for the HCl^{-*}. This

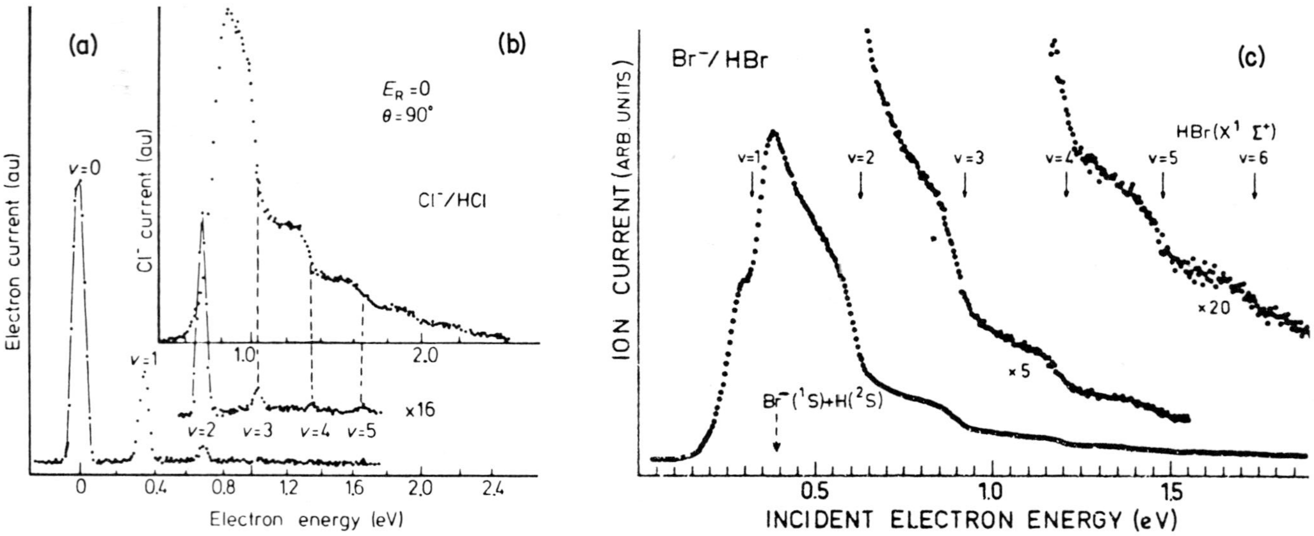

Fig. 8. Very low-energy scattered-electron current (a) and Cl^- ion current (b) formed by

$$HCl + e(\varepsilon) \longrightarrow HCl^{-*} \begin{array}{c} \longrightarrow Cl^- + H \\ \longrightarrow HCl^* + e' \end{array}$$

as a function of ε. Both the ions and the electrons are those at a 90° scattering angle. Only Cl^- ions formed with kinetic energy E_k between 0 and 20 meV are detected. [From Azria et al. (1980).] (c) Br^- ion current formed by $HBr + e(\varepsilon) \rightarrow Br^- + H$ as function of ε. The arrows indicate the energy position of the vibrational levels of the HBr neutral molecule. The dashed arrow shows the location of the thermochemical threshold. [From Abouf and Teillet-Billy (1980a).] [Part (a) is reproduced from the *Journal of Physics* with the permission of the Institute of Physics. Part (b) is reproduced from *Chemical Physics Letters* with the permission of North-Holland Publishing Company.]

interpretation is supported by the recent data of Azria et al. (1980) on both the Cl⁻ and the very low-energy scattered-electron current as a function of the incident electron energy. In Fig. 8b is shown the constant-ion-energy spectrum of Cl⁻ and in Fig. 8a the intensity of the "zero-energy" scattered electrons (residual energy $E_R = 0$ eV) measured at a scattering angle of 90° as a function of the incident electron energy. The structure in the Cl⁻ cross section is observed as decreasing steps which correspond in energy positions to the vibrational levels of the $^1\Sigma^+$ ground state of HCl; concomitantly, as each vibrational excitation energy is reached, scattered electrons are detected which have lost essentially all of their energy, presumably to vibrational excitation, demonstrating the competition between autodetachment and dissociative attachment of the $^2\Sigma^+$ shape resonance of HCl⁻*.

Similar well-defined structure has been observed by Abouaf and Teillet-Billy (1980a) for Br⁻ from HBr below ~ 2 eV (Fig. 8c) and less-resolved structure for F⁻ from HF in the range ~ 2.3 to 3.5 eV which was again attributed to the opening of new decay channels for the HX⁻* species. The onset for Br⁻(^{1}S) + H(^{2}S) was found (Fig. 8c) to be below the thermochemical value of 0.390 ± 0.009 eV. This apparent lowering of the observed onset was attributed by Abouaf and Teillet-Billy (1980a) to the large contribution of high rotational levels of the HBr ($v = 0$) in accordance with the earlier interpretation by Christophorou et al. (1968) of a similar observation for Cl⁻ from HCl (see Section V on dissociative attachment to "hot" molecules).

The cross section for (30) was found by Azria et al. (1980) to exhibit in the energy range ~ 6 to 11 eV two maxima at 7.1 and 9.3 eV which they ascribed, respectively, to the repulsive $^2\Sigma^+$ and $^2\Pi$ negative-ion states. The second peak showed structure on the high-energy side which was interpreted as the interaction of the repulsive $^2\Pi$ state of HCl⁻ with Feshbach resonances. Similar studies of HF by Abouaf and Teillet-Billy (1980b) in the range 10–15 eV led to the finding of one Feshbach resonance in the range 12.8–14 eV. A series of peaks was observed in $\sigma_{DA}(\varepsilon)$ in the same energy region corresponding to F⁻(^{1}S) + H* ($n = 2, 3, 4, 5$) products.

More recently Le Coat et al. (1982) measured the differential cross sections and angular distributions of H⁻ produced by dissociative attachment to HBr in the energy range 5–11 eV. The ground state of H⁻ is ^{1}S and that of Br is ^{2}P; the two P states of Br are split by spin-orbit coupling into the two levels $^2P_{3/2}$ and $^2P_{1/2}$. Thus the dissociation limits for the HBr⁻* resonance are H⁻(^{1}S) + Br($^2P_{3/2}$) (lower) and H⁻(^{1}S) + Br($^2P_{1/2}$) (higher). Indeed, Le Coat et al. observed such spin-orbit structure in dissociative attachment; i.e., they found that single HBr⁻ states dissociate through two exit channels. The observed H⁻ yields associated with both the Br($^2P_{3/2}$) and Br($^2P_{1/2}$) limits exhibited two peaks: one in the energy range 5–7 eV and another in the energy range 7–10 eV. The angular distributions for the former (5–7 eV) were characteristic of a $^2\Sigma$ and for the latter (7–10 eV) of a $^2\Pi$ state of HBr$^-$. Spin-orbit interaction splits the $^2\Pi$ state of HBr$^-$ into $^2\Pi_{3/2}$ and $^2\Pi_{1/2}$ states (i.e., the states with total symmetry $\Omega = \frac{1}{2}$, where Ω is the component of the total electronic angular momentum on the internuclear axis). The three $^2\Sigma_{1/2}$, $^2\Pi_{3/2}$, and $^2\Pi_{1/2}$ states will be adiabatically correlated, respectively, to the dissociation limits H$^-$(^{1}S) + Br(^{2}P$_{3/2}$), H$^-$(^{1}S) + Br(^{2}P$_{3/2}$), and H$^-$(^{1}S) + Br(^{2}P$_{1/2}$). However, these states did not account for the cross section measurements of Le Coat et al., and the authors invoked nonadiabatic coupling between these states to explain their results.

The aforementioned results and interpretations are still the subject of much discussion. For example, Nesbet (1977) interpreted the structure in $\sigma_{DA}(\varepsilon)$ for Cl$^-$/HCl as Wigner cusp structures enhanced by virtual states connected with each vibrational level of HCl, and Taylor et al. (1977) argued that the structure can be understood as the result of predissociative interference between two low-lying $^2\Sigma^+$ states of HCl$^-$ indicated by their calculations. Taylor et al. (1977) and Goldstein et al. (1978) used the stabilization method and configuration interaction and computed six resonant states for HCl$^-$ below ~ 10 eV. Two of these dissociate into H$^-$ + Cl, one into Cl$^-$ + H, and three ("dipole" states) into H + Cl + e (the dipole states are generated by the permanent dipole moment of HCl). Taylor et al. rationalized the known experimental observations in HCl on the basis of those six states. More experimental and theoretical work is anticipated.

The early work summarized in Christophorou (1971) showed that the cross sections for electron attachment to the hydrogen halide molecules vary substantially from halogen to halogen (see also Table II). With the exception of this work only scattered measurements have been made of the absolute electron attachment rate constants for these molecules. Trainor and Boness (1978) reported an attachment rate constant k_a of 8×10^{-10} cm^3 s^{-1} for HBr at 300 K and $\varepsilon \simeq 0.6$ eV, and Miller and Gould (1978) reported attachment rate constants for HCl in the range 1.7×10^{-13} to 1.0×10^{-12} cm^3 s^{-1} for flames at temperatures between 1730 and 2475 K. The existing data on the attachment rate constants $(k_a)_{th}$ at thermal energies are quite uncertain. For example, for HBr the data of Christophorou et al. (1968) indicate that $k_a(\langle\varepsilon\rangle)$ increases with decreasing $\langle\varepsilon\rangle$ at least down to ~ 0.2 eV, where $k_a \simeq 10^{-8}$ cm^3 s^{-1}, while Mothes et al. (1972) reported that the $(k_a)_{th}$ for HBr is 9.6×10^{-11} cm^3 s^{-1}. It seems that either the cross section near zero energy is much smaller than at a few tenths of an electron volt above thermal and/or that experimental conditions, gas purity, and complex pressure dependences of the attachment processes at thermal energies are responsible for these discrepancies. Electron attachment processes other than those leading to dissociative attachment have been indicated by the work of Christophorou et al. (1968) at near thermal energy which most probably involve more than one H(D)X molecule (see also Davidow and Armstrong, 1968). No parent negative ion of any of the H(D)X molecules has been observed. Recently, however, Spence et al. (1982) reported experimental observation of long-lived ($\geq 10^{-4}$ s) HI$^-$ (and H$_2$I$^-$) ions using a mass spectrometer and a Penning ionization source. No explicit mechanism of the production was given. The authors concluded that the electron affinity of HI is ≥ 0.007 eV.

c. N_2

Athough the stable N$^-$ ion does not exist (e.g., see Hotop and Lineberger, 1975) the unstable N^{-*} ion can exist and has been detected. Thus Mazeau et al. (1978) observed N^{-*}(^{3}P) in dissociative attachment to N$_2$ and NO by observing the slow electrons emitted when the short-lived N^{-*}(^{3}P) ion decays to the N(^{4}S) ground state. The presumed process for N$_2$ is (Mazeau et al., 1978; Spence and Burrow, 1979)

$$e + N_2(X\,^1\Sigma_g^+) \longrightarrow N_2^{-*}(A\,^2\Pi_u) \longrightarrow N_2^*(X\,^1\Sigma_g^+ \text{ or } A\,^3\Sigma_u^+) + e$$

$$\longrightarrow N(^4S) + N^{-*}(^3P) \quad (31a)$$

$$\longrightarrow N(^4S) + e \quad (31b)$$

Spence and Burrow (1979) reported that the total cross section for (31) has a vertical onset and a peak value of 2.5×10^{-18} cm^2 ($\pm 50\%$), just above the dissociation energy for N$_2$ at 9.76 eV with the autodetached electron [in (31b)] having 0.09 eV energy. Similarly, Huetz et al. (1980) reported the integral cross section for (31) to be $(0.8 \pm 0.4) \times 10^{-18}$ cm^2 at 10.07 eV. At higher energies, Hiraoka et al. (1977) observed long-lived N^{-*} formed in ^{1}D or ^{1}S states.

d. O_2

Dissociative attachment to O$_2$ has been extensively studied and reviewed (e.g., see Christophorou, 1971, 1978b; Massey, 1976). The O$^-$ from O$_2$ is produced in the energy range 4.4–10 eV via the process

$$O_2(X\,^3\Sigma_g^-) + e \longrightarrow O_2^{-*}(^2\Pi_u) \longrightarrow O^-(^2P) + O(^3P) \quad (32)$$

The cross section for (32) shows a single structureless peak at 6.7 eV (with a peak value of $\sim 1.4 \times 10^{-18}$ cm^2), suggesting that only a single repulsive state is responsible for the production of O$^-$ in this energy range, in spite of the fact that other NISs are known to exist (Michels and Harris, 1971). Consistent with this are the angular distribution measurements for O$^-$ ions from O$_2$ produced by electrons having energies between 5.75 and 8.40 eV which could be explained simply in terms of Eq. (32) (Van Brunt and Kieffer, 1970).

The onset, absolute magnitude, and width of $\sigma_{DA}(\varepsilon)$ for (32) depend on the temperature T; the former is lowered and the latter two are increased as T increases (see Section V).

e. CO

Christophorou (1971) summarized and discussed the available information on dissociative attachment to CO prior to 1970. Three dissociative attachment processes have been observed below 13 eV, two leading to O^- and one to C^-. The dominant process leading to O^- is

$$CO(^1\Sigma^+) + e \longrightarrow CO^{-*}(^2\Pi) \longrightarrow O^-(^2P) + C(^3P) \tag{33}$$

where the products O^- and C are in their ground states. The $\sigma_{DA}(\varepsilon)$ for (33) peaks very close to the threshold (9.62 eV) and has a maximum value of $\sim 2 \times 10^{-19}$ cm^2 (see Christophorou, 1971). The NIS involved has been characterized as a Π state based on the experimental data of Hall et al. (1977) on the energy and angular distributions of O^- from CO in the energy range of (33). The NIS is bound in the Franck–Condon region, which accounts for the vertical onset behavior of $\sigma_{DA}(\varepsilon)$.

The second process leading to O^- is

$$CO(^1\Sigma^+) + e \longrightarrow CO^{-*}(^2\Pi) \longrightarrow O^-(^2P) + C^*(^1D) \tag{34}$$

The cross section for (34) peaks at threshold (10.88 eV) and Chantry (1968) reported a peak cross section value of 9.5×10^{-21} cm^2. The work of Hall et al. (1977) indicated that the NIS responsible for (34) has a Π symmetry and is bound in the Franck–Condon region.

A third and much weaker process leading to C^- was found by Stamatovic and Schulz (1970b). The $\sigma_{DA}(\varepsilon)$ for this process showed an onset at 10.20 ± 0.04 eV and peaks at 10.4 and 10.9 eV. Stamatovic and Schulz (1970b) estimated the cross section at the first peak to be $\sim 6 \times 10^{-23}$ cm^2.

f. NO

Dissociative electron attachment to NO producing O^- has been discussed and reviewed by Christophorou (1971). In general the cross section for O^- formation begins at ~ 6.6 eV, extends to ~ 12 eV, shows reproducible structure with peaks at 8.1 and 8.6 eV, and has a peak cross section value at 8.1 eV of 1.12×10^{-18} cm^2. Although the dissociative attachment process is generally accepted to be

$$NO(X\,^2\Pi) + e \longrightarrow NO^{-*}(?) \longrightarrow O^-(^2P) + N^*(^2D) \tag{35}$$

there is still considerable uncertainty as to the nature and the number of the negative-ion states of NO involved. Thus Chantry (1968) was unable to resolve the origin of the double-peak structure of $\sigma_{DA}(\varepsilon)$ for O^-/NO; however, he established that if more than one NIS is involved, these separate NISs must be associated with the same dissociation limit $O^-(^2P) + N^*(^2D)$. On the other hand, Locht and Momigny (1970) interpreted the structure in $\sigma_{DA}(\varepsilon)$ in terms of three possible NISs. They argued that the cross section below 8.5 eV is accounted for primarily by two NISs, to which they assigned the symmetries $^3\Pi$ and $^1\Pi$; the region above 8.5 eV is due primarily to a single state of undetermined symmetry. However, measurements of the angular distribution of O^- from NO produced by dissociative attachment to NO in the energy range 8–11 eV by Van Brunt and Kieffer (1974) do not support the symmetry assignments of Locht and Momigny (1970). The measured O^-/NO angular distributions of Van Brunt and Kieffer are shown in Fig. 9; they are observed

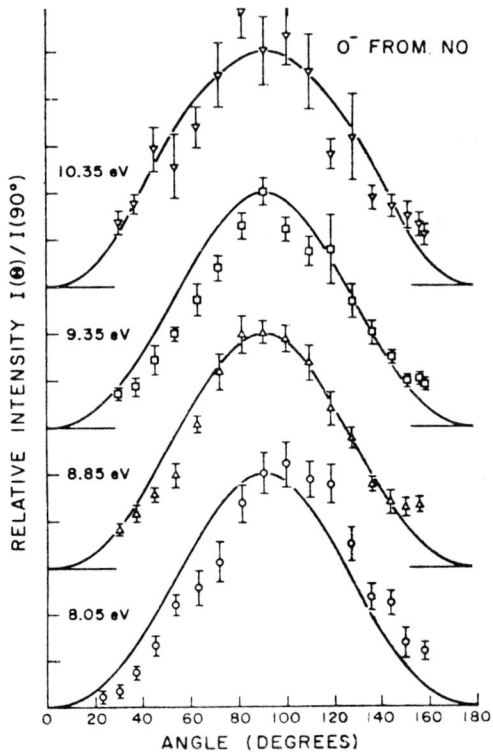

Fig. 9. Measured angular distributions of O^- from NO at the indicated electron energies. The solid curves represent least-squares fits to the data obtained using a model described in Van Brunt (1974). [From Van Brunt and Kieffer (1974). Reproduced from *Physical Review*.]

TABLE III

Thermal-Electron Attachment Rate Constant $(k_a)_{th}$
for the Halogen Molecules

Halogen molecule	$(k_a)_{th}$ ($cm^3\,s^{-1}$)	T (K)	Method	Reference
F_2	$(3.1 \pm 1.2) \times 10^{-9}$	350	FAG[a]	b
	$(4.6 \pm 1.2) \times 10^{-9}$	600	FAG	b
	7.5×10^{-9}	~500	FAG	c
		(see also Fig. 10)		
Cl_2	$(2.8 \pm 0.4) \times 10^{-10}$	300	PS[d]	e
	3.1×10^{-10}	293	ECR[f]	g
	1.1×10^{-9}	300	ECR	h
	$(3.7 \pm 1.7) \times 10^{-9}$	350	FAG	b
	(see footnote i)			
Br_2	$(0.82 \pm 0.08) \times 10^{-12}$	296	MC[j]	k
	$(1.0 \pm 0.9) \times 10^{-11}$	350	FAG	b
	$(1.34 \pm 0.34) \times 10^{-10}$	300	PS	e
	(see footnote l)			
I_2	$(1.36 \pm 0.28) \times 10^{-10}$	300	PS	e
	1.8×10^{-10}	295	MC	m
	$(0.9$ to $4.2) \times 10^{-10}$	253–467[n]	MC	o
	(see footnote p)			

[a] Flowing afterglow technique.
[b] Sides *et al.* (1976).
[c] Sides and Tiernan (1974).
[d] Pulse sampling technique using 90% Ar/10% CH_4 as buffer gas.
[e] Ayala *et al.* (1981a).
[f] Electron cyclotron resonance technique (He, Ar, Ar/CO_2, and Ar/NO were used as buffer gases).
[g] Christodoulides *et al.* (1975a) (Ar was used as buffer gas).
[h] Schultes *et al.* (1975) (Ar/NO was used as buffer gas).
[i] Rokni *et al.* (1979) used an electron-beam controlled-discharge method and obtained an attachment rate constant for Cl_2 in Ar at ~5.2 eV equal to ~2.9×10^{-10} and ~$2 \times 10^{-10}\,cm^3\,s^{-1}$ for $T = 573$ and 300 K, respectively. Similarly, they estimated the $k_a(Cl_2)$ to be ~$1.6 \times 10^{-10}\,cm^3\,s^{-1}$ at ~1 eV and $T = 523$ K using N_2 as buffer gas.
[j] Microwave cavity technique.
[k] Truby (1971).
[l] Trainor and Boness (1978) reported an attachment rate constant of $1 \times 10^{-10}\,cm^3\,s^{-1}$ at 0.8 eV from high-pressure electron-beam-sustained gas-discharge experiments on mixtures of Br_2 with N_2.
[m] Truby (1968).
[n] The rate constant increased from the lower to the higher value with increasing T from 253 to 467 K.
[o] Truby (1969).
[p] Brooks *et al.* (1979) also studied electron attachment to I_2 in N_2 from 308 to 383 K; electron attachment generally increased as T was increased. A theoretical calculation by Birtwistle and Modinos (1978) also found the rate constant to increase from ~$10^{-10}\,cm^3\,s^{-1}$ at 250 K to ~$2 \times 10^{-10}\,cm^3\,s^{-1}$ at ~300 K.

to be anisotropic and of the form that is nearly $\sin^2 \theta$ at all energies (Fig. 9). These data are consistent with the final repulsive negative-ion state(s) of NO involved in (35) being either Σ or Δ states. Although the results do not exclude the possibility of two or more states being involved, Van Brunt and Kieffer conjectured that the simplest explanation may be that (35) proceeds via a single Σ (or Δ) state and that since the neutral fragment is always produced in the first excited state, one might expect the first excited NIS of NO to be involved (possibly $^3\Sigma^-$).

g. F_2, Cl_2, Br_2, I_2

There has been considerable interest recently in the electron attachment properties of the halogen molecules, motivated largely by gas laser applications. In Table III are listed recent measurements of the thermal electron attachment rate constant $(k_a)_{th}$ for the halogen molecules. The data are rather uncertain, complicated perhaps by the dependence of $(k_a)_{th}$ on T, ε, and gas purity. It seems, however, that the $(k_a)_{th}$ for F_2 is larger than for the heavier halogens. Measurements of k_a at energies higher than thermal have also been made. Some of these are given in the footnotes to Table III and are plotted in Fig. 10 for F_2. The comparison of the k_a vs. $\langle \varepsilon \rangle$ data for F_2 in Fig. 10, although

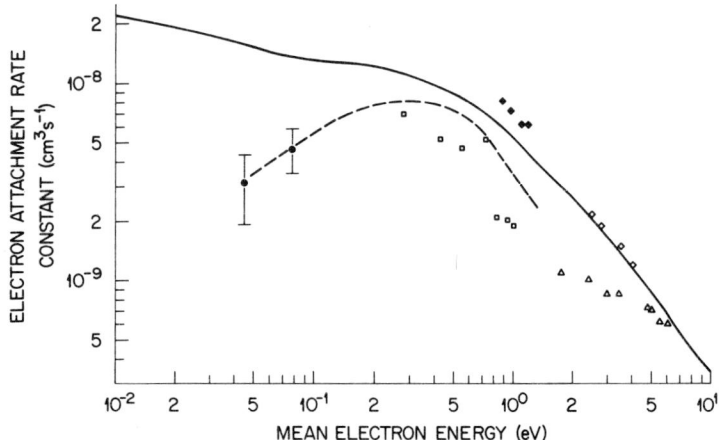

Fig. 10. Electron attachment rate constant vs. mean electron energy $\langle \varepsilon \rangle$ for F_2. The $\langle \varepsilon \rangle$ values were generally computed at each E/N via the solution of the Boltzmann equation for the respective buffer gas used: (●) Sides *et al.* (1976) (buffer gas Ar), (□) Chen *et al.* (1977) (buffer gas N_2), (△) Nygaard *et al.* (1978) (buffer gas He), (◆) Schneider and Brau (1978) (buffer gas N_2), (◇) Schneider and Brau (1978) (buffer gas Ar), (---) calculation by Hall (1978) for N_2 buffer gas; the calculated k_a values for the $v = 2$ level were found to be ~ 5 times larger than those for the $v = 0$ level of F_2; (—) calculation by Chantry (1982).

convenient and useful, is rather complicated by the fact that even at the same $\langle \varepsilon \rangle$, $f(\varepsilon, E/N)$ and thus k_a are not quite the same for the different buffer gases (He, Ar, N_2) employed in these measurements. The differences, however, are not expected to be large (see Section III.B).

A recent study of dissociative attachment to F_2, Cl_2, Br_2, and I_2 has been made by Tam and Wong (1978) using an electron-impact mass spectrometer. They observed three atomic negative-ion peaks for Cl_2, Br_2, and I_2, but only one for F_2 (Fig. 11; inset). A common feature in these mass spectra is a sharp peak at zero energy for all halogens X_2. The systematic trends (Table IV) in the energies of the three maxima of the relative $\sigma_{DA}(\varepsilon)$ of Cl_2, Br_2, and I_2 are worth noting. These maxima can be interpreted as due to the process

$$X_2(^1\Sigma_g^+) + e \longrightarrow X_2^{-*}(\text{NIS}) \longrightarrow X(^2P_{3/2,1/2}) + X^-(^1S_0) \qquad (36)$$

i.e., dissociative attachment involving the ground state $^1\Sigma_g^+$ of the X_2 molecules (in $v = 0$ and perhaps $v = 1$) via the four [$^2\Sigma_u^+$, $^2\Pi_g$, $^2\Pi_u$, $^2\Sigma_g^+$; see Gilbert and Wahl (1971) and Tam and Wong (1978) for calculated potential-energy curves for Cl_2 and Cl_2^-] NISs of X_2^- which are correlated with the dissociation limit $X(^2P_{3/2,1/2}) + X^-(^1S_0)$. Thus, Tam and Wong (1978) attributed the sharp peak at zero energy for F_2 to the shape resonance of symmetry $^2\Sigma_u^+$. For the other three halogens they interpreted the observed three maxima in their $\sigma_{DA}(\varepsilon)$ data as due to core-excited NISs of $^2\Pi_g$, $^2\Pi_u$, and $^2\Sigma_g^+$ symmetries (see Table IV).

Dissociative attachment cross sections as a function of ε have become available for F_2, Cl_2, and Br_2 (Fig. 11). In Table IV we give the NIS assignments of Kurepa and co-workers, which differ from those of Tam and Wong (1978) but which agree with recent assignments made on the basis of Cl^- angular distribution measurements by Azria et al. (1982). The assignments of Kurepa et al. and Azria et al. are in agreement with the calculations of Gilbert and Wahl (1971) (see also Peyerimhoff and Buenker, 1981). It is worth noting that although the work of Tam and Wong (1978) for F^-/F_2 showed only one maximum at ~ 0.0 eV, an earlier study by DeCorpo et al. (1970) indicated structure in the relative cross section of F^-/F_2 at ~ 3 and ~ 6 eV. It is worth noting also that recently Bardsley and Wadehra (1983) calculated a dissociative attachment cross section for ground state F_2 molecules which, while agreeing well with the experimental cross section (see Fig. 11) between ~ 0.1 and 1 eV, shows no zero energy peak.

h. Other Diatomics

Negative ions produced by dissociative attachment to diatomic molecules of the form XY with X = Cs, Li, K, Na, Rb and Y = F, Cl, Br, I have been studied by Ebinghaus (1964).

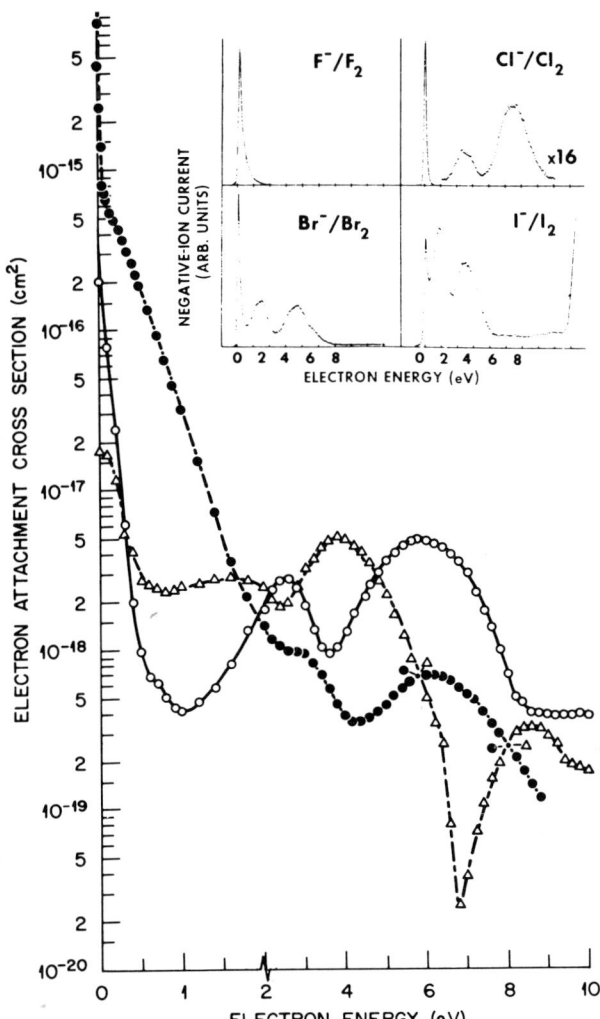

Fig. 11. Total attachment cross section vs. electron energy for F_2 (●---)(Chantry, 1978), Cl_2 (—○)(Kurepa and Belić, 1978), and Br_2 (---△)(Kurepa *et al.*, 1981b); note that the energy scale is expanded below 2 eV. *Inset*: Negative-ion current vs. ε for F^-/F_2, Cl^-Cl_2, Br^-/Br_2, and I^-/I_2. [From Tam and Wong (1978). Reproduced with permission from the *Journal of Chemical Physics*.]

TABLE IV

Maxima in the $\sigma_{DA}(\varepsilon)$ of the Halogens and Assigned Symmetries of the Corresponding NISs

Ion/halogen	Energy of maxima in $\sigma_{DA}(\varepsilon)^a$ (eV)				Assigned symmetries of corresponding NIS			
	TW[b]	KB[c]	KBB[d]	AAT[e]	TW	KB	KBB	AAT
F^-/F_2	0.09				$^2\Sigma_u^+$			
Cl^-/Cl_2	0.03	0.0		0	$^2\Pi_g$	$^2\Sigma_u^+$		$^2\Sigma_u^+$
	2.5	2.5		2.5	$^2\Pi_u$	$^2\Pi_g$		$^2\Pi_g$
	5.5	5.75		5.5	$^2\Sigma_g^+$	$^2\Pi_u$		$^2\Pi_u^f$
		9.7		not observed		$^2\Sigma_g^+$		
Br^-/Br_2	0.07		0.0		$^2\Pi_g$		$^2\Sigma_u^+$	
			0.50				$^2\Pi_{g3/2}$	
	1.4		1.45		$^2\Pi_u$		$^2\Pi_{g1/2}$	
	3.7		3.75		$^2\Sigma_g^+$		$^2\Pi_{u3/2}$	
			5.30				$^2\Pi_{u1/2}$	
			8.50				$^2\Sigma_g^+$	
I^-/I_2	0.05				$^2\Pi_g$			
	0.9				$^2\Pi_u$			
	2.5				$^2\Sigma_g^+$			

[a] See Fig. 11.
[b] Tam and Wong (1978).
[c] Kurepa and Belić (1978).
[d] Kurepa et al. (1981b).
[e] Azria et al. (1982).
[f] The results of Azria et al. (1982) do not exclude a small contribution of the $^2\Sigma_g^+$ state of Cl_2^- to the Cl^- formation in the low-energy side of this process.

2. Triatomic Molecules

a. H_2O, D_2O

Electron attachment to H_2O and D_2O prior to 1970 has been summarized and discussed by Christophorou (1971). Subsequently, measurements have been made of the σ_{DA} of the dissociation products and of the energy and angular distributions of the fragment negative ions. In Fig. 12 we summarize the available data on σ_{DA} (see also Tables II and V). Clearly as detection sensitivity improved in recent years, all three ions $H^-(D^-)$, $OH^-(OD^-)$, and O^- have been observed in the decomposition channels of three well-defined

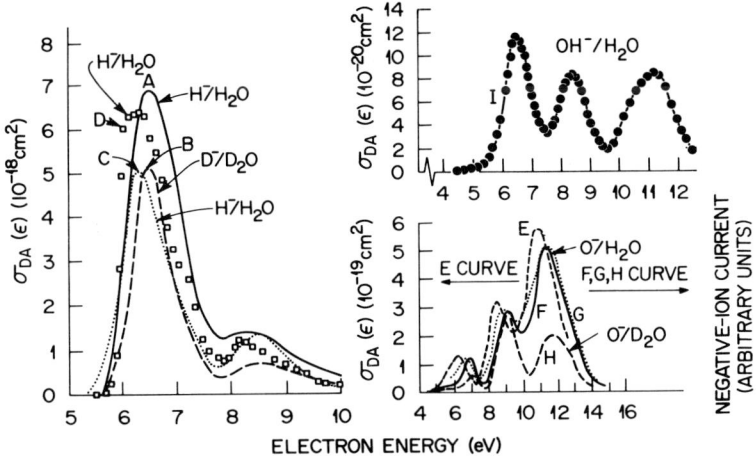

Fig. 12. Dissociative attachment cross section as a function of electron energy for H^-, O^-, and OH^- from H_2O and D^-, and O^- from D_2O (A, B, F, H—Compton and Christophorou, 1967; C—Buchel'nikova, 1959; G—Dorman, 1966; D, E, I—Melton, 1972).

NISs of $H_2O(D_2O)$ at 6.5, 8.6, and 11.8 eV (see Table V). Although the energy dependence of the cross section for the three fragments (H^-, OH^-, O^-) varies, the magnitude of the cross section decreases in the order H^-, O^-, OH^- [see discussion by Claydon et al. (1971)]. The cross sections for H_2O and D_2O also show isotopic dependences (Table II, Section IV.B) which Compton and Christophorou (1967) employed to obtain an estimate ($\sim 2 \times 10^{-14}$ s) for the τ_a of the first (6.5 eV) NIS of H_2O and D_2O.

The nature of the NISs responsible for the three dissociative attachment processes has been clarified by recent theoretical and especially recent experimental studies on the angular distribution of the negative ion fragments produced in the decay of each NIS of $H_2O(D_2O)$ (see also Chapters 1, 2, and 5). The angular distribution of the negative ions produced via the decay of NISs is directly determined by the symmetry of the states and the number of partial waves involved. Since usually one or two partial waves are involved, the form of the angular distribution of the negative ion fragments can be simple and can be interpreted to give information about the symmetry of the decaying NIS. Trajmar and Hall (1974) measured the angular distributions for $H^-(D^-)$ from $H_2O(D_2O)$ originating from the 6.5-eV process. The angular distribution of H^- (see Fig. 13a) (and D^-) peaked at 100°, which is consistent with a $^2B_1$ symmetry for the NIS at 6.5 eV. The subsequent work of Azria et al. (1979) on H_2S is also consistent with this assignment. More recently Belić et al. (1981) measured the angular distribution of H^- (and D^-) from

TABLE V

Positions $(\varepsilon_{max})^a$ of the Cross Section Maxima for the Production of Fragment Negative Ions from H_2O and D_2O^b

Ion	ε_{max}(eV)	
	H_2O	D_2O
$H^-(D^-)$	$6.5,^c\ 6.4^d$	6.5^c
	$8.6,^c\ 8.6^d$	8.6^c
	11.8^e	
O^-	$6.9,^c\ 6.6,^d\ 7.0^f$	$7.0^{c,f}$
	$8.9,^c\ 8.4,^d\ 9.1^f$	$9.0,^c\ 9.1^f$
	$11.4,^c\ 11.2,^d\ 11.8^f$	$11.8^{c,f}$
$OH^-(OD^-)$	$6.4,^d\ 7.0^f$	7.0^f
	$8.4,^d\ 9.1^f$	9.1^f
	$11.2,^d\ 11.8^f$	11.8^f

[a] Uncertainty $\sim \pm 0.1$ eV.
[b] See cross section data in Table II and Fig. 12.
[c] Compton and Christophorou (1967).
[d] Melton (1972).
[e] Belić et al. (1981).
[f] Jungen et al. (1979).

$H_2O(D_2O)$ for all three processes at 6.5, 8.6, and 11.8 eV. Their results for the 6.5-eV process are similar to those of Trajmar and Hall and are shown along with those for the 8.6- and 11.8-eV processes in Figs. 13a, 13b, and 13c, respectively. The angular distributions indicate the symmetries of the three NISs at 6.5, 8.6, and 11.8 eV to be, respectively, $^2B_1$, $^2A_1$, and $^2B_2$. Calculations by Jungen et al. (1979) indicated the same symmetries; Jungen et al. (1979) interpreted the three processes as Feshbach resonances.

The distributions of the excess energy available in dissociative attachment among the different degrees of freedom of the fragments depends on the conformation and lifetime of the NIS. Trajmar and Hall (1974) measured the energy distributions of the H^-/H_2O (and D^-/D_2O) ions produced in the 6.5-eV process and found that the OH(OD) fragments are rotationally and, in large part, vibrationally excited although most of the dissociation energy is carried away by the fragments as kinetic energy. Belić et al. (1981) measured the kinetic-energy distributions of H^-/H_2O (and D^-/D_2O) for all three processes at 6.5, 8.6, and 11.8 eV. The $H^-(D^-)$ kinetic-energy distributions for all three processes showed that most of the dissociation energy is in the form

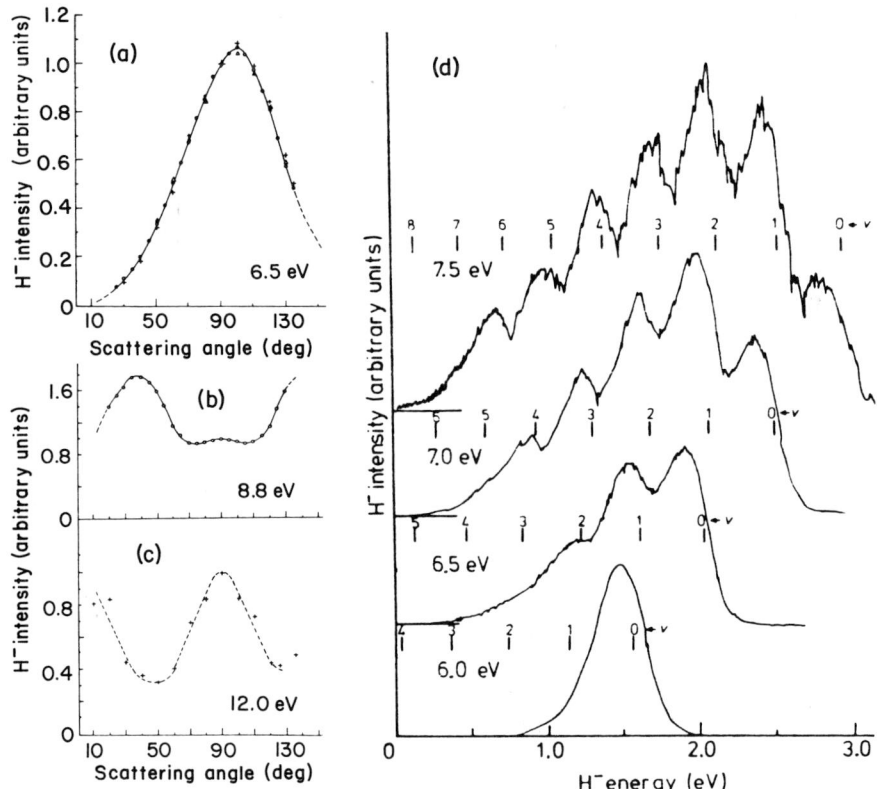

Fig. 13. (a) Angular distribution of $H^-(D^-)$ from $H_2O(D_2O)$ at 6.5 eV: (+) H_2O, ($\triangle$) D_2O, ($\bigcirc$) H_2O (Trajmar and Hall, 1974). Angular distribution of H^- from H_2O at 8.8 eV (b) and 12.0 eV (c). (d) Energy distribution of H^- ions produced in H_2O at $\varepsilon = 6.0$, 6.5, 7.0, and 7.5 eV and 90° scattering angle. The energies corresponding to excitation of the vibrational level v of OH are indicated in the figure (see text). [From Belić et al. (1981). Reproduced from the *Journal of Physics* with the permission of the Institute of Physics.]

of translational energy and that the distribution for all three processes is similiar and the same for H_2O and D_2O. This similarity would indicate that the potential-energy surfaces and lifetimes of the H_2O^- states are similar, in support of the calculations of Jungen et al. (1979). The energy distribution of H^- from H_2O produced via the 6.5-eV process is shown in Fig. 13d. As is indicated by the vibrational thresholds, the structure corresponds to vibrational excitation of the OH fragment. At 6 eV all the excess energy is in the form of translational energy and the OH fragment in the $v = 0$ level. Higher vibrational levels of OH are populated as the incident electron energy increases and at 7.5 eV excitation of up to the $v = 6$ level of OH can be seen. However, the experimental data show that the larger part of the available energy always goes into kinetic energy of the fragments. Belić et al. considered the broadness of the observed peaks—and the shift of the peak positions from the vibrational thresholds—as evidence for rotational excitation of OH. Similar, but weaker, structure was observed for the 6.5-eV process in D_2O.

For the H^-/H_2O process at 8.6 eV, Belić et al. obtained results similar to the H^-/H_2O process at 6.5 eV. However, for the H^-/H_2O process at 11.8 eV, the H^- signal was too weak to allow such studies and the OH fragment was produced in its first excited electronic state $A\,^2\Sigma^+$.

We can, then, summarize the three dissociative attachment processes for $H_2O(D_2O)$ producing H^- (D^-) as follows:

$$H_2O(X\,^1A_1) + e \xrightarrow{6.5\ eV} H_2O^{-*}(^2B_1) \longrightarrow H^-(^1S) + OH(^2\Pi) \tag{37}$$

$$\xrightarrow{8.6\ eV} H_2O^{-*}(^2A_1) \longrightarrow H^-(^1S) + OH(^2\Pi) \tag{38}$$

$$\xrightarrow{11.8\ eV} H_2O^{-*}(^2B_2) \longrightarrow H^-(^1S) + OH(A\,^2\Sigma^+) \tag{39}$$

b. H_2S

In Fig. 14 are shown the dissociative attachment cross section data of Azria et al. (1972). The first peak (onset 1.45 eV; maximum 2.2 eV) is essentially due to HS^- ions with a small contribution of S^- ions. The second peak (onset 4.25 eV; maximum 5.35 eV) and the third peak (see Fig. 14) are essentially due to H^- ions with small contributions from S^- ions. The fourth peak is essentially all S^- ions. Interesting isotope effects have been observed in the

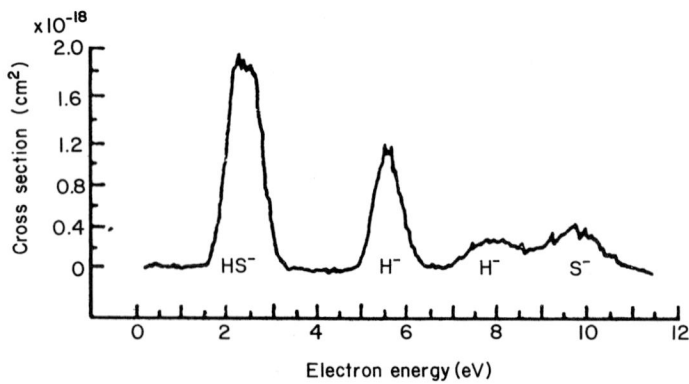

Fig. 14. Cross section for dissociative attachment anions for H_2S. [From Azria et al. (1972). Reproduced with permission from the *Journal of Chemical Physics*.]

cross sections of all ions (see Table II). Thus at 2.2 eV the cross section ratio $HS^-(H_2S)/DS^-(D_2S)$ was reported by Fiquet-Fayard et al. (1972) to be 25, at 5.35 eV the cross section ratio $H^-(H_2S)/D^-(D_2S)$ was reported by Tronc et al. (1973) to be 1.33, and at 9.5 eV the cross section ratio $S^-(H_2S)/S^-(D_2S)$ was reported by Azria et al. (1972) to be 2.7. (See discussion in Tronc et al., 1973; Fiquet-Fayard et al., 1972; and Section IV.B.)

More recently Azria et al. (1979) studied the energy and angular distributions of H^- produced via the two dissociative attachment processes in the energy range 4.5–9.0 eV. The kinetic-energy distributions for the two resonant processes indicated that both involve NISs of H_2S^- which are repulsive in the Franck–Condon region and which are correlated with the $H^-(^1S) + HS(^2\Pi)$ limit. Although in the range (around 7.5 eV) of the second resonance for H^-, the HS fragments are in their ground vibrational state, for the first process leading to H^- (4.8–6 eV) the HS fragments are weakly vibrationally excited and the fraction of H^- ions associated with the vibrationally ($v = 1, v = 2$) excited HS fragments increases with increasing energy; for both processes the HS fragments are rotationally excited. This behavior is analogous to that discussed for the case of H_2O.

An analysis of the observed angular distributions for H^- associated with $HS_{v=0}$ led Azria et al. (1979) to suggest that the first and second NISs of H_2S leading to H^- have, respectively, the symmetries $^2B_1$ and $^2A_1$.

c. CO_2

Dissociative attachment to CO_2 has been studied extensively with a variety of techniques and CO_2 has served, in many ways, as a prototype of linear triatomics with bent negative-ion configurations (Christophorou, 1971; Massey, 1976). The early work on $\sigma_{DA}(\varepsilon)$ for production of O^-/CO_2 has been summarized in Fig. 15. Two well-separated resonances exist with peaks at 4.4 and 8.2 eV having cross sections at their respective maxima of $\sim 1.4 \times 10^{-19}$ and $\sim 4.5 \times 10^{-19}$ cm^2. In addition, another well-defined O^-/CO_2 peak has been observed at 13.0 eV (Rapp and Briglia, 1965; Chantry, 1972). Theoretical calculations of the negative-ion states of CO_2 by Claydon et al. (1970) aid the understanding of these—and other (see below)—dissociative attachment processes in CO_2. Claydon et al. (1970) considered the energies of the CO_2^- states as a function of C—O separation assuming a bond angle of 180°, i.e., a linear CO_2^-. The results of their calculation are shown in Fig. 16 and allow examination of the onsets and peak positions of the observed dissociative attachment resonances in spite of the fact that the energy of a triatomic system depends on both the geometry and the atomic separation. This is because the initial capture transition proceeds according to the Franck–Condon principle, and since the ground-state neutral CO_2 molecule is linear such transition will occur at CO_2^- states possessing a nearly linear configuration.

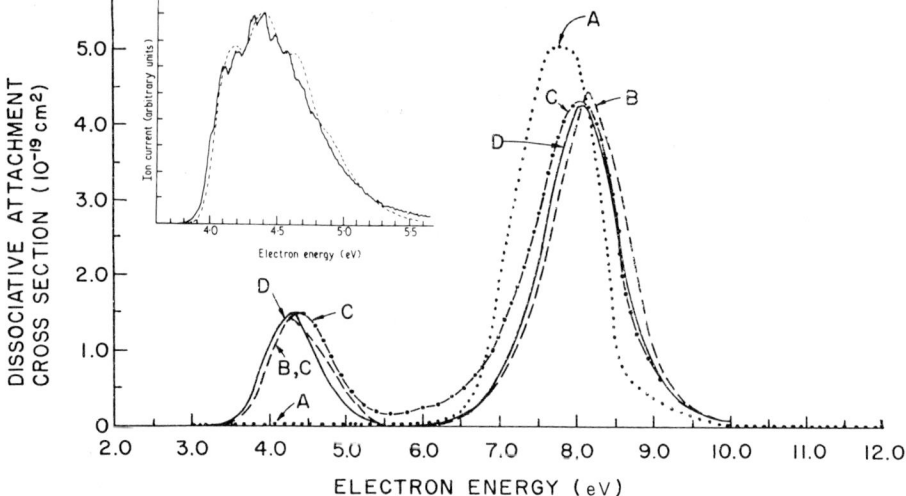

Fig. 15. Dissociative attachment cross section for the production of O^- from CO_2 as a function of incident electron energy (A—Craggs and Tozer, 1960; B—Schulz, 1962; C—Asundi et al., 1963; D—Rapp and Briglia, 1965). [From Christophorou (1971). Reproduced by permission of John Wiley & Sons, Ltd.] *Inset*: Relative cross section for the 4.4-eV O^-/CO_2 process showing distinct structure (see text). The two curves, (—) Abouaf et al. (1976), (---) Stamatovic and Schulz (1973), were normalized to the maximum. [From Abouaf et al. (1976). Reproduced from the *Journal of Physics* with permission of the Institute of Physics.]

i. The 4.4-eV Process. The O^-/CO_2 resonance at 4.4 eV has been the subject of extensive recent investigations and is generally ascribed to the process

$$CO_2(X\,^1\Sigma_g^+) + e \longrightarrow CO_2^{-*}(^2\Pi_u) \longrightarrow O^-(^2P) + CO(X\,^1\Sigma^+) \quad (40)$$

From measurements of the dependence of the O^-/CO_2 cross section on temperature, Stamatovic and Schulz (1973) (see also Spence and Schulz, 1969) established the onset for (40) at 3.97 ± 0.04 eV, which is consistent with the 3.988-eV value determined from accepted molecular constants. Similarly, Abouaf et al. (1976) obtained for this onset the value of 3.98 ± 0.05 eV. The low cross section for (40) and the rather sharp increases in σ_{DA} with ε close to the threshold led to the suggestion that this is a vertical-onset capture process. As can be seen from Fig. 16, the $^2\Pi_u$ state of CO_2 is attractive. Abouaf et al. (1976) found the onset to be quasivertical and believed that (40) may be considered a vertical-onset process.

The early measurements of the cross section for (40) (Fig. 15)—due principally to a limited incident beam energy resolution—showed no structure in the $\sigma_{DA}(\varepsilon)$ for this process. The improved energy resolution in the

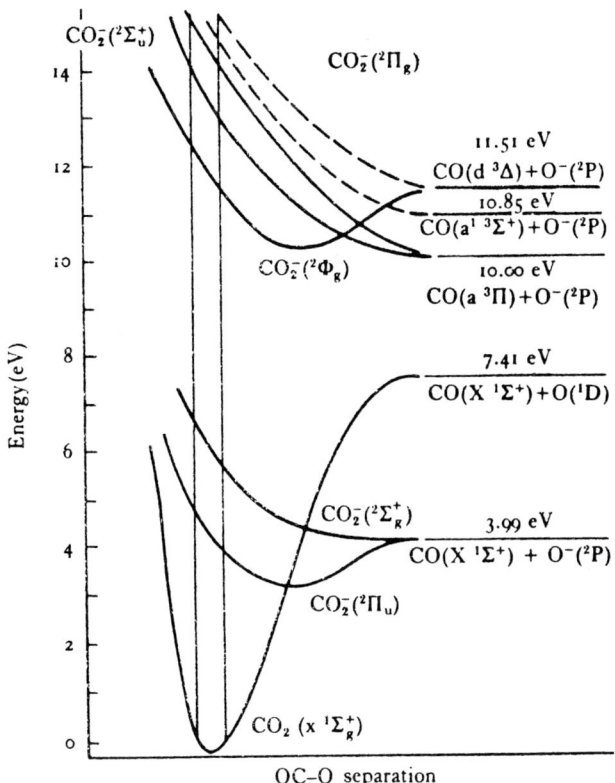

Fig. 16. Potential-energy curves for linear configurations of CO_2^- involved in dissociative attachment to the linear $X\,^1\Sigma_g^+$ ground state of CO_2. [From Claydon et al. (1970). Reproduced with permission from the *Journal of Chemical Physics*.]

experiments of Stamatovic and Schulz (1973) and Abouaf et al. (1976), however, showed the cross section for (40) to exhibit vibrational structure (see inset in Fig. 15). Stamatovic and Schulz (1973) attributed the structure to the formation of vibrationally excited CO fragments, the measured cross section being the sum of the partial dissociative attachment cross sections whose thresholds differ by a vibrational quantum of the CO fragment. On the other hand, Abouaf et al. (1976) pointed out that although their cross section data (Fig. 15, inset) agree in overall shape with those of Stamatovic and Schulz, they show more structure with spacings ($=0.12 \pm 0.02$ eV) which do not correspond to the vibrational levels of the CO fragment. The smaller spacings led Abouaf et al. to suggest that the observed structure may be connected with the capture rather than with the decay step of (40), i.e., with the vibrational structure of the CO_2^- above the $CO + O^-$ dissociation limit rather than with the CO fragment produced in the dissociative decay of CO_2^-. It would seem that both possibilities can contribute to the observed structure.

Measurements of the kinetic energy of O^- from CO_2 produced via (40) have been made by Chantry (1972) and are shown in Fig. 17a for four incident energies 3.8, 4.3, 4.8, and 5.3 eV. The distribution has a single peak close to zero energy and a high-energy tail whose extent increases with increasing ε. The most probable kinetic energy E_0 of O^- in reaction (40) is

$$E_0 = (1 - \beta)[\varepsilon - D(CO\text{---}O) + EA(O) - E_n^*], \qquad (41)$$

where $D(CO\text{---}O) = 5.45$ eV, $EA(O) = 1.465$ eV, E_n^* is the excitation energy of the products, and β is the ratio ($= 16/44$) of the mass of O^- to that of CO_2. With $E_n^* = 0$, Eq. (41) becomes

$$E_0 = 0.636(\varepsilon - 3.99)(\text{eV}). \qquad (42)$$

The fact that Eq. (42) predicts higher E_0 values than are experimentally observed (Fig. 17a) clearly shows that there is a high probability that most of the excess energy of reaction (40) is absorbed as vibrational/rotational

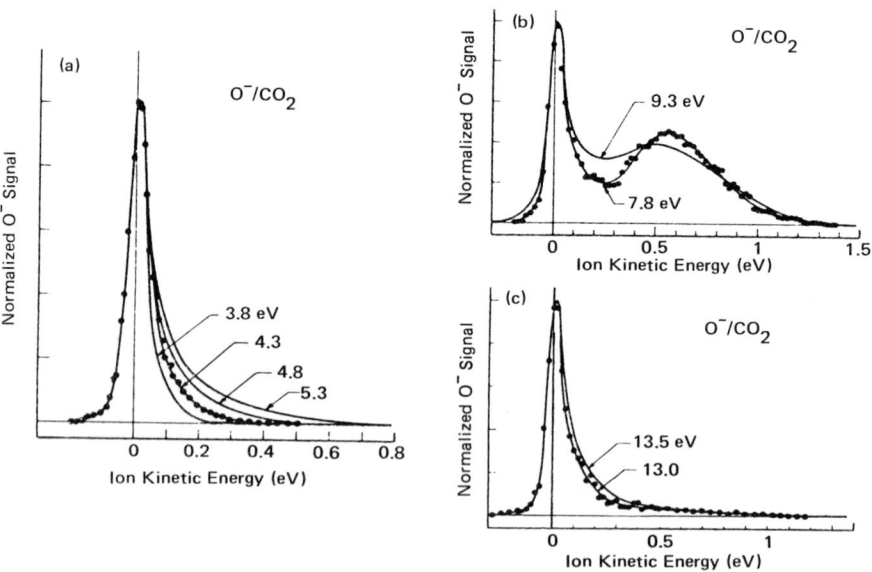

Fig. 17. Kinetic-energy distributions of O^-/CO_2 at the indicated incident electron energies typical of (a) the 4.4-eV process, (b) the 8.2-eV process, and (c) the 13.0-eV process. All curves were normalized to the same peak height. [From Chantry (1972). Reproduced with permission from the *Journal of Chemical Physics.*]

excitation of the CO fragment. Equation (42), moreover, predicts that the threshold for O^- ($E_0 = 0$) for (40) is 3.99 eV, which is higher than the observed onset of 3.85 eV. This reflects the dependence of the threshold energy on T (Section V); measurements of the onset value between 300 and 950 K by Spence and Schulz (1969) extrapolate nicely at 0 K to 3.97 eV.

ii. The 8.2-eV Process. Dissociative attachment at this energy has been ascribed to the reaction

$$CO_2(X\ ^1\Sigma_g^+) + e \longrightarrow CO_2^{-*}(^2\Sigma_g^+) \longrightarrow O^-(^2P) + CO(X\ ^1\Sigma^+) \quad (43)$$

The O^- kinetic-energy distributions for (43) are shown in Fig. 17b and indicate again that the CO fragment has appreciable internal energy. However, in addition to the sharp peak close to zero another broad maximum is observed at ~ 0.6 eV which was attributed by Chantry (1972) to the shape of the potential-energy surface of the negative-ion state.

iii The 13-eV Process. Claydon et al. (1970) ascribed the dissociative attachment process at this energy to

$$CO_2(X\ ^1\Sigma_g^+) + e \longrightarrow CO_2^{-*}(^2\Phi_g) \longrightarrow O^-(^2P) + CO^*(d\ ^3\Delta\ \text{or}\ a\ ^3\Pi) \quad (44)$$

Chantry (1972) found the peak cross section for (44) to be $\sim 6 \times 10^{-21}$ cm^2 at 300 K and $\sim 2 \times 10^{-20}$ cm^2 at 1000 K. The O^- kinetic-energy distributions for (44) are shown in Fig. 17c and, similarly to reactions (40) and (43), the CO fragment possesses significant amounts of internal energy.

In addition, Spence and Schulz (1974) used monoenergetic electron beams and mass spectrometry to study the production of O_2^- and C^- from CO_2 by dissociative attachment. They found that the $\sigma_{DA}(\varepsilon)$ for O_2^- was rather small ($\sim 10^{-24}$ cm^2) and showed two peaks at 11.3 $\pm$ 0.2 and 12.9 $\pm$ 0.2 eV, which they interpreted as a manifestation of the Renner–Teller effect.* To produce O_2^- by dissociative attachment to a linear CO_2 molecule, the CO_2^- must first bend and bring the two O atoms into close proximity; formation of O_2^- from CO_2^- requires excitation of a bending mode. Thus an electronic state which is degenerate in linear CO_2^- may couple with the bending mode, producing two nondegenerate vibronic states of CO_2^-. The cross section for C^- showed three peaks at 16.0 $\pm$ 0.2, 17.0 $\pm$ 0.2, and 18.7 $\pm$ 0.1 eV. The magnitude of σ_{DA} at 18.7 eV was estimated to be $\sim 2 \times 10^{-21}$ cm^2. It can be seen from Fig. 16 that a number of CO_2^- states exist between 15 and 20 eV.

---

* Linear triatomic molecules or ions which have a nonzero component of the orbital angular momentum along the molecular axis have degenerate electronic states which split into two when the molecule or ion is bent. The splitting results from the interaction of electronic and vibrational angular momenta. The removal of the degeneracy by vibronic interaction is known as the Renner–Teller effect (Herzberg and Teller, 1933; Renner, 1934).

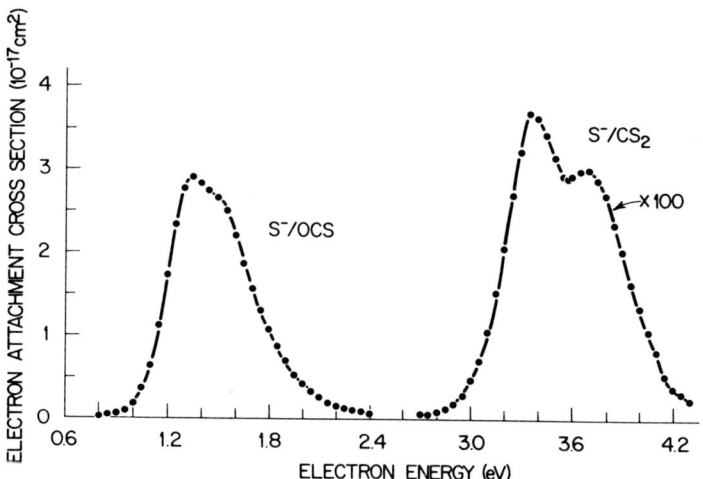

Fig. 18. Dissociative attachment cross section for the formation of S^- from OCS and CS_2 [data of Ziesel et al. (1975b); see text].

Finally, electron attachment coefficients for CO_2 have been reported by Alger and Rees (1976) in the E/N range between 60 and 152×10^{-17} V cm^2.

d. OCS; CS$_2$

Dillard and Franklin (1968) found that the principle negative ions produced by dissociative attachment to OCS (carbonyl sulfide) are S^- and O^- and those for CS_2 (carbon disulfide) are S^- and CS^-.[†] Hubin-Franskin et al. (1976) observed the onset for S^-/OCS at 1.05 eV and the peak at 1.4 eV; the onset for O^-/OCS was found at 3.9 eV and its cross section showed well-defined maxima at 4.2, 9.6, and 11.0 eV and weaker structure at 7.2 and 8.1 eV.

The cross section $\sigma_{DA}(\varepsilon)$ for S^- produced by dissociative attachment to OCS and CS_2 has been investigated by Ziesel et al. (1975b) using monoenergetic electron beams. In Fig. 18 are shown the $\sigma_{DA}(\varepsilon)$ for S^- from OCS and CS_2. The relative $\sigma_{DA}(\varepsilon)$ of Ziesel et al. were put on an absolute scale using the value 2.9×10^{-17} cm^2 they reported for the peak of the S^-/OCS cross section at 1.35 eV and the value 3.7×10^{-19} cm^2 they reported for the peak of the S^-/CS$_2$ cross section at 3.35 eV.

The threshold for S^-/OCS was found by Ziesel et al. to be 1.07 eV (i.e., close to the dissociation limit), suggesting that the potential-energy surface of OCS^- is attractive in a part of the Franck–Condon region. The threshold for

---

[†] They also observed CS_2^- presumably produced by charge transfer.

S^-/CS_2 was at 3.09 eV, which is ~0.5 eV higher than the dissociation limit. The delayed onset for S^-/CS_2 and the rather steep threshold of the cross section led Ziesel et al. to suggest that the second maximum in the $\sigma_{DA}(\varepsilon)$ for S^-/CS_2 at 3.65 eV may be due to the presence of a hump in the CS_2^- potential-energy surface. The peak in the $\sigma_{DA}(\varepsilon)$ for S^-/OCS was explained by Ziesel et al. as arising from the presence of CO fragments in a vibrationally ($v = 1$) excited state.

In analogy to CO_2 the 3.35-eV dissociative attachment process for CS_2 was viewed by Ziesel et al. as proceeding via the $^2\Pi_u$ state of CS_2^- with the dissociation limit $S^-(^2P) + CS(X\,^1\Sigma_g^+)$.

e. N_2O

The neutral N_2O molecule, as the neutral CO_2 molecule, is linear, while N_2O^-, as CO_2^-, is bent; this difference between the geometries of the neutral- and the negative-ion species influences greatly the rates of dissociative attachment. Christophorou (1971) and Massey (1976) reviewed the processes and cross sections for both dissociative and nondissociative electron attachment to N_2O. Briefly, below ~4 eV two states of N_2O^- are involved which dissociate into $O^- + N_2$. The $\sigma_{DA}(\varepsilon)$ due to the first state stretches from 0 to <2 eV and is strongly dependent on temperature, whereas the $\sigma_{DA}(\varepsilon)$ due to the second state peaks at 2.25 eV and is independent of T (see Fig. 35a in Section V).

The strong T-dependent dissociative attachment process involves the lowest (ground) state of N_2O^- and is due to capture of thermal and near-thermal energy electrons by vibrationally/rotationally excited N_2O^* molecules. According to Chantry (1969) this strong dependence of the cross section on temperature is principally due to the excitation of the bending mode of vibration, and arises from the dependence on bond angle of the separation energy of the electronic ground states of N_2O and N_2O^-. Bardsley (1969) suggested a $^2\Pi$ symmetry for this state, which splits into A and B states in the bent configuration.

The temperature-independent process peaking at 2.25 eV is ascribed to dissociative attachment via the second N_2O^- state. Bardsley (1969) suggested that this state is linear with $^2\Sigma^+$ symmetry. The kinetic energy of O^- from this state was found by Chantry (1969) to be 0.38 eV and independent of the incident electron energy in the range 1.5–2.5 eV. This result has been confirmed by Tronc et al. (1977b), who found that O^- has an energy distribution extending from 0 to 1 eV in the form of a broad peak with a maximum at 0.38 eV. Tronc et al. also studied the angular distribution of O^- produced by dissociative attachment to N_2O. They found that for incident electrons of 1.9, 2.25, and 2.9 eV energy and for scattering angles from 30° to

130°, the angular distribution of O^- does not depend on incident energy and is almost symmetrical with respect to 90°. Analysis of these data showed that the contribution of the s wave to the process is important.

For a discussion of other attachment processes—possibly production N_2O^-—at thermal energies, see Christophorou (1971), Caledonia (1975), and Massey (1976).

f. NO_2

The most recent study of dissociative attachment to NO_2 is that of Abouaf et al. (1976). Three fragment negative ions were observed: O^-, O_2^-, and NO^-, in decreasing order of abundance. The authors found the onsets for these ions to be, respectively, 1.61, 4.03, and 3.11 eV and consistent with the thermochemical onsets for the processes

$$e^- + N_2O \longrightarrow O^-(^2P) + NO(^2\Pi) \quad \text{(1.650-eV limit)} \tag{45}$$

$$\longrightarrow O_2^-(^2\Pi_g) + N(^4S) \quad \text{(4.065-eV limit)} \tag{46}$$

$$\longrightarrow NO^-(X\,^3\Sigma^-) + O(^3P) \quad \text{(3.091-eV limit)} \tag{47}$$

Earlier studies (e.g., Stockdale et al., 1969) using the retarding potential difference technique showed onsets lower than the thermochemical values, possibly owing to inferior electron energy resolution. The O^- ion peaked at 1.8 and 3.5 eV, the O_2^- ion at 4.3 eV, and the NO^- ion at 3.2 eV.

g. SO_2

Mass spectrometric studies of dissociative attachment to SO_2 identified O^- as the most abundant fragment anion with peaks at ~ 5 and 8 eV (Kraus, 1961; Dorman, 1966; Harland and Thynne, 1970; Rallis and Goodings, 1971; Lifshitz et al., 1973). In addition to O^-, Kraus (1961) observed SO^- and Dorman (1966) reported observation of O^-, SO^-, and S^- with intensities decreasing in that order and peaks at ~ 5 and 8 eV for all three ions. Two more recent electron beam mass spectrometric studies of SO_2 have been made (Čadež et al., 1983; Orient and Srivastava, 1983) where all dissociative attachment anions (O^-, S^-, SO^-) have been observed. Čadež et al. (1983) found that the total electron attachment cross section has maxima at 4.7 and 7.4 eV with peak cross-section values, respectively, equal to 6.21×10^{-18} cm^2 and 2.53×10^{-18} cm^2. On the other hand, Orient and Srivastava (1983) used a cross target SO_2 molecule-electron beam geometry, and a relative flow technique (Srivastava et al., 1975) to determine the absolute cross sections, and reported cross-section values for the production of O^-, S^-, and SO^- from SO_2 with an estimated error of $\pm 20\%$. The peak energies and peak cross-section values reported by Orient and Srivastava (1983) are: O^-/SO_2 [4.30 eV

(8.08×10^{-18} cm^2), 7.1 eV (2.68×10^{-18} cm^2)]; S$^-$/SO$_2$ [4.0 eV (0.313×10^{-18} cm^2), 7.5 eV (0.036×10^{-18} cm^2), 8.9 eV (0.030×10^{-18} cm^2)]; SO$^-$/SO$_2$ [4.7 eV (10.98×10^{-18} cm^2), 7.5 eV (0.51×10^{-18} cm^2]. The total dissociative attachment cross section determined from these partial cross sections has maxima at 4.5 and 7.1 eV which are not too dissimilar from those of Čadež et al. (1983). However, the total cross-section value at the first maximum ($\sim 17.6 \times 10^{-18}$ cm^2) is 2.8 times larger than that reported by Čadež et al. (see also Fig. 19). Also, while the work of Čadež et al. shows a ratio of the peak cross-section values at 4.5 and 7.1 eV equal to ~ 2.5, the work of Orient and Srivastava gives a ratio of ~ 5.7.

Two swarm studies revealed SO$_2^-$ formation at low energies ($\lesssim 0.5$ eV) and dissociative attachment at higher energies (>2 eV). Thus Rademacher et al. (1975) studied SO$_2$ in high-pressure buffer gases Ar, N$_2$, and C$_2$H$_4$. In Fig. 19 (inset) are shown the rate constants for total electron attachment to SO$_2$ in Ar. The mean energy scale was established from the known $f(\varepsilon, E/N)$ in Ar (see Table I). In the mean energy range >2 eV the attachment is due to dissociative

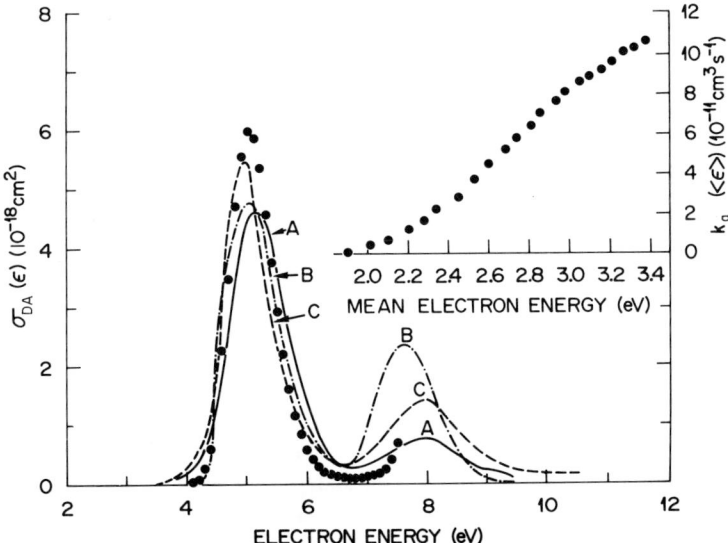

Fig. 19. Electron attachment cross sections as a function of electron energy for SO$_2$ obtained by Rademacher et al. (1975) from swarm-beam and swarm unfolding analyses (see text). (A) Cross sections obtained by a swarm-beam analysis using the $k_a(\langle\varepsilon\rangle)$ shown in the inset and the relative cross sections for O$^-$ + SO$^-$ from SO$_2$ of Kraus (1961). (B) and (C) Similar cross sections obtained using, respectively, the relative cross sections for O$^-$ from SO$_2$ of Harland and Thynne (1970) and Lifshitz et al. (1973). The points (●) are the swarm-unfolded cross section data using the $k_a(\langle\varepsilon\rangle)$ in the inset and the $f(\varepsilon, E/N)$ for Ar.

processes. Rademacher et al. applied the swarm–beam technique (Christophorou et al., 1965) using their $k_a(\langle\varepsilon\rangle)$ data and the shape of the negative-ion yields as reported by various authors (see Fig. 19 caption) and obtained the total dissociative attachment cross sections shown in Fig. 19. They also analyzed their $k_a(\langle\varepsilon\rangle)$ data using the swarm-unfolding technique (Christophorou et al., 1971c); the resultant total cross sections are also shown in Fig. 19. The total dissociative attachment cross section shows a maximum at 5 eV (principally due to O^-) and a secondary maximum at ~ 8 eV. At 5 eV the cross section is $\sim 5.5 \times 10^{-18}$ cm^2.

Lakdawala and Moruzzi (1981) studied electron attachment to SO_2 in a pulsed drift tube experiment and reported electron attachment coefficients for E/N values from 3.03 to 242.4×10^{-17} V cm^2. In accord with the work of Rademacher et al. (1975), they found the attachment coefficients below $\sim 40 \times 10^{-17}$ V cm^2 to be pressure dependent (owing to SO_2^- formation) and above this value to be pressure independent (owing to the pressure-independent dissociative attachment processes).

h. Other Triatomics

Mothes et al. (1972) reported for the thermal ($T \simeq 300$ K) attachment rate constant $(k_a)_{th}$ of HCN, ClCN, BrCN, and ICN the values of 9.1×10^{-11}, 1.1×10^{-10}, 1.8×10^{-7}, and 2.1×10^{-9} cm^3 s^{-1}, respectively. Similarly, Wecker et al. (1981) reported for the $(k_a)_{th}$ of ClO_2 and Cl_2O, respectively, the values (1.5 ± 0.5) and $(2.6 \pm 0.5) \times 10^{-10}$ cm^3 s^{-1}. The $(k_a)_{th}$ of NOCl was reported by Schultes et al. (1975) to be 6.9×10^{-10} cm^3 s^{-1}. Finally, Wiegand and Boedeker (1982) found that the cross section for Br^- formation from $HgBr_2$ (a compound of interest for mercury halide lasers) has a threshold at 3.1 eV and a peak value of 1×10^{-17} cm^2 at ~ 3.6 eV. These authors also reported data on k_a vs. E/N for $HgBr_2$/Xe mixtures.

3. Complex Polyatomic Molecules

In the case of polyatomic molecules, many types of fragment negative ions can be produced and dissociative attachment processes may not be fast. The dissociation time would depend on the distribution of the internal energy in the transient intermediate, the structural rearrangements following electron capture, as well as the orbital in which the electron is captured. The modes of decomposition and the nature of the fragment negative ions—and the internal energy of their excitation—depend rather strongly on the details of molecular structure. Four types of dissociative attachment fragment negative ions may be identified in polyatomic molecules (Christophorou, 1980a): directly cleaved (I), complementary (II), multiple (III), and rearrangement (IV). These can be stable or unstable, subject to autoionization (V) and/or autodissociation

(VI) [see Eq. (6)]. All six types of fragment negative ions seem to be produced, for example, when slow electrons collide with 2-C_4F_8 (perfluorobutene-2) (Fig. 20): directly cleaved (e.g., F^-), complementary (e.g., F^-, $C_4F_7^-$), rearrangement (e.g., $C_3F_5^-$), multiple (e.g., F^- and $C_2F_3^-$ at ~ 5 eV, $C_4F_7^{-*}$ and $C_4F_6^{-*}$ at ~ 0.0 eV), metastable autodetaching ($C_4F_7^{-*}$, $C_4F_6^{-*}$, and $C_3F_5^{-*}$), and, perhaps, also metastable autodissociating ($C_3F_5^{-*}$). The autodetachment lifetimes of the fragments $C_4F_7^{-*}$, $C_4F_6^{-*}$ at ~ 0.0 eV and $C_3F_5^{-*}$ at ~ 2.3 eV were found to be 7, 17, and 70 μs, respectively; that of the parent ion $C_4F_8^{-*}$ was found to be 10 μs at ~ 0.0 eV (Sauers et al., 1979; Christophorou, 1980a).

The multiplicity of negative-ion states and the production of a variety of negative ion fragments with large cross sections clearly show the fragility of metastable polyatomic negative ions and the effectiveness of slow (including thermal) electrons to cause extensive and often multiple molecular decompositions tantamount to what Christophorou (1980a) termed a molecular "explosion." This and the delicate dependence of the type and abundance of negative ions on the details of molecular structure can, perhaps, be illustrated by the work of Johnson et al. (1977) on the negative ions produced in collisions of slow ($\lesssim 3$ eV) electrons with chloroethanes and chloroethylenes. For these chlorocarbons, the transient anions M^{-*} fragment (even when they are formed by the capture of electrons with virtually zero energy) through a

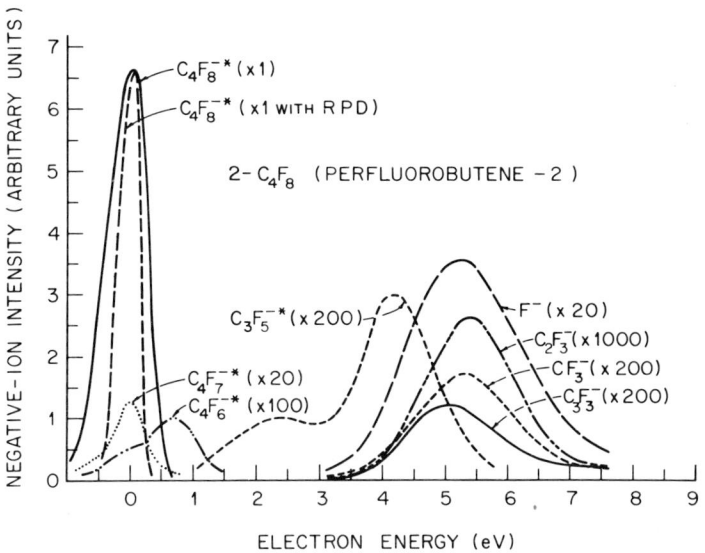

Fig. 20. Negative ions produced by low-energy electron impact on perfluorobutene-2 (2-C_4F_8). [From Sauers et al. (1979). Reproduced with permission from the Journal of Chemical Physics.]

multiplicity of channels. Depending on M^{-*}, all or some of the decomposition channels shown below were found (Johnson et al., 1977) to be possible:

$$e + M \rightarrow M^{-*} \begin{cases} M^{(*)} + e^{(')} \\ (M - Cl) + Cl^- \\ Cl + (M - Cl)^- \\ (M - 2Cl) + Cl_2^- \\ M^{-*}(\text{long-lived}, > 10^{-6} \text{ s}) \end{cases}$$

The most abundant fragment anion for these and similar chlorocarbons in electron-beam studies is Cl^- (e.g., see Johnson et al., 1977; Illenberger et al., 1978, 1979; Verhaart et al., 1978, 1980). The relative cross section for Cl^- production of Johnson et al. (1977) shown in Fig. 21 indicates the existence of five maxima below ~ 2.0 eV for the chloroethylenes (and also for the chloroethanes). It seems that these are not due to vibrational excitation, but

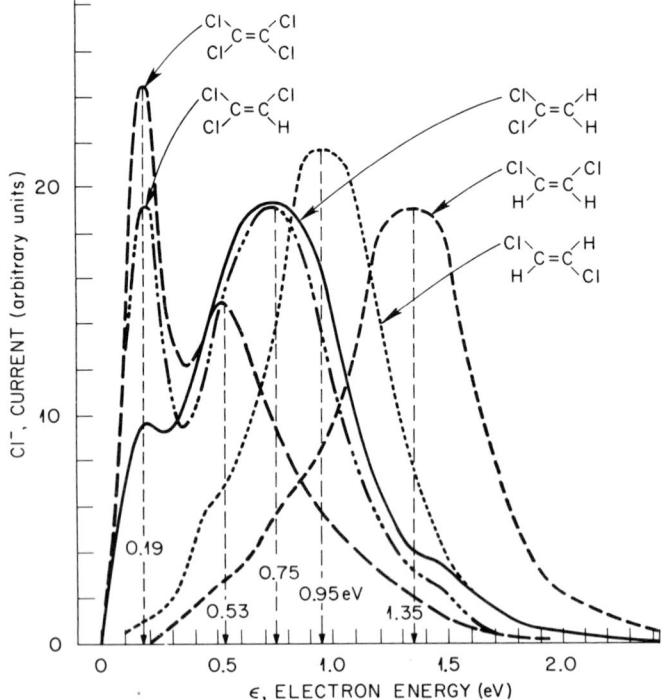

Fig. 21. Comparison of the relative cross section for the production of Cl^- for various chloroethylenes. [From Johnson et al. (1977). Reproduced with permission from the *Journal of Chemical Physics.*]

rather due to separate negative-ion states associated (Johnson et al., 1977) with orbitals dominated by the p orbitals of the Cl atom. Their positions are relatively insensitive, and their cross sections very sensitive to the details of the molecular structure. Although for the molecules in Fig. 21 dissociative attachment processes are expected to be fast ($\ll 10^{-6}$ s), for $C_2Cl_4^{-*}$ at ~ 0.0 eV (and for a number of other autodissociating long-lived parent negative ions) both the autodissociation and the autodetachment processes are slow, as a result of vibrational redistribution of internal energy (Christophorou, 1978a; Section VII).

The decomposition mechanisms responsible for specific fragment negative ions in complex molecules and the amount of translational and internal energy in the fragments depends on the decaying negative-ion state and the amount and distribution of the internal energy in the excited negative ion. In spite of recent work in this area, much knowledge is still needed concerning the patterns of multiple fragmentation of molecular negative-ion states and their relation to molecular structure. Of particular interest in this connection are the dissociative attachment studies of Franklin and co-workers (Franklin et al., 1967; DeCorpo and Franklin, 1971; DeCorpo et al., 1971; Harland et al., 1973; Wang et al., 1974a,b; Harland and Franklin, 1974; Bennett et al., 1974; Carter, 1976; Pabst et al., 1976, 1977a,b; Franklin, 1976) of a large number of complex molecules employing time-of-flight mass spectrometry. From measurements of the translational energy of dissociative attachment negative ions Franklin and co-workers have shown that for a transient polyatomic molecular anion with N vibrational degrees of freedom only a fraction α is effective in absorbing the excess energy immediately prior to dissociation, so that the average total translational energy $\bar{\varepsilon}_t$ (in the center of mass) of the fragments is

$$\bar{\varepsilon}_t = E_{v,t}^*/\alpha N. \tag{48}$$

In Eq. (48), $E_{v,t}^*$ is the total excess energy in the negative-ion intermediate, and αN may be assumed to be the number of active vibrational modes. Franklin and co-workers reported values of α ranging from 0.35 to 0.57 with an average of ~ 0.40. This was taken to indicate that the effective number of vibrational degrees of freedom in the metastable intermediate is less than the available number N. Although this may in part be attributed to a dissociation time which is much shorter than the time for complete internal energy equilibration, it is also apparent that the effective number of vibrational degrees of freedom—and hence the value of α—must depend on the spatial distribution of the wave function describing the orbit occupied by the captured electron (Christophorou, 1980a). Franklin's findings on α values less than one are in accord with independent work on the decomposition of transient negative ions not by autodissociation but by autodetachment

(Christophorou, 1978a; Christophorou et al., 1977b). In the latter studies basically the same model for statistical distribution of $E_{v,t}^*$ was used, and although the transient negative ion considered ($C_6F_6^{-*}$) lives for ~ 12 μs (Christophorou, 1978a), the effective number of degrees of freedom was found to be about a third of the available N. In this connection, a Monte Carlo sampling method calculation by Goursaud et al. (1978) showed that the partitioning of excess energy of dissociation in triatomic negative ions favors the kinetic energy if the potential-energy surface of the NIS is repulsive and favors the internal energy if it is attractive.

a. Halocarbons

Studies of the dissociative attachment patterns, identification of the product ions, and their yields for halocarbons have received considerable attention in recent years partly as a result of the presence of these compounds in the environment—and hence the need to follow their reaction pathways in the atmosphere—and partly as a result of their biological significance and potential use (by themselves or in gas mixtures) for gaseous dielectrics purposes (see Chapter 5 of Volume 2). Electron attachment data on freons prior to 1970 have been summarized, among others, by Christophorou (1971). Although many electron beam–mass spectrometric studies identifying the decomposition products and their relative abundances have been made since [e.g., see references to the work of Franklin and co-workers earlier in this section; Johnson et al. (1977); Illenberger et al. (1978, 1979); Verhaart et al. (1978, 1980); Table X later in this section], only limited measurements have been made of the absolute electron attachment rate constants and cross sections as a function of electron energy.

In Fig. 22a the total attachment rate constants $k_a(\langle\varepsilon\rangle)$ of Christodoulides and Christophorou (1971) and in Fig. 22b the $\sigma_{DA}(\varepsilon)$ for the production of Br$^-$ of Christophorou et al. (1971a) are shown for brominated alkanes of the form n-$C_NH_{2N+1}Br$. Since mass spectrometric studies (Christophorou et al., 1971a) have shown that the dominant negative ion for these halocarbons below 2 eV is Br$^-$, the $k_a(\langle\varepsilon\rangle)$ in Fig. 22a are due to the production of this ion.

In Fig. 23a the data of McCorkle et al. (1980, 1982a) on $k_a(\langle\varepsilon\rangle)$ are presented for CCl_4, CCl_3F, CCl_2F_2, and $CClF_3$. The $k_a(\langle\varepsilon\rangle)$ functions for the last three molecules were used by McCorkle et al. (1980) to obtain—through the swarm-unfolding technique (Section III.B)—the total attachment cross sections as a function of electron energy shown in Fig. 23b. The total attachment cross section of Pejčev et al. (1979) for CCl_2F_2 measured in a beam experiment is also shown in Fig. 23b. Electron-beam studies (Johnson et al., 1977; Illenberger et al., 1978, 1979; Verhaart et al., 1978, 1980) have revealed that although a multiplicity of dissociative attachment fragment anions are

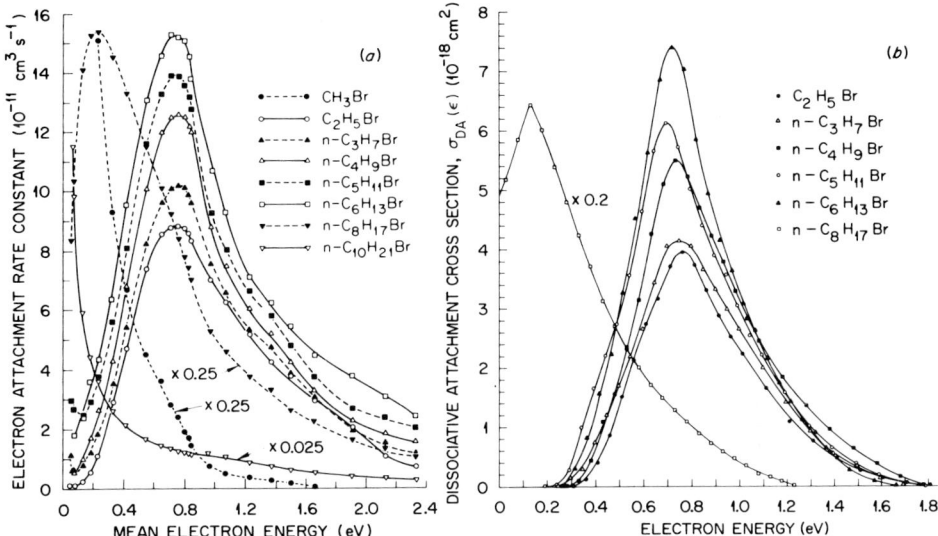

Fig. 22. (a) $k_a(\langle\varepsilon\rangle)$ (Christodoulides and Christophorou, 1971) and (b) $\sigma_a(\varepsilon)$ for Br^- formation (Christophorou et al., 1971a) for brominated hydrocarbons of the form $n\text{-}C_NH_{2N+1}Br$.

produced, the major ion is Cl^-. The total attachment rates and cross sections, therefore, are principally due to the production of Cl^-. Similar data are shown in Fig. 24 for chlorohydro- and chlorofluoroethanes.

The data in Fig. 23 show that the magnitude of $\sigma_a(\varepsilon)$ increases with increasing number of Cl atoms in the molecule, becoming very large for multiply chlorinated hydrocarbons (for CCl_4, CCl_3F, and $1,1,1\text{-}C_2Cl_3F_3$, for example, the k_a are near the theoretical maximum s-wave capture), demonstrating the extreme fragility of halocarbons toward slow electrons.* These and the beam data also indicate the existence of a large number of NISs at subexcitation energies via which such decompositions proceed and demonstrate the extreme and delicate dependence of the positions, the cross sections, and the decomposition products of the NISs on the details of molecule structure. As a rule the magnitude of $\sigma_a(\varepsilon)$ increases and the positions of the maxima in $\sigma_a(\varepsilon)$ shift toward lower energy with increasing Cl substitution. The effect of F substitution for H, as well as the effect of the number and relative positions of the Cl atoms in the molecule on the energies and corresponding cross sections (or rate constants) of the NISs is also apparent from the data in

---

* This molecular decomposition needs to be considered when such compounds are used as dielectrics in view of its possible effect on the long-range stability of the dielectric.

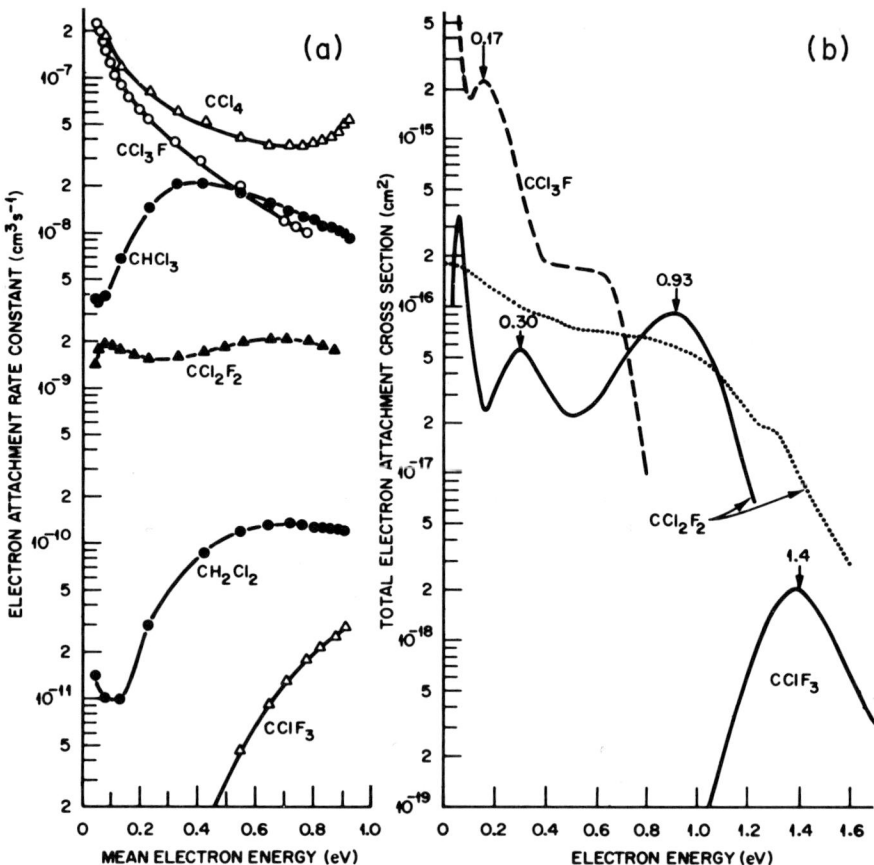

Fig. 23. (a) Electron attachment rate constant vs. mean electron energy: CCl_4, $CHCl_3$, CH_2Cl_2 (Blaunstein and Christophorou, 1968; Christodoulides and Christophorou, 1971); CCl_3F (McCorkle et al., 1982a); CCl_2F_2, $CClF_3$ (McCorkle et al., 1980). (b) Total electron attachment cross section vs. electron energy: CCl_3F (McCorkle et al., 1982a), CCl_2F_2, $CClF_3$ (McCorkle et al., 1980); (···) CCl_2F_2 (Pejčev et al., 1979).

Figs. 23 and 24. For a given number of Cl atoms in the molecule, replacement of H by F lowers the position of the resonance maxima and increases the magnitude of k_a [see McCorkle et al. (1982b); Christophorou (1981), and also Section VII]. Interestingly, the 1,1,1 or 1,1 isomers (Fig. 24) capture slow electrons much more efficiently than the 1,1,2 or 1,2 isomers, suggesting that Cl^- is more easily produced when all Cl atoms are associated with the same C atom—and the position of the resonance maxima lie at lower energies for the former compared to the latter.

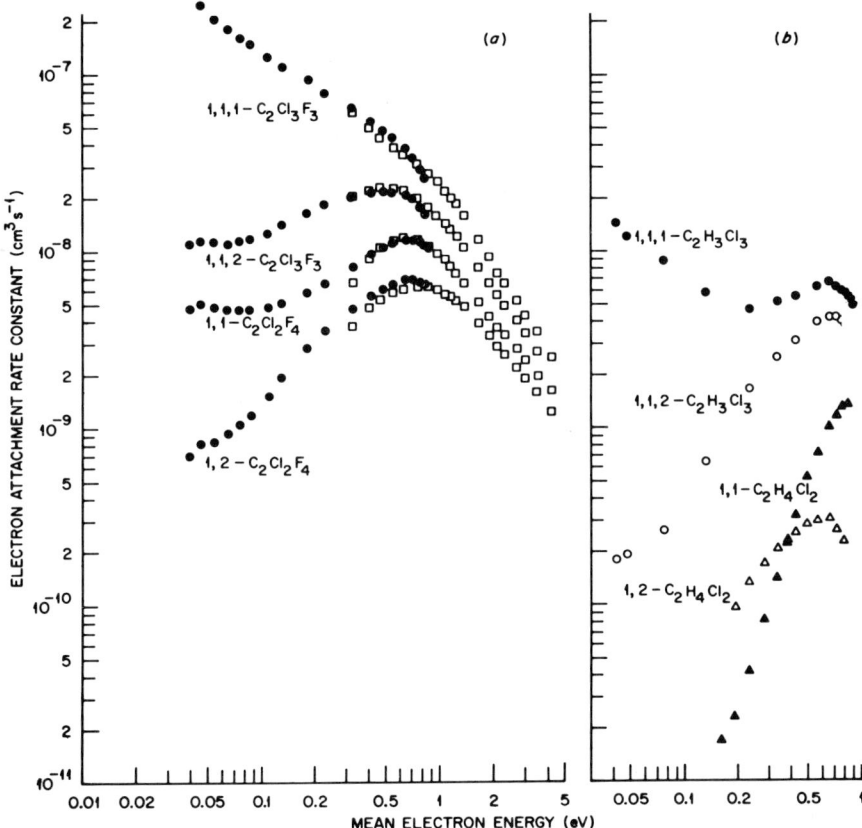

Fig. 24. Total electron attachment rate constant as a function of mean electron energy for chlorofluoroethanes (a) and chlorohydroethanes (b) (Christophorou *et al.*, 1981a; McCorkle *et al.*, 1982b).

Thermal (~ 300 K) attachment rate constants for halocarbons are given in Table VI. The $(k_a)_{th}$ increases in the order $F < Cl < Br < I$ and with the number of the halogens in the molecule.

b. Perfluorocarbons

This group of molecules has been extensively studied recently in view of their basic significance in understanding the effects of molecular structure on the NISs of polyatomic molecules [e.g., see Chapter 5, Frazier *et al.* (1978); Christophorou (1980a, 1981); Chiu *et al.* (1979)] and also in view of the potential use of certain perfluorocarbon aliphatics (in the pure form or in gas mixtures) as gaseous dielectrics [e.g., Christophorou (1980b, 1982); Chapter 5

TABLE VI

Thermal ($\sim$300 K) Electron Attachment Rate Constants for Halocarbons

Halocarbon	Rate constant (cm^3 s^{-1})
CH_3Cl	$<1.9 \times 10^{-15}$,[a] $<5 \times 10^{-15}$,[b] $<10^{-13}$ [c]
CH_3Br	3.6×10^{-12},[d] 7×10^{-12},[a] 7×10^{-9} [e]
CH_3I	7×10^{-8} [f]
CH_2Cl_2	$(6.5 \pm 1.1) \times 10^{-13}$,[g] 4.6×10^{-12},[h] 4.7×10^{-12},[i] 4.8×10^{-12} [c]
$CHCl_3$	1.3×10^{-9},[h] 2.2×10^{-9},[j] $(2.3 \pm 0.2) \times 10^{-9}$,[g] 2.4×10^{-9},[c] 3.9×10^{-9} [k]
CCl_4	2.5×10^{-7},[l] 2.8×10^{-7},[e,f,k] 3.6×10^{-7},[m] 4.1×10^{-7},[d,j,n] $(4.4 \pm 0.5) \times 10^{-7}$ [g]
CF_4	$<10^{-16}$,[i] $\leq 10^{-16}$,[o] $<3.1 \times 10^{-13}$ [m]
CF_3Cl	5.2×10^{-14},[i] 7×10^{-14},[o] $<3.1 \times 10^{-13}$,[m] 2×10^{-13} [p]
CF_2Cl_2	7×10^{-10},[o] 8.3×10^{-10},[d] 1.2×10^{-9},[p] 1.3×10^{-9},[q] 1.9×10^{-9},[f] 2.2×10^{-9} [r]
$CFCl_3$	1×10^{-7},[o] 1.2×10^{-7},[f] 2.4×10^{-7},[p] 2.37×10^{-7} [w]
CHF_3	3.6×10^{-14},[s] 4.6×10^{-14},[i] $<6.2 \times 10^{-14}$ [m]
CHF_2Cl	$<1.6 \times 10^{-13}$,[m] $<3.3 \times 10^{-13}$ [s]
$CHFCl_2$	1.5×10^{-12} [s]
1,1-$C_2H_4Cl_2$	2.1×10^{-11} [t]
1,2-$C_2H_4Cl_2$	3.2×10^{-11} [t]
1,1,1-$C_2H_3Cl_3$	1.5×10^{-8},[t,u] 1.6×10^{-8} [k]
1,1,2-$C_2H_3Cl_3$	1.5×10^{-10},[k] 1.8×10^{-10} [t,u]
1,1-$C_2F_4Cl_2$	4.8×10^{-9} [t,u]
1,2-$C_2F_4Cl_2$	7×10^{-10} [t,u]
1,1,1-$C_2F_3Cl_3$	2.8×10^{-7} [t,u]
1,1,2-$C_2F_3Cl_3$	1.1×10^{-8} [t,u]
C_2F_3Cl	6.6×10^{-12} [v]

[a] Bansal and Fessenden (1972).
[b] Christodoulides et al. (1975a).
[c] Schultes et al. (1975).
[d] Mothes et al. (1972).
[e] Christodoulides and Christophorou (1971).
[f] Christophorou (1976).
[g] Ayala et al. (1981b).
[h] Christodoulides et al. (1975b).
[i] Fessenden and Bansal (1970).
[j] Warman and Sauer (1971).
[k] Blaunstein and Christophorou (1968).
[l] Bouby et al. (1965).
[m] Davis et al. (1973).
[n] Mothes and Schindler (1971).
[o] Schumacher et al. (1978).
[p] McCorkle et al. (1980, 1982a).
[q] Bansal and Fessenden (1973).
[r] Christophorou et al. (1974).
[s] Christodoulides et al. (1978).
[t] Christophorou et al. (1981b).
[u] McCorkle et al. (1982b).
[v] Christodoulides et al. (1976).
[w] Crompton et al. (1982.)

TABLE VII

Effect of Perfluorination on the Energies and Lifetimes of the NISs and the $(k_a)_{th}$ of Hydrocarbons

Molecule	Formula	Position of lowest observed NIS (eV)[a]	EA (eV)	τ_a (μs)	$(k_a)_{th}$ (cm^3 s^{-1})
1,3-butadiene	H$_2$C=CH—CH=CH$_2$	-0.62[b]		$\ll 1$[c]	Very small
1,3-perfluorobutadiene	F$_2$C=CF—CF=CF$_2$	d	>0[e]	7[f]	$(1.26, 1.32) \times 10^{-7}$[f]
Cyclohexene	c-C$_6$H$_{10}$	-2.07[b]		$\ll 1$[c]	Very small
Perfluorocyclohexene	c-C$_6$F$_{10}$	d	$\geq 1.4 \pm 0.3$[g]	106, 113[f]	$(3.13, 3.89) \times 10^{-7}$[f]
Benzene	C$_6$H$_6$	-1.35[h]	$<0(?)$[g,i]	$<10^{-12}$[i]	$\geq 10^{-12}$[i]
Perfluorobenzene	C$_6$F$_6$	-0.42[h,j]	$\geq 1.8 \pm 0.3$[g]	12, 13.3[f]	$(1.02, 1.06) \times 10^{-7}$[f]

[a] Vertical position of the lowest observed NIS.
[b] Burrow and Jordan (1975).
[c] Most likely $<10^{-14}$ s on the basis of electron scattering experiments.
[d] The $\sigma_a(\varepsilon)$ shows at least three maxima below ~ 1 eV and is the largest at ~ 0.0 eV (Christodoulides et al., 1979; Pai et al., 1979).
[e] Based on the EA values of similar compounds (Chapter 6 of Volume 2).
[f] From Table VIII.
[g] Table II (Chapter 6 of Volume 2).
[h] Lowest π-NIS (Frazier et al., 1978).
[i] See discussion in Christophorou and Goans (1974) and Christophorou (1976).
[j] This is the lowest π-NIS and not the lowest negative-ion state of C$_6$F$_6$, which is a σ rather than a π state (see Frazier et al., 1978), as is indicated by the large positive EA, large τ_a, and large $\sigma_a(\varepsilon)$.

of Volume 2]. As was discussed in the preceding paragraph, the effect of substitution of an H atom by a halogen atom in such molecules is in general a lowering of the energy of the lowest NIS(s). This is dramatized in the case of total replacement of the H atoms in the molecule by fluorine atoms in both the aliphatic and the aromatic hydrocarbons. In Table VII are listed three examples—similar behavior is exhibited by other aliphatic hydrocarbons with four or more carbon atoms and aromatics (Christophorou, 1978a, 1980a, 1981)—where perfluorination lowers the lowest NIS such that the EA of the molecule becomes positive. This has a profound effect on the magnitude of the attachment cross section and the autodetachment lifetime of the parent anions. Thus, although the hydrogenated forms do not attach low-energy electrons, the perfluorinated molecules have very large electron attachment cross sections and the lifetimes of the parent anions are $> 10^{-6}$ s, i.e., orders of magnitude longer than those ($< 10^{-12}$ s) of the NISs of the nonfluorinated analogs. The cross sections for parent negative-ion formation are the largest at thermal energies (close to the maximum s-wave capture cross section), as can be seen from the examples in Fig. 25 [see also Fig. 31 in Chapter 5 of Volume 2 and data on other perfluorocarbons in Christodoulides et al. (1979), Pai et al. (1979), Christodoulides and Christophorou (1979)]. Thermal (~ 300 K) attachment rate constants $(k_a)_{th}$ and parent negative-ion lifetimes τ_a at ~ 0.0 eV for a number of perfluorocarbons are given in Table VIII (see also Christophorou, 1978a).

It is interesting to note the structure in the total electron attachment cross sections below ~ 1 eV in Fig. 25, which indicates the existence of three NISs in this energy range. Since at thermal and epithermal energies the ions formed are long-lived and the measured attachment rate constants showed no pressure dependence, the data in Fig. 25 are taken to give a true measure of the rate of formation of the lowest NIS of these molecules (Christodoulides et al., 1979; Pai et al., 1979). Although for those and other perfluorocarbons at thermal energies the largest contribution to the magnitude of $\sigma_a(\varepsilon)$ comes from the formation of parent anions, at higher energy $\sigma_a(\varepsilon)$ contains varied contributions from dissociative attachment fragments which depend on the details of molecular structure (e.g., see Sauers et al., 1979). For example, the polyatomic molecule 1,3-C_4F_6—which is nonplanar with two ethylene groups lying in different planes—is more easily decomposed (Sauers et al., 1979) by dissociative attachment than its isomers 2-C_4F_6 and c-C_4F_6. This is clearly shown in Fig. 26, where the fragment and parent ions are shown for 1,3-C_4F_6 and 2-C_4F_6. The lifetimes of 1,3-$C_4F_6^{-*}$ and 2-$C_4F_6^{-*}$ at thermal energies are (Table VIII) 7 and 9 μs, respectively.

Dissociative attachment to a variety of aliphatic perfluorocarbons (e.g., MacNeil and Thynne, 1969a; Lifshitz and Grajower, 1969, 1972; Lifshitz et al., 1970; Thynne and MacNeil, 1970; Harland and Thynne, 1972a; Sauers et al.,

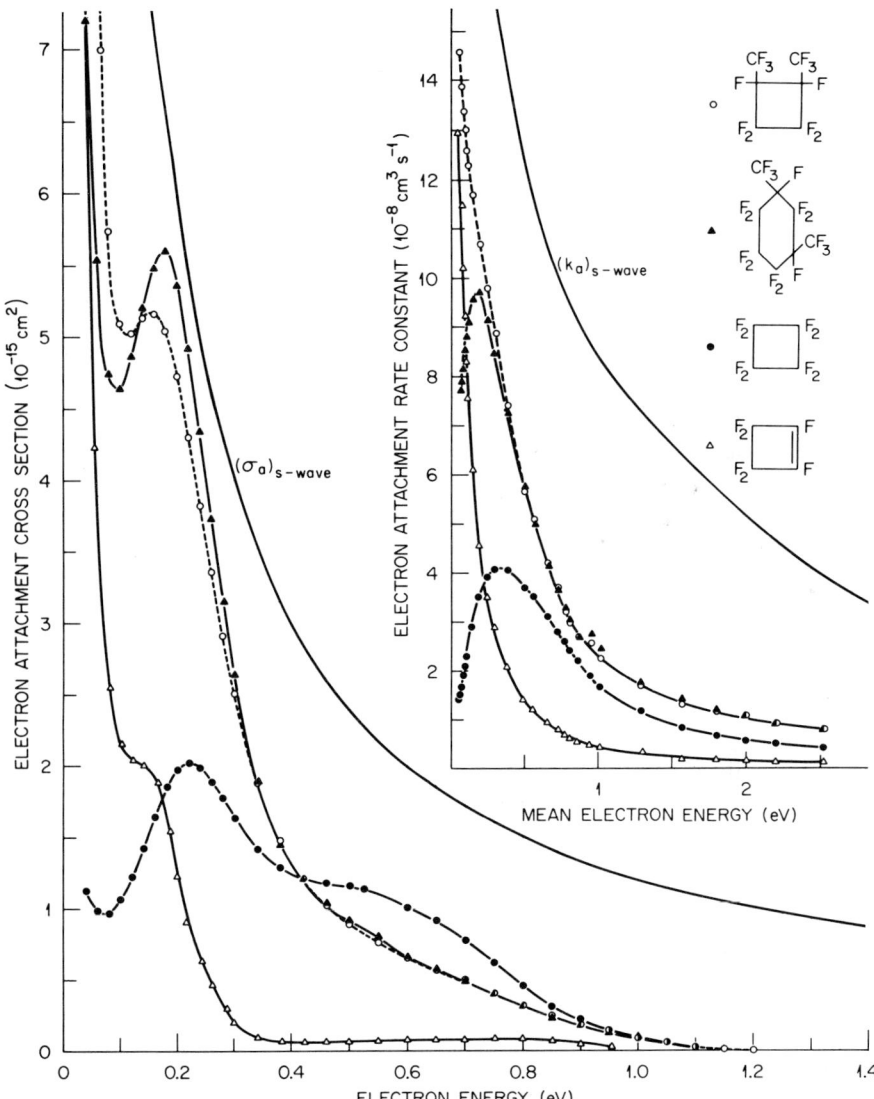

Fig. 25. Total electron attachment rate constants and cross sections for aliphatic perfluorocarbons (data of Christodoulides *et al.*, 1979; Pai *et al.*, 1979). $(\sigma_a)_{s\text{-wave}}$ and $(k_a)_{s\text{-wave}}$ are, respectively, the maximum s-wave capture cross section $(\pi \lambda^2)$ and maximum s-wave capture rate constant $[=(2/m)^{1/2}(\pi\hbar^2/2m)\int_0^\infty \varepsilon^{-1/2} f(\varepsilon, \langle\varepsilon\rangle)d\varepsilon$ where $f(\varepsilon, \langle\varepsilon\rangle)$ are the distribution functions in N_2 (up to $\langle\varepsilon\rangle \approx 0.8$ eV) and Ar $(\langle\varepsilon\rangle \gtrsim 0.8$ eV)] (see Christodoulides *et al.*, 1979).

TABLE VIII

Thermal (~300 K) Attachment Rate Constants and Autodetachment Lifetimes (at ~0.0 eV) of Parent Negative Ions of Perfluorocarbons

Perfluorocarbon	Parent negative ion	Thermal attachment rate constant $(10^{-7}\text{ cm}^3\text{ s}^{-1})$	Autodetachment lifetime of parent negative ion[a] (μs)
Perfluoro-n-butane	$C_4F_{10}^{-*}$	[b]	$12.7,^{c,d}\ 21^{e,f}$
Perfluoro-n-pentane	$C_5F_{12}^{-*}$	[b]	$34.3,^d\ 49^{f,g,h}$
Perfluoro-n-hexane	$C_6F_{14}^{-*}$	[b]	$43.9,^d\ 53^{f,g,h}$
Perfluorobutyne-2	$2\text{-}C_4F_6^{-*}$	$0.55^{i,j}$	$9,^k\ 16.3^d$
Perfluorobutadiene	$1,3\text{-}C_4F_6^{-*}$	$1.26,^j\ 1.32^i$	7^k
Perfluorocyclobutene	$c\text{-}C_4F_6^{-*}$	$1.4,^l\ 1.43,^j\ 1.52^i$	$6,^k\ 6.9,^m\ 11.2^d$
Perfluorobutene-2	$2\text{-}C_4F_8^{-*}$	$0.47,^j\ 0.49^{i,t}$	$10,^k\ 30.6^d$
Perfluorocyclobutane	$c\text{-}C_4F_8^{-*}$	$0.09,^n\ 0.11,^o\ 0.12,^{i,l}\ 0.13^j$	$10,^k\ 12,^m\ 14.8,^p\ 200^{q,r}$
Perfluorocyclopentene	$c\text{-}C_5F_8^{-*}$	$1.2,^o\ 3.9^{i,s}$	$26.2,^d\ 50^m$
Perfluorocyclohexene	$c\text{-}C_6F_{10}^{-*}$	$3.13,^o\ 3.92,^j\ 3.99^s$	$106,^d\ 113^m$
Perfluorocyclohexane	$c\text{-}C_6F_{12}^{-*}$	$1.50^{i,s}$	$236,^d\ 450^m$
Perfluoro-1-heptene	$1\text{-}C_7F_{14}^{-*}$	$0.04^{i,t}$	
Perfluoromethylcyclohexane	$c\text{-}C_7F_{14}^{-*}$	$0.4,^o\ 0.52,^{j,t}\ 0.8,^u\ 0.98^v$	$757,^d\ 793^m$
Perfluoro-1,3-dimethylcyclohexane	$c\text{-}C_8F_{16}^{-*}$	$0.74^{j,s}$	
Perfluorobenzene	$C_6F_6^{-*}$	$1.02,^w\ 1.06^o$	$12,^{m,r}\ 13.3^d$
Perfluorotoluene	$C_7F_8^{-*}$	$2.42,^o\ 2.74,^j\ 2.83^s$	$12,^d\ 37.8^d$

[a] These values are for electron energies at the peak of the negative-ion resonance, which is usually at ~0.0 eV. In some cases the peak of the parent negative-ion cross section was at higher energies than thermal; these are identified accordingly in the table.
[b] Hunter and Christophorou (1983a); see Fig. 27.
[c] Harland and Thynne (1973).

[d] Thynne (1972).
[e] Spyrou et al. (1983).
[f] Parent negative-ion cross section maximum at 0.6 eV.
[g] Christophorou et al. (1982).
[h] Lifetime decreases with increasing electron energy.
[i] Christodoulides et al. (1979).
[j] Christophorou et al. (1981b).
[k] Sauers et al. (1979).
[l] Bansal and Fessenden (1973).
[m] Naff et al. (1968).
[n] Christodoulides et al. (1974).
[o] Davis et al. (1973).
[p] Harland and Thynne (1972b).
[q] Henis and Mabie (1970).
[r] See discussion in Christophorou (1978a) concerning the differences in τ_a as measured using time-of-flight mass spectrometers and ion cyclotron resonance techniques.
[s] Pai et al. (1979).
[t] Christodoulides and Christophorou (1979).
[u] Chen et al. (1968).
[v] Mahan and Young (1966).
[w] Gant and Christophorou (1976).

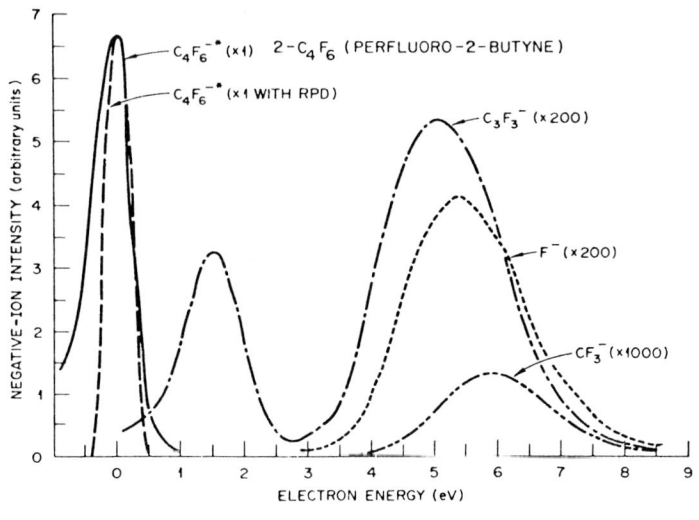

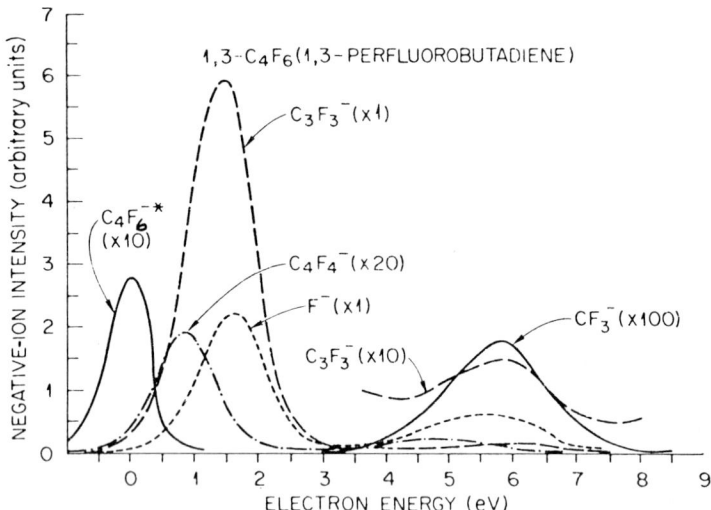

Fig. 26. Parent and fragment negative ions from perfluoro-2-butyne and 1,3-perfluorobutadiene upon electron impact as a function of electron energy. [From Sauers *et al.* (1979).]

1979; Heni *et al.*, 1982; Spyrou *et al.*, 1983) clearly indicated the importance of configurational and geometrical molecular rearrangements on the kind and extent of anionic decomposition.

A recent electron-beam study (employing time-of-flight mass spectrometry) of dissociative and nondissociative attachment to the perfluoroalkane molecules n-$C_N F_{2N+2}$ ($N = 1$–6) by Spyrou et al. (1983) has shown that F^- is the predominant ion for $N < 3$. For n-$C_4 F_{10}$ a weak parent negative ion was observed, but again the predominant ion was F^-. For n-$C_5 F_{12}$ and n-$C_6 F_{14}$ the parent negative ions were the most abundant with relative cross sections peaking at 0.6 eV and autodetachment lifetimes from 10 to 100 μs depending on the molecular size and the electron energy. In addition to the parent negative ion and the F^- ion, fragment negative ions of the form $C_N F_{2N+1}^-$, $C_N F_{2N}^-$, and $C_N F_{2N-1}^-$ ($N = 1$–6) have been detected. This work has shown also that anionic decompositions proceed through a number of NISs whose energy positions shift toward lower energies with increasing N (i.e., with increasing molecular size).

The attachment rate constants for perfluoroalkanes have been measured by Hunter and Christophorou (1983a) in high-pressure (0.133–1.6 MPa) buffer gases of N_2 and Ar over a mean energy range of 0.04–4.8 eV. These data are presented in Fig. 27. Similarly to the beam results of Spyrou et al. (1983), the $k_a(\langle \varepsilon \rangle)$ functions shift toward lower energies and increase in magnitude with increasing molecular size. The pressure dependences of the $k_a(\langle \varepsilon \rangle)$ observed by Hunter and Christophorou (1983a) reflect nicely the magnitudes of the lifetimes of the NISs involved and the kind (dissociative versus nondissociative) of electron attachment processes. Thus the $k_a(\langle \varepsilon \rangle)$ for CF_4 and $C_2 F_6$, for which no parent anions are known to form as their extremely short-lived NISs dissociate rapidly into, respectively, F^- and CF_3^- and F^-, CF_3^-, and $C_2 F_5^-$, were found to be pressure independent. The $k_a(\langle \varepsilon \rangle)$ for $C_3 F_8$ was found to be very strongly dependent on total pressure, indicating that the parent anion has a moderately short lifetime ($\sim 10^{-10}$ s; Hunter and Christophorou, 1983a) and can be stabilized by collision with another molecule—but is too short-lived to be observed in the beam study. For the high total pressures employed by Hunter and Christophorou, the $k_a(\langle \varepsilon \rangle)$ showed progressively less dependence on total pressure with increasing molecular size, i.e., in the order $C_3 F_8$, n-$C_4 F_{10}$, and n-$C_5 F_{12}$, and they became pressure independent for n-$C_6 F_{14}$. It thus appears that at least three carbon atoms are necessary for a perfluoroalkane molecule to have a positive EA and to be, therefore, capable of forming long-lived NISs at low energies. This observation may also hold for double-bonded perfluorocarbons since no parent negative ions of the *isolated* molecules of $C_2 F_4$ (or 1-$C_3 F_6$) have been observed in electron-impact studies. The $C_2 F_4^-$ anion was observed, however, in solid solution and its EPR (electron paramagnetic resonance) spectrum was found to be consistent with electron capture into a σ orbital (McNeil et al., 1977). It was reported recently also that for $C_2 H_4$, fluorination—contrary to the cases just discussed—destabilizes the π^* anion of $C_2 F_4$ with respect to that of $C_2 H_4$ (Chiu et al., 1979).

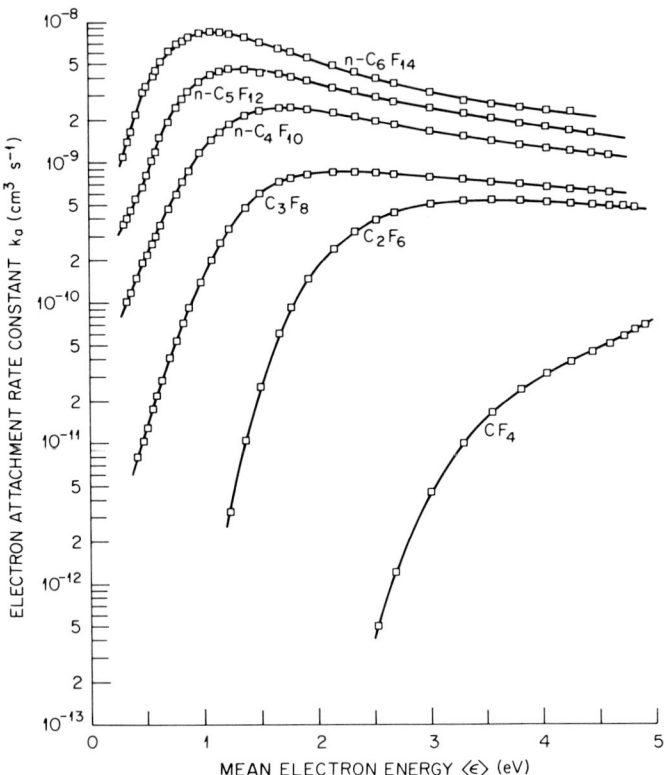

Fig. 27. Total electron attachment rate constants as a function of mean electron energy for perfluoroalkanes. When the attachment rate constants were increasing with buffer gas pressure (see text), their values extrapolated to infinite gas pressure are the ones plotted in the figure. [From Hunter and Christophorou (1983a).]

c. *Other Complex Polyatomic Molecules*

Foremost among the complex polyatomic molecules which have been extensively studied in recent years is sulfur hexafluoride (SF_6), principally because of its wide use as a gas dielectric. In Fig. 28 are shown the total and partial attachment cross sections as a function of electron energy and the total attachment rate constant as a function of the mean electron energy for this molecule.* Thermal values of the attachment rate constant are listed in Table IX. At room temperature SF_6^{-*} is the dominant ion formed at

---

* The large attachment cross section at thermal energies and its sharp decrease with increasing energy above thermal for this and other (e.g., CCl_4, perfluorocarbons) polyatomic molecules led to their use as detectors of slow electrons.

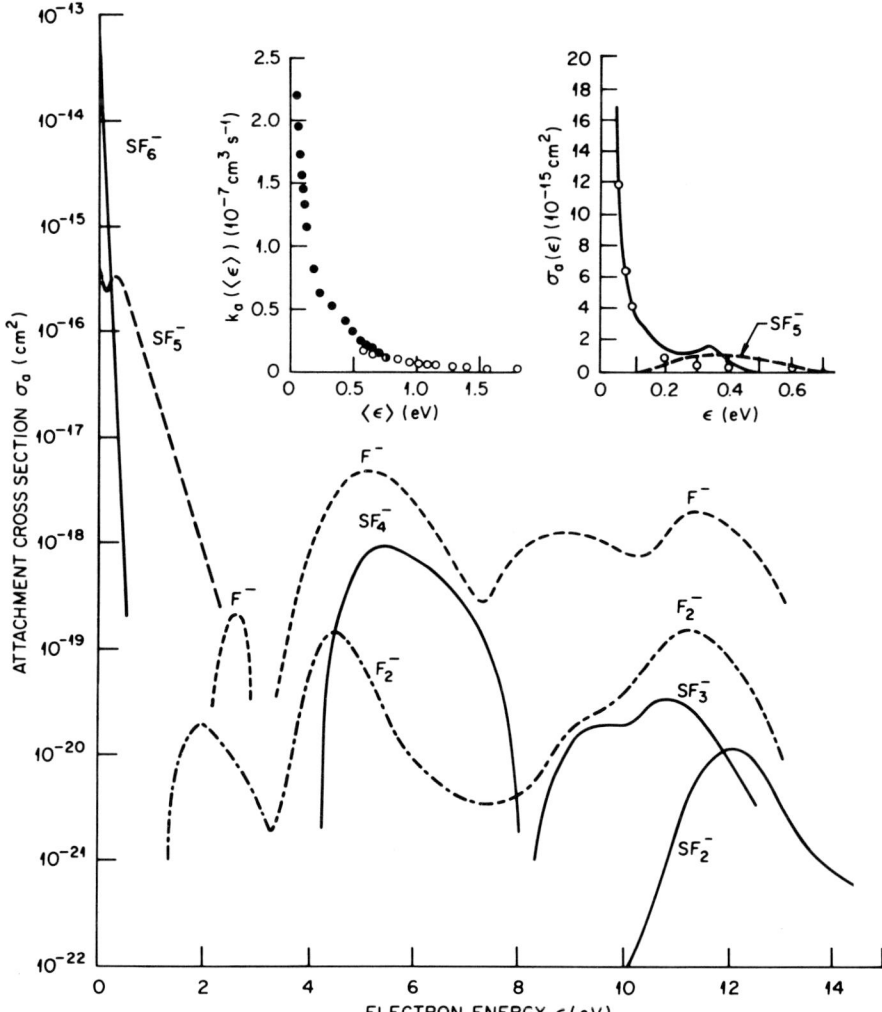

Fig. 28. Cross section for parent and fragment negative ions produced by electron impact on SF_6 (Kline et al., 1979. Reproduced with permission from the *Journal of Applied Physics*.] *Inset*: $k_a(\langle\varepsilon\rangle)$ measured in N_2 (●, Christophorou et al., 1971b) and Ar (○, Gant and Christophorou, 1976). $\sigma_a(\varepsilon)$: (—) Swarm-unfolded cross section determined by McCorkle et al. (1980) using the $k_a(\langle\varepsilon\rangle)$ measured in N_2 (Christophorou et al., 1971b). (○) $\sigma_a(\varepsilon)$ for the production of SF_6^- and SF_5^- of Kline et al. (1979). (---) Cross section for SF_5^- from SF_6 (Christophorou et al., 1971b, 1972).

TABLE IX

Thermal (~ 300 K) Attachment Rate Constants $(k_a)_{th}$
for Other Polyatomic Molecules

Molecule	$(k_a)_{th}$ $(cm^3 s^{-1})$	Reference
BF_3	$<1.5 \times 10^{-11}$	Stockdale et al. (1972)
BCl_3	2.8×10^{-9}	Stockdale et al. (1972)
ClF_3	3.6×10^{-10}	Christodoulides et al. (1976)
NF_3	$(2.1 \pm 0.8) \times 10^{-11 a}$	Sides and Tiernan (1977)
	2.4×10^{-11}	Mothes et al. (1972), Mothes and Schindler (1971)
	$(1.6 \pm 0.6) \times 10^{-9 b}$	Shaw and Jones (1977)
SF_4	$(1.5 \pm 0.5) \times 10^{-9}$	Babcock and Streit (1982)
SF_6^c	2.1×10^{-7}	Davis and Nelson (1970)
	2.21×10^{-7}	Fehsenfeld (1970)
	2.2×10^{-7}	Fessenden and Bansal (1970)
	$(2.28 \pm 0.07) \times 10^{-7}$	Crompton et al. (1982)
	2.41×10^{-7}	Chen et al. (1968)
	2.49×10^{-7}	Christophorou et al. (1981b)
	2.6×10^{-7}	Mothes and Schindler (1971)
	$2.7; 2.8 \times 10^{-7}$	Christophorou et al. (1971b)
	$(2.8 \pm 0.3) \times 10^{-7}$	Ayala et al. (1981b)
	3.1×10^{-7}	Mahan and Young (1966)
$V(CO)_6$	0.6×10^{-7}	George and Beauchamp (1982)[d]
$Cr(CO)_6$	3.0×10^{-7}	George and Beauchamp (1982)[d]
$Fe(CO)_5$	2.0×10^{-7}	George and Beauchamp (1982)[d]
$Ni(CO)_4$	2.0×10^{-7}	George and Beauchamp (1982)[d]
$Mo(CO)_6$	1.3×10^{-7}	George and Beauchamp (1982)[d]
$W(CO)_6$	1.2×10^{-7}	George and Beauchamp (1982)[d]

[a] Attributed to F^- ions.
[b] Total; for both F^- and NF_2^- ions.
[c] The various methods used to measure $(k_a)_{th}$ for SF_6 employed SF_6 mixtures with a variety of carrier gases (see original references).
[d] These authors reported that the dominant anions resulting from dissociative attachment to the $M(CO)_n$ molecules are $M(CO)_{n-1}^-$.

thermal energies [the τ_a of SF_6^{-*} is $>10^{-5}$ s; see Christophorou (1978a)]. The SF_5^- is much weaker and its cross section peaks at ~ 0.37 eV. Although at room temperature the intensity of SF_5^- at thermal energies is very small, it increases sharply with increasing T (see Section V). At higher energies other fragment anions are observed (Fig. 28).

In Fig. 29 total dissociative attachment cross sections are shown for two isotopic pairs of molecules, namely NH_3 and ND_3 and CH_4 and CD_4, and

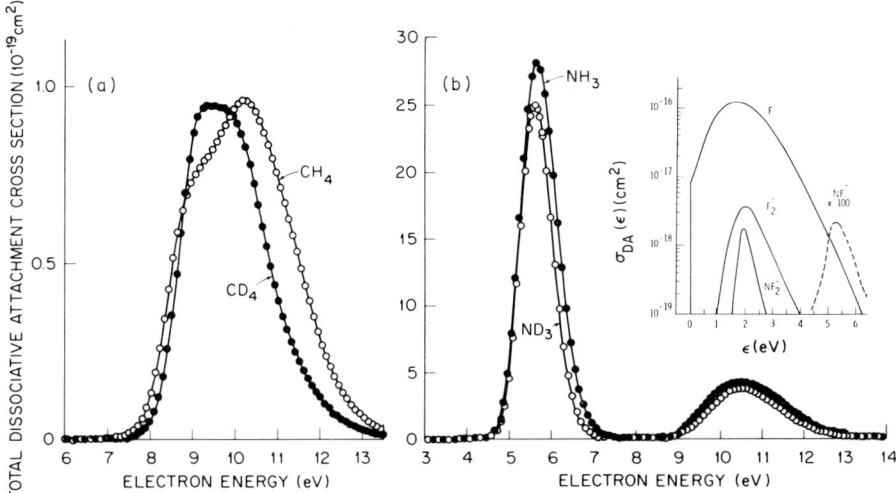

Fig. 29. Total dissociative attachment cross section as a function of electron energy: (a) CH_4 and CD_4 (Sharp and Dowell, 1967), (b) NH_3 and ND_3 [Sharp and Dowell (1969); the cross section scale was established by multiplying the reported ratio of the total dissociative attachment cross section to the total ionization cross section by 2.3×10^{-16} cm^2, the total ionization cross section value at 85 eV]. *Inset*: F^- from NF_3 [$T = 365$ K, Chantry (1982, private communication]; F^- is the dominant fragment negative ion; weaker fragment negative ions (F_2^-, NF_2^-, and NF^-) were also observed.

also for F^- from NF_3. Total attachment rate constants as a function of mean electron energy for three fluorinated ethers, two of which [$(CF_3)_2O$ and $(CF_3)_2S$)] are of interest as possible constituents in diffuse discharge switch gas mixtures owing to their desirable electron attaching properties (see Chapter 5 of Volume 2) are presented in Fig. 30. Values of $(k_a)_{th}$ for some other polyatomic molecules are listed in Table IX.

Numerous other studies of dissociative attachment to polyatomic molecules have been made. These usually identified the nature of the fragment negative ions, their energy dependence, energy onset and maxima, and often their relative abundances. In Table X are given a number of these studies [see also Dillard (1973) and, for benzene derivatives, Christophorou et al. (1977a)].

D. Dependence of the Magnitude of the Dissociative Attachment Cross Section on the Resonance Energy

The experimental data discussed earlier in this section show that the energy position of the dissociating negative-ion state affects critically the magnitude of the dissociative attachment cross section. The peak cross section $\sigma_{DA}(\varepsilon_{max})$

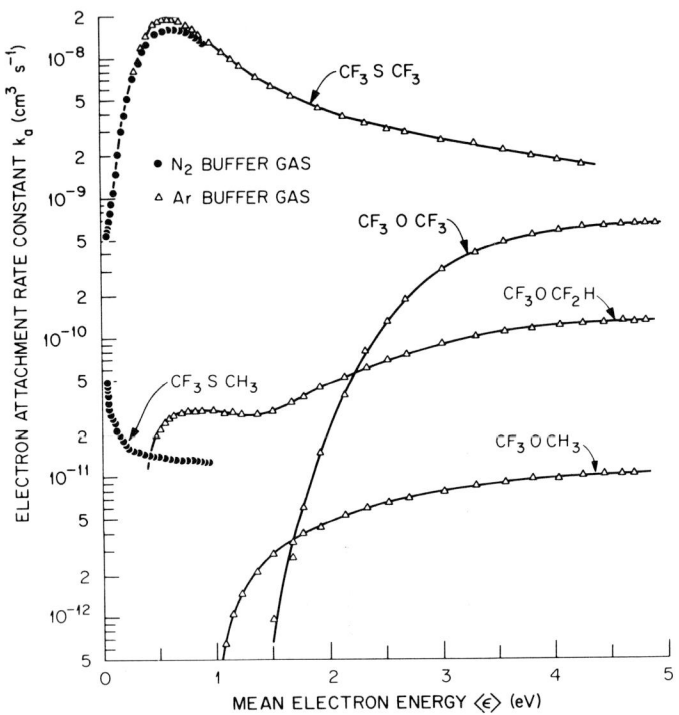

Fig. 30. Total electron attachment rate constant as a function of mean electron energy for CF_3SCF_3, CF_3SCH_3, CF_3OCF_3, CF_3OCF_2H, and CF_3OCH_3, measured at room temperature in the buffer gases of N_2 (solid points) and Ar (open points). [From Hunter and Christophorou (1983b).]

at the resonance energy ε_{max} varies by about ten orders of magnitude, from $\sim 10^{-24}$ to $\gtrsim 10^{-14}$ cm^2 as ε_{max} varies from ~ 15 to 0.04 eV (thermal energy). This behavior of $\sigma_{DA}(\varepsilon_{max})$ was first pointed out and discussed by Christophorou and Stockdale (1968). Christophorou and Stockdale (1968) and Christophorou (1971) plotted $\sigma_{DA}(\varepsilon_{max})$ as a function of ε_{max} and observed that when ε_{max} lies below the energy ε_N of the electronic states of the neutral molecule and dissociation is fast (small autoionization, purely repulsive negative ion potential-energy curve/surface in the Franck–Condon region) $\sigma_{DA}(\varepsilon_{max})$ varies roughly as ε_{max}^{-1}, reflecting the ε^{-1} dependence of σ_c [Eq. (24)]. When, however, $\varepsilon_{max} \geq \varepsilon_N$, $\sigma_{DA}(\varepsilon_{max})$ is a much stronger decreasing function of ε_{max} and this was attributed to increased autoionization. For vertical-onset cases [e.g., $H^-(D^-)/H_2(D_2)$ at 3.75 eV; O^-/CO at 9.62 eV; O^-/CO_2 at 4.4 eV] $\sigma_{DA}(\varepsilon_{max})$ is obviously very much smaller than would be expected owing to the small survival probability (large autoionization) characteristic of

TABLE X

Dissociative Attachment Studies of Other Complex
Polyatomic Molecules

Reference	Polyatomic molecule(s) studied
Bennett et al. (1974)	$TiCl_4, TiBr_4, TiI_4$
Brion (1969)	SF_6, SeF_6, TeF_6
Chantry (1976)	CeI_3
Christophorou et al. (1966); Naff et al. (1971); Hadjiantoniou et al. (1973a)	Substituted benzenes
Compton (1977)	UF_6
Compton and Bouby (1967)	$(CHO)_2, (CH_3CO)_2$
Compton and Cooper (1977)	$C_{12}H_4N_4(TCNQ)^a$
Compton and Stockdale (1976)	$Fe(CO)_5, Ni(CO)_4$
Cooper and Compton (1973)	$C_4H_2O_3, C_4H_4O_3, C_4H_6O_3, C_8H_4O_3,$ $C_5H_6O_3, C_6H_6O_3, C_4H_6O_4,$ $C_{10}H_2O_6, C_4H_5O_2N, C_4H_3O_2N$
Cooper and Compton (1974)	$C_4F_4O_3, C_5F_6O_3$
DeCorpo and Franklin (1971)	$CCl_4, CBr_4, CI_4, CHI_3, BF_3$
DeCorpo et al. (1971)	$CF_4, CCl_4, BF_3, CH_3SH, C_2H_5ONO_2,$ $CH_3NH_2, CH_2Cl_2, CH_2Br_2, CH_2I_2$
Di Domenico and Franklin (1972)	CH_3NO_2
Di Domenico et al. (1972)	$c\text{-}C_5H_6$
Dillard et al. (1975)	Substituted benzils
Dorman (1966)	$H_2O, CO_2, N_2O, SO_2, CS_2, NH_3, S_2Cl_2,$ $CCl_4, SiCl_4, CH_3CHO, (CH_3)_2CO,$ $(C_2H_5)_2CO, C_2H_5OH, CH_3COCH_2Cl,$ $CH_2BrCH_2OH, 1,1,2,2\text{-}C_2H_2Cl_4$
Ebinghaus (1964)	MX and M_2X_2 (M = Li, Na, K, Rb, Cs; X = F, Cl, Br, I)
Franklin et al. (1974)	$As_4, GeF_4, PbF_4, SiF_4, SnF_4, CF_4$
Harland and Franklin (1974)	$NF_3, BF_3, CF_4, C_2F_6, C_3F_8, c\text{-}C_4F_8$
Harland and Thynne (1969)	SF_5Cl
Harland and Thynne (1970)	$(CF_3)_2CO$
Harland and Thynne (1971)	SF_4
Harland and Thynne (1975)	$CF_3CHO, C_2F_5CHO, n\text{-}C_3F_7CHO$
Harland et al. (1972)	GeF_4
Hildenbrand (1977)	WF_5, WF_6, WOF_4
Jäger and Henglein (1967)	$CH_3NO_2, C(NO_2)_4, C_2H_5NO_2, C_2H_5ONO,$ $C_2H_5ONO_2$
Jäger and Henglein (1968)	$SiCl_4, (CH_3)_3SiCl, (CH_3)_2SiCl_2,$ $CH_3SiCl_3, CH_2CHSiCl_3, C_6H_5SiCl_3$
Khvostenko et al. (1973)	$(CH_3)_2CO, C_4H_4S, C_5H_5N, C_6H_6$
MacNeil and Thynne (1969a)	$C_2F_6, (CF_3)_2CO$
MacNeil and Thynne (1969b)	CS_2, OCS
MacNeil and Thynne (1969c)	SO_2, CF_2O
MacNeil and Thynne (1970a)	BF_3, PF_3

TABLE X (cont.)

Reference	Polyatomic molecule(s) studied
MacNeil and Thynne (1970b)	CF_4, SiF_4
MacNeil and Thynne (1972)	$(CF_3O)_2$
Naff et al. (1972)	$CO(NH_2)_2$ (urea), CH_3COCH_2Cl
Pabst et al. (1976)	$AsF_3, AsCl_3, AsBr_3$
Pabst et al. (1977a)	$SnCl_4, SnBr_4, SnI_4$
Pabst et al. (1977b)	$SiCl_4, SiBr_4, GeCl_4, GeBr_4$
Spyrou et al. (1983)	$n\text{-}C_NF_{2N+2}$ ($N = 1\text{-}6$), $i\text{-}C_4F_{10}$
Stockdale et al. (1970)	$SeF_6, TeF_6, MoF_6, ReF_6, UF_6$
Stockdale et al. (1972)	BF_3, BCl_3
Stockdale et al. (1974)	$CH_3X (X = NO_2, CN, I, Br)$
Thynne (1969)	NF_3
Thynne and Harland (1973a)	WF_6
Thynne and Harland (1973b)	C_2F_5CN, C_3F_7CN
Von Trepka and Neuert (1963)	$CH_4, C_2H_6, C_3H_8, n\text{-}C_4H_{10}, C_2H_4,$ $i\text{-}C_4H_{10}, C_2H_2, C_3H_6, i\text{-}C_4H_8,$ $1\text{-}C_4H_8, CH_3OH, C_2H_5OH, n\text{-}C_3H_7OH,$ $i\text{-}C_3H_7OH, C_6H_6$
Wang et al. (1973; 1974b)	CF_4, SiF_4
Wang et al. (1974a)	GeF_4

[a] TCNQ = 7,7,8,8-tetracyanoquinodimethane.

these cases. The dissociative attachment data that have appeared since fully support these conclusions. To illustrate further this behavior of σ_{DA}, Christophorou (1978c) plotted representative $\sigma_{DA}(\varepsilon)$ functions for a number of molecules in the energy range from 0 to ~ 16 eV (Fig. 31). The salient features of the behavior of σ_{DA} are clearly borne out: the magnitude of σ_{DA} decreases with increasing energy position of the NIS; as $\varepsilon_{max} \to$ thermal energy, $\sigma_{DA} \to \pi \lambda^2$ where $\pi \lambda^2$ is the s-wave capture cross section ($\lambda = 2\pi\lambda$ is the electron de Broglie wavelength).

V. Dissociative Electron Attachment to "Hot" Molecules (Effects of Temperature on Dissociative Electron Attachment)

The effects of temperature on the various electron attachment processes are of intrinsic value, of basic significance in determining thermochemical data from electron attachment studies, and of interest to many applied areas (employing temperatures higher than ambient) where electron densities are crucially affected by negative-ion formation (see Chapter 5 of Volume 2). The effects of temperature on the magnitude and energy dependence of the cross

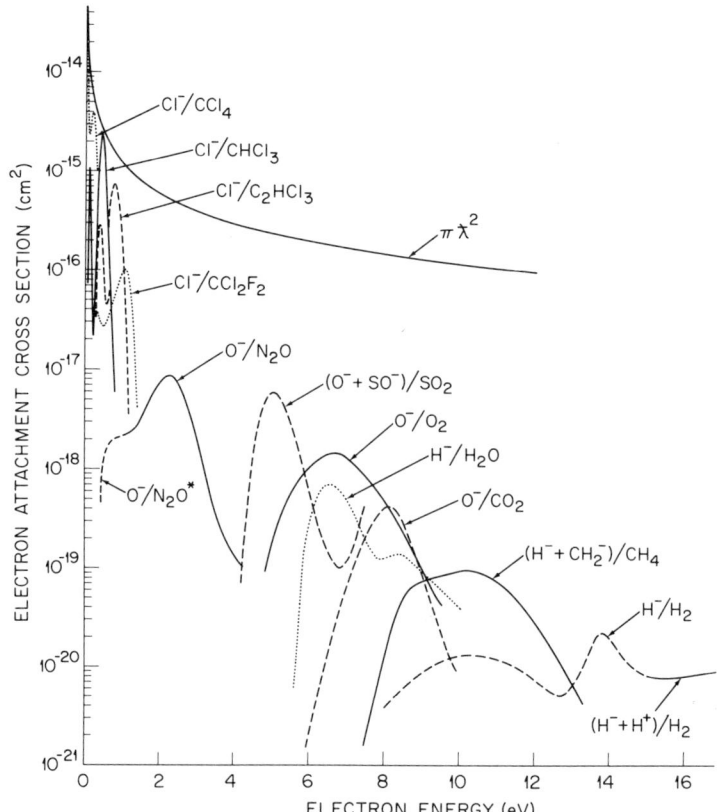

Fig. 31. Dissociative attachment cross sections as a function of electron energy for a number of molecules. Some of the plotted $\sigma_{DA}(\varepsilon)$ were deduced from swarm experiments and are thus "total" cross sections. They are identified with the specific ions as shown because these were the most abundant in mass spectrometric studies. Some of the molecules shown have other resonances which were not plotted for convenience of display. O^-/N_2O^* denotes dissociative attachment from vibrationally excited N_2O molecules, and $(H^- + H^+)/H_2$ denotes ion pair formation from H_2. [From Christophorou (1978c).]

section for specific and total negative-ion production have been investigated by both the electron-beam (Hickam and Berg, 1958; Fite and Brackmann, 1963; Fite et al., 1965; Henderson et al., 1969; Spence and Schulz, 1969, 1973; Chantry, 1969; Chen and Chantry, 1970, 1972, 1979; Allan and Wong, 1978, 1981) and the electron swarm (Compton et al., 1966; Wentworth et al., 1967, 1969; Chaney and Christophorou, 1969; Fehsenfeld, 1970; Warman and Sauer, 1971; Shimamori and Fessenden, 1979; McCorkle et al., 1982b; Hunter et al., 1983) methods; the former extended to $T \simeq 3000$ K and the latter have been restricted to $T \lesssim 500$ K.

A. Diatomic Molecules

Let us, then, first refer to the experimental results of Fite and co-workers in Fig. 32 on the temperature dependence of dissociative attachment to O_2 producing O^-. In the range of ~ 4–10 eV, O^- is produced from O_2 by resonance dissociative attachment. The cross section $\sigma_{DA}(\varepsilon)$ for this process shows (Fig. 32a; 300 K) a single, structureless peak at 6.7 eV, suggesting that only one repulsive (in the Franck–Condon region) state ($^2\Pi_u$) is responsible for this process, which can be described as

$$e + O_2(X\ ^3\Sigma_g^-) \longrightarrow O_2^-(^2\Pi_u) \longrightarrow O^-(^2P) + O(^3P) \tag{49}$$

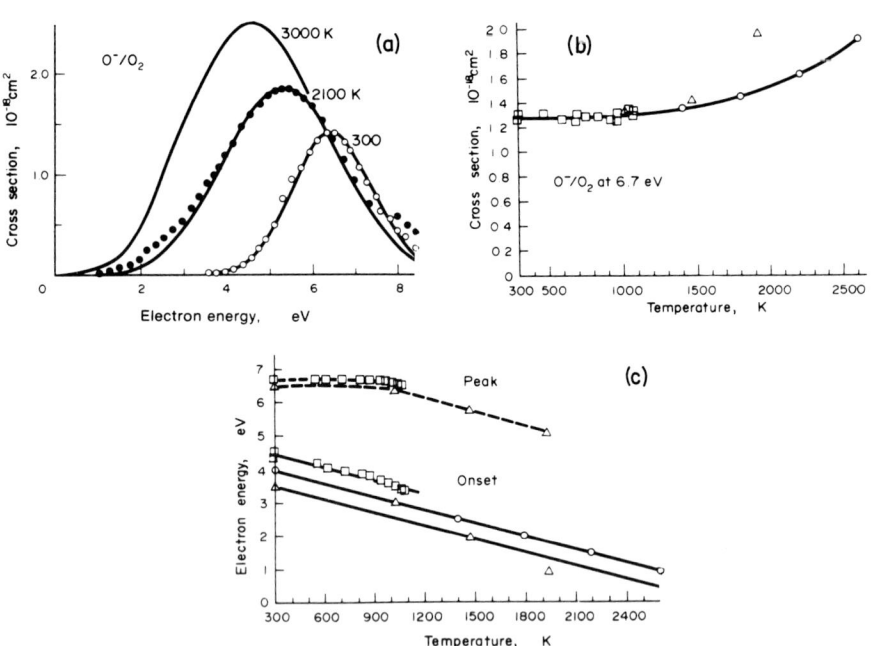

Fig. 32 (a) Temperature dependence of $\sigma_{DA}(\varepsilon)$ for O^- from O_2. The solid curves are values calculated by O'Malley (1967a); the points are the experimental results of Fite and Brackmann (1963). [From O'Malley (1967a). Reproduced with permission from the *Physical Review*.] (b) Temperature dependence of the peak value of $\sigma_{DA}(\varepsilon)$ for O^- from O_2: (-□-) (Spence and Schulz, 1969; experiment), (△) (Henderson *et al.*, 1969; experiment), (-○-) (O'Malley, 1967a; theory). [Reproduced with permission from the *Physical Review*.] (c) Temperature dependence of the energies of the onset and the maximum of $\sigma_{DA}(\varepsilon)$ for O^- from O_2: (□) (Spence and Schulz, 1969; experiment), (△) (Fite and Brackmann, 1963; experiment), (-○-) (O'Malley, 1967a; theory). [From Spence and Schulz (1969).]

6 Electron Attachment Processes

It is seen (Fig. 32a) that as T increases the threshold energy decreases (it actually shifts from ~ 4 eV at ~ 300 K to ~ 1.2 eV at 1970 K), the energy position of the resonance maximum decreases (Fig. 32c), while the magnitude of the cross section (Fig. 32b) and the resonance width increase (Fig. 32a). These data have been explained theoretically by O'Malley (1967a), who assumed that the direct effect of T on O_2 is to produce a Maxwellian distribution of vibrational (v) and rotational (j) states. The effective cross section $\sigma_{DA}(T, \varepsilon)$ for dissociative attachment, then, is the Boltzmann average of the cross section $\sigma_{DA}^{v,j}(\varepsilon)$ from each of the individual states, i.e.,

$$\sigma_{DA}(T, \varepsilon) = \sum_{v=v_{min}}^{\infty} \sum_{j=j_{min}}^{\infty} Ne^{-(E_v + E_j)/kT} \sigma_{DA}^{v,j}(\varepsilon) \tag{50}$$

where

$$N = \left(\sum_{v=v_{min}}^{\infty} \sum_{j=j_{min}}^{\infty} e^{-(E_v + E_j)/kT} \right)^{-1}$$

and v_{min} and j_{min} are subject to the energy threshold requirement $\varepsilon + E_v + E_j \geq E_{threshold} = D(O - O) - EA(O) = 5.08 - 1.46 = 3.62$ eV. The cross section $\sigma_{DA}^{v,j}(\varepsilon)$ is (O'Malley, 1966, 1967a)

$$\sigma_{DA}^{v,j}(\varepsilon) = \frac{4\pi^2 g}{k^2} \frac{\Gamma_{aX}}{\Gamma_d} \left| \tilde{\chi}_v \left(R_\varepsilon - i \frac{\Gamma_a}{\Gamma_d} \right) \right|^2 e^{-\rho} \tag{51}$$

where $\tilde{\chi}_v$ is the initial vibrational wave function, k is the electron wave number, $\Gamma_{a,X}$ is the partial autodetachment width for the state X, and the rest of the symbols are as defined earlier. In his treatment O'Malley considered the effect of rotational states to be negligible, and the excellent agreement of his predictions on the threshold, magnitude, width, and energy position of the O^-/O_2 resonance with the experimental results (Fig. 32) justifies this assumption. As T increases, higher vibrational levels are populated, the internuclear distances increase (and hence the Franck–Condon region broadens) significantly, and although even at 2000 K there is only a limited amount of vibrational excitation, there is a profound change in the survival factor $e^{-\rho}$ in Eq. (51) which dominates the temperature dependence of the cross section.

The insignificant effect of rotational excitation seems to be at variance with earlier theoretical predictions (Chen and Peacher, 1967) of a strong effect of rotational excitation on the yield of H^-/H_2. Recent electron-beam experiments (Allan and Wong, 1978) yielded $\sigma_{DA}(\varepsilon)$ from individual vibrational and rotational states of H_2 (and D_2) and along with theoretical work (Wadehra and Bardsley, 1978; Bardsley and Wadehra, 1979) clarified and quantified the role of vibrational and rotational excitation on the magnitude of $\sigma_{DA}(\varepsilon)$ for the production of $H^-(D^-)$ from $H_2(D_2)$ in the energy range 1–5 eV via the extremely short-lived $^2\Sigma_u^+$ shape resonance. In Fig. 33a the cross section (in relative units) for H$^-$ from H$_2$ is shown for 300 and 1400 K. At 300 K the well-established vertical onset is observed at 3.75 eV. At 1400 K the H$^-$ from H$_2$ curve shows contributions from vibrational states up to $v = 4$ and from rotational states up to $j = 7$. When the intensities of the measured peaks are compared with the corresponding state populations at 1400 K, a drastic increase in $\sigma_{DA}(\varepsilon)$ with vibrational and rotational quantum is evident. Thus the

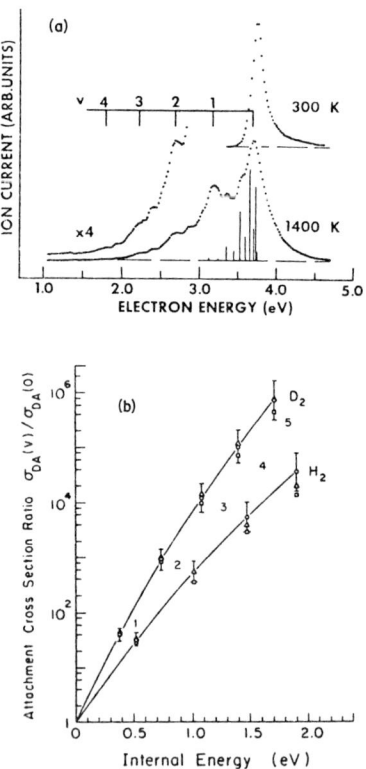

Fig. 33. (a) Threshold region of $\sigma_{DA}(\varepsilon)$ for H$^-$ from H$_2$ at 300 and 1400 K. Note that in the 1400-K curve, peak intensities from excited molecules are much larger than expected from the vibrational population (e.g., 1.4% for $v = 1$) and rotational population (vertical bars), reflecting a drastic increase of the cross section with vibrational and rotational quanta. The vertical lines ($v = 0$–4) indicate expected peak positions for rotational/vibrational profiles at 1400 K. [From Allan and Wong (1978). Reproduced with permission from the *Physical Review Letters*.] (b) Internal-state dependence of threshold dissociative attachment cross sections in H$_2$ and D$_2$ via the $^2\Sigma_u^+$ resonance: (○) (Allan and Wong, 1978; experiment), (□) (Wadehra and Bardsley, 1978; theory), (△) (Bardsley and Wadehra, 1979; theory). [From Bardsley and Wadehra (1979). Reproduced with permission from the *Physical Review*.]

threshold cross section for H^- from H_2 in the $^2\Sigma_u^+$ resonance region between 1–5 eV was reported by Allan and Wong (1978) to increase by four orders of magnitude from $v = 0$ to $v = 4$ and by a factor of 5 from $j = 0$ to $j = 7$. Similar observations were made for D^- from D_2. In Fig. 33b are plotted experimental and theoretical results on the internal-energy dependence of σ_{DA} for H_2 and D_2. For a given amount of internal energy, vibrational excitation of D_2 has a larger effect on σ_{DA} than for H_2. The results of recent calculations employing resonance scattering theory with semiempirical parameters (Wadehra and Bardsley, 1978; Bardsley and Wadehra, 1979) on the attachment cross sections near threshold for several rotational and vibrational states of H_2 and D_2 are compared with experimental data in Fig. 33b and are listed in Table XI for some values of v and j. Theory and experiment are in comfortable agreement and both show that although the effect of rotational excitation is not as high as that of vibrational excitation and not as significant as suggested earlier (Chen and Peacher, 1967), it is, nonetheless, considerable.

Additional examples of the effects of T on $\sigma_{DA}(\varepsilon)$ have been reported by Allan and Wong (1981) for the formation of Cl^- and F^-, respectively, from the polar diatomic molecules HCl(DCl) and HF. Similarly to the case of H_2, the $\sigma_{DA}(\varepsilon)$ for Cl^- from HCl and F^- from HF was found to be very sensitive to nuclear motion, increasing with vibrational excitation. Furthermore, Allan and Wong observed a pronounced structure in $\sigma_{DA}(\varepsilon)$ for Cl^- production due to rotationally excited HCl molecules. The electric dipole moment (1.1 D) of

TABLE XI

Calculated Dissociative Attachment Cross Sections Near Threshold for Various Vibrational/Rotational Levels of H_2 and D_2[a]

		H_2		D_2	
v	j	ε (eV)	$\sigma_{DA}^{v,j}$ (cm^2)	ε (eV)	$\sigma_{DA}^{v,j}$ (cm^2)
0	0	3.73	1.6×10^{-21}	3.83	3.0×10^{-24}
0	1	3.73	1.7×10^{-21}	3.80	3.3×10^{-24}
0	5	3.53	3.7×10^{-21}	3.70	5.7×10^{-24}
0	10	3.13	2.2×10^{-20}	3.43	2.2×10^{-23}
0	20	1.63	5.5×10^{-18}	2.55	2.5×10^{-21}
1	0	3.23	5.5×10^{-20}	3.45	1.5×10^{-22}
2	0	2.73	8.0×10^{-19}	3.08	3.3×10^{-21}
4	0	1.85	3.2×10^{-17}	2.43	3.6×10^{-19}
8	0	0.40	3.5×10^{-16}	1.25	9.6×10^{-17}
12	0			0.30	3.7×10^{-16}

[a] Bardsley and Wadehra (1979).

HCl introduces a long-range interaction which is absent in H_2. Figure 34a shows the results of Allan and Wong (1981) on the energy dependence of Cl^- from HCl in the energy region of 0–2 eV at four temperatures between 300 and 1180 K. At 300 K, a steep rise with a maximum at 0.82 eV is seen, as in earlier studies (Christophorou et al., 1968; Azria et al., 1974), corresponding to the

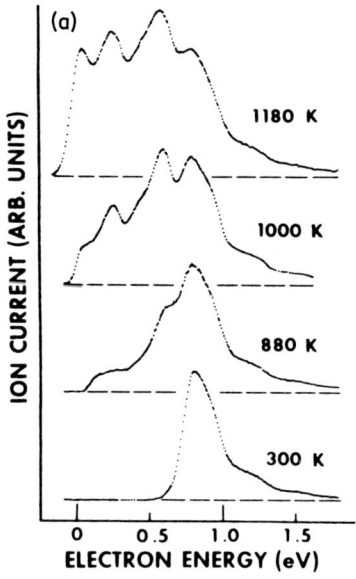

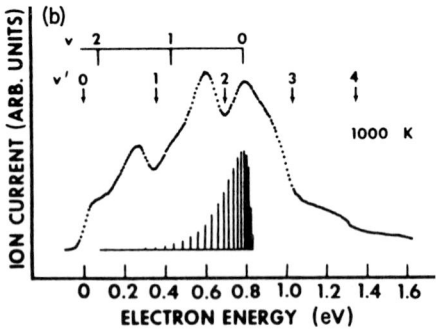

Fig. 34 (a) Ion current for Cl^- from HCl at various T (the four spectra have approximatley the same vertical scales). (b) Ion current for Cl^- from HCl in the threshold region at 1000 K. The vertical bars under the curve give the rotational population and their threshold positions. The expected peak positions from vibrationally excited HCl are marked by v and the vibrational-level energies of neutral HCl by v'. The accuracy of the energy scale is ± 40 meV. [From Allan and Wong (1981).]

TABLE XII

Vibrational Enhancement in Threshold Cross Section of Dissociative Attachment[a]

	HCl	DCl	HF
$\sigma_{v=1}/\sigma_{v=0}$	38	32	21
$\sigma_{v=2}/\sigma_{v=0}$	880	580	300

[a] The quoted experimental errors for the ratios listed are $\pm 30\%$ for $v = 1$ and $\pm 50\%$ for $v = 2$, from Allan and Wong (1981).

ground dissociation limit of $Cl^- + H$. The structure seen above the peak occurs at ~ 1.03 and 1.35 eV, at which energies the vibrational channels $v' = 3$ and 4 become open. As T increases additional structure in the Cl^- production is observed at lower energies owing to rotationally and vibrationally excited HCl. The detailed spectrum shown in Fig. 34b at 1000 K helps to clarify the processes. Thus the two peaks centered ~ 0.2 eV below the expected onsets for Cl^- production from $v = 0$ and $v = 1$ states of HCl arise from the increase of the cross section with rotational quantum (these are shaped by the decrease in σ_{DA} at the opening of each vibrational channel, v' in Fig. 34b) of HCl. The increase in σ_{DA} due to vibrationally excited HCl, v in Fig. 34b, shows up as shoulders in the cross section curve. From the vibrational population ratios and the Cl^- and F^- signal ratios measured by Allan and Wong (1981) at various T, the ratios $\sigma_{DA}(v = 1)/\sigma_{DA}(v = 0)$ and $\sigma_{DA}(v = 2)/\sigma_{DA}(v = 0)$ were determined and are listed in Table XII.

More recently Bardsley and Wadehra (1983) calculated dissociative attachment cross sections for HCl and DCl in several initial rotational and vibrational states. Their cross sections for electron attachment to HCl in specific vibrational states averaged over a thermal distribution of rotational states increased dramatically and shifted to lower energies with increasing vibrational quantum number. Thus the cross section for $v = 0$ peaked at ~ 0.8 eV and had a peak cross-section value of $\sim 10^{-17}$ cm^2, whereas for $v = 3$ the cross section peaked at ~ 0.0 eV and had a peak cross section value of $\sim 10^{-14}$ cm^2 (see Bardsley and Wadehra, 1983, for details).

B. Polyatomic Molecules

The pronounced effects of temperature on $\sigma_{DA}(\varepsilon)$ have been demonstrated also for triatomic and polyatomic molecules. In Fig. 35a is shown the effect of T on the production of O^- from N_2O below ~ 4 eV. The production of O^- from N_2O is very sensitive to T in one region (close to thermal and epithermal

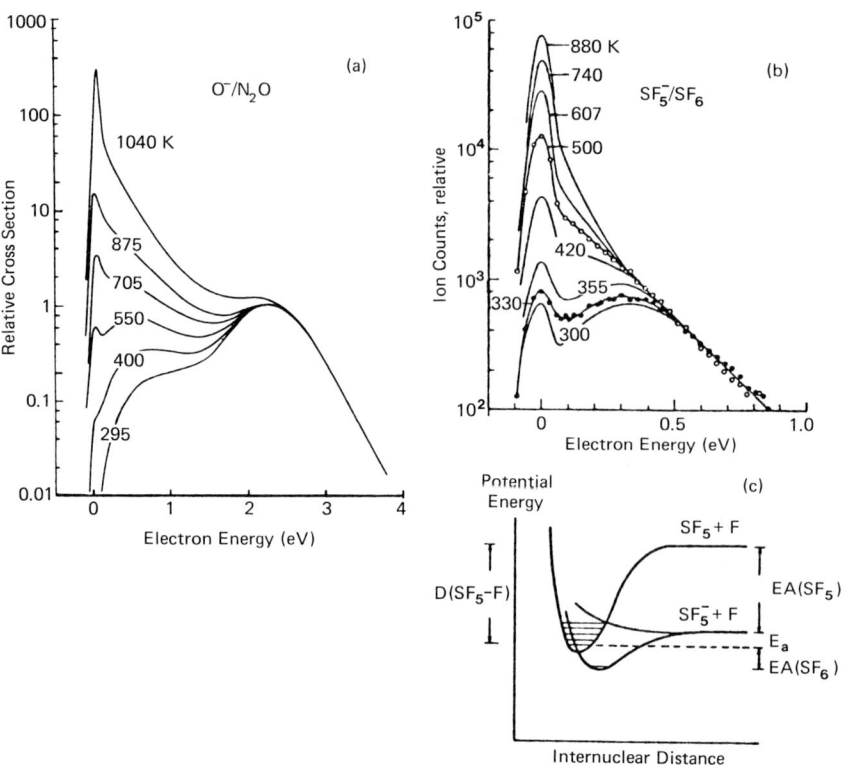

Fig. 35. (a) Dependence of the relative cross section for formation of O⁻ from N₂O as a function of electron energy on temperature. For all temperatures the curves were normalized at 2.25 eV; they coincide at higher energies and they differ drastically at lower energies as T changes. [From Chantry (1969). Reproduced with permission from the *Journal of Chemical Physics*.] (b) SF_5^- from SF_6 vs. ε at various T (Chen and Chantry, 1979). (c) Schematic potential-energy diagram illustrating the origin of the two peaks in the SF_5^-/SF_6 current (see text) (Chen and Chantry, 1979). [Parts (b) and (c) reproduced with permission from the *Journal of Chemical Physics*.]

energies) and insensitive in another ($\gtrsim 2.3$ eV). It appears that two states of N_2O^- are involved in the production of O⁻ from N_2O below ~4 eV. The strongly temperature-sensitive portion of the cross section involves the lowest (ground) state of N_2O^- and is due to excitation of the bending mode of vibration (Chantry, 1969); it is thought to arise from the dependence on bond angle of the energy separation of the electronic ground states of N_2O and N_2O^- (the potential energy of the lowest N_2O^- state depends significantly on bond angle). The temperature-independent peak at 2.25 eV is ascribed to dissociative attachment via the second N_2O^- state. These findings, along with those on the temperature dependence of dissociative attachment to O_2 and O⁻ from CO_2 (Spence and Schulz, 1969; Chantry, 1972), demonstrate the significance of temperature studies in determinations of *true* onsets for specific product ions and the deduction therefrom of thermochemical data.

In Fig. 35b are shown the results of Chen and Chantry (1979) on the formation of SF_5^- from SF_6. Two peaks in the SF_5^- vs. ε curve are seen: one at ~ 0.0 eV and the other at ~ 0.38 eV. The ~ 0.38-eV peak is rather broad and temperature insensitive. The ~ 0.0-eV peak is extremely sensitive to T. A semilogarithmic plot of the relative SF_5^- signal at the zero-energy peaks vs. T^{-1} yielded an activation energy of 0.2 eV. The two peaks can be understood on the basis of Fig. 35c. The zero-energy peak is the result of attachment of near-zero-energy electrons to those vibrational/rotational states of SF_6 at energies ≥ 0.2 eV above the ground state and the 0.38-eV peak is the result of vertical Franck–Condon transitions of SF_6 to the repulsive negative-ion curve after capturing a ~ 0.38-eV electron. It is observed that, although the cross section for SF_5^- from SF_6 at ~ 0.0 eV increases dramatically with increasing T, the *total* electron attachment cross section is T independent (see Fig. 36a). This implies that at ~ 0.0 eV the formation of SF_6^- (see Section VII) should decrease with increasing T. Indeed such a decrease has been reported

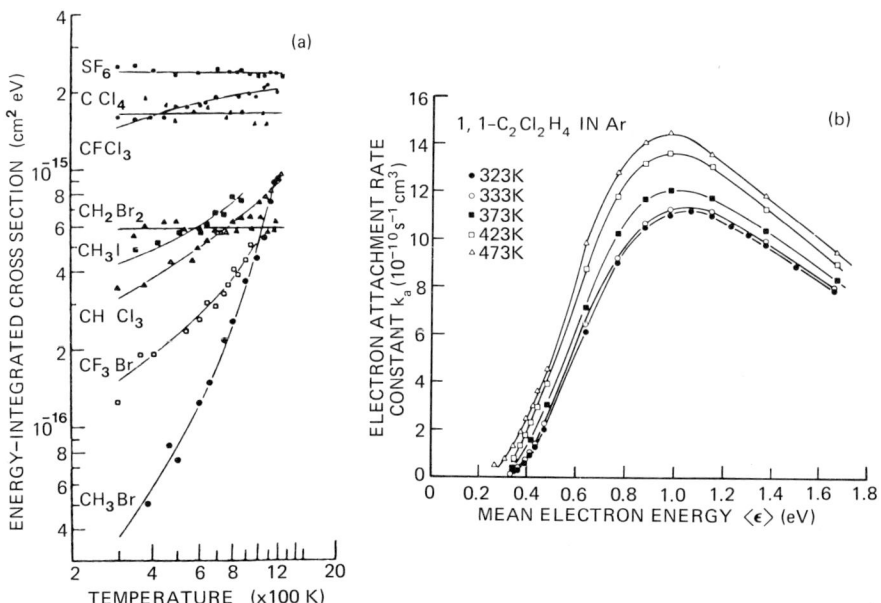

Fig. 36. (a) Energy-integrated attachment cross sections as a function of temperature for various halocarbons and SF_6 [From Spence and Schulz, 1973. Reproduced with permission from the *Journal of Chemical Physics*.] (b) Total electron attachment rate constant as a function of mean electron energy for 1,1-$C_2Cl_2H_4$ in Ar at several temperatures [Christophorou et al., 1980; McCorkle et al., (1982b).]

(Hickam and Berg, 1958) (see also Section VII.C on a large "negative" temperature effect, i.e., a decrease of the total attachment rate constant with increasing T).

In Fig. 36a is shown the variation with T of the energy-integrated total attachment cross section for various polyatomic molecules studied by Spence and Schulz (1973). Three observations can be made with respect to Fig. 36a: (i) the spread in the magnitude of the total attachment cross section seen at room temperature decreases as the temperature increases, (ii) the smaller the magnitude of the attachment cross section at room temperature the larger seems to be its temperature dependence, and (iii) the attachment cross section for certain polyatomic molecules approaches the theoretical maximum value (see Fig. 25 and Section VII) as the temperature increases.

Finally, in Fig. 36b are shown the results (Christophorou et al., 1980; McCorkle et al., 1982b) on the T-dependence of the total electron attachment rate constant as a function of $\langle \varepsilon \rangle$ for $1,1\text{-}C_2Cl_2H_4$. When the $\sigma_a(\varepsilon, T)$ were deconvoluted from these data, structure was indicated which could arise from vertical transitions to a repulsive negative-ion state from the $v = 0$ and $v = 1$ levels of the neutral molecule.

VI. Dissociative Electron Attachment to Electronically Excited Molecules

To our knowledge the only existing measurements on electron attachment to electronically excited molecules [reaction (10)] are those on the dissociative attachment to the O_2 molecule excited in the lowest (metastable) $a\,^1\Delta_g$ state, which lies 0.98 eV above the ground state, $X\,^3\Sigma_g^-$, of O_2. Burrow (1973) found the dissociative attachment cross section (at its maximum; located 5.5 eV above a $^1\Delta_g$ or 6.5 eV above $X\,^3\Sigma_g^-$) for the reaction

$$e + O_2(a\,^1\Delta_g) \longrightarrow O_2^-(^2\Pi_u) \longrightarrow O^-(^2P) + O(^3P) \quad [3.64\text{-eV limit}] \quad (52)$$

to be $(4.6 \pm 1.3) \times 10^{-18}$ cm^2, i.e., 3.5 ± 1 times larger than the peak cross section value of 1.3×10^{-18} cm^2 (at 6.5 eV) from the ground state $X\,^3\Sigma_g^-$, viz.,

$$e + O_2(X\,^3\Sigma_g^-) \longrightarrow O_2^-(^2\Pi_u) \longrightarrow O^-(^2P) + O(^3P) \quad [3.64\text{-eV limit}] \quad (53)$$

More recently Belić and Hall (1981) reported for reaction (52) a peak cross section of 3.8×10^{-18} cm^2. These authors also reported peak cross section values of 1.8×10^{-18} cm^2 and $< 10^{-20}$ cm^2 at 7.5 and 8.5 eV above a $^1\Delta_g$, respectively, for the dissociative attachment reactions

$$e + O_2(a\,^1\Delta_g) \longrightarrow O_2^-(^2\Sigma_g^+) \longrightarrow O^-(^2P) + O(^1D) \quad [5.61\text{-eV limit}] \quad (54)$$

$$e + O_2(a\,^1\Delta_g) \longrightarrow O_2^-(^2\Sigma_g^+?) \longrightarrow O^-(^2P) + O(^1S) \quad [7.89\text{-eV limit}] \quad (55)$$

6 Electron Attachment Processes

Calculations by Bottcher and Buckley (1979) indicated that the cross section for dissociative attachment to the metastable c $^3\Pi_u$ state of H_2, viz.,

$$e + H_2(c\ ^3\Pi_u) \longrightarrow H_2^-(^2\Pi_u) \longrightarrow H^-(1s^2) + H(2p) \tag{56}$$

is substantially larger ($\gtrsim 10^{-18}$ cm^2) than for dissociative attachment to the low vibrational levels of the ground electronic state $X\ ^1\Sigma_g^+$ of H_2. The possibility that σ_{DA} is considerably larger for excited- than for ground-state molecules is most interesting. It raises the possibility that in hydrogenic plasmas metastable $H_2(c\ ^3\Pi_u)$ molecules may play an important role in forming negative ions by capturing slow electrons, when substantial numbers of such metastables are produced.

Dissociative attachment processes for electronically excited molecules naturally have lower energy onsets and should be significant in cases where the population of low-lying metastable states is high.

VII. Molecular Parent Negative Ions

For a parent negative ion to form [reaction (7), Section II] the electron affinity of the molecule must be positive. If the excess energy of the ion is not removed quickly, however, it will be destructed by autodetachment within an average time τ_a, even though its EA is positive. If the autodetachment lifetime τ_a is $\gtrsim 10^{-6}$ s the transient parent anion can be detected and its τ_a measured directly (e.g., by using conventional time-of-flight mass spectrometers). On the other hand, for the total pressures normally employed in swarm experiments the time between collisions of the transient parent anion and a stabilizing body (usually a buffer gas molecule) is much smaller than τ_a, and thus such ions are completely stabilized; swarm studies then can provide $\sigma_c(\varepsilon)$ for these ions. If the parent anion is moderately short-lived ($10^{-12} \lesssim \tau_a < 10^{-6}$ s), its complete stabilization by collision requires high pressure in a swarm experiment. In this section we discuss long-lived parent negative ions and moderately short-lived parent negative ions.

A. Long-Lived Parent Negative Ions

A variety of polyatomic structures have been found (see Christophorou, 1971, 1978a) to capture thermal and epithermal electrons in the gas phase via a nuclear-excited Feshbach resonance mechanism (Section II) and to form long-lived ($\tau_a > 10^{-6}$ s) parent negative ions. Such structures include benzene derivatives with highly electron withdrawing substituents (e.g., $-NO_2$,

—CN, —CHO); —CN-(poly)substituted organic molecules [e.g., $C_2(CN)_4$]; a variety of perfluorinated organic compounds with π or with only σ orbitals (e.g., see Table VIII) and other multiply halogenated molecules for which dissociative attachment is either not energetically possible or not too fast (e.g., C_2Cl_4); C=O-containing organic molecules [e.g., $(CHO)_2$, $(CF_3)_2CO$]; strained molecules (e.g., azulene, cyclooctatetraene); higher aromatic hydrocarbons (e.g., anthracene, 1,2-benzanthracene); and miscellaneous organic [e.g., $(C_5H_5)_2Co$] and inorganic [e.g., SF_6, SF_4, MF_6 (M = Se, W, Re, U)] polyatomics. Hadjiantoniou et al. (1973b) also reached the conclusion that organic molecules containing the groups —COCO—, —COCH(OH)—, —COOH, and =CHCHO capture thermal electrons and form long-lived parent negative ions.

For all observed long-lived parent negative ions which are formed via a nuclear-excited Feshbach resonance mechanism, the EA is positive (>0 eV), the number of vibrational degrees of freedom N is quite large, and the electron attachment cross section $\sigma_a(\varepsilon)$ is substantial. A positive EA is necessary to facilitate strong binding for the attached electron and a large N is required to provide an effective distribution of the metastable ion's excess energy amongst its vibrational degrees of freedom to delay autodetachment.

1. Lifetimes

The data summarized by Christophorou (1978a)* show that the τ_a is crucially affected by the details of the molecular structure and by the nature of the functional groups present in the molecule.[†] This is clearly illustrated by the values of τ_a for the parent anions of NO_2-containing disubstituted benzene derivatives in Table XIII. Christophorou and co-workers studied over 40 nitrobenzenes containing in addition to the NO_2 group another substituent X around the benzene periphery and investigated the effects of the nature and the position of X on τ_a. These compounds attach strongly thermal and epithermal electrons and form long-lived parent negative ions (unless a fast dissociative attachment process effectively competes), the electron being captured into a delocalized π orbital (Johnson et al., 1975; Christophorou, 1978a). The lifetimes of the parent anions of these compounds show an

---

* See this reference for a discussion of the methods employed to determine τ_a and for a discussion of long-lived *fragment* anions.

† Unfavorable Franck–Condon factors between the initial (unstable ion) and final (neutral molecule and electron) states can allow considerable time to elapse before the ion's excess energy is concentrated in a mode which ejects the electron. Long-lived NISs can also result from electronic state and symmetry selection rules as well as from distinct structural characteristics—such as the molecular geometry and the geometrical changes that often occur concomitantly with electron capture.

TABLE XIII

Lifetimes of Long-Lived Parent Negative Ions of NO_2-Containing Disubstituted Benzene Derivatives[a]

Compound	Substituent X	Lifetime (10^{-6} s)		
		$[o\text{-}XC_6H_4NO_2]^{-*}$	$[m\text{-}XC_6H_4NO_2]^{-*}$	$[p\text{-}XC_6H_4NO_2]^{-*}$
Nitroanisole	OCH_3	16	52	10
Nitrobromobenzene	Br	18	21	10
Nitrofluorobenzene	F	17	28	10
Nitrophenol	OH	460[b,c]	31[c]	14[c]
Nitrochlorobenzene	Cl	17	47[d]	14
Nitrotoluene	CH_3	13[e]	19[e]	14[e]
Nitroaniline	NH_2	46[c]	21[c]	15[c]
Nitrobenzene	H	—	—	18[d]
Nitrothiophenol	SH			23
Nitrobenzaldehyde	CHO	395	205	47
Nitrobenzoic acid	COOH	200	338	142
Nitro-$\alpha\alpha\alpha$-trifluorotoluene	CF_3	200	187[b]	143
Nitroacetophenone	$COCH_3$	189	310[b]	196
Nitrobenzonitrile	CN	209	315[b]	205
Dinitrobenzene	NO_2	463	537	421

[a] Unless otherwise indicated the lifetime data listed are from Johnson et al. (1975).
[b] Lifetime decreases with increasing electron energy.
[c] Data of Hadjiantoniou et al. (1973a).
[d] Data of Naff et al. (1971).
[e] Data of Christophorou et al. (1973a).

increase with increasing EA (Johnson et al., 1975; Christophorou, 1978a) and depend on the π-electron donor–acceptor properties of the substituent groups (NO_2 and X): strong electron withdrawing substituents X (NO_2 is a strongly electron withdrawing group as well) greatly increase τ_a, while τ_a is little affected by π-electron donating substituents. Actually, CNDO-2 molecular orbital calculations by Johnson et al. (1975) on nitrobenzenes with two substituents, NO_2 and X, in the *para* position to avoid intramolecular complexing between them, showed that when X is an electron acceptor, the magnitude of τ_a correlates with the amount of π-electron charge withdrawn from the ring, while τ_a is little affected by the amount of π-electron charge donated to the ring by X when X is an electron donor. The large values of τ_a for the *ortho* isomers of nitrophenol, nitroaniline, and nitrobenzaldehyde indicate the sensitivity of τ_a to the intramolecular interaction between the substituents (X and NO_2) at the *ortho* position (Hadjiantoniou et al., 1973a).

The τ_a is expected to increase with increasing N and some authors (e.g., Naff et al., 1968) claim to have observed a quantitative dependence of τ_a on N. A comprehensive analysis of all the available data by Christophorou (1978a) showed that the dependence of τ_a on N is rather complicated. Christophorou (1978a) found—as did Naff et al. (1968)—that $\ln \tau_a$ increases with N for particular groups of molecules. However, this dependence varies greatly from one group of molecules to another and within a particular group of molecules the $\ln \tau_a$ vs. N function shows a leveling off as N increases, perhaps due, in part, to the increase in $\sigma_a(\varepsilon)$ with increasing molecular size. Besides $\sigma_a(\varepsilon)$ and N other molecular properties (e.g., EA; see also the last footnote) are known to influence τ_a. The work of Christophorou et al. (1977b) on $C_6F_6^{-*}$ showed also that only a fraction ($\sim \frac{1}{3}$ for $C_6F_6^{-*}$) of the available vibrational degrees of freedom may be involved in the sharing of the excess energy of the anion.

Parent negative-ion lifetimes are expected (see Section VII.A.3) to depend on the internal energy of the anion and thus should decrease with increasing energy ε of the captured electron. The larger the internal energy of the metastable anion the larger is the rate of autodetachment. In the preponderance of the cases studied so far, time-of-flight mass spectrometers with RPD sources were employed and the measured lifetimes were not found to vary with ε. This can be attributed to the large widths of the electron-beam pulses employed and the precipitous decline of $\sigma_a(\varepsilon)$ above thermal energy so that even though the mean energy of the electron pulse is increased, experimentally one observes predominantly ions produced by capture of electrons whose energy coincides with the position of the cross section maximum. In certain such experiments, however, the τ_a was found to decrease with ε. The first experimental observation of the dependence of τ_a on ε was reported by Collins et al. (1970) for 1,4-naphthoquinone and *p*-benzoquinone. Subsequently, similar observations were made for a number of other anions

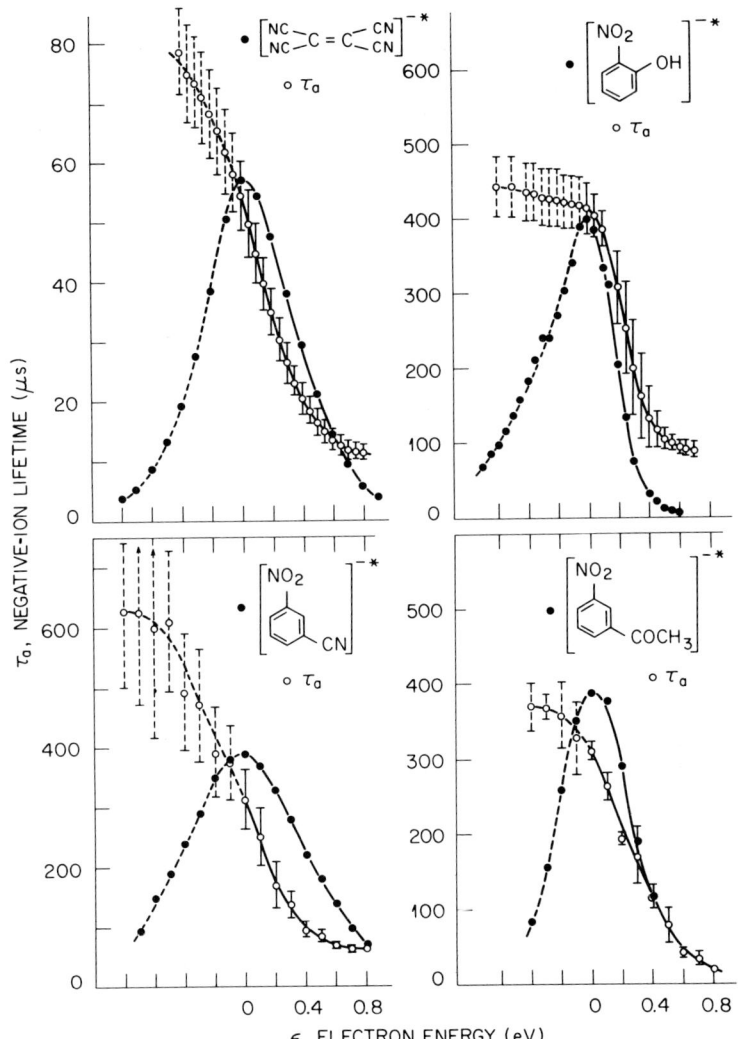

Fig. 37. Negative-ion current (●) in arbitrary units and parent negative-ion lifetime (○) as a function of electron energy for $C_2(CN)_4$, o-$C_6H_4NO_2OH$, m-$C_6H_4NO_2CN$, and m-$C_6H_4NO_2COCH_3$. [From Christophorou (1978a).]

(e.g., Christophorou *et al.*, 1973b; Cooper and Compton, 1973; Johnson *et al.*, 1975; Syprou *et al.*, 1983) and some examples are shown in Fig. 37. The use of highly monoenergetic electron beams is therefore indicated as a further investigation of the variation of τ_a with ε.

2. Attachment Rate Constants and Cross Sections

The rate constants and cross sections of the long-lived parent anions of complex polyatomic molecules are usually very large at thermal and near-thermal (<0.5 eV) energies [see Christophorou (1971, 1978a), Figs. 25 and 28, and Table VIII], approaching their maximum values for s-wave capture. The $\sigma_{DA}(\varepsilon)$ are largest at thermal energies and decline precipitously thereafter* [see Fig. 25, Christophorou et al. (1971b), Gant and Christophorou (1976), Christodoulides et al. (1979), Christodoulides and Christophorou (1979), Pai et al. (1979)]; the shapes of the relative cross sections in most beam studies are instrumental [see Fig. 37 and Christophorou (1978a)]. In certain cases, however, the cross sections for long-lived parent anions peak at energies higher than thermal [e.g., at 0.7 eV for $C_6H_5COCOC_6H_5^{-*}$ (Christophorou et al., 1973b); at 0.6 eV for $n\text{-}C_4F_{10}^{-*}$, $n\text{-}C_5F_{12}^{-*}$, and $n\text{-}C_6F_{14}^{-*}$ (Spyrou et al., 1983)], but still, below 1 eV. Thermal attachment rate constants for fluorocarbons were given in Table VIII and for SF_6 in Table IX.

3. A Theoretical Treatment

The observed large cross sections and long lifetimes for complex polyatomic molecules have been reconciled (Compton et al., 1966; Klots, 1967; Christophorou et al., 1973b; 1977b) by assuming that the former results from a strong transition between the neutral molecule and the metastable anion, and the latter from a distribution of the anion's excess energy (comprised primarily of the molecule's electron affinity and the captured electron's kinetic energy) among its many internal degrees of freedom which delays the time required for the metastable anion to return to a configuration which will lead to autodetachment. The latter assumption explicitly implies that intramolecular relaxation in the ion is fast and that all forms of energy input are equivalent.

Using the above hypothesis Compton et al. (1966), Klots (1967), and Christophorou et al. (1973b, 1977b) related the rates for the capture and the autodetachment processes through the principle of detailed balance, and with the aid of unimolecular reaction-rate theory (see Robinson and Holbrook, 1972; Forst, 1973) derived expressions relating $\tau_a, \sigma_a(\varepsilon)$, EA, N, and v (the vibrational frequencies of the molecule). In particular, Christophorou et al. (1973b, 1977b) considered the reaction

$$e(\varepsilon_i) + AX(\varepsilon_t + \varepsilon_z) \xrightarrow{\sigma_a(\varepsilon_i)} AX^{-*}(\varepsilon_t + \varepsilon_z' + \varepsilon_i + EA)$$
$$\text{(a)} \hspace{4cm} \text{(b)}$$
$$\underset{\sigma_a^{(*)}(\varepsilon_f)}{\overset{\tau_a^{-1}}{\rightleftarrows}} AX^{(*)}[\varepsilon_t + \varepsilon_z + (\varepsilon_i - \varepsilon_f)] + e(\varepsilon_f) \quad (57)$$
$$\text{(c)}$$

---

* These properties—large cross sections at ~ 0.0 eV with a sharp decline thereafter—make some of these compounds [e.g., $c\text{-}C_6F_{10}$, for which $(k_a)_{th}$ ($=3.9 \times 10^{-7}$ cm^3 s^{-1}) is the largest known to date] ideal for detection of "zero"-energy electrons.

where ε_t is the translational energy (assumed to be zero in their analysis) of the neutral *polyatomic* molecule (AX) and the negative ion, ε_i and ε_f are, respectively, the energies of the incident and autodetached electrons, $\varepsilon_z \, (= \frac{1}{2} \Sigma_{i=1}^N h\nu_i)$ and $\varepsilon_z' \, (= \frac{1}{2} \Sigma_{i=1}^N h\nu_i')$ are, respectively, the zero-point energies of the neutral molecule and the anion, ν_i and ν_i' are, respectively, the neutral molecule's and the ion's *i*th fundamental vibrational frequency, and the rest of the symbols are as they have been defined earlier. We note that the attachment cross section (the one actually measured experimentally) for (a) → (b) is $\sigma_a(\varepsilon_i)$, while the attachment cross section for (c) → (b) is $\sigma_a^{(*)}(\varepsilon_f)$.

The rates for (b) and (c) [Eq. (57)] are related through the principle of detailed balance as

$$\tau_a^{-1} = (\rho^0/\rho^-) v_f \sigma_a^{(*)}(v_f) \tag{58}$$

where ρ^0 is the number of states per unit energy per unit volume for the final products $AX^{(*)}$ and $e(\varepsilon_f)$, ρ^- is the number of states per unit energy for the negative ion AX^{-*}, and v_f is the velocity of the autodetached electron. Christophorou et al. (1973b) assumed that $v_f \sigma_a^{(*)}(v_f) = v_i \sigma_a(v_i)$ and $v_i' = v_i$, viewed the molecule as a set of weakly coupled harmonic oscillators, assumed that the ion's internal energy is shared *equally* between the system's N vibrational degrees of freedom ($N = 3n - 6$ for a nonlinear molecule with n atoms), and used for ρ^- the continuous-energy density expression of Whitten and Rabinovitch (1963)

$$\rho^- = [\varepsilon_T + (1 - \beta\omega')\varepsilon_z']^{N-1} \bigg/ \Gamma(N) \prod_{i=1}^N h\nu_i \tag{59}$$

In Eq. (59) ε_T is the total internal energy of the metastable ion in excess of ε_z', $\Gamma(N)$ is a gamma function of N, $\Pi_{i=1}^N h\nu_i$ is the product of all vibrational energies $h\nu_i$ (assumed to be the same for the ion and the neutral molecule), and $1 - \beta\omega'$ is a correction factor described by Whitten and Rabinovitch (1963) which is to be evaluated at each value of $\varepsilon_T/\varepsilon_z'$ (see also Christophorou, 1978a).

The density of states of the neutral molecule to which the metastable negative ion decays, ρ^M, and the density of states of the products [neutral molecule + $e(\varepsilon_f)$] to which the metastable negative ion decays, ρ^0, were assumed to be given, respectively, by

$$\rho^M = [(\varepsilon_i - \varepsilon_f) + (1 - \beta\omega)\varepsilon_z]^{N-1} \bigg/ \Gamma(N) \prod_{i=1}^N h\nu_i \tag{60}$$

and

$$\rho^0 = \rho^e N^M$$
$$= \int_0^{\varepsilon_i} m^2 v_f / \pi^2 \hbar^3 \left\{ [\varepsilon_i - \varepsilon_f + (1 - \beta\omega)\varepsilon_z]^{N-1} \bigg/ \Gamma(N) \prod_{i=1}^N h\nu_i \right\} d\varepsilon_f. \tag{61}$$

In the above equations, $1 - \beta\omega$ is a correction factor for the neutral molecule analogous to that, $1 - \beta\omega'$, for the negative ion. It should be observed that the density of states ρ^0 was taken equal to the density of states of the ejected electron ($= m^2 v_f / \pi^2 \hbar^3$) multiplied by the number of molecular states N^M. From Eqs. (58)–(61), we have (Christophorou et al., 1973b)

$$\tau_a^{-1} = \frac{\Gamma(N) \prod_{i=1}^{N} h\nu_i}{(\varepsilon_i + \text{EA} + a' \varepsilon_z)^{N-1}} \left(\frac{2\varepsilon_i}{m}\right)^{1/2} \sigma_a(\varepsilon_i)$$

$$\times \int_0^{\varepsilon_i} \frac{m^2}{\pi^2 \hbar^3} \left(\frac{2\varepsilon_f}{m}\right)^{1/2} \frac{(\varepsilon_i - \varepsilon_f + a\varepsilon_z)^{N-1}}{\Gamma(N) \prod_{i=1}^{N} h\nu_i} d\varepsilon_f \qquad (62)$$

where a', a, and ε are $1 - \beta\omega'$, $1 - \beta\omega$, and $\tfrac{1}{2} mv^2$, respectively.

If we assume a twofold degeneracy for the negative ion and multiply and divide Eq. (62) by $(\varepsilon_i + a\varepsilon_z)^{N+1/2}$ we have

$$\tau_a^{-1} = \frac{\sigma_a(\varepsilon_i) \varepsilon_i^{1/2}}{(\varepsilon_i + \text{EA} + a' \varepsilon_z)^{N-1}} \frac{m}{\pi^2 \hbar^3} (\varepsilon_i + a\varepsilon_z)^{(N+1/2)} I \qquad (63)$$

where $I = \int_0^A X^{1/2} (1-X)^{N-1} dX$, $A = \varepsilon_i / (\varepsilon_i + a\varepsilon_z)$, and $X = \varepsilon_f / (\varepsilon_i + a\varepsilon_z)$. The quantity I can be evaluated numerically for any value of A. If the assumptions that $\sigma_a(v_i) = \sigma_a^{(*)}(v_f)$ and $v_i' = v_i$ are removed, we have

$$\tau_a^{-1} = \frac{\Gamma(N) \prod_{i=1}^{N} h\nu_i'}{(\varepsilon_i + \text{EA} + a' \varepsilon_z')^{N-1}} \frac{m}{\pi^2 \hbar^3}$$

$$\times \int_{\varepsilon_f = 0}^{\varepsilon_i} \frac{(\varepsilon_i - \varepsilon_f + a\varepsilon_z)^{N-1}}{\Gamma(N) \prod_{i=1}^{N} h\nu_i} \varepsilon_f \sigma_a^{(*)}(\varepsilon_f) d\varepsilon_f. \qquad (64)$$

The assumptions that the molecules are initially in their ground vibrational states, that the density of states of the neutral molecule and the negative ion can be represented by similar expressions, and that the excess energy in both the negative ion and the neutral molecule is shared equally among the available vibrational degrees of freedom implicit in the derivation of the preceding expressions are difficult to assess quantitatively. This treatment also restricts itself to the intermediate state (i.e., the transient negative ion) and does not consider in detail the entrance (initial electron capture process) and the exit (final autodetaching process) steps, both of which depend strongly on selection rules and the symmetry of the state(s) involved. Nevertheless, Eq. (64) is a rather important one; it relates $\tau_a(\varepsilon)$, $\sigma_a(\varepsilon)$, N, $h\nu_i$, $h\nu_i'$, EA, and also the energies ε_i and ε_f of the incoming and outgoing electrons. It allows a qualitative understanding of long-lived negative ions as well as of the energy dependence of τ_a on ε_i (see Compton et al., 1966; Klots, 1967; Christophorou et al., 1973b, 1977b; Johnson et al., 1975; Christophorou, 1978a). A more comprehensive theoretical understanding is, however, needed [see a relevant disscussion in Ivanov and Ponomarev (1977)].

While long-lived parent negative ions formed by electron capture in the field of the *ground electronic* state of complex polyatomic molecules are abundant, long-lived parent negative ions formed by electron capture in the field of an *excited electronic* state are rare. In the case of atoms, He^{-*} [(1s2s2p) $^4\text{P}_j$] seems to be a distinct such example (Sweetman, 1960). In the case of molecules, the long-lived p-benzoquinone parent anion observed (Christophorou et al., 1969a; Collins et al., 1970) at ~ 2.0 eV and attributed to electron capture in the field of the lowest triplet state of the molecule seems to be a distinct such example [see discussion in Christophorou (1978a)] for molecules.

B. Moderately Short-Lived Parent Negative Ions

1. Effect of the Density and Nature of the Gaseous Medium on Electron Attachment

Electron attachment to molecules—and other processes and molecular properties involving separated charges—are dramatically influenced by both the density and the nature of the medium in which the electron–molecule interaction takes place. In spite of the basic and applied significance of the understanding of these environmental influences, our knowledge of the transition from the gasesous (isolated-molecule) to the condensed-phase behavior is still incomplete. Recent "interphase studies" [Christophorou (1976), Chapter 4 of Volume 2] provide a new insight as to the effects of the density and the nature of the environment on the fundamental electron–molecule interaction processes at densities intermediate to those corresponding to low-pressure gases and liquids and on the gradual transition from "isolated-molecule" to "condensed-phase" behavior. In this section we focus on electron attachment to molecules in high ($\lesssim$ few hundred Torr) and very high (to $\sim 40{,}000$ Torr) pressure gases.

It has long been recognized that of the electron attachment processes referred to in Section II, the nondissociative electron attachment process involving a transient ion AX^{-*} which requires collisional stabilization is the one (though not the only one) to show a strong dependence on the medium. Consider, then, the AX molecule to be embedded in a gaseous medium M of density N_M. The apparent rate constant k_a of attachment of electrons to AX forming AX^- via the simple process

$$\text{AX} + \text{e} \xrightarrow{k_c} \text{AX}^{-*} \text{(vibr.)} \begin{array}{c} \xrightarrow{\tau_a^{-1}} \text{AX}^{(*)} + \text{e}^{(\prime)} \quad (65\text{a}) \\ \xrightarrow{k_{st} N_M} \text{AX}^- + \text{energy} \quad (65\text{b}) \end{array}$$

would be

$$k_a = k_c[k_{st}N_M/(\tau_a^{-1} + k_{st}N_M)]. \tag{66}$$

The dependence of k_a on N_M is a function of the relative magnitudes of τ_a^{-1} and $k_{st}N_M$. If $\tau_a^{-1} \ll k_{st}N_M$, k_a is independent of N_M. When, however, $\tau_a^{-1} \gg k_{st}N_M$, $k_a \propto N_M$ if other processes do not enter as N_M increases (see below).

For the long-lived parent anions discussed in the preceding section, the long τ_a, large σ_a, and the position of the cross section maxima at ~ 0.0 eV minimize the effects of the gaseous medium on them (Christophorou, 1972, 1976). The effect of the nature and density of the gaseous medium, however, is most profound when the nondissociative attachment process leads to moderately short-lived parent anions. To illustrate these effects we shall focus on the O_2 molecule.

2. Nondissociative Electron Attachment to O_2

Nondissociative electron attachment to O_2 has been the subject of many studies for over half a century and has been reviewed by many authors (e.g., Caledonia, 1975; Christophorou, 1976, 1978b; Massey, 1976; Hatano and Shimamori, 1981). The ground electronic state of O_2 is the triplet $X\,^3\Sigma_g^-$, the lowest NIS has the configuration $X\,^2\Pi_g$, and the electron affinity of O_2 is 0.44 eV (Chapter 6 of Volume 2); the O_2^{-*} formed at low energies is moderately short-lived (see later this section) and hence it can be stabilized collisionally in a high-pressure swarm experiment. In Fig. 38 are shown approximate potential-energy curves for O_2 and O_2^- on the basis of which the formation of O_2^- from O_2 below ~ 1 eV proceeds as

$$O_2(X\,^3\Sigma_g^-; v=0) + e \xrightarrow{k_c} O_2^{-*}(X\,^2\Pi_g; v' \geq 4) \begin{array}{c} \xrightarrow{\tau_a^{-1}} O_2^{(*)} + e^{(')} \quad (67a) \\ \xrightarrow[(+M)]{k_{st}N_M} O_2^-(X\,^2\Pi_g; v' < 4) + M + \text{energy} \quad (67b) \end{array}$$

i.e., capture of the electron into vibrational levels (fourth, fifth, etc.) of O_2^- lying above the $v=0$ level of O_2. The lowest v' level of O_2^- above the $v=0$ of O_2 is at 0.082 eV and thus a peak in the energy dependence of the measured attachment rate constant and cross section is expected, and indeed found, at this energy.

The two-step mechanism of reaction (67) (Bloch and Bradbury, 1935; Herzenberg, 1969) is often (when the time between collisions of O_2^{-*} and M is much longer than the autodetachment lifetime of O_2^{-*}) written as a one-step process:

$$e + O_2 + M \xrightarrow{k_{aM}} O_2^- + M + \text{energy} \tag{68}$$

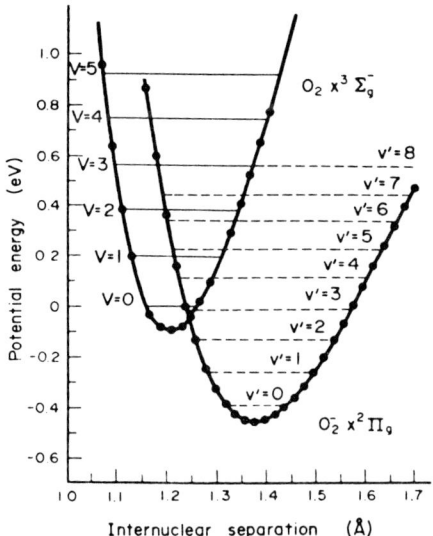

Fig. 38. Approximate potential-energy curves for O_2 and O_2^- (Boness and Schulz, 1970). The O_2^- curve should have its minimum at 1.341 Å (Celotta et al., 1972). [Reproduced with permission from the *Physical Review*.]

where k_{aM} is the three-body attachment coefficient. Although the role of collisional detachment ($O_2^{-*} + M \rightarrow O_2 + M + e$) is not yet clear, reactions (67) [or (68)] explain satisfactorily most of the experimental observations at room temperature and low densities (say $\lesssim 100$ Torr). However, at higher gas densities and certain types of M neither (67) nor (68) can explain the experimental results (see below).

Three-body attachment rates as a function of $\langle \varepsilon \rangle$ and M ($= O_2, N_2, C_2H_4, C_2H_6$) obtained from swarm experiments at low pressures are shown in Fig. 39a. In agreement with process (67), they show a broad maximum at ~ 0.09 eV, but no structure due to capture into higher vibrational levels ($v' > 4$) is resolved. Such structure, which occurs at the positions of the vibrational levels $v' \geq 4$ of $O_2^-(X\,^2\Pi_g)$, is clearly seen, however, (Fig. 39b) in the three-body attachment rate constants as a function of ε in pure O_2 ($2 \times 10^{14} \leq N_{O_2} \leq 10^{16}$ molecules cm^{-3}) obtained (Spence and Schulz, 1972) from electron-beam (~ 100-meV half-width) experiments. Similar structure has been observed independently in the swarm-unfolded cross sections for formation of O_2^- in N_2 (McCorkle et al., 1972).

In Fig. 39b are shown also the results of a theoretical calculation of the effective rate constant for reaction (67) ($M = O_2$) by Chapman and Herzenberg (see Spence and Schulz, 1972). Chapman and Herzenberg (see

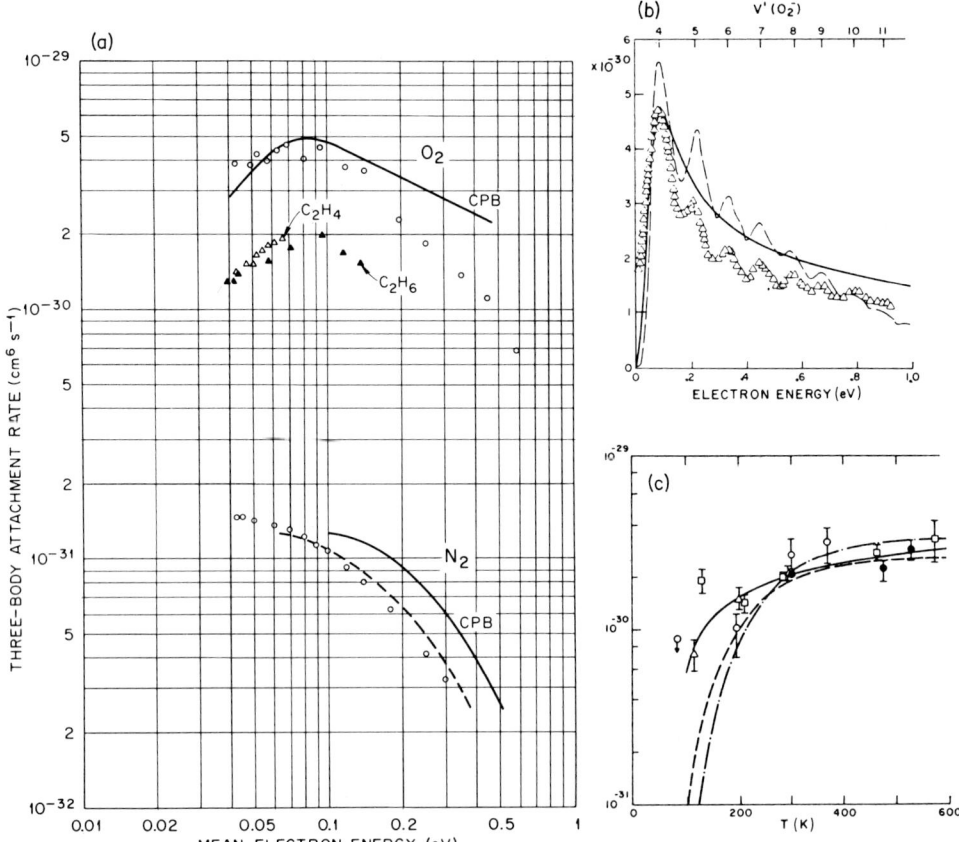

Fig. 39. (a) Three-body attachment rates as a function of mean electron energy for reaction (68); $M = O_2$: (○) (McCorkle et al., 1972), (—) (Chanin et al., 1962); $M = C_2H_4$: (△) (Goans and Christophorou, 1974); $M = C_2H_6$: (▲) (Goans and Christophorou, 1974); $M = N_2$: (○) (McCorkle et al., 1972), (—) (Chanin et al., 1962); the broken line represents the data of Chanin et al. (1962) replotted by McCorkle et al. (1972) using a different energy scale. [From Christophorou (1978b). Reprinted with permission from *Rodiat. Phys. Chem.* **12**, 1978, Pergamon Press, Ltd.] (b) k_{aO_2} vs. electron energy for the reaction $e + O_2 + O_2 \xrightarrow{k_{aO_2}} O_2^- + O_2$. (—) Swarm data of Pack and Phelps (1966a), (---) beam data of Spence and Schulz (1972) normalized to the swarm data, (△) results calculated by Chapman and Herzenberg (1972) (see Spence and Schulz, 1972). [From Caledonia (1975).] (c) k_{aO_2} vs. T for the reaction $e + O_2 + O_2 \xrightarrow{k_{aO_2}} O_2^- + O_2$: (□) Van Lint et al. (1971), (△) Truby (1972), (○) Pack and Phelps (1966b), (●) Chanin et al. (1962). Empirical fits: (—) (Truby, 1972), (---) (Phelps, 1969), (-··-) (Caledonia, 1975). [From Caledonia (1975).] [Parts (b) and (c) are reprinted with permission from Caledonia (1975). Copyright 1975, American Chemical Society.]

Herzenberg, 1969) employed the Breit–Wigner (1936) formalism for the cross section $\sigma_c^n(a, \varepsilon)$ for capture of an electron by O_2 initially in the state a forming a negative ion in the nth vibrational level (Eq. (67)], viz.,

$$\sigma_c^n(a, \varepsilon) = \pi \lambdabar^2 (C^n/g) \{\Gamma_a^n \Gamma^n / [(\varepsilon - E_n)^2 + (\tfrac{1}{2}\Gamma^n)^2]\}. \quad (69)$$

In Eq. (69) C^n is the number of degenerate resonances corresponding to the nth level of the anion, g is the total number of spin states of the electron and the molecule, $\lambdabar$ is the electron de Broglie wavelength, Γ_a^n is the partial width for the resonance from state a to n, Γ^n is the total width from all states a, and E_n is the energy at which the resonance occurs. Consequently, the total *capture* rate constant k_c for molecules in the initial state a is

$$k_c = \sum_n \int \sigma_c^n(a, \varepsilon) \, v f_M(\varepsilon) \, d\varepsilon, \quad (70)$$

where v is the electron velocity and $f_M(\varepsilon)$ is a Maxwellian distribution function. If now the autodetachment lifetime of the resonance, $\tau_a^n = \hbar/\Gamma^n$, is shorter than the stabilization time τ_{st}, the percentage of the metastable anions in state n which are stabilized is τ_a^n/τ_{st}. To determine τ_{st} for O_2^{-*} in pure O_2, Herzenberg (1969) used the Langevin expression for the classical cross section for spiraling collisions between O_2^{-*} and O_2, viz.,

$$\tau_{st} = (2\pi e N p)^{-1} (M_r/\alpha)^{1/2}, \quad (71)$$

where p is the probability of stabilization per collision of O_2^{-*} and O_2, N is the number density, M_r is the reduced mass, and α is the static polarizability of O_2.

The total *attachment* rate constant for O_2 initially in state a (for $\tau_a^n < \tau_{st}$) is

$$k_{aO_2} = 2\pi p e \left(\frac{\alpha}{M_r}\right)^{1/2} \left(\sum_n \tau_a^n \int \sigma_c^n(a, \varepsilon) v f_M(\varepsilon) \, d\varepsilon \right). \quad (72)$$

When the width of $f_M(\varepsilon)$ is broad compared to Γ^n, $v f_M(\varepsilon)$ can be taken out of the integral and Eq. (72) becomes

$$k_{aO_2} = \frac{4\pi^3 \hbar^2 e}{m} \left(\frac{\alpha}{M_r}\right)^{1/2} p \left(\sum_n f_M(E_n) \frac{C^n}{g} \lambdabar(E_n) \frac{\Gamma_a^n}{\Gamma^n}\right). \quad (73)$$

For the lowest resonance ($v' = 4$) $\Gamma_a^n/\Gamma^n = 1$. Herzenberg used the vibrational-excitation cross sections of Hake and Phelps (1967) for O_2 and obtained $\Gamma_{a0}^n \simeq 2 \times 10^{-6}$ eV ($a0$ denotes the $v = 0$ level of O_2) for the $v' = 4$ level of O_2^-. An *ab initio* calculation by Koike and Watanabe (1973) gave a width of 4×10^{-6} eV for the same resonance. In comparing the results of their calculation with those of Spence and Schulz, Chapman and Herzenberg replaced $f_M(\varepsilon)$ in Eq. (72) with a Gaussian distribution of 110 meV half-width to approximate the beam energy resolution in the Spence and Schulz experiment. They also considered only transitions from the $v = 0$ of O_2 and took $p = 1$. Their result (Fig. 39b) is in agreement with both Spence and Schulz's beam experiment and the Pack and Phelps (1966a) swarm data.

The temperature dependence of the three-body attachment coefficient for O_2 has been investigated by several workers. In Fig. 39c are shown those data for pure O_2, as they have been summarized by Caledonia (1975). The sources of the experimental results are given in the figure caption. The solid curve was obtained by Truby (1972) by fitting his data at 300, 200, and 113 K to an Arrhenius plot. The dashed line is an estimation of the T dependence of k_{aO_2} (see figure caption) by Phelps (1969) and the dash-dot line was calculated (see figure caption) by Caledonia (1975) from Eq. (73) using $E_n = 0.079$ eV, $p = 1$, $C^n = 4$, and $g = 6$ (see Herzenberg, 1969). Both the Phelps and the Caledonia estimates are lower than the experimental results below ~ 300 K (see discussion later in this section as to a possible reason).

Thermal ($T \simeq 290$–300 K) values of k_{aM} [reaction (68)] are summarized in Table XIV. Although these data are not yet fully understood, it is apparent that for the rare gases, where the only mode of energy transfer from O_2^{-*} to M is translational, k_{aM} is small. Polyatomic molecules with large vibrational and rotational states are more effective stabilizers, as are polar molecules.

Electron attachment to O_2 at higher (>100 Torr) pressures, however, is complex. The results obtained by the various groups and their interpretation are still the subject of much and inconclusive discussion. Let us first briefly review the results of the Oak Ridge National Laboratory group. Christophorou and co-workers studied electron attachment to O_2 in N_2, C_2H_4, and C_2H_6 at total pressures ranging from ~ 200 Torr to, respectively, 27,500, 17,000, and 17,500 Torr. Figure 40a dramatizes the increase of the electron attachment rate $(\alpha w)_0$ as a function of $\langle \varepsilon \rangle$ with increasing N_2 pressure P_{N_2} over the entire pressure range covered.* No saturation of the dependence of $(\alpha w)_0$ on P_{N_2} is evident. The cross section unfolded (McCorkle et al., 1972) from the $(\alpha w)_0$ vs. $\langle \varepsilon \rangle$ functions in Fig. 40a for $P_{N_2} < 1000$ Torr showed a gradual shift toward lower energies as P_{N_2} increased. That the increase of $(\alpha w)_0$ with P_{N_2} does not saturate but takes off just when the simple Bloch–Bradbury mechanism would have predicted such a saturation, is clearly seen in Fig. 40b where $(\alpha w)_0$ for $\langle \varepsilon \rangle \simeq 0.05$ eV is plotted as a function of P'_{N_2} (compressibility-corrected pressure). A similar but stronger behavior is seen for C_2H_6. Only the data for C_2H_4 (see Fig. 40b) exhibit the expected saturation behavior.

The $(\alpha w)_0$ vs. $\langle \varepsilon \rangle$ data for O_2 in N_2 were interpreted (Christophorou, 1972; Goans and Christophorou, 1974) in terms of a model whereby N_2 is assumed

---

* In Fig. 40 $(\alpha w)_0$ represents the measured attachment rate when $P_{O_2} \to 0$ Torr (α is the electron attachment coefficient and w is the electron swarm drift velocity). The units of $(\alpha w)_0$ are $\text{Torr}^{-1} \text{s}^{-1}$; they can be transformed into $\text{cm}^3 \text{s}^{-1}$ by dividing by 3.24×10^{16}.

TABLE XIV

Three-Body Attachment Rate Constants for the Reaction
$e + O_2 + M \xrightarrow{k_{aM}} O_2^- + M + \text{energy}$ ($T \simeq 290\text{--}300$ K)

M	k_{aM} (10^{-30} cm^6 s^{-1})
He	0.033,[a] ~0.03,[b] 0.07[c]
Ne	0.023[a]
Ar	0.05[a]
Kr	0.05[a]
Xe	0.085[a]
H_2	0.48[d]
D_2	0.14[d]
N_2	~0.06,[b] 0.085,[e] 0.09,[f] 0.10,[g] 0.11,[c] 0.15,[h,i] 0.26[j]
O_2	1.7,[j] 2.0,[k] 2.1,[c,l,m] 2.15,[g] 2.2,[d,f,n] 2.3,[e,i] 2.6,[o] 2.8[b]
CH_4	0.34[a]
C_2H_4	~1.5,[p] 1.7,[q] 2.0,[r] 2.3,[o] 2.5,[s] ~3,[a] 3.4[t]
C_2H_6	1.3,[u] ~1.5,[p] 1.7[a]
C_3H_8	3.2,[u] 3.3[a]
n-C_4H_{10}	4.2,[f] 4.5,[u] ~5[a]
n-C_5H_{12}	7.9[a]
neo-C_5H_{12}	7,[r] 8.0[a]
n-C_6H_{14}	8.1[a]
C_6H_6	8.5,[s] 18[q]
CO	1.31[f]
CO_2	~3,[o,q,s,v,w] 3.2,[r] 3.23[g]
H_2O	14,[s,t,v] 15.2[q]
H_2S	9,[s] 10[q]
NH_3	6.8,[s] 7.5[q]
CH_3OH	8.8,[s] 9.6,[q] 11[a]
C_2H_5OH	18[a]
CH_3COCH_3	27,[s] >35[q]

[a] Shimamori and Hatano (1977).
[b] Chanin et al. (1962).
[c] Van Lint et al. (1960).
[d] Shimamori and Hatano (1976a).
[e] Shimamori and Hatano (1976b).
[f] Shimamori and Fessenden (1981).
[g] Crompton et al. (1979).
[h] McCorkle et al. (1972).
[i] Hackam and Lennon (1965).
[j] Young et al. (1963).
[k] Pack and Phelps (1966a).
[l] Nelson and Davis (1971).
[m] Truby (1972).
[n] Warman et al. (1971).
[o] Hurst and Bortner (1959).
[p] Goans and Christophorou (1974).
[q] Bouby and Abgrall (1967).
[r] Kokaku et al. (1979).
[s] Bouby et al. (1970).
[t] Stockdale et al. (1967).
[u] Kokaku et al. (1980).
[v] Pack and Phelps (1966b).
[w] Zastawny (1974).

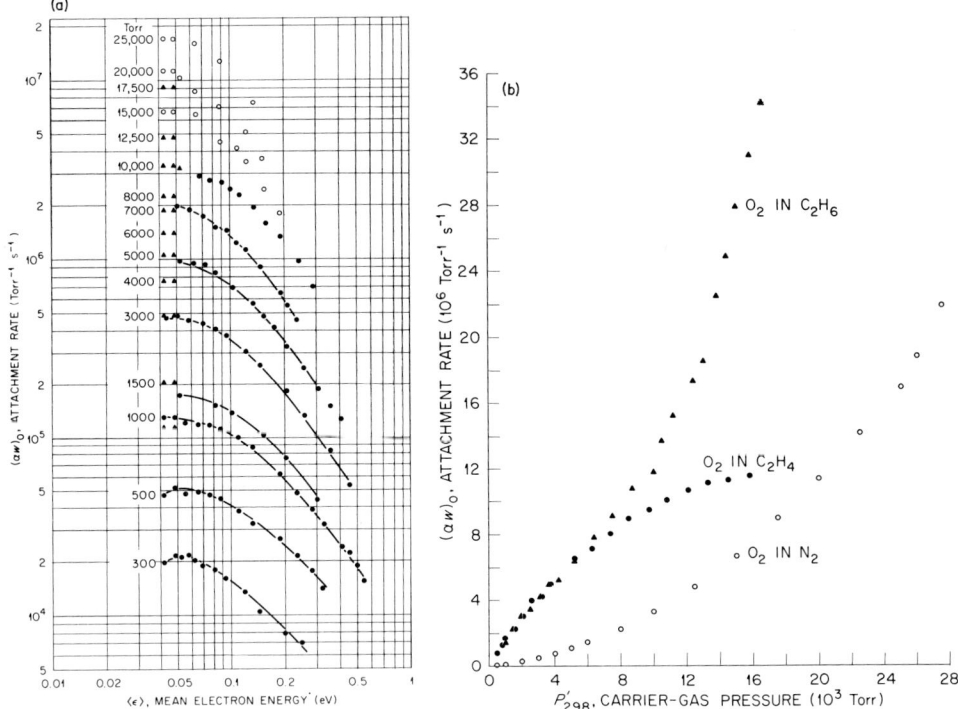

Fig. 40 (a) Attachment rate constant $(\alpha w)_0$ as a function of mean electron energy $\langle \varepsilon \rangle$ for O_2 in N_2, for N_2 pressures between 300 and 25,000 Torr: (●) (McCorkle et al., 1972), (▲) and (○) (Goans and Christophorou, 1974). (b) Attachment rate constant $(\alpha w)_0$ for O_2 in N_2 (○), C_2H_4 (●), and C_2H_6 (▲) as a function of the pressure P'_{298} of the gaseous medium corrected for compressibility. The data plotted are for E/P'_{298} values equal to 0.03 V cm^{-1} Torr^{-1} for N_2 and 0.1 V cm^{-1} Torr^{-1} for C_2H_4 and C_2H_6. The E/P values correspond to a mean electron energy of ~ 0.05 eV. [From Goans and Christophorou (1974).] $(\alpha w)_0$ represents the measured attachment rate constant for $P_{O_2} \to 0$ Torr. [Reproduced with permission from the *Journal of Chemical Physics*.]

(i) to act as a stabilizing third body in distant collisions and (ii) to be involved in close, "sticky," collisions which seriously perturb the O_2^- potential-energy curves [as suggested by the changes in $\sigma_a(\varepsilon)$ with N_2 pressure], i.e., via the two mechanisms

(i)
$$e + O_2 \xrightarrow{k_c} O_2^{-*}$$
$$O_2^{-*} \xrightarrow{\tau_a^{-1}} O_2 + e$$
$$O_2^{-*} + N_2 \xrightarrow{k_{st}N_{N_2}} O_2^- + N_2 + \text{energy}$$
$$\xrightarrow{k_d N_{N_2}} O_2 + N_2 + e$$

6 Electron Attachment Processes

(ii)
$$e + O_2 + N_2 \xrightarrow{k_c N_{N_2}} [O_2^{-*} - N_2]$$
$$[O_2^{-*} - N_2] \xrightarrow{\tau_a'^{-1}} O_2 + N_2 + e$$
$$[O_2^{-*} - N_2] + N_2 \xrightarrow{k'_{st} N_{N_2}} O_2^- + 2N_2 + \text{energy}$$
$$\xrightarrow{k'_d N_{N_2}} O_2 + 2N_2 + e$$

In mechanism (ii) the formation of a transient complex $[O_2^{-*} - N_2]$ was postulated which can either autoionize or lead to O_2^- upon collision with a second N_2 molecule. Good agreement was found between the predicted dependence of the measured attachment rate constant and the experimental results over the entire pressure range covered in these experiments.

The dramatic increase of $(\alpha w)_0$ with $P'_{C_2H_6}$ (Fig. 40b) could not be modeled and was attributed to molecular clusters involving four or more C_2H_6 molecules. Only the $(\alpha w)_0$ vs. $P'_{C_2H_4}$ data for C_2H_4 have been found to be consistent with a simple two-step (Bloch–Bradbury) process, viz.,

$$e + O_2 \xrightarrow{k_c} O_2^{-*}$$
$$O_2^{-*} \xrightarrow{\tau_a^{-1}} O_2 + e$$
$$O_2^{-*} + C_2H_4 \xrightarrow{k_{st} N_{C_2H_4}} O_2^- + C_2H_4 + \text{energy}$$

which predicts

$$\frac{1}{(\alpha w)_0} = \frac{1}{k_c} + \frac{\tau_a^{-1}}{k_c k_{st} P'_{C_2H_4}}. \tag{74}$$

From plots of the experimental data in the manner suggested by Eq. (74), Goans and Christophorou (1974) obtained $k_c = 2.33 \times 10^7$ Torr^{-1} s^{-1} = 7.2×10^{-10} cm^3 s^{-1} and $\tau_a^{-1}/k_{st} = 10{,}700$ Torr. The latter quantity was assumed to give the pressure at which the rate of autodetachment of O_2^{-*}, τ_a^{-1}, is equal to the rate of stabilization of O_2^{-*}, k_{st}, via collisions with C_2H_4 molecules. Goans and Christophorou denoted by N_c the number density of C_2H_4 at which $k_{st} = k_{coll} p = \tau_a^{-1}$, where k_{coll} is the collision rate of O_2^{-*} and C_2H_4 at N_c and p is the probability of stabilization of O_2^{-*} at each O_2^{-*}–C_2H_4 collision, and estimated the collision time τ_{coll} from

$$\tau_{coll} = (2\pi N_c)^{-1}(M_r/e^2\alpha)^{1/2}. \tag{75}$$

In Eq. (75) M_r is the reduced mass of O_2^{-*} and C_2H_4 and α is the static polarizability of C_2H_4. An estimate of 2×10^{-12} s was obtained for $\tau_a(O_2^{-*})$ from $\tau_a(O_2^{-*}) = \tau_{coll}/p$ by setting $p = 1$. Since $p \le 1$ this estimate is a lower limit [see discussion in Christophorou (1976, 1978a)]. This lifetime value is considered to be representative of $O_2^{-*}(X\,^2\Pi_g, v' = 4)$ since the attachment rates in Fig. 40b are for $\langle \varepsilon \rangle \simeq 0.05$ eV and the electron scattering experiments position the $v' = 4$ level of $O_2^-(X\,^2\Pi_g)$ at ~ 0.08 eV. The 2×10^{-12}-s lifetime for $O_2^{-*}(X\,^2\Pi_g, v'=4)$ is considerably lower than that of $\sim 10^{-10}$ s obtained from other swarm studies (Chanin et al., 1962; Shimamori and Hatano, 1977) at much lower pressures. Theoretical estimates using electron scattering cross sections gave lifetimes for $O_2^{-*}(X\,^2\Pi_g, v'=4)$ equal to $\sim 3 \times 10^{-10}$ (Herzenberg, 1969), 1.65×10^{-10} (Koike and Watanabe, 1973), 0.72×10^{-10} (Koike, 1973, 1975), and 0.88×10^{-10} s (Parlant and Fiquet-Fayard, 1976). It seems (see Christophorou, 1978a,b) that the difference between the lifetime values obtained from the high-pressure swarm data on the one hand and those obtained from low-pressure swarm studies and theory on the other is to a large part due to the higher rate constant k_c used in the former. It should be noted that all of the theoretical calculations were made, obviously, on isolated O_2^{-*}, while in the high-pressure studies the medium may affect the O_2^{-*} lifetime.

Following the work of Christophorou and co-workers, Hatano, Shimamori, Fessenden, and their associates made extensive measurements of *thermal* electron attachment to O_2 in various gaseous media ($P < 800$ Torr, usually below ~ 300 Torr) using a microwave conductivity technique combined with pulse radiolysis (Shimamori and Hatano, 1976a,b; 1977; Kokaku et al., 1979, 1980; Shimamori and Fessenden, 1981). This work has been reviewed by Hatano and Shimamori (1981) and has further demonstrated the large and delicate effects of the medium on electron attachment to O_2. Thus, Shimamori and Hatano (1977) found that thermal electron attachment to O_2 followed a Bloch–Bradbury mechanism in many binary gas mixtures at pressures below a few hundred Torr. They, however, found the rate constant for formation of O_2^{-*}, k_c, equal to $(4.8 \pm 0.6) \times 10^{-11}$ cm^3 s^{-1}, which is much lower than that $(7.2 \times 10^{-10}$ cm^3 s$^{-1})$ determined by Goans and Christophorou (1974) at higher pressures. Their lower value led to a lower estimate ($\sim 10^{-10}$ s) for the $O_2^{-*}(v'=4)$ lifetime. Subsequently, Kokaku et al. (1979) studied *thermal* electron attachment to O_2 in C_2H_4, CO_2, and neopentane to pressures of $\lesssim 850$ Torr and Kokaku et al. (1980) to O_2 in C_2H_6, C_3H_8, and n-C_4H_{10}. In each O_2/M mixture they found that the effective rate constant for electron attachment continued to increase with increasing density of M, exceeding by a sizeable amount that predicted by a Bloch–Bradbury mechanism.* The excess attachment was attributed to electron capture by preexisting van der Waals complexes ($O_2 \cdot M$) at high pressures. They proposed that electron attachment in these mixtures proceeds through the Bloch–Bradbury mechanism [reaction (67)] and in addition by

---

* A deviation from the three-body pressure dependence of the attachment rate was also observed by Grünberg (1969, 1978) for pure O_2 and O_2/CO_2 mixtures at pressures below one atmosphere.

6 Electron Attachment Processes

$$(O_2 \cdot M) + e \longrightarrow (O_2 \cdot M)^{-*} \longrightarrow O_2 + M + e \quad (76)$$

$$(O_2 \cdot M)^{-*} \begin{array}{c} \xrightarrow{(+O_2)} O_2^- + M + O_2 \\ \xrightarrow{(+M)} O_2^- + M + M \end{array} \quad (77)$$

and concluded that the rate constant for initial attachment to the van der Waals molecules $(O_2 \cdot M)$ is much larger than that $[k_c,$ Eq. (67)] to initial attachment to O_2. An extrapolation of their data for O_2/C_2H_6 mixtures to higher pressures predicted well the measurements of Goans and Christophorou (1974).

In a more recent publication, Shimamori and Fessenden (1981) measured the temperature dependence of the three-body attachment coefficient in pure $O_2(^{16}O_2$ and $^{18}O_2)$, O_2/N_2, and O_2/CO mixtures ($P_{O_2} < 10$, $P_{N_2} < 60$, $P_{CO} < 40$ Torr) from ~ 78 to 330 K. These results are shown in Fig. 41. The increases in the rate constants below certain values of T were taken as evidence for electron attachment to van der Waals molecules: $(O_2)_2, (O_2 \cdot N_2)$,

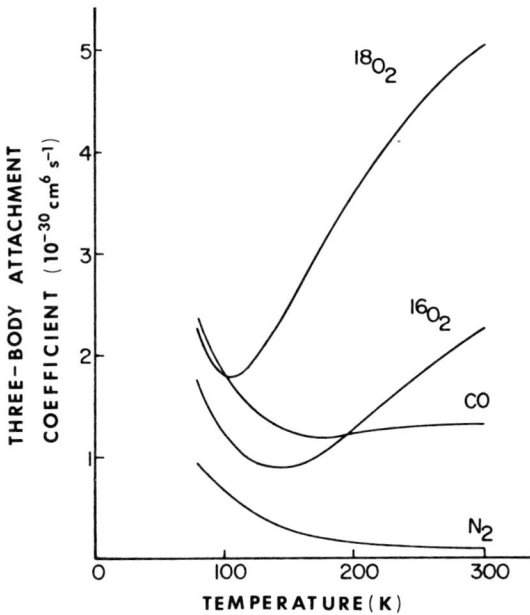

Fig. 41. Temperature dependence of the three-body attachment coefficient for pure $^{16}O_2$ and $^{18}O_2$ and for O_2/CO and O_2/N_2 mixtures. [From Shimamori and Fessenden (1981). Reproduced with permission from the *Journal of Chemical Physics*.]

and ($O_2 \cdot CO$). The authors argued that the increases in the rate constants with decreasing T follow closely their predicted increases in the concentrations of van der Waals molecules with decreasing T. For pure O_2 they estimated that the rate constant for electron attachment to $(O_2)_2$ is nearly two orders of magnitude larger than for O_2 itself and that even at room temperature about 10% of the measured rate constant is due to van der Waals dimers; the differences between the $^{16}O_2$ and $^{18}O_2$ were attributed to the lowering of the resonance energy for $^{18}O_2$ due to its lower vibrational frequencies. The monotonic increase of the three-body attachment coefficient for O_2/N_2 mixtures with decreasing T led to the remarkable suggestion that virtually all of the electron attaching in O_2/N_2 mixtures over the entire T range (78–330 K) is done by van der Waals $O_2 \cdot N_2$ dimers. This work, then, suggests that van der Waals molecules are important in electron attachment processes not only at very high pressures but also at low pressures.

Finally, McMahon (1981a,b, 1982) attempted to model the observed dependences of the rate constants of electron attachment to O_2 in various gaseous media using a statistical model and three-, four-, and more-body electron attachment processes.

Although a satisfactory understanding of nondissociative electron attachment to O_2 in dense gases is not yet in sight, the experimental results dramatize the profound effects of the nature and the density of the medium on electron attachment to this and other molecules and the need for more experimental and theoretical work.

3. Nondissociative Electron Attachment to Other Molecules

For a survey of nondissociative electron attachment studies and three-body attachment coefficients for other molecules (Br_2, I_2, NO, CO_2, H_2O, N_2O, NO_2, O_3, SO_2) see Caledonia (1975). See also Christophorou (1976) for a discussion of electron attachment to O_2, SO_2, C_6H_6, and C_2H_5Br in high-pressure gases ("quasiliquids").

C. Effects of Temperature on Nondissociative Electron Attachment

In Section V we discussed the effects of temperature on dissociative electron attachment to molecules, and in Fig. 39c we presented data on the effect of T on the three-body attachment coefficient for O_2. A weak dependence of the effective three-body attachment coefficient for NO has also been reported in the range 200–500 K (McFarland et al., 1972; Parkes and Sugden, (1972). Interestingly, the *total* rate of the electron attachment to SF_6 has been found to be independent of T by Compton et al. (1966) in the range 300–420 K, by Fehsenfeld (1970) in the range 293–523 K, and by Spence and Schulz (1973) in the range 300 to ~ 1200 K (see Fig. 36a), whereas the rates of formation of SF_6^- and SF_5^- at near zero energy were found to be very much dependent on T, SF_5^- *increasing* (Chen and Chantry, 1970, 1979) and SF_6^- *decreasing* (Hickam and Berg, 1958) as T increases. These observations are interpreted as being due to the competition between parent ion formation and fragmentation of SF_6^{-*}. As T increases SF_5^- increases owing to electron attachment to hot SF_6^* molecules (see Section V) and the SF_6^- decreases because of increased autodetachment.

Recently Hunter et al. (1983) and McCorkle et al. (1983a) observed (Fig. 42a) a *large negative* (attachment decreasing dramatically with increasing T) temperature effect for nondissociative electron attachment to perfluoropropylene (1-C_3F_6). The attachment rate constant for this molecule measured in mixtures with either N_2 or Ar showed a very large dependence on the pressure of either buffer gas (see Fig. 43) and also on the partial pressure of 1-C_3F_6 itself (Hunter et al., 1982). From the observed pressure and temperature dependences, and from relevant data on electron scattering and mass spectrometric studies, Hunter et al. (1983) proposed that electron attachment to this molecule below ~ 2 eV proceeds as

$$C_3F_6 + e \longrightarrow C_3F_6^{-*} \xrightarrow{(+C_3F_6)} (C_3F_6)_2^{-*} \xrightarrow{(+M)} (C_3F_6)_2^- + \text{energy} \qquad (78)$$

with "decay" arrows above $C_3F_6^{-*}$ and $(C_3F_6)_2^{-*}$.

The dramatic decrease in electron attachment with increasing T for $T > 300$ K (Fig. 42a) has been attributed to a decrease in the lifetime of the intermediate species $C_3F_6^{-*}$ [and perhaps $(C_3F_6)_2^{-*}$] due to its low EA. In a moderately high-pressure (~ 225 Torr) quadrupole mass spectrometric study, Hunter et al. (1983) observed $(C_3F_6)_2^-$ ions and ruled out their formation via direct electron attachment to stable dimers $(C_3F_6)_2$.

Additional measurements by McCorkle et al. (1983a,b) on the temperature dependence of the $k_a(\langle \varepsilon \rangle)$ of 1-C_3F_6 in the buffer gas N_2 (Fig. 42b) have shown that while the $k_a(\langle \varepsilon \rangle)$ decreases with increasing T for $T \gtrsim 300$ K, it actually increases with increasing T for small values of $\langle \varepsilon \rangle$. McCorkle et al. (1983a) indicated that the $k_a(\langle \varepsilon \rangle)$ may be affected by changes in T in two opposite ways: (1) decrease with T because the probability of destruction of 1-$C_3F_6^{-*}$ by autodetachment increases with T, and (2) increase with T because the rate constant for formation of 1-$C_3F_6^{-*}$ [first step in reaction (78)] may increase when increases in T make the transition to the negative ion state energetically more favorable (this, of course, is negligible when $\langle \varepsilon \rangle \gg kT$). McCorkle et al. (1983a) thus suggested that when $\langle \varepsilon \rangle \gg kT$ the T-dependence of $k_a(\langle \varepsilon \rangle)$ is dominated by the T-dependence of the 1-$C_3F_6^{-*}$ autodetachment rate constant, and $k_a(\langle \varepsilon \rangle)$ increases with decreasing T as long as T is

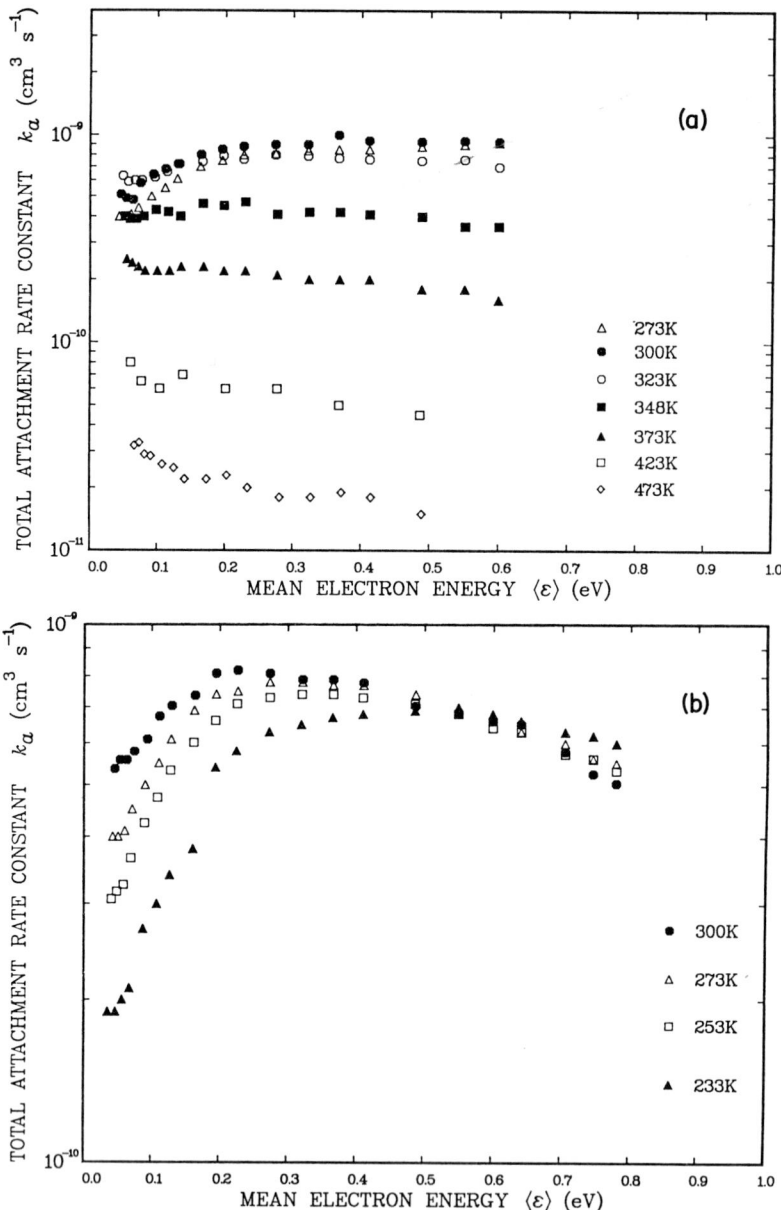

Fig. 42. Electron attachment rate constant vs. mean electron energy for 1-C_3F_6 in N_2 at various gas temperatures. (a) With total gas number density = 1.3×10^{20} cm^{-3}, $N_{C_3F_6} = 1.62 \times 10^{15}$ cm^{-3} at temperatures represented by △ 273 K, ● 300 K, ○ 323 K ■ 348 K, ▲ 373 K, □ 423 K, ◇ 473 K; data from McCorkle *et al.* (1983a). (b) With total gas number density = 6.5×10^{19} cm^{-3}, $N_{C_3F_6} = 1.62 \times 10^{15}$ cm^{-3} at temperatures represented by ● 300 K, △ 273 K, □ 253 K, ▲ 233 K; data from McCorkle *et al.* (1983b). [See text and McCorkle *et al.* (1983a) for discussion.]

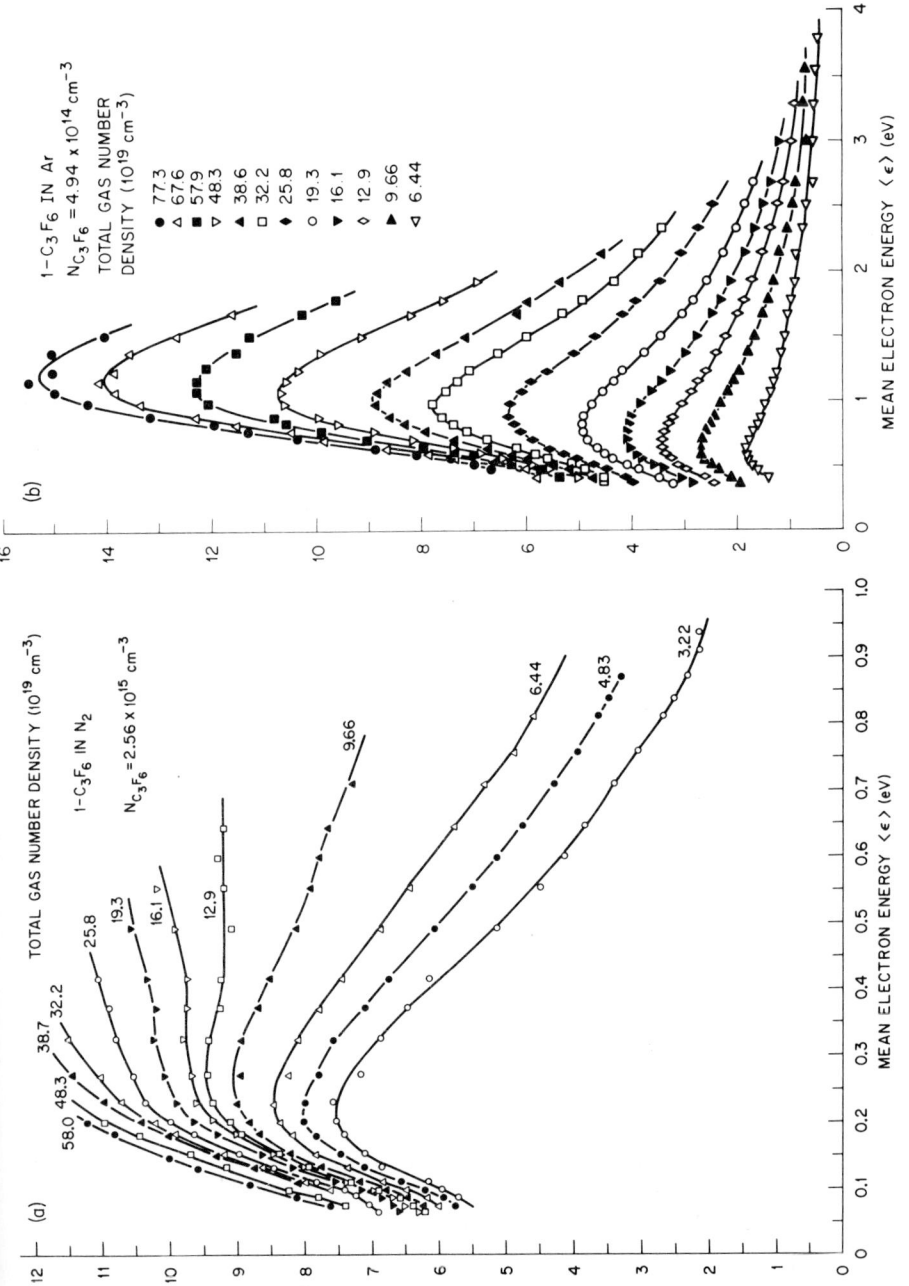

Fig. 43. Electron attachment rate constant for perfluoropropylene (1-C_3F_6) vs. mean electron energy for fixed 1-C_3F_6 gas density and various buffer gas densities in N_2 (a) and Ar (b). [From Hunter et al. (1983).]

high enough for autodetachment to be significant. On the other hand, for small values of $\langle \varepsilon \rangle$, $k_a(\langle \varepsilon \rangle)$ may increase with T when the increase with T of the rate constant for formation of $1\text{-}C_3F_6^{-*}$ exceeds that for autodetachment of $1\text{-}C_3F_6^{-*}$.

D. Electron Attachment to Molecular Clusters

Recently electron-beam experiments with sonic nozzle expansions have been employed to increase molecular gas densities well above those in the collision chambers of such instruments. The fast adiabatic expansion results in greatly reduced molecular rotational, vibrational, and translational energies and can lead to strong molecular clustering (Wegener, 1974). Electron attachment to the van der Waals polymers so produced can then be studied. Such studies were reported for a number of molecules including CO_2 (Klots and Compton, 1977, 1978a), N_2O (Klots and Compton, 1978a), and H_2O (Klots and Compton, 1978b; Klots, 1979; Armbruster et al., 1981).

For CO_2 Klots and Compton found that the most intense cluster anion had the general formula $O^-(CO_2)_n$ ($n = 0$–6). They also detected long-lived anions with charge to mass ratios corresponding to $(CO_2)_n^-$ ($2 \leq n \leq 6$). Some of the smaller ions may be produced by dissociative attachment to larger clusters. For N_2O they observed $O^-(N_2O)_n$ ($0 \leq n \leq 30$), while for H_2O the most prominent ions observed constituted a homologous series with $OH^-(H_2O)_n$. In similar experiments Armbruster et al. (1981) reported observation of $(H_2O)_{11}^-$, which they took as the free-hydrated electron. In a similar study of H_2O/CO_2 mixtures, Klots (1979) observed stable hydrates of the CO_2^- ion with the formula $(H_2O)_x(CO_2)_y^-$ ($x, y \geq 1$); the simplest ion, $H_2OCO_2^-$, showed no evidence of autodissociation.

Undoubtedly, this technique will provide valuable information for the understanding of electron attachment processes in the "interphase region" between low-density and condensed media.

E. Radiative Attachment to Molecules

Although the radiative attachment process has been investigated for a number of atoms (e.g., see Branscomb, 1962; Rothe, 1969; Christophorou, 1971; Massey, 1976; Mital et al., 1977), radiative attachment data for molecules are virtually nonexistent. Radiative attachment cross sections from laboratory measurements of photodetachment cross sections have been made for some species, such as O_2 (Burch et al., 1958, 1959; Branscomb, 1962) and OH (Branscomb, 1966). The rate coefficients ($T = 300\text{ K}$) for the processes $e + O_2 \rightarrow O_2^- + h\nu$ and $e + OH \rightarrow OH^- + h\nu$ have been reported to be, respectively, equal to $10^{-19 \pm 1}$ and $10^{-15 \pm 1}$ cm^3 s^{-1} by the above authors.

However, indirect radiative attachment cross section determinations based on photodetachment data (the inverse process of radiative attachment) are beset by uncertainties. In the case of O_2, photodetachment data near threshold are lacking and the equilibrium separation and vibrational level spacings between O_2 and O_2^- are not the same; in the case of OH^-, uncertainty results from rotational structure in the photodetachment cross section. Direct laboratory measurements of radiative attachment cross sections for low-energy electrons in molecular gases are indicated.

It should, finally, be noted that when anions of polyatomic molecules are formed by capture of near-zero-energy electrons, radiative stabilization of the transient anions could become important if their autodetachment lifetimes are sufficiently long (Section VII.A). Evidence for this has been obtained by Foster and Beauchamp (1975), who studied SF_6^{-*} using an ion cyclotron resonance technique and long (~ 10 ms) observation times.

VIII. Negative Ions Formed by Ion Pair Processes and by Collisions of Molecules with Ground-State and Rydberg Atoms

These processes have been discussed in Section II.D. In this section we present and discuss recent experimental results.

A. Ion Pair Formation

As we mentioned in Section II.D.2, free-electron capture by molecules via the ion pair process [reaction (11)] is nonresonant, and is thus not limited in energy range above threshold as are the resonance electron attachment processes. This can be seen from the early data of Rapp and Briglia (1965) on the total cross section for negative-ion production as a function of electron energy for O_2, CO, CO_2, and N_2O. Apparently, in recent years the ion pair process has not been studied as extensively as the resonance electron attachment processes at low energies. Among the scattered studies of this process are those of Kurepa and Belić (1978) on Cl_2 and Kurepa et al. (1981b) on Br_2. These studies were motivated by the need to understand the dissipative processes in systems of interest to gas lasers and gas discharges. The data of these investigators on the ion pair formation cross section $\sigma_{IP}(\varepsilon)$ for Br_2 and the total ion cross section for Cl_2 are shown in Fig. 44. The threshold for the ion pair process with the positive ion in its ground state is 11.87 eV for Cl_2 and 10.41 eV for Br_2. In the region of the first peak dissociative attachment processes contribute to the measured cross sections. The structure in the

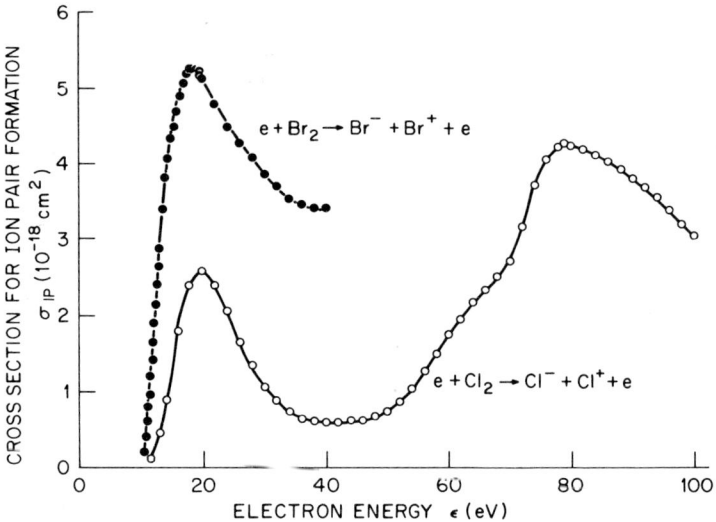

Fig. 44. Cross section σ_{IP} for ion pair formation as a function of electron energy ε for Br_2 (Kurepa et al., 1981b) and total ion formation as a function of ε for Cl_2 (Kurepa and Belić, 1978). In the figure the cross section for Cl_2 is identified by the ion pair process since with the exception of the peak at ~ 20 eV this process predominates. The estimated relative error is ± 0.20 for Cl_2 and ± 0.05 for Br_2.

$Cl^- + Cl^+$ cross section above 40 eV could be associated with excited states of Cl_2^- and the product Cl^+ in various states of excitation.

Ion pair production by electron impact on H_2 and CO close to threshold has been investigated by Köllmann (1975).

There are three other processes which lead to ion pair formation. These involve collisions of molecules XY with fast neutral atoms,

$$A_{fast} + XY \longrightarrow A^+ + XY^- \qquad (79)$$

collisions of XY with Rydberg atoms A^{**},

$$A^{**} + XY \longrightarrow A + X^+ + Y^- \qquad (80)$$

and collisions of XY with photons,

$$h\nu + XY \longrightarrow X^+ + Y^- \qquad (81)$$

Processes (79) and (80) represent negative-ion formation by capture of bound electrons (Section II.D.5) and are discussed in parts B and C of this section. Process (81), referred to by Massey (1976) as polar photodissociation, is outside the scope of the present chapter.

B. Capture of Bound Electrons by Molecules in Collision with Ground-State Neutral Atoms

Experimentally the study of these processes (Section II.D.5) has become possible because of the development of new techniques for producing neutral atomic beams with kinetic energies in the electron volt range. Energetic beams of neutral atoms (mostly Li, Na, K, and Cs) were used to strike neutral target molecules in crossed-orthogonal beam arrangements. The resulting anionic products were mass analyzed and their intensity—and at times their kinetic energy and angular dependence—was measured as a function of the projectile's kinetic energy. Often also the internal energy of excitation of the products was studied and information as to the states of the transient molecular anions leading to the observed anions was obtained. Such experiments not only provided knowledge on negative ions which supplements that from direct electron-impact studies, but also yielded extensive data on electron affinities of molecules and on bond strengths.

1. Examples of Negative-Ion Studies Involving Collisions of Ground-State Atoms with Molecules

In this section we give selected results on reactions (12a) and (12b) for diatomic, triatomic, and polyatomic molecules to illustrate the complementary nature of bound and free-electron capture processes.

a. Reaction (12a)

One of the most significant aspects of this reaction is that it can lead to the production of anions not observed in direct electron–molecule collisions. The XY^- ions can be produced in their ground vibrational state compared to electron-impact studies where the transient anions are initially formed with substantial internal energy ($\geq EA$) and can thus undergo fast autodetachment and/or autodissociation. This is in accord with abundant earlier observations (see Christophorou, 1971) of negative ions in ion–molecule reactions which are not normally observed in direct electron-impact studies. They relate to the fact that in neutral atom-molecule (or ion–molecule) collisions anions are formed at lower states of internal excitation for which the anion is stable or less unstable with respect to autodetachment and/or autodissociation than in direct electron-impact reactions. Also consistent with this is the finding that a number of metastable parent and fragment anions formed via (12a) and (12b) are longer lived than when formed in direct electron attachment reactions.

The observation of parent anions via (12a) not observed in *direct* electron-impact studies can be illustrated by the work of Tang et al. (1974) and Compton et al. (1975) on negative ions formed by fast-alkali-atom impact on the triatomic molecules CO_2, CS_2, and COS. These studies have shown that in addition to fragment anions, the parent ions CO_2^-, CS_2^-, and COS^- are produced. No CO_2^- or COS^- ions were reported in any electron-impact study, and the observation of CS_2^- reported in some electron-impact experiments is unclear (it may be the product of secondary charge transfer reactions rather than a direct electron-impact product). However, CO_2^- was produced (Paulson, 1970) from reactions of O^-, O_2^-, and NO^- with CO_2 and also by low-energy electron impact on several cyclic anhydrides (Cooper and Compton, 1973) and on maleic and succinic anhydrides (Compton et al., 1974) where the O—C—O angle in the molecule is near 120°, i.e., close to that of CO_2^-.

The significance of studies of reaction (12a) in obtaining information on molecular negative-ion states can be illustrated by referring to the work of Kimura and Lacmann (1980) on NO_2 and Mochizuki and Lacmann (1976) on O_2. Kimura and Lacmann (1980) analyzed the kinetic-energy and scattering angle distributions of the product K^+ ions in the reaction

$$K_{fast}^0 + NO_2 \longrightarrow K^+ + NO_2^- \qquad (82)$$

and from the loss in translational energy they obtained information on the excitation of the NO_2^- ion. The energy loss spectrum for (82) in the forward direction is shown in Fig. 45a for collision energies of 3.5–30.8 eV in the center of mass (c.m.) system.

The main peak around 2.5 eV corresponds to the production of the ground electronic state $^1A_1$ of NO_2^- and the smaller peak located at 4–6 eV was attributed to an excited state of NO_2^-, possibly $^3B_1$. Bond bending during the collision was found to have a stronger influence on (82) than bond stretching. In a similar study, Mochizuki and Lacmann (1976) measured both the angle and energy differential cross sections for the reaction

$$K_{fast}^0 + O_2 \longrightarrow K^+ + O_2^- \qquad (83)$$

monitoring K^+ for laboratory alkali collision energies between 0.75 and 30 eV. Their energy loss spectrum for K^+ in the forward direction (see Fig. 45b) shows two peaks at 4.6 and 5.1 eV. The 4.6-eV peak indicates a maximum transition probability into the 4th to 6th vibrational levels of the ground state $O_2^-(^2\Pi_g)$ [i.e., $4.6 - I(K) (=4.34) \simeq 0.26$ eV above the ground state of O_2]. The 5.1-eV peak was attributed to an electronically excited state of O_2^- ($^4\Sigma_u^-$ or $^2\Pi_u$?) located ~ 1.1 eV above the ground state of O_2^- or ~ 0.7 eV above the ground state of O_2. As the deflection angle was increased, the energy loss spectrum was found to shift toward lower energy. Reaction (83) was studied by Stockdale and Warmack (1980) using K (and Cs) with alkali laboratory collision energies from threshold to ~ 100 eV. Their energy loss distributions for K energies equal to 17.2 and 21.6 eV showed the principal

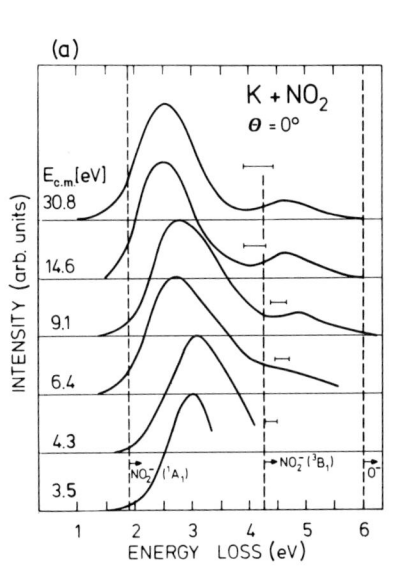
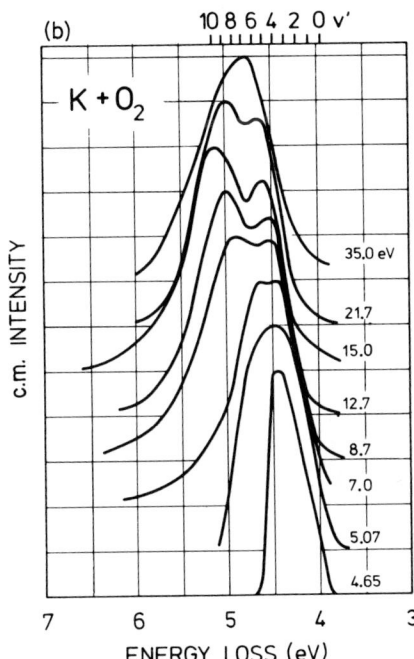

Fig. 45 (a) Normalized relative intensity of K^+ ions ($d^2\sigma/dE\,d\Omega$) from the reaction $K + NO_2 \rightarrow K^+ + NO_2^-$ in the direction of the incoming beam ($\theta = 0°$) as a function of energy loss for different collision energies $E_{c.m.}$ (all values in c.m. system). Minimum energy loss values for NO_2^- and O^- formation are indicated by broken lines. The final energy resolution δE is shown by a horizontal error bar on each curve. [From Kimura and Lacmann (1980). Reproduced from *Chemical Physics Letters* with permission of North-Holland Publishing Company.] (b) Intensity of K^+ ($d^2\sigma/d\Omega\,dE$) ions, normalized to their maximum ion yield, as a function of the energy loss for different colliding energies in the direction of the incoming K beam (all values in c.m. system). [From Mochizuki and Lacmann (1976). Reproduced with permission from the *Journal of Chemical Physics*.]

peak at 5.1 eV but did not clearly indicate the structure at 4.6 eV. The authors pointed out, however, that this might have been due to their inferior neutral-beam energy resolution [see Stockdale and Warmack (1980) and Kleyn *et al.* (1978) for further discussion and results on reaction (83)].

b. Reaction (12b)

Energy loss in this reaction provided—as in reaction (12a)—information on the negative-ion states of the target molecules in a manner analogous to dissociative attachment processes. For example, Okada (1979) measured the energy loss spectra for the forward-scattered product K^+ ions in collisions of fast (15–60 eV lab.) K atoms with HCl and observed two loss peaks in the c.m. system at ~ 6.0 and 8.0 eV. He attributed the former to the production of Cl^- and the latter to H^- in the reaction

$$K^0_{fast} + HCl \begin{cases} \rightarrow K^+ + Cl^- + H & (84a) \\ \rightarrow K^+ + H^- + Cl & (84b) \end{cases}$$

The threshold energy for production of K^+ in (84a) is $\sim 5.1 \pm 0.2$ eV, which locates [since $I(K) = 4.34$ eV] the product negative ions at ~ 0.8 eV above the ground state of HCl, in agreement with dissociative attachment studies (Christophorou et al., 1968). A weaker peak was observed in the energy loss spectrum at ~ 3.7 eV above the ground state of HCl which was attributed to the production of H^-. A weak H^- signal was reported in electron-impact studies (Azria et al., 1974). No HCl^- was observed. Similar studies on $K^0_{fast} + Cl_2$ by Mochizuki and Lacmann (1977) indicated that Cl^- is produced through several electronic states of Cl_2^-.

A number of other studies involving triatomic targets have also been made. Thus Stockdale and Reinhardt (1980) measured positive- and negative-ion energy spectra in the forward direction in collisions of K^0_{fast} with H_2S and D_2S from near threshold energies to ~ 100 eV. Unlike the case of electron impact (Azria et al., 1972, 1979) *no* $H^-(D^-)$ or S^- ions—but rather only $HS^-(DS^-)$—were observed in the collision $K^0_{fast} + H_2S(D_2S)$, i.e.,

$$K^0_{fast} + H_2S(D_2S) \longrightarrow K^+ + HS^-(DS^-) + H(D) \qquad (85)$$

The threshold for $HS^-(DS^-)$ production was found to correspond to the $HS^-(DS^-) + H(D)$ limit of the $^2A_1$ state of $H_2S^-(D_2S^-)$ at 1.55 eV, which, in turn, is the same to within ± 0.2 eV as that of the dissociative attachment limit of the $^2A_1$ $H_2S^-(D_2S^-)$ state. Since production of negative ions via reactions (12a,b) is known *not* to be limited by Franck–Condon-type transitions, these authors used reaction (85) to probe the nature of the ground $^2A_1$ state of H_2S^-. In agreement with electron-impact studies (Fiquet-Fayard et al., 1972), they found that any minimum in the $^2A_1$ state must be shallow.

The absence of $H^-(D^-)$ or S^- ions in reaction (85) is similar to that of H_2O and D_2O. The ionic products observed (Warmack et al., 1978) in collisions of Cs^0_{fast} and K^0_{fast} on H_2O and D_2O over the laboratory coordinate system energy range from 10 to 200 eV and within the angular range of $0°$ to $\sim 140°$ were K^+, Cs^+, and $OH^-(OD^-)$. No O^-, $H^-(D^-)$ ions were found (if present their intensity would be $\lesssim 1\%$ that of OD^-), although in dissociative attachment studies O^- and especially $H^-(D^-)$ are the dominant ions observed (Section IV.C, Fig. 12). The absence of $H^-(D^-)$ and O^- in alkali-atom impact is not yet clearly understood; distortion or mixing of the H_2O states by the presence of the alkali-atom ion may be a possible cause. The positive- and negative-ion thresholds are equal to within ~ 0.1 eV and correspond to the $H(^2S) + OH^-(^1\Sigma)$ limit of H_2O^- at 3.28 eV. Only the ground $^2A_1$ state of H_2O^- correlates to this limit, and it seems likely that the $^2A_1$ state (or a version distorted by the presence of A^+) is involved in the production of $OH^-(OD^-)$. It is interesting to note that $OH^-(OD^-)$ from $H_2O(D_2O)$ is formed in several energy groups in contrast to $HS^-(DS^-)$ from $H_2S(D_2S)$, where only a single low-energy peak was observed. Evidently, as suggested by Stockdale and Reinhardt (1980), with H_2S electron transfer and subsequent dissociation of $H_2S^-(D_2S^-)$ take place so that little momentum is acquired in the collision with K^0 and any internal energy in H_2S^- is largely transferred into kinetic energy of H.

Neutral alkali atom–aliphatic halocarbon reactions revealed most useful features, which nicely complement our knowledge from electron-impact studies. Branching ratios for fragment anions as a function of alkali-atom energy have been determined by several authors (e.g., Tang et al., 1974, 1976; Rothe et al., 1974) and demonstrated—as did electron-impact studies—the fragility of halocarbons toward low-energy electrons (free and bound alike) as well as the multiplicity of the ensuing decompositions of the transient anions. Figure 46 shows one such example for $K + CCl_2F_2$. Alkali atom–halomethane collisions also showed the existence of parent anions such as CF_3I^- (Rothe et al., 1974; Tang et al., 1976; Compton et al., 1978), CF_3Br^- (Compton et al., 1978), $CFCl_3^-$, and $CF_2Cl_2^-$ (Dispert and Lacmann, 1978) which were not observed in direct electron-impact studies. Similarly the ion $CH_3NO_2^-$ was observed (Compton et al., 1978), but not the CH_3CN^- ion in spite of efforts (Compton et al., 1978; Warmack et al., 1980) to detect it in view of its large (3.92 D) dipole moment. It seems that both the attachment cross section and the EA of CH_3CN are small and the transient molecular anion is subject to rapid autodetachment and/or autodissociation into CN^-.

In addition, studies of neutral alkali–halocarbon reactions demonstrated—in agreement with electron-impact studies—that the atomic halogen anions (Cl^-, Br^-, I^-) are by far the most abundant in bound-electron-capture collisions and that the relative yields are strong functions of the structure of the halocarbon. Some differences, however, exist both in nature and relative abundance of the fragment anions as induced by bound- and free-electron capture. Thus for $CFCl_3$ Dispert and Lacmann (1978) reported, in decreasing order of intensity, Cl^-, $CFCl_2^-$, Cl_2^-, F^-, $CFCl_3^-$, FCl^-, and CCl_2^-, while the electron-impact studies of Illenberger et al. (1979) showed—in decreasing order of intensity—Cl^-, F^-, Cl_2^-, and CCl_3^- and no $CFCl_2^-$, $CFCl_3^-$, CCl_2^-, or FCl^-. Similarly, for $CHCl_3$ Dispert and Lacmann (1978) reported Cl^-, Cl_2^-, CCl_2^-, $CHCl_2^-$, and CCl_3^-, while the electron-impact study of this molecule (Johnson et al., 1977) showed only Cl^-, $CHCl_2^-$, and Cl_2^-. In neither study was $CHCl_3^-$ observed.

2. Determination of Electron Affinities and Bond Dissociation Energies

Use of reaction (12a) [and (12b)]—with A being, as a rule, an alkali atom—has been made to determine EA(XY) or EA(Y); or, if EA(Y) is known, the dissociation energy D(XY) of XY. The threshold energy for (12a) is I(A) − EA(XY) and for (12b) is I(A) + D(XY) − EA(Y). Thus from relative or absolute cross section measurements for reactions (12a,b) as a function of the relative impact energy down to the energy threshold, values of electron affinities and bond strengths can be obtained assuming that the threshold energy can be accurately established. For such determinations, the threshold portion of the cross section is usually convoluted taking into account the velocity distribution of the alkali-atom beam and the thermal velocity of the target molecules—both of which blur the onset energy—in a manner analogous to that adopted (Chantry, 1971) for electron–molecule collisions (e.g., see Nalley et al., 1973; Hubers and Los, 1975; Mathur et al., 1976; Dispert and Lacmann, 1978). An example is shown in Fig. 46, where the solid line for Cl^- is an assumed linearly increasing trial cross section for Cl^- which is convoluted with the relative velocity distribution of the reactant. Two other factors need to be considered in this connection, namely the internal energy of the target and the energy state of the products. Although the latter are thought

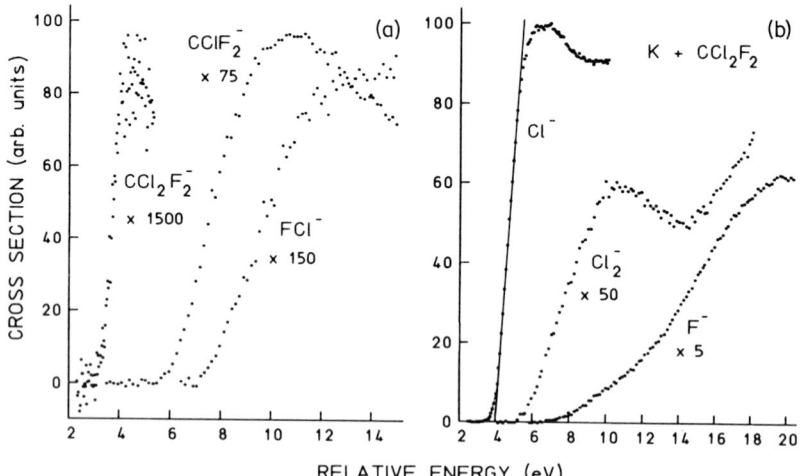

Fig. 46. Relative cross sections from the reaction $K + CCl_2F_2$ for the production of (a) $CCl_2F_2^-$, $CClF_2^-$, and FCl^-; (b) Cl_2^-, Cl^-, and F^-. The solid curve is an assumed linearly increasing trial cross section for Cl^- formation which is convoluted with the relative velocity distribution of the reactants. [From Dispert and Lacmann (1978).]

to be produced in their relaxed states, the former requires proper consideration and measurements as a function of T. Electron affinities obtained from such studies can be found in Table II of Chapter 6 of Volume 2, and some bond dissociation energies can be found in Mathur et al. (1976) and Dispert and Lacmann (1978).

Finally, many studies in this area dealt with metal hexafluoride molecules XF_6 (X = S, Se, Te, Mo, W, U) [e.g., Annis and Datz (1977, 1978), Dispert and Lacmann (1977), Mathur et al. (1977), Hubers and Los (1975); see also other references quoted by these authors). Interest in this class of molecules was kindled in part by their significance in chemical lasers and isotope separation schemes. As can be seen from Table II (Chapter 6 of Volume 2) the EA of some hexafluoride compounds exceeds the $I(A)$ of the alkali atoms [e.g., the EA of MoF_6, WF_6, and UF_6 exceeds $I(K)$ or $I(Cs)$] and reaction (12a) is exoergic. In such studies care must be taken to account properly for the effect of two other processes: one dealing with the alkali beam itself and the other with the reaction products. With regard to the first, Mathur et al. (1977), among others, pointed out that in addition to the reaction

$$A + MF_6 \longrightarrow A^+ + MF_6^- \tag{86}$$

in some cases exoergic reactions of MF_6 with dimeric alkalis (A_2) can be the cause of the observed ion pairs even when process (86) is also exoergic. With regard to the second, Annis and Datz (1977) found that in reactions of K and Cs with MF_6 (M = S, Se, Te, Mo, W, U) at thermal energies exchange reactions forming AF (and possibly MAF_5) predominate over (86). Annis and Datz pointed out that atom abstraction reactions of the form

$$A + MF_6 \longrightarrow AF + MF_5 \tag{87}$$

are exoergic by 0.39–2.69 eV depending on M.

C. Capture of Bound Electrons by Molecules in Collision with Rydberg Atoms

A number of experimental studies of reactions (14d) and (14e) have been made and their results compared to those of reactions (15a) and (15b). The experimental studies fall into two types. Those using tunable dye lasers to study collisions of single well-characterized Rydberg atoms with molecules and those where electron-impact excitation is utilized to create Rydberg atoms with an unknown distribution of n, l states. The upper limit of n values in the latter type of studies is determined by residual electric fields in the apparatus which will ionize Rydberg atoms with sufficiently high n, while the lower limit of n is determined by the radiative lifetime which must be sufficiently long for collisions to occur. In Table XV are listed experimental studies of reactions (14d) and (14e).

The first absolute measurements of the rate constants and cross sections for reaction (14d) have been made by West et al. (1976) for the reaction

$$\text{Xe}^{**}(nf) + \text{SF}_6 \xrightarrow{k_{ab}} \text{Xe}^+ + \text{SF}_6^- \qquad (88)$$

where k_{ab} is the bound-electron attachment rate constant. These studies were followed by those of Foltz et al. (1977) and Hildebrandt et al. (1978), who studied reaction (14d) for SF_6, $c-C_7F_{14}$, and C_6F_6 and reaction (14e) for

TABLE XV

Experimental Studies of Reactions (14d) and (14e)

Rydberg atom	Target molecule	Observed products	Reference
Unknown Rydberg-state distribution[a]			
R^{**}	SF_6	R^+, SF_6^-	Hotop and Niehaus (1967)
R^{**}	CH_3CN, CD_3CN	R^+, CH_3CN^-, CD_3CN^-	Sugiura and Arakawa (1970)
He^{**}	SF_6	He^+, SF_6^-	Klots (1977)
Ar^{**}	SF_6	Ar^+, SF_6^-, SF_5^-	Astruc et al. (1979)
Ar^{**} [b]	SF_6	Ar^+, SF_6^-	Dimicoli and Botter (1981b)
	C_6F_6	$Ar^+, C_6F_6^-$	
Ar^{**} [c]	CCl_4	Ar^+, Cl^-	Dimicoli and Botter (1981a)
	CCl_3F	Ar^+, Cl^-	
	CH_3I	Ar^+, I^-	
	SF_6	Ar^+, SF_6^-	
	C_6F_6	$Ar^+, C_6F_6^-$	
Single-Rydberg state			
$Xe^{**}(nf)$ $25 \leq n \leq 40$	SF_6	Xe^+, SF_6^-	West et al. (1976)
$Xe^{**}(nf)$ $25 \leq n \leq 40$	SF_6	Xe^+, SF_6^-	Foltz et al. (1977)
	CCl_4	Xe^+, Cl^-	
	CCl_3F	Xe^+, Cl^-	
$Xe^{**}(nf)$ $26 \leq n \leq 39$	$c\text{-}C_7F_{14}$	$Xe^+, c\text{-}C_7F_{14}^-$	Hildebrandt et al. (1978)
	C_6F_6	$Xe^+, C_6F_6^-$	
	CH_3Br	Xe^+, Br^-	
	CH_3I	Xe^+, I^-	
$Kr^{**}(np)$	$HCl + SF_6$	SF_6^-	Matsuzawa and Chupka (1977)
	$HF + SF_6$	SF_6^-	

[a] Evidence for reactions $Ar^{**}, Kr^{**} + CH_3NO_2 \rightarrow CH_3NO_2^- + Ar^+, Kr^+$, and $Ar^{**} + CH_3I \rightarrow Ar^+ + CH_3 + I^-$ has been obtained also by Stockdale et al. (1974).

[b] For selected ranges of n using field ionization.

[c] The rate constants and cross sections were measured for three Ar^{**}-beam velocities.

CCl_4, CCl_3F, and CH_3I, e.g.,

$$Xe^{**}(nf) + CCl_4 \xrightarrow{k_{dab}} Xe^+ + CCl_3 + Cl^- \tag{89}$$

where k_{dab} is the bound-electron dissociative attachment rate constant. The results of these workers for SF_6 and $c\text{-}C_7F_{14}$ (reaction 14d) and CCl_4 and CCl_3F [reaction (14e)] are plotted in Fig. 47 where they are compared with the

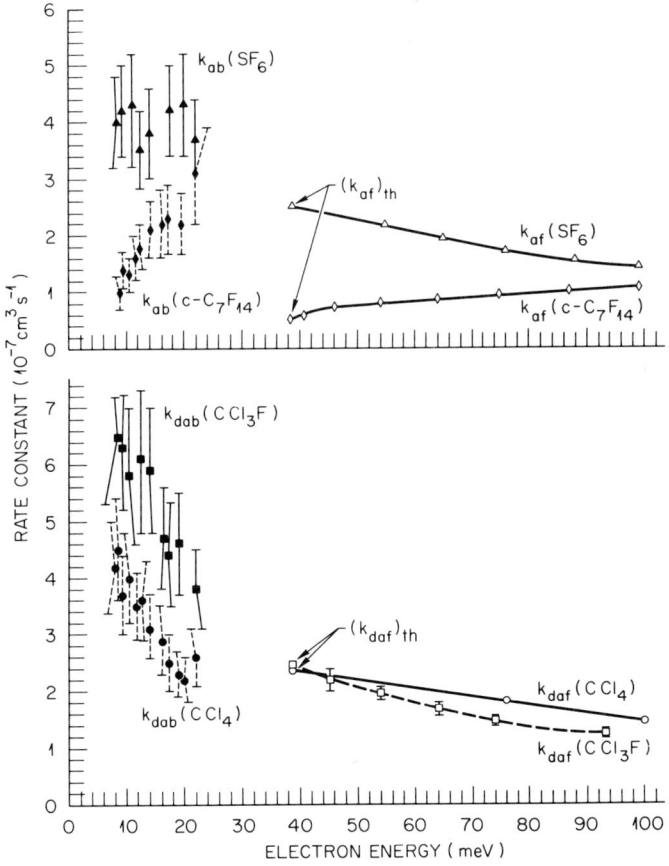

Fig. 47. Comparison of free- and bound-electron attachment rate constants for SF_6 and $c\text{-}C_7F_{14}$ and CCl_3F and CCl_4 (see text). The energy scale for k_{af} and k_{daf} refers to the mean electron energy determined from known electron energy distributions (Section III); the energy scale for k_{ab} and k_{dab} was determined from R_H/n^2 (Section II.D.5). The plotted data are from the following sources: k_{ab}—SF_6 (Foltz et al., 1977), $c\text{-}C_7F_{14}$ (Hildebrandt et al., 1978); k_{dab}—CCl_4, CCl_3F (Foltz et al., 1977); k_{af}—SF_6 (Christophorou et al., 1971b), $c\text{-}C_7F_{14}$ (Christodoulides and Christophorou, 1979); k_{daf}—CCl_4 (Christodoulides and Christophorou, 1971); CCl_3F (McCorkle et al., 1980, 1982a); $(k_{af})_{th}$ ($T \simeq 298$ K)—SF_6, $c\text{-}C_7F_{14}$ (Christophorou et al., 1981b); $(k_{daf})_{th}$ ($T \simeq 298$ K)—CCl_4 (Christophorou et al., 1981b); CCl_3F (McCorkle et al., 1980, 1982a).

corresponding free-electron capture reactions, e.g.,

$$e_f + SF_6 \xrightarrow{k_{af}} SF_6^{-*} \tag{90}$$

$$e_f + CCl_4 \xrightarrow{k_{daf}} CCl_3 + Cl^- \tag{91}$$

In the "free"-electron model of Matsuzawa $k_{ab} \simeq k_{af}$ and $k_{dab} \simeq k_{daf}$ if the electron velocity distributions are the same. However, this is not the case. The mean electron energies in free-electron capture studies ($T \simeq 300$ K) are ≥ 0.038 eV, while in the bound-electron capture studies they are between 0.009 and 0.022 eV. Furthermore, the Rydberg-state velocity spread is narrower than Maxwellian (Rundel, 1980), and this is a major obstacle in comparisons of bound- and free-electron capture data, although efforts have been made to take this difference in energy spread into consideration (West et al., 1976; Foltz et al., 1977; Rundel, 1980).

In addition to differences in energy distributions, in the case of reaction (14d) differences may arise as a result of stabilization of the parent anion (e.g., SF_6^{-*}) by collision with the Rydberg core—i.e., the free-electron picture may not be entirely valid. Indeed Dimicoli and Botter (1981b) attempted to determine the role of the ionic core on both σ_{ab} and τ_a for SF_6^{-*} and $C_6F_6^{-*}$, and in a similar effort Astruc et al. (1979) concluded that the influence of the ionic core on the lifetime of SF_6^{-*} cannot be neglected. In this connection Dimicoli and Botter (1981a,b) found a rather large decrease in k_{ab} and σ_{ab} for SF_6 and in k_{dab} and σ_{dab} for CCl_4 and CCl_3F with increasing Ar** velocity, indicating that the ionic core Ar^+ plays a role in the reaction. Interestingly, Astruc et al. (1979) found that in collisions of Ar** ($n \geq 20$) with SF_6 the intensity of SF_5^- increases and that of SF_6^- decreases with increasing temperature, as in the T dependence of free-electron capture by SF_6 producing these very same anions (see Section V).

In spite of the as yet not fully assessed role of the ionic core on the bound-electron capture reactions, the differences in the electron energy distributions in bound- and free-electron experiments, and the lack of overlap between bound- and free-electron capture measurements, the data on k_{ab} and k_{af}, and k_{dab} and k_{daf} [Fig. 47; for C_6F_6 the thermal ($T = 298$ K) value of k_{af} is 1×10^{-7} cm^3 s^{-1} (Gant and Christophorou, 1976) and k_{ab} at 20 meV is 0.3×10^{-7} cm^3 s^{-1} (Hildebrandt et al., 1978)] seem to be in reasonable agreement, lending support to the "free"-electron model. The experimental data also show that in the energy range below ~ 25 meV—most difficult to probe in free-electron experiments—the bound-electron studies using highly excited Rydberg atoms have a most important function to play. Undoubtedly more results will be forthcoming which will clarify further the basic physics of bound-electron capture reactions (14d) and (14e) and their relation to the respective free-electron capture processes.

IX. Doubly Charged Negative Ions

Dinegative ions should be stable in the gas phase if the additional binding energy due to the second electron exceeds the Coulomb repulsion between the mononegative ion and the second electron. The autodetachment lifetimes of reported doubly charged negative ions have been reviewed by Christophorou (1978a) and an extensive discussion of doubly charged negative ions has been given by Kiser (1979). Although Kiser contends firmly that long-lived ($>10^{-6}$ s) doubly charged atomic, diatomic, and polyatomic negative ions exist, the existence of at least some of those species is still uncertain and the results of some authors on long-lived doubly charged atomic anions have not been reproduced by others (e.g., see discussion in Christophorou, 1978a; Kiser, 1979; Spence et al., 1981; and Hird and Ali, 1981). Apparently, the observations are crucially influenced by the ion-source type and conditions.

One of the first doubly charged anions reported was H^{2-}. The H^{2-} ion is the first in the three-electron isoelectronic series H^{2-}, He^-, Li, Be^+, B^{2+}, etc. The formation of an extremely short-lived ($\sim 10^{-15}$ s) H^{2-} has been associated with the structure observed at 14.2 eV in the cross section for the reaction $e + H^- \rightarrow H + 2e$ (Walton et al., 1970, 1971), and also at 17.2 eV (Peart and Dolder, 1973). The resonance in the cross section at 14.2 eV was assigned to the configuration $(2s)^2(2p)(^2P^0)$ and that at 17.2 eV largely to the configuration $(2p)^3(^2P^0; {}^2D^0)$ of H^{2-} (Taylor and Thomas, 1972; Thomas, 1974). Reports of the existence of longer-lived ($\sim 2 \times 10^{-18}$ s) H^{2-} and D^{2-} are still in question [see references and discussion in Christophorou (1978a) and Kiser (1979)]. An extremely short-lived (5×10^{-16} s) doubly charged O^{2-} has been predicted at 5.38 eV by Herrick and Stillinger (1975).

Stuckey and Kiser (1966) first reported the existence of long-lived ($\sim 10^{-4}$ s) doubly charged negative ions O^{2-}, F^{2-}, Cl^{2-}, and Br^{2-}. Since then other such species have been reported [see references in Christophorou (1978a) and Kiser (1979)] and include I^{2-}, Te^{2-}, P^{2-}, As^{2-}, Sb^{2-}, and Bi^{2-}.

Little is known of the mechanisms of formation and properties of these doubly charged anions. Kiser (1979) argues that since it is known that long-lived doubly negative ions have been observed only in the energy region in which singly charged negative ions are produced, the latter must be the precursors of the former. Although at this stage it is not possible to specify the excited states of either the doubly charged anions or the precursor mononegative ions, it seems certain that excited states of both must be involved to allow for the relatively long lifetimes observed. It is indeed of great interest to identify the atomic configurations that allow for the formation of such negative-ion states and the processes that lead to their production. It is also of importance to consider whether beam intensities can be high enough to seriously warrant consideration of their use in high-energy accelerators.

Reports of doubly charged parent and fragment anions of complex molecules have also appeared. Thus Dougherty (1969) reported doubly charged parent anions of benzo(cd)pyrene-6-one, Bowie and Stapleton (1976) of nitrobenzoic acid, cyanobenzoic acid, and $p\text{-}NO_2\text{-}C_6H_4\text{-}(CH_2)_n\text{-}CO_2R$ with R = H or CH_3 (the most abundant dinegative ion occurred for $n = 3$ or 4), and Stapleton and Bowie (1977) (see also Kiser, 1979) of SF_6. Doubly charged fragment molecular anions such as CN^{2-} from CF_3CN and C_2H_5CN (Kiser, 1979), SF_5^{2-} from SF_6 (Stapleton and Bowie, 1977; Kiser, 1979), and $(M-H)^{2-}$ from organic acids, $(NO_2)^{2-}$ from nitrobenzenes, and $(M-CH_3)^{2-}$ from p-nitroanisole and p-nitrobenzyl methyl ether (Bowie and Stapleton, 1976) have been reported. Although no lifetimes were given for these dinegative species, the experimental arrangements employed for their detection require them to be long-lived ($> 10^{-6}$ s).

Certainly, both the experimental and the theoretical information on doubly charged anions is incomplete; such species remain a challenge to both theory and experiment.

The dinegative species discussed in this section are for the gas phase. In the condensed phase (liquids, solutions, and solids) multiply-charged anions can be stabilized by the surrounding medium, and many such anions have been reported. Hydration of anions in the gas phase could also lead to species able to support more than one electron.

Acknowledgments

We wish to thank Spyros Spyrou and Scott R. Hunter for reading the manuscript and Dinesh Naik for help with the references.

References

Aberth, W., Schnitzer, R., and Anbar, M. (1975). *Phys. Rev. Lett.* **34**, 1600–1603.
Abouaf, R., and Teillet-Billy, D. (1977). *J. Phys. B* **10**, 2261–2268.
Abouaf, R., and Teillet-Billy, D. (1980a). *Chem. Phys. Lett.* **73**, 106–109.
Abouaf, R., and Teillet-Billy, D. (1980b). *J. Phys. B* **13**, L275–L279.
Abouaf, R., Paineau, R., and Fiquet-Fayard, F. (1976). *J. Phys. B* **9**, 303–314.
Ajello, J. M., and Chutjian, A. (1979). *J. Chem. Phys.* **71**, 1079–1087.
Alger, S. R., and Rees, J. A. (1976). *J. Phys. D* **9**, 2359–2367.
Allan, M., and Wong, S. F. (1978). *Phys. Rev. Lett.* **41**, 1791–1794.
Allan, M., and Wong, S. F. (1981). *J. Chem. Phys.* **74**, 1687–1691.
Annis, B. K., and Datz, S. (1977). *J. Chem. Phys.* **66**, 4468–4477.
Annis, B. K., and Datz, S. (1978). *J. Chem. Phys.* **69**, 2553–2561.
Armbruster, M., Haberland, H., and Schindler, H. G. (1981). In "Electron and Ion Swarms" (L. G. Christophorou, ed.), pp. 203–207. Pergamon, New York.
Astruc, J. P., Barbé, R., and Schermann, J. P. (1979). *J. Phys. B* **12**, L377–L381.
Asundi, R. K., Craggs, J. D., and Kurepa, M. V. (1963). *Proc. Phys. Soc. London* **82**, 967–978.
Ayala, J. A., Wentworth, W. E., and Chen, E. C. M. (1981a). *J. Phys. Chem.* **85**, 768–777.

Ayala, J. A., Wentworth, W. E., and Chen, E. C. M. (1981b). *J. Phys. Chem.* **85**, 3989–3994.
Azria, R., Tronc., M., and Goursaud, S. (1972). *J. Chem. Phys.* **56**, 4234–4235.
Azria, R., Roussier, L., Paineau, R., and Tronc, M. (1974). *Rev. Phys. Appl.* **9**, 469–473.
Azria, R., Le Coat, Y., Lefevre, G., and Simon, D. (1979). *J. Phys. B* **12**, 679–687.
Azria, R., Le Coat, Y., Simon, D., and Tronc, M. (1980). *J. Phys. B* **13**, 1909–1918.
Azria, R., Abouaf, R., and Teillet-Billy, D. (1982). *J. Phys. B* **15**, L569–L574.
Babcock, L. M., and Streit, G. E. (1982). *J. Phys. Chem.* **86**, 1240–1242.
Baede, A. P. M. (1975). *Adv. Chem. Phys.* **30**, 463–535.
Bansal, K. M., and Fessenden, R. W. (1972). *Chem. Phys. Lett.* **15**, 21–23.
Bansal, K. M., and Fessenden, R. W. (1973). *J. Chem. Phys.* **59**, 1760–1768.
Bardsley, J. N. (1969). *J. Chem. Phys.* **51**, 3384–3389.
Bardsley, J. N., and Mandl, F. (1968). *Rep. Prog. Phys.* **31**, 471–531.
Bardsley, J. N., and Wadehra, J. M. (1979). *Phys. Rev.* **20**, 1398–1405.
Bardsley, J. N., and Wadehra, J. M. (1983). *J. Chem. Phys.* **78**, 7227–7234.
Bardsley, J. N., Herzenberg, A., and Mandl, F. (1964). *In* "Atomic Collision Processes" (M. R. C. McDowell, ed.), pp. 415–427. North-Holland, Amsterdam.
Bardsley, J. N., Herzenberg, A., and Mandl, F. (1966). *Proc. Phys. Soc. London* **89**, 321–340.
Beauchamp, J. L. (1976). *J. Chem. Phys.* **64**, 929–935.
Belić, D. S., and Hall, R. I. (1981). *J. Phys. B* **14**, 365–373.
Belić, D. S., Landau, M., and Hall, R. I. (1981). *J. Phys. B* **14**, 175–190.
Bennett, S. L., Pabst, R. E., Margrave, J. L., and Franklin, J. L. (1974). *Int. J. Mass Spectrom. Ion Phys.* **15**, 451–461.
Berry, R. S., and Leach, S. (1981). *Adv. Electron. Electron Phys.* **57**, 137–182.
Birtwistle, D. T., and Modinos, A. (1978). *J. Phys. B* **11**, 2949–2955.
Blaunstein, R. P., and Christophorou, L. G. (1968). *J. Chem. Phys.* **49**, 1526–1531.
Bloch, F., and Bradbury, N. E. (1935). *Phys. Rev.* **48**, 689–695.
Boness, M. J. W., and Schulz, G. J. (1970). *Phys. Rev. A* **2**, 2182–2186.
Bortner, T. E., and Hurst, G. S. (1958). *Health Phys.* **1**, 39–45.
Bottcher, C., and Buckley, B. D. (1979). *J. Phys. B* **12**, L497–L500.
Bouby, L., and Abgrall, H. (1967). *Proc. 5th Int. Conf. Phys. Electron. At. Coll., Leningrad, USSR 1967*, pp. 584–585.
Bouby, L., Fiquet-Fayard, F., and Abgrall, H. (1965). *C. R. Hebd. Seances Acad. Sci.* **261**, 4059–4062.
Bouby, L., Fiquet-Fayard, F., and Le Coat, Y. (1970). *Int. J. Mass Spectrom. Ion Phys.* **3**, 439–454.
Bowie, J. H., and Stapleton, B. J. (1976). *J. Am. Chem. Soc.* **98**, 6480–6843.
Branscomb, L. M. (1962). *In* "Atomic and Molecular Processes" (D. R. Bates, ed.), Chapter 4. Academic Press, New York.
Branscomb, L. M. (1966) *Phys. Rev.* **148**, 11–18.
Breit, G., and Wigner, E. (1936). *Phys. Rev.* **49**, 519–531.
Brion, C. E. (1969). *Int. J. Mass Spectrom. Ion Phys.* **3**, 197–202.
Brooks, H. L., Hunter, S. R., and Nygaard, K. J. (1979). *J. Chem. Phys.* **71**, 1870–1873.
Buchel'nikova, I. S. (1959). *Sov. Phys. JETP Eng. Transl.* **8**, 783–791. [(1958). *J. Exp. Theor. Phys. USSR* **35**, 1119–1130.]
Burch, D. S., Smith, S. J., and Branscomb, L. M. (1958). *Phys. Rev.* **112**, 171–175.
Burch, D. S., Smith, S. J., and Branscomb, L. M. (1959). *Phys. Rev.* **114**, 1652.
Burrow, P. D. (1973). *J. Chem. Phys.* **59**, 4922–4931.
Burrow, P. D., and Jordan, K. D. (1975). *Chem. Phys. Lett.* **36**, 594–598.
Caledonia, G. E. (1975). *Chem. Rev.* **75**, 333–351.
Čadež, I. M., Pejčev, V. M., and Kurepa, M. V. (1983). *J. Phys. D* **16**, 305–314.
Caledonia, G. E. (1975). *Chem. Rev.* **75**, 333–351.
Carter, D. E. (1976). *J. Chem. Phys.* **65**, 2584–2591.

Cavalleri, G. (1969). *Phys. Rev.* **179**, 186–202.
Celotta, R. J., Bennett, R. A., Hall, J. L., Siegel, M. W., and Levine, J. (1972). *Phys. Rev. A* **6**, 631–642.
Cermac, V., and Herman, Z. (1964). *Collect. Czech. Chem. Commun.* **29**, 953–959.
Chaney, E. L., and Christophorou, L. G. (1969). *J. Chem. Phys.* **51**, 883–892.
Chanin, L. M., Phelps, A. V., and Biondi, M. A. (1962). *Phys. Rev.* **128**, 219–230.
Chantry, P. J. (1968). *Phys. Rev.* **172**, 125–136.
Chantry, P. J. (1969). *J. Chem. Phys.* **51**, 3369–3379.
Chantry, P. J. (1971). *J. Chem. Phys.* **55**, 2746–2759.
Chantry, P. J. (1972). *J. Chem. Phys.* **57**, 3180–3186.
Chantry, P. J. (1976). *J. Chem. Phys.* **65**, 4412–4420.
Chantry, P. J. (1978). "Attachment Measurements in Halogen Bearing Molecules." Westinghouse R&D Document No. 78-9C6-ATTACH-R1, unpublished.
Chantry, P. J. (1982) *In* "Applied Atomic Collision Physics," (H. S. W. Massey, E. W. Mc Daniel, and B. Bederson, eds.) Volume 3, "Gas Lasers," (E. W. Mc Daniel, and W. L. Nighan, eds.) Academic Press, New York, pp. 35–70.
Chapman, C. J., and Herzenberg, A. (1972). See Spence and Schulz, 1972.
Chen, C. L., and Chantry, P. J. (1970). *Bull. Am. Phys. Soc.* **15**, 418–419.
Chen, C. L., and Chantry, P. J. (1972). *Bull. Am. Phys. Soc.* **17**, 1133.
Chen, C. L., and Chantry, P. J. (1979). *J. Chem. Phys.* **71**, 3897–3907.
Chen, H.-L., Center, R. E., Trainor, D. W., and Fyfe, W. I. (1977). *Appl. Phys. Lett.* **30**, 99–101.
Chen, J. C. Y. (1966). *Phys. Rev.* **148**, 66–73.
Chen, J. C. Y., and Peacher, J. L. (1967). *Phys. Rev.* **163**, 103–111.
Chen, U. C., George, R. D., and Wentworth, W. E. (1968). *J. Chem. Phys.* **49**, 1973–1974.
Chiu, N. S., Burrow, P. D., and Jordan, K. D. (1979). *Chem. Phys. Lett.* **68**, 121–126.
Christodoulides, A. A., and Christophorou, L. G. (1971). *J. Chem. Phys.* **54**, 4691–4705.
Christodoulides, A. A., and Christophorou, L. G. (1979). *Chem. Phys. Lett.* **61**, 553–557.
Christodoulides, A. A., Schultes, E., Schumacher, R., and Schindler, R. N. (1974). *Z. Naturforsch.* **29a**, 389–399.
Christodoulides, A. A., Schumacher, R., and Schindler, R. N. (1975a). *J. Phys. Chem.* **79**, 1904–1911.
Christodoulides, A. A., Schumacher, R., and Schindler, R. N. (1975b). *Z. Naturforsch.* **30a**, 811–814.
Christodoulides, A. A., Schumacher, R., and Schindler, R. N. (1976). *Ber. Bunsenges. Phys. Chem.* **80**, 965–969.
Christodoulides, A. A., Schumacher, R., and Schindler, R. N. (1978). *Int. J. Chem. Kinet.* **10**, 1215–1223.
Christodoulides, A. A., Christophorou, L. G., Pai, R. Y., and Tung, C. M. (1979). *J. Chem. Phys.* **70**, 1156–1168.
Christophorou, L. G. (1971) "Atomic and Molecular Radiation Physics." Wiley (Interscience), New York.
Christophorou, L. G. (1972). *J. Phys. Chem.* **76**, 3730–3734.
Christophorou, L. G. (1976). *Chem. Rev.* **76**, 409–423.
Christophorou, L. G. (1978a). *Adv. Electron. Electron Phys.* **46**, 55–129.
Christophorou, L. G. (1978b). *Radiat. Phys. Chem.* **12**, 19–34.
Christophorou, L. G. (1978c). *Proc. XIIIth Int. Conf. on Phenomena in Ionized Gases, Berlin, GDR, September 11-17, 1977*, pp. 51–72. VEB Export-Import, Leipzig.
Christophorou, L. G. (1980a). *Environ. Health Perspect.* **36**, 3–32.
Christophorou, L. G. (1980b). Gaseous Dielectrics II. *Proceedings of the Second International Symposium on Gaseous Dielectrics, Knoxville, Tennessee, March 9-13, 1980*. Pergamon, New York.

Christophorou, L. G. (1981). In "Photon, Electron, and Ion Probes of Polymer Structure and Properties" (D. W. Dwight, T. J. Fabish, and H. R. Thomas, eds.), pp. 11–34. American Chemical Society, Washington, D. C.

Christophorou, L. G. (1982). Gaseous Dielectrics III. *Proceedings of the Third International Symposium on Gaseous Dielectrics, Knoxville, Tennessee*, March 7–11, 1982. Pergamon, New York.

Christophorou, L. G., and Goans, R. E. (1974). *J. Chem. Phys.* **60**, 4244–4250.

Christophorou, L. G., and Stockdale, J. A. D. (1968). *J. Chem. Phys.* **48**, 1956–1960.

Christophorou, L. G., Compton, R. N., Hurst, G. S., and Reinhardt, P. W. (1965). *J. Chem. Phys.* **43**, 4273–4281.

Christophorou, L. G., Compton, R. N., Hurst, G. S., and Reinhardt, P. W. (1966). *J. Chem. Phys.* **45**, 536–547.

Christophorou, L. G., Compton, R. N., and Dickson, H. W. (1968). *J. Chem. Phys.* **48**, 1949–1955.

Christophorou, L. G., Carter, J. G., and Christodoulides, A. A. (1969a). *Chem. Phys. Lett.* **3**, 237–240.

Christophorou, L. G., Chaney, E. L., and Christodoulides, A. A. (1969b). *Chem. Phys. Lett.* **3**, 363–366.

Christophorou, L. G., Carter, J. G., Collins, P. M., and Christodoulides, A. A. (1971a). *J. Chem. Phys.* **54**, 4706–4714.

Christophorou, L. G., McCorkle, D. L., and Carter, J. G. (1971b). *J. Chem. Phys.* **54**, 253–260.

Christophorou, L. G., McCorkle, D. L., and Anderson, V. E. (1971c). *J. Phys. B* **4**, 1163–1172.

Christophorou, L. G., McCorkle, D. L., and Carter, J. G. (1972). *J. Chem. Phys.* **57**, 2228.

Christophorou, L. G., Carter, J. G., Chaney, E. L., and Collins, P. M. (1973a). In "Proc. IVth Int. Congress Radiat. Res. July 1970, Physics and Chemistry" (J. F. Duplan and A. Chapiro, eds.), Vol. 1, pp. 145–157. Gordon and Breach, London.

Christophorou, L. G., Hadjiantoniou, A., and Carter, J. G. (1973b). *J. Chem. Soc. Faraday Trans. II* **69**, 1713–1722.

Christophorou, L. G., McCorkle, D. L., and Pittman, D. (1974). *J. Chem. Phys.* **60**, 1183–1184.

Christophorou, L. G., Grant, M. W., and McCorkle, D. L. (1977a). In "Advances in Chemical Physics" (I. Prigogine and S. A. Rice, eds.), Vol. 36, pp. 413–520. Wiley, New York.

Christophorou, L. G., Gant, K. S., and Anderson, V. E., (1977b). *J. Chem. Soc. Faraday Trans. II* **73**, 804–811.

Christophorou, L. G., James, D. R., Pai, R. Y., Mathis, R. A., Sauers, I., Smith, D. H., Frees, L., Pace, M. O., Bouldin, D. W., Chan, C. C., Fatheddin, A., and Maughan, D. (1980). Oak Ridge National Laboratory Report ORNL/TM-7405.

Christophorou, L. G., James, D. R., Pai, R. Y., Mathis, R. A., Sauers, I., Smith, D. H., Frees, L. C., Hunter, S. R., Pace, M. O., Bouldin, D. W., Spyrou, S. M., Fatheddin, A. (1981a). Oak Ridge National Laboratory Report ORNL/TM-7862.

Christophorou, L. G., Mathis, R. A., James, D. R., and McCorkle, D. L. (1981b). *J. Phys. D* **14**, 1889–1901.

Christophorou, L. G., James, D. R., Pai, R. Y., Mathis, R. A., Sauers, I., Frees, L. C., Hunter, S. R., Pace, M. O., Bouldin, D. W., Spyrou, S. M., Fatheddin, A., and Hanna, M. C. (1982). Oak Ridge National Laboratory Report ORNL/TM-8158.

Chutjian, A. (1981). *Phys. Rev. Lett.* **46**, 1511–1514.

Claydon, C. R., Segal, G. A., and Taylor, H. S. (1970). *J. Chem. Phys.* **52**, 3387–3398.

Claydon, C. R., Segal, G. A., and Taylor, H. S. (1971). *J. Chem. Phys.* **54**, 3799–3816.

Collins, P. M., Christophorou, L. G., Chaney, E. L., and Carter, J. G. (1970). *Chem. Phys. Lett.* **4**, 646–650.

Compton, R. N. (1977). *J. Chem. Phys.* **66**, 4478–4485.

Compton, R. N., and Bouby, L. (1967). *C. R. Hebd. Seances Acad. Sci. Ser. C* **264**, 1153–1156.

Compton, R. N., and Christophorou, L. G. (1967). *Phys. Rev.* **154**, 110–116.

Compton, R. N., and Cooper, C. D. (1977). *J. Chem. Phys.* **66**, 4325–4329.
Compton, R. N., and Stockdale, J. A. D. (1976). *Int. J. Mass Spectrom. Ion Phys.* **22**, 47–55.
Compton, R. N., Christophorou, L. G., Hurst, G. S., and Reinhardt, P. W. (1966). *J. Chem. Phys.* **45**, 4634–4639.
Compton, R. N., Reinhardt, P. W., and Cooper, C. D. (1974). *J. Chem. Phys.* **60**, 2953–2957.
Compton, R. N., Reinhardt, P. W., and Cooper, C. D. (1975). *J. Chem. Phys.* **63**, 3821–3827.
Compton, R. N., Reinhardt, P. W., and Cooper, C. D. (1978). *J. Chem. Phys.* **68**, 4360–4367.
Cooper, C. D., and Compton, R. N. (1973). *J. Chem. Phys.* **59**, 3550–3565.
Cooper, C. D., and Compton, R. N. (1974). *J. Chem. Phys.* **60**, 2424–2429.
Craggs, J. D., and Tozer, B. A. (1960). *Proc. R. Soc. London, Ser. A* **254**, 229–241.
Crawford, O. H., and Garrett, W. R. (1977). *J. Chem. Phys.* **66**, 4968–4970.
Crompton, R. W., Hegerberg, R., and Skullerud, H. R. (1979). *Proc. Int. Seminar on Swarm Experiments in Atomic Collision Research, Tokyo*, p. 18.
Crompton, R. W., Hegerberg, R., and Skullerud, H. R. (1980). *J. Phys. B* **13**, L455–L459.
Crompton, R. W., Haddad, G. N., Hegerberg, R., and Robertson, A. G. (1982). *J. Phys. B* **15**, L483–L484.
Davidow, R. S., and Armstrong, D. A. (1968). *J. Chem. Phys.* **48**, 1235–1241.
Davis, F. J., and Nelson, D. R. (1970). *Chem. Phys. Lett.* **6**, 277–278.
Davis, F. J., Compton, R. N., and Nelson, D. R. (1973). *J. Chem. Phys.* **59**, 2324–2329.
DeCorpo, J. J., and Franklin, J. L., (1971). *J. Chem. Phys.* **54**, 1885–1888.
DeCorpo, J. J., Steiger, R. P., Franklin, J. L., and Margrave, J. L. (1970). *J. Chem. Phys.* **53**, 936–938.
DeCorpo, J. J., Bafus, D. A., and Franklin, J. L. (1971). *J. Chem. Phys.* **54**, 1592–1597.
Di Domenico, A., and Franklin, J. L. (1972). *Int. J. Mass Spectrom. Ion Phys.* **9**, 171–184.
Di Domenico, A., Harland, P. W., and Franklin, J. L. (1972). *J. Chem. Phys.* **56**, 5299–5307.
Dillard, J. G. (1973). *Chem. Rev.* **73**, 589–643.
Dillard, J. G., and Franklin, J. L. (1968). *J. Chem. Phys.* **48**, 2349–2352.
Dillard, J. G., O'Toole, J. H., and Ogliaruso, M. A. (1975). *Org. Mass Spectrom.* **10**, 728–736.
Dimicoli, I., and Botter, R. (1981a). *J. Chem. Phys.* **74**, 2346–2354.
Dimicoli, I., and Botter, R. (1981b). *J. Chem. Phys.* **74**, 2355–2360.
Dispert, H., and Lacmann, K. (1977). *Chem. Phys. Lett.* **45**, 311–315.
Dispert, H., and Lacmann, K. (1978). *Int. J. Mass Spectrom. Ion Phys.* **28**, 49–67.
Dorman, F. H. (1966). *J. Chem. Phys.* **44**, 3856–3863.
Dougherty, R. C. (1969). *J. Chem. Phys.* **50**, 1896–1897.
Dowell, J. T., and Sharp, T. E. (1968). *Phys. Rev.* **167**, 124–127.
Dutton, J. (1975). *J. Phys. Chem. Ref. Data* **4**, 577–856.
Ebinghaus, H. (1964). *Z. Naturforsch.* **19a**, 727–732.
Fehsenfeld, F. C. (1970). *J. Chem. Phys.* **53**, 2000–2004.
Feneuille, S., and Jacquinot, P. (1981). *In* "Advances in Atomic and Molecular Physics" (D. Bates and B. Bederson, eds.), Vol. 17, pp. 99–166. Academic Press, New York.
Fessenden, R. W., and Bansal, K. M. (1970). *J. Chem. Phys.* **53**, 3468–3473.
Fiquet-Fayard, F. (1974). *J. Phys. B* **7**, 810–816.
Fiquet-Fayard, F., Ziesel, J. P., Azria, R., Tronc, M., and Chiari, J. (1972). *J. Chem. Phys.* **56**, 2540–2548.
Fite, W. L., and Brackmann, R. T. (1963). *Proc. Sixth Int. Conf. on the Ionization Phen. in Gases, Paris* **1**, 21–27.
Fite, W. L., Brackmann, R. T., and Henderson, W. R. (1965). *Proc. Fourth Int. Conf. on the Phys. of Electron. and Atomic Coll.* New York, pp. 100–103.
Foltz, G. W., Latimer, C. J., Hildebrandt, G. F., Kellert, F. G., Smith, K. A., West, W. P., Dunning, F. B., and Stebbings, R. F. (1977). *J. Chem. Phys.* **67**, 1352–1359.

Forst, W. (1973). "Theory of Unimolecular Reactions." Academic Press, New York.
Foster, M. S., and Beauchamp, J. L. (1975). *Chem. Phys. Lett.* **31**, 482–486.
Fox, R. E., Hickam, W. M., Grove, D. J., and Kjeldaas, T., Jr. (1955). *Rev. Sci. Instrum.* **26**, 1101–1107.
Franklin, J. L. (1976). *Science* **193**, 725–732.
Franklin, J. L., and Harland, P. W. (1974). *Annu. Rev. Phys. Chem.* **25**, 485–526.
Franklin, J. L., Hierl, P. M., and Whan, D. A. (1967). *J. Chem. Phys.* **47**, 3148–3153.
Franklin, J. L., Wang, J. L.-F., Bennett, S. L., Harland, P. W., and Margrave, J. M. (1974) *Adv. Mass Spectrom.* **6**, 319–325.
Frazier, J. R., Christophorou, L. G., Carter, J. G., and Schweinler, H. C. (1978). *J. Chem. Phys.* **69**, 3807–3818.
Gant, K. S., and Christophorou, L. G. (1976). *J. Chem. Phys.* **65**, 2977–2981.
Garrett, W. R. (1970). *Chem. Phys. Lett.* **5**, 393–397.
Garrett, W. R. (1972). *Mol. Phys.* **24**, 465–487.
Garrett, W. R. (1978). *J. Chem. Phys.* **69**, 2621–2626.
George, P. M., and Beauchamp, J. L. (1982). *J. Chem. Phys.* **76**, 2959–2964.
Gilbert, T. L., and Wahl, A. C. (1971). *J. Chem. Phys.* **55**, 5247–5261.
Goans, R. E., and Christophorou, L. G. (1974). *J. Chem. Phys.* **60**, 1036–1045.
Goldstein, E., Segal, G. A., and Wetmore, R. W. (1978). *J. Chem. Phys.* **68**, 271–279.
Goursaud, S., Sizun, M., and Fiquet-Fayard, F. (1978). *J. Chem. Phys.* **68**, 4310–4319.
Grünberg, R., (1969). *Z. Naturforsch.* **24a**, 1039–1048.
Grünberg, R., (1978). *Z. Naturforsch.* **33a**, 1346–1360.
Hackam, R., and Lennon, J. J. (1965). *Proc. Phys. Soc. London* **86**, 123–131.
Hadjiantoniou, A., Christophorou, L. G., and Carter, J. G. (1973a). *J. Chem. Soc. Faraday Trans. II* **69**, 1691–1703.
Hadjiantoniou, A., Christophorou, L. G., and Carter, J. G. (1973b). *J. Chem. Soc. Faraday Trans. II* **69**, 1704–1712.
Hake, R. D., and Phelps, A. V. (1967). *Phys. Rev.* **158**, 70–84.
Hall, R. I., Joyez, G., Mazeau, J., Reinhardt, J., and Schermann, C. (1973). *J. Phys. Orsay Fr.* **34**, 827–843.
Hall, R. I., Cadez, I., Schermann, C., and Tronc, M. (1977). *Phys. Rev. A* **15**, 599–610.
Hall, R. J. (1978). *J. Chem. Phys.* **68**, 1803–1807.
Harland, P. W., and Franklin, J. L. (1974). *J. Chem. Phys.* **61**, 1621–1636.
Harland, P. W., and Thynne, J. C. J. (1969). *J. Phys. Chem.* **73**, 4031–4035.
Harland, P. W., and Thynne, J. C. J. (1970). *J. Phys. Chem.* **74**, 52–59.
Harland, P. W., and Thynne, J. C. J. (1971). *J. Phys. Chem.* **75**, 3517–3523.
Harland, P. W., and Thynne, J. C. J. (1972a). *Int. J. Mass Spectrom. Ion Phys.* **9**, 253–266.
Harland, P. W., and Thynne, J. C. J. (1972b). *Int. J. Mass Spectrom. Ion Phys.* **10**, 11–23.
Harland, P. W., and Thynne, J. C. J. (1973). *Int. J. Mass Spectrom. Ion Phys.* **11**, 445–454.
Harland, P. W., and Thynne, J. C. J. (1975). *Int. J. Mass Spectrom. Ion Phys.* **18**, 73–86.
Harland, P. W., Cradock, S., and Thynne, J. C. J. (1972). *Int. J. Mass Spectrom. Ion Phys.* **10**, 169–195.
Harland, P. W., Franklin, J. L., and Carter, D. E. (1973). *J. Chem. Phys.* **58**, 1430–1437.
Hatano, Y., and Shimamori, H. (1981). *In* "Electron and Ion Swarms" (L. G. Christophorou, ed.), pp. 103–116. Pergamon Press, New York.
Heideman, H. G. M., Kuyatt, C. E., and Chamberlain, G. E. (1966). *J. Chem. Phys.* **44**, 355–358.
Henderson, W. R., Fite, W. L., and Brackmann, R. T. (1969). *Phys. Rev.* **183**, 157–166.
Heni, M., Illenberger, E., Baumgartel, H., and Suzer, S. (1982). *Chem. Phys. Lett.* **87**, 244–248.
Henis, J. M. S., and Mabie, C. A. (1970). *J. Chem. Phys.* **53**, 2999–3013.
Herrick, D. R., and Stillinger, F. H. (1975). *J. Chem. Phys.* **62**, 4360–4365.

Herzberg, G., and Teller, E. (1933). *Z. Phys. Chem. Ser. B* **21**, 410–446.
Herzenberg, A. (1969). *J. Chem. Phys.* **51**, 4942–4950.
Hickam, W. M., and Berg, D. (1958). *J. Chem. Phys.* **29**, 517–523.
Hildebrandt, G. F., Kellert, F. G., Dunning, F. B., Smith, K. A., and Stebbings, R. F. (1978). *J. Chem. Phys.* **68**, 1349–1354.
Hildenbrand, D. L. (1977). *Int. J. Mass Spectrom. Ion Phys.* **25**, 121–128.
Hiraoka, H., Nesbet, R. K., and Welsh, L. W., Jr. (1977). *Phys. Rev. Lett.* **39**, 130–133.
Hird, B., and Ali, S. P. (1981). *J. Chem. Phys.* **74**, 3620–3621.
Holstein, T. (1951). *Phys. Rev.* **84**, 1073.
Hotop, H., and Lineberger, W. C. (1975). *J. Phys. Chem. Ref. Data* **4**, 539–576.
Hotop, H., and Niehaus, A. (1967). *J. Chem. Phys.* **47**, 2506–2507.
Hotop, H., Patterson, T. A., and Lineberger, W. C. (1974). *J. Chem. Phys.* **60**, 1806–1812.
Hubers, M. M., and Los, J. (1975). *Chem. Phys.* **10**, 235–259.
Hubin-Franskin, M.-J., Katihabwa, J., and Collin, J. E. (1976). *Int. J. Mass Spectrom. Ion Phys.* **20**, 285–293.
Huetz, H., Gresteau, F., and Mazeau, J. (1980). *J. Phys. B* **13**, 3275–3284.
Hunter, S. R., and Christophorou, L. G. (1983a). *J. Chem. Phys.* (submitted).
Hunter, S. R., and Christophorou, L. G. (1983b). *J. Chem. Phys.* (in preparation).
Hunter, S. R., Christophorou, L. G., James, D. R., and Mathis, R. A. (1982). In "Gaseous Dielectrics III" (L. G. Christophorou, ed.), pp. 7–20. Pergamon Press, New York.
Hunter, S. R., Christophorou, L. G., McCorkle, D. L., Sauers, I., Ellis, H. W., and James, D. R. (1983). *J. Phys. D*, **16**, 573–580.
Hurst, G. S., and Bortner, T. E. (1959). *Phys. Rev.* **114**, 116–120.
Illenberger, E., Scheunemann, H.-U., and Baumgartel, H. (1978). *Ber. Bunsenges. Phys. Chem.* **82**, 1154–1158.
Illenberger, E., Scheunemann, H.-U., and Baumgartel, H. (1979). *Chem. Phys.* **37**, 21–31.
Itikawa, Y. (1978). *Phys. Rep.* **46**, 117–164.
Ivanov, A. I., and Ponomarev, O. A. (1977). *Bull. Acad. Sci. USSR Div. Chem. Sci. Engl. Transl.* **25**, 2323–2328. [(1976). *Izv. Akad. Nauk SSSR Ser. Khim.* **25**, 2493–2499.]
Jäger, K., and Henglein, A. (1967). *Z. Naturforsch.* **22a**, 700–704.
Jäger, K., and Henglein, A. (1968). *Z. Naturforsch.* **23a**, 1122–1127.
Johnson, J. P., McCorkle, D. L., Christophorou, L. G., and Carter, J. G. (1975). *J. Chem. Soc. Faraday Trans. II* **71**, 1742–1751.
Johnson, J. P., Christophorou, L. G., and Carter, J. G. (1977). *J. Chem. Phys.* **67**, 2196–2215.
Jordan, K. D., and Burrow, P. D. (1978). *Acc. Chem. Res.* **11**, 341–348.
Jordan, K. D., and Wendoloski, J. J. (1977). *Chem. Phys.* **21**, 145–154.
Jordan, K. D., Griffing, K. M., Kenney, J., Andersen, E. L., and Simons, J. (1976). *J. Chem. Phys.* **64**, 4730–4740.
Jungen, M., Vogt, J., and Staemmler, V. (1979). *Chem. Phys.* **37**, 49–55.
Khvostenko, V. I., Furlei, I. I., Aminev, I. K., and Mazunov, V. A. (1973). *High Energy Chem. Engl. Transl.* **7**, 474–478; [(1973). *Khim. Vys. Energii* **7**, 537–542.]
Kimura, M., and Lacmann, K. (1980). *Chem. Phys. Lett.* **70**, 41–44.
Kiser, R. W. (1979). In "Instrumental Inorganic Chemistry." In "Topics in Current Chemistry" (F. L. Boschke, ed.). Vol. 85, pp. 89–158. Springer-Verlag, Berlin.
Kleyn, A. W., Hubers, M. M., and Los, J. (1978). *Chem. Phys.* **34**, 55–63.
Kline, L. E., Davies, D. K., Chen, C. L., and Chantry, P. J. (1979). *J. Appl. Phys.* **50**, 6789–6796.
Klots, C. E. (1967). *J. Chem. Phys.* **46**, 1197–1199.
Klots, C. E. (1977). *J. Chem. Phys.* **66**, 5240–5241.
Klots, C. E. (1979). *J. Chem. Phys.* **71**, 4172.

Klots, C. E., and Compton, R. N. (1977). *J. Chem. Phys.* **67**, 1779–1780.
Klots, C. E., and Compton, R. N. (1978a). *J. Chem. Phys.* **69**, 1636–1643.
Klots, C. E., and Compton, R. N. (1978b). *J. Chem. Phys.* **69**, 1644–1647.
Koike, F. (1973). *J. Phys. Soc. Jpn.* **35**, 1166–1170.
Koike, F. (1975). *J. Phys. Soc. Jpn.* **39**, 1590–1595.
Koike, F., and Watanabe T. (1973). *J. Phys. Soc. Jpn.* **34**, 1022–1028.
Kokaku, Y., Hatano, Y., Shimamori, H., and Fessenden, R. W. (1979). *J. Chem. Phys.* **71**, 4883–4887.
Kokaku, Y., Toriumi, M., and Hatano, Y. (1980). *J. Chem. Phys.* **73**, 6167–6168.
Köllmann, K. (1975). *J. Chem. Phys.* **63**, 1314–1316.
Kraus, K. (1961). *Z. Naturforsch.* **16a**, 1378–1385.
Kurepa, M. V., and Belić, D. S. (1978). *J. Phys. B* **11**, 3719–3729.
Kurepa, M. V., Babić, D. S., and Belić, D. S. (1981a). *Chem. Phys.* **59**, 125–136.
Kurepa, M. V., Babić, D. S., and Belić, D. S. (1981b). *J. Phys. B* **14**, 375–384.
Kuyatt, C. E., Simpson, J. A., and Mielczarek, S. R. (1966). *J. Chem. Phys.* **44**, 437–439.
Lakdawala, V. K., and Moruzzi, J. L. (1981). *J. Phys. D* **14**, 2015–2026.
Lane, N. F. (1980). *Rev. Mod. Phys.* **52**, 29–119.
Le Coat, Y., Azria, R., and Tronc, M. (1982). *J. Phys. B* **15**, 1569–1579.
Lifshitz, C., and Grajower, R. (1969). *Int. J. Mass Spectrom. Ion Phys.* **3**, 211–219.
Lifshitz, C., and Grajower, R. (1972). *Int. J. Mass Spectrom. Ion Phys.* **10**, 25–37.
Lifshitz, C., Peers, A. M., Grajower, R., and Weiss, M. (1970). *J. Chem. Phys.* **53**, 4605–4619.
Lifshitz, C., Agam, J., Weinberg A., Kantor, D., Shainok, U., and Peres, M. (1973). *Int. J. Mass Spectrom. Ion Phys.* **11**, 243–253.
Locht, R., and Momigny, J. (1970). *Int. J. Mass Spectrom. Ion Phys.* **4**, 379–391.
Lozier, W. W. (1934). *Phys. Rev.* **46**, 268–276.
McCorkle, D. L., Christophorou, L. G., and Anderson, V. E. (1972). *J. Phys. B* **5**, 1211–1220.
McCorkle, D. L., Christodoulides, A. A., Christophorou, L. G., and Szamrej, I. (1980). *J. Chem. Phys.* **72**, 4049–4057.
McCorkle, D. L., Christodoulides, A. A., Christophorou, L. G., and Szamrej, I. (1982a). *J. Chem. Phys.* **76**, 753–754.
McCorkle, D. L., Szamrej, I., and Christophorou, L. G. (1982b). *J. Chem. Phys.*, **77**, 5542–5548.
McCorkle, D. L., Christophorou, L. G., and Hunter, S. R. (1983a). *In Proc. 3rd Intern. Swarm Seminar, Innsbruck, Austria, August 3-5, 1983* (Lindinger, W., Villinger, H., and Federer, W., eds.), pp. 37–41.
McCorkle, D. L., Christodoulides, A. A., and Christophorou, L. G. (1983b). (unpublished results).
McFarland, M., Dunkin, D. B., Fehsenfeld, F. C., Schmeltekopf, A. L., and Ferguson, E. E. (1972). *J. Chem. Phys.* **56**, 2358–2364.
McMahon, D. R. A. (1981a). *Chem. Phys.* **63**, 95–111.
McMahon, D. R. A. (1981b). *In* "Electron and Ion Swarms" (L. G. Christophorou, ed.), pp. 117–126. Pergamon Press, New York.
McMahon, D. R. A. (1982). *Chem. Phys.* **66**, 67–84.
McMahon, T. B., and Beauchamp, J. L. (1972). *Rev. Sci. Instrum.* **43**, 509–512.
McNeil, R. I., Shiotani, M., Williams, F., and Yim, M. B. (1977). *Chem. Phys. Lett.* **51**, 433–437.
MacNeil, K. A. G., and Thynne, J. C. J. (1969a). *Int. J. Mass Spectrom. Ion Phys.* **2**, 1–13.
MacNeil, K. A. G., and Thynne, J. C. J. (1969b). *J. Phys. Chem.* **73**, 2960–2964.
MacNeil, K. A. G., and Thynne, J. C. J. (1969c). *Int. J. Mass Spectrom. Ion Phys.* **3**, 35–46.
MacNeil, K. A. G., and Thynne, J. C. J. (1970a). *J. Phys. Chem.* **74**, 2257–2262.
MacNeil, K. A. G., and Thynne, J. C. J. (1970b). *Int. J. Mass Spectrom. Ion Phys.* **3**, 455–464.
MacNeil, K. A. G., and Thynne, J. C. J. (1972). *Int J. Mass Spectrom. Ion Phys.* **9**, 135–140.
Mahan, B. H., and Young, C. E. (1966). *J. Chem. Phys.* **44**, 2192–2196.

Massey, H. S. W., (1976). "*Negative Ions.*" Cambridge University Press, London.
Mathur, B. P., Rothe, E. W., Tang, S. Y., and Reck, G. P. (1976). *J. Chem. Phys.* **65**, 565–569.
Mathur, B. P., Rothe, E. W., and Reck, G. P. (1977). *J. Chem. Phys.* **67**, 377–381.
Matsuzawa, M. (1972a). *J. Phys. Soc. Jpn.* **32**, 1088–1094.
Matsuzawa, M. (1972b). *J. Phys. Soc. Jpn.* **33**, 1108–1119.
Matsuzawa, M. (1975). *J. Phys. B* **8**, 2114–2122.
Matsuzawa, M. (1980). *In* "Electronic and Atomic Collisions, Proceedings of the XIth Int. Conf. on the Phys. of Electron. and Atom. Collisions, Kyoto, 29 Aug.-4 Sept., 1979, Invited Papers and Progress Reports" (N. Oda and K. Takayanagi, eds.), pp. 493–506. North Holland, Amsterdam.
Matsuzawa, M., and Chupka, W. A. (1977). *Chem. Phys. Lett.* **50**, 373–376.
Mazeau, J., Gresteau, F., Hall, R. I., and Huetz, A. (1978). *J. Phys. B* **11**, L557–L560.
Melton, C. E. (1972). *J. Chem. Phys.* **57**, 4218–4225.
Michels, H. H., and Harris, F. E. (1971). *In* "Electronic and Atomic Collisions, Abstracts of the VIIth Int. Conf. on the Physics of Electronic and Atomic Collisions, Amsterdam, 26–30 July, 1971", Vol. II, pp. 1170–1171. North-Holland, Amsterdam.
Miller, W. J., and Gould, R. K. (1978) *J. Chem. Phys.* **68**, 3542–3547.
Mital, H. P., Chandra, S., and Narain, U. (1977). *J. Phys. Soc. Jpn.* **42**, 1282–1286.
Mochizuki, T., and Lacmann, K. (1976). *J. Chem. Phys.* **65**, 3257–3265.
Mochizuki, T., and Lacmann, K. (1977). *Chem. Phys. Lett.* **49**, 604–607.
Mothes, K. G., and Schindler, R. N. (1971). *Ber. Bunsenges. Phys. Chem.* **75**, 938–945.
Mothes, K. G., Schultes, E., and Schindler, R. N. (1972). *J. Phys. Chem.* **76**, 3758–3764.
Naff, W. T., Cooper, C. D., and Compton, R. N. (1968). *J. Chem. Phys.* **49**, 2784–2788.
Naff, W. T., Compton, R. N., and Cooper, C. D. (1971). *J. Chem. Phys.* **54**, 212–222.
Naff, W. T., Compton, R. N., and Cooper, C. D. (1972). *J. Chem. Phys.* **57**, 1303–1307.
Nalley, S. J., Compton, R. N., Schweinler, H. C., and Anderson, V. E. (1973). *J. Chem. Phys.* **59**, 4125–4139.
Nelson, D. R., and Davis, F. J. (1971). *Bull. Am. Phys. Soc.* **16**, 217.
Nenner, I., and Schulz, G. J. (1975). *J. Chem. Phys.* **62**, 1747–1758.
Nesbet, R. K. (1977). *J. Phys. B* **10**, L739–L743.
Nygaard, K. J., Hunter, S. R., Fletcher, J., and Foltyn, S. R. (1978). *Appl. Phys. Lett.* **32**, 351–353.
Odom, R. W., Smith, D. L., and Futrell, J. H. (1975). *J. Phys. B* **8**, 1349–1366.
Okada, S. (1979). *Chem. Phys. Lett.* **61**, 245–248.
O'Malley, T. F. (1966). *Phys. Rev.* **150**, 14–29.
O'Malley, T. F. (1967a). *Phys. Rev.* **155**, 59–63.
O'Malley, T. F. (1967b). *J. Chem. Phys.* **47**, 5457–5458.
O'Malley, T. F., and Taylor, H. S. (1968). *Phys. Rev.* **176**, 207–221.
Orient, O. J., and Srivastava, S. K. (1983). *J. Chem. Phys.* **78**, 2949–2952.
Pabst, R. E., Bennett, S. L., Margrave, J. L., and Franklin, J. L. (1976). *J. Chem. Phys.* **65**, 1550–1560.
Pabst, R. E., Perry, D. L., Margrave, J. L., and Franklin, J. L. (1977a). *Int. J. Mass Spectrom. Ion Phys.* **24**, 323–333.
Pabst, R. E., Margrave, J. L., and Franklin, J. L. (1977b). *Int. J. Mass Spectrom. Ion Phys.* **25**, 361–374.
Pack, J. L., and Phelps, A. V. (1966a). *J. Chem. Phys.* **44**, 1870–1883.
Pack, J. L., and Phelps, A. V. (1966b). *J. Chem. Phys.* **45**, 4316–4329.
Pai, R. Y., Christophorou, L. G., and Christodoulides, A. A. (1979). *J. Chem. Phys.* **70**, 1169–1176.
Parkes, D. A., and Sugden, T. M. (1972). *J. Chem. Soc. Faraday Trans. II* **68**, 600–614.
Parlant, G., and Fiquet-Fayard, F. (1976). *J. Phys. B* **9**, 1617–1628.

Paulson, J. F. (1970). *J. Chem. Phys.* **52**, 963–964.
Peart, B., and Dolder, K. T. (1973). *J. Phys. B* **6**, 1497–1502.
Pejčev, V. M., Kurepa, M. V., and Čadež, I. M. (1979). *Chem. Phys. Lett.* **63**, 301–304.
Peyerimhoff, S. D., and Buenker, R. J. (1981). *Chem. Phys.* **57**, 279–296.
Phelps, A. V. (1969). *Can. J. Chem.* **47**, 1783–1793.
Puckett, L. J., Kregel, M. D., and Teague, M. W. (1971). *Phys. Rev. A* **4**, 1659–1666.
Rademacher, J., Christophorou, L. G., and Blaunstein, R. P. (1975). *J. Chem. Soc. Faraday Trans. II* **71**, 1212–1226.
Rallis, D. A., and Goodings, J. M. (1971). *Can. J. Chem.* **49**, 1571–1574.
Rapp, D., and Briglia, D. D. (1965). *J. Chem. Phys.* **43**, 1480–1489.
Rapp, D., Sharp, T. E., and Briglia, D. D. (1965). *Phys. Rev. Lett.* **14**, 533–535.
Renner, R. (1934). *Z. Phys.* **92**, 172–193.
Robinson, P. J., and Holbrook, K. A. (1972). "Unimolecular Reactions." Wiley (Interscience), New York.
Rokni, M., Jacob, J. H., and Mangano, J. A. (1979). *Appl. Phys. Lett.* **34**, 187–189.
Rothe, D. E. (1969). *Phys. Rev.* **177**, 93–99.
Rothe, E. W., Tang, S. Y., and Reck, G. P. (1974). *Chem. Phys. Lett.* **26**, 434–436.
Rundel, R. D. (1980). In "Electronic and Atomic Collisions, Proceedings of the XIth Int. Conf. on the Phys. of Electron. and Atom. Collisions, Kyoto, 29 Aug.-4 Sept. 1979, Invited Papers and Progress Reports" (N. Oda and K. Takayanagi, eds.), pp. 481–491. North Holland, Amsterdam.
Sanche, L., and Schulz, G. J. (1973). *J. Chem. Phys.* **58**, 479–493.
Sauers, I., Christophorou, L. G., and Carter, J. G. (1979). *J. Chem. Phys.* **71**, 3016–3024.
Schermann, C., Cadez, I., Delon, P., Tronc, M., and Hall, R. I. (1978). *J. Phys. E* **11**, 746–747.
Scheunemann, H.-U., Illenberger, E., and Baumgärtel, H. (1980). *Ber. Bunsenges. Phys. Chem.* **84**, 580–585.
Schneider, B. I., and Brau, C. A. (1978). *Appl. Phys. Lett.* **33**, 569–571.
Schnitzer, R., and Anbar, M. (1976). *J. Chem. Phys.* **65**, 1117–1123.
Schultes, E., Christodoulides, A. A., and Schindler, R. N. (1975). *Chem. Phys.* **8**, 354–365.
Schulz, G. J. (1959). *Phys. Rev.* **113**, 816–819.
Schulz, G. J. (1960). *J. Chem. Phys.* **33**, 1661–1665.
Schulz, G. J. (1962). *Phys. Rev.* **128**, 178–186.
Schulz, G. J. (1973). *Rev. Mod. Phys.* **45**, 378–486.
Schulz, G. J., and Asundi, R. K. (1965). *Phys. Rev. Lett.* **15**, 946–949.
Schulz, G. J., and Asundi, R. K. (1967). *Phys. Rev.* **158**, 25–29.
Schumacher, R., Sprünken, H.-R., Christodoulides, A. A., and Schindler, R. N. (1978). *J. Phys. Chem.* **82**, 2248–2252.
Sharp, T. E., and Dowell, J. T. (1967). *J. Chem. Phys.* **46**, 1530–1531.
Sharp, T. E., and Dowell, J. T. (1969). *J. Chem. Phys.* **50**, 3024–3035.
Shaw, M. J., and Jones, J. D. C. (1977). *Appl. Phys.* **14**, 393–398.
Shimamori, H., and Fessenden, R. W. (1978). *J. Chem. Phys.* **68**, 2757–2766.
Shimamori, H., and Fessenden, R. W. (1979). *J. Chem. Phys.* **70**, 1137–1141.
Shimamori, H., and Fessenden, R. W. (1981). *J. Chem. Phys.* **74**, 453–466.
Shimamori, H., and Hatano, Y. (1976a). *Chem. Phys. Lett.* **38**, 242–247.
Shimamori, H., and Hatano, Y. (1976b). *Chem. Phys.* **12**, 439–445.
Shimamori, H., and Hatano, Y. (1977). *Chem. Phys.* **21**, 187–201.
Sides, G. D., and Tiernan, T. O. (1974). ARL Report No. 74–0105; see Sides *et al.*, 1976.
Sides, G. D., and Tiernan, T. O. (1977). *J. Chem. Phys.* **67**, 2382–2384.
Sides, G. D., Tiernan, T. O., and Hanrahan, R. J. (1976). *J. Chem. Phys.* **65**, 1966–1975.
Smirnov, B. M. (1982). "Negative Ions." McGraw Hill, New York.

Spence, D. (1975). In "The Physics of Electronic and Atomic Collisions" (J. S. Risley and R. Geballe, eds.), pp. 241–255. University of Washington Press, Seattle.
Spence, D., and Burrow, P. D. (1979). J. Phys. B **12**, L179–L184.
Spence, D., and Schulz, G. J. (1969). Phys. Rev. **188**, 280–287.
Spence, D., and Schulz, G. J. (1972). Phys. Rev. A **5**, 724–732.
Spence, D., and Schulz, G. J. (1973). J. Chem. Phys. **58**, 1800–1803.
Spence, D., and Schulz, G. J. (1974). J. Chem. Phys. **60**, 216–220.
Spence, D., Chupka, W. A., and Stevens, C. M. (1981). Argonne National Laboratory Report ANL-79-65, Part 1, p. 63–67.
Spence, D., Chupka, W. A., and Stevens, C. M. (1982). J. Chem. Phys. **76**, 2759–2761.
Spyrou, S., Sauers, I., and Christophorou, L. G. (1983). J. Chem. Phys. **78**, 7200–7216.
Srivastava, S. K., Chutjian, A., and Trajmar, S. (1975). J. Chem. Phys. **63**, 2659–2665.
Stamatovic, A., and Schulz, G. J. (1970a). Rev. Sci. Instrum. **41**, 423–427.
Stamatovic, A., and Schulz, G. J. (1970b). J. Chem. Phys. **53**, 2663–2667.
Stamatovic, A., and Schulz, G. J. (1973). Phys. Rev. A **7**, 589–592.
Stapleton, B. J., and Bowie, J. H. (1977). Aust. J. Chem. **30**, 417–420.
Stebbings, R. F. (1976). Science **193**, 537–542.
Stockdale, J. A. D., and Reinhardt, P. W. (1980). Chem. Phys. Lett. **71**, 274–276.
Stockdale, J. A. D., and Warmack, R. J. (1980). J. Chem. Phys. **72**, 5673–5678.
Stockdale, J. A. D., Christophorou, L. G., and Hurst, G. S. (1967). J. Chem. Phys. **47**, 3267–3270.
Stockdale, J. A. D., Compton, R. N., Hurst, G. S., and Reinhardt, P. W. (1969). J. Chem. Phys. **50**, 2176–2180.
Stockdale, J. A. D., Compton, R. N., and Schweinler, H. C. (1970). J. Chem. Phys. **53**, 1502–1507.
Stockdale, J. A. D., Nelson, D. R., Davis, F. J., and Compton, R. N. (1972). J. Chem. Phys. **56**, 3336–3341.
Stockdale, J. A. D., Davis, F. J., Compton, R. N., and Klots, C. E. (1974). J. Chem. Phys. **60**, 4279–4285.
Stuckey, W. K., and Kiser, R. W. (1966). Nature **211**, 963–964.
Sugiura, T., and Arakawa, K. (1970). In "Recent Developments in Mass Spectroscopy. Proceedings of the International Conference on Mass Spectroscopy" (K. Ogata and T. Hayakawa, eds.), pp. 848–853. University of Tokyo Press, Tokyo, 1970.
Sweetman, D. R. (1960). Proc. Phys. Soc. London **76**, 998–1000.
Tam, W.-C., and Wong, S. F. (1978). J. Chem. Phys. **68**, 5626–5630.
Tang, S. Y., Rothe, E. W., and Reck, G. P. (1974). J. Chem. Phys. **61**, 2592–2595.
Tang, S. Y., Mathur, B. P., Rothe, E. W., and Reck, G. P. (1976). J. Chem. Phys. **64**, 1270–1275.
Taylor, H. S., and Thomas, L. D. (1972). Phys. Rev. Lett. **28**, 1091–1092.
Taylor, H. S., Nazaroff, G. V., and Golebiewski, A., (1966). J. Chem. Phys. **45**, 2872–2888.
Taylor, H. S., Goldstein, E., and Segal, G. A. (1977). J. Phys. B **10**, 2253–2259.
Thomas, L. D. (1974). J. Phys. B **7**, L97–L98.
Thynne, J. C. J. (1969). J. Phys. Chem. **73**, 1586–1588.
Thynne, J. C. J. (1972). In "Dynamic Mass Spectrometry" (D. Price, ed.), Vol. 3, pp. 67–97. Heyden, London.
Thynne, J. C. J., and Harland, P. W. (1973a). Int. J. Mass Spectrom. Ion Phys. **11**, 137–147.
Thynne, J. C. J., and Harland, P. W. (1973b). Int. J. Mass Spectrom. Ion Phys. **11**, 399–408.
Thynne, J. C. J., and MacNeil, K. A. G. (1970). Int. J. Mass Spectrom. Ion Phys. **5**, 329–335.
Tozer, B. A. (1958). J. Electron. Control **4**, 149–159.
Trainor, D. W., and Boness, M. J. W. (1978). Appl. Phys. Lett. **32**, 604–606.
Trajmar, S., and Hall, R. I. (1974). J. Phys. B **7**, L458–L461.
Tronc, M., Goursaud, S., Azria, R., and Fiquet-Fayard, F. (1973). J. Phys. Orsay Fr. **34**, 381–388.
Tronc, M., Fiquet-Fayard, F., Schermann, C., and Hall, R. I. (1977a). J. Phys. B **10**, 305–321.

Tronc, M., Fiquet-Fayard, F., Schermann, C., and Hall, R. I. (1977b). *J. Phys. B* **10**, L459–L462.
Tronc, M., Hall, R. I., Schermann, C., and Taylor, H. S. (1979). *J. Phys. B* **12**, L279–L282.
Truby, F. K. (1968). *Phys. Rev.* **172**, 24–30.
Truby, F. K. (1969). *Phys. Rev.* **188**, 508–512.
Truby, F. K. (1971). *Phys. Rev. A* **4**, 613–618.
Truby, F. K. (1972). *Phys. Rev. A* **6**, 671–676.
Turner, J. E. (1977). *Am. J. Phys.* **45**, 758–766.
Van Brunt, R. J. (1974). *J. Chem. Phys.* **60**, 3064–3070.
Van Brunt, R. J., and Kieffer, L. J. (1970). *Phys. Rev. A* **2**, 1899–1905.
Van Brunt, R. J., and Kieffer, L. J. (1974). *Phys. Rev. A* **10**, 1633–1637.
Van Lint, V. A. J., Wikner, E. G., and Trueblood., D. L. (1960). *Bull. Am. Phys. Soc.* **5**, 122–123.
Van Lint, V. A. J., Colwell, J. F., and Vroom, D. A. (1971). *Air Force Weapons Lab. Report AFWL* TR-70-115.
Verhaart, G. J., Van der Hart, W. J., and Brongersma, H. H. (1978). *Chem. Phys.* **34**, 161–167.
Verhaart, G. J., Van Sprang, H. A., and Brongersma, H. H. (1980). *Chem. Phys.* **51**, 389–398.
Von Trepka, L., and Neuert, H. (1963). *Z. Naturforsch.* **18a**, 1295–1303.
Wadehra, J. M., and Bardsley, J. N. (1978). *Phys. Rev. Lett.* **41**, 1795–1798.
Walton, D. S., Peart, B., and Dolder, K. T. (1970). *J. Phys. B* **3**, L148–L150.
Walton, D. S., Peart, B., and Dolder, K. T. (1971). *J. Phys. B* **4**, 1343–1348.
Wang, J. L.-F., Margrave, J. L., and Franklin, J. L. (1973). *J. Chem. Phys.* **58**, 5417–5421.
Wang, J. L.-F., Margrave, J. L., and Franklin, J. L. (1974a). *J. Chem. Phys.* **60**, 2158–2162.
Wang, J. L.-F., Margrave, J. L., and Franklin, J. L. (1974b). *J. Chem. Phys.* **61**, 1357–1360.
Warmack, R. J., Stockdale, J. A. D., and Compton, R. N. (1978). *J. Chem. Phys.* **68**, 916–925.
Warmack, R. J., Stockdale, J. A. D., and Schweinler, H. C. (1980). *J. Chem. Phys.* **72**, 5930–5934.
Warman, J. M., and Sauer, M. C., Jr. (1971). *Int. J. Radiat. Phys. Chem.* **3**, 273–282.
Warman, J. M., Bansal, K. M., and Fessenden, R. W. (1971). *Chem. Phys. Lett.* **12**, 211–213.
Wecker, D., Christodoulides, A. A., and Schindler, R. N. (1981). *Int. J. Mass Spectrom. Ion Phys.* **38**, 391–406.
Wegener, P. P. (1974). "Molecular Beams and Low Density Gasdynamics." Dekker, New York.
Weller, C. S., and Biondi, M. A. (1968). *Phys. Rev.* **172**, 198–206.
Wentworth, W. E., and Becker, R. S. (1962). *J. Am. Chem. Soc.* **84**, 4263–4266.
Wentworth, W. E., Becker, R. S., and Tung, R. (1967). *J. Phys. Chem.* **71**, 1652–1665.
Wentworth, W. E., George, R., and Keith, H. (1969). *J. Chem. Phys.* **51**, 1791–1801.
West, W. P., Foltz, G. W., Dunning, F. B., Latimer, C. J., and Stebbings, R. F. (1976). *Phys. Rev. Lett.* **36**, 854–858.
Whitten, G. Z., and Rabinovitch, B. S. (1963). *J. Chem. Phys.* **38**, 2466–2473.
Wiegand, W. J., and Boedeker, L. R. (1982). *Appl. Phys. Lett.* **40**, 223–227.
Young, B. G., Johnson, A. W., and Carruthers, J. A. (1963). *Can. J. Phys.* **41**, 625–631.
Zastawny, A. (1974). *Acta Phys. Pol. A* **46**, 39–45.
Zembekov, A. A. (1971). *Chem. Phys. Lett.* **11**, 415–416.
Ziesel, J. P., Nenner, I., and Schulz, G. J. (1975a). *J. Chem. Phys.* **63**, 1943–1949.
Ziesel, J. P., Schulz, G. J., and Milhaud, J. (1975b). *J. Chem. Phys.* **62**, 1936–1940.

7 Electron Detachment Processes

R. L. Champion and L. D. Doverspike
Department of Physics
The College of William and Mary
Williamsburg, Virginia 23185

I.	Introduction	619
II.	Nomenclature and Experimental Techniques	621
III.	Atomic Reactants	623
IV.	Molecular Targets	639
	A. Halogen Ions + N_2, CO, CO_2, O_2	640
	B. $H^-(D^-)$ + N_2, CO, CO_2, O_2	646
V.	Detachment Accompanied by Rearrangement	656
	A. $H^-(D^-)$ + H_2, D_2	658
	B. Halogen Ions + H_2, D_2, HD	660
	C. I^- + Methyl Halides	670
VI.	Molecular Negative Ions	673
VII.	Detachment Rate Constants	677
VIII.	Summary	678
	References	679

I. Introduction

The collisional detachment of negative ions by atomic and molecular targets and the reverse processes of electron attachment and three-body recombination are important in many areas of physics and chemistry. These processes have posed challenging problems for many years in the field of atomic and molecular collision physics. Electron attachment and recombination mechanisms are the subjects of Chapter 6 of Volume 1 and Chapter 2 of Volume 2.

The purpose of this review is to survey the recent activity in the field of collisional electron detachment of negative ions, with particular emphasis on electron detachment by molecular targets. Those interested in the more general field of negative ions and their properties may wish to refer to the texts by Massey (1976) and Smirnov (1982) and to Chapter 6 of this volume.

Additionally, a recent review which focuses primarily on the subject of collisional detachment by atomic targets has been given by Champion (1982).

Interest in gas-phase negative ions and their collisional properties stems from the fact that the majority of the elements in the periodic table and an incredible variety of molecules and radicals form stable negative ions. The binding energy of an "additional" electron to the neutral atom or molecule is termed the electron affinity of the species, and for atoms many of these electron affinities have been determined in high-precision experiments in which a laser is used to photodetach the negative ions. Reviews of the subject of atomic electron affinities have been presented by Hotop and Lineberger (1975), by Smirnov (1982), and is also covered in Chapter 6 of Volume 2.

The definition of the electron affinity of a molecule is not so clear-cut as for an atom owing to the vibrational and rotational degrees of freedom and, for polyatomic molecules, configurational degrees of freedom of the molecular negative ion and the neutral parent molecule. For molecules, the "vertical" electron affinity is taken to be the minimum energy necessary to detach the electron from the ground vibrational state of the negative molecular ion *without* a change in the nuclear coordinates. The minimum energy separation between the negative ion and neutral parent (which may have different nuclear coordinates) is often called simply the electron affinity: this latter quantity is less than the vertical electron affinity. Recent reviews on the subject of molecular electron affinities have been given by Janousek and Brauman (1979), by Franklin and Harland (1974), and in Chapter 6 of Volume 2.

There are several areas of applied interest in which negative ions play an important role. They all involve ionized gases and include such diverse fields as ionospheric chemistry, the development of intense negative ion beams, and high-power switching circuits, flame chemistry, and magnetohydrodynamics where the plasma conductivity is greatly affected by the negative-ion concentration. In all of these environments, the principal method whereby negative ions are destroyed is that of collisional detachment.

Collisions of negative ions with neutral atoms or molecules can lead to detachment of the loosely bound electron in several distinct ways; these may be enumerated as follows:

Direct detachment, in which there are no excited intermediate or final states of the reactants:

$$X^- + B \longrightarrow X + B + e \tag{1}$$

Detachment with excitation, in which excited states of either the target or negative ion (or its neutral parent) are involved:

$$X^- + B \longrightarrow (X^-)^* + B \longrightarrow X + B + e \tag{2}$$

(excitation to autodetaching levels),

$$X^- + B \longrightarrow X + B^* + e \text{ or } X^* + B + e \tag{3}$$

(excitation of a neutral product), and

$$X^- + B \longrightarrow X + (B^-)^* \longrightarrow X + B + e \tag{4}$$

(charge transfer to a temporary negative-ion state of the target).

Detachment with bonding, in which the neutral reactants can form bound systems:

$$X^- + BC \longrightarrow XB + C + e \tag{5}$$

(reactive collision with detachment), and

$$X^- + B \longrightarrow XB + e \tag{6}$$

(associative detachment).

Direct detachment (1) is important (and often the dominant detachment channel) at all collision energies above threshold, whereas (2) represents only a few percent of the total detachment cross section, and then only at collision energies above a few kilo-electron-volts. Detachment with excitation (3) and detachment via charge transfer (4) may be important (and, in fact, indistinguishable) at all collision energies for molecular targets. Electron detachment with concomitant chemical bonding, (5) and (6), is important for selected reactants, but only for collision energies below several tens of electron volts.

Detailed experimental studies of electron detachment by molecular targets are just beginning to surface in the literature. In this review (or perhaps, preview) we will present examples and results from some of these studies with illustrative examples of each of the mechanisms discussed above. Owing to the limited number of studies completed to date, this presentation is undoubtedly far from definitive.

To introduce the subject of collisional electron detachment, we will first briefly discuss the collisional dynamics for a few selected atomic reactants: it is only for such atomic reactants that both theory and experiment exist. With this fundamental understanding, we will then move on to collisional detachment for reactants which involve molecular targets or molecular negative ions.

II. Nomenclature and Experimental Techniques

The notation that we shall employ to refer to the total electron detachment cross section is $\sigma_{-10}(E)$, where $(-1, 0)$ refer to the charge states of the incoming and outgoing atomic or molecular projectiles, respectively. The collision energy E refers to the relative energy (energy in the center of mass frame of reference) unless specifically stated otherwise.

The differential cross section for a given collision energy is usually denoted by $\sigma(\theta)$ [rather than $\sigma(\theta, E)$] where

$$\sigma_{\text{total}}(E) = 2\pi \int \sigma(\theta)\sin\theta \, d\theta$$

for cases of azimuthal symmetry. In addition, some further symbol must be assigned to $\sigma(\theta)$ to identify the product channel. For example, $\sigma_e(\theta)$ refers to elastic differential cross sections and $\sigma_n(\theta)$ refers to the differential cross section for neutrals which are the products of collisional detachment.

It has become common practice for experimentalists to report their results for differential cross section measurements with "reduced" coordinates:

$$\tau \equiv E\theta, \qquad \rho \equiv \theta\sigma(\theta)\sin\theta$$

This representation is attractive in that results at various energies can be compared conveniently to each other. Structure in $\sigma(\theta)$ which is due to some particular feature of the internuclear potential (e.g., curve crossing) occurs at an approximately fixed value of τ. Moreover, the magnitude of $E \cdot \theta$ is approximately independent of the coordinate system—i.e., $E \cdot \theta \simeq E_{\text{lab}}\theta_{\text{lab}}$. The usefulness and efficacy of these reduced coordinates are discussed by Smith et al. (1967).

Experiments have been performed in which the absolute total detachment cross section, the angular differential and doubly differential (i.e., both the angular and product energy spectra and obtained) cross sections have been measured. There have been experiments in which energy and sometimes angular spectra of the detached electrons have been measured and others in which the light emitted from the products of collisional detachment has been monitored. For several negative ion–molecule systems that have been studied, the mechanisms that complement electron detachment (such as dissociative charge transfer) have also been investigated.

Several methods have been used to measure total detachment cross sections. For lower collision energies, it is convenient to detect the detached electrons by using weak electrostatic fields often combined with magnetostatic fields to steer the free electrons to a collector plate. This can be done without seriously disturbing the trajectories of the heavier negative ions, and examples of this technique can be found in the work of Hasted (1952), Roche and Goodyear (1969), Hummer et al. (1960), Smith et al. (1978), and Bydin and Dukel'skii (1957). Bailey and Mahadevan (1970) used a radio-frequency electron trap to distinguish detached electrons from slow, heavy product negative ions.

For collision energies at which large-angle elastic scattering is not too important, the total electron loss cross section can be determined by an attenuation technique in which the intensity of a transmitted ion beam is observed as the target gas pressure is varied. Such an approach has been used by Risley and Geballe (1974) and Bennett et al. (1975). Meyer (1980) has employed a slight variation of the above in which the collision path length is varied (rather than the pressure) in an attenuation experiment in which the neutralization cross sections for negative ion–alkali reactants were determined.

For both the attenuation and electron-trap techniques, it is generally not possible to separate electron detachment from detachment with ionization. For example, in an electron trap which detects all of the electrons that are collision products, the experiment measures

$$\sigma(E) = \sigma_{-10}(E) + 2\sigma_{-11}(E),$$

where $\sigma_{-11}(E)$ is the cross section for detachment with ionization. At low collision energies the second term is usually negligible, however.

At higher collision energies, the fast-neutral collision products can be measured directly to determine $\sigma_{-10}(E)$ and $\sigma_{-11}'(E)$ as has been done, for example, by Geddes et al. (1980).

Differential cross sections have been measured with the time-of-flight method (e.g., Fayeton et al., 1978; Cheung and Datz, 1980), by using electrostatic energy analysis (e.g., Lam et al., 1974), and with position-sensitive channel-plate detectors (e.g., deVreugd et al., 1979). In order to investigate the role of excited states in collisional detachment, it is necessary to measure doubly differential cross sections, and the time-of-flight apparatus is especially well suited to this purpose.

Cross sections for the production of negatively charged ions which are products of negative ion–molecule collisions have been determined in experiments both with and without product mass analysis (e.g., Paulson, 1970; Lifshitz et al., 1978; Huq et al., 1982b). These product ions are normally identified by capitalizing upon the fact that their laboratory kinetic energies are small when compared to that of the reactant negative ion, enabling electrostatic trapping of the product ion. The low-energy detached electrons can be separated from the low-energy product ions with an appropriate magnetic field configuration within the electrostatic trap (Huq et al., 1982b).

All of these methods being used in negative ion–molecule investigations have been employed previously in positive ion–molecule experiments.

III. Atomic Reactants

Of all the negative ion–atom systems, those which involve collisions of H^- and D^- with the rare gases have been studied most extensively. Because of the relative simplicity of the $H^- +$ He and $H^- +$ Ne systems, they have also been the subject of several theoretical studies. In order to discuss the dynamics of collisional detachment, it is instructive to first examine the experimental observations and theories for $H^-(D^-) +$ He and $H^-(D^-) +$ Ne for collision energies near the threshold for detachment. For these collision energies (say, under several hundred electron volts), direct detachment (1) is by far the dominant mechanism among the various detachment processes, (1)–(6).

The qualitative behavior of the total detachment cross section at low collision energies was first discussed by Mason and Vanderslice (1958), and it is easily understood if one refers to the potential-energy curves for HeH^- and HeH. The results of recent calculations for these potentials (Olson and Liu, 1980) are given in Fig. 1. At large internuclear separations, the HeH^- electronic energy lies 0.75 eV (i.e., the electron affinity of H) below the HeH curve. In a collision, as the negative ion approaches the target atom, the electronic energy of the HeH^- system rises until it crosses or merges with the continuum of states representing neutral atoms and a free electron. If the curves cross at $R = R_x$, as in Fig. 1, then for $R < R_x$, HeH^- can no longer be regarded as a stable state, but perhaps as a quasibound resonance. It can be

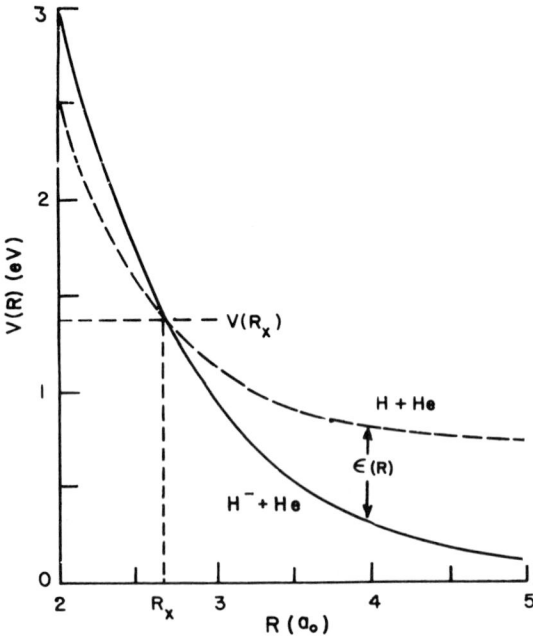

Fig. 1. Internuclear potentials for the ground states of HeH and HeH^-. [From a calculation by Olson and Liu (1980).]

7 Electron Detachment Processes

anticipated that detachment will occur during a collision when the separation of the negative ion and the atom is less than (or close to) the crossing radius R_x.

For these collision energies the de Broglie wavelength associated with the nuclear motion is considerably smaller than the molecular size, and the motion of the nuclei may be discussed within the framework of classical mechanics. Moreover, for low collision energies the nuclear velocities are considerably smaller than electron velocities and molecular calculations based upon a near-adiabatic approximation should be appropriate for calculating the internuclear potential.

The cross section for collisional detachment may be expressed

$$\sigma_{-10}(E) = \int_0^\infty 2\pi b P_d(b, E) \, db \tag{7}$$

where b is the classical impact parameter and $P_d(b, E)$ is the detachment probability which depends upon b and E. Several different theoretical approaches have been employed to describe $P_d(b, E)$ and these will be discussed later. It may be useful, however, to first consider the simplest (but not too realistic) assumption for $P_d(b, E)$, which is

$$P_d(b, E) = 0, \quad b > b_x,$$
$$= 1, \quad b \le b_x, \tag{8}$$

where b_x is the impact parameter for which the classical turning point of the nuclear motion is R_x, i.e., where

$$b_x^2/R_x^2 + V(R_x)/E = 1. \tag{9}$$

In this case,

$$\sigma_{-10}(E) = \pi b_x^2 = \pi R_x^2 [1 - V(R_x)/E]. \tag{10}$$

Thus, the threshold energy will be $V(R_x)$ as indicated in Fig. 1 and the high-energy limit for $\sigma_{-10}(E)$ will be approximately πR_x^2.

An improvement on the above step-function model for $P_d(b, E)$ is one in which the direct detachment is described by a complex potential model (see Lam et al., 1974, for example). In this model, the decay of the negative-ion state is governed simply by a width $\Gamma(R)$, which is generally zero for $R \gtrsim R_x$. The lifetime of the unstable HeH⁻ state is proportional to $1/\Gamma(R)$ and the detachment probability, in this model, is given by

$$P_d(b, E) = \int \Gamma[R(t)] \, dt$$
$$= \int \frac{\Gamma(R) \, dR}{v(R)}, \tag{11}$$

where $v(R)$ is the relative nuclear velocity. Equation (11) predicts an isotope effect for $H^-(D^-) + He$ since, for a given relative collision energy, the heavier projectile (D^-) spends more time in the region $R < R_x$ and is hence more likely to detach. This simple model predicts the correct size of the isotope effect observed for $H^-(D^-) + He$ (Champion et al., 1976; Huq et al., 1982a), which can be seen in the plot of the measured $\sigma_{-10}(E)$ exhibited in Fig. 2.

Another approach to the problem of collisional detachment has been taken by Gauyacq (1979, 1980a), who calculated $P_d(b, E)$ for the HeH^- system by using an extension of a formalism known as the "zero-radius model," which was previously discussed by Devdarianni (1973) and Demkov (1964, 1980). In this model, detachment can occur for R near R_x, which is in the region in which

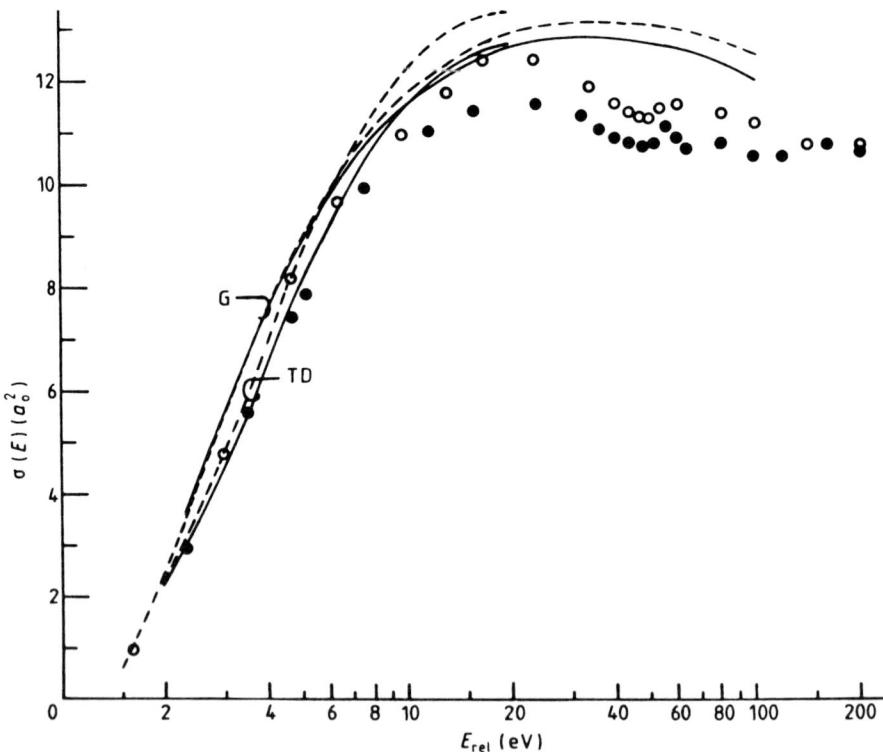

Fig. 2. Total electron detachment cross sections for H^- and D^- on He. The open circles are for D^- and the full circles are for H^-. The theoretical calculations of Gauyacq (1980a) are the full and broken curves labeled G which extend up to 100 eV. The theoretical calculations of Taylor and Delos (1982) are the full and broken curves labeled TD which extend up to 20 eV. All full curves refer to H^- and broken curves refer to D^-. [From Huq et al. (1982a). Reproduced with permission from the *Journal of Physics*.]

7 Electron Detachment Processes

the binding energy $\varepsilon(R)$ of the extra electron is very small. [$\varepsilon(R)$ is the difference between the neutral and ionic potentials for $R > R_x$.] For small values of $\varepsilon(R)$, the wavelength of the outer electron becomes much larger than the size of the HeH molecular core (which is on the order of R_x) and the probability that the electron is outside this core becomes large. The detachment problem is then addressed by dividing the space into two regions; the core and the outer region where the electron is essentially in a field-free region with energy $\varepsilon(R)$. In this outer region, the electron wave function (assumed to be an s-state) satisfies the free-particle Schrödinger equation

$$\left(\frac{\hbar^2}{2m}\frac{\partial^2}{\partial r^2} + i\hbar\frac{\partial}{\partial t}\right)\psi = 0,$$

with initial condition $r\psi \sim e^{-kr}$, where $k = [2\varepsilon(R)]^{1/2}$. The logarithmic derivative of the wave function is specified on the boundary between the two regions, which is allowed to be at arbitrarily small r:

$$\frac{1}{r\psi}\frac{\partial(r\psi)}{\partial r}\bigg|_{r\to 0} = f[R(t)] \qquad (12)$$

and where the time dependence of the boundary condition is due to the motion of the nuclei. For $R > R_x$, $f[R(t)]$ is negative and equal to $-[2\varepsilon(R)]^{1/2}$. For $R < R_x$, $f[R(t)]$ changes sign and is not so easily obtained as for $R > R_x$. Consequently, calculations for $H^- + He$ based on the zero-radius model use a linear extrapolation to approximate $f[R(t)]$ for $R < R_x$. The potential-energy curves (such as those of Fig. 1) can be used to find $R(t)$ and $\varepsilon[R(t)]$ for a given impact parameter and Eq. (12) can be numerically integrated to find $\psi(r, t)$. For a large value of t, $t = T$, the collision partners will have separated and the survival probability [i.e., the complement of the detachment probability $P_d(b)$] can be found by projecting the electron wave function $\psi(r, T)$ onto the bound eigenfunction. The result of such a procedure for $D^- + He$ at $E = 16$ eV is seen in Fig. 3. The calculation for $P_d(b)$ indicates that a substantial fraction of $\sigma_{-10}(E)$ comes from impact parameters larger than b_x, a feature which is not present in a complex potential model. The calculated detachment cross section for $D^- + He$ which is based upon the above zero-radius approximation is given in Fig. 2, and the agreement between the experiment and the calculation is seen to be quite good.

The isotope effects observed in the $H^- + He$ experiment are not completely accounted for by this calculation: the calculation predicts about 50% of the observed difference in $\sigma_{-10}(E)$ for $H^-(D^-) + He$. Gauyacq (1980a) suggests that the zero-radius model inherently contains two types of isotope effects which are in opposition and, as a consequence, may result in $\sigma_{-10}(E)$ exhibiting the behavior of either the $H^-(D^-) + He$ *or* the inverse. Alternatively, there may be no discernible isotope effect. These two opposing effects at a

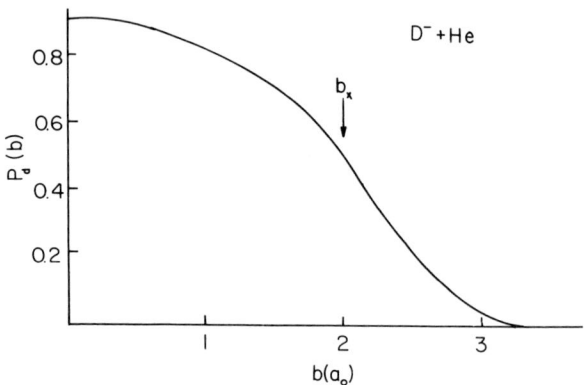

Fig. 3. Detachment probability as a function of impact parameter b for $E_{rel} = 16$ eV; b_x is the impact parameter for which the distance of closest approach is R_x. [From Gauyacq (1980a). Reproduced with permission from the *Journal of Physics*.]

given relative energy are due to (i) the lengthening amount of time spent in the region $R < R_x$ for the heavier isotope (the same effect as for the local complex potential model) and (ii) the velocity dependence of "dynamical transitions" for $R \gtrsim R_x$, in which detachment is more probable for the higher-velocity reactants.

In order to illustrate the features of the zero-range model, it is interesting to follow the temporal behavior of the electron wave function as the collision proceeds. Figure 4 illustrates this behavior for the $H^- + He$ system for a relative collision energy of 200 eV. Prior to the collision (negative time on Fig. 4), the wave function is a straight line on the semilogarithmic plot, indicative of the bound state. At $t = 0$ {the point at which $f[R(t)]$ changes sign}, the wave packet has begun to spread appreciably. For large positive times the wave function appears to be split into two distinct parts; that which remains near the origin and represents the survival probability, and an outgoing wave packet which represents the outgoing (detached) electron.

A different approach to the theory of these processes has been developed by Taylor and Delos (1982). They expand the electronic wave function as

$$\psi = C_0(t)\phi_0 + \int C_E(t)\phi_E \rho(E) dE \tag{13}$$

where ϕ_0 is the state in which the electron is bound to the molecule; ϕ_E is a state in which the electron is free, with kinetic energy E; and $\rho(E)$ is the density of states in this continuum. Thus $C_0(t)$ is the probability amplitude for finding the electron bound; $|C_0(-\infty)|^2 = 1$, and $1 - |C_0(\infty)|^2$ is the probability of electron detachment.

7 Electron Detachment Processes

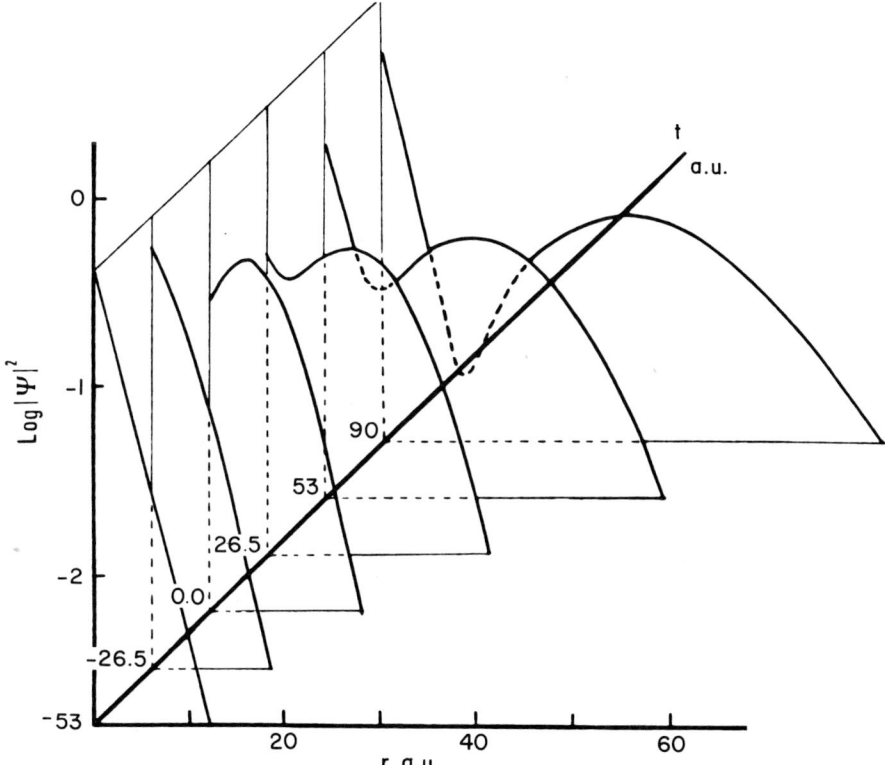

Fig. 4. Logarithm of the probability of finding the "extra" electron at a distance r plotted as a function of time. Negative (positive) times are pre-(post-)collision. The system is $H^- + He$ and the relative collision energy is 200 eV. The impact parameter is $1a_0$. [From Gauyacq (1980a). Reproduced with permission from the *Journal of Physics*.]

Taylor and Delos assume that the states ϕ_0, ϕ_E can be constructed in such a way that nonadiabatic couplings among them are negligible, and detachment occurs because of electrostatic couplings (i.e., matrix elements V_{0E} of the electronic Hamiltonian). Furthermore, they assume that transitions only occur between the bound state and the continuum (and vice versa), and that direct continuum–continuum transitions are insignificant.

From these assumptions, they derive coupled equations that must be satisfied by the coefficients $C_0(t), C_E(t)$:

$$i\hbar \frac{d}{dt} C_0(t) = V_{\text{ion}}(t) C_0(t) + \int V_{0E}(t) C_E(t) \rho(E) \, dE, \qquad (14)$$

$$i\hbar \frac{d}{dt} C_E(t) = [V_{\text{neutral}}(t) + E] C_E(t) + V_{E0} C_0(t).$$

The problem then reduces to solving this non-denumerably-infinite set of coupled equations.

Neglecting the time dependence of $V_{E0}(t)$, and approximating $V_{ion}(t) - V_{neutral}(t)$ by a quadratic function of time [which for the $H^- - $He system is fitted to the calculation of Olson and Liu (1980)], Taylor and Delos show that these equations can be solved, and proceed to derive a rather complicated formula for the survival probability. Their results can be interpreted by using Fig. 5. Transitions from the bound state to the free state with electron energy E occur primarily at the crossing between these states, and can occur either on the incoming or the outgoing part of the trajectory. These two contributions add coherently, leading to a possible interference pattern in the kinetic-energy spectrum of the detached electrons. In addition, transitions into the final state (of energy E) can also occur through detachment to an intermediate state (of energy $\hat{E}$ in Fig. 5) on the incoming part of the trajectory, then reattachment into the bound state followed by detachment into the final state on the outgoing part of the trajectory.

Results of their calculation of the total detachment cross section for $H^-(D^-)$ on He are also shown in Fig. 2. Good agreement with experimental

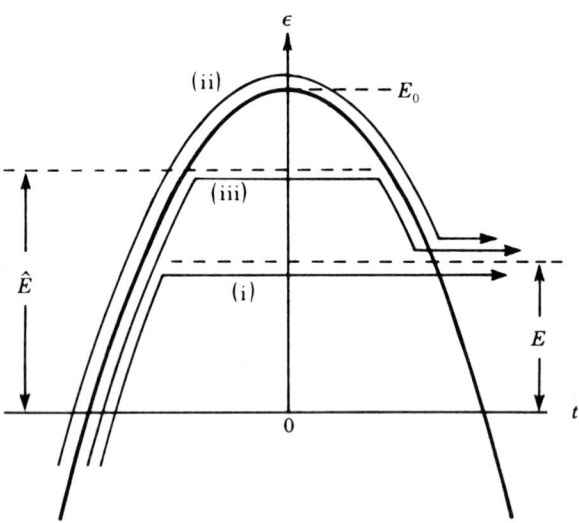

Fig. 5. Schematic illustration of various paths which may lead to detachment. The heavy parabola represents the energy difference between ionic and neutral states as a function of time, i.e., $E(t)$. For $E(t) > 0$, the discrete state is moving through the continuum. The classical turning point is represented by $t = 0$. The energy of the detached electron is E. [From Taylor and Delos (1982). Reproduced with permission from the *Proceedings of the Royal Society*.]

7 Electron Detachment Processes

results is obtained. It is also interesting that the calculations of Gauyacq and of Taylor and Delos agree quite well with each other, even though they are based on what appear to be very different assumptions about the dynamics of these processes.

Herzenberg and Ojha (1979) have investigated the $H^- + He$ problem at slightly higher energies using an approach closely related to the zero-range approximation already discussed. They argue that detachment occurs when, as the particles approach, the binding energy of the extra electron becomes sufficiently small that the Born–Oppenheimer approximation is no longer valid and the electron "forgets" that it started off in a bound state of H^-. This "forgetting-point" R_f is defined as that value of R for which

$$\sum_{i \neq 0} |a_i[R_f(t)]|^2 \simeq 0.5 \tag{15}$$

where the a_i are the expansion coefficients for the electron wave function

$$\psi(r, t) = \sum_i a_i[R(t)]\psi_i[r, R(t)] \exp \frac{-i}{\hbar} \int_{-\infty}^{t} E_i[R(\tau)] \, d\tau \tag{16}$$

that satisfy the boundary conditions

$$a_0(-\infty) = 1, \quad a_i(-\infty) = 0, \quad i \neq 0.$$

In Eq. (16), the ψ_i and E_i are eigenfunctions and eigenvalues of the electronic Hamiltonian at fixed R. As in the results of Gauyacq (1980a) discussed earlier, there is substantial detachment due to "undercrossing transitions" since R_f is found to be considerably larger than the crossing radius. For a given impact parameter, R_f increases with increasing energy because the B–O breakdown occurs earlier at higher energies. Thus, this model predicts that $\sigma_{-10}(E)$ will rise for $E \gtrsim 200$ eV, as is found to be the case by Risley and Geballe (1974). A local complex potential model is not compatible with such a rise.

For collision energies which exceed several kilo-electron-volts, the projectile (H^-) velocity is comparable to that of the electron and the adiabatic picture of the collision is not appropriate. For $E \gtrsim 5$ keV, $\sigma_{-10}(E)$ has been observed to decrease with increasing energy [see data compilation in Risley (1980), for example] and this energy range has been treated theoretically in an impulse approximation in which the target scatters the essentially free electron associated with the H^- projectile (Lopantseva and Firsov, 1966; Dewangan and Walters, 1978). This high-energy theory also predicts a decreasing detachment cross section and is in reasonable agreement with the experimental observations.

Now let us turn to the problem of collisional detachment of $H^-(D^-)$ by Ne. The experimental results for $\sigma_{-10}(E)$ are seen in Fig. 6; it is clear that these results are in contradistinction to those for $H^-(D^-) + He$. First, the isotope

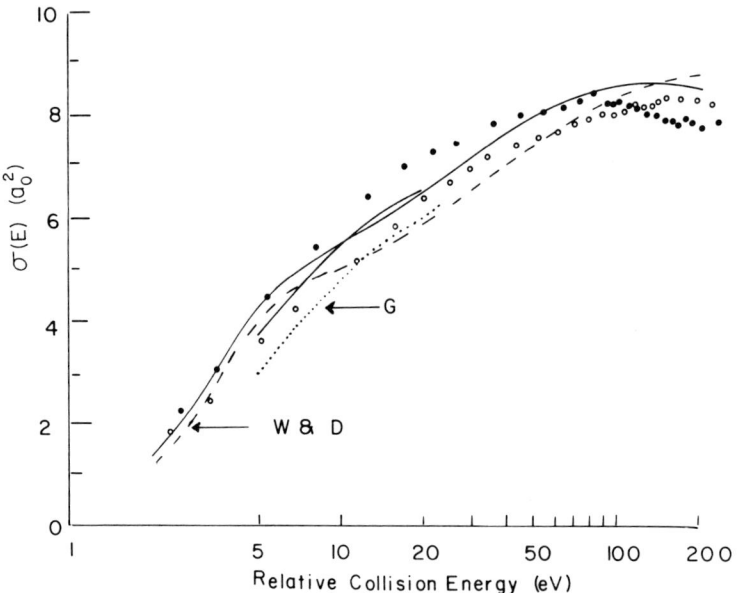

Fig. 6. Total detachment cross sections for $H^-(D^-)$ + Ne. The solid (open) circles are the experimental points for $H^-(D^-)$. The theoretical curves from Gauyacq (1980b) are labeled G and those of Wang and Delos (1983) are labeled WD. Full curves refer to H^- and broken curves refer to D^-.

effect for the Ne target is the reverse of that observed for He and, second, the magnitude of $\sigma_{-10}(E)$ for Ne is considerably smaller than that for the He target. Internuclear potentials for NeH and NeH$^-$ (Olson and Liu, 1980) are presented in Fig. 7. In contrast to HeH and HeH$^-$, the ionic and neutral curves do not clearly cross, but rather merge in the vicinity of $R = 2.2a_0$. Olson and Liu argue that this basic difference in the internuclear potentials accounts for the different isotope effects observed for the He and Ne targets. More specifically, a complex potential approach for $H^-(D^-)$ + Ne is inappropriate because no crossing into the continuum occurs and it should be replaced with a collision dynamics more descriptive of "charge transfer to a continuum." For cases similar to that for NeH$^-$ where the potential curves for charge transfer states may be close together [see, e.g., Melius and Goddard (1974) or Wijnaendts van Resandt et al. (1978) for LiNa$^+$], the probability for charge transfer increases with an increase in collision velocity. Thus, any model based upon such an assumption will result in the isotope effects for $\sigma_{-10}(E)$ observed in Fig. 6.

Calculations of $\sigma_{-10}(E)$ for $H^-(D^-)$ + Ne, which are based upon the zero-range potential approximation discussed previously, have been reported by

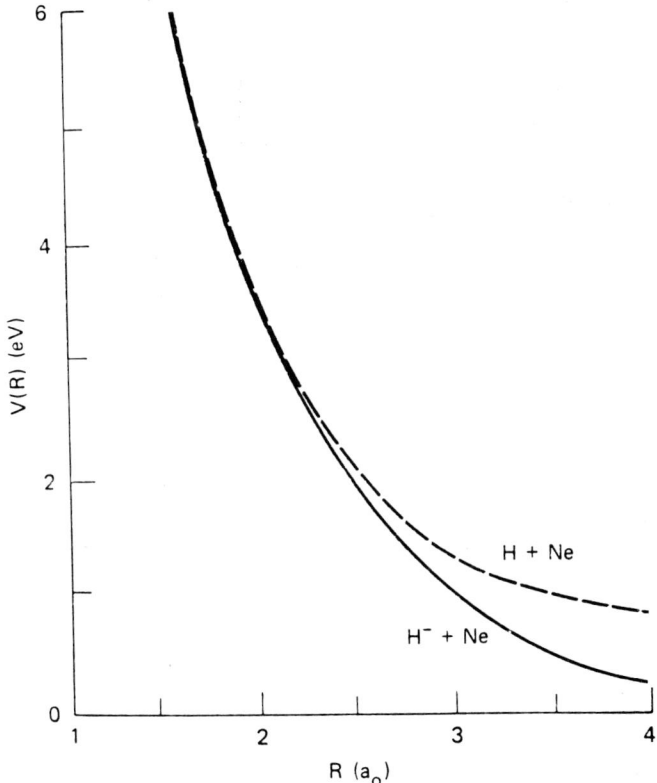

Fig. 7. Internuclear potentials for the ground states of NeH and NeH$^-$. [From a calculation by Olson and Liu (1980).]

Gauyacq (1980b). The magnitude of the cross sections, as well as the isotope effect observed for these reactants, are very nicely reproduced by the calculation (see Fig. 6), thus illustrating the importance of dynamical effects in the detachment process.

Wang and Delos (1983) have also calculated $\sigma_{-10}(E)$ for H$^-$(D$^-$) + Ne for $E \lesssim 200$ eV. The starting point of their calculation was Eq. (14), utilizing potential curves V_{ion} and $V_{neutral}$ close to those calculated by Gauyacq, and an assumed form for $V_{E0}(R)$:

$$V_{E0}(R) \simeq A \exp(-0.66R). \tag{17}$$

Equations (14) were then solved by a first-order approximation: with

$$C_0(t) \simeq \exp\left(-\frac{i}{\hbar}\int_0^t V_{ion}(t')\,dt'\right), \tag{18}$$

the equations for $C_E(t)$ are easily integrated, and the total probability of detachment is

$$P_d = \int_0^\infty |C_E(\infty)|^2 \rho(E)\,dE. \tag{19}$$

Cross sections were calculated from these formulas, and the results are shown in Fig. 6. There is very good agreement between theory and experiment.

The observations and calculations that have been presented for low-energy electron detachment for the relatively simple systems of $H^-(D^-) + He$, Ne suggest that there are two somewhat distinct mechanisms for direct detachment: (i) curve crossing of a discrete (incoming) state into a continuum (representative of the outgoing state), and (ii) a dynamical transition in which the motion of the nuclei is used to promote the "extra" electron into the continuum. The former mechanism is characterized by sharp, well-defined thresholds, an isotope effect in which the slower isotope (for a given relative energy) has a larger detachment cross section, and a detachment cross section which is fairly large. The latter mechanism is characterized by small cross sections, fuzzy thresholds, and an isotope effect which is the reverse of that stated above. It is obvious that dynamical coupling, (ii), may always accompany detachment due to curve crossing. These same ideas will be carried over to the discussion of electron detachment by molecular targets where one must substitute the idea of potential surfaces for internuclear potential curves.

Another class of reactants in which the near-threshold behavior of $\sigma_{-10}(E)$ has been studied in great detail includes halogen anions and rare-gas targets (Champion and Doverspike, 1976a; Fayeton et al., 1978; Smith et al., 1978; Haywood et al., 1981a; Huq et al., 1982a). These systems are exemplified by what appears to be detachment by curve crossing with sharp thresholds (at about twice the electron affinity of the halogen atom) and fairly large detachment cross sections. The experimental results for $\sigma_{-10}(E)$ for several halogen anion–rare gas reactants can be seen in Fig. 8.

The internuclear potentials for $ArCl^-$ and $HeBr^-$ have been calculated by Olson and Liu (1978, 1979) and the results for $ArCl^-$ (and ArCl) are shown in Fig. 9. The calculated potential parameters R_x (the value of R for which the neutral and ionic curves cross) and $V(R_x)$ are in very good agreement with those values deduced from fitting a complex potential to low-energy experimental observations of $\sigma_{-10}(E)$ and the elastic differential scattering cross section $\sigma_e(\theta)$. Results for $\sigma_e(\theta)$ for $Cl^- + Ar$ are presented in Fig. 10 for several collision energies. The onset of detachment is clearly evident as the sudden decrease in the reduced differential cross section $\rho(\theta)$ for reduced scattering angles $\tau \gtrsim 1.2$ keV deg. The solid lines in the figure are calculations of elastic differential cross sections based upon a complex potential model.

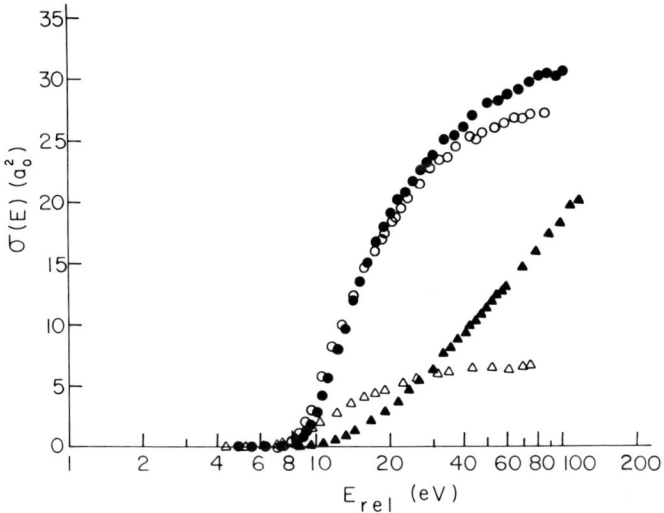

Fig. 8. Total detachment cross sections for various halogen anion rare-gas systems: (●) $Br^- + Ar$, (○) $Cl^- + Ar$, (▲) $I^- + Ar$, (△) $F^- + He$.

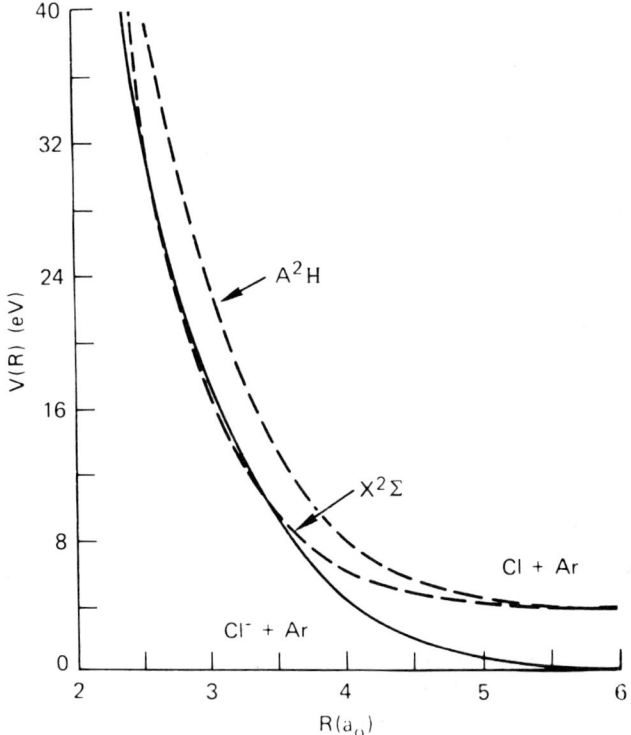

Fig. 9. Internuclear potentials for the ground electronic states of $ArCl^-$ and $ArCl$. [From Olson and Liu (1978). Reproduced with permission from the *Physical Review*.]

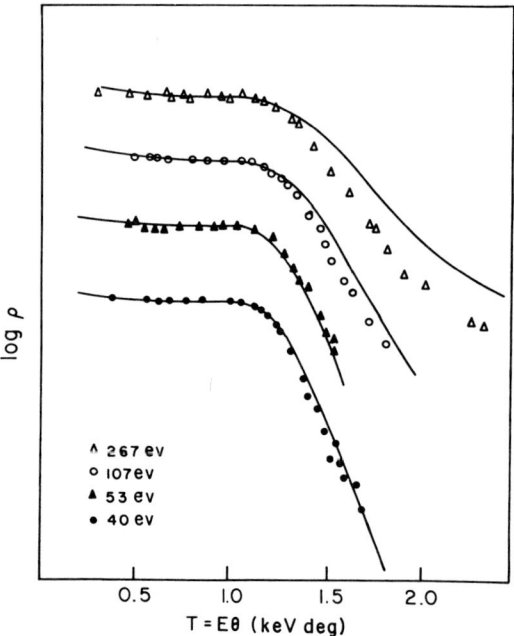

Fig. 10. Differential elastic scattering cross section for $Cl^- + Ar$ at various relative collision energies. The lowest-energy result is from Champion and Doverspike (1976a) and the three highest-energy results are from Fayeton et al. (1978). The solid line is a calculation based upon a local complex potential model. It is clear that the calculation fails to reproduce the experimental observations for $E \gtrsim 100$ eV.

Fayeton et al. (1978) demonstrated that a single complex potential could not describe the experimental results over a large energy range; this is clearly evident in Fig. 10, where it can be seen that a potential which reproduces the experimental results for $E \lesssim 100$ eV fails to do so for $E \gtrsim 100$ eV.

DeVreugd et al. (1982) have further illustrated this point in studies at higher collision energies (500–3000 eV) in which both $\sigma_n(\theta)$ and the energy spectra of the detached electrons have been measured. These experiments illustrated that the $P_d(b, E)$ increased with increasing velocity and that detachment was localized in a certain $\tau(b)$ interval. These results at high collision energies were explained by assuming that rotational coupling was the principal mechanism which determined the discrete–continuum interaction.

Collision mechanisms other than direct detachment begin to become important for energies larger than several hundred electron volts. These mechanisms have been studied in some detail by Esaulov et al. (1978) over the energy range 80–2000 eV and by Risley et al. (1978) for the range 1–6 keV. The former experiments measured the doubly differential cross sections for the neutral hydrogen atoms which resulted from detachment. These measurements were able to separate direct detachment (1) from the detachment/excitation channels (2) through (4). An example of an energy loss spectrum for the H coming from collisions of H$^-$ with He is given in Fig. 11. It is clear that there is considerable production of excited-state hydrogen at this particular value of the scattering angle, as indicated by the relative size of the peak labeled "B" in the figure. The peak labeled "A" represents direct detachment where no excited states of the reactants are involved. In principle, the shape of spectrum A is a reflection of the energy spectrum of the detached electrons. There are, however, broadening effects since the ejected electron carries momentum and can be ejected in any direction (e.g., nearly isotropic for s-waves), thus giving a range of H-atom velocities for a given electron energy.

Risley *et al.* (1978) have measured the absolute cross sections for the production of excited H atoms in the detachment process by observing the intensities of the various Lyman lines (forbidden transitions are induced by external fields) from the detachment products. They found that the H(np) production accounts for about 11% of the total detachment cross section at 5 keV. The probabilities of H(2s) production are considerably smaller than those for H(2p) formation.

The energy spectra of the detached electrons resulting from detachment in H$^-$ + rare gas collisions have been measured by Risley (1973) and such spectra have been determined indirectly from the H-velocity spectra in time-of-flight experiments by Esaulov *et al.* (1978). In both cases, there are broadening effects which cause some difficulties in the determination of the

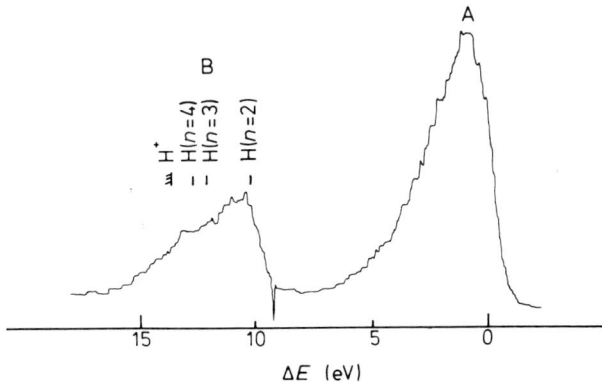

Fig. 11. Energy loss spectrum of neutral hydrogen atoms scattered in H$^-$-He collisions at $E_{lab} = 730$ eV, $\theta_{lab} = 0.6°$. The zero of the energy loss scale refers to a hydrogen atom having lost 0.75 eV, its electron affinity. [From Esaulov *et al.* (1978). Reproduced with permission from the *Journal of Physics.*]

energy spectrum of the detached electrons with respect to the motion of the parent negative ion. Nevertheless, there are several salient features which have been deduced from the experiments: (i) the maximum in the detachment spectrum occurs for electron energies less than 1 eV and the distribution function is asymmetric, possessing a high-energy tail; (ii) the maximum and width of the distribution increase as the collision energy is increased; and (iii) the spectra depend only slightly upon scattering angle. These features are all essentially reproduced by the previously discussed calculations of Herzenberg and Ojha (1979) and Gauyacq (1980a).

If one looks beyond the low-energy maximum in the spectra of detached electrons (i.e., at electrons with higher kinetic energies), discrete lines are seen to be superimposed on the continuum background of electrons from direct detachment. The electrons that comprise the rather sharp lines in the spectra come from autodetaching levels of the hydrogen negative ion and have been the subject of a detailed study by Risley *et al.* (1974). In addition to exciting the projectile to its autodetaching levels, the electron energy spectra indicate that autodetachment from the negative-ion resonances of the target—which are propulated by a charge transfer (4)—may also be important in the autodetachment mechanism. This latter process will be seen to be very important for molecular targets. An energy spectrum of the higher-energy detached electrons for the case of 3-keV H^- incident upon Ar is shown in Fig. 12.

The total cross section for H^- + He collisions for excitation of H^- to the $1s^2$ autodetaching state is found to be rather flat for $0.2 \lesssim E \lesssim 10$ keV at about 0.03 Å^2, while excitation to the $(2p^2) + (2s2p)$ levels decreases from 0.04

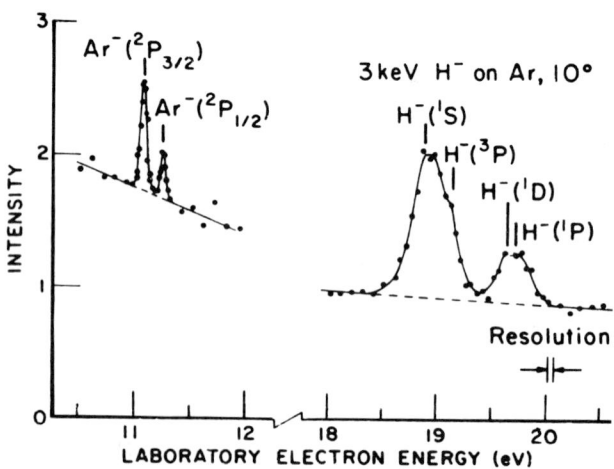

Fig. 12. Electron energy spectrum from 3-keV H^- on Ar observed at 10°. [From Risley *et al.* (1974). Reproduced with permission from the *Physical Review.*]

to about 0.01 Å² over the same range (Risley, 1973). Thus, for a collision energy of 5 keV, detachment via excitation to autodetaching states of H^- constitutes about 1% of the total detachment cross section.

IV. Molecular Targets

Studies of collisions between atomic negative ions and atomic and molecular targets have demonstrated that electron detachment is usually the dominant inelastic product channel. The collisional dynamics is much more complicated when molecular targets are involved, however, because one has to deal with a number of molecular and negative-molecular-ion potential surfaces and their interactions. With the possible exception of H_3^- (Michels and Paulson 1981), there is currently very little detailed information about surfaces, in particular the negative-molecular-ion surfaces.

The experimental studies to date indicate that for molecular targets, electron detachment may proceed by any of several distinct mechanisms. The most common detachment mechanism is perhaps direct detachment, which was introduced in the discussion for atomic targets. Detachment occurs in this case after the loosely bound electron of the negative ion is promoted into the continuum by either a curve (surface) crossing mechanism or by a coupling of the nuclear motion of the collision partners to the electronic motion. Typical examples of this process have already been discussed in detail for the atomic systems $H^- + (He, Ne)$.

A variety of experiments designed to probe various aspects of electron detachment in negative ion–molecule collisions have been completed recently, and these studies have confirmed that several distinct scattering channels are operative in the detachment process. In addition to direct detachment, another mechanism involves an initial charge transfer to a shape resonance of the molecular target followed by a rapid decay of the negative molecular ion [Eq. (4)], which likely leaves the target molecule vibrationally excited. Evidence for this type of process has been found in the kinetic energy spectra of the detached electrons (Risley, 1977) as well as in the time-of-flight spectra of the fast-neutral products of collisional detachment (Annis *et al.*, 1980; Tuan and Esaulov, 1982a).

Another mechanism of electron detachment by molecular targets is one in which reactive scattering (i.e., molecular rearrangement) accompanies detachment, as given in Eq. (5). In those cases where the product molecule can form a stable negative ion, direct competition between the ion–molecule reaction channel $A^- + BC \rightarrow AB^- + C$ and detachment is possible.

Finally, at the lowest collision energies, i.e., from thermal energies to several electron volts, there is always the possibility of associative detachment for selected reactants as in Eq. (6).

In general the relative importance of processes (4)–(6) increases as the collision energy is lowered. Associative detachment is only important at low collision energies, $E \lesssim 1$ eV, and this particular process will not be reviewed in this report. Reactive scattering accompanied by detachment [i.e., (5)] has been observed for collision energies in the electron volt range. For higher collision energies, i.e., $100 < E \lesssim 1000$ eV, both direct detachment and detachment via charge transfer have been observed to be of similar magnitude for some systems studied thus far. For example, the time-of-flight studies of the collisional detachment of H^- by N_2 by Tuan and Esaulov (1982a) indicate that the total cross sections for direct detachment and detachment via charge transfer are approximately the same at $E \simeq 500$ eV.

For purposes of clarification, the ensuing discussion will break up the negative ion–molecule systems that have been the subject of recent experimental investigations into groups for which electron detachment appears to be dominated by direct detachment and one or more of the reaction channels (4)–(6) discussed above. A unique characterization, however, is generally not possible since in most cases several channels may be operative simultaneously with branching ratios which depend upon the collision energy. A few specific systems will be discussed at length in order to illustrate the rich variety of physics that can be involved in the detachment of electrons from negative ions in collisions with molecules.

A. Halogen Ions + N_2, CO, CO_2, O_2

Fairly extensive studies of electron detachment in which rearrangement processes appear to play only a minor role have been reported. For collision energies near the threshold for detachment, the negative ion–molecule systems which have received considerable attention are halogen negative ions in collisions with various molecular targets (Doverspike *et al.*, 1980; Huq *et al.*, 1982a). Some measurements of $\sigma_{-10}(E)$ for halogen ions and several molecular targets are shown in Figs. 13–15.

All these cross sections display the same general dependence on the relative collision energy. With the exception of O_2, all rise rather rapidly from thresholds which are considerably greater than the electron affinities of the incident negative ions and continue to increase more slowly at higher energies. The cross sections vary smoothly with energy, are very similar to those observed for rare-gas targets, and offer no evidence that a detachment mechanism other than direct detachment is involved in the collisional dynamics.

However, to see that this is not the case, let us examine the system $Cl^- + N_2$ in more detail. For these reactants measurements of $\sigma_n(\theta)$ (the neutral differential cross section integrated over all product velocities) have been made

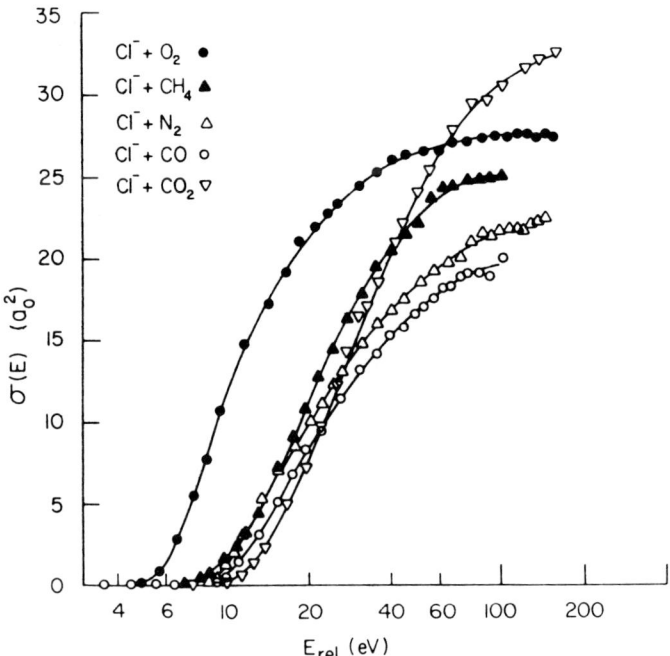

Fig. 13. Electron detachment cross sections for Cl^- molecule systems. [From Doverspike *et al.* (1980). Reproduced with permission from the *Physical Review*.]

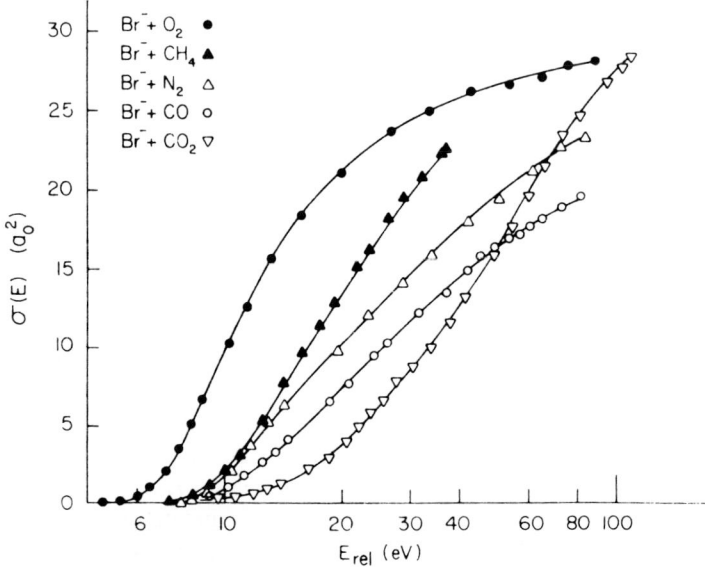

Fig. 14. Electron detachment cross sections for Br^-—molecule systems. [From Doverspike *et al.* (1980). Reproduced with permission from the *Physical Review*.]

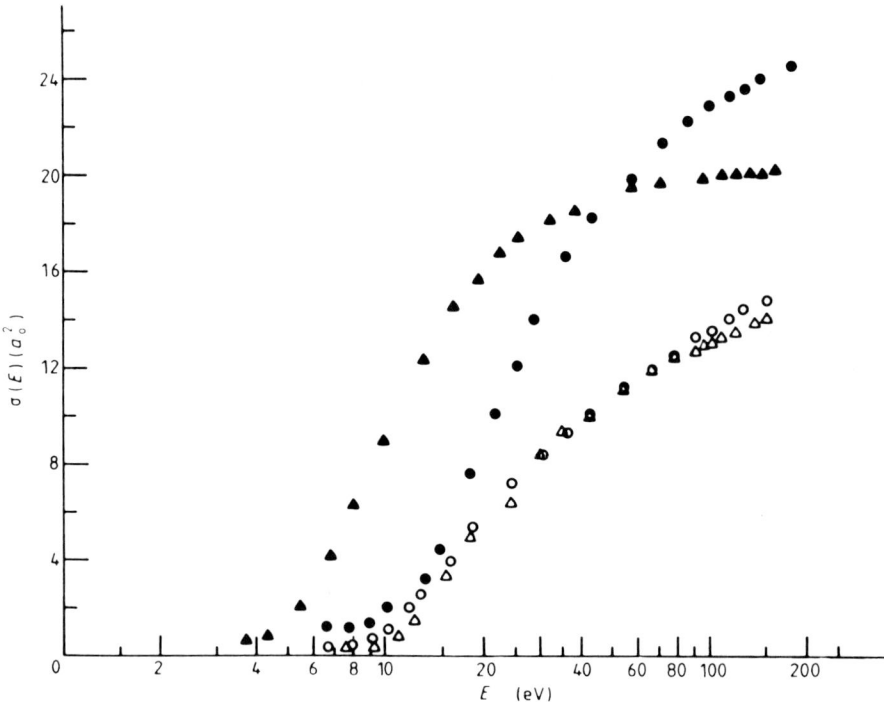

Fig. 15. Electron detachment cross sections for F^-—molecule systems: ($\triangle$) CO, ($\bigcirc$) N_2, ($\bullet$) CO_2, ($\blacktriangle$) O_2 [From Huq et al. (1982a).]

down to relatively low energies (Champion, 1980). As may be seen in Fig. 16, for $E = 44$ eV, the angular distribution for the neutral chlorine atoms exhibits two distinct maxima, one at a τ value ($\tau = E\theta$) of $\tau_1 \simeq 700$ eV deg and the other at $\tau_2 \simeq 900$ eV deg. As E is increased to about 80 eV, the τ_1 process becomes dominant, while the τ_2 process appears to dominate the scattering at lower energies. This behavior suggests that there are two distinct mechanisms involved in the neutralization of Cl^-.

In a further study of this system, Annis et al. (1980) have measured the time-of-flight spectra for product chlorine atoms, and those experiments give strong evidence that the resonances of N_2^- [the $^2\Pi_g$ and a'; see Schultz (1973) for a discussion of molecular-negative-ion resonances] are very important in the detachment process. Their results for a Cl^- energy of 200 eV are exhibited in Fig. 17 for several scattering angles. The two groups of chlorine atoms observed in the time-of-flight spectra have (most probable) energy losses Q

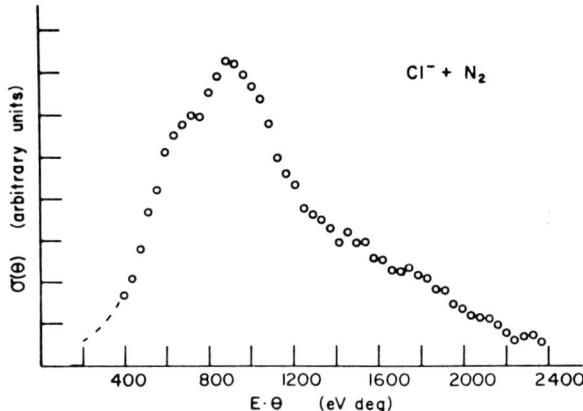

Fig. 16. Differential cross section for neutral chlorine atoms which are products of collisional detachment from N_2 at $E_{rel} = 44$ eV. [From Champion (1980).]

which are compatible with

$$Cl^- + N_2 \longrightarrow Cl + N_2^-(^2\Pi_g)$$
$$\longrightarrow Cl + N_2^-(a')$$

The bottom panel of Fig. 17 also contains a spectrum for the argon target for which there is only direct detachment; i.e., no appreciable target or projectile excitation is observed. For the negative ion–molecule experiments, one cannot ascertain whether (a) the detached electron carries away the excess energy or (b) the target, N_2, is left rather highly excited as a result of the detachment. The structure of the spectra and the angular dependence of the amplitudes of the two peaks as seen in Fig. 17 strongly suggest that direct detachment (at least as understood in ion–atom collisions) is perhaps a minor contributor to detachment for these reactants.

The $\sigma_{-10}(E)$ for $Cl^- + N_2$ shows a threshold for detachment at $E \simeq 7.6$ eV, which does not rule out the possibility that the $^2\Pi_g$ state of N_2^- is involved in detachment at collision energies even down to threshold. Recent time-of-flight studies of Tuan and Esaulov (1982a) suggest that the role of the resonance states of N_2 in the detachment of Cl^- is insignificant for collision energies near 1 keV. Clearly, more experiments, particularly at low collision energies, are needed to clarify the role of negative-ion resonance states in such collisional detachment.

In a recent study (Vedder et al., 1981), it was demonstrated that there are considerable differences between the inelastic (but nondetaching) scattering of

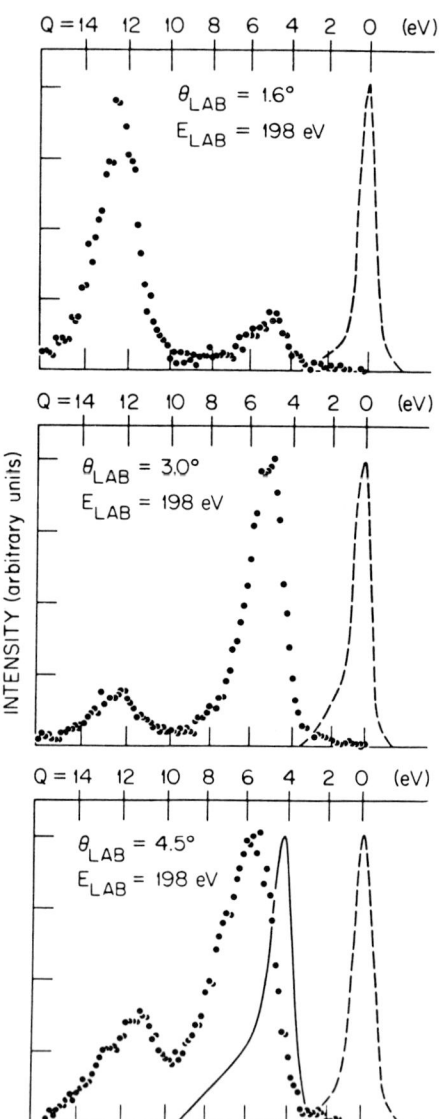

Fig. 17. Time-of-flight energy loss spectra of chlorine formed by detachment in Cl$^-$ (E_{lab} = 198 eV) + N$_2$ collisions. Also shown are spectra for elastic scattering from Ar (dashed lines) and collisional detachment from Ar (solid lines). [From Annis *et al.* (1980). Reproduced with permission of *Physical Review Letters.*]

Cl^- by N_2 and CO_2 and that of the isoelectronic systems $K^+ + N_2$, CO_2. These results lend strong support to the hypothesis that the known negative-ion resonance states of CO_2 are responsible for significant target excitation in nondetaching collisions with Cl^-, but their role is not as evident in the case of N_2, where the situation is clouded by the competing direct excitation process. Thus it appears that the resonance states of the target molecule may be involved in most aspects of the scattering of negative ions by molecules. Although $\sigma_{-10}(E)$ for $Cl^- + CO$ shows no obvious structure, other experiments indicate that at least two distinct channels are responsible for detachment. Angular distributions of Cl atoms produced in collisions of Cl^- with CO show several prominent maxima which are apparent in the spectra shown in Fig. 18. Measurements at energies above and below those shown in Fig. 18 indicate that the process at $\tau \simeq 225$ eV deg becomes dominant at higher energies, while the τ process at 900 eV deg is favored at the lower energies.

Time-of-flight measurements of Cl for $Cl^- + CO$ have illustrated (Annis and Datz, 1982) that detachment is associated with two different energy loss mechanisms. The low-loss channel is commensurate with charge transfer to the $^2\Pi$ resonance of CO^-, while the high-loss channel with $Q \simeq -11.5$ eV is

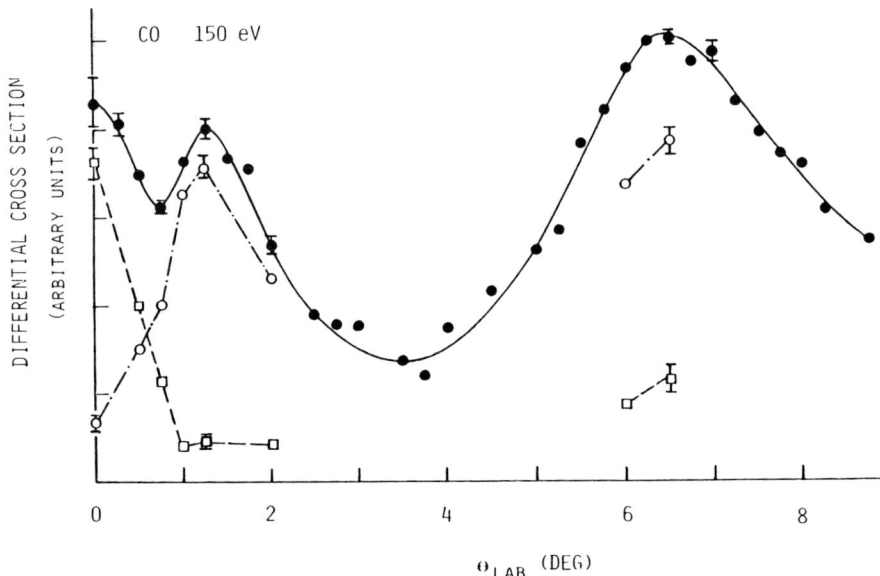

Fig. 18. Differential cross section for Cl from $Cl^- + CO$ at 150 eV (lab). The open circles and squares indicate the relative contributions of the low-loss and high-loss channels (with endothermicities of about 5 and 12 eV, respectively). [From Annis (1982, private communication).]

in an energy range where no resonances exist for CO. The high-loss channel appears to involve the excitation of the A $^1\Pi$ state of CO accompanied by detachment and is dominant at small scattering angles.

Of all the systems studied for the Cl$^-$ projectile, the $\sigma_{-10}(E)$ exhibiting the lowest threshold is for the O_2 target. Neutral time-of-flight studies (Annis and Datz, 1982) for Cl$^-$ + O_2 indicate that detachment occurs solely by a direct process; there are no indications of other channels contributing to the process. Similar measurements by Tuan and Esaulov (1982a) on F$^-$ + N_2 show that detachment in this system occurs only via direct detachment. To date, time-of-flight studies have not been reported for the other systems shown in Figs. 13–15.

Although numerous reaction channels are energetically possible for some of these systems, the experimental studies indicate that electron detachment is by far the most dominant inelastic process in the energy range covered by these studies. Vogt and Opiela (1975) have measured the total charge transfer cross section for the process Cl$^-$ + $O_2 \rightarrow O_2^-$ + Cl for laboratory energies from 20 eV to 5 keV and report that the cross section remains quite small over the entire range, rising slowly from 0.2 to $0.7a_0^2$. Tiernan et al. (1971) and Vogt et al. (1976) have measured the energy dependence of the charge transfer cross section for Cl$^-$ + O_2 in the threshold region, but report only relative cross section values. However, both indicate that the cross section shows a threshold of about 3.2 eV and reaches its maximum at roughly 5 eV. It may be that the observed low thresholds for detachment in the systems (Cl$^-$, F$^-$ + O_2) are due to the fact that the corresponding potential surfaces are attractive. This would give a low-energy threshold for charge transfer possibly leading to autodetachment if v > 3, as in

$$F^- + O_2 \longrightarrow O_2^-(v) + F \longrightarrow O_2 + F + e$$

as well as a low threshold for direct electron detachment.

Finally, Dimov and Roslyakov (1972) have measured electron loss cross sections between 0.3 and 3 keV for the systems (Cl$^-$, Br$^-$) + (O_2, CO_2). The energy for the measurements presented in Figs. 13 and 14 just overlaps their results, and in the region of overlap, they are 45% lower in all cases. The reason for this disagreement is not clear.

We will return later to examine the results for the collisional detachment of halogen anions by other molecular targets (viz., H_2 and D_2) where reactive scattering accompanies detachment.

B. H$^-$(D$^-$) + N_2, CO, CO_2, O_2

We turn now to the negative ion that perhaps has been studied most extensively, namely H$^-$. Several measurements of $\sigma_{-10}(E)$ for molecular targets have been reported and those results published prior to 1974 have been reviewed in some detail by Risley and Geballe (1974) and Risley (1974). Many recent experiments which have focused upon the collisional detchment of H^- have also examined the effects of isotopic substitution (by substituting D^- for H^-) upon the various cross sections.

The near-threshold results (Huq et al., 1983b) for $\sigma_{-10}(E)$ for $H^-(D^-) + N_2$ are shown in Fig. 19. An interesting feature of these cross sections is the appearance of a "dual" isotope effect when the total detachment cross section is exhibited as a function of E.

At the higher relative collision energies, the reactants with the higher relative collision velocity exhibit the larger detachment cross section, whereas the trend is just the opposite at low relative collision energies. It is likely that this difference is due to the fact that there are different mechanisms which dominate the detachment in the high- and low-energy regions. The magnitude of the low-energy isotope effect varies from about 5 to 10% over the energy range 10–40 eV and is consistent with a description of electron detachment that involves the crossing, or merging, of the discrete reactant state (which represents the interaction potential of the negative ion with N_2) with the continuum of product states (representing $H + N_2$ along with a free electron

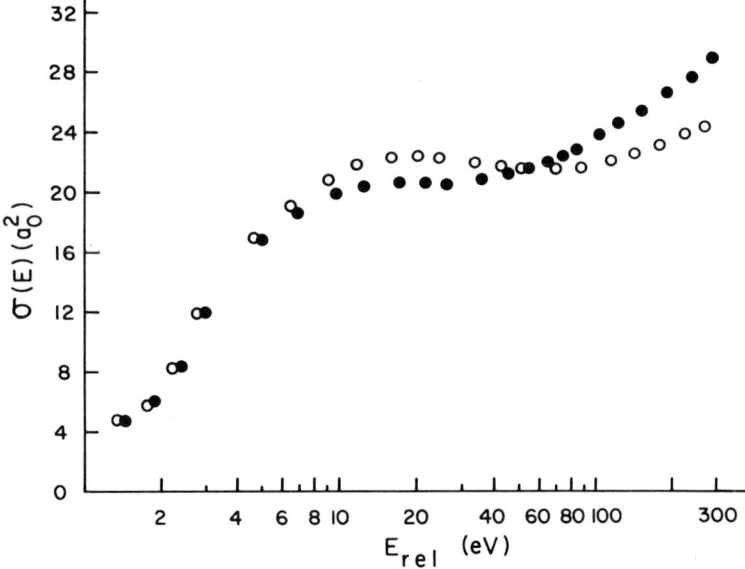

Fig. 19. Total electron detachment cross sections for H^- and D^- on N_2 as a function of relative collision energy. The solid circles are for H^- and the open circles are for D^-. [From Huq et al. (1983b). Reproduced with permission from the *Physical Review*.]

of arbitrary energy). According to this description, for a given E, both D^- and H^- will follow the same trajectories but with different velocities (the isotopic masses being different) and hence the time spent by D^- in the continuum is larger than that of H^-, resulting in a larger detachment cross section for D^-.

At higher relative collision energies (i.e., $E > 50$ eV), the isotope effect reverses its character, where the faster reactants give the larger detachment cross section. In addition, it is found that the detachment cross sections in this energy region (i.e., $E > 50$ eV) increase with relative collision energy and, more importantly, scale with relative collision velocity: for the same relative collision velocity, the cross sections for the isotopic doublet are approximately the same. This can be seen very clearly in Fig. 20, where the electron detachment cross sections for both isotopes are plotted as a function of relative collision velocity. For higher collision energies, this velocity scaling for $H^-(D^-) + N_2$ has been investigated by Risley (1974) and has been found to be valid for energies up to at least 10 keV. It is thus clear that there are two distinct regions of energy: at low collision energy, detachment cross sections for both isotopes are found to be the same at identical collision energies,

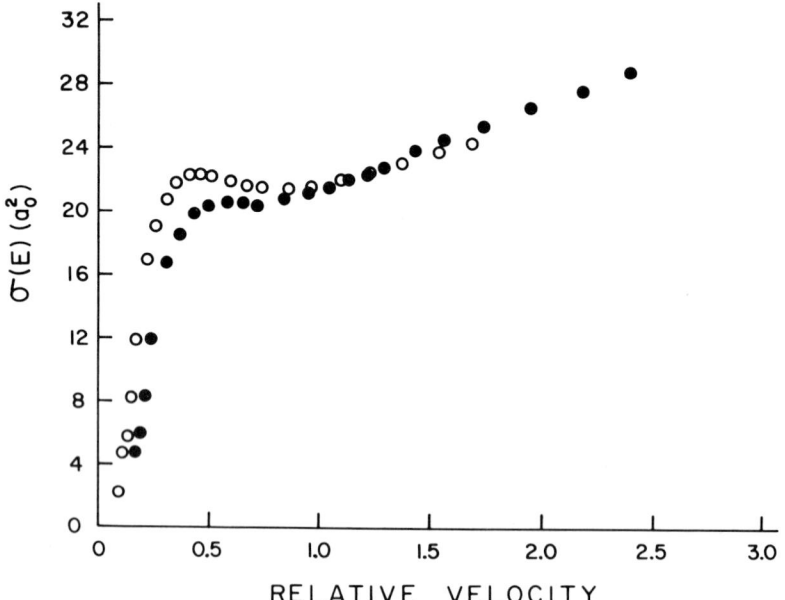

Fig. 20. Total electron detachment cross sections for H^- and D^- on N_2 given as a function of relative collision velocity, which is expressed in units of 10^7 cm/s. The full circles are for H^- and the open circles are for D^-. [From Huq et al. (1983b). Reproduced with permission from the Physical Review.]

whereas at high collision energy, detachment cross sections are found to scale with relative collision velocity. As will be discussed later, this is essentially true for all the $H^-(D^-)$–molecule systems studied to date.

The kinetic-energy spectra for electrons arising from the collisional detachment of H^- by N_2 (Risley, 1977) revealed that the $^2\Pi_g$ shape resonance of N_2^- was involved in the detachment mechanism, at least for $E \gtrsim 1$ keV. The evidence for this may be clearly seen in Fig. 21, where kinetic-energy distributions of the detached electrons show regular oscillations associated with $N_2^-[^2\Pi_g(v')] \rightarrow N_2[^1\Sigma_g(v)]$ transitions. The oscillatory structure appears to be superimposed upon a rather broad distribution typical of direct

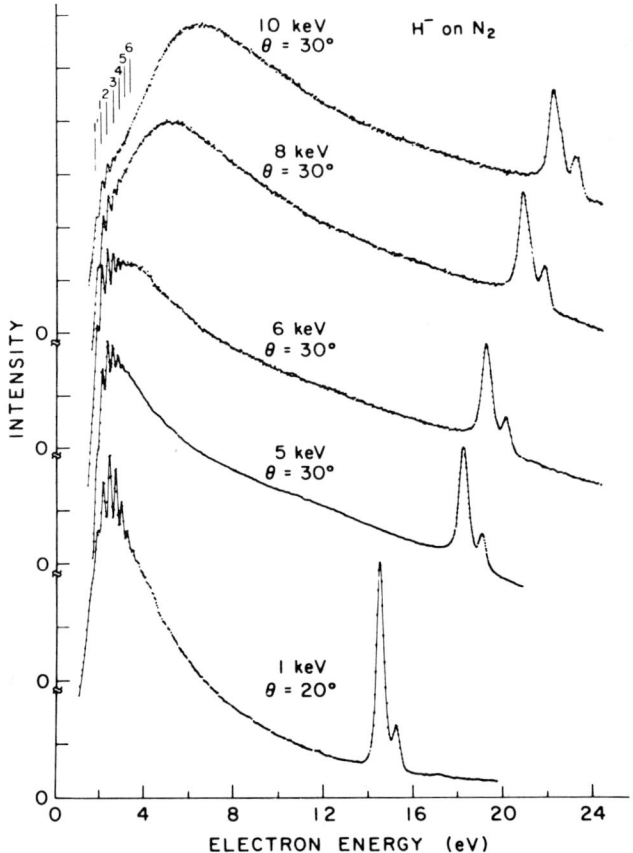

Fig. 21. Energy distribution of secondary electrons produced in collisions of H^- with N_2. H^- lines are shifted in the laboratory reference frame because of reaction kinematics. [From Risley (1977). Reproduced with permission from the *Physical Review*.]

detachment by atomic targets, suggesting that detachment via charge transfer may be minor compared to direct detachment. It is, however, difficult to delineate the two in the spectrum because of the inherent resolution of the experiment.

Tuan and Esaulov (1982a) have further investigated the role of detachment via charge exchange to a shape resonance for $H^- + N_2$ (for $200 \lesssim E \lesssim 1000$ eV) by examining the energy loss spectra for neutral hydrogen atoms which are products of detachment.

An example of an energy loss spectra for $E \sim 1$ keV is seen in Fig. 22. The three peaks in the spectra are attributed to direct detachment, detachment via charge transfer, and detachment with simultaneous electronic excitation of N_2. By integrating over all product scattering angles, it is estimated that roughly 40% of the total electron detachment cross section is due to detachment via the $^2\Pi_g$ shape resonance of N_2^-. These data unambiguously demonstrate the possible role of these shape resonances in the dynamics of detachment by molecular targets.

A theoretical study of the $H^- + N_2$ system has been reported by Tuan *et al.* (1982) in which electron scattering data (i.e., $e + N_2$) are used to calculate the

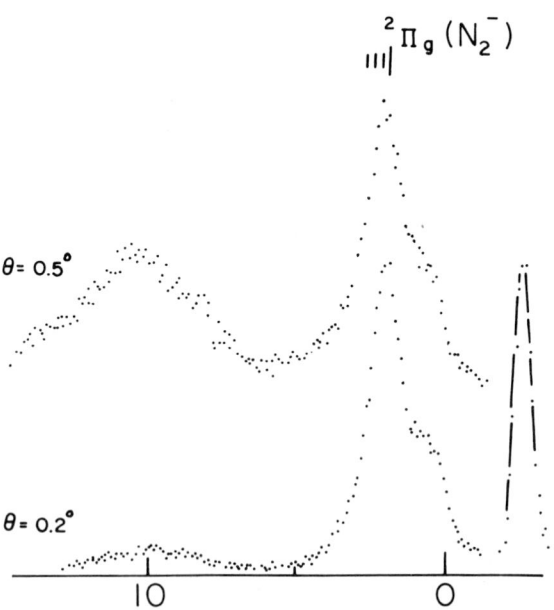

Fig. 22. Energy loss spectra for H atoms from $H^- + N_2$ collisions at 980 eV. The abcissa is the energy loss in eV and the zero of the abcissa refers to ground-state products. The location of the $^2\Pi_g$ state of N_2^- is indicated, as is the energy profile of the incident H^- beam. [From Tuan and Esaulov (1982a). Reproduced with permission from *Journal of Physics*.]

scattering amplitude (taken as a Breit–Wigner resonance amplitude) for the shape-resonance-assisted detachment channel. Not only do the calculations reproduce the correct energy loss spectra (similar to that in Fig. 22), but they also correctly indicate the relative importance of the $^2\Pi_g$ state in the detachment process, viz., $\sim 40\%$ of $\sigma_{-10}(E)$ for $E \sim 1$ keV. Such calculations appear not to have been extended to low collision energies ($E \lesssim 1$ keV).

Finally, it is reasonable to suggest that the different scaling behaviors seen in $\sigma_{-10}(E)$—with velocity scaling for $E \gtrsim 50$ eV—may be due to the fact that direct detachment is the unique detachment mechanism for low collision energies. Owing to a lack of either low-energy data or a theoretical calculation, this is speculative.

Total electron detachment cross sections for $H^-(D^-) + CO$ show striking similarities to those just discussed for N_2. For $E \gtrsim 50$ eV, $\sigma_{-10}(E)$ scales approximately with the collision velocity whereas $\sigma_{-10}(E)$ scales with E for $E \lesssim 50$ eV with no discernible isotope effect at low energies. This can be clearly seen in Fig. 23, where the cross sections are presented for $E \lesssim 300$ eV (Huq et al., 1983b). CO is similar to N_2 in that there is a low-lying shape resonance ($^2\Pi$, $E \sim 1.5$ eV) for the negative molecular ion. Thus it is quite natural to

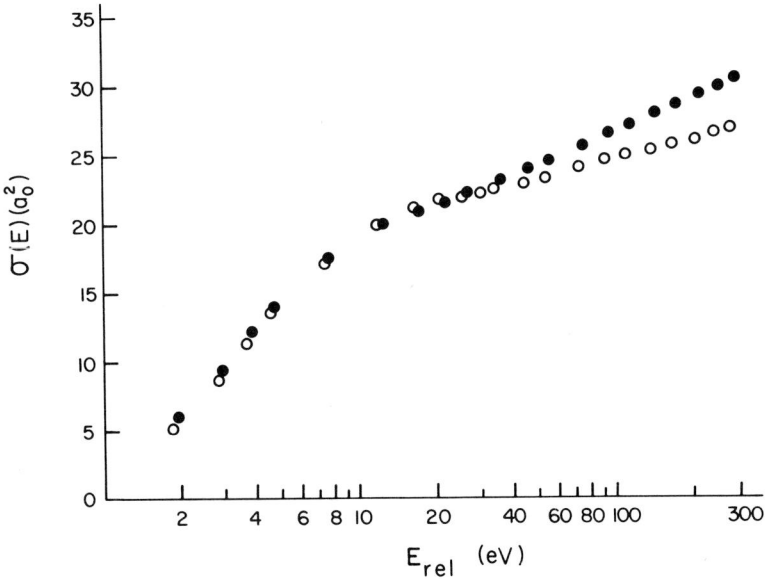

Fig. 23. Cross sections for electron detachment for collisions of H^- and D^- with CO. The solid circles are for H^- and the open circles are for D^-. [From Huq et al. (1983b). Reproduced with permission from the *Physical Review*.]

expect that this resonance might also be involved in the detachment mechanism. Studies by Tuan and Esaulov (1982a) similar to those presented in Fig. 22 indicate that the $^2\Pi$ resonance of CO^- is indeed important in the collisional detachment of H^- by CO. Specifically, their work indicates that direct detachment and detachment via the shape resonance are comparable at $E \simeq 500$ eV. For the CO target, no evidence for dissociative charge transfer:

$$H^- + CO \longrightarrow O^- + C + H$$
$$\longrightarrow O + C^- + H$$

was found for $2 < E \lesssim 300$ eV (Huq et al., 1983b).

$\sigma_{-10}(E)$ for the $H^-(D^-) + CO_2$ system is different from that for any system presented thus far in that structure is observed in the detachment cross section. A plot of $\sigma_{-10}(E)$ is given in Fig. 24. The cross sections given in this figure scale very well with collision velocity at the higher energies. For $7 \lesssim E \lesssim 30$ eV, a strong isotope effect is observed in $\sigma_{-10}(E)$, the D^- projectile giving the larger cross section. In a separate experiment, Huq et al. (1983b) have been able to measure the total cross sections for the production of "slow" negative product ions, as in

$$H^- + CO_2 \longrightarrow O^- + CO + H \tag{20}$$

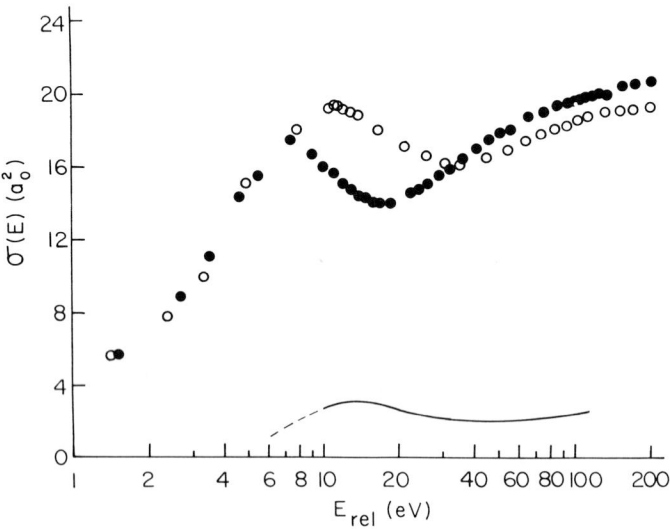

Fig. 24. Total cross sections for electron detachment in collisions of H^- (solid circles) and D^- (open circles) with CO_2. The lowest curve is the cross section for producing slow negative product ions [see eq. (20)] from $H^- + CO_2$. [From Huq et al. (1983b). Reproduced with permission from the *Physical Review*.]

This cross section, denoted $\sigma_1(E)$, is determined by electrostatically trapping all low-energy product ions regardless of the product-ion mass. Since these channels may compete with electron detachment it is of interest to know, e.g., if the structure in $\sigma_{-10}(E)$ is due to competing reactions such as (20). The results for $\sigma_1(E)$ for $H^-(D^-) + CO_2$ are also shown in Fig. 24. It was found that the only product ion contributing to $\sigma_1(E)$ in the vicinity of the maximum ($E \sim 13$ eV) in $\sigma_1(E)$ is O^-, whereas $OH^-(OD^-)$ comprises approximately 15% of the signal at $E \sim 7$ eV. The maxima in $\sigma_1(E)$ observed in the vincinity of $E \simeq 13$ eV may be correlated to the local minima in $\sigma_{-10}(E)$ at the same energy because of competition among the channels responsible for the production of O^- and free electrons. For these low collision energies, it may be that the resonance states of CO_2^- (see Claydon et al., 1970; Chantry, 1972; Chantrell et al., 1982, for example) are involved in the detachment or dissociative attachment channels and are responsible for the structure observed in $\sigma_{-10}(E)$ and $\sigma_1(E)$ presented in Fig. 24.

The $^2\Pi_u$ resonance of CO_2^- is definitely involved in the neutralization of H^- by CO_2 at $E = 500$ eV, as can be seen in the energy loss spectra for the product hydrogen atoms which are shown in Fig. 25 (Tuan and Esaulov, 1982b). There are several salient features of these spectra: first, the $^2\Pi_u$ resonance channel becomes relatively more important (compared to direct detachment) as the scattering angle is increased and the impact parameter is

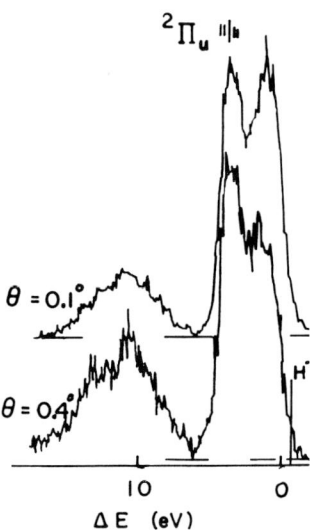

Fig. 25. Energy loss spectra from H atoms from $H^- + CO_2$ collisions at 500 eV. The zero of the abcissa refers to ground-state products. The location of the $^2\Pi_u$ state of CO_2^- is indicated. [From Tuan and Esaulov (1982b). Reproduced with permission from the *Journal of Physics*.]

presumable decreased, and second, there appears to be a significant amount of detachment with concomitant target (or projectile) electronic excitation. This can be seen in Fig. 25 as the broad peak centered about an 11-eV energy loss. Recent studies of electron scattering on CO_2 by Chantrell *et al.* (1982) indicated that there were several core-excited resonances of CO_2^- with energies in the range of 10–13 eV. These resonances, although not indicated on Fig. 25, can not be ruled out as the source of the broad peak observed at $\Delta E \simeq -11$ eV.

By integrating over all scattering angles, Tuan and Esaulov (1982b) estimate that this broad peak constitutes 40% of $\sigma_{-10}(E)$ at $E \simeq 500$ eV. Thus, in the collisional detachment of H^- by CO_2, direct detachment may not be the principal mechanism of producing free electrons, at least at 500 eV.

The results of several different measurements for the total electron detachment cross sections for $H^-(D^-) + O_2$ are shown in Fig. 26, for $E \lesssim 300$ eV.

For $H^- + O_2$, there are several mechanisms in addition to direct detachment which can result in the production of free electrons. The charge exchange reaction for the formation $O_2^-(v' = 0)$ is endothermic by 0.31 eV. However,

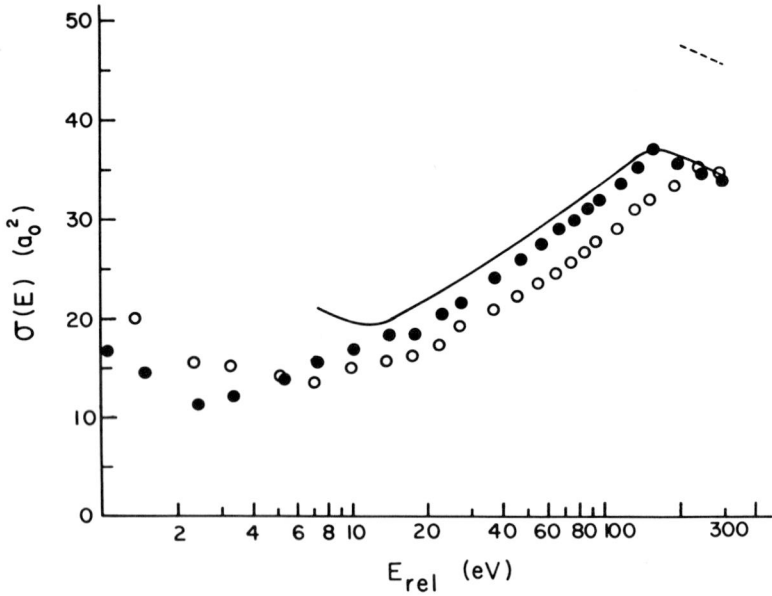

Fig. 26. $\sigma_{-10}(E)$ for H^- (solid circles) and D^- (open circles) on O_2. The solid curve is representative of the results of Bailey and Mahadevan (1970) for $H^- + O_2$ and the dotted line is representative of the results from Risley and Geballe (1974). [From Huq *et al.* (1983b). Reproduced with permission from the *Physical Review*.]

O_2^- may be produced in a vibrationally stable (i.e., $v' \leq 3$) or unstable (i.e., $v' > 3$) state. For $v' > 3$, O_2^- will autodetach to give free electrons and the neutral O_2 molecule. These electrons will then contribute to the observed detachment cross sections. It is impossible to give a quantitative estimate of the cross section for free-electron production from such a process since there is no way to distinguish these electrons from those produced from direct detachment.

Previous experimental studies have demonstrated that, for selected reactants, associative detachment may compete with direct collisional detachment and charge transfer at very low collision energies (Roche and Goodyear, 1969; Commer and Schultz, 1974). Thus, associative detachment ($H^- + O_2 \to HO_2 + e$) may also be an important source of detached electrons at low collision energies. The apparent increase in the measured detachment cross section $\sigma_{-10}(E)$, observed in Fig. 26 as the energy is decreased below 3 eV is probably due to the onset of associative detachment (Dunkin et al., 1970).

For high energies, there is a descrepancy between the magnitudes of the measured cross sections, but the rather unusual decrease in $\sigma_{-10}(E)$ as E is increased beyond 150 eV is reported in all three measurements presented in Fig. 26.

For these reactants, the charge transfer cross section [with $O_2^-(v')$ having $v' \leq 3$] undoubtedly competes with electron detachment at low collision energies, and the measured cross sections for charge transfer are presented in Fig. 27, where they are displayed as a function of collision velocity. The striking feature in Fig. 27 is the extent to which the oscillating charge transfer cross sections for H^- and D^- "scale" with the collision velocity.

In an attempt to understand the mechanism(s) responsible for such behavior, it has been suggested (Huq et al., 1983b) that the basic dynamics for the charge transfer can be described within the framework of a two-state problem in which the initial and final states involved in the process are given by

$$H^- + O_2(v = 0) \longrightarrow H + O_2^-(v' = 3)$$

The asymptotic energy difference ΔE between these two states is given approximately by the electron affinity of the hydrogen atom (0.75 eV). In a two-state approximation, the total cross section contains an oscillatory term which depends upon the collision velocity (Wu and Ohmura, 1962),

$$\sigma(v) \sim M(v)\sin^2(\Delta Ez/2v). \qquad (21)$$

In Eq. (21), z/v represents the average time spent in the region where transitions can take place and $M(v)$ represents the coupling of the initial state ($H^- + O_2$) to the final state ($H + O_2^-$). By inserting $z = 7a_0$ into Eq. (21), the two maxima and one minimum of Fig. 27 can be well-reproduced.

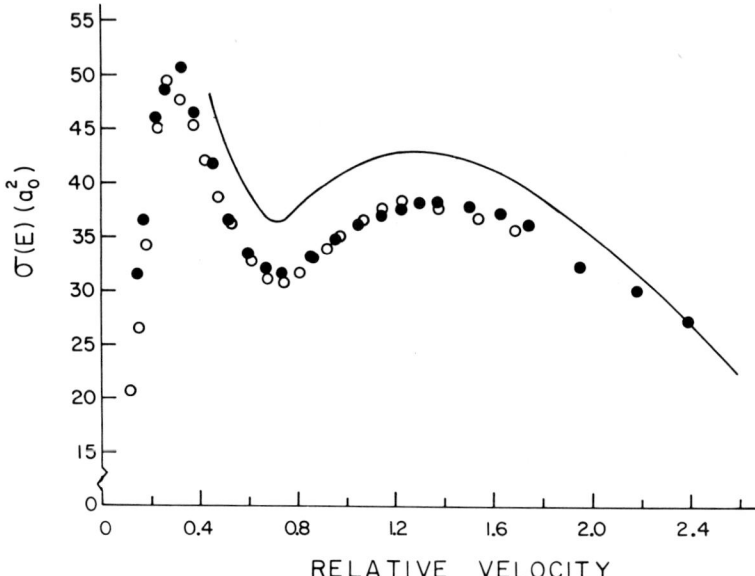

Fig. 27. Cross sections for charge transfer in collisions of H$^-$ (solid circles) and D$^-$ (open circles) with O$_2$, as a function of collision velocity (in units of 10^7 cm/s). The solid line is representative of the results from Bailey and Mahadevan (1970). [From Huq et al. (1983b). Reproduced with permission from the *Physical Review*.]

Moreover, the total cross section, which should be on the order of $\pi(\frac{1}{2}z)^2$, is observed to be of about this magnitude. The isotope effect which is observed is also compatible with this rather simple model.

Although this analysis and description of the production of O$_2^-$ ions is plausible, it neglects the electron detachment channel and is obviously an oversimplified description of the reaction dynamics.

Thus we find once again that the negative-ion states of the target molecule play a vital role in the dynamics of negative ion–molecule collisions. For H$^-$(D$^-$) + O$_2$, $\sigma_{-10}(E)$ is not the dominant neutralization mechanism for $E \lesssim 300$ eV. We will see further examples of detachment relegated to an inferior role in systems where rearrangement channels can compete with electron detachment.

V. Detachment Accompanied by Rearrangement

For many systems, electron detachment by molecular targets cannot be treated in isolation. This is due to the fact that some type of rearrangement or

7 Electron Detachment Processes

reactive collision channel often accompanies (and sometimes dominates) electron detachment, especially at low collision energies. Unfortunately, our current understanding of reactive scattering in ion–molecule systems which involve negative ions is not as extensive as it is for reactants which include positive ions. This is due, in part, to the rather limited number of experimental studies performed in the field of negative ion–molecule collisions. Moreover, from a theoretical point of view, reactive scattering in negative ion–molecule collisions is more complicated than for positive ions because of the existence of the additional product channel of electron detachment. The manner in which electron detachment "competes" with conventional reactive scattering in collisions of negative ions with molecules is not well understood.

In what follows, no attempt will be made to be exhaustive. Only a few studies which shed some light upon the relative importance of electron detachment in such collisions will be considered.

A number of collision systems which illustrate the competition between electron detachment and rearrangement processes (such as reactive and ion exchange collisions) have been the subject of recent investigations. For example, the total cross sections for both electron detachment and rearrangement collisions,

$$A^- + BC \longrightarrow e + \text{products} \tag{22}$$

$$A^- + BC \longrightarrow AB + C^- \tag{23}$$

have been studied for a few systems for collision energies under several hundred electron volts.

The collisional dynamics for (22) and (23), for reactions of O^- with D_2, have been discussed by Herbst et al. (1979) in terms of a "trajectory surface leaking model" (Preston and Cohen, 1976) in which detachment is described by a complex potential (Miller, 1970). Other than this, it appears that there are no detailed theoretical studies of detachment accompanied by reactive scattering in collisions of negative ions with molecules.

A complex potential model has been used by Goursaud et al. (1976, 1978) to describe dissociative attachment processes for collisions of electrons with triatomic molecules, $e^- + ABC \rightarrow A^- + BC$. In particular, they have examined the effect of Γ [see Eq. (11)] upon the kinetic-energy distributions of the dissociation fragments and point out that isotope effects are caused by preferential autodetachment in the case of heavier isotopes (and hence slower dissociation fragments) in isotopic molecules. These $e^- + ABC$ collisions may be regarded as reactive "half-collisions" and hence the above theoretical studies are somewhat relevant to the study of reactive scattering in negative ion–molecule collisions.

A. $H^-(D^-) - H_2, D_2$

Electron detachment in collisions of H^- with H_2 (and isotopic variations) has been studied extensively. These studies have covered the range of collision energies from about one electron volt to several million electron volts. A thorough review of detachment in $H^- + H_2$ collisions has been given by Risley and Geballe (1974).

$\sigma_{-10}(E)$ has been measured recently (Huq et al., 1983a) for these isotopic systems. The measurements emphasize the detachment threshold region, and examples of these measurements are shown in Fig. 28, where the detachment cross sections for $H^-(D^-)$ on D_2 are plotted as functions of the relative collision velocity. The cross sections show a threshold in the vicinity of 1 eV, scale with collision energy up to approximately 10 eV, and scale well with collision velocity at higher energies. Similar isotopic studies by Risley (1974) show that velocity scaling of these cross sections continues up to at least 10 keV.

The H_2 molecule is known to possess a broad ($\Gamma \simeq 8$ eV) shape resonance for the formation of H_2^- at about 2 eV above the ground state of the neutral molecule. Tuan and Esaulov (1982a) have measured the energy loss spectra of hydrogen atoms produced in collisions of H^- with D_2 at 420 eV. The spectra show a most probable energy loss well below 2 eV and give no indication of

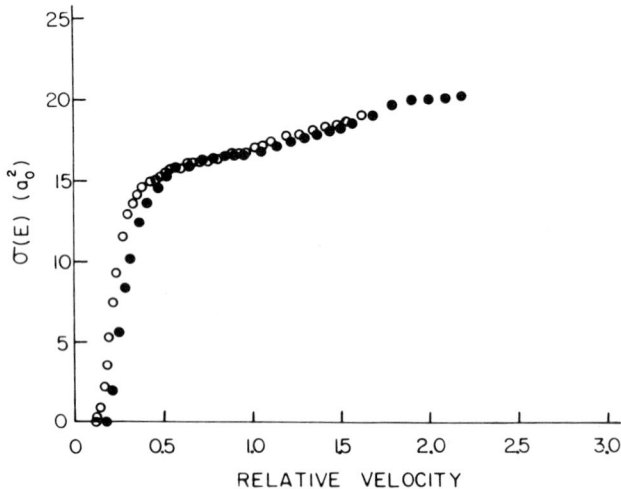

Fig. 28. $\sigma_{-10}(E)$ for collisions of H^- (solid circles) and D^- (open circles) with D_2, as a function of collision velocity (in units of 10^7 cm/s). [From Huq et al. (1983a). Reproduced with permission from the *Physical Review*.]

the involvement of the broad, low-lying $^2\Sigma_u$ resonance of D_2^- in the detachment process. Thus, detachment in these systems is probably direct detachment, governed by an appropriate coupling of the ground electronic state of the relevant molecular and negative-ion potential surfaces, similar to that already discussed for atomic reactants.

Michels and Paulson (1981) have carried out detailed *ab initio* calculations of the potential-energy surface for $H^- + H_2$. In order to define the lowest singlet potential-energy surface for H_3^-, calculations were performed for both triangular (C_{2v}) and linear geometries. These calculations indicated that the H_3^- surface lies below the ground state of the congruent H_3 surface for both configurations (Michels, 1982, private communication). If these surfaces are found not to cross for intermediate orientations (other than C_{2v} and linear), then it is clear that the mechanism of direct detachment is dynamic coupling, as in the case of $H^- + Ne$. The observation that the $^2\Sigma_u$ resonance of H_2^- is not involved in the detachment is compatible with the "noncrossing" results.

Although the H_3^- surface seems to lie below that for H_3, ion exchange reactions such as

$$H^- + D_2 \longrightarrow D^- + HD$$

can occur and the calculations reported by Michels and Paulson indicate that the minimum-energy reaction path for this ion exchange occurs for a linear configuration with a barrier height of about 0.7 eV. For C_{2v} symmetry, these same calculations indicate that the minimum-energy reaction path leads to dissociation,

$$H^- + H_2 \longrightarrow H^- + H + H$$

rather than ion exchange. It is clear for $H^- + H_2$, the ion exchange and detachment product channels are both open at low collision energies. The cross sections for rearrangement have been reported by Michels and Paulson (1981) and an example of these measurements may be seen in Fig. 29. They exhibit thresholds of approximately 1 eV, quickly reach a maximum between 2 and 3 eV, and decrease rapidly thereafter. The most striking feature of these cross sections are the large isotope effects, which appear to be larger than any found to date for such abstraction reactions. The thresholds exhibited in Fig. 29 are about the same as those for $\sigma_{-10}(E)$ seen in Fig. 28. Thus any theoretical treatment of low-energy collisional detachment for $H^- + H_2$ must include both ion exchange and detachment as product channels. The $H^- + H_2$ system and its isotopic analogs can serve as the prototype for any theoretical attack on these complicated scattering systems. Considerable theoretical work has been done on the H_3^- and H_3 surfaces and there now exists a considerable amount of experimental data of sufficient quality to test the details of any dynamical model for these reactants.

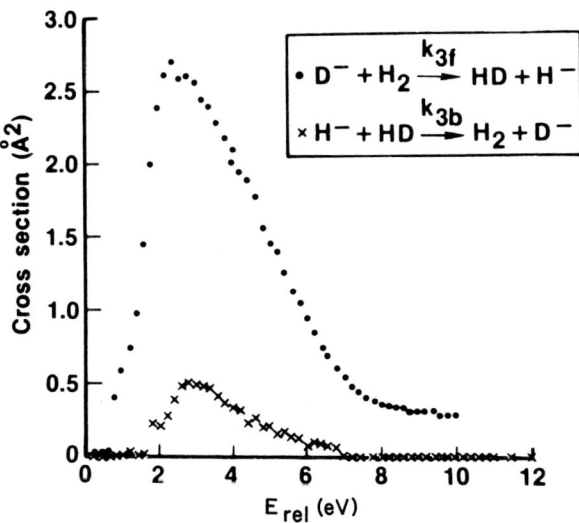

Fig. 29. Ion exchange cross sections for $D^- + H_2$ and $H^- + HD$, as a function of relative collision energy, [From Michels and Paulson (1981). Reproduced with the permission of Plenum Press from "Potential Energy Surface and Dynamics Calculations (D. G. Truhlar, ed.), 1981."]

B. Halogen Ions + H_2, D_2, HD

Measurements of the total cross sections for (22) and (23) have been reported for the above systems (Huq *et al.*, 1982b). The cross sections for electron and H^- (or D^-) production [the latter termed $\sigma_I(E)$] may result from a number of product channels. For the $F^- + H_2$ system, they are

$$F^- + H_2 \longrightarrow \begin{cases} HF + H^- & (1.28 \text{ eV}) \quad (24) \\ F + H + H^- & (7.12 \text{ eV}) \quad (25) \\ HF + H + e & (2.03 \text{ eV}) \quad (26) \\ F + H_2 + e & (3.40 \text{ eV}) \quad (27) \\ F + H + H + e & (7.87 \text{ eV}) \quad (28) \end{cases}$$

The endothermicities for ground-state reactants and products are listed along with each product channel. The endothermicities for the analogous reactions with D_2 and HD targets are slightly different [except for (27)] owing to differences in the zero-point energies of the various deuterated molecules. The experimental results for $\sigma_I(E)$ and $\sigma_{-10}(E)$ for $F^- + D_2$ are given in Fig. 30.

At low energies $\sigma_I(E)$ can be observed to be an order of magnitude larger than $\sigma_{-10}(E)$ and exhibits an onset compatible with Eq. (24). For $E < 7$ eV,

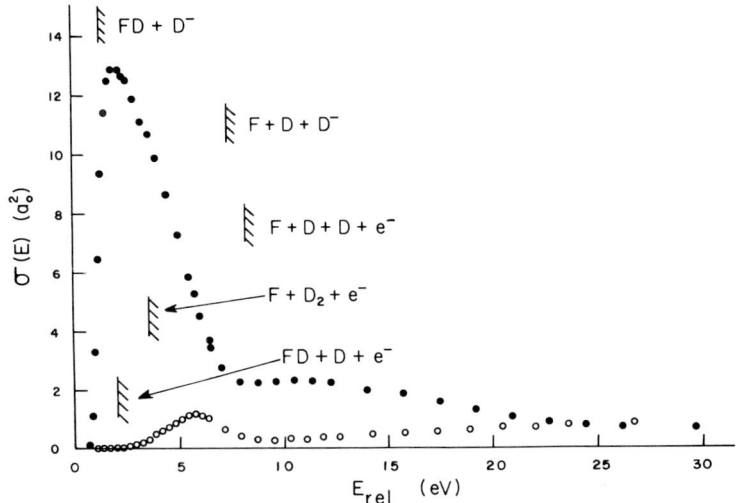

Fig. 30. Cross sections for D^- production (solid circles) and electron detachment (open circles) for $F^- + D_2$. The thresholds from Eqs. (24)–(28) are indicated. [From Huq et al. (1982b). Reproduced with permission from the *Journal of Chemical Physics*.]

energetic consideration dictate that the sole mechanism for D^- production is, in fact, reaction (24). It should be noted that the general shape and magnitude of $\sigma_1(E)$ for $E < 10$ eV strongly resembles the results of trajectory calculations by Muckerman (1972) for the reactive scattering of "fast" F atoms in

$$F + HD \longrightarrow HF(DF) + D(H)$$

$\sigma_{-10}(E)$ is seen to be surprisingly small over the entire energy range of the experiment. For low energies, electron detachment may accompany reactive scattering according to the following scheme:

$$F^- + H_2 \longrightarrow (F\text{---}H\text{---}H)^- \begin{cases} HF + H^- \\ HF + H + e \end{cases} \qquad (29)$$

This suggestion is supported by the bell-shaped form of $\sigma_{-10}(E)$, which is similar to that of $\sigma_1(E)$. For $E \gtrsim 10$ eV, the dominant contribution to $\sigma_{-10}(E)$ is probably from direct detachment as in (27).

For $E \gtrsim 10$ eV, the $H^-(D^-)$ production may arise from either (24) or (25). The trajectory calculations of Muckerman (1972) for $F + HD$ show a substantial tailing of the cross section for the reactive scattering for relative collision energies up to about 30 eV. Moreover, these same calculations also indicate a rather large ($\sim 8a_0^2$) and flat cross section for collision-induced dissociation (CID), $F + HD \rightarrow F + H + D$, which is predicted to be the dominant channel at high energies. It is plausible that in the negative ion–molecule system the rather flat behavior of $\sigma_1(E)$ for $E \gtrsim 10$ eV is also indicative of the CID channel (25). It should be emphasized that these comparisons of the trajectory calculations for the neutral reactants to the experimental observations for the corresponding ionic reactants is made solely because of their suprising similarities and not because one expects such comparisons to be valid *a priori*.

The effects of isotopic substitution upon $\sigma_1(E)$ can be seen in Fig. 31 where $\sigma_1(E)$ for H_2, HD, and D_2 targets are presented. For HD the $\sigma_1(E)$ measurement does not distinguish between H^- and D^- products.

A substantial isotope effect is observed in the region of the maximum of $\sigma_1(E)$. At low energies ($E \lesssim 10$ eV) the magnitude of the cross section increases in the order $D_2 < HD < H_2$. This ordering is not preserved, however, for $E \gtrsim 10$ eV, where the cross section for the HD target behaves somewhat differently than those for H_2 and D_2. The behavior of the isotope effect with energy can be seen more clearly if one examines the ratios of the cross sections for the various isotopic targets. These ratios can be defined as

$$R_{24} = \sigma_1(H_2)/\sigma_1(D_2), \qquad \text{etc.,}$$

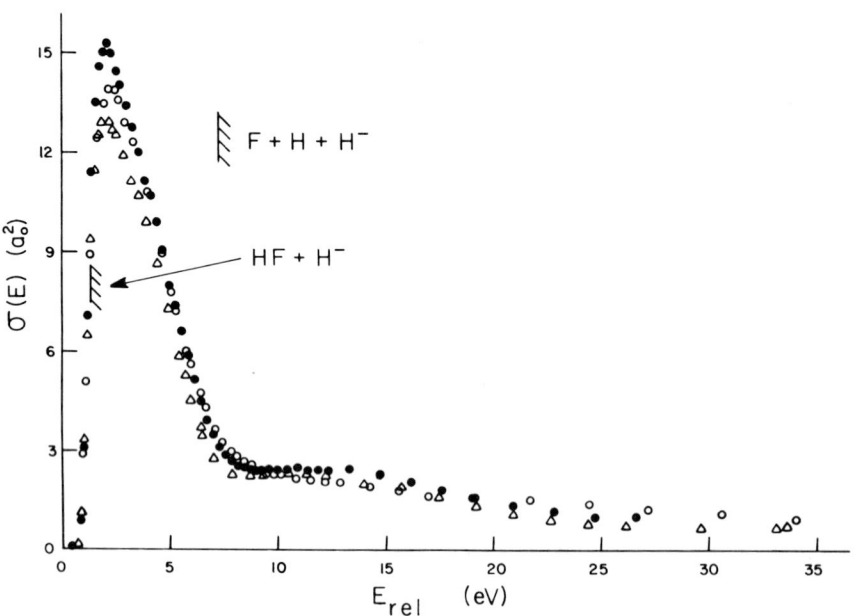

Fig. 31. Cross sections for H^- or D^- production are given for $F^- + H_2$ (solid circles), $F^- +$ HD (open circles), and $F^- + D_2$ (triangles). [From Huq *et al.* (1982b). Reproduced with permission from the *Journal of Chemical Physics*.]

7 Electron Detachment Processes

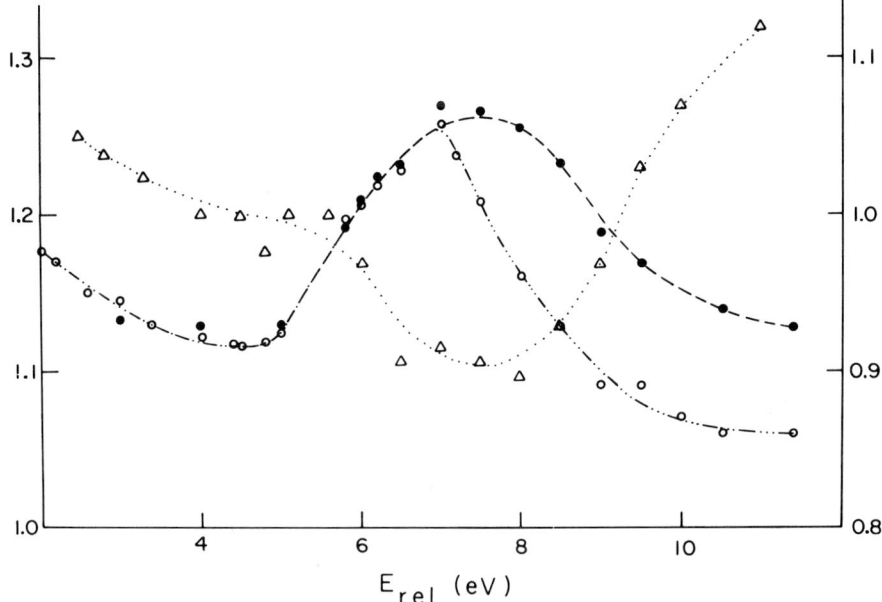

Fig. 32. Ratios of the cross sections for $H^-(D^-)$ production from $F^- + H_2$, HD, and D_2 as a function of relative collision energy. The solid circles are for R_{24}, open circles for $\mathscr{R}_{24}$, and the triangles are for R_{23}. The scales for R_{24} and $\mathscr{R}_{24}$ are indicated on the left of the figure; that for R_{23} is indicated on the right. All of the experimental points have been joined by a smooth curve. [From Huq et al. (1982b). Reproduced with permission from the *Journal of Chemical Physics*.]

where the subscripts on R refer to the masses of H_2 and D_2, respectively. The ratios $R_{24}(E)$ and $R_{23}(E)$ are given in Fig. 32. The interesting feature of $R_{24}(E)$ is that this ratio decreases slowly as the energy is increased from threshold and then displays a local maximum at an energy of about 7 eV. In contrast to $R_{24}(E)$, the plot of $R_{23}(E)$ shows a minimum at this energy. These features at about 7 eV are observed in a region which is near the onset of the CID channel and where the cross section for electron detachment maximizes. As may be seen in Fig. 32, a plot of the ratio of the summed cross sections, $\mathscr{R}_{24} \equiv [\sigma_I(H_2) + \sigma_{-10}(H_2)]/[\sigma_I(D_2) + \sigma_{-10}(D_2)]$, displays a similar feature in the same region. This suggests that the origin of this structure lies mainly in the reactive scattering dynamics.

$\sigma_{-10}(E)$ for all three targets are shown in Fig. 33 along with indications of the various thresholds for the production of free electrons. It is obvious that electron detachment is governed by two distinct mechanisms. For $E \gtrsim 10$ eV, $\sigma_{-10}(E)$ increases in the order $D_2 < HD < H_2$. Thus for a given E, the reactants with the higher collision velocity possess the larger $\sigma_{-10}(E)$.

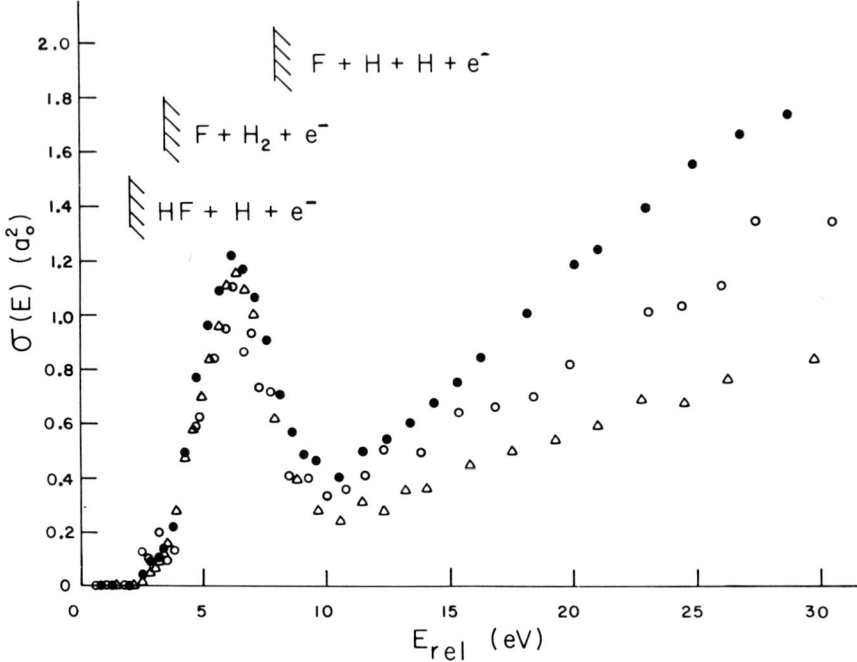

Fig. 33. $\sigma_{-10}(E)$ for $F^- + H_2$ (solid circles), $F^- + HD$ (open circles), and $F^- + D_2$ (triangles). [From Huq et al. (1982b).]

Moreover, in this energy range, the three detachment cross sections scale approximately with collision velocity; i.e., they are the same when compared at the same velocity. As discussed earlier, this behavior of $\sigma_{-10}(E)$ has been observed for many systems and, along with a small detachment cross section, it is indicative of electron detachment via dynamic coupling. It is possible that the negative-ion resonance states of H_2^- are involved in the detachment for $E \gtrsim 10$ eV, but there have been no studies reported for these systems which could substantiate this suggestion.

For $E \lesssim 10$ eV, detachment by the different isotopes does not scale with collision velocity but is rather a universal function of the relative collision energy with no significant isotope effect. The mechanism for detachment in this low-energy regime must be due to some type of surface crossing which is not available for the direct electron detachment channel represented by (27). These observations are compatible with the earlier suggestion that low-energy electron detachment occurs by (26) as a companion to reactive scattering indicated in (24).

To illustrate the above idea with a model, one can assume that there exists a region in configuration space associated with reactive scattering where the

7 Electron Detachment Processes

(FH—H$^-$) surface lies above that for (FH—H). Such a region will be unstable with respect to detachment and one can attempt to describe this detachment in terms of the decay of a quasistationary state of width Γ as was done in the trajectory leaking model (Herbst et al., 1979) and the work on dissociative attachment (Goursaud et al., 1976, 1978) mentioned earlier. If one assumes that the total reactive cross section is given by Eq. (29), i.e., $\sigma_R(E) = \sigma_e(E) + \sigma_1(E)$, then (within the framework of a quasistationary state or complex potential model) the low-energy electron detachment cross section may be written as

$$\sigma_{-10}(E) = \sigma_R(E)P_d(E),$$

where $P_d(E)$ is the average detachment probability for a given E. Since it is known that $P_d(E)$ is small, one can write

$$P_d(E) \simeq 1 - e^{-\bar{\Gamma}\Delta t/\hbar} \simeq \bar{\Gamma}\Delta t/\hbar,$$

where Δt is the time the product H$^-$ (or D$^-$) spends in the aforementioned unstable region and $\bar{\Gamma}$ is the suitably averaged value for the autodetachment width. Thus

$$\frac{\sigma_{-10}(E, H_2)}{\sigma_{-10}(E, D_2)} \simeq \mathscr{R}_{24}(E)\frac{\bar{\Gamma}(H_2)\Delta t(H_2)}{\bar{\Gamma}(D_2)\Delta t(D_2)}.$$

For $E \simeq 6$ eV, the above ratio of the detachment cross sections is observed to be about 1.1. If we approximate $\Delta t(H_2)/\Delta t(D_2)$ by the square root of the ratio of the reduced masses of the products (H$^-$ + HF and D$^-$ + DF) and take $\mathscr{R}_{24}(6\text{ eV})$ from Fig. 32, then

$$\bar{\Gamma}(H_2)/\bar{\Gamma}(D_2) \simeq 1.25.$$

Such a result is not unreasonable since the impact parameters that lead to reactive scattering in (29) for the H$_2$ target are undoubtedly different from those that lead to reactive scattering for the D$_2$ target [this is presumably the reason that $\mathscr{R}_{24}(E) > 1$]. Consequently, one would not expect the ratio $\bar{\Gamma}(H_2)/\bar{\Gamma}(D_2)$, which is an average over all reactive impact parameters and molecular orientations, to be unity. Although this analysis generates reasonable numbers, it is certainly an oversimplified description of the reaction dynamics. It is more than likely, however, that any future theory of these molecular collisions will incorporate some of the above quasistationary-state arguments.

The reactive cross sections $\sigma_1(E)$ show very sharp onsets in the vicinity of 1 eV. One question of interest is whether a barrier exists for these reactions which, in the case of H$_2$, has an asymptotic threshold for ground-state reactants of 1.28 eV. In order to explore this question it is necessary to correct the results of Fig. 31 for the effects of broadening which were due primarily to

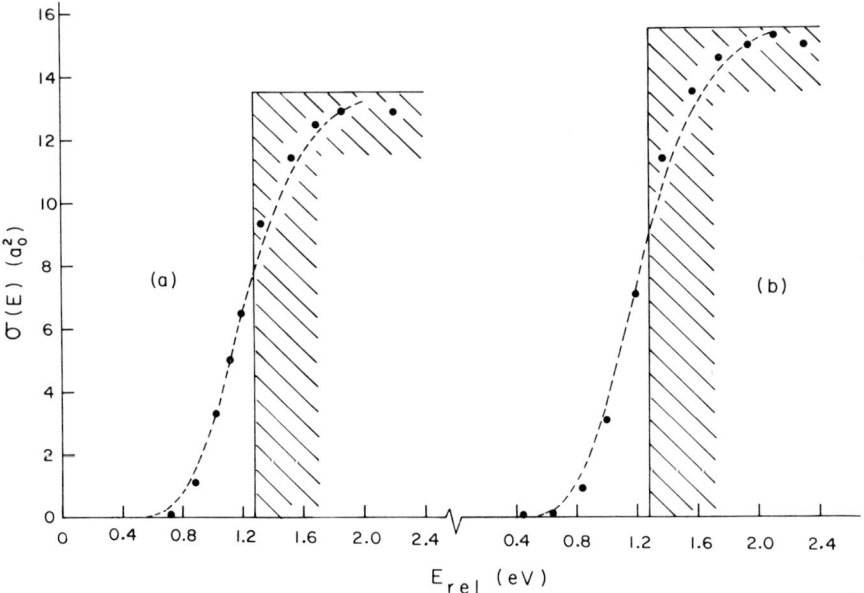

Fig. 34. Cross sections for $H^-(D^-)$ production for (a) $F^- + D_2$ and (b) $F^- + H_2$ for the near-threshold region. The dashed line is the result of a convolution (Chantry 1971) of the step functions shown. [From Huq et al. (1982b). Reproduced with permission from the *Journal of Chemical Physics*.]

the thermal motion of the target gas (at 300 K). The near-threshold results for $\sigma_1(E)$ for both H_2 and D_2 are presented in Fig. 34, along with the results of a convolution (Chantry, 1971) which assumes a step-function cross section (the most extreme case) form for reaction, a target-gas temperature of 300 K, and threshold of 1.28 eV. The convolutions are in excellent agreement with the experimental observations. Thus if a barrier to reactive scattering exists, it can be no larger than approximately 100 meV.

Studies such as those just discussed for F^- and the hydrogenic molecules have also been carried out with Cl^- (Huq et al., 1982b). The several channels that have been studied, along with their ground-state endothermicities for the H_2 target, are

$$Cl^- + H_2 \longrightarrow \begin{cases} HCl + H^- & (2.91 \text{ eV}) \quad (30) \\ Cl + H + H^- & (7.34 \text{ eV}) \quad (31) \\ \cdot HCl + H + e & (3.66 \text{ eV}) \quad (32) \\ Cl + H_2 + e & (3.60 \text{ eV}) \quad (33) \\ Cl + H + H + e & (8.09 \text{ eV}) \quad (34) \end{cases}$$

7 Electron Detachment Processes

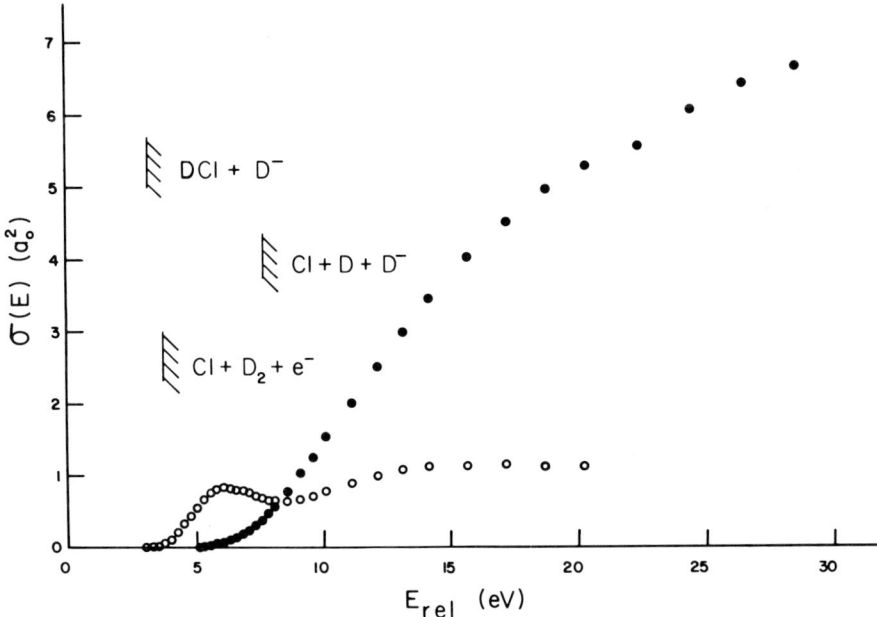

Fig. 35. $\sigma_{-10}(E)$ for Cl + D$_2$ (solid circles) and cross section for D$^-$ production (open circles). [From Huq et al. (1982b). Reproduced with permission from the *Journal of Chemical Physics*.]

The measured cross sections, $\sigma_1(E)$ and $\sigma_{-10}(E)$, are shown in Fig. 35 for the D$_2$ target along with some indicated thresholds of the processes above. As in the case for the F$^-$ projectile, $\sigma_1(E)$ displays a local maximum at about twice the threshold energy for (30), i.e., at about 6 eV. However, in contrast to F$^-$, the magnitude of $\sigma_1(E)$ at low energies is about an order of magnitude smaller than the detachment cross section. At higher enrgies ($E > 8$ eV), a broad plateau is observed for $\sigma_1(E)$ which could be related to the CID channel (31). In this region the magnitude of the cross section is of the same order as for F$^-$, viz., $1.5a_0^2$. In contrast to the F$^-$ + D$_2$ case, $\sigma_{-10}(E)$ is found to be large and is the dominant process for energies above 8 eV. An interesting question is—what particular features of the relevant potential surfaces cause such drastic differences in $\sigma_{-10}(E)$ for Cl$^-$ when compared to that for F$^-$?

Measurements of $\sigma_1(E)$ for all three isotopic targets are presented in Fig. 36. The cross sections are seen to be a universal function of E for $E \lesssim 10$ eV and show larger isotope effects than those found for the F$^-$ projectile. In this same energy range the magnitude of $\sigma_1(E)$ increases in the order D$_2$ < H$_2$ < HD, which reflects the ordering of the thresholds for (H$^-$, D$^-$) production; i.e., the lowest threshold gives the largest cross section.

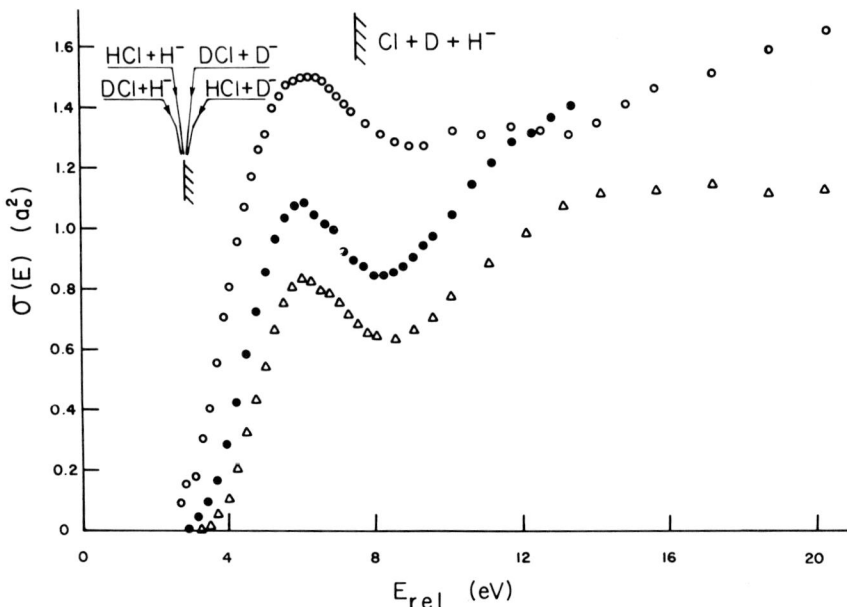

Fig. 36. Isotope effect for H^- and D^- production for $Cl^- + H_2$ (solid circles), $Cl^- + HD$ (open circles), and $Cl^- + D_2$ triangles). [From Huq et al. (1982b). Reproduced with permission from the *Journal of Chemical Physics*.]

The results for $\sigma_{-10}(E)$ are shown in Fig. 37 for all three targets. Bydin and Dukel'skii (1957) have measured the detachment cross section for $Cl^- + H_2$ for laboratory collision energies between approximately 200 eV and 2 keV and there is excellent agreement between the two experiments where they overlap. The isotope effects in $\sigma_{-10}(E)$ are not consistent with a simple model for detachment based upon either a quasistationary-state or dynamic coupling type description, since according to both, $\sigma_{-10}(E)$ for the HD target should lie between those of H_2 and D_2. This is contrary to the observations. Such bahavior suggests that detachment is probably affected by some other inelastic channel(s), which in this case may be reactive scattering as in (32). Thus, in contrast to the F^- case, detachment in Cl^- collisions appears to occur as a companion of reactive scattering even at relatively high energies.

There is additional experimental evidence in support of this conclusion. Cheung and Datz (1980) studied the electron detachment of Cl^- in collisions with H_2 and D_2 in the energy range commensurate with the studies just presented. Using time-of-flight techniques to determine the energy loss spectra of chlorine atoms produced by detachment, they were able to resolve four distinct detachment channels characterized by different energy losses. Their

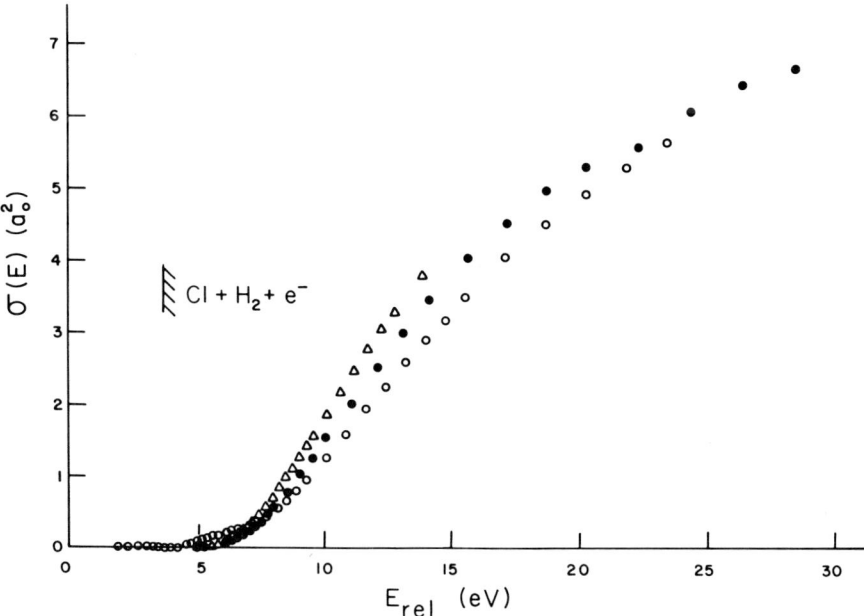

Fig. 37. $\sigma_{-10}(E)$ for $Cl^- + H_2$ (triangles), $Cl^- + HD$ (open circles), and $Cl^- + D_2$ (solid circles). [From Huq et al. (1982b). Reproduced with permission from the *Journal of Chemical Physics*.]

energy loss measurements associated with the three principal channels are plotted as a function of relative collision energy in Fig. 38. The loss spectra indicate that three different detachment mechanisms are operative throughout the whole energy range investigated. The channel labeled A is the direct detachment process already discussed; channel B is the result of dissociation of the target molecule and probably leads to $H^-(D^-)$ production or detachment. Cheung and Datz point out that the channel labeled C is consistent with the formation of an H_2^- in the $^2\Sigma_g^+$ state via electron transfers which involve Franck–Condon transitions from ground-state H_2. Since this state crosses the $(^3\Sigma_u^+)$ H_2 molecular potential curve the H_2^- can decay into either a pair of H atoms plus a free electron or into an H atom and an H^- ion.

Relative cross sections obtained for channels A–C are shown in Fig. 39. These measurements establish that the simple detachment channel becomes dominant at the higher energies, but that all channels are important in this energy range. It should be pointed out that the energy dependence of the sum of these individual cross sections is in general agreement with the detachment cross section measurements in Fig. 37. Based upon the observations of

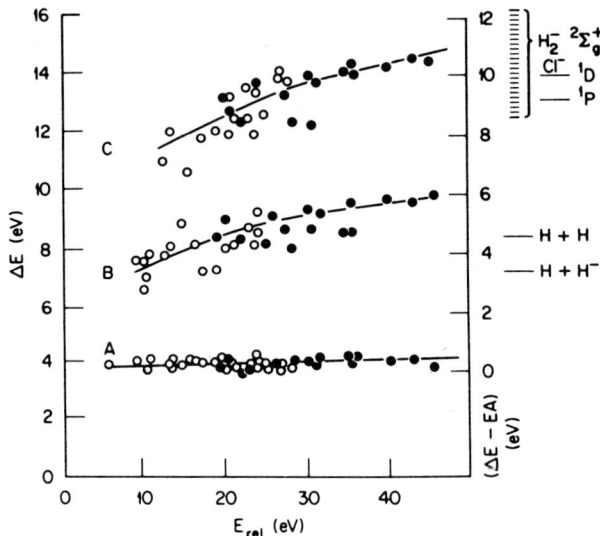

Fig. 38. Energy losses associated with channels A, B, and C as a function of collision energy: (○) $Cl^- + H_2$, (●) $Cl^- + D_2$. A second scale on the right-hand side gives the energy loss after subtracting the Cl electron affinity (3.6 eV). Losses associated with possible exit channel excitations are also indicated. [From Cheung and Datz (1980). Reproduced with permission from the *Journal of Chemical Physics*.]

Cheung and Datz, it is now clear why no single mechanism is capable of explaining the isotope effects in $\sigma_{-10}(E)$ as outlined above.

Calculations of potential surfaces for these systems are not currently available. The many and varied features found in the various cross sections should provide adequate tests for trajectory calculations and models for collisional detachment.

C. I^- + Methyl Halides

As a final example of a system in which detachment and reactive channels compete, it is interesting to look at the gas-phase analogs for a class of reactions which are termed "S_N2" (substitution–nucleophilic, type 2). The reaction,

$$I^- + CH_3Cl \longrightarrow CH_3I + Cl^-$$

is an example of such an S_N2 reaction. There are several basic ideas about these S_N2 reactions which are deduced from observations (see e.g., Morrison and Boyd, 1973). First, it is known that the reaction occurs by "backside attack" in

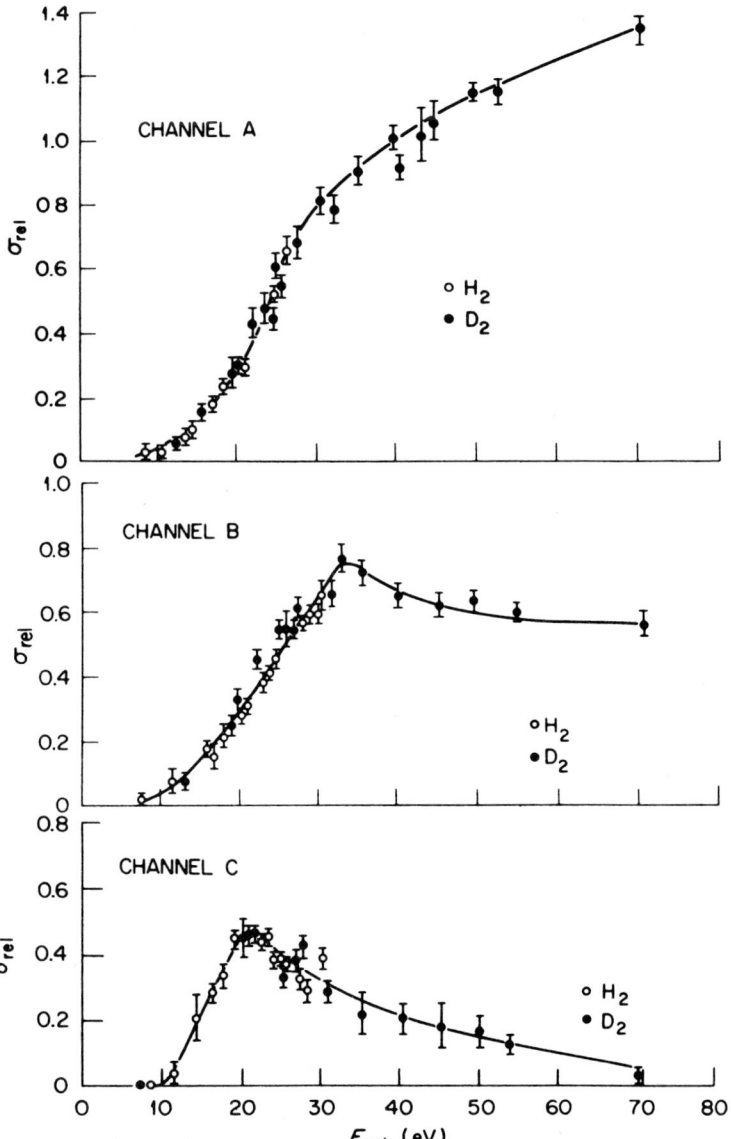

Fig. 39. Relative cross sections for channels A, B, and C for $Cl^- + H_2, D_2$. The cross section units on each panel are the same. [From Cheung and Datz (1980). Reproduced with permission from the *Journal of Chemical Physics*.]

the preceding reaction, as in

$$I^- + \overset{H}{\underset{H}{\overset{H\cdots}{C}}}-Cl \longrightarrow I-\overset{H}{\underset{H}{\overset{\cdots H}{C}}} + Cl^-$$

forming a mirror image of the original methy halide. The second idea is that when the reactions occur in typical organic solvents, heavier "attacking" or "leaving" groups give a higher probability of reaction. Electron detachment is not a consideration in the solution chemistry for S_N2 reactions.

Measurements for $\sigma_{-10}(E)$ for several S_N2 reactants are given in Fig. 40. For both cases, $\sigma_{-10}(E)$ rises from a threshold which is several electron volts above the electron affinity of iodine and continues to increase smoothly with no discernible structure up to the highest energies of the experiment. The cross sections for producing slow negatively charged product ions [$\sigma_l(E)$, as discussed earlier] are also shown in the figure. It is clear that these (S_N2?) channels dominate electron detachment for $E \lesssim 20$ eV. If the product ions which comprise $\sigma_l(E)$ are Br$^-$ and Cl$^-$ (in Figs. 40a and 40b, respectively), then the magnitudes of $\sigma_1(E)$ are in accord with the previously mentioned rule of

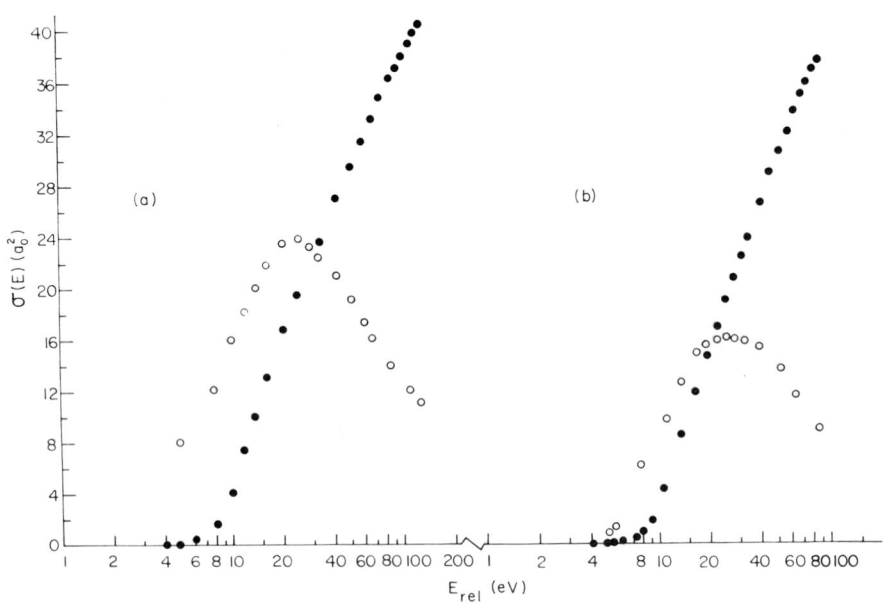

Fig. 40. Cross sections for electron detachment (solid circles) and slow-ion production (open circles) for I$^-$ + methylhalide: (a) I$^-$ + CH$_3$Br, (b) I$^-$ + CH$_3$Cl.

7 Electron Detachment Processes

organic chemistry stating that the heavier "leaving" anion should have the higher reaction probability.

No additional detailed information, such as differential cross sections, product-ion identification, etc., seems to be available for these reactants. It will be of interest to see if such reactions exhibit strong backward peaking in the center of mass system, as is the case for the somewhat similar $K + CH_3I$ reaction (Rulis and Bernstein, 1972).

VI. Molecular Negative Ions

There have been relatively few detachment studies involving molecular negative ions: the most extensive studies are those associated with O_2^-. Detachment studies have been reported for O_2^- and several target gases by Hasted and Smith (1956), Roche and Goodyear (1969), Bailey and Mahadevan (1970), and Wynn et al. (1970). All the investigations emphasized collision energies below several hundred volts. Higher-energy measurements have been reported by Doering (1964). The striking difference in the behavior of the detachment cross sections in the threshold region for collisions of O_2^- with O_2 and He can be seen in Figs. 41 and 42. The threshold for $\sigma_{-10}(E)$ appears to be zero in the case of He, but about 3.5 eV for the O_2 target. The electron

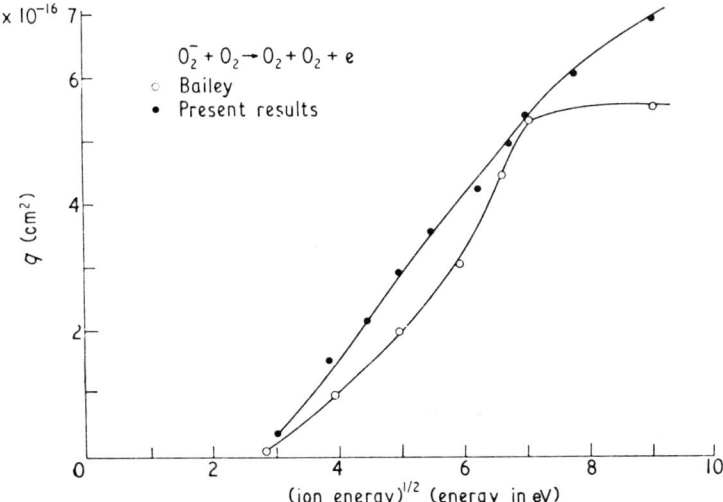

Fig. 41. $\sigma_{-10}(E)$ for $O_2^- + O_2$ as a function of the square root of the laboratory energy. The open circles are from Bailey and Mahadevan (1970) and the solid circles are from Roche and Goodyear (1969).

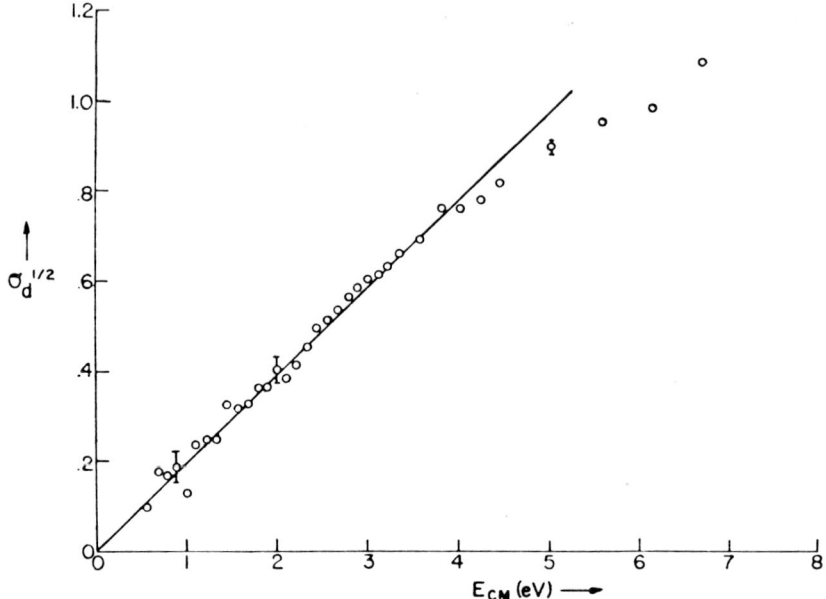

Fig. 42. Square root of $\sigma_{-10}(E)$ (10^{-8} cm) for $O_2^- + $ He. The straight line is a fit to the data. [From Wynn et al. (1970). Reproduced with permission from the *Journal of Chemical Physics*.]

affinity of O_2 is approximately 0.44 eV, and it was suggested by Wynn et al. (1970) that excited vibrational states of O_2^- in the primary ion beam are responsible for the low threshold observed with He.

For the oxygen target, it could be that an avoided crossing of the relevant potential surfaces for $O_2^- + O_2$ and $O_2 + O_2$ explains the high threshold. However, one cannot rule out the possibility that the apparent high threshold is the result of depleting the detachment channel by charge transfer. The charge transfer cross section was measured by Bailey and Mahadevan: it exhibited a resonant type ($Q = 0$) shape and was still increasing as collision energies were decreased down to 2 eV. They found charge transfer to be the dominant channel for collision energies up to 300 eV. Detachment cross sections for O_2^- and numerous gas targets have been reported by Bennet et al. (1975) and Hasted and Smith (1956) at higher collision energies.

As another example of systems which possess a potential barrier to detachment, Wynn et al. (1970) found the threshold for detachment of OH^- in He to be several electron volts above the electron affinity of OH (~ 1.8 eV).

One of the few collision systems involving a polyatomic negative ion that has been studied experimentally deserves mention. The collisional detachment and dissociation cross sections for collisions of UF_6^- with the rare gases have

7 Electron Detachment Processes

been reported by Haywood et al. (1981b) for laboratory energies below 500 eV. Experiments in which the kinetic-energy and angular distributions of the product fragment ions produced in collisions of UF_6^- with Ar, Xe, and selected molecules have been measured by Annis and Stockdale (1981) for collision energies up to 200 eV (lab).

The total cross section measurements showed that $\sigma_{-10}(E)$ is extremely small over the whole energy range, reaching only a few percent of that for collision-induced dissociation. The principal dissociation channels were those that lead to the products $UF_5 + F^-$ and $UF_5^- + F$, the production of F^- being the dominant channel. The energetic thresholds for F^- production were determined for the rare-gas targets, all giving a value of approximately 2 eV, which is well below the currently accepted value of 4.7 eV for ground-state reactants and products. The experimental thresholds would be expected to lie below the actual threshold owing to Doppler broadening effects (Chantry, 1971). Moreover, the UF_6^- is produced by chemical desorption from a hot surface and therefore will possess a significant amount of internal energy if it is in thermal equilibrium with the surface before desorption. However, these effects cannot account for the observed low thresholds, implying that the UF_6^- ion beam is formed with an average internal energy which is in excess of that predicted if it is in thermal equilibrium with the surface.

Identical experiments were carried out for the molecular targets N_2 and SF_6 to see if there were substantive differences when compared to the atomic-target experiments: the molecular-target results are shown in Fig. 43 for $\sigma_{-10}(E)$ and the F^- production cross sections. In all respects, however, the results show essentially the same behavior as the atomic systems: in particular the near-threshold behavior is almost identical for both targets.

In order to investigate the possible role that internal energy plays in the collisional decomposition of UF_6^-, the above experiments were repeated using UF_6^- beams produced on a carbon filament whose temperature varied from 225 to 1540°C. The square root of the F^- production cross section measured for an argon target and for filament temperatures of 330 and 1320°C is shown in Fig. 44. The cross section for the lower temperature is shifted approximately 1 eV (to higher collision energies) with respect to the high-temperature results. The 1-eV shift ($\sim 15k\triangle T$) is reasonable since this is approximately the anticipated change in the average internal energy of desorbing UF_6^- ions if thermal equilibration is assumed. However, the low-temperature threshold results for F^- continue to indicate an onset in the neighborhood of 2 eV. Hence, the reason for the apparently low threshold is yet to be understood.

The general conclusion of these collision studies involving UF_6^- is that the target serves merely as a spectator in the collisional excitation of UF_6^-. Although larger targets provide a larger cross section for excitation of the

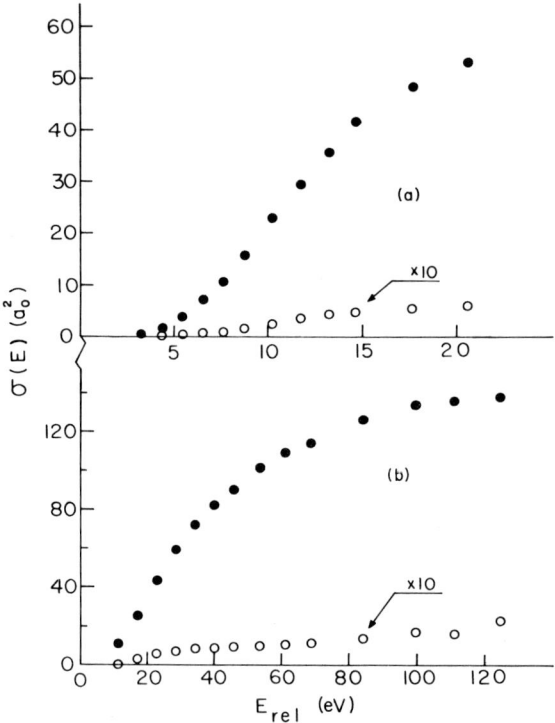

Fig. 43. Cross sections for electron detachment (open circles) and for the production of F^- (solid circles) for collisions of UF_6^- with N_2 (top panel) and SF_6. The electron detachment cross sections have been multiplied by ten for clarity.

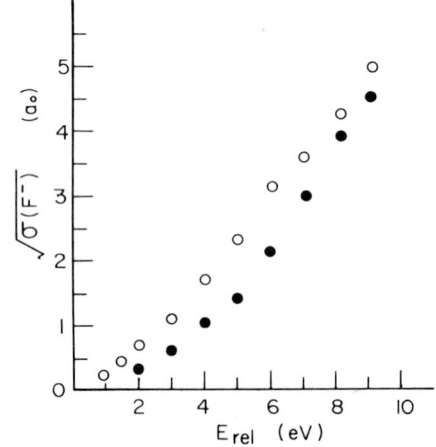

Fig. 44. Behavior of the near-threshold cross section for the production of F^- in collisions of UF_6^- with Ar. The open circles are for UF_6^- which desorbs from a carbon filament at 1320°C and the solid circles are for 330°C.

7 Electron Detachment Processes

negative ion, the subsequent decomposition of the excited negative ion is governed by statistical arguments that are independent of which target provides the collisional excitation. A statistical model (Haywood et al., 1981b) for these collisions was used with success to predict the observed branching ratios for the collision products, including the low yield of detached electrons.

VII. Detachment Rate Constants

There are numerous areas of chemistry and physics where the rate constants, $k(T)$, for collisional detachment and the reverse process, three-body attachment, are important. The detachment rate constants are defined in terms of the detachment cross section

$$k(T) = \frac{8\pi\mu}{(2\pi\mu kT)^{1.5}} \int_0^\infty E\sigma_{-10}(E) e^{-E/kT} \, dE. \tag{35}$$

For $T \simeq 10^4$ K, $k(T)$ is extremely sensitive to the details of the detachment cross sections in the threshold region. Implicit in the definition of $k(T)$ above is the assumption that all degrees of freedom of the reactants are in thermal equilibrium. For the collisional detachment of atomic negative ions by atoms, the determination of the rate constant from low-energy beam experiments in which the detachment cross sections are measured is straightforward, since there are no low-lying excited states of either reactant and the equipartition assumption is essentially correct.

The only measurements of $\sigma_{-10}(E)$ which have sufficient detail around the threshold energy so that meaningful calculations of $k(T)$ can be made are for the halogen–negative ion rare-gas systems (Smith et al., 1978; Haywood et al., 1981a). Even for these systems only an upper limit to $k(T)$ can be ascertained owing to the fact that kT is small with respect to the threshold energy for temperatures of interest (in most applications) and therefore leads to very small detachment rate constants for these systems. This upper limit to $k(T)$ is primarily dictated by the minimum detachment cross section which can be determined in any experimental measurement.

Experiments which use "flowing afterglow" or "flow-drift-tube" techniques (McFarland et al., 1972) can sample the energy range 0.04 to ~ 3 eV, but it appears that there are no measurements for $k(T)$ for direct detachment of atomic negative ions by atomic targets. There have been, however, numerous flow-tube studies of associative detachment, ion–molecule reactions, and other processes for molecular reactants; the resulting rate constants have been compiled in an exhaustive study by Albritton (1978).

Shock-tube techniques have been utilized to measure detachment rate constants for reactants involving halogen–negative ion and rare-gas targets

(Mandl, 1976a,b; Mandl et al., 1970). The shock-tube results for $k(T)$ are consistently several orders of magnitude higher than the rate constants determined from Eq. (35) and beam measurements of $\sigma_{-10}(E)$ (Champion and Doverspike, 1976b; Mandl, 1976b; Smith et al., 1978); Haywood, 1981a). There is not as yet any explanation for the rather serious disagreement between the two results (Berry, 1980).

Shui and Keck (1973) developed a phase-space theory for collisional detachment and applied their technique to systems that include the halogen negative ions. In order for their calculations of $k(T)$ to be compatible with the results of shock-tube measurements, it was necessary to assume that the threshold for detachment occurred at essentially the electron affinity of the halogen, which is contrary to experimental observations.

For molecular targets, deducing rate constants from the measured detachment cross sections is questionable since the assumption of equipartition is not fulfilled. In the beam experiments for molecular targets, the translational and internal energies of the reactants are obviously not equal. In such circumstances, there is no general technique for converting such cross section measurements into rate constants which are based upon the principle of thermodynamic equilibrium. This problem in conjunction with drift-tube measurements has been discussed in some detail by Albritton et al. (1977).

Nevertheless, rate constants as defined by Eq. (35) have been determined (Doverspike et al., 1980) for those molecular targets for which unambiguous detachment thresholds have been determined. Because of the rather large detachment thresholds exhibited by these systems, such rate constants are quite small, again in contrast to the rate constants determined in shock-tube experiments. There appear to be no reasonable arguments about non-equilibrium conditions in the beam results which could significantly reduce the disagreement between the two sets of experiments.

VIII. Summary

We have pointed out various pathways to collisional electron detachment with illustrations from specific systems. The principal focus of this discussion has been upon collisional detachment by molecular targets for collision energies where the molecular properties of the reactants and products are important. A brief review of collisional detachment by atomic targets has been presented to serve as an introduction to the subject.

The various detachment mechanisms which have been discussed include direct detachment, detachment via charge transfer to a shape resonance of the molecular (target) negative ion, and detachment accompanied by various chemical or rearrangement reactions. Where possible, the effects of isotopic

7 Electron Detachment Processes

substitution, for either the negative ion or the molecular target, have been examined in some detail. This has led to some insights concerning the mechanisms and dynamics of collisional detachment by molecular targets. Nevertheless, our current understanding of negative ion–molecular interactions remains somewhat elementary.

Acknowledgments

It is a pleasure for the authors to express their gratitude to M. S. Huq, who was instrumental in the acquisition of much of the data presented in this discussion, and to J. B. Delos for many helpful conversations. The support of the Division of Chemical Sciences, Office of Basic Energy Sciences of the Department of Energy is also gratefully acknowledged.

References

Albritton, D. L. (1978). *At. Data Nucl. Data Tables* **22**, 1.
Albritton, D. L., Dotan, I., Lindinger, W., McFarland, M., Tellinghuisen, J., and Feshenfeld, F. (1977). *J. Chem. Phys.* **66**, 410.
Annis, B. K., and Datz, S. (1982). "Electronic and Atomic Collisions" (S. Datz, ed.), XIIth ICPEAC, p. 115. North Holland, Amsterdam.
Annis, B. K., and Stockdale, J. A. D. (1981). *J. Chem. Phys.* **73**, 297.
Annis, B. K., Datz, S., Champion, R. L., and Doverspike, L. D. (1980). *Phys. Rev. Lett.* **45**, 1554.
Bailey, T. L., and Mahadevan, P. (1970). *J. Chem. Phys.* **52**, 179.
Bennett, R. A., Moseley, J. T., and Peterson, J. R. (1975). *J. Chem. Phys.* **62**, 2223.
Berry, R. S. (1980). *J. Chim. Phys.* **77**, 759.
Bydin, Yu. F., and Dukel'skii, V. M. (1957). *Sov. Phys. JETP* **4**, 474.
Champion, R. L. (1980). "Electronic and Atomic Collisions" (N. Oda and K. Takayanagi, ed.), XIIth ICPEAC, p. 635. North Holland, Amsterdam.
Champion, R. L. (1982). *Adv. Electron. Electron Phys.* **58**, 143.
Champion, R. L., and Doverspike, L. D. (1976a). *Phys. Rev. A* **13**, 609.
Champion, R. L., and Doverspike, L. D. (1976b). *J. Chem. Phys.* **65**, 2482.
Champion, R. L., and Doverspike, L. D., and Lam, S. K. (1976). *Phys. Rev. A* **13**, 617.
Chantrell, S. J., Field, D., and Williams, P. I. (1982). *J. Phys. B* **15**, 309.
Chantry, P. J. (1971). *J. Chem. Phys.* **55**, 2746.
Chantry, P. J. (1972). *J. Chem. Phys.* **57**, 3180.
Cheung, J. T., and Datz, S. (1980). *J. Chem. Phys.* **73**, 3159.
Claydon, C. R., Segal, G. A., and Taylor, H. S. (1970). *J. Chem. Phys.* **52**, 3387.
Commer, J., and Schultz, G. J. (1974). *Phys. Rev. A* **10**, 2100.
Demkov, Yu. N. (1964). *Sov. Phys. JETP* **19**, 762.
Demkov, Yu. N. (1980). "Electronic and Atomic Co-lisions" (N. Oda and K. Takayahagi, ed.), IXth ICPEAC, p. 645. North Holland, Amsterdam.
Devdarianni, A. Z., (1973). *Sov. Phys. Tech. Phys.* **18**, 255.
deVreugd, C., Wijnaendts van Resandt, R. W., Los, J., Smith, B. T., and Champion, R. L. (1979). *Chem. Phys.* **42**, 305.
deVreugd, C., Wijnaendts van Resandt, R. W., Delos, J. B., and Los, J. (1982). *Chem. Phys.* **68**, 261, 275.

Dewangan, J. P., and Walters, H. R. J. (1978). *J. Phys. B* **11**, 3983.
Dimov, G. I., and Roslyakov, G. V. (1972). *Sov. Phys. Tech. Phys.* **17**, 90.
Doering, J. P. (1964). *J. Chem. Phys.* **41**, 1164.
Doverspike, L. D., Smith, B. T., and Champion, R. L. (1980). *Phys. Rev. A* **22**, 393.
Dunkin, D. B., Fehsenfeld, F. C., and Ferguson, E. E. (1970). *J. Chem. Phys.* **53**, 987.
Esaulov, V., Dhuicq, D., and Gauyacq, J. P. (1978). *J. Phys. B* **11**, 1049.
Fayeton, J., Dhuicq, and Barat, M. (1978). *J. Phys. B* **11**, 1267.
Franklin, J. L., and Harland, P. W. (1974). *Annu. Rev. Phys. Chem.* **25**, 485.
Gauyacq, J. P. (1979). *J. Phys. B* **13**, L387.
Gauyacq, J. P. (1980a). *J. Phys. B* **13**, 4417.
Gauyacq, J. P. (1980b). *J. Phys. B* **13**, L501.
Geddes, J., Hill, J., Sha, M. B., Goffe, T. V., and Gilbody, H. B. (1980). *J. Phys. B* **13**, 319.
Goursaud, S., Sizun, M., and Fiquet-Fayard, F. (1976). *J. Chem. Phys.* **65**, 5453.
Goursaud, S., Sizun, M., and Fiquet-Fayard, F. (1978). *J. Chem. Phys.* **68**, 4310.
Hasted, J. B. (1952). *Proc. R. Soc. London A* **212**, 235.
Hasted, J. B., and Smith, R. A. (1956). *Proc. R. Soc. London A* **235**, 349.
Haywood, S. E., Bowen, D. J., Champion, R. L., and Doverspike, L. D. (1981a). *J. Phys. B* **14**, 261.
Haywood, S. E., Doverspike, L. D., Champion, R. L., Herbst, E., Annis, B. K., and Datz, S. (1981b). *J. Chem. Phys.* **74**, 2845.
Herbst, E., Payne, L. G., Champion, R. L., and Doverspike, L. D. (1979). *Chem. Phys.* **42**, 413.
Herzenberg, A., and Ojha, P. (1979). *Phys. Rev. A* **20**, 1905.
Hotop, H., and Lineberger, W. C. (1975). *J. Phys. Chem. Ref. Data* **4**, 539.
Hummer, D. G., Stebbings, R. F., and Fite, W. L. (1960). *Phys. Rev.* **119**, 668.
Huq, M. S., Doverspike, L. D., Champion, R. L., and Esaulov, V. A. (1982a). *J. Phys. B* **15**, 951.
Huq, M. S., Fraedrich, D. S., Doverspike, L. D., and Champion, R. L. (1982b). *J. Chem. Phys.* **76**, 4952.
Huq, M. S., Doverspike, L. D., and Champion, R. L., (1983a). *Phys. Rev. A* **27**, 2831.
Huq, M. S., Doverspike, L. D., and Champion, R. L. (1983b). *Phys. Rev. A* **27**, 785.
Janousek, B. K., and Brauman, J. I. (1979). *Gas Phase Ion Chem.* **2**, 53–86.
Lam, S. K., Delos, J. B., Champion, R. L., and Doverspike, L. D. (1974). *Phys. Rev. A* **9**, 1828.
Lifshitz, C., Wu, R. L. C., Tiernan, T. O., and Terwillinger, D. T. (1978). *J. Chem. Phys.* **68**, 247.
Lopantseva, G. B., and Firsov, O. B. (1966). *Sov. Phys. JETP* **23**, 648.
McFarland, M., Dunkin, D. B., Fehsenfeld, F. C., Schmeltekopf, A. L., and Ferguson, E. E. (1972). *J. Chem. Phys.* **56**, 2358.
Mandl, A. (1976a). *J. Chem. Phys.* **64**, 903.
Mandl, A. (1976b). *J. Chem. Phys.* **65**, 2484.
Mandl, A., Kivel, B., and Evans, E. W. (1970). *J. Chem. Phys.* **53**, 2363.
Mason, E. A., and Vanderslice, I. E. (1958). *J. Chem. Phys.* **28**, 253.
Massey, H. S. W. (1976). "Negative Ions," 3rd ed. Cambridge Univ. Press.
Melius, C. F., and Goddard, W. A. (1974). *Phys. Rev. A* **10**, 1541.
Meyer, F. W. (1980). *J. Phys. B* **13**, 3823.
Michels, H. H., and Paulson, J. F. (1981). "Potential Energy Surfaces and Dynamics Calculations" (D. G. Truhlar, ed.), p. 535 ff. Plenum, New York.
Miller, W. H. (1970). *J. Chem. Phys.* **52**, 3563.
Morrison, R. T., and Boyd, R. N. (1973). "Organic Chemistry." Allyn and Bacon, Rockleigh, New Jersey.
Muckerman, J. T. (1972). *J. Chem. Phys.* **56**, 2997; *J. Chem. Phys.* **57**, 3388.
Olson, R. E., and Liu, B. (1978). *Phys. Rev. A* **17**, 1568.
Olson, R. E., and Liu, B. (1979). *Phys. Rev.* **20**, 1344.
Olson, R. E., and Liu, B. (1980). *Phys. Rev. A* **22**, 1389.

Paulson, J. F. (1970). *J. Chem. Phys.* **52**, 959.
Preston, R. K., and Cohen, J. S. (1976). *J. Chem. Phys.* **65**, 1589.
Risley, J. S. (1973). Electron detachment in collisions of H$^-$ with atomic and molecular targets, Ph.D. Thesis, University of Washington, Seattle. (Univ. Microfilm, Ann Arbor Mich.)
Risley, J. S. (1974). *Phys. Rev. A* **10**, 731.
Risley, J. S. (1977). *Phys. Rev. A* **16**, 2346.
Risley, J. S. (1980). *Electron. At. Collisions, Proc. Int. Conf. 11th*, p. 619.
Risley, J. S., and Geballe, R. (1974). *Phys. Rev. A* **9**, 2485.
Risley, J. S., Edwards, A. K., and Geballe, R. (1974). *Phys. Rev. A* **9**, 115.
Risley, J. S., deHeer, F. J., and Kerkdijk, C. D. (1978). *J. Phys. B* **11**, 1783.
Roche, A. E., and Goodyear, C. C. (1969). *J. Phys. B* **2**, 191.
Rulis, A. M., and Bernstein, R. B. (1972). *J. Chem. Phys.* **57**, 5497.
Schultz, G. J. (1973). *Rev. Mod. Phys.* **45**, 423.
Shui, V. H., and Keck, J. C. (1973). *J. Chem. Phys.* **59**, 5242.
Smirnov, B. M. (1982). "Negative Ions." McGraw-Hill, New York.
Smith, F. T., Marchi, R. P., Aberth, W., Lorents, D. C., and Heinz, O. (1967). *Phys. Rev.* **161**, 31.
Smith, B. T., Edwards, W. R., Doverspike, L. D., and Champion, R. L. (1978). *Phys. Rev. A* **18**, 945.
Taylor, R., and Delos, J. B. (1982). *Proc. R. Soc. A* **379**, 179, 209.
Tiernan, T. O., Hughes, B. M., and Lifshitz, C. (1971). *J. Chem. Phys.* **55**, 5692.
Tuan, V. N., and Esaulov, V. A. (1982a). *J. Phys. B* **15**, L95.
Tuan, V. N., and Esaulov, V. A. (1982b). *Intl. Symp. Phys. of Ionized Gases*, Dubrovnik.
Tuan, V. N., Gauyacq, J. P., and Herzenberg, A. (1982). *Intl. Symp. Phys. of Ionized Gases*, Dubrovnik.
Vedder, M., Doverspike, L. D., and Champion, R. L. (1981). *Phys. Rev. A* **24** 615.
Vogt, D., and Opiela, K. H. (1975). *Phys Lett. A* **54**, 331.
Vogt, D., Mischke, J., and Dreves, W. (1976). *J. Phys. E* **9**, 38.
Wang, T. S., and Delos, J. B. (1983). *J. Chem. Phys.* **79**, 4306.
Wijnaendts van Resandt, R. W., deVreugd, C., Champion, R. L., and Los, J. (1978). *Chem. Phys.* **29**, 151.
Wu, T. Y., and Ohmura, T. (1962). "Quantum Theory of Scattering", p. 234. Prentice-Hall, Englewood Cliffs, New Jersey.
Wynn, M. J., Martin, J. D., and Bailey, T. L. (1970). *J. Chem. Phys.* **52**, 191.

Index

A

Absorption potential, 72
Activation energy, 299
Adiabatic nuclei approximation, 46, 50, 73
Airglow, 346, 347, 363, 367, 371, 377, 381, 382, 391
Alcohols, negative ion resonances of, 452, 453
Aldehydes, negative ion resonances of, 457, 460, 461
Alkali atom (beams), 292, 493–495, 595–601
Alkanes
    mass spectra of, 264, 271
    negative ion resonances of, 454–458
Angular distribution
    coefficients, 414
    correlations, 191, 309, 412
    of scattered electrons, *see* specific molecules
    via resonance states, 412
Appearance potential, 275, 293, 296–299, 312, 313
Ar
    as buffer gas, 498, 499
        mean energy versus E/N, 499
    collisions of, with UF_6^-, 675, 676
    detachment cross section (total) in collisions of, with Br^-, Cl^-, I^-, 635, 636
    differential excitation cross section of, 198
    double ionization cross section of, 299
    momentum transfer cross section of, 124
ArCl, $ArCl^-$, internuclear potential of, 634, 635
Associative detachment, *see* Electron detachment
Associative ionization, 354

Atmospheric processes, 226–231
Auger decay, 257
Auger electrons, 257
Auger transitions, 422
Aurora, 127, 132, 135, 194, 230, 231, 371, 373, 377
Aurora boreallis, 347, 354, 375
Aurora processes, 157, 194, 373, 377
Aurora substorms, 336, 341, 362, 372
Autocorrelation diagram, 257
Autocorrelation spectrum, 258
Autodetachment, 404, 468, 569, 638, *see also* Negative ions; Negative ion states; Negative ion resonances
Autoionization, 255, 256, 296, 309, 339, 404, 421
    of doubly excited states, 421
    of negative ions, *see* Negative ions
    of neutrals, *see* Positive ions
    via bound states, 255

B

Balmer-alpha emission, 340, 342, 362
BCl_3, thermal attachment rate constant of, 554
Benzene
    negative ion states of, 460–464
    transmission spectrum of, 462, 467
Bethe–Born approximation, 157, 279–282
Beulter–Fano parameter, 407
Beulter–Fano profile, 190, 193, 407, 421
BF_3
    ionization of, by electrons, 304
    thermal electron attachment rate constant, 554

Bloch–Bradbury mechanism, 578, 582, 586, 587
Bond dissociation energies, 600
Bonham–Ochkur–Rudge approximation, 83, 84
"Boomerang" model, 410, 449
Born approximation, 42, 44, 45, 47, 60–62, 80, 81, 84–89, 102, 119, 165, 195, 197, 200, 253, 279, 354
Born operator, 165
Born–Oppenheimer approximation, 65, 66, 82, 83, 166, 197, 416, 482, 488, 504, 631
Born-type approximations, 81, 84, 197, 200
Bound-electron capture, 493–495, 595–604
Br_2
 dissociative electron attachment of, 518–522
  cross section, 521
  thermal rate constant, 518
  three-body rate constant, 588
 ion-pair formation of, 593, 594
 ionization of, by electrons, 303
 negative ion states of, 448, 520–522
Branching ratio, 339, 347, 368, 374, 378, 379, 382–384, 422, 599, 677
 technique, 338, 349
Brominated alkanes, dissociative electron attachment, 540, 541
Breit–Wigner factor, 412
Breit–Wigner formula, 410, 581
Breit–Wigner resonance amplitude, 651

C

Cameron bands, 351, 395
Cascade excitations, 388
Cascade transitions, 385, 386
1-C_3F_6, electron attachment of, 589–592
C_2H_2
 close-coupling electron scattering calculation of, 92
 elastic electron scattering of, 34
C_2H_4, C_2D_4
 dissociation of, by electrons, 396
 elastic electron scattering of, 34
 excitation by electrons
  total scattering, 188
  vibrational, 179
 negative ion states of, 456, 457
C_2H_6, C_3H_8
 dissociation of, by electrons, 396
 elastic electron scattering of, 34
 negative ion states of, 457
$CClF_3$
 dissociation of, by electrons, 396
 dissociative electron attachment of, 540–542
  thermal rate constant, 544
 excitation of, by electrons, total scattering, 188
CCl_2F_2
 dissociation of, by electrons, 396
 dissociative electron attachment of, 540–542
  cross section, 542
  rate constant versus mean energy, 542
  thermal rate constant, 544
 elastic electron scattering of, 35
 excitation of, by electrons
  total scattering, 188
  vibrational, 180
 ionization of, by electrons, 304
 negative ions in collisions with alkali atoms, 599, 600
CCl_3F
 dissociation of, by electrons, 396
 dissociative electron attachment of, 540–542
  cross section, 542
  rate constant versus mean energy, 542
  thermal rate constant, 544
 elastic electron scattering of, 35
 excitation of, by electrons
  total scattering, 188
  vibrational excitation, 180
 negative ions in collision
  with alkali atoms, 599, 600
  with Rydberg atoms, 602–604
CCl_4
 collisions with Rydberg atoms, 602–604
 dissociative electron attachment of, 540–542
  rate constant versus mean energy, 542
  thermal rate constant, 544
 elastic electron scattering of, 35
 excitation of, by electrons, total scattering, 188
 ionization of, by electrons, 304
  total cross section, 305
Centrifugal barrier, 274, 416, 456
Centrifugal potential, 405
CF_4
 dissociation of, by electrons, 396
 dissociative electron attachment of, rate constant, 544, 552
 electron scattering of, total cross section, 188
 ionization of, by electrons, 304

Index   685

total cross section, 305
CH_4
  dissociation of, by electrons, 396
  dissociative electron attachment of
    cross section, 506, 555
    isotope effect, 506, 508, 555
  electron detachment, cross section for Br^-, Cl^-, 641
  electron scattering of
    calculations, 92, 98
    elastic, 34
      differential cross section, 37, 38, 115
      integral cross section, 40
  excitation of, by electrons, cross section
    for electronic, 186
    for total scattering, 188
    for vibrational, 179
  ionization of, by electrons, 304
    cross section
      single differential, 307
      total, 305
  negative ion states of, 454–457
  transmission spectrum of, 455
CH_3X (X = Br, Cl)
  dissociative electron attachment of
    rate constant versus mean energy, 541
    thermal rate constant, 544
  electron detachment, cross section in collision with I^-, 672
$CHCl_3$
  dissociative electron attachment of
    rate constant versus mean energy, 542
    thermal rate constant, 544
  ionization of, by electrons, 304
    cross section, 305
  negative ions of, in collision with alkali atoms, 599
Charge transfer
  Cl^- and F^- on O_2, 646
  electron detachment via, 621
  neutralization technique, 295, 296
Chloroethylenes, electron attachment of, 538, 539
Chlorofluoroethanes, electron attachment rate constant of, 543, 544
Chlorohydroethanes, electron attachment rate constant of, 543, 544
Clamped nuclei approximation, 49
Clastogram, 261, 262
Cl_2
  dissociative electron attachment of, 518–522
    cross section, 521

thermal rate constant, 518
ion-pair formation of, 593, 594
ionization of, by electrons, cross section, 303
negative ion states of, 520–522
  Feshbach resonances, 448
Cl_2O, thermal electron attachment rate constant of, 536
ClF_3, thermal electron attachment rate constant of, 554
ClX (X = O_2, NO), thermal electron attachment rate constant of, 536
Close-coupling calculation, 45, see also specific species
Close-coupling equations, 55–59, 63, 69
Close-coupling method, 201, 202
Cluster (molecular)
  electron attachment of, 493, 585–588, 592
  ionization of, by electrons, 305, 306
  multiply charged ions, 272
  positive ions, 272, 275
  van der Waals, 272, 305, 585–588, 592
CNX (X = Br, Cl, H, I), thermal electron attachment rate constant of, 536
CO
  coincidence spectrum of, 257
  dissociative electron attachment of, 516
    angular distribution of O^-, 516
  dissociative excitation of, by electrons, 351, 390–396, 440, 441
    Cameron bands, 351, 395
    cross section, 392
    emission cross section, 380
    rotational temperature, 395
    UV spectrum, 351
  elastic electron scattering
    calculations, 91, 94, 98, 99
    differential cross section, 30, 37, 38, 113
    integral cross section, 40, 114, 124
  electron detachment, cross section
    for Br^-, 641
    for Cl^-, 641, 645
    for H^- (D^-), 651, 652
  electron energy-loss spectrum, 256
  excitation of, by electrons, cross section
    for electronic, 184, 185, 210, 225
    for rotational, 36
    for total, 188
    for vibrational, 176, 177, 394, 442
  ionization of, by electrons, cross section, 301, 303
  negative ion resonances of, 422, 441–443, 516

CO_2
    dissociative electron attachment of, 527–532
        cross section, 528, 531
        effect of temperature on, 528, 531, 566
        kinetic energy of O^-, 530, 531
        potential energy curve for CO_2^-, 529
        vibrational excitation of fragments, 529
    dissociative excitation of, by electrons, 351, 390–396
        cross section for OI multiplet, 392, 394
        Doppler width of OI transition, 393
        emission cross section, 380
        total fragment kinetic energy, 393
    electron detachment, cross section
        for Br^-, Cl^-, 641
        for H^- (D^-), 652–654
    electron scattering
        calculations, 92, 98
        elastic cross section
            differential, 37, 38
            integral, 40
            integral momentum, 124
        excitation of, by electrons
            electronic, 211
            total scattering, 188, 189, 449
            vibrational, 178, 196, 205, 206
        ionization of, by electrons, 303
        negative ion resonances of, 422, 449, 450, 527–531
        negative ions of, in collision with alkali atoms, 596
Coefficient functions, 45, 51
Coincidence measurements, 191
Coincidence spectrum, 257
Coincidence techniques, 190, 256, 278
Collision complex, 53, 67
Collision cross section, 120, 158
Collision frequency, 120, 127, 132
Collisional detachment, *see* Electron detachment
Complex potential model, 625, 627, 657, 665
Convolution method, 294
Core-ion model, 342–344, 346, 371, 397
Correlation
    angular, 191, 260, 309
    auto, 257, 258
    diagram, 257, 459
    electron–electron, 258–260
    radial, 260
    rules, 255, 267
    studies of positive ions, 266
Coulomb explosion, 271

Coulomb potential, 404
Cracking patterns, 266
Cross section, definitions
    collisional detachment, *see* Electron detachment
    ionization
        counting cross section, 281
        double differential, 280, 306, 308
        macroscopic, 288
        partial, 281
        single differential, 280, 306, 307
        total, 281
        triple differential, 306, 308
    methods of normalization, 24–27
        by relative flow technique, 25, 26
        by theory in Born limit, 25
        by total scattering cross section, 26, 27
    scattering
        absolute elastic, 170
        Born differential, 165, 280
        differential excitation, 2, 3, 89, 159, 160, 168
        double differential, 16, 159
        elastic differential, 3
        inelastic momentum transfer, 168
        integral, 2, 27, 122, 160
        integral elastic, 3, 122, 169
        integral excitation, 2, 89, 167
        integral resonant, 410
        measured differential, 3, 159
        momentum transfer, 2, 3, 89, 121, 122, 160
        partial, 407
        reduced differential, 80, 81, 117, 118
        reduced momentum transfer, 81
        total, 3, 27, 160, 181, 187–189
        total excitation, 166, 169
        uncertainties, 170, 171
Crossed-beam techniques, 595
CS_2
    dissociative electron attachment of, 532, 533
    negative ions of, in collision with alkali atoms, 596
    negative ion states of, 532, 533
CsX (X = Cl, F)
    differential cross section of, for rotational excitation (CsF), 80
    elastic electron scattering of, 31, 32
    electron scattering calculations of, 91, 94, 96–98
    total electron scattering cross section of, 187, 188

Index 687

Curve crossing, 634

D

Deflection method, 311
Degrees of freedom
  active, 539
  vibrational, 66, 69, 256, 277, 539, 572, 575
Density of states, 575
Derivative technique, 294, 426
Detailed balance, 574, 575
Dielectronic recombination, 404
Dissociation
  of molecules, see Dissociative excitation
  of negative ions, see Dissociative attachment
  of positive ions, see Positive ions
  resonance-assisted, 421
  by tunneling, 273
Dissociation barrier, 274
Dissociative electron attachment, see also Electron attachment; Negative ions; specific molecules
  angular distribution of products of, 501, 502, 509, 513, 522–527
  data
    for Br_2, 518–522
    for Cl_2, 518–522
    for CO, 516
    for D_2, 506, 509–511
    for F_2, 518–522
    for HD, 509–511
    for H_2, 506, 509–511
    for I_2, 518–522
    for N_2, 515
    for NO, 516, 517, 519
    for O_2, 515, 516
    for hydrogen halides, 506, 511–515
    for CO_2, 527–531
    for CS_2, 532, 533
    for D_2O, 506, 522–526
    for ethers, 555, 556
    for halocarbons, 538, 540–544
    for H_2O, 506, 522–526
    for H_2S, 506, 526, 527
    for N_2O, 533, 534
    for NO_2, 534
    for OCS, 532, 533
    for SO_2, 534–536
    for CD_4, 506, 555
    for CH_4, 506, 555
    for ND_3, 506, 555

    for NF_3, 555
    for NH_3, 506, 555
    for perfluorocarbons, 537, 546, 550–552
    for other polyatomics, 557, 558
    for SF_6, 552–554
    for thioethers, 555, 556
  dependence of, on resonance energy, 555–559
  effects of temperature on, 491, 558–568
  data
    for D_2, 561–563
    for DCl, 563–565
    for halocarbons, 567, 568
    for HCl, 563–565
    for HF, 563–565
    for H_2, 561–563
    for O_2, 560, 561
    for CO_2, 528, 531, 566
    for N_2O, 565, 566
    for SF_6, 566–568, 589
  to electronically excited molecules, 568, 569
    H_2, 569
    O_2, 568
  high-resolution studies of, 501–503
  identification of products of, 500
  isotope effects of, 505–511, 527
  kinetic energy of fragment anions of, 490, 501, 522, 524–531
  spin–orbit structure, 513, 514
  theoretical expression, 504
  vertical onset, 489, 528, 556
Dissociative excitation of molecules, by electrons, 335
  core-ion model of, 342–346, 371, 397
  data
    for CO, CO_2, 390–396
    for N_2, 344, 365–376
    for O_2, 376–390
  emission cross sections, 369, 370, 380
  experimental techniques, 346–363
  kinetic energy of fragments (atoms), 339–346, 356, 357, 374, 386, 387, 392, 393
  kinetic energy spectra of ions, 342–345, 357
  models, 338–346
  repulsive energy curves, 358, 359
Dissociative ionization, 257, 258, 277, 301, 312–314, 349, 350, 366, 397, see also Ionization
  recombination, 336, 391

Distorted wave approximation, 46, 59, 201
Doppler broadening, 16, 336, 340, 362, 392, 428, 435
　linewidth, 336, 340, 362, 363, 377, 386, 392, 393
Double escape, 260
Double hemispherical analyzer, 294

E

Earth's atmosphere, 128, 132, 362
Effective scattering
　path length, 16, 20, 169
　volume, 20, 27, 169
Eikonal approximation, 86–88, 201
Eikonal phase shift, 86
Electron affinity, definition, 484–486, 600, 601, 620
　adiabatic, 446
　vertical, 620
Electron attachment, *see also* Dissociative electron attachment; Negative ions; Negative ion states
　buffer gases for, 498
　cross section, *see also* specific molecules
　　maximum, 547
　to dipolar molecules, 492
　dissociative, 487–491, 504
　effect of medium on, 577–588
　effect of temperature on, 580–582, 588–592
　to electronically excited molecules, 491, 568, 569
　frequency, 497
　to ground-state molecules, 487–491
　to "hot" molecules, 491, 558–568
　to molecular clusters, 493, 585, 592
　nondissociative, 487–490, 569–592
　nonresonance, 491, 492
　potential energy diagrams, 488–490
　radiative, 592, 593
　rate coefficient for, 497
　rate constant for, 497, 581
　　maximum, 547, 558, 559
　　monoenergetic, 500
　　thermal values, 497, 500
　resonance processes, 479, 486–491
　techniques for, 495–504
　three-body rate constant for, 579
　vertical onset behavior of cross section, 489, 528

Electon beam technique, 424, 496
Electron capture, *see* Negative ions
　rate constant, 581
Electron collimator, 9
Electron collisions in laser fields, 191, 192
Electron detachment, collisional, 619
　accompanied by rearrangement, 656, 657
　data
　　for $Cl^- + H_2$, HD, D_2, 666–671
　　for $F^- + H_2$, HD, D_2, 660–666
　　for $H^- (D^-) + H_2(D_2)$, 658–660
　　for $I^- +$ methyl halides, 670, 671
　associative, 621, 639, 640, 648, 655
　attenuation techniques, 622, 623
　autodetaching levels, 621, 638, 665
　with bonding, 621
　via charge transfer, 621, 646, 650, 655, 656
　　through resonances, 621, 639, 643, 645, 651, 656
　complex potential model of, 625, 657, 665
　cross sections, *see* specific molecules
　　angular differential, 622
　　collisional, 625
　　differential, 622, 636, 643, 645
　　direct, 639, 646, 647
　　doubly differential, 622, 636, 637
　　elastic differential, 622, 634, 635
　　with excitation, 620, 621, 643, 669
　　reactive, 639, 640, 665
　　total, 621, 622, 626, 632, 634, 635, 641, 642, 647–652
　direct, 620, 621
　dual isotope effect, 647
　via dynamic coupling, 664
　energy spectra of detached electrons, 630, 637, 638, 649, 650, 653
　electron-trap techniques, 623
　with excitation, 620, 621
　high-energy theory, 631
　impulse approximation, 631
　isotope effects, 626, 627, 632–634, 647, 648, 651, 652, 656–659, 662–664, 667, 668
　probability, 625, 627, 628, 634
　quasibound resonance, 624
　rate constants, 677, 678
　reactive scattering, 621, 639, 640, 657, 665
　rotational coupling, 636
　statistical model, 677
　threshold region, 624, 634, 666
　time of flight, 623, 644, 645

total cross section
  for Br^-, Cl^-, I^- + Ar, 634, 635
  for Br^-, Cl^- + N_2, O_2, CO, CO_2, CH_4, 641
  for F^- + He, 635
  for F^- + molecule, 642
  for H^- (D^-) + He, Ne, 623, 624, 626, 632
  for H^- (D^-) + N_2, O_2, CO, CO_2, 647–649, 651–654, 656
  for O_2^- + O_2, 673, 674
  for UF_6^- + Ar, Xe, N_2, SF_6, 675, 676
  zero-radius model, 626–628
Electron–electron
  correlation, 258–260
  spectroscopy, 306–310
Electron energy analyzer, 14
Electron energy calibration, 23, 435
Electron energy distribution, 7, 161
  difference technique, 294
  functions, 424, 497, 498
  loss spectroscopy, 21
  loss spectrum, 21, 161, 258, see also individual species
  resolution, 10, 11, 172, 429, 432
  selectors, 6, 9–14, 23, 293
Electron escape process, 260
Electron exchange, 163, 193, 280
  polarization, 46, 77
Electron impact
  excitation, see individual molecules
  mass spectrometers, 501, 502
  spectrometers, 5, 6, 21, 425–432, 501–503
  spectroscopy, 160–166
Electron monochromator, 5, 6, 293
Electron scattering, elastic
  ab initio treatment, 63, 71
  complete lab-frame treatment, 65, 71
  cross sections, see Cross section
  isotope effects, 434
  by laser-excited species, 192
  by rigid rotor, 55, 67
  by vibrating rotor, 62, 69
Electron source, 6–9
Electron superelastic scattering, 192
Electron swarm technique, 121, 424, 497
Electron transmission
  function, 422
  spectroscopy, 166, 423–428
  spectrum, see individual molecules
Electron trap techniques, 623

Electronegativity, 465
Electronic excitation by electrons, 174, 181, 182–189, 197–202, 207–226, see also specific molecules
  data
    on CO, 184, 185, 210, 225
    on F_2, 211, 226
    on H_2, 182, 183, 207–209, 212–217
    on N_2, 183, 184, 198, 209, 210, 217–224
    on NO, 185
    on O_2, 185, 211, 225
    on CO_2, 211
    on SO_2, 186
    on CH_4, 186
Electrophilicity, 422
Electrostatic coupling, 629
Electrostatic effects, 63, 71
Electrostatic energy analysis, 311
Electrostatic lense, 9, 428
Electrostatic selectors, 9, 11, 294
Electrostatic trap, 623
Emission cross sections, 295
Escape model, 260
Ethers, dissociative electron attachment, 555, 556
Exchange distorted wave approximation, 65
Exchange effects, 63, 71
Exchange polarization, 77
Exchange potential, 449
Exchange scattering, 309
Excimer, 195, 236–240
Exciplex, 236–240

F

Fano plot, 281, 282
Feshbach resonance, see Negative ion resonances
F_2
  dissociative electron attachment of, 518–522
    cross section, 521
    rate constant versus mean energy, 519
    thermal rate constant, 518, 519
  electron scattering calculations of, 98
  excitation of, by electrons, electronic, 211, 226
  ionization of, by electrons, 303
  negative ion states of, 448, 520–522
    Feshbach resonances, 448
Fixed nuclei approximation, 41, 42, 50

Fluorescence technique, 295
Fourier analysis, 450
Fourier transform, 162, 349
Fractional energy loss per collision, 122, 195
Fragment ions, 265–268, 536–538
   angular distribution of, 306, 313, 314
   even electron, 265, 268
   kinetic energy of, 310–314
   multiply charged, 268–272
   odd electron, 265, 268
   primary, 265
   secondary, 265
Fragmentation patterns
   for negative ions, 536–540
   for positive ions, 262, 265–268
Frame transformation, 53, 67
   theory, 78–81
Franck–Condon double overlap, 441
Franck–Condon factor, 166, 297, 393, 408, 409, 434, 446
Franck–Condon overlap, 253, 409, 421, 441, 504
Franck–Condon principle, 252, 253, 266, 311, 312, 339, 488
Franck–Condon region, 276, 311, 339, 358, 489
Franck–Condon transitions, 277, 504
Free-electron model, 495, 604
Functional groups, 266, 570

G

Gas breakdown, 233–235
Gas discharge, 194, 231, 233, 336, 339, 346, 365, 366, 371, 390
Gas lasers, 137, 157, 171, 233, 236–241, 337, 404
Glauber approximation, 81, 86–88, 119, 201

H

Halocarbons, *see also* Perfluorocarbons; specific molecules
   dissociation of, by electrons, 396
   electron attachment of, 540–552
      dissociative, 538
         cross section, 541, 542
         effect of temperature, 567, 568
         rate constants, 540–552
      nondissociative, 537–539, 545–552

ionization of, by electrons, 304, 305
negative ion states of, 458–460, 538, 540–543
negative ions in collision
   with alkali atoms, 599–601
   with Rydberg atoms, 602–604
HBr (DBr)
   dissociation of, by electrons, 396
   dissociative electron attachment of, 511–514
      angular distribution of Br^-, 513
      cross section, 506, 512
      isotope effect, 506
      thermal rate constant, 514
   elastic electron scattering of, 31
   excitation by electrons, vibrational, 177
   negative ion states, 511–514
HCl (DCl)
   dissociation of, by electrons, 396
   dissociative electron attachment of, 511–515
      cross section, 506, 512
      effect of temperature on, 563–565
      isotope effect on, 506, 511
      thermal rate constant, 514
   elastic differential cross section of, 116
   electron scattering calculations of, 91, 94, 98
   excitation of, by electrons, vibrational, 177, 417–419
   negative ion states of, 447, 448, 511–515
      dipole resonances, 447, 514
      Feshbach resonances, 448
   negative ions in collision with alkali atoms, 598
   potential energy curves of, 419
HCN
   elastic electron scattering of, 32
   integral elastic scattering cross section of, 40
   thermal attachment rate constant of, 536
He
   differential elastic scattering cross section of, 38
   electron detachment, in collision
      with F^-, 635
      with H^- (D^-), 623–631, 637
      with O_2^-, 673, 674
   negative ion resonance of, 22
He_2^{2+}, potential energy diagram of, 269
HeH; HeH^-, potential energy curves of, 624
Hemispherical analyzers, 13, 14, 428
HF (DF)

close coupling calculation of, for elastic
    scattering, 92
dissociative electron attachment of, 511, 513
    effect of temperature on, 563–565
elastic electron scattering of, 30
excitation of, by electrons, cross section for
    vibrational, 177
negative ion states of, 447, 448
    dipole resonances, 447
    Feshbach resonances, 447, 448
$HgBr_2$
    electron attachment of, 536
    ionization of, by electrons, 304
H_2^+
    excitation of, by electrons, 208
    ionization potential of, 253
    potential energy curves of, 253, 254
H_2 (HD, D_2)
    dissociation of, by electrons, 342, 352, 353,
        396, 397
    dissociative electron attachment of
        cross section, 506, 509, 510
        effect of temperature on, 561–563
        to electronically excited H_2, 569
        isotope effect on, 506, 508–510
    electron detachment, in collisions with H^-
        (D^-), Cl^-, F^-, 658–671
    electron energy-loss spectrum of, 22
    electron scattering (elastic)
        calculations, 90, 92–95, 97, 99
        differential cross section, 37, 38, 51, 104,
            105, 108
        integral, 40, 52, 107
        momentum transfer, 124
    electron thermalization in, 125, 126, 232
    excitation of, by electrons
        cross section
            differential, 203
            integral, 187, 204
            vibrational, 175, 203–205
        electronic
            differential, 182, 183, 208, 209,
                212–217
            integral, 182, 183, 208, 209, 213–217
        rotational
            differential, 52, 106
            integral, 60, 103, 107
        total scattering, 103, 187
    ionization of, by electrons, 300
        angular distribution of H^+, 314
        cross section (partial) for H_2, 264

kinetic energy distribution of H^+, 312
ion exchange, in collision with H^-, D^-,
    659, 660
ion production, in collision with Cl^-, F^-,
    660–670
negative ion states of, resonances, 432–434,
    482, 483, 509–511
potential energy curves
    of H_2, H_2^-, 433
    of H_2^{2+}, H_2^+, 254
VUV emission in electron impact excitation,
    352
H_3^-, 639, 659
H_2O (D_2O)
    dissociation of, by electrons, 396
    dissociative electron attachment of, 452,
        522–526
        angular distribution of products, 522–526
        cross section, 506, 523
        energy distribution of products, 522–525
        isotope effect on, 506
        vibrational excitation of products, 524,
            525
    elastic scattering calculations of, 93, 96, 98
    excitation of H_2O, by electrons
        cross section
            for rotational, 36
            for total scattering, 189
            for vibrational, 187
    ionization of, by electrons, 303
    momentum transfer cross section of, 124
    negative ion states of, 452, 453, 523–526
    negative ions, in collision of H_2O, with
        alkali atoms, 598, 599
    potential energy curves of H_2O^-, 452
$H_2S(D_2S)$
    dissociation of, by electrons, 396
    dissociative electron attachment of, 526, 527
        angular distribution of products, 527
        cross section, 506, 526
        energy distribution of products, 527
        isotope effect on, 506, 527
    elastic electron scattering, 32
    excitation by electrons
        total scattering, 189
        vibrational, 178
    negative ion states of, 452, 453, 526, 527
    negative ions in collision with alkali atoms,
        598, 599
    potential energy diagram of H_2S^+, 274
HI (DI)

dissociative attachment
cross section, 506, 511
isotope effect, 506
long-lived anion of, 515
Hydrocarbons
aliphatic, negative ion resonances, 423, 454–457
alkenes, negative ion resonances, 423, 456–458
alternant, negative ion resonances, 458
aromatic
negative ion resonances, 423, 458, 460–465
parent anions, 570
heterocyclic, negative ion resonances, 464, 468–471
nitrobenzenes
autodetachment lifetimes, 569–573
negative ion states, 464
substituted benzenes
autodetachment lifetimes, 569–573
negative ion resonances, 464–468
resonance energies, 464, 468
Hydrogenic atoms, 493
Hydrogenic orbitals, 494
Hydrogenic plasmas, 569

I

I_2
dissociative electron attachment of, 518–522
thermal rate constant, 518
negative ion states of, 520–522
Feshbach resonances, 448
Impact parameter, 625
model, 197, 199
Impulse approximation, 631
Independent atom model, 84
Inelastič collision frequency, 235
Inelastic loss factor, 234, 235
Inner shell
electron, 190, 257
excitation, 190
ionization, *see* Ionization
resonances, *see* Negative ion resonances
Instrument(al) function, 19, 168
Internuclear potential, 624, 633–635
Ion exchange, 659, 660
Ion kinetic energy spectroscopy, 306, 310–313
Ionosphere, 127, 129, 130, 134, 135, 194, 226, 228, 230

Ion-pair formation, 276, 313, 491, 492, 593, 594
Ion-pair processes, 594
Ion-pair results
for Br_2, 593, 594
for Cl_2, 593, 594
Ions, *see* Negative ions; Positive ions
Ionization, by electrons
angular
correlation, 260, 309
distribution of fragments, 306, 314
ejected electrons, 307, 308
asymmetric kinematics, 308, 309
auto, *see* Autoionization
binary peak, 309
classical theories, 283
cross section data
for BF_3, 304
for Br_2, 303
for CH_4, 304, 305, 307
for Cl- and F-methanes, 304, 305
for clusters, 305, 306
for Cl_2, 303
for CO, 301, 303
for CO_2, 303
for D_2, 300
for F_2, 303
for H_2, 300
for H_2O, 303
for $HgBr_2$, 304
for N_2, 298, 300–302
for NH_3, 304
for NO, 301
for N_2O, 303
for NO_2, 299, 303
for O_2, 297, 301
for O_3, 303
for PH_3, 304
for SF_6, 305
for SO_2, 303
for SO_3, 304
for UF_6, 305
cross sections
additivity of atomic, 285
counting, 281
differential, 280
double differential, 306, 308
gross (total) 281
partial, 264, 281, 284, 301
single differential, 280, 306, 307
triple differential, 306, 308

direct, 255
dissociative, 257, 277, 301, 312–314
double, 257, 258, 282, 298, 299, 313
double potential, 269
efficiency, 270, 288
empirical formulas, 282–285
energy-loss spectrum, 307, 308
of excited molecules, 283, 295
indirect, see Autoionization
inner shell, 257, 258, 295
isotope effect in dissociative, 312
kinetic ion spectrum, 312
mechanisms, 252–260
momentum distribution, 256
multiple, 268–272, 286, 298
partial, 284, 286
positron impact, 287
processes, 252
quantal approximations, 283
recoil effects, 306
recoil peak, 309
semiclassical theories, 283
semiempirical theories, 283–285
stepwise, 295
symmetric kinematics, 309
techniques for, 287–296
threshold-electron impact, 258
threshold behavior, law, 286, 287, 296, 298, 299

K

K_2, total electron scattering cross section, 187
Ketones, negative ion resonances, 460
Kinematics
asymmetric, in ionization, 308, 309
of fragment ions, 310
symmetric, in ionization, 309
Kinetic shift, 299
KI, electron scattering
calculations, 91, 92, 97
differential elastic cross section, 117
total cross section, 188
Kr, differential elastic scattering cross section, 37, 38

L

Lasers, see Gas lasers
Legendre polynomials, 74, 412, 416

LiF, electron scattering calculations of, 91, 94
Li_2, electron scattering of, total, 189
Limit theorem, 161, 162, 164, 165
Line parameter, 407
Line shape, 386, 422
Line width, 337, 340, 362, 384
Lyman alpha, 352, 353

M

Magic angle, 309
Magnetic analyzer, 293
Magnetic field analysis, 311
Mass spectra, 258, 262–264, 266–268
Mass spectrometry, 290–293
atom-beam technique, 292
collection efficiency, 290
cycloidal, 293
deflection method, 292
discrimination effects, 290, 291
ion extraction efficiency, 293
magnetic, 501
Nier-type ion source, 291
quadrupole, 293
sector-type, 291
trapped-ion, 296
$M(CO)_n$ molecules (M = V, Cr, Fe, Ni, Mo, W), thermal attachment rate constant, 554
Mean free path, 120
Metal hexafluorides, 601
Metastable decomposition, 272
Metastable fragments, 356, 359, 360
Metastable ion
negative, see Negative ions
peak, 271, 272
positive, 272–275, 310, see also Positive ions
states, 255, 256
transition, 271, 273, 275
Molecular beam geometry, 15, 18, 19
Molecular flow, 18
Molecular target, 17–20
Mott–Möller formula, 284
Multiply charged ions, see Negative ions; Positive ions

N

N(atomic)
cross section for electron impact excitation, 377

partial term diagram, 368
Na$_2$, electron scattering, total cross section of, 187
Ne
  differential elastic scattering cross section, 38
  electron detachment in collision with H$^-$ (D$^-$), 624, 631–634
    total cross section, 632
  momentum transfer cross section, 124
Negative ions, *see also* Electron attachment; Negative ion resonances
  autodetachment lifetimes, 452, 479, 480, 482, 484, 486, 503, 569–577
    measurement of, 486, 503
  dipolar, 492
  doubly charged, 605, 606
  extremely short-lived, 452, 486, 509
  fragment, *see* Fragmentation patterns; Dissociative electron attachment
  long-lived, 486, 515, 537–539, 545–552, 569–577
  moderately short-lived, 486, 578, 586, 587
  modes of production, 479–495
  survival probability, 505, 508
  theoretical treatment of long-lived, 574–577
Negative ion resonances, 173, 174, 403, 405, 479–484, *see also* Negative ions; Negative ion states
  above first ionization limit, 421, 422
  angular distributions, 412–416, 434
  barrier, 405, 412, 480
  in collisional detachment, *see* Electron detachment
  core excited, 479, 480, 482
  core-excited shape, 174, 410, 479, 480, 483
  cross section, 406, 410
  data
    for aldehydes, 457, 460, 461
    for aliphatic hydrocarbons, 454–458, 538
    for alkenes, 456–458
    for aromatic hydrocarbons, 460–465
      heterocyclic compounds, 464, 460–471
      substituted benzenes, 464–468
    for C$_n$H$_{2n+1}$OH, 452, 453
    for CO, 422, 441–443, 516
    for CO$_2$, 422, 449, 450, 527–531
    for D$_2$, 432–434, 509–511
    for H$_2$, 432–434, 509–511
    for H$_2$O, 452, 453, 523–526
    for H$_2$S, 526, 527
    for halocarbons, 458, 459, 540–543
    for halogens, 448, 519–522
    for hydrogen halides, 447, 448, 511–514
    for ketones, 460
    for N$_2$, 411, 422, 435–441, 515
    for N$_2$O, 422, 450, 453
    for NO, 422, 442, 445–447, 517–519
    for NO$_2$, 450, 451, 534
    for O$_2$, 442, 444, 447
    for OCS, 450, 532, 533
    for perfluorocarbons, 458–460, 464, 467, 537, 545, 547, 550
    for SF$_6$, 453, 454
    for SO$_2$, 451, 534–536
  decay of, 405, 420
  dipole, 447
  energy, 406, 559
  experimental techniques, 424–432
  Feshbach, 404, 405, 435, 437, 441, 442, 446, 447, 479–483, 569, 570
  inner shell, 422, 447
  lifetimes, *see* Negative ions; Negative ion states
  orbitals, 413
  shape, 405, 410, 442, 479–482
  state radius, 410
  in theoretical chemistry, 422, 423
  theory of, 405–423, 479–484
  threshold behavior of, 416–420
  types, 479–484
  width of, 408, 410, 479, 504, 505
  window, 408
Negative ion states, *see also* Electron attachment; Negative ions; Negative ion resonances
  "grandparent," 405, 446
    model, 442, 446, 447
  lifetimes, 452, 468, 479–484, 486, 503–505, 509, 515, 545, 548, 570–577
  "parent," 405, 446
NeH, NeH$^-$, internuclear potential of, 633
NeKr, NeKr^{2+}, potential energy diagram of, 270
NF$_3$
  dissociative electron attachment of, 555
  thermal attachment rate constant of, 554
NH$_3$
  dissociation of, by electrons, 396
  dissociative electron attachment of, 506, 554, 555

Index 695

isotope effect on NH_3, ND_3, 506, 555
dissociative excitation of, 345, 346
electron scattering of, 33, 93, 96
ionization of, by electrons, 304

N_2
  as buffer gas in swarm studies, 498, 499
  dissociation of, by electrons, 339, 344, 349, 356, 357, 364–376
    cross section
      dissociative excitation, 369, 370, 376, 377
      dissociative ionization, 367
      emission, 369, 370, 374
      metastable decay, 357
      simultaneous excitation and ionization, 371
      total, 367, 375
    translational energy of N^+, 344
    translational energy of Rydberg N atoms, 344
  dissociative electron attachment of, 515
  dissociative ionization of, 349, 350, 366, 367
  elastic electron scattering
    differential cross section, 37–39, 109–112, 123
    integral cross section, 40, 109–112
    momentum transfer cross section, 123, 124
  electron detachment, cross section in collision
    with Br^-, 641
    with Cl^-, 641–643
    with F^-, 642
    with H^- (D^-), 647, 648
    with UF_6^-, 676
  electron energy-loss spectra of, 162, 163, 173
  electron scattering calculations of, 90, 92–94, 96–98
  excitation by electrons, cross sections
    differential, 183, 198, 206, 218–221
    electronic, 183, 184, 209, 210, 217–224
    integral, 112, 221–224
    rotational, 36, 131, 435, 437
    total, 188, 189, 436
    vibrational, 176, 206, 411, 438–441
  inelastic electron scattering of, by metastable N_2, 338
  ionization of, by electrons, cross section, 300, 301
    angular dependence, 313
    near threshold, 298
    partial, 302
  mean electron energy versus E/N, 499
  metastable, 356, 357
  negative ion resonances of, 39, 410, 411, 422, 435–441, 483, 515
    core excited, 410
    Feshbach, 435, 437
    inner-shell, 422
    shape, 39, 436, 438
  potential energy curves, 439

N_2O
  dissociative electron attachment of, 533, 534
    effect of temperature on, 533, 565, 566
    three-body rate constant, 588
  excitation of, by electrons
    elastic, 33
    total scattering, 189
  ionization of, by electrons, 303
  negative ion states of, 422, 450, 533
    inner-shell resonances, 422, 450
    shape resonances, 450

NO
  dissociation of, by electrons, 396
  dissociative electron attachment of, 516, 517, 519
    angular distribution of O^-, 517
    effect of temperature on, 588
    three-body rate constant of, 588
  elastic electron scattering of, 30
  excitation of, by electrons
    electronic, 185
    total scattering, 188
  ionization of, by electrons, 301
  negative ion states of, 422, 442, 445–447, 517, 519
    Feshbach resonances, 442, 446
    inner-shell resonances, 422, 447

NO_2
  dissociative electron attachment of, 534
    three-body rate constant, 588
  ionization of, by electrons, 299, 303
  negative ion states of, 450, 451, 534
  negative ions of, in collisions with alkali atoms, 596
  transmission spectrum of, 451

Nondissociative electron attachment, *see* Electron attachment; Negative ions; Negative ion resonances; Negative ion states

O

O(atomic)
  excitation of, by electrons, 384, 389
  formation of, in electron impact with O_2, 385
  term diagram of, 378, 383
  triplet transition probabilities of, 379
Ochkur–Rudge model, 197, 198, 212–224
OCS
  dissociation of, by electrons, 396
  dissociative electron attachment of, 532, 533
  electron scattering of, 188
  negative ion states of, 450, 532, 533
    shape resonances, 450
  negative ions of, in collision with alkali atoms, 596
Odd-nitrogen chemistry, 336, 354, 365, 372
O_2
  charge transfer, in collision
    with Cl^-, F^-, 646
    with H^- (D^-), 656
  dissociative electron attachment, 515
    angular distribution of O^-, 515
    effect of temperature on, 560, 561
    to electronically excited O_2, 568
    to ground-state O_2, 515
  dissociative excitation of, by electrons, 339, 340, 343, 376–390
    branching ratios, 379, 382, 383
    cross sections for excited atoms, 385–390
    emission cross section, 380
    fine structure levels, 382
    kinetic energy of fragments, 343, 386
  electron detachment, cross section
    for Br^-, 641
    for Cl^-, 641
    for H^- (D^-), 654
    for O_2^-, 673
  electron energy-loss spectrum of, 164, 229
  electron scattering
    calculations, 98, 99
    differential elastic cross section, 37, 38
    integral elastic cross section, 40
    momentum transfer cross section, 124
  excitation of, by electrons, cross section
    for electronic, 185, 211, 225
    for total scattering, 188
    for vibrational, 176·
  ionization of, by electrons
    cross section, 301
    threshold behavior, 297
  negative ion states of, 442, 444, 447, 515
  negative ions of, in collision with alkali atoms, 596, 597
  nondissociative electron attachment of, 578–588
    effect of temperature on, 580, 582, 587
    lifetime of O_2^{-*}, 585, 586
    potential energy diagrams of, 579
    rate constant of, 584
    three-body rate constant of, 579, 580, 583
    radiative attachment of, 592, 593
  Schumann–Runge continuum, 163, 230, 342, 381
O_3
  ionization of, by electrons, 303
  three-body rate constant of, 588
Optical absorption spectrum, 161, 162
Optical cross section measurements, 347 354
Optical excitation function, 167, 170
Optical mass spectrometer, 364, 376
Optical pumping effects, 377, 383
Optical selection rules, 163, 165
Orbital
  highest-occupied, 423
  lowest-unoccupied, 423
  π, 423
  Rydberg, 422, 446
  valence, 422
Oscillator strength
  apparent generalized, 165
  differential optical, 280, 282, 308
  generalized, 164, 169, 181, 190, 279, 280, 308, 373
  optical, 169, 190
  partial dipole, 309
Overlap integral, 253, 408

P

Pairing theorem, 458
Paraffins, mass spectra of, 263, 268
Partial wave
  cross section, 191
  formula, 406
  theory, 412
Penetration factor, 410
Perfluorocarbons, *see also* Halocarbons; specific species
  electron attachment of, 537, 543–552
    attachment rate constant versus mean energy, 547, 552

dissociative, 537, 546, 550–552
lifetimes of parent anions, 545, 546, 548
nondissociative, 537, 545, 552
thermal rate constant, 545, 548
negative ion states of, 458–460, 464–467, 537, 545–547, 550, 551
negative ions of, in collision with Rydberg atoms, 602–604
PH_3
clastogram, 262
ionization of, by electrons, 304
Phase shift, 405, 418, 419, 425
nonresonant, 406
resonance, 406
Photodissociative absorption, 336
Photoelectron spectroscopy, 423
Photon-simulation experiments, 309, 310
Photoion–photoelectron coincidence spectroscopy, 278
Pierce element, 7
Planetary atmospheres, 127, 135, 136, 157, 335–337, 339, 341, 349, 362, 365, 366, 390
Plasma, ionized, 2, 127, 231, 234
Polar molecules
close-coupling calculations of, for electron scattering, 45
differential elastic cross section of, 116, 117
momentum transfer cross section of, 39, 41
negative ion of, 492
states, 514
reduced differential scattering cross section of, 118
rotational excitation of, 45, 102, 181
threshold resonance behavior of, 416–420
vibrational excitation of, by electrons, 195–197
Polarization potential, 72, 78
adiabatic, 66, 77
local adiabatic, 46
local exchange, 78
nonadiabatic, 46, 66
Polarized orbital approximation, 77
Population inversion, 158, 236
Positive ions, *see also* Ionization by electrons
doubly charged, 268–272, 274, 312, 313
even, 261, 265, 268
excited, 252
fragment, *see also* Fragment ions, 265–268
angular distribution, 306, 309, 314
even-electron, 261, 265, 268

kinematics, 308–314
odd-electron, 265, 268
primary, 265
rearrangement, 266, 275, 276
secondary, 265
metastable, 272–275
multiply charged, 252, 268–272, 298, 313
odd, 261, 265, 268
parent, 261–265
state-selected, 278
triply charged, 272
Potential
centrifugal, 405, 416, 456, 480
local, 73, 84
local exchange, 54, 64, 75, 76, 78
muffin tin, 89
nonlocal exchange, 64, 84
optical, 45
Potential barrier, 405, 429, 674
Potential energy curves, diagrams, *see* specific species
Predissociation, 256, 273, 274, 339, 374, 375
Profile parameter, 407
Pseudo-Jahn–Teller systems, 465

Q

Quantum defect, 448
Quasi-equilibrium theory, 255, 267, 274, 277–279
degrees of freedom in, 277
rate expression, 278

R

Radiative attachment, 592, 593
Radiative entrapment processes, 340, 362, 366, 377, 383
Radiative lifetime, 339, 347, 356, 357
Ramsauer–Townsend minimum, 454, 455
Rare gas monohalides, 236–241
dimers of, 270
Reaction zone, 260, 286
Recoil effects, 306, 361
Recoil method, 166, 167
Recoil peak, 309
Recombination, dielectronic, 404
dissociative, 336
Reflection coefficient, 417

Relief plot, 259, 260
Rener–Teller effect, 531
Resonance-assisted dissociation, 421
Resonance electron attachment, see Electron attachment; Negative ions; Negative ion resonances
Resonance radiation, 390
Resonance spectroscopy, 423, 425
Resonances, see Negative ion resonances; Negative ion states
Retarding potential analysis, 310
 analyzer, 167
 difference method, 10, 23, 293
 energy selector, 9–11
Rotational excitation, see specific molecules
 temperature, 347
Rydberg atom, 337, 340–346, 360, 371, 495, 601–604
 electron, 343, 405
 molecule, 343
 orbital, 422, 466
 states, 493–495

S

Schumann–Runge continuum, 163, 230, 342, 381
Screening constant, 447
Selection rules, 164–166, 174, 181
 optical, 165, 253
Separated atom
 approximation of, 84, 88
 distorted-wave theories of, 88
SF_4, thermal attachment rate constant of, 554
SF_6
 collision of, with UF_6^-, 675, 676
 differential elastic scattering cross section of, 37, 38
 electron attachment of, 552–554
  dissociative, 553
  effect of temperature on, 566–568, 589
  nondissociative
   effect of temperature, 567, 568, 589
   lifetime of SF_6^{-*}, 554
   thermal rate constant, 554
   total cross section, 553
   total rate constant, 553
 electron energy loss spectrum of, 258
 electron scattering cross section of, 35, 188, 189
 as an electron scavenging gas, 429, 430, 496

integral elastic scattering cross section of, 40
 ion autocorrelation spectrum of, 258
 ionization of, by electrons, 305
 negative ion states of, 452, 453, 553
 negative ions of, in collisions
  with alkali atoms, 601
  with Rydberg atoms, 602–604
Shake-off process, 422
Shape resonance, see Negative-ion resonances
Shift
 competitive, 299
 kinetic, 299
 thermal, 299
SiH_4
 dissociation of, by electrons, 396
 ionization of, by electrons, 305
SO_2
 dissociation of, by electrons, 396
 elastic electron scattering of, 33
 electron attachment
  dissociative, 534–536
  nondissociative, 535, 536, 588
 excitation of, by electrons, cross section
  for electronic, 186
  for total scattering, 189
  for vibrational, 179
 ionization of, by electrons, 303
 negative ion states of, 451, 534–536
SO_3
 ionization of, by electrons, 304
Spectral purity, 338, 350, 351, 396
Spin exchange, 156, 174
Spin-selected electrons, 193
Static exchange approximation, 71–74, 76, 102
Stevenson's rule, 267
Superexcited states, 190, 256
Supersonic beams, 18
Supersonic nozzles, 592
Survival factor, 561
 probability, 505, 508, 627, 628, 630
Swarm-beam technique, 499
Swarm-unfolding technique, 499, 500
Symmetry, molecular, 412
 operations, 165
Synchrotron light, 337, 353

T

Terrestrial atmosphere, 365, 382, 391
Thioethers, dissociative electron attachment, 555, 556

Index

Thomas–Kuhn–Reiche sum rule, 190
Threshold
  electron spectrum, 421, 429, 503
  photoelectron technique, 503
Threshold laws, 286, 287, 293, 296, 298
Threshold peak in vibrational excitation, 417, 418
Threshold resonances, 174
Time-of-flight analysis, 311
Time-of-flight method, 623
Time-of-flight monochromator, 431
Time-of-flight spectra, 341, 343
Time-of-flight technique for dissociation studies, 347, 354–360
Total ionization apparatus, 288, 289, 496
  condensor tube, 289
  path length corrections, 289
Transmission method, 166, 167, 423, 428
Trapped-electron method, 168, 420, 429, 430, 496, 623
Trapped-ion mass spectrometry, 296
Trochoidal monochromator, 293, 425–427, 501
Tunneling, 273, 274

U

UF_6
  elastic electron scattering cross section of, 35, 37, 38
  ionization of, by electrons, 305
  negative ions of, in collisions with alkali atoms, 601
Unimolecular decomposition, 255, 275, 278, 574

V

Vertical attachment energy, 484–486
Vertical detachment energy, 484–486
Vertical electron affinity, 419, 484
Vertical ionization energy, 459
Vertical onset, 489, 528, 556
Vertical transition, 253, 408, 409, 488–490
Vibrational excitation by electrons, *see also* specific molecules, 172, 174–180, 195–197, 202–207, 418, 419
  direct, 174
  indirect, 174
  nonresonant, 194–196, 205
  resonant, 172–174
  threshold peaks in, 417
Volume correction factor, 20

W

Wannier point, 260
Wehnelt cylinder, 8, 9

X

Xe, collisions with UF_6^-, 675

Z

Zero momentum transfer, 161, 165, 169, 181, 190
Zero radius model, 626–628